**Privileged Chiral Ligands
and Catalysts**

*Edited by
Qi-Lin Zhou*

Related Titles

Blaser, H.-U., Federsel, H.-J. (eds.)

Asymmetric Catalysis on Industrial Scale

Challenges, Approaches and Solutions

Second Edition
2010
ISBN: 978-3-527-32489-7

Bandini, M., Umani-Ronchi, A. (eds.)

Catalytic Asymmetric Friedel-Crafts Alkylations

2009
ISBN: 978-3-527-32380-7

Tao, J., Lin, G.-Q., Liese, A.

Biocatalysis for the Pharmaceutical Industry

Discovery, Development, and Manufacturing

2009
ISBN: 978-0-470-82314-9

Crabtree, R. H. (ed.)

Handbook of Green Chemistry - Green Catalysis

3 Volume Set
2009
ISBN: 978-3-527-31577-2

van Santen, R. A., Sautet, P. (eds.)

Computational Methods in Catalysis and Materials Science

An Introduction for Scientists and Engineers

2009
ISBN: 978-3-527-32032-5

Carreira, E. M., Kvaerno, L.

Classics in Stereoselective Synthesis

2009
ISBN: 978-3-527-29966-9

Ding, K., Uozumi, Y. (eds.)

Handbook of Asymmetric Heterogeneous Catalysis

2008
ISBN: 978-3-527-31913-8

Rothenberg, G.

Catalysis

Concepts and Green Applications

2008
ISBN: 978-3-527-31824-7

Morokuma, K., Musaev, D. (eds.)

Computational Modeling for Homogeneous and Enzymatic Catalysis

A Knowledge-Base for Designing Efficient Catalysts

2008
ISBN: 978-3-527-31843-8

Edited by Qi-Lin Zhou

Privileged Chiral Ligands and Catalysts

WILEY-VCH Verlag GmbH & Co. KGaA

The Editor

Prof. Qi-Lin Zhou
Nankai University
Institute of Elemento-Organic Chemistry
94 Weijin Road
Tianjin 300071
China

■ All books published by **Wiley-VCH** are carefully produced. Nevertheless, authors, editors, and publisher do not warrant the information contained in these books, including this book, to be free of errors. Readers are advised to keep in mind that statements, data, illustrations, procedural details or other items may inadvertently be inaccurate.

Library of Congress Card No.: applied for

British Library Cataloguing-in-Publication Data
A catalogue record for this book is available from the British Library.

Bibliographic information published by the Deutsche Nationalbibliothek
The Deutsche Nationalbibliothek lists this publication in the Deutsche Nationalbibliografie; detailed bibliographic data are available on the Internet at <http://dnb.d-nb.de>.

© 2011 Wiley-VCH Verlag & Co. KGaA, Boschstr. 12, 69469 Weinheim, Germany

All rights reserved (including those of translation into other languages). No part of this book may be reproduced in any form – by photoprinting, microfilm, or any other means – nor transmitted or translated into a machine language without written permission from the publishers. Registered names, trademarks, etc. used in this book, even when not specifically marked as such, are not to be considered unprotected by law.

Cover Design   Grafik-Design, Schulz Fußgönheim
Typesetting   MPS Limited, a Macmillan Company, Chennai
Printing and Binding   Fabulous Printers Pte Ltd, Singapore

Printed in Singapore
Printed on acid-free paper

ISBN: 978-3-527-32704-1

Contents

Preface *XV*
List of Contributors *XIX*

1   BINAP *1*
    Takeshi Ohkuma and Nobuhito Kurono
1.1   Introduction: Structural Consideration *1*
1.2   Hydrogenation of Olefins *3*
1.3   Hydrogenation of Ketones *6*
1.3.1   Functionalized Ketones *6*
1.3.2   Simple Ketones *9*
1.4   Isomerization of Allylamines and Allylalcohols *13*
1.5   Hydroboration, Hydrosilylation, Hydroacylation, and Hydroamination *14*
1.6   Allylic Alkylation *18*
1.7   Heck Reaction *18*
1.7.1   Intramolecular Reaction *18*
1.7.2   Intermolecular Reaction *19*
1.8   Aldol and Mannich-Type Reactions *21*
1.8.1   Aldol Reaction *21*
1.8.2   Mannich-Type Reaction *23*
1.9   Nucleophilic Additions to Carbonyl and Imino Compounds *24*
1.9.1   Allylation *24*
1.9.2   Alkenylation and Arylation *25*
1.9.3   Dienylation *26*
1.9.4   Cyanation *27*
1.10   α-Substitution Reactions of Carbonyl Compounds *28*
1.10.1   Fluorination and Amination *28*
1.10.2   Arylation and Orthoester Alkylation *29*
1.11   Michael-Type Reactions *30*

Privileged Chiral Ligands and Catalysts. Edited by Qi-Lin Zhou
Copyright © 2011 WILEY-VCH Verlag GmbH & Co. KGaA, Weinheim
ISBN: 978-3-527-32704-1

1.11.1	Michael Reaction 30	
1.11.2	Aza-Michael Reaction 32	
1.12	Conjugate Additions Using Organoboron and Grignard Reagents 32	
1.13	Diels–Alder Reaction 35	
1.14	Ene Reaction 38	
1.15	Cyclization 38	
1.15.1	Intramolecular Reactions of Enynes 38	
1.15.2	[3 + 2] and [5 + 2] Cycloaddition Reactions 39	
1.15.3	[2 + 2 + 2] Cycloaddition Reactions 42	
1.15.4	Pauson–Khand Type Reactions 43	
1.16	Ring-Opening Reactions 45	
1.17	Concluding Remarks 45	
2	**Bisphosphacycles — From DuPhos and BPE to a Diverse Set of Broadly Applied Ligands** 55	
	Weicheng Zhang and Xumu Zhang	
2.1	Introduction 55	
2.2	Development of Bisphosphacycle Ligands 55	
2.2.1	Structural Features of DuPhos and BPE 55	
2.2.2	Strategies of Ligand Design 58	
2.3	Applications of Bisphosphacycle Ligands 65	
2.3.1	Asymmetric Hydrogenation 65	
2.3.2	Asymmetric Hydroformylation 75	
2.3.3	Asymmetric Hydrosilylation 77	
2.3.4	Asymmetric Hydroacylation 77	
2.3.5	Asymmetric Cycloisomerization, Cycloaddition, and Cyclization 78	
2.3.6	Asymmetric Phosphination 81	
2.3.7	Asymmetric Nucleophilic Addition to Ketones and Ketimines 82	
2.3.8	Asymmetric Conjugate Addition 85	
2.3.9	Miscellaneous Reactions 85	
2.4	Concluding Remarks 87	
3	**Josiphos Ligands: From Discovery to Technical Applications** 93	
	Hans-Ulrich Blaser, Benoît Pugin, Felix Spindler, Esteban Mejía, and Antonio Togni	
3.1	Introduction and Background 93	
3.2	Discovery and Development of the Josiphos Ligand Family 94	
3.3	Why Are Josiphos Ligands So Effective? 97	

3.3.1	General Considerations	97
3.3.2	Structural Aspects of Transition Metal Complexes Containing Josiphos and Josiphos-Like Ligands	99
3.4	Catalytic Profile of the Josiphos Ligand Family	104
3.4.1	Enantioselective Reductions of C=C, C=O and C=N Bonds	104
3.4.1.1	Enantioselective Hydrogenation of C=C Bonds	104
3.4.1.2	Copper-Catalyzed Reduction of Activated C=C bonds with PMHS (Conjugate Reduction)	110
3.4.1.3	Enantioselective Hydrogenation of C=O Bonds	111
3.4.1.4	Enantioselective Hydrogenation of C=N Bonds	114
3.4.2	Enantioselective Hydrofunctionalizations	118
3.4.2.1	Hydroboration	118
3.4.2.2	Hydroamination and Hydrophosphonation	119
3.4.2.3	Hydrocarboxylation	120
3.4.3	Enantioselective C–C Bond Forming Reactions	120
3.4.3.1	Allylic Alkylation	120
3.4.3.2	Michael Addition	121
3.4.3.3	Heck Reaction	122
3.4.3.4	Miscellaneous C–C Reactions	123
3.4.4	Miscellaneous Enantioselective Reactions	125
3.4.4.1	Isomerization of Allylamines	125
3.4.4.2	Ring-Opening of Oxabicycles	125
3.4.4.3	Allylic Substitution	126
3.4.5	Application in Non-Enantioselective Reactions	127
3.5	Concluding Remarks	127
4	**Chiral Spiro Ligands**	**137**
	Shou-Fei Zhu and Qi-Lin Zhou	
4.1	Introduction	137
4.2	Preparation of Chiral Spiro Ligands	139
4.3	Asymmetric Hydrogenation	144
4.3.1	Hydrogenation of Functionalized Olefins	144
4.3.1.1	Hydrogenation of Enamides	144
4.3.1.2	Hydrogenation of Enamines	146
4.3.1.3	Hydrogenation of α,β-Unsaturated Acids	146
4.3.2	Hydrogenation of Ketones and Aldehydes	151
4.3.2.1	Hydrogenation of Simple Ketones	151
4.3.2.2	Hydrogenation of Racemic 2-Substituted Ketones via DKR	151
4.3.2.3	DKR Hydrogenation of Racemic 2-Substituted Aldehydes	153
4.3.3	Hydrogenation of Imines	154
4.3.4	Hydrogenation of 2-Substituted Quinolines	154
4.4	Asymmetric Carbon–Carbon Bond Forming Reaction	155

4.4.1	Rhodium-Catalyzed Arylation of Carbonyl Compounds and Imines *155*
4.4.2	Palladium-Catalyzed Umpolung Allylation of Aldehydes *158*
4.4.3	Copper-Catalyzed Conjugate Addition Reaction *158*
4.4.4	Copper-Catalyzed Ring-Opening Reaction with Grignard Reagents *158*
4.4.5	Nickel-Catalyzed Three-Component Coupling Reaction *159*
4.4.6	Nickel-Catalyzed Hydrovinylation Reaction *161*
4.4.7	Rhodium-Catalyzed Hydrosilylation/Cyclization Reaction *161*
4.4.8	Palladium-Catalyzed Asymmetric Oxidative Cyclization *162*
4.4.9	Gold-Catalyzed Ring Expanding Cycloisomerization *163*
4.5	Asymmetric Carbon–Heteroatom Bond Forming Reaction *163*
4.5.1	Palladium-Catalyzed Hydrosilylation *163*
4.5.2	Palladium-Catalyzed Wacker-Type Oxidative Cyclization Reaction *163*
4.5.3	Copper-Catalyzed Carbene Insertion into X–H Bonds *164*
4.5.4	Allene-Based Allylic Cyclization Reactions *166*
4.6	Conclusion *167*
5	**Chiral Bisoxazoline Ligands** *171*
	Levi M. Stanley and Mukund P. Sibi
5.1	Introduction *171*
5.2	Enantioselective Carbon–Carbon Bond Formation *176*
5.2.1	Addition of Carbon Nucleophiles to C=O and C=N Bonds *176*
5.2.1.1	Aldol Reactions *176*
5.2.1.2	Mannich-Type Reactions *177*
5.2.1.3	Nitroaldol (Henry) Reactions *179*
5.2.1.4	Nitro-Mannich (Aza-Henry) Reactions *181*
5.2.1.5	Addition of Activated Carbon Nucleophiles to Carbonyl Electrophiles *182*
5.2.1.6	Addition of Activated Carbon Nucleophiles to Imines *183*
5.2.1.7	Ene Reactions *184*
5.2.1.8	Friedel–Crafts Reactions of Aromatic Compounds with C=O and C=N Bonds *184*
5.2.2	1,4-Addition of Carbon Nucleophiles to α,β-Unsaturated Acceptors *186*
5.2.3	Reactions of Radicals Alpha to Carbonyls *190*
5.2.4	Cyclization Reactions *191*
5.2.5	Rearrangement Reactions *191*
5.3	Enantioselective Carbon–Heteroatom Bond Formation *193*
5.3.1	1,4-Addition of Heteroatom Nucleophiles to α,β-Unsaturated Acceptors *193*
5.3.1.1	1,4-Addition of Nitrogen Nucleophiles *193*

5.3.1.2	1,4-Addition of Sulfur and Oxygen Nucleophiles	195
5.3.1.3	1,4-Addition of Boron Nucleophiles	195
5.3.2	Allylic Functionalization Reactions	196
5.3.3	α-Heteroatom Functionalization of Carbonyl Compounds	196
5.3.3.1	Amination	196
5.3.3.2	Oxygenation	197
5.3.3.3	Halogenation	197
5.3.4	X–H Insertion Reactions (X = O, N, S)	198
5.3.5	Cyclization Reactions	199
5.3.5.1	Carbonylative Cyclization	199
5.3.5.2	Wacker-Type Cyclizations	199
5.3.5.3	Hydroamination	201
5.3.6	Kinetic Resolution and Desymmetrization Reactions	201
5.3.6.1	Kinetic Resolution	201
5.3.6.2	Desymmetrization	201
5.4	Enantioselective Cycloaddition Reactions	202
5.4.1	Carbo-Diels–Alder Cycloadditions	202
5.4.2	Hetero-Diels–Alder Cycloadditions	204
5.4.3	Cyclopropanations	205
5.4.4	Aziridination	208
5.4.5	1,3-Dipolar Cycloadditions	209
5.4.6	Additional Cycloaddition Reactions	211
5.5	Conclusions	212
6	**PHOX Ligands**	**221**
	Cory C. Bausch and Andreas Pfaltz	
6.1	Introduction	221
6.2	Synthesis of PHOX Ligands	222
6.3	Nucleophilic Allylic Substitution	224
6.3.1	Palladium-Catalyzed Allylic Substitution	224
6.3.2	Tungsten- and Iridium-Catalyzed Allylic Substitution	229
6.3.3	Allylic Substitution in Total Synthesis	230
6.4	Decarboxylative Tsuji Allylations	231
6.4.1	Method Development	231
6.4.2	Application to Fluorinated Derivatives	234
6.4.3	Applications in Total Synthesis	235
6.5	Heck Reaction	237
6.5.1	Intermolecular Heck Reaction	237
6.5.2	Intramolecular Heck Reaction	238
6.6	Hydrogenation	240
6.6.1	Hydrogenation of Imines	240
6.6.2	Hydrogenation of Trisubstituted Olefins	240
6.6.3	Hydrogenation of Tetrasubstituted Olefins	243
6.6.4	Hydrogenation of Vinyl Phosphonates	243

6.6.5	Hydrogenation of α,β-Unsaturated Ketones	244
6.6.6	Hydrogenation of Ketones	244
6.6.7	Transfer Hydrogenation of Ketones	246
6.7	Cycloadditions	246
6.7.1	[3 + 2] Cycloadditions	246
6.7.2	Diels–Alder Reactions	247
6.8	Miscellaneous Reactions	248
6.8.1	Hydrosilylations	248
6.8.2	Pauson–Khand Reaction	248
6.8.3	Decarboxylative Protonation	249
6.8.4	Sigmatropic Rearrangements	250
6.8.5	Desymmetrization Reactions	251
6.8.6	Asymmetric Arylations	253
6.9	Conclusion	253

7	**Chiral Salen Complexes** 257	
	Wen-Zhen Zhang and Xiao-Bing Lu	
7.1	Introduction	257
7.2	Synthesis of Chiral Salen Complexes	257
7.3	Structural Properties of Chiral Salen Complexes	259
7.4	Asymmetric Reactions Catalyzed by Chiral Salen Complexes	262
7.4.1	Asymmetric Epoxidation	262
7.4.2	Asymmetric Ring-Opening of Epoxides	266
7.4.2.1	Desymmetrization of Meso-Epoxides	266
7.4.2.2	Kinetic Resolution of Racemic Epoxides	269
7.4.2.3	Enantioselective Addition of Carbon Dioxide to Propylene Oxide	271
7.4.2.4	Asymmetric Alternating Copolymerization of Racemic Epoxides and Carbon Dioxide	272
7.4.2.5	Enantioselective Homopolymerization of Epoxides	273
7.4.3	Asymmetric Cyclopropanation	274
7.4.4	Asymmetric Conjugate Addition Reaction	277
7.4.5	Asymmetric Diels–Alder Reaction	281
7.4.6	Asymmetric Cyanohydrin Synthesis	284
7.4.7	Miscellaneous Reactions	287
7.5	Conclusion and Outlook	289

8	**BINOL** 295	
	Masakatsu Shibasaki and Shigeki Matsunaga	
8.1	Introduction	295
8.2	Applications in Reduction and Oxidation	296
8.3	Metal/BINOL Chiral Lewis Acid Catalysts in Asymmetric C–C Bond Forming Reactions	300

8.3.1	Group IV Metal/BINOL Lewis Acid Catalysts	300
8.3.2	Group XIII Metal/BINOL Lewis Acid Catalysts	304
8.3.3	Rare Earth Metal/BINOL Lewis Acid Catalysts	307
8.4	Acid/Base Bifunctional Metal/BINOL Catalysts	308
8.4.1	Rare Earth Metal/Alkali Metal/BINOL Catalysts	308
8.4.2	Group XIII Metal/Alkali Metal/BINOL Catalysts	312
8.4.3	Other Metal/BINOL Complexes as Acid/Base Bifunctional Catalysts	316
8.4.4	Lewis Acid/Lewis Base Bifunctional Aluminium-Catalyst	321
8.5	BINOL in Organocatalysis	324
8.6	Summary	329

9 TADDOLate Ligands 333
Hélène Pellissier

9.1	Introduction	333
9.2	Nucleophilic Additions to C=O Double Bonds	334
9.2.1	Organozinc Additions to Aldehydes	334
9.2.2	Allylations	335
9.2.3	Aldol-Type Reactions	336
9.2.4	Miscellaneous Reactions	338
9.3	Nucleophilic Conjugate Additions to Electron-Deficient C=C Double Bonds	339
9.4	Nucleophilic Substitutions	342
9.4.1	Allylic Substitutions	342
9.4.2	α-Halogenations of Carbonyl Compounds	344
9.4.3	Miscellaneous Substitutions	345
9.5	Cycloaddition Reactions	345
9.5.1	Diels–Alder reactions	346
9.5.2	Hetero-Diels–Alder Reactions	347
9.5.3	Miscellaneous Cycloadditions	348
9.6	Oxidation and Reduction Reactions	349
9.7	Miscellaneous Reactions	351
9.8	Conclusions	354

10 Cinchona Alkaloids 361
Hongming Li, Yonggang Chen and Deng Li

10.1	Introduction	361
10.2	Metal Catalysis	363
10.3	Phase-Transfer Catalysis	367
10.3.1	Asymmetric Alkylations	367
10.3.2	Asymmetric Conjugate Additions	368
10.3.3	Asymmetric Aldol Reactions	368
10.3.4	Examples of Recent Applications	369
10.4	Nucleophilic Catalysis	370

10.4.1	Asymmetric Reactions with Ketenes	370
10.4.2	Asymmetric Morita–Baylis–Hillman Reactions	373
10.4.3	Asymmetric Cyanation of Simple Ketones	374
10.4.4	Recent Applications of Nucleophilic Catalysis by Cinchona Alkaloids	375
10.4.4.1	Asymmetric Conjugate Additions	375
10.4.4.2	Asymmetric Electrophilic Halogenations of Olefins	376
10.5	Base Catalysis	377
10.5.1	Asymmetric Alcoholysis of Cyclic Anhydrides	377
10.5.2	Conjugate Additions	380
10.5.3	Asymmetric Mannich and Aldol Reactions	381
10.6	Cooperative and Multifunctional Catalysis	382
10.6.1	Acid–Base Cooperative Catalysis	382
10.6.1.1	Asymmetric Conjugate Additions	382
10.6.1.2	Asymmetric 1,2-Additions to Carbonyls	386
10.6.1.3	Asymmetric 1,2-Additions to Imines	390
10.6.1.4	Asymmetric Friedel–Crafts Reactions	391
10.6.1.5	Asymmetric Diels–Alder Reactions	393
10.6.1.6	Asymmetric Fragmentation	394
10.6.2	Base–Iminium Cooperative Catalysis	396
10.6.2.1	Asymmetric Conjugate Additions	396
10.6.2.2	Asymmetric Fridel–Crafts Additions	398
10.6.2.3	Asymmetric Diels–Alder Reactions	399
10.6.2.4	Semipinacol-Type 1,2-Carbon Migration	399
10.6.3	Multifunctional Cooperative Catalysis	400
10.6.3.1	Tandem Conjugate Addition–Protonation Reactions	400
10.6.3.2	Catalytic Asymmetric Peroxidations	400
10.7	Conclusion	404
11	**Proline Derivatives**	**409**
	Shilei Zhang and Wei Wang	
11.1	Introduction	409
11.2	Proline as Organocatalyst	410
11.2.1	Aldol Reactions	410
11.2.1.1	Intermolecular Aldol Reactions	410
11.2.1.2	Intramolecular Aldol Reactions	412
11.2.1.3	Synthesis of Carbohydrates by Proline-Catalyzed Aldol Reactions	413
11.2.2	Mannich Reactions Catalyzed by Proline	414
11.2.3	Michael Addition Reactions Catalyzed by Proline	415
11.2.4	Morita–Baylis–Hillman (MBH) Reactions Catalyzed by Proline	416
11.2.5	α-Amination, α-Aminoxylation, and α-Alkylation of Carbonyl Compounds Catalyzed by Proline	417

11.2.6	Cascade/One-Pot Reactions Catalyzed by Proline	418
11.3	Proline Analogs as Organocatalysts	419
11.3.1	4-Hydroxyproline as Organocatalyst	419
11.3.2	Other Proline Analogs as Organocatalysts	421
11.4	5-Pyrrolidin-2-yltetrazole as Organocatalyst	422
11.5	Pyrrolidine-Based Sulfonamides as Organocatalysts	424
11.6	Pyrrolidine-Based Amides as Organocatalysts	425
11.7	Pyrrolidine Diamine Catalysts	427
11.8	Diarylprolinols or Diarylprolinol Ether Catalysts	429
11.8.1	Aldol Reactions, Mannich Reactions, and Other α-Functionalizations of Aldehydes Catalyzed by Diarylprolinols or Diarylprolinol Silyl Ethers	429
11.8.2	Michael Addition Reactions Catalyzed by Diarylprolinols or Diarylprolinol Silyl Ethers.	430
11.8.2.1	Michael Additions through an Enamine Pathway	430
11.8.2.2	Michael Additions through an Iminium Mechanism	430
11.8.3	Cycloaddition Reactions Catalyzed by Diarylprolinols or Diarylprolinol Silyl Ethers	433
11.8.4	Cascade Reactions Catalyzed by Diarylprolinol Silyl Ethers	435
11.8.4.1	Three-Membered Rings Formed by a [1 + 2] Strategy	435
11.8.4.2	Five-Membered Rings Formed by a [3 + 2] Strategy	436
11.8.4.3	Six-Membered Rings Formed by a [4 + 2] Strategy	437
11.8.4.4	Six-Membered Rings Formed by a [3 + 3] Strategy	437
11.8.4.5	Six-Membered Rings Formed by a [2 + 2 + 2] Strategy	438
11.8.4.6	Other Cascade Reactions	439
11.9	Concluding Remarks	439

Index 447

Preface

Catalytic asymmetric synthesis has been one of the most active research areas in modern chemistry. Asymmetric catalyses with enzymes, chiral metal complexes, and chiral organic molecules have emerged as successful and powerful tools in asymmetric synthesis. Among the three catalytic asymmetric processes, artificial metal complex catalysis and organocatalysis have only a very short history compared to traditional biocatalysis but are now a predominant part of asymmetric synthesis in both research and application. The development of efficient synthetic chiral catalysts, including chiral metal complex catalysts modified with various chiral ligands and chiral organo-molecule catalysts, is at the center of research in asymmetric catalysis. Although numerous chiral ligands as well as chiral catalysts have been reported in past decades, only a handful of them, rooted in a very few core structures, can be regarded as truly successful in demonstrating proficiency in various mechanistically unrelated reactions. Researchers have designated chiral catalysts showing good enantioselectivity over a wide range of different reactions as "privileged chiral catalysts," a term coined by Jacobsen. The essential feature that makes a catalyst "privileged" is its scaffold (core structure). To understand the relationship between the structure of a "privileged" catalyst and its catalytic features in reactions is the key to opening the door to designing more efficient catalysts. Furthermore, a deep insight into the structural characteristics of the most successful catalysts so far reported will facilitate the selection of appropriate catalysts in developing new asymmetric processes. However, available books on asymmetric synthesis have focused predominantly on asymmetric reactions, making it difficult to perceive the suite of chiral catalysts in terms of structural characteristics and catalytic abilities. This book, *Privileged Chiral Ligands and Catalysts*, tells the stories of these ligands and catalysts from the core structure point of view, a rarity in previous books. This book is a timely overview of a few popularly used chiral ligands and catalysts, focused on their structural aspects and the relationship between the structure of catalysts and their success in catalytic operations.

It is not the goal, and it would be an almost impossible undertaking, to provide a comprehensive book on chiral ligands and catalysts. To illustrate clearly the key points of "privileged" chiral ligands and catalysts in a 400-page book we have selected eleven ligands and catalysts as examples to discuss in detail, namely,

BINAP, DuPhos, Josiphos, spiro ligands, BOX, PHOX, Salen complexes, BINOL, TADDOL, cinchona alkaloids, and proline, rather than examining all of the high-profile candidates. Among the eleven, BINAP, DuPhos, Josiphos, spiro ligands, BOX, and PHOX are chiral ligands in metal catalysts; Salen complexes are chiral metal catalysts, and cinchona alkaloids and proline are generally used as organocatalysts. BINOL and TADDOL were used as chiral ligands in Lewis acid catalysts in earlier studies but recently they have also been used as organocatalysts in various reactions. The editor, based solely on personal taste, has sought to arrange the presentation of the eleven ligands or catalysts by starting with ligands, then addressing metal catalysts and organocatalysts. Although the selection is subjective we believe that the important ligands and catalysts in the field of asymmetric synthesis are included and that the general aspects of ligand and catalyst design will thus be fully exhibited through these eleven examples.

The eleven selected ligands or catalysts are independent of each other and so, as a result, each chapter in this book provides an individual overview of each one. Although the authors responsible for each chapter were given sufficient freedom to organize their material, we encouraged them to provide a short discussion of the family of ligands or catalysts to which the individual ligand or catalyst belongs. It is beneficial to readers to see the full spectrum of the ligands or catalysts rooted on the same scaffolds. Each chapter also emphasizes the chiral-inducing models of metal catalysts or organocatalysts to illustrate the transfer of chirality from catalysts to substrates in different reactions. The most successful applications, especially the latest identified reactions of these ligands or catalysts, have been described to support their designation as "privileged" catalytic properties. In contrast, well-known classic applications are discussed only briefly.

The editor believes that the authors have described the most important features of these specific ligands or catalysts discussed in this book. The reader can find the design principle of the chiral ligands and catalysts readily in each chapter. Because many common principles have been considered during the development of the most successful ligands or catalysts, some overlap of these principles inevitably occurs among the chapters. For instance, the features of high chemical robustness and ease of modification can be found in all eleven selected ligands or catalysts. Further, it is generally accepted that the high scaffold rigidity of the ligand or catalyst plays a crucial role in making it "privileged." Accordingly, the reader may readily notice that almost all of the successful ligands or catalysts contain five- or six-membered rings. In addition to the structural properties of each ligand or catalyst, an easily available starting material is also important. At least three selected ligands and catalysts – proline, cinchona alkaloids, and TADDOL – are derived directly from a "chiral pool." The chiral moiety of the ligands BOX and PHOX are chiral amino alcohols, derived from natural amino acids. The coordinating atom is another important aspect of ligand design, and the most successful ligands all have phosphorus or nitrogen as the coordinating atom. In addition, the dentate number of chiral ligands and the chelating ring size of catalysts are crucial features for obtaining satisfactory chiral induction.

It is our hope that the new descriptive model in this book will lead readers to constructive thinking about what makes chiral catalysts "privileged" and encourage more creative work for the development of "privileged ligands and catalysts." If this book is helpful to our colleagues in the chemistry community in their design of chiral ligands or catalysts, selection of appropriate catalysts in their chiral synthesis, or other aspects of their research we believe our goal will have been met.

I am deeply indebted to all chapter authors for their significant contributions to the book. I am grateful to Dr. Elke Maase of Wiley-VCH, who initiated the project of editing this book, and to Lesley Belfit for her support during the editing process. I also thank my colleague Dr. Shou-Fei Zhu for his constructive suggestions in editing this book.

Qi-Lin Zhou
Nankai University
Tianjin, China

List of Contributors

Cory Bausch
University of Basel
Department of Chemistry
St. Johanns-Ring 19
4056 Basel
Switzerland

Hans-Ulrich Blaser
Solvias AG
P.O. Box
4002 Basel
Switzerland

Yonggang Chen
Brandeis University
Department of Chemistry
415 South St., Waltham
MA 024543
USA

Li Deng
Brandeis University
Department of Chemistry
415 South St., Waltham
MA 024543
USA

Nobuhito Kurono
Hokkaido University, Graduate School of Engineering
Division of Chemical Process Engineering, Laboratory of Organic Synthesis
Sapporo 060- 8628
Japan

Hongming Li
Brandeis University
Department of Chemistry
415 South St., Waltham
MA 024543
USA

Xiao-Bing Lu
Dalian University of Technology
State Key Laboratory of Fine Chemicals
No. 2 Linggong Road
Dalian
Liaoning 116024
China

Privileged Chiral Ligands and Catalysts. Edited by Qi-Lin Zhou
Copyright © 2011 WILEY-VCH Verlag GmbH & Co. KGaA, Weinheim
ISBN: 978-3-527-32704-1

Shigeki Matsunaga
The University of Tokyo
Graduate School of Pharmaceutical
Sciences
Tokyo 113-0033
Japan

Esteban Mejía
ETH Zürich
Laboratorium für Anorganische Chemie
Wolfgang-Pauli-Str. 10
8093 Zürich
Switzerland

Takeshi Ohkuma
Hokkaido University, Graduate School
of Engineering
Division of Chemical Process
Engineering, Laboratory of Organic
Synthesis
Sapporo 060- 8628
Japan

Hélène Pellissier
Université Paul Cézanne –
Aix-Marseille III
Institut Sciences Moléculaires de
Marseille
Avenue Esc. Normandie-Niemen
13397 Marseille
France

Andreas Pfaltz
University of Basel
Department of Chemistry
St. Johanns-Ring 19
4056 Basel
Switzerland

Benoît Pugin
Solvias AG
P.O. Box
4002 Basel
Switzerland

Masakatsu Shibasaki
The University of Tokyo
Graduate School of Pharmaceutical
Sciences
Tokyo 113-0033
Japan

Mukund P. Sibi
North Dakota State University
Department of Chemistry and
Molecular Biology
1231 Albrecht Boulevard
Fargo
USA

Felix Spindler
Solvias AG
P.O. Box
4002 Basel
Switzerland

Levi M. Stanley
University of Illinois at
Urbana-Champaign
Department of Chemistry
Urbana
USA

Antonio Togni
ETH Zürich
Laboratorium für Anorganische Chemie
Wolfgang-Pauli-Str. 10
8093 Zürich
Switzerland

Wei Wang
University of New Mexico
Department of Chemistry and
Chemical Biology
Albuquerque
NM 87131-0001
USA

and

Chinese Academy of Sciences
Shanghai Institute of Materia Medica
Shanghai 201203
China

Shilei Zhang
University of New Mexico
Department of Chemistry and
Chemical Biology
Albuquerque
NM 87131-0001
USA

Weicheng Zhang
Nankai University
College of Pharmacy
94 Weijin Road
Tianjin 300071
China

Wen-Zhen Zhang
Dalian University of Technology
State Key Laboratory of Fine
Chemicals
No. 2 Linggong Road
Dalian
Liaoning 116024
China

Xumu Zhang
Rutgers, The State University of
New Jersey
Department of Chemistry and
Chemical Biology
610 Taylor Road
Piscataway
NJ 08854
USA

Qi-Lin Zhou
Nankai University
Institute of Elemento-organic
Chemistry
94 Weijin Road
Tianjin 300071
China

Shou-Fei Zhu
Nankai University
Institute of Elemento-organic
Chemistry
94 Weijin Road
Tianjin 300071
China

1
BINAP

Takeshi Ohkuma and Nobuhito Kurono

1.1
Introduction: Structural Consideration

BINAP (2,2′-diphenylphosphino-1,1′-binaphthyl), which was devised by Ryoji Noyori (winner of the Nobel Prize in Chemistry 2001), is typical among chiral diphosphine ligands [1–3]. BINAP chemistry has contributed notably the development of the field of asymmetric catalysis [1, 4]. This ligand with transition metallic elements forms C_2-symmetric chelate complexes. Figure 1.1 indicates the chiral structure created by an (R)-BINAP–transition metal complex. The naphthalene rings of BINAP are omitted in the side view (right-hand side) for clarity. As illustrated in the top view, the axial-chirality information of the binaphthyl backbone is transferred through the P-phenyl rings to the four coordination sites shown by □ and ■. The coordination sites □ placed in the P^1–M–P^2 plane are sterically influenced by the "equatorial" phenyl rings, whereas the out-of-plane coordination sites, ■, are affected by the "axial" phenyl groups (side view). Consequently, the two kinds of quadrant of the chiral structure (first and third versus second and forth in the side view) are clearly discriminated spatially, where the second and fourth quadrants are sterically crowded, while the first and third ones are open for approach of substrates and reagents. This chiral structure realizes excellent enantiodifferentiation in various asymmetric catalytic reactions. The flexibility of the binaphthyl backbone appears to enable a wide scope of substrate.

The great success of BINAP chemistry has encouraged researchers to develop BINAP derivatives and related chiral biaryl diphosphines [5]. Figure 1.2 illustrates representative examples. As shown in Figure 1.1, substitution manner of P-aryl rings of BINAP ligands obviously affects the chiral structure of metal complexes. TolBINAP, which has P-4-tolyl groups, shows similar enantioselective features to those of BINAP, although the solubility of the metal complexes in organic solvents is increased [6]. XylBINAP and DTBM-BINAP bearing 3,5-dialkyl groups on the P-phenyl rings give better enantioselectivity than that with the original ligand in some cases [6]. The bulkier P-aryl groups seem to make the contrast of congestion in the chiral structure clearer. Diphosphines with a relatively small P^1–M–P^2 angle in the complexes, that is, MeO-BIPHEP [7], SEGPHOS [8], SYNPHOS

Privileged Chiral Ligands and Catalysts. Edited by Qi-Lin Zhou
Copyright © 2011 WILEY-VCH Verlag GmbH & Co. KGaA, Weinheim
ISBN: 978-3-527-32704-1

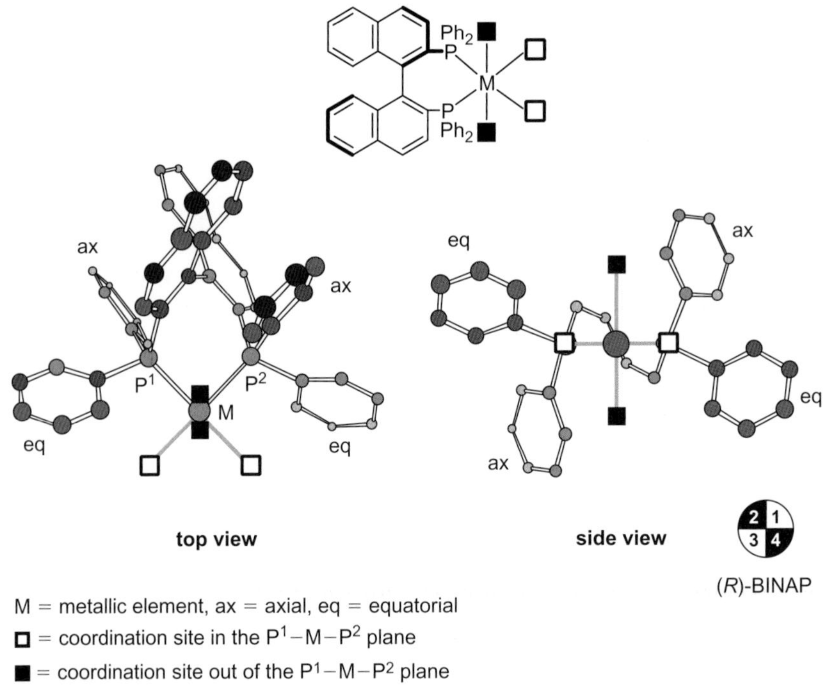

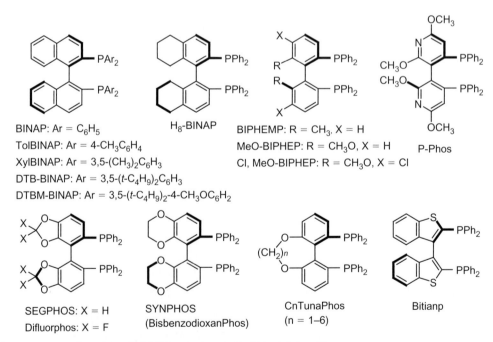

Figure 1.1 Molecular models of an (R)-BINAP–transition metal complex.

M = metallic element, ax = axial, eq = equatorial
□ = coordination site in the P^1–M–P^2 plane
■ = coordination site out of the P^1–M–P^2 plane

BINAP: Ar = C_6H_5
TolBINAP: Ar = 4-$CH_3C_6H_4$
XylBINAP: Ar = 3,5-$(CH_3)_2C_6H_3$
DTB-BINAP: Ar = 3,5-$(t-C_4H_9)_2C_6H_3$
DTBM-BINAP: Ar = 3,5-$(t-C_4H_9)_2$-4-$CH_3OC_6H_2$

H_8-BINAP

BIPHEMP: R = CH_3, X = H
MeO-BIPHEP: R = CH_3O, X = H
Cl, MeO-BIPHEP: R = CH_3O, X = Cl

P-Phos

SEGPHOS: X = H
Difluorphos: X = F

SYNPHOS
(BisbenzodioxanPhos)

CnTunaPhos
(n = 1–6)

Bitianp

Figure 1.2 (R)-BINAP and selected chiral biaryl diphosphines.

(BisbenzodioxanPhos) [9], P-Phos [10], and Difluorphos [11], place the "equatorial" phenyl groups in forward regions, providing highly contrasted chiral structures. The chiral structures are varied by the size of the P^1–M–P^2 angle. CnTunaphos ($n = 1$–6) can control the angle by changing the number of CH_2 moieties [12]. H_8-BINAP [13] and BIPHEMP [14], which are alkylated biphenyl diphosphines, exhibit some unique stereoselective characters. Heteroaromatic biaryl ligands, Bitianp [15] and P-Phos, as well as fluorinated diphosphine, Difluorphos, are expected to add some electronic perturbation in the catalytic systems.

In this chapter we introduce typical, but not comprehensive, asymmetric reactions catalyzed by the BINAP–metal complexes, achieving excellent enantioselectivity. Mechanistic considerations for some reactions are commented on with molecular models.

1.2
Hydrogenation of Olefins

In 1980, highly enantioselective hydrogenation of α-(acylamino)acrylic acids and esters catalyzed by the cationic BINAP–Rh(I) complexes was reported [16–18]. For example, (Z)-α-(benzamido)cinnamic acid is hydrogenated with [Rh{(R)-binap}(CH$_3$OH)$_2$]ClO$_4$ to afford (S)-N-benzoylphenylalanine in 100% enantiomeric excess (ee) and 97% yield [substrate/catalyst molar ratio (S/C) = 100, 4 atm H$_2$, room temperature), 48 h, in C$_2$H$_5$OH] (Scheme 1.1). In terms of enantioselectivity this hydrogenation appears to be excellent; however, very careful choice of reaction parameters, such as low substrate concentration and low hydrogen pressure, is required [19–22]. The scope of the olefinic substrates is insufficiently wide.

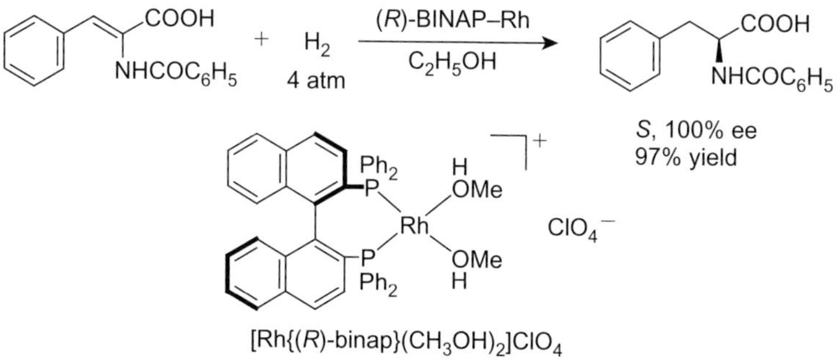

Scheme 1.1 Hydrogenation of (Z)-α-(benzamido)cinnamic acid with a BINAP–Rh catalyst.

BINAP–Ru(II) catalysis resolved the above problems. Methyl (Z)-α-(acetamido)cinnamate is hydrogenated in the presence of Ru(OCOCH$_3$)$_2$[(R)-binap] (S/C = 200) in CH$_3$OH (1 atm H$_2$, 30 °C, 24 h) to give methyl (R)-α-(acetamido)cinnamate in 92% ee and 100% yield (Scheme 1.2) [17, 23, 24]. Various olefinic substrates,

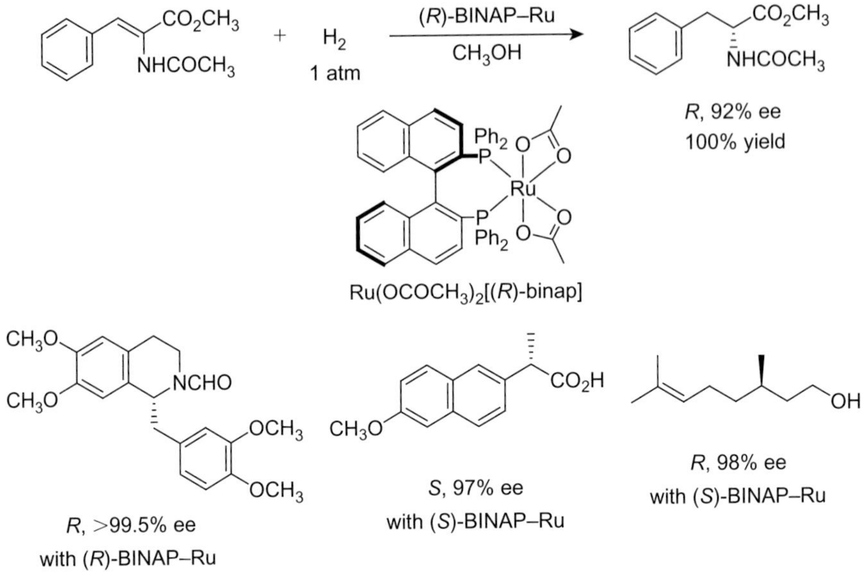

Scheme 1.2 Hydrogenation of functionalized olefins catalyzed by BINAP–Ru complexes.

including enamides, α,β- and β,γ-unsaturated carboxylic acids, and allylic and homoallylic alcohols, are converted into the desired products in high ee [25]. About 300 tons per year of optically active citronellol is produced by this hydrogenation [26]. The citronellol synthesis, by the use of a Ru(II) catalyst with a MeO-BIPHEP derivative, is applied to the production of vitamin E [27]. The H_8-BINAP–Ru (II) catalyst reduces α,β-unsaturated carboxylic acids with even higher enantioselectivity [28].

Figure 1.3 illustrates a mechanism for the hydrogenation of methyl (Z)-α-acetamidocinnamate catalyzed by the BINAP–Ru(II) complex [19, 20, 23]. Precatalyst Ru(OCOCH$_3$)$_2$[(R)-binap] [(R)-**1**] was converted into the RuH(OCOCH$_3$) complex **2**, which is an active species, under a H$_2$ atmosphere with release of CH$_3$CO$_2$H. The enamide substrate reversibly coordinates to the Ru center in bidentate fashion, forming **3**. Migratory insertion gives **4**, followed by Ru–C bond cleavage largely by H$_2$, but also by CH$_3$OH solvent to some extent, resulting in the chiral product and regenerating the catalytic species **2**. The stereochemistry of the product is determined at the irreversible step (**4**→**2**). Because the reactivities of the two diastereomers of **4** are similar, the enantioselectivity of the product corresponds well to the relative stability (population) of the diastereomeric enamide–RuH(OCOCH$_3$) intermediates [not transition states (TSs)], Si-**3** and Re-**3** (Figure 1.4). The Si-**3** is more favored over the diastereomeric isomer Re-**3**, because the latter suffers nonbonded repulsion between an equatorial phenyl ring of the (R)-BINAP and the methoxycarbonyl group of substrate. Therefore, the major (favored) intermediate Si-**3** is converted into the (R) hydrogenation product via **4**.

1.2 Hydrogenation of Olefins

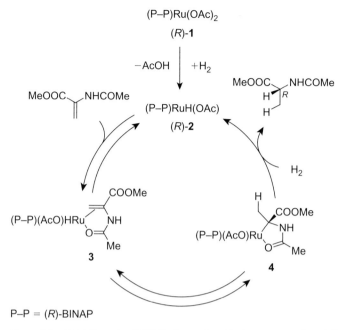

P–P = (R)-BINAP

Figure 1.3 Catalytic cycle of BINAP–Ru catalyzed hydrogenation of methyl (Z)-α-acetamidocinnamate. (For clarity the β-substituents in the substrates are omitted.)

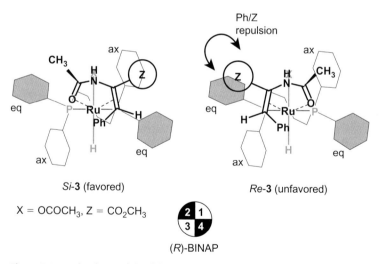

Figure 1.4 Molecular models of diastereomeric (R)-BINAP/enamide Ru complexes **3** (not transition state).

1.3
Hydrogenation of Ketones

1.3.1
Functionalized Ketones

Ru(OCOCH$_3$)$_2$(binap) is feebly active for the hydrogenation of ketones, although it shows remarkable catalytic efficiency for the reaction of functionalized olefins. This problem is resolved simply by replacing the carboxylate ligands with halides. For instance, β-keto esters (R = alkyl) are hydrogenated with RuCl$_2$(binap) (polymeric form; S/C = 2000) in CH$_3$OH (100 atm H$_2$, 30 °C, 36 h) to give the β-hydroxy esters in >99% ee quantitatively (Scheme 1.3) [29, 30]. A turnover number (TON) of 10 000 is achieved in the best cases. Several related complexes exhibit comparable catalytic efficiency, including RuCl$_2$(binap)(dmf)$_n$ (oligomeric form) [31], [RuCl(binap)(arene)]Cl [6, 32], [NH$_2$(C$_2$H$_5$)$_2$][{RuCl(binap)}$_2$(μ-Cl)$_3$] [24, 33], and other *in situ* prepared halogen-containing BINAP–Ru complexes [34]. A range of α-, β-, and γ-hetero substituted ketones as well as difunctionalized ketones and diketones is converted into the chiral alcohols in high ee (Scheme 1.3) [17, 35, 36]. Ruthenium(II) complexes with biaryldiphosphines that have smaller dihedral angles (MeO-BIPHEP [37], P-Phos [10], SEGPHOS [8], and SYNPHOS [9]) hydrogenate an aromatic β-keto ester (R = C$_6$H$_5$) in high stereoselectivity [38]. For the reaction of an analogue with trifluoromethyl group (R = CF$_3$), the Difluorphos–Ru(II) complex exhibits fine enantioselectivity [11]. The electronic deficient character of this ligand is supposed to be important.

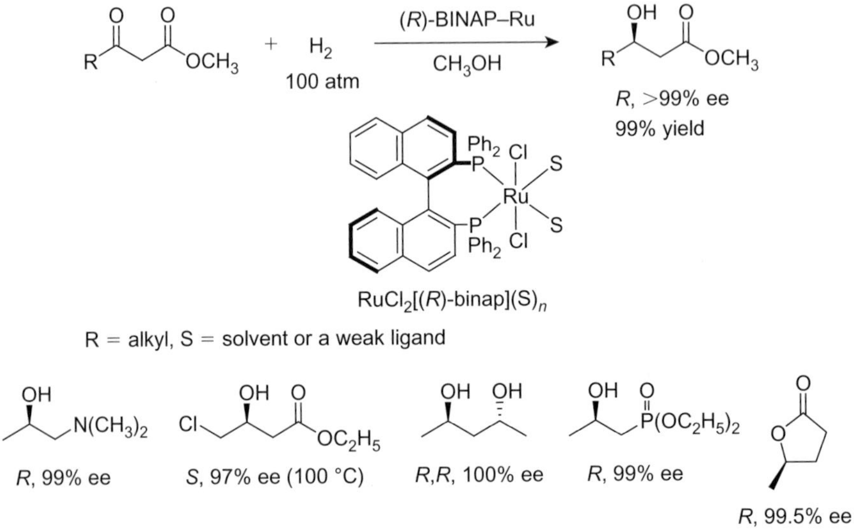

Scheme 1.3 Hydrogenation of functionalized ketones catalyzed by BINAP–Ru complexes.

Figure 1.5 shows a plausible catalytic cycle of hydrogenation of β-keto esters mediated by BINAP–Ru(II) complexes [19, 20]. The *(R)*-BINAP–RuCl$_2$ precatalyst [*(R)*-**5**] and H$_2$ generate the active catalytic RuHCl species **6** with release of HCl. The β-keto ester reversibly coordinates to the Ru center of **6**, forming the σ-type chelate complex **7**. Protonation at the carbonyl oxygen of the substrate increases electrophilic ability of the carbonyl carbon, and induces a change in chelate fashion from σ to π. Therefore, the subsequent migration of hydride on the Ru to the keto-ester carbonyl carbon occurs smoothly, resulting in the hydroxy-ester complex **8**. Replacement of the *(R)* hydroxy ester with solvent molecules gives the cationic species **9**. Heterolytic cleavage of H$_2$ by **9** regenerates **6** along with proton. Enantioface selection of the substrate occurs in the first irreversible hydride migration step, **7**→**8**. Protonation of the carbonyl oxygen with a strong acid (HCl) in **7** is crucial for conversion into **8**. Hydrogenation with the BINAP–Ru(OCOCH$_3$)$_2$ complex releases CH$_3$CO$_2$H instead of HCl, however, the acidity of CH$_3$CO$_2$H is insufficient to activate the keto-ester substrate.

Figure 1.6 illustrates molecular models of diastereomeric TSs, *Si*-**10** and *Re*-**10**, in the hydride transfer step (**7**→**8** in Figure 1.5) in the hydrogenation of β-keto esters catalyzed by the *(R)*-BINAP–Ru(II) complex [19, 20]. The protonated

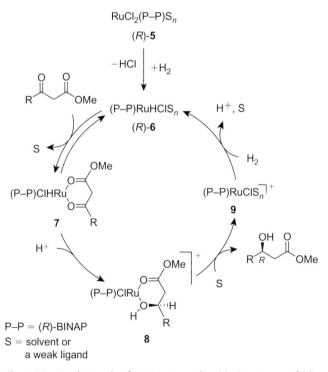

Figure 1.5 Catalytic cycle of BINAP–Ru catalyzed hydrogenation of β-keto esters.

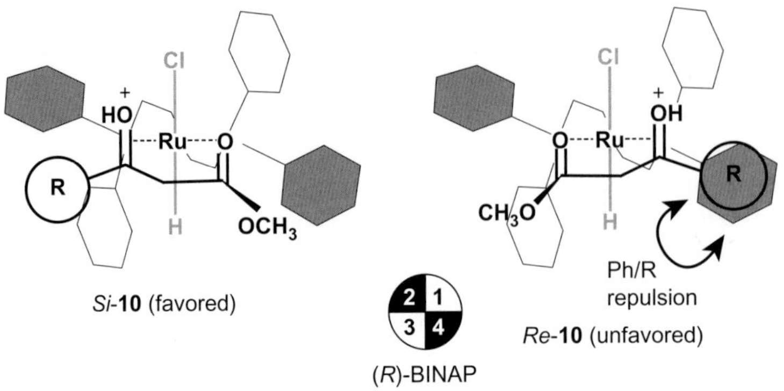

Figure 1.6 Molecular models of diastereomeric transition states in *(R)*-BINAP–Ru catalyzed hydrogenation of β-keto esters.

carbonyl moiety of substrate, C=O⁺H, coordinates parallel to the H–Ru bond in a π fashion. Thus, the "R" group connecting to the carbonyl is placed close to the *P*-phenyl group of the *(R)*-BINAP ligand. The TS *Re*-**10** producing the *(S)* alcohol is disfavored due to the notable R/Ph repulsion in the crowded region of the fourth quadrant. Therefore, the *(R)* product is predominantly obtained via the TS *Si*-**10**.

BINAP–Ru(II) catalyzed hydrogenation of functionalized ketones is used for the production of useful chiral compounds in the chemical industry. Scheme 1.4 shows an example for the synthesis of carbapenems, a class of β-lactam antibiotics (>100 tons per year) [1, 17]. The racemic α-substituted β-keto ester is hydrogenated in the presence of BINAP–Ru catalyst with a stereo-mutation at the α position [39]. The chiral structure of the BINAP complex as well as intramolecular asymmetric induction of the substrate efficiently control the contiguous two chiral centers of the product. Therefore, the (2*S*,3*R*) alcoholic product is obtained

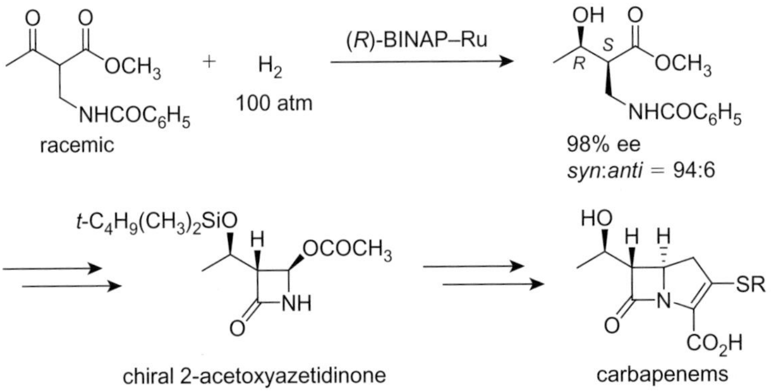

Scheme 1.4 Synthesis of a carbapenem intermediate by BINAP–Ru catalyzed hydrogenation of an α-substituted β-keto ester.

selectively among four possible stereoisomers through the dynamic kinetic resolution [40]. The use of DTBM-SEGPHOS–Ru(II) complex for this reaction affords exclusively the (2S,3R) product [8].

1.3.2
Simple Ketones

The BINAP–RuCl$_2$ catalyst hydrogenates functionalized ketones with high enantioselectivity through TSs stabilized by the substrate–Ru chelate structure (Figure 1.6). However, this catalyst is totally inert in hydrogenation of simple (unfunctionalized) ketones, because the chelate-stabilization in the TS is not available. High reactivity and enantioselectivity for this reaction is achieved by the use of a combined catalyst system of RuCl$_2$(xylbinap)(daipen) and alkaline base or RuH(η^1-BH$_4$)(xylbinap)(dpen) with or without a base (Scheme 1.5) [41, 42]. For example, hydrogenation of acetophenone with RuH(η^1-BH$_4$)[(S)-xylbinap][(S,S)-dpen] at an S/C of 100 000 (8 atm H$_2$, 45 °C) in t-C$_4$H$_9$OK (0.014 M, 400 equiv to Ru) containing 2-propanol is completed in 45 min to afford (R)-1-phenylethnol in 99% ee. The average turnover frequency (TOF$_{av}$ = TON min^{-1}) is about 2200. The ee

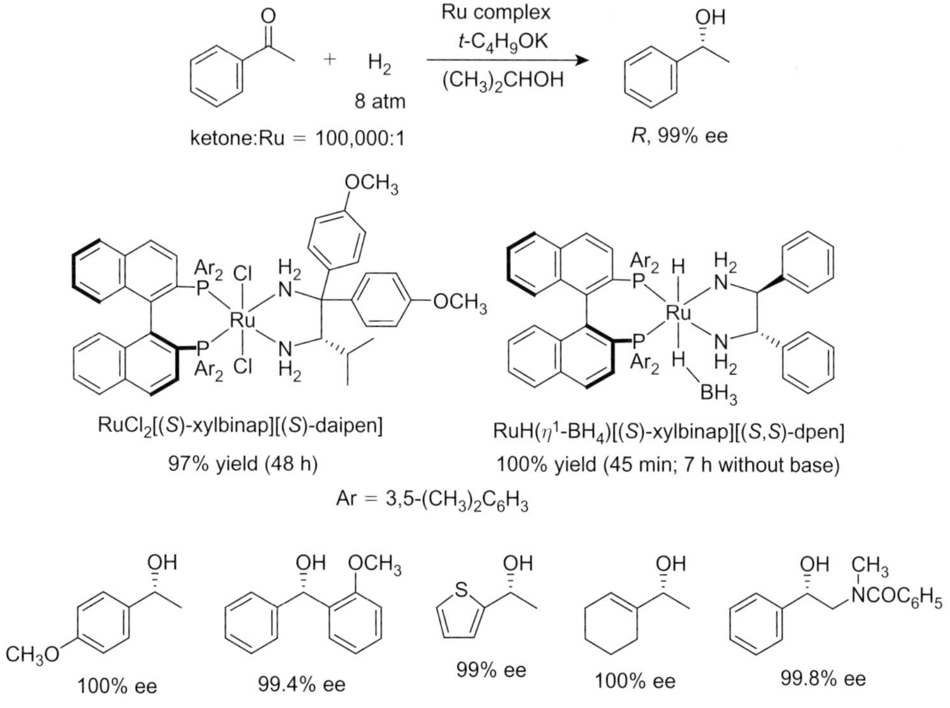

Scheme 1.5 Hydrogenation of simple ketones catalyzed by XylBINAP/1,2-dimaine–Ru complexes.

value is decreased to 80–82% when BINAP or TolBINAP is used instead of XylBINAP [43]. A series of simple ketones, including alkyl aryl ketones, unsymmetrical benzophenones, hetero-aromatic ketones, α,β-unsaturated ketones, α-amino ketones, and some aliphatic ketones, is hydrogenated in high enantioselectivity [44, 45]. The analogous catalyst systems using chiral biaryl diphosphines with P-3,5-xylyl groups exhibit similar efficiency [30, 46].

Figure 1.7 illustrates a proposed catalytic cycle for the hydrogenation of simple ketones with the (S)-TolBINAP/(S,S)-DPEN–Ru(II) complex [42, 47]. The precatalyst **11** (X, Y = Cl, Cl or η^1-BH$_4$, H) is converted into the cationic complex **12** in 2-propanol under reductive conditions with or without base. Species **12** and H$_2$ reversibly forms the molecular hydrogen complex **13**, followed by deprotonation to

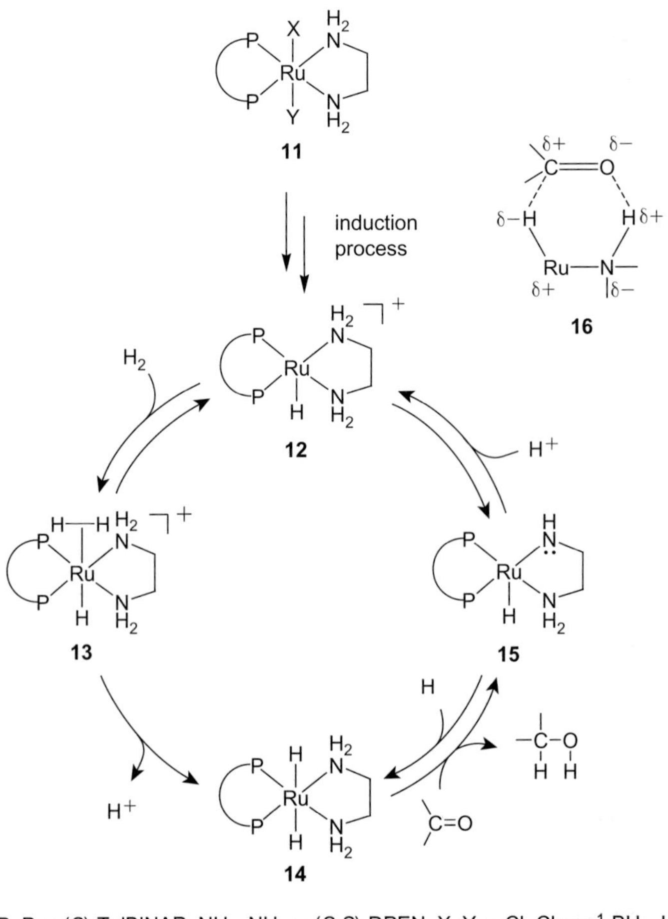

P–P = (S)-TolBINAP, NH$_2$–NH$_2$ = (S,S)-DPEN, X, Y = Cl, Cl or η^1-BH$_4$, H

Figure 1.7 Catalytic cycle of TolBINAP/DPEN–Ru catalyzed hydrogenation of simple ketones.

give the active RuH$_2$ complex **14**. A ketone is readily reduced by **14**, affording the alcoholic product and the 16-electron Ru–amide complex **15**. The cationic complex **12** is promptly regenerated by protonation of **15** in an alcoholic medium; **15** is partly converted into **14** by the addition of H$_2$. The reaction of RuH$_2$ species **14** and a ketonic substrate proceeds thorough the pericyclic six-membered TS **16**. The H$^{\delta-}$–Ru$^{\delta+}$–N$^{\delta-}$–H$^{\delta+}$ quadrupole on the catalyst and the C$^{\delta+}$=O$^{\delta-}$ dipole of the substrate effectively interact to reduce the activation energy. Therefore, presence of an "NH" moiety on the diamine ligand is crucial to achieve high catalytic activity.

The mode of enantioface-selection is explained by the use of TS molecular models shown in Figure 1.8 [42, 47]. These are completely different from the TSs in the hydrogenation of β-keto esters catalyzed by the BINAP–Ru(II) complex without diamine ligand (Figure 1.6). The (S)-TolBINAP/(S,S)-DPEN–RuH$_2$ complex **14** has a C_2-symmetric structure, in which the skewed five-membered DPEN chelate-ring appropriately arranges the amino protons. The axially directed proton, H$_{ax}$, is more reactive than the equatorial one (H$_{eq}$), because the H$^{\delta-}$–Ru$^{\delta+}$–N$^{\delta-}$–H$_{ax}$$^{\delta+}$ quadrupole forming a smaller dihedral angle preferentially interacts with the C$^{\delta+}$=O$^{\delta-}$ dipole of the ketone. Acetophenone approaches the reaction site

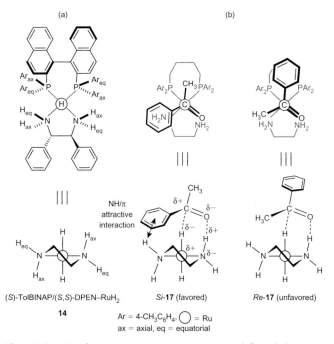

Figure 1.8 (S)-TolBINAP/(S,S)-DPEN–RuH$_2$ species (left) and diastereomeric transition states in the hydrogenation of acetophenone. (For clarity the equatorially oriented phenyl substituents in the DPEN ligands are omitted in the molecular models.)

with the Si-face (Si-**17**) or Re-face (Re-**17**) forming the pericyclic TS. The TS Re-**17** suffers significant nonbonded repulsive interaction between the phenyl ring of substrate and the aromatic groups of the (S)-TolBINAP. Therefore, the reaction selectively proceeds via the TS Si-**17** to give the (R) product. The secondary NH/π attractive interaction between the NH$_{eq}$ and the phenyl of acetophenone appears to further stabilize the TS Si-**17**. This interpretation is supported by the result that the catalyst with the sterically more demanding XylBINAP instead of TolBINAP exhibits higher enantioselectivity (Scheme 1.5).

The chiral environment of the BINAP/diamine–Ru(II) complexes is readily modified by changing the combination of the diphosphine and diamine ligands (Figure 1.9). For example, the (S)-TolBINAP/(R)-DMAPEN–Ru(II) catalyst hydrogenates phenylglyoxal diethyl acetal and (E)-chalcone to afford the corresponding alcohols in 96% and 97% ee [48]. The same products in only 36% and 45% ee, respectively, are obtained by the use of the (S)-XylBINAP/(S)-DAIPEN–Ru(II) catalyst, which gives the highest ee (99%) of product in the reaction of acetophenone (Scheme 1.5). The combination of (S)-TolBINAP and (R)-IPHAN, a chiral 1,4-diamine, achieves high reactivity and enantioselectivity in the hydrogenation of 1-tetralones, a class of cyclic aromatic ketones, and substituted 2-cyclohexenones [49]. With the use of TolBINAP/PICA–Ru(II) catalyst, sterically

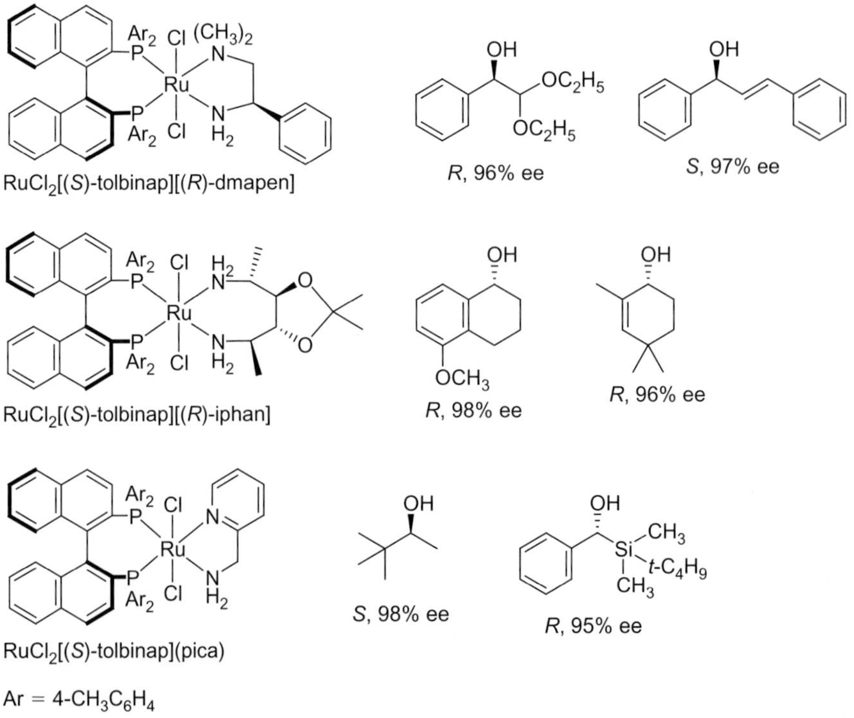

Figure 1.9 TolBINAP/amine ligand–Ru complexes and alcoholic products.

1.4 Isomerization of Allylamines and Allylalcohols

very crowded *tert*-alkyl ketones and acyl silanes are smoothly hydrogenated to quantitatively afford the chiral products in high ee [50]. Custom catalyst design notably expands the scope of substrates. This method is utilized in chemical industries on a productive scale [51, 52].

1.4
Isomerization of Allylamines and Allylalcohols

Asymmetric isomerization of allylamines is a simple and atom-economical reaction by which to obtain optically active enamines [53]. Scheme 1.6 shows a representative example for the BINAP–Rh(I) catalyzed isomerization of diethylgeranylamine to citronellal diethylenamine. [Rh{(S)-binap}(cod)]ClO$_4$ promotes the reaction in THF (60–120 °C) to afford the chiral enamine in 96–99% ee [54]. The TON approaches 8000 within 7 h. The catalyst can be recycled, so that the total chiral-multiplication number reaches 400 000. [Rh(binap)$_2$]ClO$_4$ acts at higher reaction temperature, but the robustness of this complex leads to its preferred practical use. More than 1500 tons of (−)-menthol are produced per year by this method [53]. Several allylamine substrates with different *N*-substituents, alkoxy alkyl groups, aromatic moieties, and cyclic skeletons are also converted into the desired compounds in high ee.

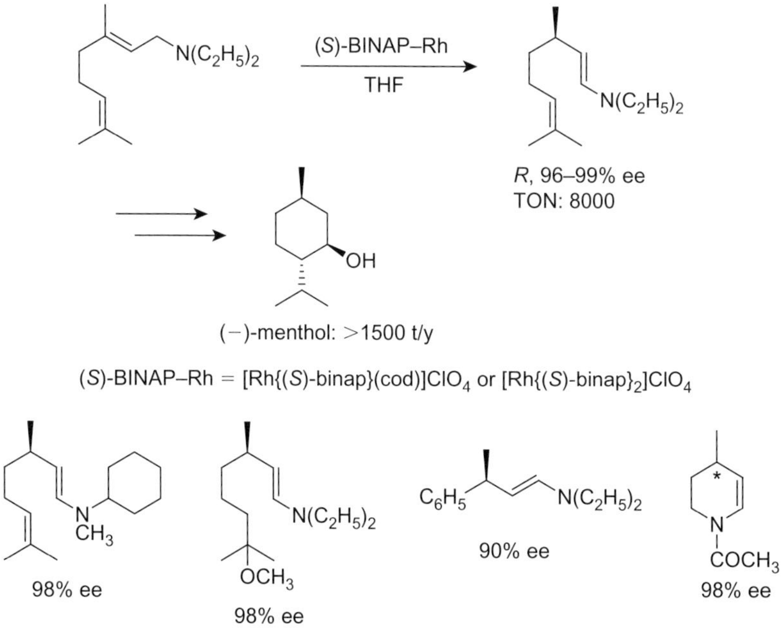

Scheme 1.6 Isomerization of allylamines catalyzed by BINAP–Rh complexes and the industrial application to the synthesis of (−)-menthol.

The BINAP–Rh(I) catalyzed isomerization of allylalcohols is less successful in terms of enantioselectivity. Nevertheless, some useful reactions are reported [53]. When racemic 4-hydroxy-2-cyclopentenone is subjected to isomerization conditions in the presence of [Rh{(R)-binap}(CH$_3$OH)$_2$]ClO$_4$ (S/C = 200, 0 °C, 14 days), the (R)-hydroxy enone in 91% ee is recovered at 72% conversion (Scheme 1.7) [55]. The relative reaction rate between the two enantiomers (k_f/k_s) is calculated to be 5. The (R) product can be optically purified by recrystallization after conversion into the *tert*-butyldimethylsilyl ether. Desymmetrization of meso allylic 1,4-enediol di-triethylsilyl ether by enantioselective isomerization is achieved by the use of [Rh{(S)-binap}(cod)]ClO$_4$ (S/C = 50, CH$_2$ClCH$_2$Cl reflux, 16 h) [56]. The desired (R)-mono-silyloxy-ketone is obtained in 97.5% ee and 85.8% yield.

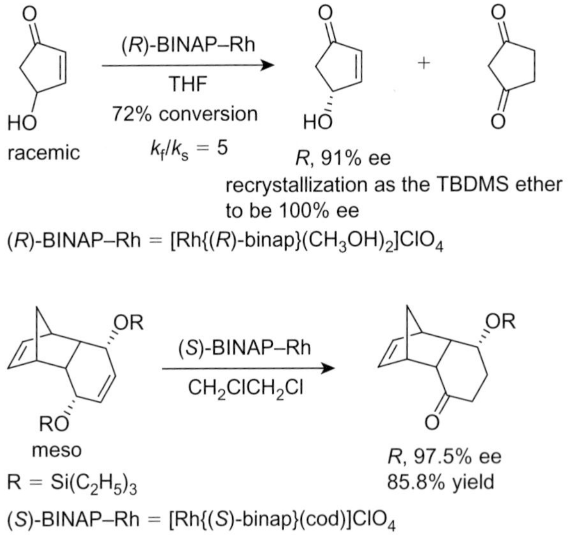

Scheme 1.7 Isomerization of allylic alcohol and allylic ether catalyzed by BINAP–Rh complexes.

1.5
Hydroboration, Hydrosilylation, Hydroacylation, and Hydroamination

Enantioselective hydroboration of olefins is a reliable procedure for the synthesis of chiral alcohols in combination with an oxidation procedure [57]. Styrene and catechol borate (1.1 equiv) react in DME with an *in situ* prepared catalyst from [Rh(cod)$_2$]BF$_4$ and (R)-BINAP (S/C = 50, −78 °C, 2 h), followed by a treatment with H$_2$O$_2$ under basic conditions, to give (R)-1-phenylethanol in 96% ee and 91% yield (Scheme 1.8) [58]. The regioisomeric 2-phenylethanol, which is a major isomer in the reaction without a catalyst, is not detected. Aromatic substituted

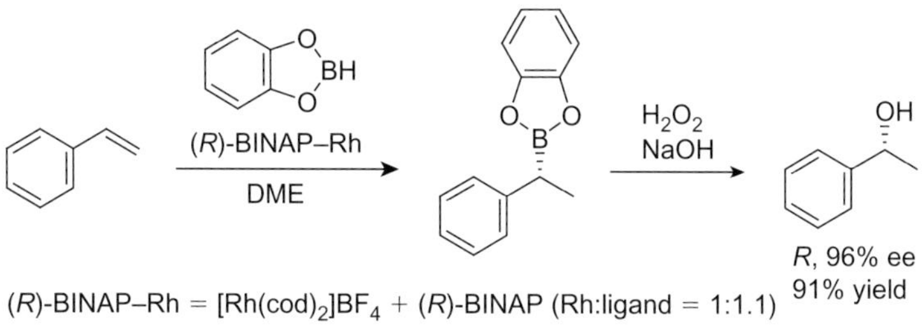

1.5 Hydroboration, Hydrosilylation, Hydroacylation, and Hydroamination

Scheme 1.8 Hydroboration of styrenes with a BINAP–Rh catalyst.

styrenes also react with high stereoselectivity. The reactivity and enantioselectivity are decreased in the hydroboration of α- or β-substituted styrenes.

The BINAP–Rh complex efficiently catalyzes asymmetric intramolecular hydrosilylation of allylic alcohol silylethers [59]. As shown in Scheme 1.9, a cinnamyl alcohol silylether (**18**) is converted into the cyclic silylether **19** in acetone with [Rh{(S)-binap}(nbd)]ClO$_4$ (S/C = 50, 25 °C, 5 min) [60]. The competitive intermolecular reaction is observed at less than 1%. Treatment of **19** under oxidative conditions afforded *(R)*-1-phenyl-1,3-propanediol in 97% ee and 75% yield.

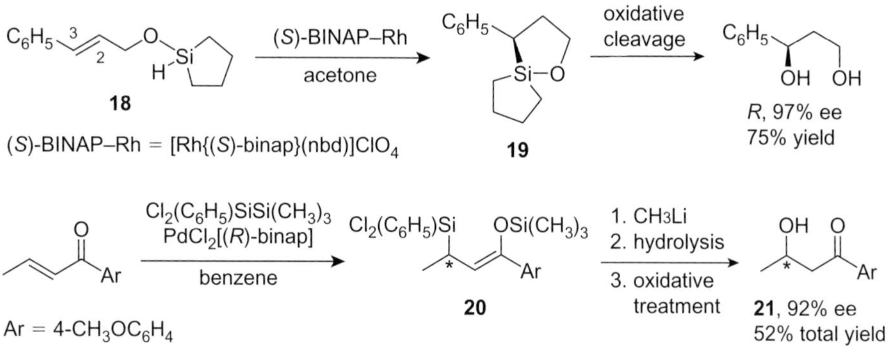

Scheme 1.9 Enantioselective hydrosilylation of olefins and disilylation of enones.

The cyclic silane structure of **18** is crucial for efficient enantioselection. Several 2- and 3-substituted compounds are converted into the desired products in high ee. Enantioselective 1,4-disilylation of 4′-methoxybenzalacetone using Cl$_2$(C$_6$H$_5$)SiSi(CH$_3$)$_3$ (1.5–2 equiv) as a reagent catalyzed by PdCl$_2$[(R)-binap] (S/C = 200, 80 °C, 0.5 h) gives the silylation compound **20**, which is converted into the β-hydroxy ketone **21** in 92% ee through treatment with CH$_3$Li, acidic hydrolysis, and oxidative cleavage of the C–Si bond [61]. The symmetric disilane reagents do not react with the enone. Aliphatic and aromatic enones react with good to high enantioselectivity.

Enantioselective hydrosilylation of α,β-unsaturated ketones catalyzed by Cu(I) complexes with TolBINAP and DTBM-SEGPHOS affords the chiral β-substituted

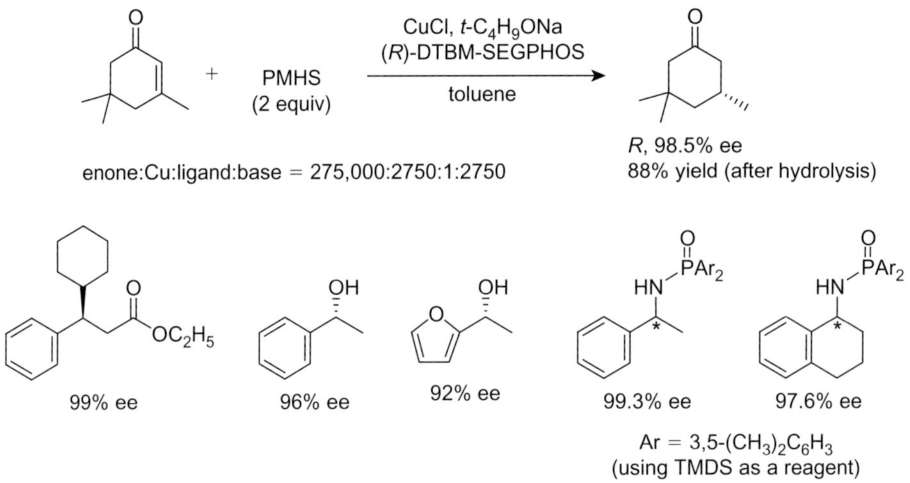

Scheme 1.10 Asymmetric hydrosilylation of enones, ketones, and imines with DTBM-SEGPHOS–Cu catalysts.

ketones in high ee. Scheme 1.10 illustrates an example using the DTBM-SEGPHOS–Cu(I) system [62]. The safe and inexpensive polymethylhydrosiloxane (PMHS) is used as a reducing agent. The reaction of isophorone is carried out with an *in situ* prepared catalyst from CuCl (S/C = 100), the *(R)* ligand (S/ligand = 275 000), and *t*-C$_4$H$_9$OK in toluene (−35 °C, 3 days) to give the *(R)* ketone in 98.5% ee and 88% yield after hydrolysis. The chiral multiplicity is remarkably high. In a similar manner, α,β-unsaturated esters [63], aromatic ketones [64], and aromatic imines [65] are also reduced with high levels of enantioselection. The use of tetramethyldisiloxane (TMDS) under modified conditions gives better stereoselectivity for the reaction of imines. Hydrosilylation of ketones with an air-stable BINAP–CuF$_2$ catalyst system is also reported [66].

Intramolecular hydroacylation of 4-*tert*-butyl-4-pentenal catalyzed by the cationic *(S)*-BINAP–Rh(I) complex (S/C = 20–25) in CH$_2$Cl$_2$ (25 °C, 2–4 h) gives *(S)*-3-*tert*-butylpentanone in >99% ee in high yield (Scheme 1.11) [67]. Desymmetrization of a meso 3,3-dialkynylpropanal **22** by hydroacylation with the TolBINAP–Rh(I) catalyst (S/C = 10, 10 °C, 2 h) gives the 2-cyclopentenone with a quaternary carbon center at the 4 position of **23** in 92% ee [68]. A keto aldehyde **24** is quantitatively converted into the seven-membered lactonic compound *(S)*-**25** in >99% ee by the DTBM-SEGPHOS–Rh(I) catalyzed hydroacylation (S/C = 20, room temperature, 2 days) [69]. A series of substrates with aliphatic and aromatic moieties reacts with high enantioselectivity.

A catalyst system consisting of [IrCl{*(S)*-binap}]$_2$ and a fluoride source (F/C = 4) effects the asymmetric reaction of norbornene and aniline without solvent (S/C = 50, 75 °C, 72 h), leading to the hydroamination product *(R)*-**26** in 95% ee and 22% yield (Scheme 1.12) [70]. Fluoride appears to act as a π-donating anionic ligand accelerating the N–H oxidative addition to Ir. Intramolecular

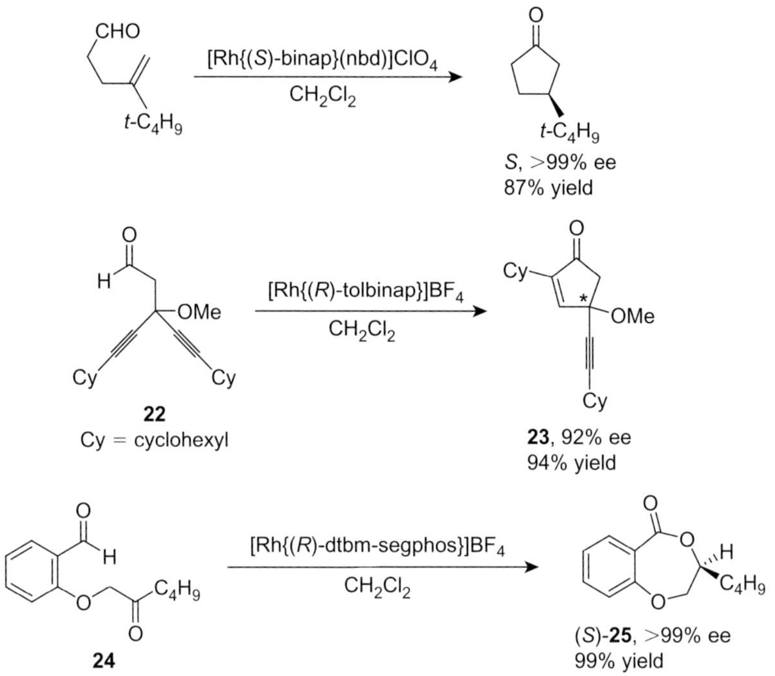

Scheme 1.11 Chiral diphosphine–Rh catalyzed intramolecular hydroacylation.

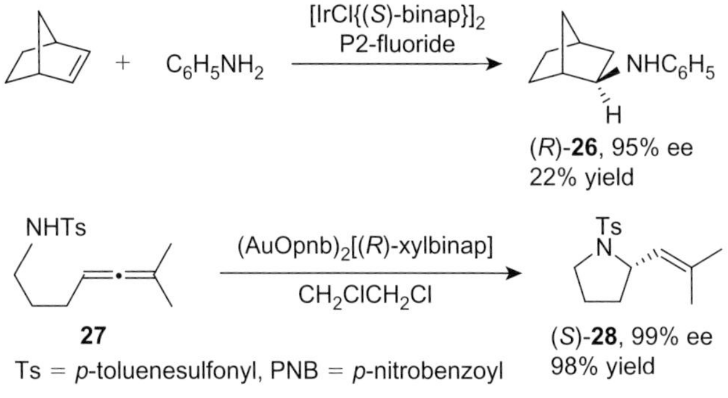

Scheme 1.12 Enantioselective hydroamination of olefins and allenes.

hydroamination of an *N*-tosyl allenic compound (**27**) is carried out in CH_2ClCH_2Cl with $(AuOpnb)_2[(R)\text{-xylbinap}]$, a unique bimetallic complex (S/C = 33, 23 °C, 15 h), to quantitatively yield the cyclic amine product *(S)*-**28** in 99% ee [71]. Choice of *p*-nitrobenzoate as an anionic ligand is crucial to achieve a high yield and ee of the product. A range of substrates with linear and cyclic alkyl substituents at the allenic moiety gives the desired products in equally high ee. The BINAP–Pd catalyzed reaction of styrenes and anilines has also been reported [72].

1.6
Allylic Alkylation

Allylic alkylation of 3-acetoxy-1,3-diphenyl-1-propene with the sodium salt of dimethyl 2-acetamidomalonate (**29a**) in the presence of a catalyst formed *in situ* from [Pd(η^3-C$_3$H$_5$)Cl]$_2$ and *(S)*-BINAP (S/C = 100, 25 °C, 120 h) affords the *(S)* adduct *(S)*-**30a** in 95% ee and 92% yield (Scheme 1.13) [73, 74]. When this reaction is conducted with the sodium salt of simple dimethyl malonate **29b** the enantioselectivity is decreased drastically. This problem is solved by using Zn(C$_2$H$_5$)$_2$ as a base to generate the malonate [75]. The desired product **30b** is obtained in 99% ee (S/C = 25, room temperature, 20 h). The reaction of a prochiral nucleophile prepared from an α-acetamide-β-keto ester **31** and *t*-C$_4$H$_9$OK with cinnamyl acetate in the presence of the *(R)*-BINAP–Pd(II) complex (S/C = 50, −30 °C, 48 h) gives the *(R)* adduct *(R)*-**32** in 95% ee [76]. The bulky γ substituent of the allylic acetate (C$_6$H$_5$ in this case) is crucial to achieve high enantioselectivity.

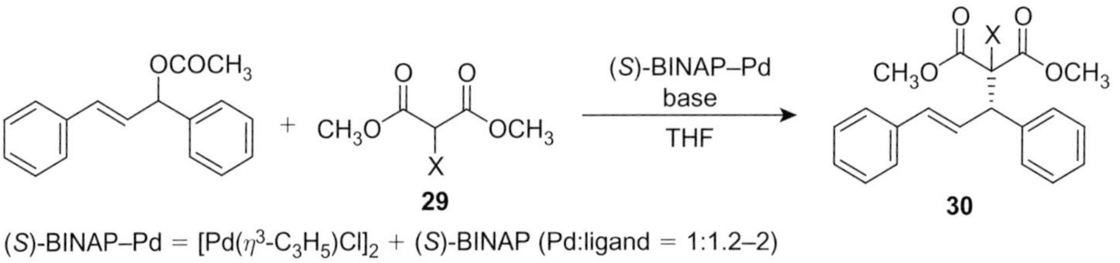

29, X	base	% ee	% yield
a, NHCOCH$_3$	NaH	95	92
b, H	Zn(C$_2$H$_5$)$_2$	99	84

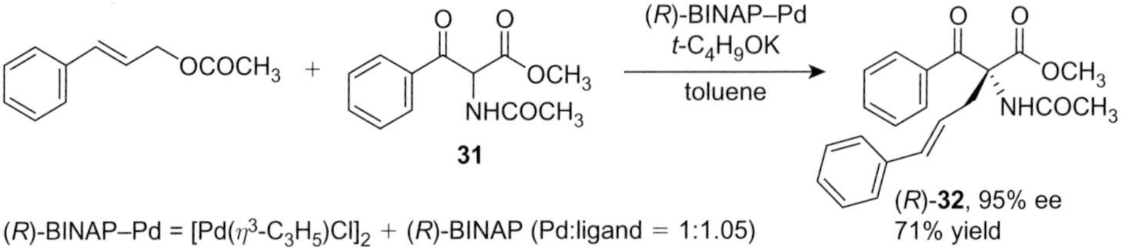

Scheme 1.13 Enantioselective allylic substitution with BINAP–Pd catalyst.

1.7
Heck Reaction

1.7.1
Intramolecular Reaction

Asymmetric Heck-type cyclization of various vinyl halides and enol triflates is catalyzed by BINAP–Pd complexes [77]. Synthetically useful chiral cyclic compounds with functionalities are produced by this reaction. For example, a prochiral enol triflate **33** is cyclized with Pd(OCOCH$_3$)$_2$, *(R)*-BINAP, and K$_2$CO$_3$ in toluene (S/C = 20, 60 °C, 27 h) to afford (S,S)-**34** in 91% ee and 60% yield (Scheme 1.14) [78]. Spirooxindols **36** are prepared from the aryl iodide **35** in up to 95% ee [79]. Both enantiomers of **36** can be selectively synthesized from the single enantiomer of the catalyst. Thus, **35** reacts with the *(R)*-BINAP–Pd catalyst in the presence of Ag$_3$PO$_4$ (2 equiv) in N-methylpyrrolidine (NMP) (S/C = 10, 60 °C, 25 h) to afford (S)-**36** in 80% ee. When the reaction is conducted with the *(R)*-BINAP–Pd catalyst and 1,2,2,6,6-pentamethylpiperidine (PMP; 5 equiv) in N,N-dimethylacetamide (DMA) (S/C = 10, 100 °C, 1.5 h), the *(R)* product is selectively obtained. These results suggest that the reaction mechanism with a cationic Pd catalyst is different from that catalyzed by the neutral species.

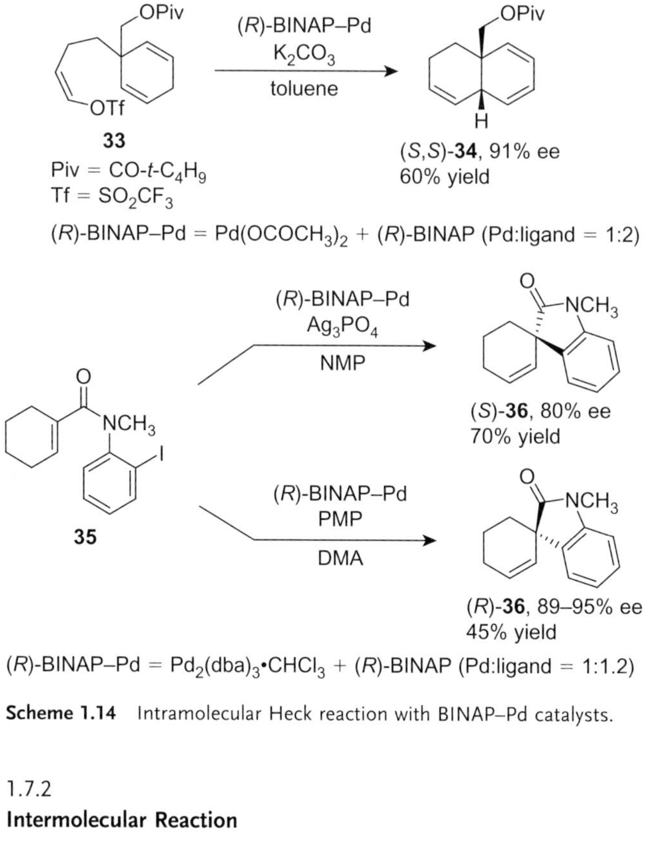

Scheme 1.14 Intramolecular Heck reaction with BINAP–Pd catalysts.

1.7.2
Intermolecular Reaction

The BINAP–Pd catalyst exhibits high enantioselectivity in the intermolecular Heck reaction [77a, d]. When phenyl triflate reacts with 2,3-dihydrofuran (5 equiv) in the presence of an *in situ* formed *(R)*-BINAP–Pd complex and 1,8-bis(dimethylamino)naphthalene (proton sponge; 3 equiv) in benzene (S/C = 33, 40 °C,

9 days), *(R)*-2-phenyl-2,3-dihydrofuran [*(R)*-**37**] in >96% ee and *(S)*-2-phenyl-2,5-dihydrofuran [*(S)*-**38**] in 17% ee are obtained in a 71 : 29 ratio (Scheme 1.15) [80]. The choice of base is important to attain high enantioselectivity. The high basicity and bulkiness of the proton sponge appears to fit with the reaction. Several substituted phenyl triflates and 2-naphthyl triflates have been used successfully under the optimized conditions. When the reaction of phenyl triflate and 2,3-dihydrofuran is carried out with the DTB-MeO-BIPHEP–Pd catalyst (DTB; see Figure 1.2) using $N(C_2H_5)(i\text{-}C_3H_7)_2$ as a base, **37** is obtained in 99% ee and 70% yield [81]. The BITIANP–Pd catalyzed reaction gives exclusively **37** in 91% ee [82].

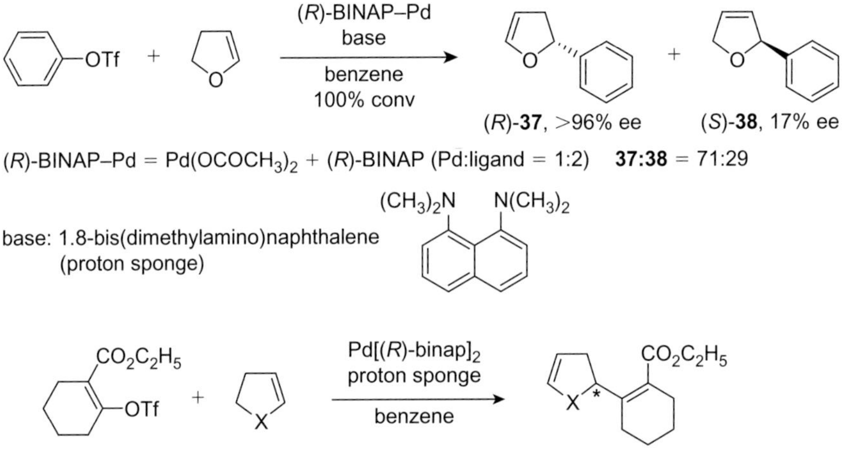

Scheme 1.15 Intermolecular Heck-type arylation and alkenylation with BINAP–Pd catalysts.

The Heck-type alkenylation of olefins is also catalyzed by the BINAP–Pd complex. The reaction of 2-ethoxycarbonyl-1-cyclohexenyl triflate (**39**) and 2,3-dihydrofuran (4 equiv) in the presence of Pd(binap)$_2$ (S/C = 33) and proton sponge (2 equiv) in benzene (40 °C, 56 h) affords the chiral product **40** (X = O) in >96% ee [83]. The use of preformed complex Pd(binap)$_2$ restrains formation of the regioisomeric products. When the reaction is conducted using 1-methoxy-carbonyl-2-pyrroline instead of dihydrofuran (S/C = 33, 60 °C, 20 h), the desired product **40** (X = NCO$_2$CH$_3$) is obtained in >99% ee and 95% yield. The BITIANP–Pd catalyst exhibits similar efficiency [84].

1.8
Aldol and Mannich-Type Reactions

1.8.1
Aldol Reaction

The asymmetric aldol reaction is a reliable method to produce synthetically useful chiral β-hydroxy carbonyl compounds [85]. Benzaldehyde and a silyl enolate of acetophenone (**41**) (1.5 equiv) react with [Pd{(*R*)-binap}(H$_2$O)$_2$](BF$_4$)$_2$ (S/C = 20) in 1,1,3,3-tetramethylurea (TMU) (0 °C, 24 h) to afford *(R)*-3-hydroxy-1,3-diphenyl-1-propanone [*(R)*-**42**] in 89% ee and 92% yield (Scheme 1.16) [86]. The structure of the BINAP–Pd(II) complex has been determined by a X-ray crystallographic analysis. A chelating acyl Pt(II) complex Pt(3,5-dtbs)[*(R)*-binap] that can be handled in the open air is activated as a catalyst by treatment with *p*-toluenesulfonic acid (acid : Pt = 1 : 1) [87]. The BINAP–Pt species catalyzes the aldol reaction of 3-phenylpropanal and a ketene silyl acetal (**43**) (1.5 equiv) with 2,6-lutidine (amine : Pt = 1 : 1) in CH$_2$Cl$_2$ (S/C = 20, −25 °C, 168 h) to give the chiral product **44** in 95% ee. Primary alkyl aldehydes react with high enantioselectivity. Aromatic

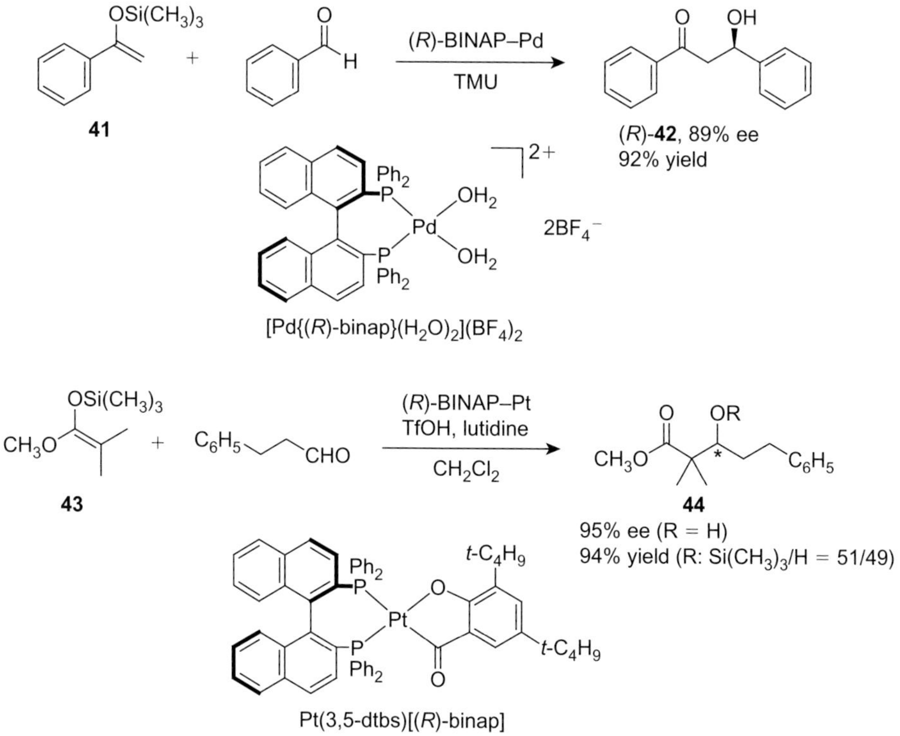

Scheme 1.16 Asymmetric aldol reaction catalyzed by BINAP–Pd and BINAP–Pt complexes.

aldehydes show higher reactivity, but the ee of the products is significantly lower. A Pt cationic complex appears to be the active species.

An *in situ* prepared complex from AgF and (*R*)-BINAP in a 1 : 1 ratio catalyzes the asymmetric aldol reaction of trimethoxysilyl enolates and aromatic aldehydes (Scheme 1.17) [88]. When the silyl enolate of *tert*-butyl ethenyl ketone **45** reacts with benzaldehyde in the presence of the BINAP–Ag catalyst in CH_3OH (S/C = 10, −78 to −20 °C, 6 h), the aldol product **46** is obtained in 97% ee (*syn* : *anti* = >99 : 1) [89]. Regardless of (*E*/*Z*) stereochemistry of the enolate the *syn* diastereo isomer is preferably formed. The reactivity is significantly decreased in the reaction with aliphatic aldehydes. The related catalyst prepared from AgOTf and BINAP promotes the aldol reaction of aldehydes with tributyltin enolates [90]. An enantioselective nitroso aldol reaction has been achieved by the use of the BINAP–Ag chemistry. The reaction of the trimethyltin enolate of cyclohexanone (**47**) and nitrosobenzene (1 equiv) with Ag(OTf)[(*R*)-tolbinap] (S/C = 10) in THF (−78 °C, 2 h) gives exclusively the (*R*) *O*-adduct (*R*)-**48** in 99% ee [91]. In contrast, the tributyltin enolate and nitrosobenzene react in the presence of a bimetallic complex $[Ag(OTf)]_2$[(*R*)-tolbinap] (S/C = 25) in ethylene glycol diethyl ether (−78 °C, 2 h) to afford selectively the *N*-adduct **49** in >99% ee. Thus, a series of *N*- and *O*-nitroso aldol products is synthesized in high chemo- and enantioselectivity.

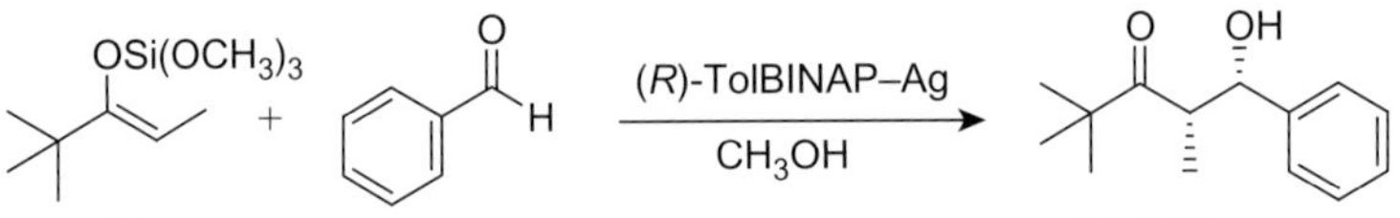

45 **46**, 97% ee
syn:*anti* = >99:1
(*R*)-TolBINAP–Ag = AgF + (*R*)-TolBINAP (Ag:ligand = 1:1) 84% yield

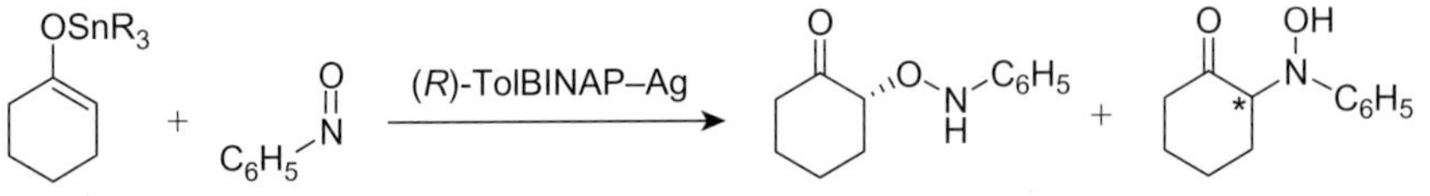

47 (*R*)-**48** **49**

R	(*R*)-TolBINAP–Ag	solvent	% yield	**48:49**	% ee
CH_3	Ag(OTf)[(*R*)-tolbinap]	THF	88	>99:1	99 (**48**)
n-C_4H_9	$[Ag(OTf)]_2$[(*R*)-tolbinap]	$C_2H_5O(CH_2)_2OC_2H_5$	95	4:96	>99 (**49**)

Scheme 1.17 BINAP–Ag catalyzed aldol reactions.

Asymmetric addition of silyl dienolates to aldehydes can be performed with the BINAP–Cu catalyst. For example, thiophene-2-carbaldehyde reacts with **50** catalyzed by a species formed *in situ* from $Cu(OTf)_2$, (*S*)-TolBINAP, and $[(n\text{-}C_4H_9)_4N](C_6H_5)_2SiF_2$ (TBAT) (S/C = 50) in THF (−78 °C, 6–8 h) to afford the (*R*)-adduct (*R*)-**51** in 95% ee almost quantitatively (Scheme 1.18) [92]. A range of aromatic and α,β-unsaturated aldehydes reacts with high enantioselectivity. The

Cu(I) dienolate prepared in this system appears to be a catalytic cycle species. Usually, the aldol reaction of ketones is difficult because of their lower reactivity than that of aldehydes. However, the reaction of methyl primary alkyl ketones and a silyl dienolate (**52**) is successfully catalyzed by a ternary system consisting of $Cu(OTf)_2$, TolBINAP, and TBAT (S/C = 10) at ambient temperature to give the cyclized product **53** in 90% ee and 81% yield [93]. Aryl methyl ketones also react with medium to high enantioselectivity.

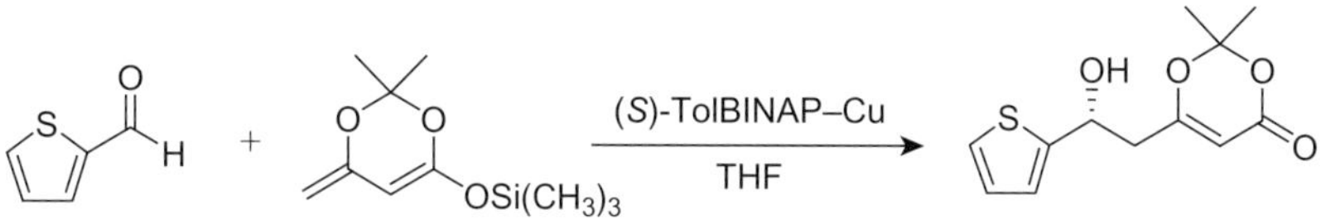

50 (*R*)-**51**, 95% ee 98% yield

(*S*)-TolBINAP–Cu = $Cu(OTf)_2$ + (*S*)-TolBINAP + TBAT (Cu:ligand:TBAT = 1:1:2)
TBAT = $[(n\text{-}C_4H_9)_4N](C_6H_5)_3SiF_2$

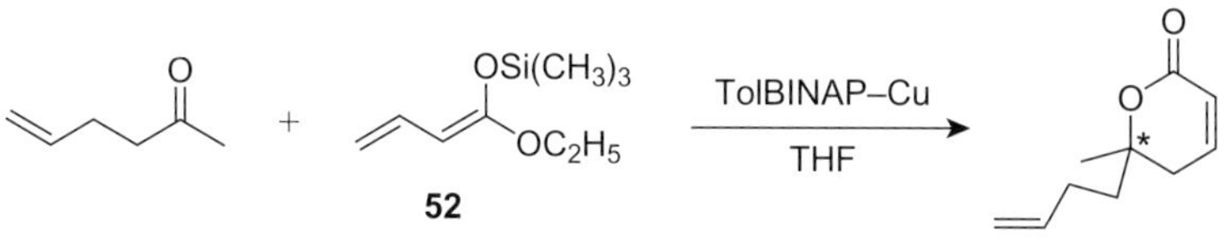

53, 90% ee
81% yield

TolBINAP–Cu = $Cu(OTf)_2$ + TolBINAP + TBAT (Cu:ligand:TBAT = 1:1.1:2)

Scheme 1.18 BINAP–Cu catalyzed aldol reactions with silyl dienolates.

1.8.2 Mannich-Type Reaction

The Mannich-type reaction of a β-keto ester **54** and *N*-Boc-protected imine of benzaldehyde **55** is catalyzed by $[Pd\{(R)\text{-binap}\}(H_2O)_2](OTf)_2$ (S/C = 20) in THF (0 °C, 5 h) to afford the adduct (*R*,*S*)-**56** (*syn* : *anti* = 88 : 12) in 99% ee and 93% yield (Scheme 1.19) [94]. The reaction appears to proceed through the cationic Pd enolate **57**, illustrated in the scheme, with release of H_2O and TfOH from the precatalyst. The bulky *tert*-butyl group of the enolate preferably locates in the first quadrant to avoid the crowded second quadrant with the equatorial *P*-phenyl ring of *(R)*-BINAP (see side view); the *tert*-butyl group then shields the *Si*-face of the enolate. Therefore, the enolate predominantly reacts with the electrophile at the *Re*-face to give the desired enantiomer of product. The chiral structure of BINAP indirectly controls the approach of the electrophile. The *in situ* formed protic acid, TfOH, is presumed to activate the imino substrate. A series of aromatic and α,β-unsaturated imines as well as α-imino esters reacts with high enantioselectivity.

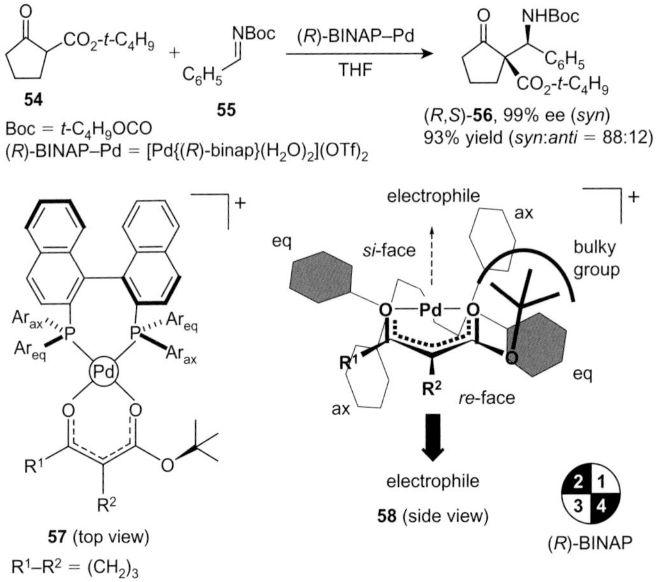

Scheme 1.19 Mannich-type reaction catalyzed by BINAP–Pd complex, and molecular models of the Pd-enolate intermediate.

1.9
Nucleophilic Additions to Carbonyl and Imino Compounds

1.9.1
Allylation

Enantioselective addition of allyltrimethoxysilane (**58**) to aldehydes is performed by the use of BINAP–Ag catalysts (Scheme 1.20) [88, 95]. When benzaldehyde reacts with **58** (1.5 equiv) in the presence of a complex formed *in situ* from AgF and *(R)*-TolBINAP (S/C = 33) in CH_3OH (−20 °C, 4 h), the homoallylic alcohol *(R)*-**59** is obtained in 94% ee and 80% yield [96]. Several aromatic and α,β-unsaturated aldehydes react with good to high enantioselectivity. The allylation of ketones is catalyzed by the chiral Ag complex. Thus, 2-chloro-2-cyclohexenone and **58** (2 equiv) react with the *(R)*-Difluorphos–AgF catalyst (S/C = 20) and CH_3OH (1 equiv) in THF (−78 °C, 12 h) to produce predominantly the 1,2-addition product **60** in 96% ee and 97% yield [97]. This reaction does not proceed in CH_3OH solution, while a stoichiometric amount of CH_3OH in THF improves the product yield. This is because the protonation of a Ag-alkoxide intermediate with CH_3OH occurs smoothly. The reaction of cyclic and acyclic aromatic ketones also proceeds with high enantioselectivity.

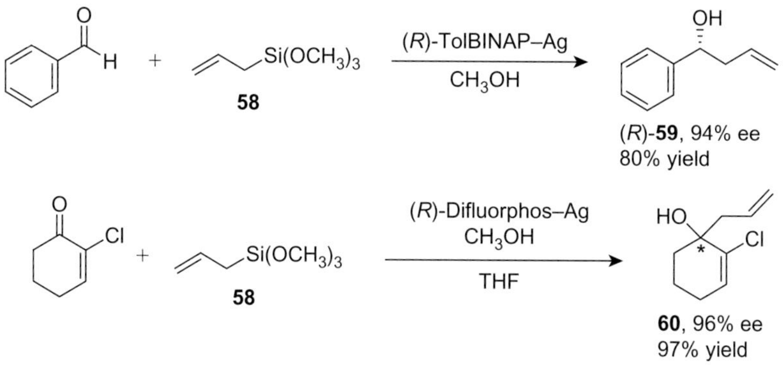

Scheme 1.20 Addition of allyltrimethoxysilane to aldehydes and ketones with chiral Ag catalysts.

1.9.2
Alkenylation and Arylation

A chiral catalyst generated *in situ* from CuF_2 and *(R)*-DTBM-SEGPHOS (ligand and also reductant) effects asymmetric vinylation and arylation to aldehydes (Scheme 1.21). The reaction of 4-chlorobenzaldehyde and trimethoxyvinylsilane (**61a**; R = CH_2=CH, X = OCH_3; 2 equiv) with the *(R)*-DTBM-SEGPHOS–Cu

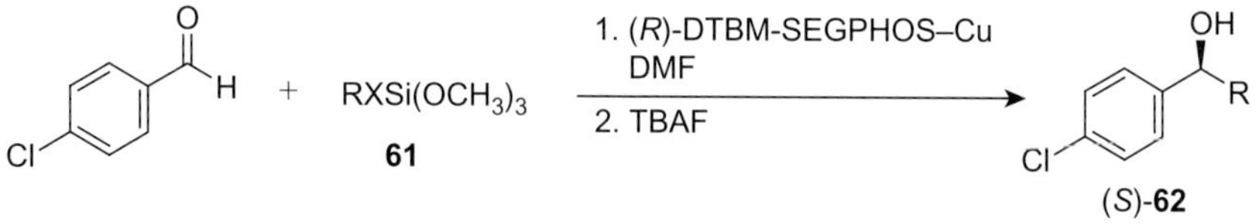

61	R	X	% yield	% ee
a	CH_2=CH	OCH_3	99	97
b	C_6H_5	C_6H_5	81	92

TBAF = tetrabutylammonium fluoride
(R)-DTBM-SEGPHOS–Cu = $CuF_2 \cdot 2H_2O$ + *(R)*-DTBM-SEGPHOS (Cu:ligand = 1:2)

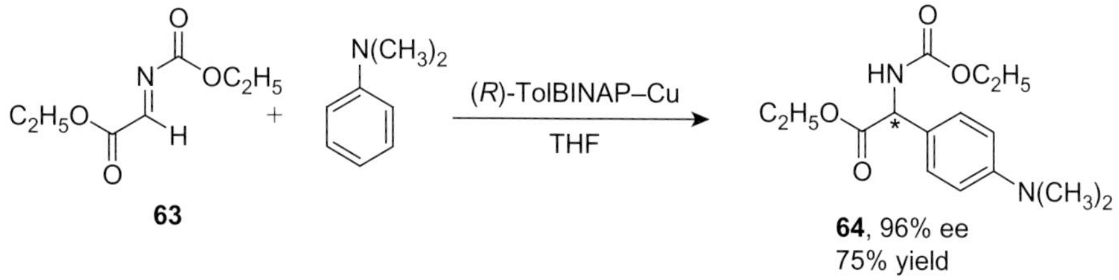

(R)-TolBINAP–Cu = $CuPF_6 \cdot 4CH_3CN$ + *(R)*-TolBINAP (Cu:ligand = 1:1.1)

Scheme 1.21 Vinylation and arylation to aldehydes and α-imino esters with chiral Cu catalysts.

catalyst (S/C = 33) in DMF (40 °C, 2 h) affords quantitatively the allylic alcohol *(S)*-**62a** (R = CH$_2$=CH) in 97% ee [98]. Several aromatic, aliphatic, and α,β-unsaturated aldehydes react with good to excellent levels of enantioselectivity. When the reaction is carried out using dimethoxydiphenylsilane (**61b**; R = X = C$_6$H$_5$) (S/C = 33, 40 °C, 1 h), the diaryl methanol *(S)*-**62b** (R = C$_6$H$_5$) is obtained in 92% ee. The asymmetric Friedel–Crafts type arylation of imines is catalyzed by a complex prepared *in situ* from CuPF$_6$ and *(R)*-TolBINAP. An N-protected α-imino ester (**63**) reacts with N,N-dimethylaniline in the presence of the *(R)*-TolBINAP–Cu catalyst (S/C = 20) in THF at −78 °C to afford the aromatic α-amino-acid derivative **64** in 96% ee [99]. Only the *para*-substituted compound is detected. A series of aniline analogues can be converted into the desired products in high regio- and enantioselectivity.

1.9.3
Dienylation

The carbonyl dienylation using acetylene and hydrogen gas is achieved with a cationic Rh catalyst (Scheme 1.22) [100]. Thus, an α-substituted aldehyde **65**, acetylene (1 atm), and H$_2$ (1 atm) react with a complex formed *in situ* from [Rh(cod)$_2$]BARF (BARF = B[3,5-(CF$_3$)$_2$C$_6$H$_3$]$_4$) and *(R)*-MeO-BIPHEP (S/C = 20) in the presence of triphenylacetic acid (7.5 mol%) and Na$_2$SO$_4$ (2 equiv) in CH$_2$ClCH$_2$Cl (25 °C, 72 h) to exclusively afford the *(Z)*-dienylation product **66** in 88% ee and 85% yield [101]. This reaction is suggested to proceed through the carbonyl insertion of a rhodacyclopentadiene that is formed by acetylene dimerization. When 6-bromopyridine-2-carboxaldehyde (**67**) reacts with an enyne **68** (2 equiv) and H$_2$ (1 atm) in the presence of the *(R)*-TolBINAP–[Rh(cod)$_2$]OTf complex (S/C = 25) and triphenylacetic acid (2 mol%) at 40 °C, the *(E)*-dienylation product **69** is obtained in 99% ee [102]. Many heterocyclic aromatic aldehydes as well as ketones are converted into the desired products in high ee.

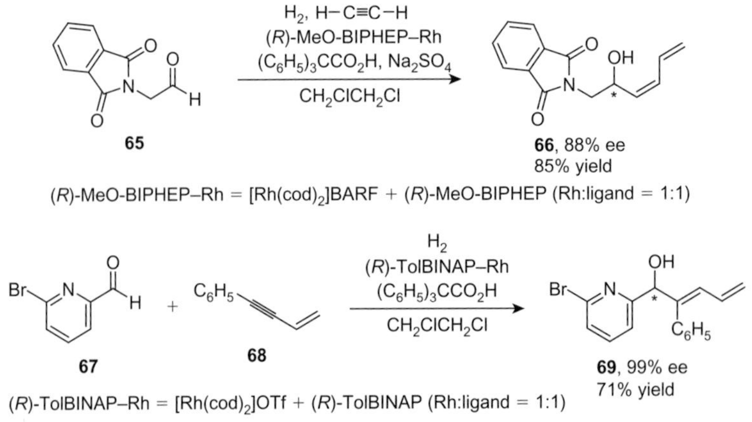

Scheme 1.22 Dienylation of aldehydes catalyzed by chiral Rh complexes.

1.9.4
Cyanation

The combined catalyst system of Ru[(S)-phgly]$_2$[(S)-binap] (phgly = phenylglycinate) and Li$_2$CO$_3$ in a 1 : 1 ratio effects asymmetric cyanosilylation of aldehydes (Scheme 1.23). The reaction of benzaldehyde and (CH$_3$)$_3$SiCN (1.2 equiv) with the Ru[(S)-phgly]$_2$[(S)-binap]–Li$_2$CO$_3$ system (S/C = 10 000) in diethyl ether (−78 °C, 12 h) affords the cyanohydrin product *(R)-70* quantitatively with 97% ee [103]. The Ru complex alone does not show substantial reactivity. The combined system can complete the reaction with an S/C of 100 000 at −40 °C to give **70** in 90% ee. A series of aromatic, heteroaromatic, and α,β-unsaturated aldehydes reacts with high enantioselectivity. Methyl benzoylformate reacts with (CH$_3$)$_3$SiCN (2 equiv) catalyzed by the Ru[(S)-phgly]$_2$[(S)-binap]–C$_6$H$_5$OLi system (S/C = 1000) in *t*-C$_4$H$_9$OCH$_3$ (−60 °C, 18 h) to give the adduct *(R)-71* quantitatively with 99% ee [104]. The reaction is completed with an S/C of 10 000 at −50 °C to afford the product in 98% ee. A range of aromatic, aliphatic, and α,β-unsaturated keto esters can be converted into the silylated products with high enantioselectivity.

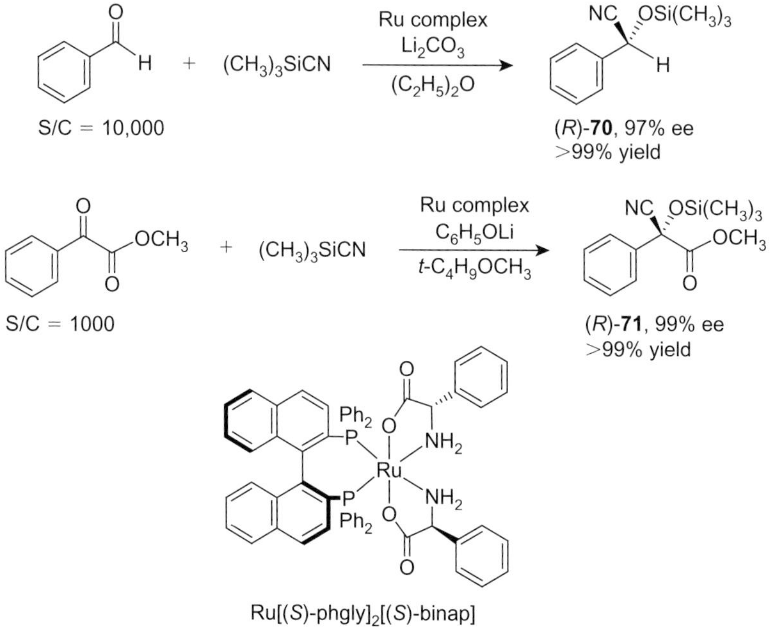

Scheme 1.23 Cyanosilylation of aldehydes and α-keto esters with a chiral Ru·Li catalyst.

1.10
α-Substitution Reactions of Carbonyl Compounds

1.10.1
Fluorination and Amination

The α-fluorination of 1,3-dicarbonyl compounds is successfully catalyzed by the XylBINAP– and DTBM-SEGPHOS–Pd complexes [105]. For example, β-keto ester **72** reacts with N-fluorobenzenesulfonimide (NFSI; 1.5 equiv) in the presence of [Pd{(R)-xylbinap}(OH)$_2$](BF$_4$)$_2$ (S/C = 40) in C$_2$H$_5$OH (−10 °C, 20 h) to afford the α-fluorinated product *(R)*-**73** in 94% ee and 91% yield (Scheme 1.24) [106]. The mode of enantioface selection is similar to that of the BINAP–Pd catalyzed Mannich-type reaction (Scheme 1.19). The more stereo-demanding XylBINAP is preferable for this reaction. C$_2$H$_5$OH is the solvent of choice, because the alcohol appears to facilitate the formation of a Pd-enolate intermediate. This chemistry is applicable to the reaction of β-keto phosphonates and α-*tert*-butoxycarbonyl lactones and lactams as well as N-(*tert*-butoxycarbonyl)oxindols [107].

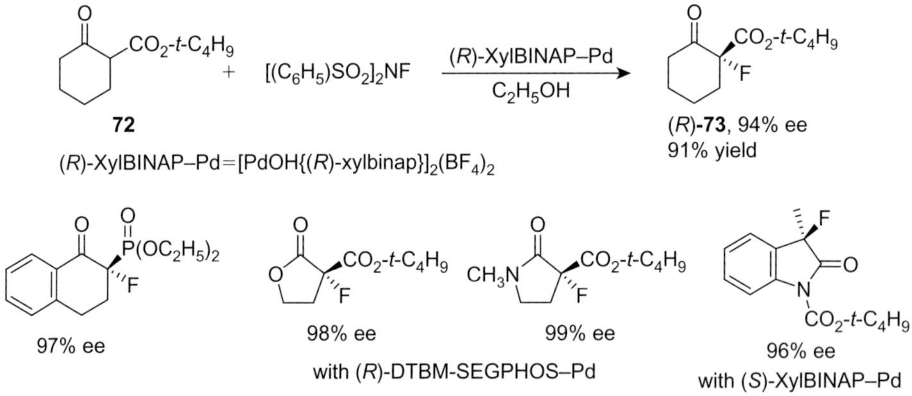

Scheme 1.24 α-Fluorination of carbonyl compounds with chiral Pd catalysts.

The BINAP–Pd catalyst effects the asymmetric reaction of β-keto esters and azodicarboxylates (Scheme 1.25). Thus, cyclic β-keto ester **74** reacts with diisopropyl

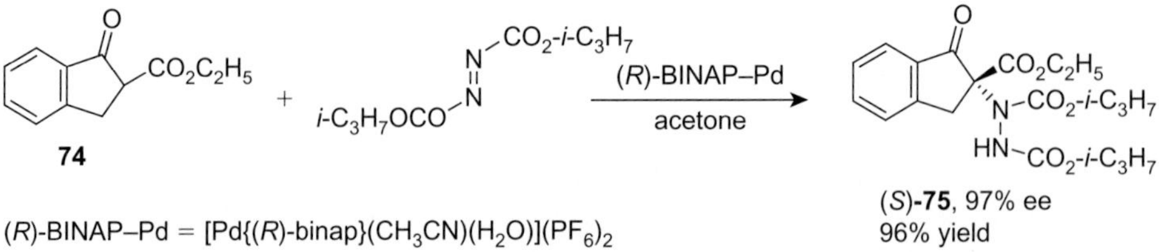

Scheme 1.25 α-Amination of β-keto esters catalyzed by BINAP–Pd complex.

azodicarboxylate using [Pd{(R)-binap}(CH$_3$CN)(H$_2$O)](PF$_6$)$_2$ (S/C = 1000) in acetone (room temperature, 70 h) to give the α adduct *(S)*-**75** in 97% ee and 96% yield [108]. The mechanism is closely related to that of α fluorination. α-Cyanoketones can also be used as substrates [109].

1.10.2
Arylation and Orthoester Alkylation

Asymmetric α-arylation of substituted carbonyl compounds has been developed for the construction of all-carbon chiral quaternary centers. The chiral catalyst formed from Pd(OCOCH$_3$)$_2$ and *(S)*-BINAP (S/C = 5) promotes the reaction of an α-methyl ketone (**76**) and 4-*tert*-butylbromobenzene (2 equiv) in the presence of *t*-C$_4$H$_9$ONa (2 equiv) at 100 °C to afford the arylation product **77** in 98% ee and 75% yield (Scheme 1.26) [110]. Some specific substrates react with high stereoselectivity. The BINAP–Ni complex effectively catalyzes asymmetric arylation of α-substituted γ-butyrolactones. For example, the reaction of α-methyl-γ-butyrolactone and chlorobenzene (1 equiv) is catalyzed by the *(S)*-BINAP–Ni(cod)$_2$ system (S/C = 20) with NaHMDS (2.3 equiv) and ZnBr$_2$ (15 mol%) in a toluene–THF mixture (60 °C, 17–20 h) to produce the α phenylation product *(S)*-**78** in >97% ee [111]. Addition of the Zn salt significantly increases the reaction rate and yield of products. A range of optically active α-alkyl α-aryl γ-butyrolactones can be synthesized by this method.

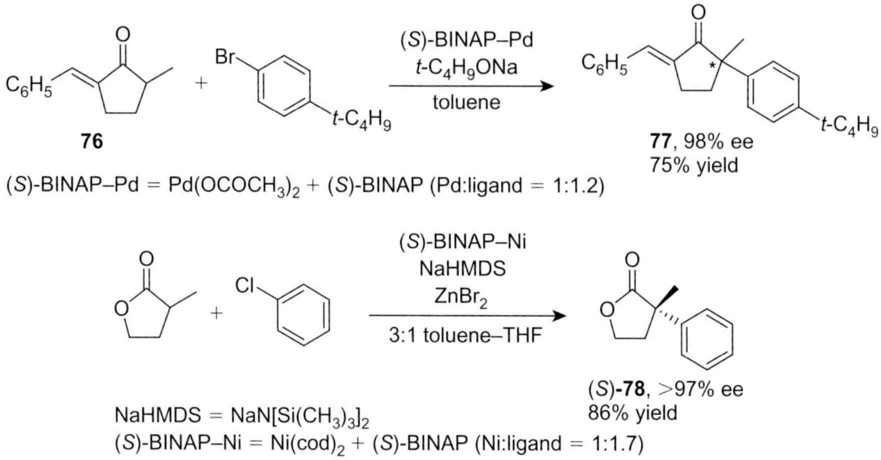

Scheme 1.26 α-Arylation of carbonyl compounds with BINAP–Pd and –Ni catalysts.

Asymmetric α alkylation of carbonyl compounds is difficult. However, the reaction of N-propionylthiazolidinethione (**79**) and trimethyl orthoformate (3 equiv) is catalyzed by [Ni{(R)-tolbinap}](OTf)$_2$ (S/C = 20) with 2,6-lutidine (3 equiv) and BF$_3$·O(C$_2$H$_5$)$_2$ (3 equiv) in CH$_2$Cl$_2$ (−78 °C, 0.5 h) to afford the α adduct *(S)*-**80** in

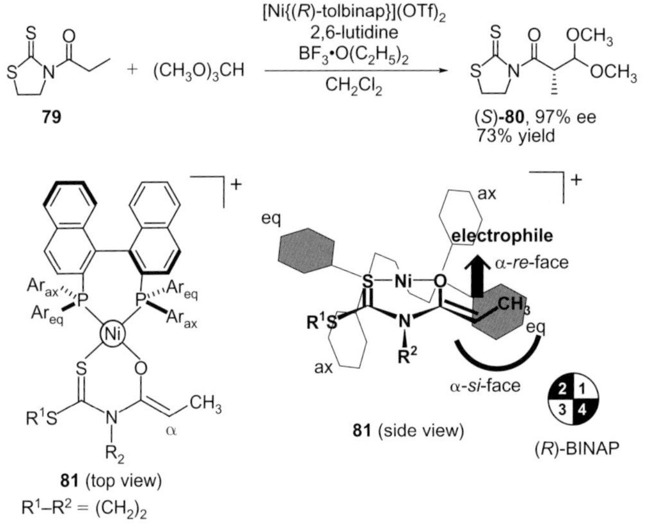

Scheme 1.27 α-Orthoester alkylation catalyzed by TolBINAP–Ni complex, and molecular models of the Ni enolate intermediate.

97% ee and 73% yield (Scheme 1.27) [112]. A series of acyl derivatives of **79** reacts with high enantioselectivity. The corresponding oxazolidinone does not give the desired product. This addition reaction is suggested to proceed via a Ni-enolate intermediate (**81**), the conformation of which is fixed by the S–Ni–O chelate structure. The electrophile ($CH_3O^+=CHOCH_3$) approaches the α-carbon of the enolate from the α-*Re*-face side or α-*Si*-face side (see the side view in Scheme 1.27). The C–C bond formation predominantly occurs at the side of α-*Re*-face [first quadrant of *(R)*-BINAP–Ni complex], because the α-*Si*-face of the enolate is covered by the equatorial *P*-phenyl ring of *(R)*-BINAP at the fourth quadrant. Thus, the *(S)* addition product is obtained in high ee.

1.11
Michael-Type Reactions

1.11.1
Michael Reaction

The asymmetric Michael reaction [113] of 1,3-dicarbonyl compounds with α,β-unsaturated ketones proceeds without addition of strong acid or base when [Pd(binap)(H$_2$O)$_2$](OTf)$_2$ is used as the catalyst (Scheme 1.28). For instance, cyclic β-keto ester **82** reacts with 3-penten-2-one using the *(R)*-BINAP–Pd aqua complex (S/C = 20) in THF (−20 °C, 24 h) to afford the conjugate addition product **83** in 89% yield (diastereomer ratio = 8 : 1) [114]. The ee value of the major diastereomer is 99%. The aqua Pd complex appears to be in equilibrium among three

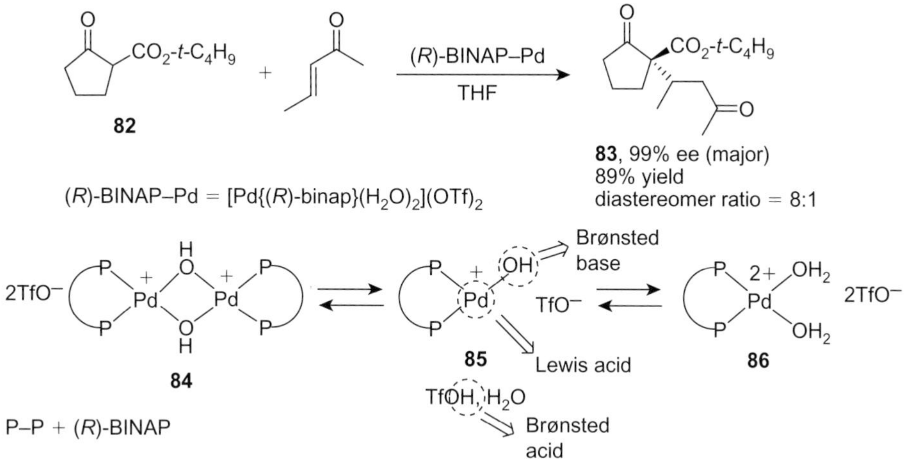

1.11 Michael-Type Reactions | 31

Scheme 1.28 Michael reaction catalyzed by BINAP–Pd complex.

species: **84** (a dimer), **85** (a hydroxo complex), and **86** (an aqua complex). Species **85**, which includes a Lewis acidic Pd center and a Brønsted basic OH, is suggested to behave as the catalyst. The acid–base bifunctionality cooperatively forms the Pd-enolate intermediate. The enone substrate is activated by a protonation with TfOH formed in this system. The enantioselective manner is similar to that shown in Section 1.8.2 (Mannich-type reaction) (Scheme 1.19).

The addition reaction of a fumarate derivative of thiazolidinethione (**87**) and *tert*-butyl 3-oxobutanoate (1.5 equiv) is catalyzed by [Ni{(S)-tolbinap}](BF$_4$)$_2$ (S/C = 10) in ethyl acetate (0 °C, 12 h) to give the Michael product **88** in 87% yield (Scheme 1.29) [115]. Treatment of **88** with DBU (0.05 equiv) affords quantitatively the dihydropyrone *(R)*-**89** in 97% ee. The nickel catalyst is presumed to activate both the thione Michael acceptor **87** and the nucleophilic β-keto ester.

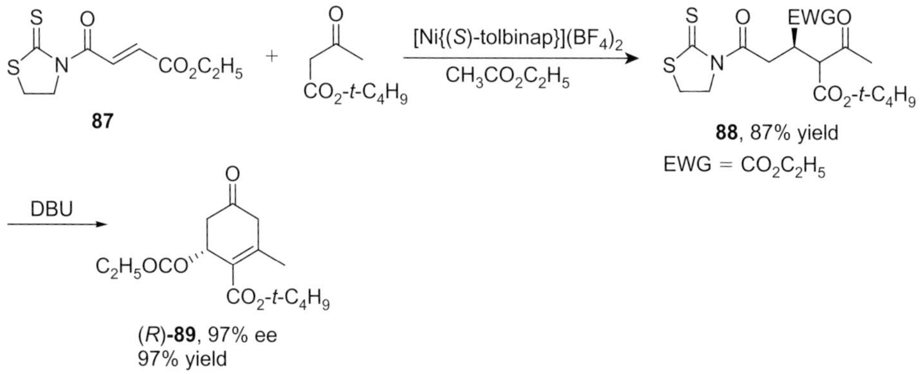

Scheme 1.29 Michael reaction catalyzed by TolBINAP–Ni complex.

1.11.2
Aza-Michael Reaction

Conjugate addition of primary aromatic amines to N-alkenoyl carbamates is catalyzed by the BINAP–Pd complex (Scheme 1.30). The reaction of 4-chloroaniline and a *tert*-butyl alkenoyl carbamate **90** (1.5 equiv) with [Pd{(R)-binap}(CH$_3$CN)](OTf)$_2$ (S/C = 50) proceeds in toluene (25 °C, 18 h) to give the amination product **91** in >99% ee quantitatively [116]. The reactivity and enantioselectivity depend on the electronic character of the anilines. Alkenoyl oxazolidinone **92** reacts with a salt of 4-methoxyaniline and TfOH (1.5 equiv) in the presence of [PdOH{(R)-binap}]$_2$(OTf)$_2$ (S/C = 50) in THF (room temperature, 12 h) to afford the β-amination product **93** in 98% ee and 92% yield [117]. A catalytic amount of the basic PdOH complex reacts with the aniline salt to generate the appropriate amount of the free amine, so that the uncatalyzed addition reaction with the aniline is inhibited. The reaction using aniline analogues substituted by electron-rich and -deficient groups also shows high enantioselectivity.

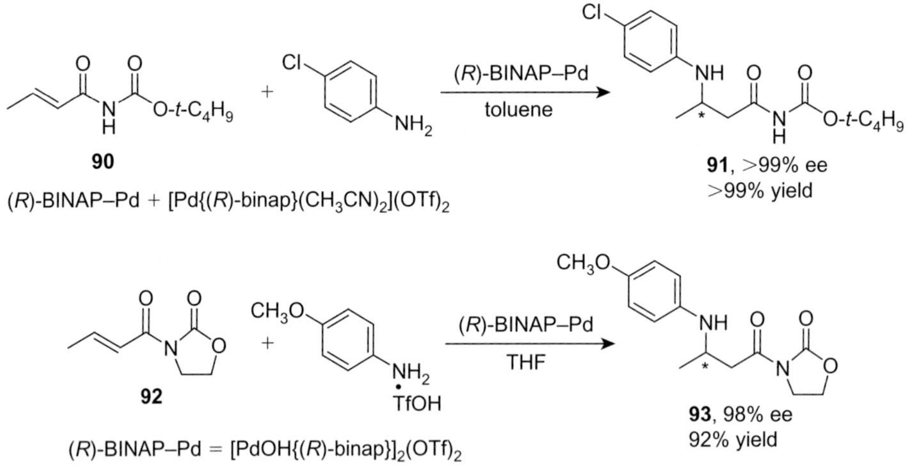

Scheme 1.30 Aza-Michael reactions catalyzed by BINAP–Pd complexes.

1.12
Conjugate Additions Using Organoboron and Grignard Reagents

Asymmetric 1,4-addition reactions of α,β-unsaturated carbonyl compounds are versatile procedures for the synthesis of useful chiral β-substituted carbonyl compounds [118, 119]. The reaction of 2-cyclohexenone and phenylboroxine (2.5 equiv) is catalyzed by [RhOH{(R)-binap}]$_2$ (S/C = 33) in a 10 : 1 mixture of dioxane and H$_2$O (35 °C, 3 h) to afford *(R)*-3-phenylcyclohexanone in 99% ee

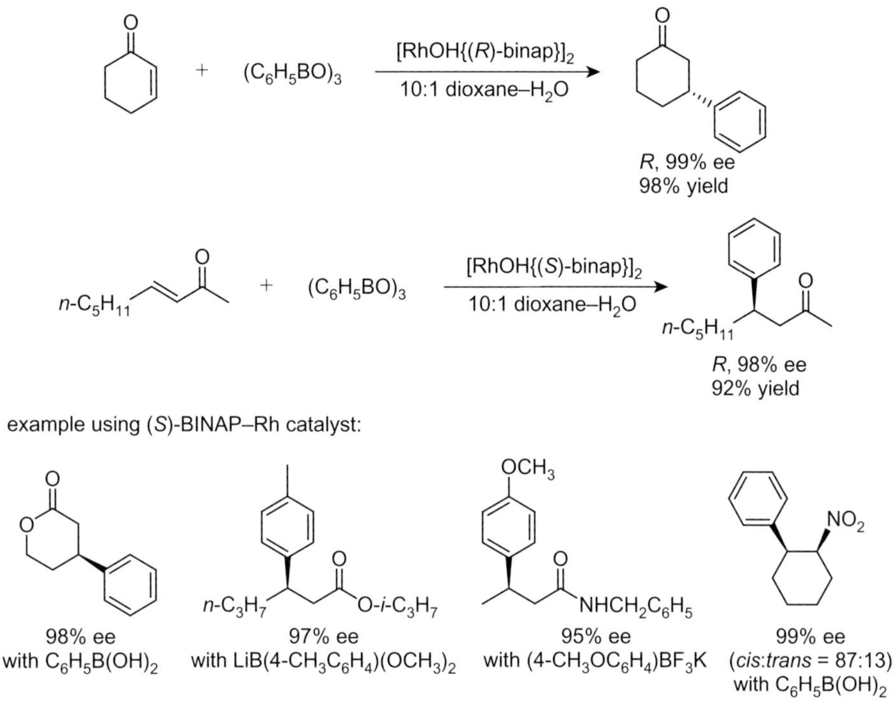

1.12 Conjugate Additions Using Organoboron and Grignard Reagents

Scheme 1.31 Conjugate addition reactions using arylboron reagents catalyzed by BINAP–Rh complexes.

quantitatively (Scheme 1.31) [120]. Under the same conditions 3-nonen-2-one, an acyclic enone, is converted into the *(R)* phenylation product in 98% ee. This method is applied to various cyclic and acyclic esters and amides as well as 1-alkenylphosphonates and nitroalkenes by appropriate use of the borane reagents, such as arylboronic acids, $LiBAr(OCH_3)_3$, and $ArBF_3K$ [121]. Arylsiloxanes are also used for this reaction [122].

Figure 1.10a illustrates a plausible reaction mechanism for the 1,4-addition of aryl boronic acids to α,β-unsaturated carbonyl compounds catalyzed by the *(R)*-BINAP–Rh complex [120]. The [Rh]OH complex **94** is converted into the [Rh]Ar species **95** by transmetallation of an aryl group from boron to rhodium. An enone substrate is inserted into the Rh–Ar bond to form an oxa-π-allyl-rhodium intermediate **96**. Hydrolysis of **96** by H_2O releases the arylation product along with regeneration of the catalyst species **94**. Therefore, an addition of H_2O in the reaction media is necessary. The configuration of the products is determined at the insertion of the enone into the Rh–Ar bond (**95→96**). At this step an enone substrate (e.g., 2-cyclohexenone) coordinates with the α-*Re*-face to avoid steric hindrance caused by an equatorial *P*-phenyl ring of the *(R)*-BINAP at the second quadrant (Figure 1.10b; **97**). Therefore, migration of an Ar group from the Rh center to the β-carbon of 2-cyclohexenone forms the β-*(R)* configuration in the intermediate **96**.

(a)

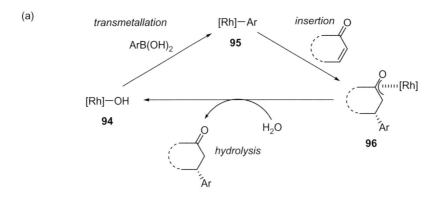

(b)

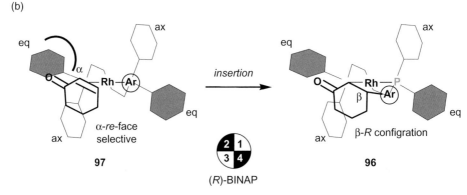

Figure 1.10 (a) Plausible 1,4-addition reaction mechanism catalyzed by (R)-BINAP–Rh complex; (b) molecular models of the stereo-determining step.

The BINAP–Rh complexes also catalyze the asymmetric 1,4-addition of alkenyl groups (Scheme 1.32). 2-Cyclohexenone reacts with an alkenylcatecholborane **98** (5 equiv) using the (S)-BINAP–Rh(acac)(CH$_2$=CH$_2$) system as a catalyst (S/C = 33) and (C$_2$H$_5$)$_3$N (10 equiv) (100 °C, 3 h) to furnish the alkenylation product (S)-**99** in 96% ee and 92% yield [123]. Addition of a base is necessary to trap acidic alkenylboronic acid and catechol generated during the reaction. The conjugate alkenylation of 2-cyclohexenone with an alkenyl Zr reagent (**100**) (1.2 equiv), which is prepared by hydrozirconation of 1-heptyne with Cp$_2$ZrHCl (room temperature, 0.5 h), can be carried out with the (S)-BINAP–[Rh(cod)(CH$_3$CN)$_2$]BF$_4$ system (S/C = 20) in an aprotic media (room temperature, 5 h) to give the adduct **99** in 99% ee and 96% yield after usual workup [124]. The enantioselectivity is somewhat decreased in the reaction of acyclic enones.

Asymmetric 1,4-addition of alkyl groups by Grignard reagents to α,β-unsaturated esters is catalyzed by the BINAP–Cu complex. For example, unsaturated ester **101** and C$_2$H$_5$MgBr (5 equiv) react with the (R)-TolBINAP–CuI system (S/C = 100) in t-C$_4$H$_9$OCH$_3$ (−40 °C, 2–3 h) to afford the β-substituted ester (S)-**102** in 95% ee and 90% yield (Scheme 1.33) [125]. The absolute configuration of

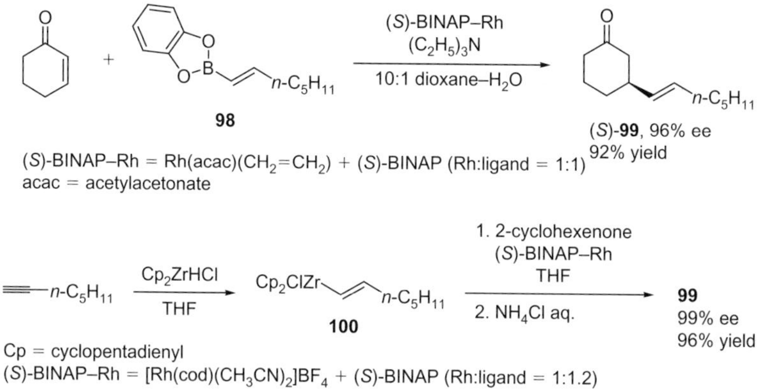

Scheme 1.32 Conjugate addition reactions using alkenyl metal reagents catalyzed by BINAP–Rh complexes.

product is reversed by using the geometrical isomer. Thus, methyl (E)-5-phenyl-2-pentenoate [(E)-103] is converted into the (R) product (R)-104 in 93% ee with the (R)-TolBINAP–Cu catalyst. On the other hand, the reaction of (Z)-103 results in the (S) adduct (S)-104 in 94% ee catalyzed by the same complex.

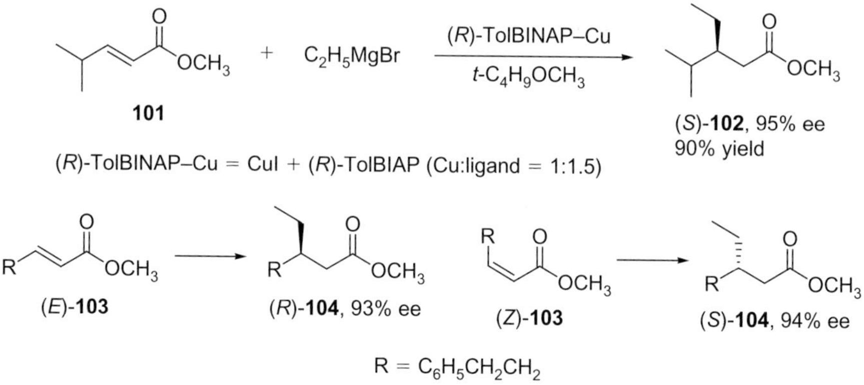

Scheme 1.33 1,4-Addition of alkyl groups using Grignard reagents catalyzed by TolBINAP–Cu complex.

1.13
Diels–Alder Reaction

A dicationic BINAP–Pd complex acts as an efficient Lewis-acidic catalyst for asymmetric Diels–Alder cyclization [126]. The reaction of N-acryloyloxazolidinone (105) and cyclopentadiene (5 equiv) with [Pd{(R)-binap}(C_6H_5CN)](BF_4)$_2$ (S/C = 10) in CH_2Cl_2 (−50 °C, 24 h) affords the cyclization product (S)-106 in

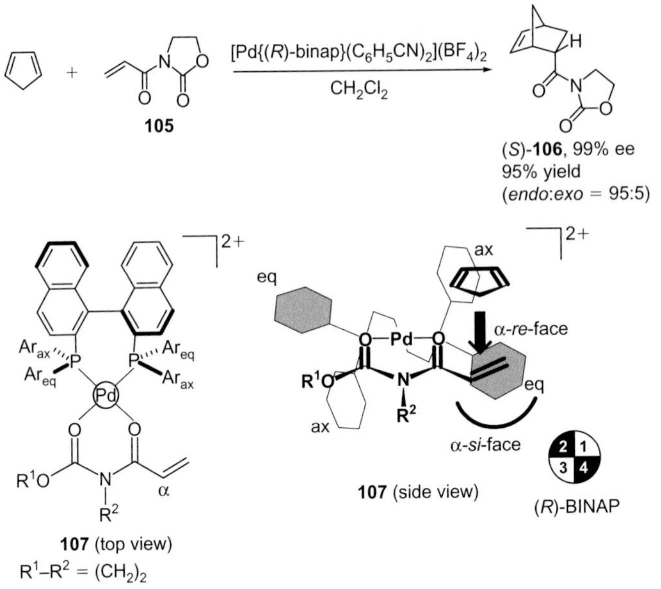

Scheme 1.34 Diels–Alder reaction catalyzed by BINAP–Pd complex, and molecular models of the reaction intermediate.

99% ee and 95% yield (*endo* : *exo* = 95 : 5) (Scheme 1.34) [127]. The enantioselectivity depends on the nature of the counter anion (77% ee with the PF_6^- complex). The reaction appears to proceed through the intermediate **107**, which has a six-membered Pd–**105** chelate-ring structure. The diene approaches the dienophile from the side of α-*Re*-face to avoid repulsive interaction with the *(R)*-BINAP's equatorial *P*-phenyl ring at the fourth quadrant (see the side view of **107**), exclusively forming the *(S)* chiral center.

The BINAP–Pd has been applied successfully to the asymmetric hetero-Diels–Alder reaction [128]. Thus, phenylglyoxal reacts with 2,3-dimethyl-1,3-butadiene (1.5 equiv) in the presence of [Pd{(S)-binap}(C$_6$H$_5$CN)](BF$_4$)$_2$ (S/C = 50) in CH$_3$Cl (0 °C, 24 h) to give the cycloadduct *(R)*-**108** in 99% ee and 67% yield (Scheme 1.35) [129]. An addition of MS-3A increases the enantioselectivity. A trace amount of H$_2$O existing in the reaction system may react with the Pd complex to generate acidic impurities, causing an achiral pathway of the reaction. The corresponding Pt complex under the same conditions also exhibits high enantioselectivity [130]. The Pd-catalyzed asymmetric 1,3-dipolar cyclization of nitrones and *N*-alkenoyloxazolidinones is also studied [131]. The asymmetric cyclization of a kind of Danishefsky's diene **(109)** and an *N*-tosyl-α-imino ester **110** (1.25 equiv) with the *(S)*-TolBINAP–CuClO$_4$·4CH$_3$CN system (S/C = 100) in THF at −78 °C produces the aza-cyclization product **111** in 96% ee and 70% yield (trans : cis = 10 : 1) [132]. The use of several other solvents decreases both diastereoselectivity and enantioselectivity. The SEGPHOS–Cu catalyzed nitroso Diels–Alder reaction proceeds in a highly diastereo- and enantioselective

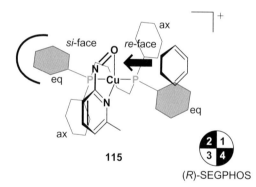

Scheme 1.35 Hetero-Diels–Alder reactions and a molecular model of the reaction intermediate with (R)-SEGPHOS–Cu catalyst.

manner. Thus, 6-methylnitrosopyridine (**113**) and a diene (**112**) (1.2 equiv) react with the *(R)*-SEGPHOS–[Cu(CH$_3$CN)$_4$]PF$_6$ system (S/C = 10) in CH$_2$Cl$_2$ (−85 to −20 °C, 6 h) to afford quantitatively the adduct **114** in 97% ee as a single diastereomer [133]. A methyl group at the 6 position of nitrosopyridine is essential to attain high enantioselectivity. The reaction is supposed to proceed through a tetrahedral intermediate (**115**) with a Cu–**113** chelate ring. The diene horizontally approaches the nitroso N=O from the *Re*-face direction at the first quadrant, because the opposite side is blocked by the equatorial *P*-phenyl ring of *(R)*-SEGPHOS at the second quadrant, resulting in **114** predominantly.

1.14
Ene Reaction

Under appropriate Lewis-acidic conditions alkenes react with imino and carbonyl compounds [134]. An *N*-toluenesulfonyl-α-imino ester (**116**) and α-methylstyrene (2 equiv) react with the *(R)*-TolBINAP–[Cu(CH$_3$CN)$_4$]ClO$_4$ system (S/C = 20) in benzotrifluoride, a polar aromatic solvent, (room temperature, 18 h) to afford the α-amino ester *(S)*-**117** in 99% ee and 92% yield (Scheme 1.36) [135]. Heterosubstituted alkenes are applicable to this reaction. A carbonyl-ene reaction is carried out with the SEGPHOS–Pd catalyst. Thus, methylenecyclopentane and ethyl trifluoropyruvate (**118**; 1.5 equiv) react, using a dicationic catalyst prepared *in situ* from PdCl$_2$[*(S)*-segphos] and AgSbF$_6$ (S/C = 20) in CH$_2$Cl$_2$ (room temperature, 15 min), to give the ene-adduct **119** in 97% ee and 84% yield [136]. Several mono- and di-substituted alkenes have been used for this reaction. The corresponding Pt complexes also exhibit high catalytic efficiency [130, 137].

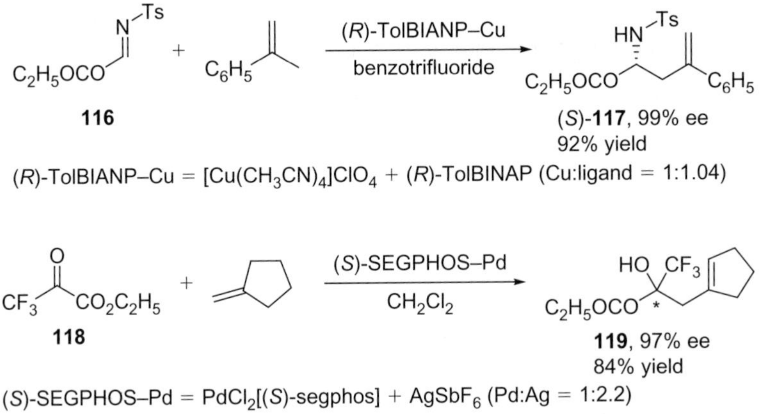

(R)-TolBIANP–Cu = [Cu(CH$_3$CN)$_4$]ClO$_4$ + *(R)*-TolBINAP (Cu:ligand = 1:1.04)

(S)-SEGPHOS–Pd = PdCl$_2$[*(S)*-segphos] + AgSbF$_6$ (Pd:Ag = 1:2.2)

Scheme 1.36 Imine- and carbonyl-ene reactions catalyzed by Cu and Pd complexes.

1.15
Cyclization

1.15.1
Intramolecular Reactions of Enynes

A SEGPHOS–Pd complex effectively catalyzes the asymmetric ene-type cyclization of a 1,6-enyne (**120**, Scheme 1.37) [138]. The reaction with the *(R)*-SEGPHOS–Pd-(OCOCF$_3$)$_2$ system (S/C = 20) proceeds in C$_6$D$_6$ (100 °C, 37 h) to give the

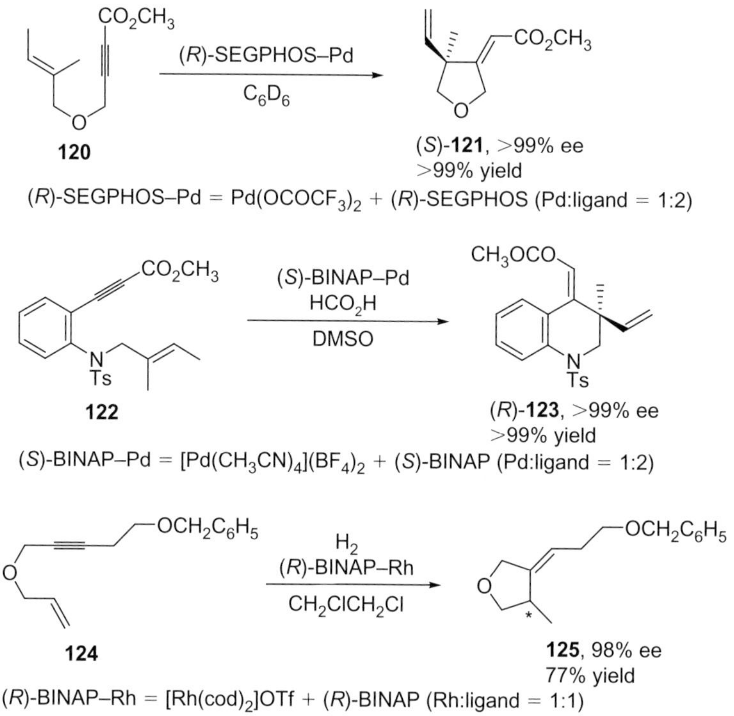

Scheme 1.37 Cyclization of 1,6- and 1,7-enynes catalyzed by chiral Pd and Rh complexes.

cyclization product *(S)*-**121** in pure form. The reactivity and enantioselectivity depend on the solvent polarity, which closely relates to the reaction mechanism. When a 1,7-enyne (**122**) reacts with the *(S)*-BINAP–[Pd(CH$_3$CN)$_4$](BF$_4$)$_2$ system (S/C = 20) and HCO$_2$H (1 equiv) in DMSO (100 °C, 3 h) the quinoline derivative *(R)*-**123** is obtained in >99% ee quantitatively [139]. The spiro compound is formed from the substrate with a cyclic alkene moiety. A BINAP–Rh complex also catalyzes the reaction of related 1,6-enynes with nearly perfect enantioselectivity [140]. Reductive cyclization of 1,6-enyne **124** with the *(R)*-BINAP–[Rh(cod)$_2$]OTf system (S/C = 20) under a H$_2$ atmosphere (1 atm) in CH$_2$ClCH$_2$Cl (25 °C, 2–3 h) affords the desired cyclization product **125** in 98% ee and 77% yield [141]. The reaction is suggested to proceed via a Rh-metallocycle intermediate. A series of carbo- and heterocyclic compounds has been synthesized by this method.

1.15.2
[3+2] and [5+2] Cycloaddition Reactions

Cyclodimerization of oxabenzonorbornadienes is catalyzed by the BINAP–Rh complex in a highly enantioselective manner (Scheme 1.38). When the reaction of **126** is carried out with [RhCl{*(R)*-binap}]$_2$ and NaBARF (S/C = 100,

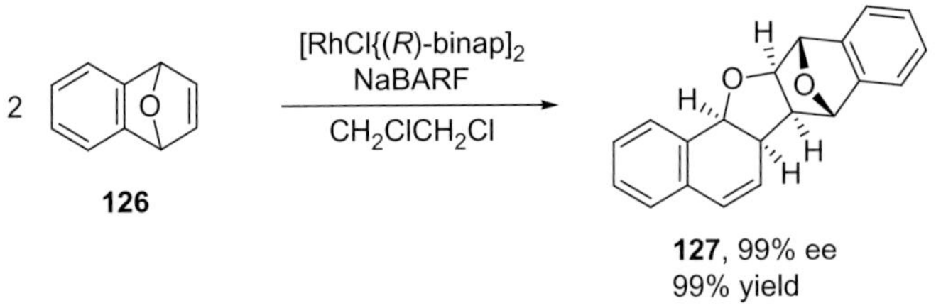

Scheme 1.38 BINAP–Rh catalyzed [3 + 2] cycloaddition reaction.

Rh : Na = 1 : 2) in CH$_2$ClCH$_2$Cl (40 °C, 1 h), the polycyclic tetrahydrofuran derivative **127** is obtained in 99% ee quantitatively [142]. The reaction of **126** and dimethyl 2-butynedioate gives the cross-cycloaddition product in high ee.

Figure 1.11a illustrates a proposed reaction mechanism of the cyclodimerization of **126** [142]. A cationic BINAP–Rh(I) species **128** is generated in this system; **128** reacts with two **126** molecules to form the rhodacyclopentane intermediate **129**, which is converted into the alkoxorhodium(III) species **130** by β-oxygen elimination. Then, reductive elimination of the sp^3 C–O bond from **130** produces the product **127** with regeneration of **128**. The stereochemistry of **127** is determined at the formation of the intermediate **129**. Figure 1.11b shows two diastereomeric molecular models of **129**. Two **126** molecules approach the Rh center via the open space of the *(R)*-BINAP–Rh species **128** (first and third quadrants) to form the favored intermediate **131**, which is then converted into the product **127**. In contrast, the diastereomeric intermediate **132** suffers notable steric repulsion between the *(R)*-BINAP's equatorial P-phenyl rings and oxabenzonorbornadiene moieties at the second and fourth quadrants of the complex.

The TolBINAP–Pd complex catalyzes the asymmetric cycloaddition of vinyloxiranes and carbodiimides (Scheme 1.39). For example, the reaction of a mixture of a vinyloxirane **133** and di(4-chlorophenyl)carbodiimide (**134**; 1 equiv) with the *(S)*-TolBINAP–Pd$_2$(dba)$_3$ · CHCl$_3$ system (S/C = 17) in THF (room temperature, 15 h) produces the 4-vinyl-1,3-oxazolidin-2-imine *(R)*-**135** in 94% ee and 95% yield [143]. Several arylcarbodiimides react with high stereoselectivity. Cyclization of an azalactone (**136**) with N-phenylmaleimide (1.5 equiv) using (AuOCOC$_6$H$_5$)$_2$[*(R)*-dtbm-segphos] (S/C = 50) in acetone (room temperature, 24 h) followed by a methyl-esterification affords the cycloadduct **137** in 98% ee and 84% yield [144]. No *endo*-isomer is observed. A range of electron-deficient alkenes can be used as dipolarophiles. The Au complex is presumed to interact with the sp^2-N of **136** but not C=C-π bond.

Intramolecular [5 + 2] cycloaddition of vinylcyclopropanes tethered with alkene moieties is catalyzed by the BINAP–Rh complex (Scheme 1.40). Thus, a cyclopropyl diene (**138**) reacts with [Rh{*(R)*-binap}]SbF$_6$ (S/C = 10) in CH$_2$ClCH$_2$Cl (70 °C, 2 days) to afford the bicyclo[5.3.0] product **139** in 99% ee and 80% yield [145]. The enantioselectivity, notably, depends on the substrate structure. Vinylcyclopropanes with malono-diester and sulfonamide groups are converted into the corresponding products in high ee.

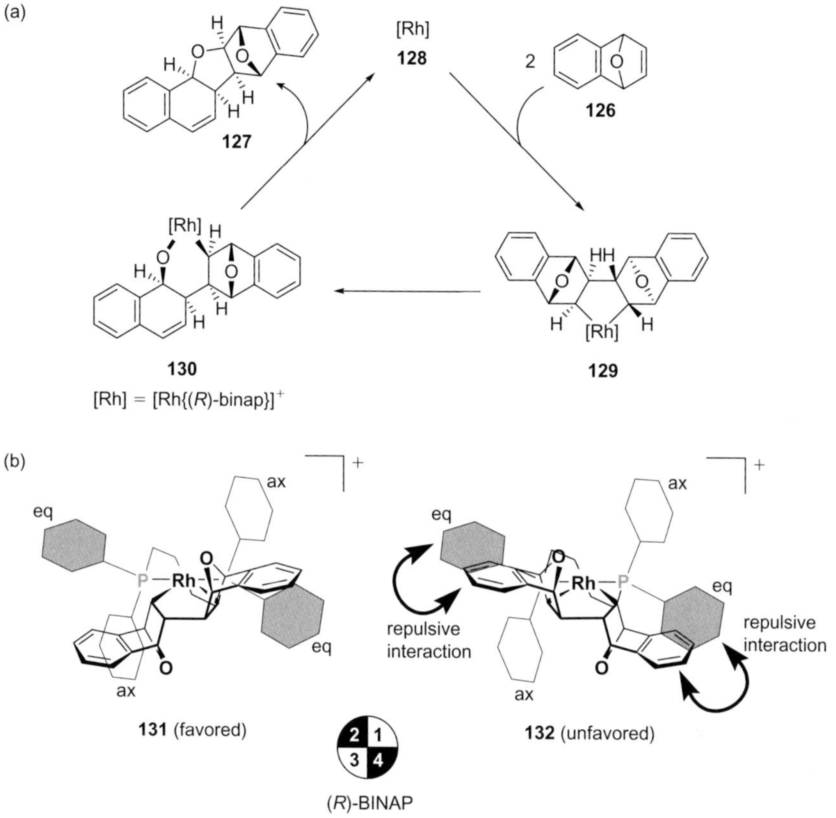
(a)
[Rh]
128
2
126
127
130
129
[Rh] = [Rh{(R)-binap}]+
(b)
eq
ax
131 (favored)
repulsive
interaction
132 (unfavored)
(R)-BINAP

Figure 1.11 (a) Plausible [3 + 2] cycloaddition reaction mechanism catalyzed by *(R)*-BINAP–Rh complex; (b) molecular models of the diastereomeric intermediates **131** and **132**.

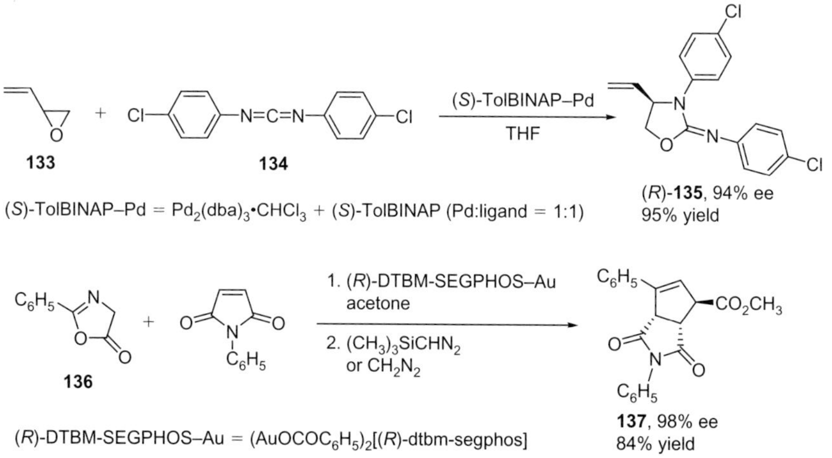
133
134
(S)-TolBINAP–Pd
THF
(R)-135, 94% ee
95% yield
(S)-TolBINAP–Pd = Pd2(dba)3•CHCl3 + (S)-TolBINAP (Pd:ligand = 1:1)
136
1. (R)-DTBM-SEGPHOS–Au
acetone
2. (CH3)3SiCHN2
or CH2N2
137, 98% ee
84% yield
(R)-DTBM-SEGPHOS–Au = (AuOCOC6H5)2[(R)-dtbm-segphos]

Scheme 1.39 [3+2] cycloaddition reactions with chiral Pd and Au catalysts.

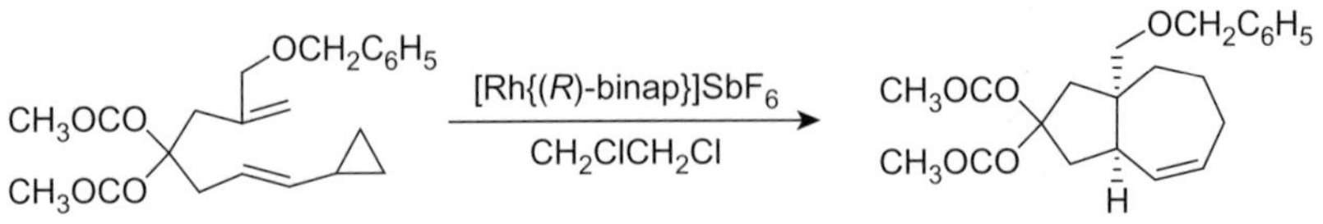

138

139, 99% ee
80% yield

Scheme 1.40 BINAP–Rh catalyzed [5+2] cycloaddition reaction.

1.15.3
[2 + 2 + 2] Cycloaddition Reactions

Asymmetric cyclotrimerization of α,ω-diynes and monoynes forms axially chiral biaryl compounds in the presence of the H_8-BINAP–Rh catalyst (Scheme 1.41) [146]. For example, unsymmetric diene **140** and 1,4-diacetoxy-2-butyne (5 equiv) react with the *(S)*-H_8-BINAP–[Rh(cod)$_2$]BF_4 system (S/C = 20) in CH_2Cl_2 (room temperature, 3 h) to afford the chiral phthalide derivative *(R)*-**141** in >99% ee and 73% yield [147]. The presence of an *ortho*-substituted aryl group in **140** is crucial to achieve high enantioselectivity. Terminal monoynes are also appropriate substrates for this reaction. This methodology has been applied to the intermolecular cross-cyclotrimerization of internal alkynes with two equivalents of dialkyl acetylenedicarboxylates. Thus, an internal alkynyl alcohol (**142**) and diethyl acetylenedicarboxylate (2 equiv) react with the *(S)*-H_8-BINAP–Rh catalyst (S/C = 20) followed by acetylation to give the biaryl product **143** in 96% ee and 75% yield [148].

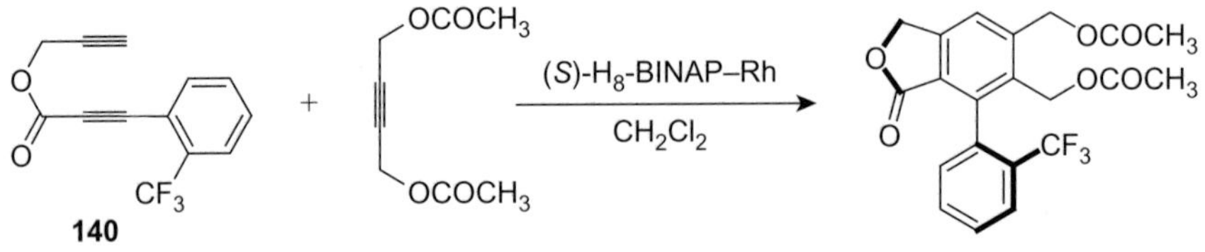

140

(R)-**141**, >99% ee
73% yield

(S)-H_8-BINAP–Rh = [Rh(cod)$_2$]BF_4 + (S)-H_8-BINAP (Rh:ligand = 1:1)

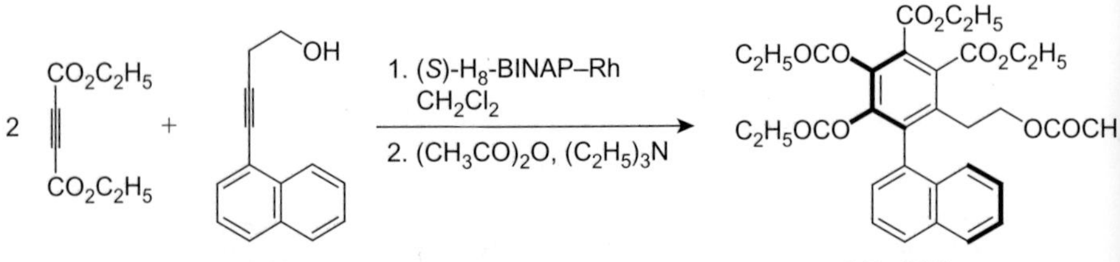

142

143, 96% ee
75% yield

(S)-H_8-BINAP–Rh = [Rh(cod)$_2$]BF_4 + (S)-H8-BINAP (Rh:ligand = 1:1)

Scheme 1.41 [2+2+2] cycloadditions of alkynes to form axially chiral biaryls.

Planar-chiral metacyclophanes can be synthesized by the H_8-BINAP–Rh catalyzed cyclotrimerization of appropriate triynes (Scheme 1.42). For instance, a triyne (144) is cyclized with the (R)-H_8-BINAP–[Rh(cod)$_2$]BF$_4$ catalyst system (S/C = 50) in CH_2Cl_2 (room temperature, 16 h) to afford the [7]–[10]metacyclophane product 145 in an optically pure form, although the chemical yield is relatively low [149]. The corresponding [7]–[10]orthocyclophane is obtained as a by-product. The asymmetric [2+2+2] cycloaddition successfully controls the central chirality of products [150, 151]. The reaction of a 1,6-enyne (146) with an ether-linkage and methyl phenylpropiolate (3 equiv) is catalyzed by the (S)-Xyl-P-Phos–[RhCl(cod)]$_2$–AgBF$_4$ system (S/C = 10) in THF (60 °C, about 2.5 h) to produce the bicyclohexadiene (S)-147 in >99% ee as a single regioisomer [150]. The steric hindrance with P-xylyl groups of Xyl-P-Phos leads to the high regioselectivity, and the narrower dihedral angle of this ligand increases the enantioselectivity. 1,6-Enynes tethered with NTs and $C(CO_2CH_3)_2$ moieties are also converted into the desired products in high stereoselectivity.

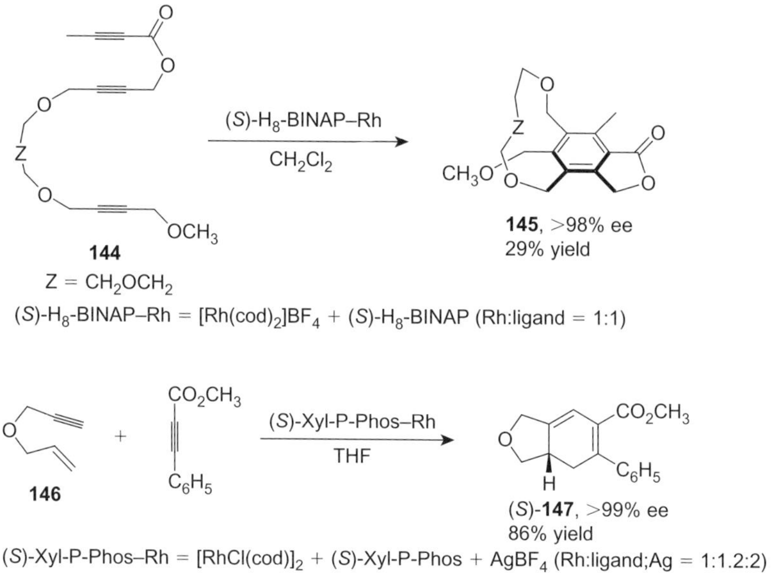

Scheme 1.42 [2+2+2] cycloadditions with control of planar and central chirality.

1.15.4
Pauson–Khand Type Reactions

Asymmetric cyclization of 1,6-enynes with CO in the presence of appropriate catalysts gives the corresponding 2-cyclopentenones in high stereoselectivity (Scheme 1.43) [152]. A Co complex prepared *in situ* from $Co_2(CO)_8$ and (S)-BINAP in a 1 : 1

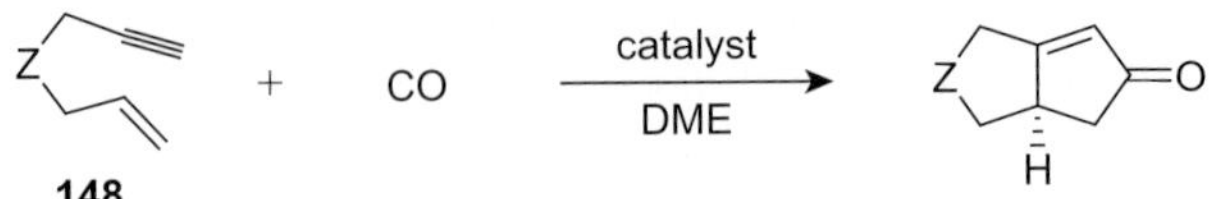

148

148	catalyst	% ee	% yield
a; Z = NTs	Co$_2$(CO)$_8$ + (S)-BINAP (Co:ligand = 2:1)	93	60
b; Z = C(CO$_2$CH$_3$)$_2$	Co$_4$(CO)$_{12}$ + (S)-TolBINAP (Co:ligand = 2:1)	96	85

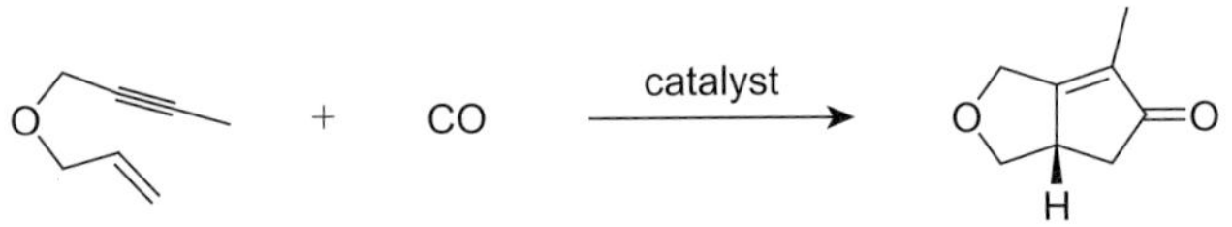

ctsylata	solvent	% ee	% yield
[RhCl(CO)$_2$]$_2$ + (S)-BINAP + AgOTf (Rh:ligand:Ag = 1:1:2)	THF	96	40
[IrCl(cod)]$_2$ + (S)-TolBINAP (Ir:ligand = 1:1)	toluene	98	61

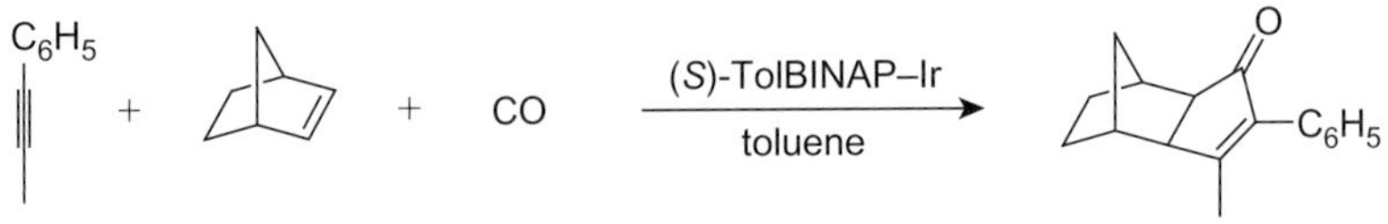

152, 93% ee
(S)-TolBINAP–Ir = [IrCl(cod)]$_2$ + (S)-TolBINAP (Ir:ligand = 1:1) 32% yield

Scheme 1.43 Pauson–Khand-type cycloaddition reactions with BINAP–metal catalysts.

ratio catalyzes the cyclocarbonylation of an enyne (**148a**, Z = NTs) in DME (S/C = 5, 1 atm CO, 85 °C, 15 h) to furnish the bicyclo[3.3.0]enone **149a** in 93% ee and 60% yield [153]. The cycloaddition of **148b** [Z = C(CO$_2$CH$_3$)$_2$] with the (S)-TolBINAP–Co$_4$(CO)$_{12}$ system (S/C = 13) gives the product **149b** in 96% ee and 85% yield under similar conditions (1.05 atm CO, 75 °C, 5 h) [154]. The bimetallic complex Co$_2$(binap)(CO)$_6$ appears to be a key intermediate for this reaction [155]. The MeO-BIPHEP–Co complex also exhibits good catalyst performance [156]. A cationic complex formed from (S)-BINAP, [RhCl(CO)$_2$]$_2$, and AgOTf (S/C = 17) effects the asymmetric cyclocarbonylation of allyl 2-butynyl ether **150** with CO (1 atm) in THF (90 °C, 5 h) to produce the adduct **151** in 96% ee and 40% yield [157]. The same reaction is catalyzed by a neutral complex consisting of [IrCl(cod)]$_2$ and (S)-TolBINAP (S/C = 5) under the toluene reflux conditions (1 atm CO, 20 h) to afford **151** in 98% ee and 61% yield [158]. This Ir system catalyzes the three-component cycloaddition of 1-phenyl-1-propyne, norbornene, and CO (S/C = 5, toluene reflux) to give the adduct **152** in 93% ee and 32% yield as a single regioisomer. Rhodium complexes bearing other chiral biaryl diphosphines, including BIPHEMP, BisbenzodioxanPhos (SYNPHOS), Difluorphos, and P-Phos also catalyze the Pauson–Khand-type reactions with high enantioselectivity [159].

1.16
Ring-Opening Reactions

Enantioselective reductive ring-opening reaction of meso oxabicyclic alkenes using DIBAL-H as a hydride source can be performed by the BINAP–Ni catalysis [160]. For example, an oxabicyclo[2.2.1] compound **153** is treated with DIBAL-H (1.1 equiv) in the presence of the *(R)*-BINAP–Ni(cod)$_2$ system (S/C = 7) in toluene (room temperature, 1 h) to afford the cyclohexenol **154** in 97% ee and 97% yield (Scheme 1.44) [161]. Slow addition of the reductant is crucial to achieve high enantioselectivity. Oxabenzonorbornadiene (**155**) and oxabicyclo[3.2.1]alkene **157** are converted into the corresponding cyclohexenol **156** and cycloheptenol **158**, respectively, in >98% ee under similar conditions [162].

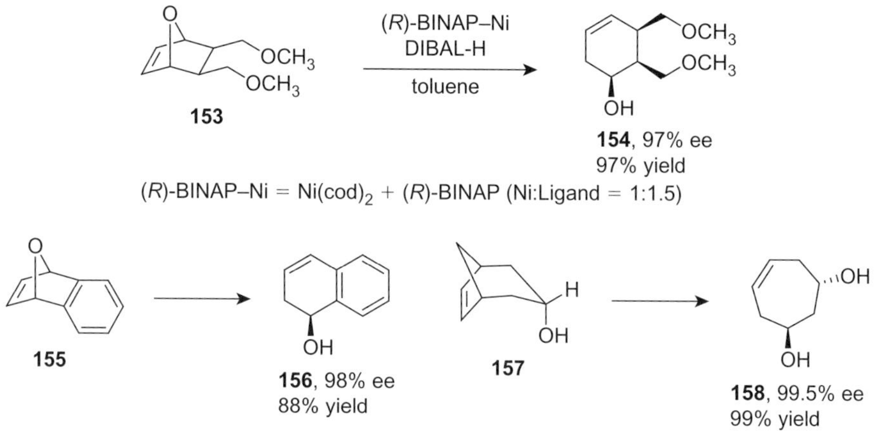

Scheme 1.44 BINAP–Ni catalyzed reductive ring-opening reactions of oxabicyclic alkenes.

1.17
Concluding Remarks

BINAP has been utilized as the typical chiral ligand in a wide variety of catalytic asymmetric reactions, including hydrogenation, isomerization, C–C bond formation, cyclization, and so on. The chiral structures of the BINAP–metal complexes, which are schematically illustrated in this text, precisely discriminate two enantiomers and enatiofaces. Owing to the clear C_2-symmetric character, the stereoselective outcome is rationally explainable and also predictable. Furthermore, the flexible binaphthyl backbone attains high generality for substrates. Recent development of BINAP analogues has notably improved the enantioselective ability in some cases. Therefore, BINAP is regarded as one of the most reliable and successful diphosphine ligands. We believe that BINAP will continue contributing to improve the chemistry of asymmetric catalysis.

References

1 Noyori, R. (1994) *Asymmetric Catalysis in Organic Synthesis*, John Wiley & Sons, Inc., New York.
2 Noyori, R. (2002) *Angew. Chem. Int. Ed.*, **41**, 2008–2022.
3 Miyashita, A., Takaya, H., Souchi, T., and Noyori, R. (1984) *Tetrahedron*, **40**, 1245–1253.
4 Kitamura, M. and Noyori, R. (2003) in *Chiral Reagents for Asymmetric Synthesis* (ed. L.A. Paquette), John Wiley & Sons, Ltd, Chichester, pp. 128–132.
5 Recent reviews: (a) MaCarthy, M. and Guiry, P.J. (2001) *Tetrahedron*, **57**, 3809–3844; (b) Tang, W. and Zhang, X. (2003) *Chem. Rev.*, **103**, 3029–3069; (c) Berthod, M., Mignani, G., Woodward, G., and Lemaire, M. (2005) *Chem. Rev.*, **105**, 1801–1836; (d) Shimizu, H., Nagasaki, I., and Saito, T. (2005) *Tetrahedron*, **61**, 5405–5432; (e) Li, Y.-M., Kwong, F.-Y., Yu, W.-Y., and Chan, A.S.C. (2007) *Coord. Chem. Rev.*, **251**, 2119–2144.
6 Mashima, K., Kusano, K., Sato, N., Matsumura, Y., Nozaki, K., Kumobayashi, H., Sayo, N., Hori, Y., Ishizaki, T., Akutagawa, S., Tandakaya, H. (1994) *J. Org. Chem.*, **59**, 3064–3076.
7 (a) Schmid, R., Foricher, J., Cereghetti, M., and Schönholzer, P. (1991) *Helv. Chim. Acta*, **74**, 370–389; (b) Schmid, R., Broger, E.A., Cereghetti, M., Crameri, Y., Foricher, J., Lalonde, M., Müller R. K., Scalone, M., Schoettel, G., and Zutter, U. (1996) *Pure Appl. Chem.*, **68**, 131–138.
8 (a) Saito, T., Yokozawa, T., Ishizaki, T., Moroi, T., Sayo, N., Miura, T., and Kumobayashi, H. (2001) *Adv. Synth. Catal.*, **343**, 264–267; (b) Shimizu, H., Nagasaki, I., Matsumura, K., Sayo, N., and Saito, T. (2007) *Acc. Chem. Res.*, **40**, 1385–1393.
9 (a) Pai, C.-C., Li, Y.-M., Zhou, Z.-Y., and Chan, A.S.C. (2002) *Tetrahedron Lett.*, **43**, 2789–2792; (b) de Paule, S.D., Jeulin, S., Ratovelomanana-Vidal, V., Genêt, J.-P., Champion, N., and Dellis, P. (2003) *Eur. J. Org. Chem.*, 1931–1941; (c) Genêt, J.-P. (2003) *Acc. Chem. Res.*, **36**, 908–918.
10 Pai, C.-C., Lin, C.-W., Lin, C.-C., Chen, C.-C., Chan, A.S.C., and Wong, W.T. (2000) *J. Am. Chem. Soc.*, **122**, 11513–11514.
11 (a) Jeulin, S., de Paule, S.D., Ratovelomanana-Vidal, V., Genêt, J.-P., Champion, N., and Dellis, P. (2004) *Angew. Chem. Int. Ed.*, **43**, 320–325; (b) Jeulin, S., de Paule, S.D., Ratovelomanana-Vidal, V., Genêt, J.-P., Champion, N., and Dellis, P. (2004) *Proc. Natl. Acad. Sci. U.S.A.*, **101**, 5799–5804.
12 Zhang, Z., Qian, H., Longmire, J., and Zhang, X. (2000) *J. Org. Chem.*, **65**, 6223–6226.
13 (a) Zhang, X., Mashima, K., Koyano, K., Sayo, N., Kumobayashi, H., Akutagawa, S., and Takaya, H. (1991) *Tetrahedron Lett.*, **32**, 7283–7286; (b) Kumobayashi, H., Miura, T., Sayo, N., Saito, T., and Zhang, X. (2001) *Synthesis*, 1055–1064.
14 Schmid, R., Coreghetti, M., Heiser, B., Schönholzer, P., and Hansen, H.-J. (1988) *Helv. Chim. Acta*, **71**, 897–929.
15 Benincori, T., Brenna, E., Sannicolò, F., Trimarco, L., Antognazza, P., Cesarotti,

E., Demartin, F., and Pilati, T. (1996) *J. Org. Chem.*, **61**, 6244–6251.

16 Miyashita, A., Yasuda, A., Takaya, H., Toriumi, K., Ito, T., Souchi, T., and Noyori, R. (1980) *J. Am. Chem. Soc.*, **102**, 7932–7934.

17 Ohkuma, T., Kitamura, M., and Noyori, R. (2000) in *Catalytic Asymmetric Synthesis*, 2nd edn (ed. I. Ojima), John Wiley & Sons, Inc., New York, pp. 1–110.

18 Brown, J.M. (1999) in *Comprehensive Asymmetric Catalysis*, Vol. 1 (eds E.N. Jacobsen, A. Pfaltz, and H. Yamamoto), Springer-Verlag, Berlin, Heidelberg, pp. 121–182.

19 Noyori, R., Kitamura, M., and Ohkuma, T. (2004) *Proc. Natl. Acad. Sci. U.S.A.*, **101**, 5356–5362.

20 Ohkuma, T., Kitamura, M., and Noyori, R. (2007) in *New Frontiers in Asymmetric Catalysis* (eds K. Mikami and M. Lautens), John Wiley & Sons, Inc., Hoboken, pp. 1–32.

21 (a) Chan, A.S.C. and Halpern, J. (1980) *J. Am. Chem. Soc.*, **102**, 838–840; (b) Brown, J.M. and Chaloner, P.A. (1980) *J. Am. Chem. Soc.*, **102**, 3040–3048; (c) Landis, C.R. and Halpern, J. (1987) *J. Am. Chem. Soc.*, **109**, 1746–1754.

22 Brown, J.M. (2007) in *The Handbook of Homogeneous Hydrogenation* Vol. 3 (eds J.G. de Vries and C.J. Elsevier), Wiley-VCH Verlag GmbH, Weinheim, pp. 1073–1103.

23 (a) Kitamura, M., Tsukamoto, M., Bessho, Y., Yoshimura, M., Kobs, U., Widhalm, M., and Noyori, R. (2002) *J. Am. Chem. Soc.*, **124**, 6649–6667; (b) Ishibashi, Y., Bessho, Y., Yoshimura, M., Tsukamoto, M., and Kitamura, M. (2005) *Angew. Chem. Int. Ed.*, **44**, 7287–7290; (c) Wiles, J.A. and Bergens, S.H. (1998) *Organometallics*, **17**, 2228–2240.

24 Ikariya, T., Ishii, Y., Kawano, H., Arai, T., Saburi, M., Yoshikawa, S., and Akutagawa, S. (1985) *J. Chem. Soc., Chem. Commun.*, 922–924.

25 (a) Noyori, R., Ohta, M., Hsiao, Y., and Kitamura, M. (1986) *J. Am. Chem. Soc.*, **108**, 7117–7119; (b) Takaya, H., Ohta, T., Sayo, N., Kumobayashi, H., Akutagawa, S., Inoue, S., Kasahara, I.,

and Noyori, R. (1987) *J. Am. Chem. Soc.*, **109**, 1596–1597; (c) Ohta, T., Takaya, H., Kitamura, M., Nagai, K., and Noyori, R. (1987) *J. Org. Chem.*, **52**, 3174–3176.

26 Blaser, H.-U., Malan, C., Pugin, B., Spindler, F., Steiner, H., and Studer M. (2003) *Adv. Synth. Catal.*, **345**, 103–151.

27 Schmidt, R. and Scalone M. (1999) in *Comprehensive Asymmetric Catalysis*, Vol. 3 (eds E.N. Jacobsen, A. Pfaltz, and H. Yamamoto), Springer-Verlag, Berlin, Heidelberg, pp. 1439–1460.

28 Zhang, X., Uemura, T., Matsumura, K., Sayo, N., Kumobayashi, H., and Takaya, H. (1994) *Synlett*, 501–503.

29 Noyori, R., Ohkuma, T., Kitamura, M., Takaya, H., Sayo, N., Kumobayashi, H., and Akutagawa, S. (1987) *J. Am. Chem. Soc.*, **109**, 5856–5858.

30 Ohkuma, T. and Noyori, R. (2007) in *The Handbook of Homogeneous Hydrogenation*, Vol. 3 (eds J.G. de Vries and C.J. Elsevier), Wiley-VCH Verlag GmbH, Weinheim, pp. 1105–1163.

31 (a) Kitamura, M., Tokunaga, M., Ohkuma, T., and Noyori, R. (1991) *Tetrahedron Lett.*, **32**, 4163–4166; (b) Kitamura, M., Tokunaga, M., Ohkuma, T., and Noyori, R. (1993) *Org. Synth.*, **71**, 1–13.

32 Mashima, K., Hino, T., and Takaya, H. (1992) *J. Chem. Soc. Dalton Trans.*, 2099–2107.

33 (a) King, S.A. and DiMichele, L. (1995) in *Catalysis of Organic Reactions* (eds M.G. Scaros and M.L. Prunier), Marcel Dekker, Inc., New York, pp. 157–166; (b) Ohta, T., Tonomura, Y., Nozaki, K., and Takaya, H. (1996) *Organometallics*, **15**, 1521–1523.

34 (a) Taber, D.F. and Silverberg, L.J. (1991) *Tetrahedron Lett.*, **32**, 4227–4230; (b) Heiser, B., Borger, E.A., and Crameri, Y. (1991) *Tetrahedron: Asymmetry*, **2**, 31–62; (c) Hoke, J.B., Hollis, L.S., and Stern, E.W. (1993) *J. Organomet. Chem.*, **455**, 193–196; (d) Genêt, J.P., Ratovelomanana-Vidal, V., Caño de Andrade, M.C., Pfister, X., Guerreiro, P., and Lenoir, J.Y. (1995) *Tetrahedron Lett.*, **36**, 4801–4804.

35 (a) Kitamura, M., Ohkuma, T., Inoue, S., Sayo, N., Kumobayashi, H., Akutagawa, S., Ohta, T., Takaya, H., and

Noyori, R. (1988) *J. Am. Chem. Soc.*, **110**, 629–631; (b) Kitamura, M., Ohkuma, T., Takaya, H., and Noyori, R. (1988) *Tetrahedron Lett.*, **29**, 1555–1556; (c) Nishi, T., Kitamura, M., Ohkuma, T., and Noyori, R. (1988) *Tetrahedron Lett.*, **29**, 6327–6330; (d) Ohkuma, T., Kitamura, M., and Noyori, R. (1990) *Tetrahedron Lett.*, **31**, 5509–5512.

36 Ohkuma, T. and Noyori, R. (1999) in *Comprehensive Asymmetric Catalysis*, Vol. 1 (eds E.N. Jacobsen, A. Pfaltz, and H. Yamamoto), Springer-Verlag, Berlin, Heidelberg, pp. 199–246.

37 Madec, J., Pfister, X., Phansavath, P., Ratovelomanana-Vidal, V., and Genêt, J.-P. (2001) *Tetrahedron*, **57**, 2563–2568.

38 Ohkuma, T. and Noyori, R. (2004) in *Comprehensive Asymmetric Catalysis*, Suppl 1 (eds E.N. Jacobsen, A. Pfaltz, and H. Yamamoto), Springer-Verlag, Berlin, Heidelberg, pp. 1–41.

39 (a) Noyori, R., Ikeda, T., Ohkuma, T., Widhalm, M., Kitamura, M., Takaya, H., Akutagawa, S., Sayo, N., Saito, T., Taketomi, T., and Kumobayashi, H. (1989) *J. Am. Chem. Soc.*, **111**, 9134–9135; (b) Kitamura, M., Ohkuma, T., Tokunaga, M., and Noyori, R. (1990) *Tetrahedron: Asymmetry*, **1**, 1–4.

40 Noyori, R., Tokunaga, M., and Kitamura, M. (1995) *Bull. Chem. Soc. Jpn.*, **68**, 36–56.

41 (a) Ohkuma, T., Koizumi, M., Doucet, H., Pham, T., Kozawa, M., Murata, K., Katayama, E., Yokozawa, T., Ikariya, T., and Noyori, R. (1998) *J. Am. Chem. Soc.*, **120**, 13529–13530; (b) Ohkuma, T., Koizumi, M., Muñiz, K., Hilt, G., Kabuto, C., and Noyori, R. (2002) *J. Am. Chem. Soc.*, **124**, 6508–6509.

42 (a) Noyori, R. and Ohkuma, T. (2001) *Angew. Chem. Int. Ed.*, **40**, 40–73; (b) Ohkuma, T. (2007) *J. Synth. Org. Chem.*, **65**, 1070–1080; (c) Ohkuma, T. (2010) *Proc. Jpn. Acad. Ser. B*, **86**, 202–219.

43 (a) Ohkuma, T., Ooka, H., Hashiguchi, S., Ikariya, T., and Noyori, R. (1995) *J. Am. Chem. Soc.*, **117**, 2675–2676; (b) Doucet, H., Ohkuma, T., Murata, K., Yokozawa, T., Kozawa, M., Katayama, E., England, A.F., Ikariya, T., and Noyori, R. (1998) *Angew. Chem. Int. Ed.*, **37**, 1703–1707.

44 (a) Ohkuma, T., Ooka, H., Ikariya, T., and Noyori, R. (1995) *J. Am. Chem. Soc.*, **117**, 10417–10418; (b) Ohkuma, T., Ooka, H., Yamakawa, M., Ikariya, T., and Noyori, R. (1996) *J. Org. Chem.*, **61**, 4872–4873; (c) Ohkuma, T., Ikehira, H., Ikariya, T., and Noyori, R. (1997) *Synlett*, 467–468; (d) Ohkuma, T., Doucet, H., Pham, T., Mikami, K., Korenaga, T., Terada, M., and Noyori, R. (1998) *J. Am. Chem. Soc.*, **120**, 1086–1087; (e) Ohkuma, T., Koizumi, M., Ikehira, H., Yokozawa, T., and Noyori, R. (2000) *Org. Lett.*, **2**, 659–662; (f) Ohkuma, T., Koizumi, M., Yoshida, M., and Noyori, R. (2000) *Org. Lett.*, **2**, 1749–1751; (g) Ohkuma, T., Ishii, D., Takeno, H., and Noyori, R. (2000) *J. Am. Chem. Soc.*, **122**, 6510–6511; (h) Ohkuma, T., Takeno, H., Honda, Y., and Noyori, R. (2001) *Adv. Synth. Catal.*, **343**, 369–375; (i) Ohkuma, T., Jing, L., and Noyori, R. (2004) *Synlett*, 1383–1386.

45 Ohkuma, T. and Noyori, R. (2004) in *Transition Metals for Organic Synthesis*, Vol. 2, 2nd edn (eds M. Beller and C. Bolm), Wiley-VCH Verlag GmbH, Weinheim, pp. 29–113.

46 (a) Wu, J., Chen, H., Kwok, W., Guo, R., Zhou, Z., Yeung, C., and Chan, A.S. C. (2002) *J. Org. Chem.*, **67**, 7908–7910; (b) Henschke, J.P., Burk, M.J., Malan, C. G., Herzberg, D., Peterson, J.A., Wildsmith, A.J., Cobley, C.J., and Casy, G. (2003) *Adv. Synth. Catal.*, **345**, 300–307.

47 (a) Sandoval, C.A., Ohkuma, T., Muñiz, K., and Noyori, R. (2003) *J. Am. Chem. Soc.*, **125**, 13490–13503; (b) Noyori, R., Sandoval, C.A., Muñiz, K., and Ohkuma, T. (2005) *Phil. Trans. R. Soc. A*, **363**, 901–912.

48 (a) Arai, N., Ooka, H., Azuma, K., Yabuuchi, T., Kurono, N., Inoue, T., and Ohkuma, T. (2007) *Org. Lett.*, **9**, 939–941; (b) Arai, N., Azuma, K., Nii, N., and Ohkuma, T. (2008) *Angew. Chem. Int. Ed.*, **47**, 7457–7460; (c) Ooka, H., Arai, N., Azuma, K., Kurono, N., and Ohkuma, T. (2008) *J. Org. Chem.*, **73**, 9084–9093.

49 Ohkuma, T., Hattori, T., Ooka, H., Inoue, T., and Noyori, R. (2004) *Org. Lett.*, **6**, 2681–2683.

50 (a) Ohkuma, T., Sandoval, C.A., Srinivasan, R., Lin, Q., Wei, Y., Muñiz, K., and Noyori, R. (2005) *J. Am. Chem. Soc.*, **127**, 8288–8289; (b) Arai, N., Suzuki, K., Sugizaki, S., Sorimachi, H., and Ohkuma, T. (2008) *Angew. Chem. Int. Ed.*, **47**, 1770–1773.

51 Tsutsumi, K., Katayama, T., Utsumi, N., Murata, K., Arai, N., Kurono, N., and Ohkuma, T. (2009) *Org. Process Res. Dev.*, **13**, 625–628.

52 Blaser, H.-U., Spindler, F., and Thommen, M. (2007) in *The Handbook of Homogeneous Hydrogenation*, Vol. 3 (eds J.G. de Vries and C.J. Elsevier), Wiley-VCH Verlag GmbH, Weinheim, pp. 1279–1324.

53 Akutagawa, S. and Tani, K. (2000) in *Catalytic Asymmetric Synthesis*, 2nd edn (ed. I. Ojima), John Wiley & Sons, Inc., New York, pp. 145–161.

54 (a) Tani, K., Yamagata, T., Akutagawa, S., Kumobayashi, H., Taketomi, T., Takaya, H., Miyashita, A., Noyori, R., and Otsuka, S. (1984) *J. Am. Chem. Soc.*, **106**, 5208–5217; (b) Inoue, S., Takaya, H., Tani, K., Otsuka, S., Sato, T., and Noyori, R. (1990) *J. Am. Chem. Soc.*, **112**, 4897–4905; (c) Tani, K., Yamagata, T., Otsuka, S., Kumobayashi, H., and Akutagawa, S. (1989) *Org. Synth.*, **67**, 33–39.

55 Kitamura, M., Manabe, K., and Noyori, R. (1987) *Tetrahedron Lett.*, **28**, 4719–4720.

56 Hiroya, K., Kurihara, Y., and Ogasawara, K. (1995) *Angew. Chem. Int. Ed. Engl.*, **34**, 2287–2289.

57 Hayashi, T. (1999) in *Comprehensive Asymmetric Catalysis*, Vol. 1 (eds E.N. Jacobsen, A. Pfaltz, and H. Yamamoto), Springer-Verlag, Berlin, Heidelberg, pp. 351–364.

58 (a) Hayashi, T., Matsumoto, Y., and Ito, Y. (1989) *J. Am. Chem. Soc.*, **111**, 3426–3428; (b) Hayashi, T., Matsumoto, Y., and Ito Y. (1991) *Tetrahedron: Asymmetry*, **2**, 601–612.

59 (a) Hayashi, T. (1999) in *Comprehensive Asymmetric Catalysis*, Vol. 1 (eds E.N. Jacobsen, A. Pfaltz, and H. Yamamoto), Springer-Verlag, Berlin, Heidelberg, pp. 319–333; (b) Nishiyama, H. and Itoh, K. (2000) in *Catalytic Asymmetric Synthesis*, 2nd edn (ed. I. Ojima), John Wiley & Sons, Inc., New York, pp. 111–143.

60 (a) Bergens, S.H., Noheda, P., Whelan, J., and Bosnich, B. (1992) *J. Am. Chem. Soc.*, **114**, 2121–2128; (b) Wang, X. and Bosnich, B. (1994) *Organometallics*, **13**, 4131–4133.

61 Hayashi, T., Matsumoto, Y., and Ito, Y. (1988) *J. Am. Chem. Soc.*, **110**, 5579–5581.

62 Lipshutz, B.H., Servesko, J.M., Papa, P.P., and Lover, A.A. (2004) *Org. Lett.*, **6**, 1273–1275.

63 (a) Appella, D.H., Moritani, Y., Shintani, R., Ferreira, E.M., and Buchwald, S.L. (1999) *J. Am. Chem. Soc.*, **121**, 9473–9474; (b) Hughes, G., Kimura, M., and Buchwald, S.L. (2003) *J. Am. Chem. Soc.*, **125**, 11253–11258; (c) Lipshutz, B.H., Servesko, J.M., and Taft, B.R. (2004) *J. Am. Chem. Soc.*, **126**, 8352–8353.

64 (a) Lipshutz, B.H., Noson, K., and Chrisman, W. (2001) *J. Am. Chem. Soc.*, **123**, 12917–12918; (b) Lipshutz, B.H., Noson, K., Chrisman, W., and Lower, A. (2003) *J. Am. Chem. Soc.*, **125**, 8779–8789.

65 Lipshutz, B.H. and Shimizu, H. (2004) *Angew. Chem. Int. Ed.*, **43**, 2228–2230.

66 Sirol, S., Courmarcel, J., Mostefai, N., and Riant, O. (2001) *Org. Lett.*, **3**, 4111–4113.

67 (a) Wu, X.-M., Funakoshi, K., and Sakai, K. (1992) *Tetrahedron Lett.*, **33**, 6331–6334; (b) Barnhart, R.W., Wang, X., Noheda, P., Bergens, S.H., Whelan, J., and Bosnich, B. (1994) *J. Am. Chem. Soc.*, **116**, 1821–1830.

68 Tanaka, K. and Fu, G.C. (2002) *J. Am. Chem. Soc.*, **124**, 10296–10297.

69 (a) Shen, Z., Kahn, H.A., and Dong, V.M. (2008) *J. Am. Chem. Soc.*, **130**, 2916–2917; (b) Shen, Z., Dornan, P.K., Woo, T.K., and Dong, V.M. (2009) *J. Am. Chem. Soc.*, **131**, 1077–1091.

70 Dorta, R., Egli, P., Zürcher, F., and Togni, A. (1997) *J. Am. Chem. Soc.*, **119**, 10857–10858.

71 LaLonde, R.L., Sherrry, B.D., Kang, E.J., and Toste, F.D. (2007) *J. Am. Chem. Soc.*, **129**, 2452–2453.

72 Kawatsura, M. and Hartwig, J.F. (2000) *J. Am. Chem. Soc.*, **122**, 9546–9547.

73 Yamaguchi, M., Shima, T., Yamagishi, T., and Hida, M. (1990) *Tetrahedron Lett.*, **31**, 5049–5052.
74 (a) Pfaltz, A. and Lautens, M. (1999) in *Comprehensive Asymmetric Catalysis*, Vol. 2 (eds E.N. Jacobsen, A. Pfaltz, and H. Yamamoto), Springer-Verlag, Berlin, Heidelberg, pp. 833–884; (b) Trost, B.M. and Lee, C. (2000) in *Catalytic Asymmetric Synthesis*, 2nd edn (ed. I. Ojima), John Wiley & Sons, Inc., New York, pp. 593–649.
75 Fuji, K., Kinoshita, N., and Tanaka, K. (1999) *Chem. Commun.*, 1895–1896.
76 Kuwano, R. and Ito, Y. (1999) *J. Am. Chem. Soc.*, **121**, 3236–3237.
77 (a) Shibasaki, M. and Vogl, E.M. (1999) in *Comprehensive Asymmetric Catalysis*, Vol. 1 (eds E.N. Jacobsen, A. Pfaltz, and H. Yamamoto), Springer-Verlag, Berlin, Heidelberg, pp. 457–487; (b) Donde, Y. and Overman, L.E. (2000) in *Catalytic Asymmetric Synthesis*, 2nd edn (ed. I. Ojima), John Wiley & Sons, Inc., New York, pp. 675–697; (c) Dounay, A.B. and Overman, L.E. (2003) *Chem. Rev.*, **103**, 2945–2963; (d) Tietze, L.F., Ila, H., and Bell, H.P. (2004) *Chem. Rev.*, **104**, 3453–3516.
78 Sato, Y., Watanabe, S., and Shibasaki, M. (1992) *Tetrahedron Lett.*, **33**, 2589–2592.
79 (a) Ashimori, A., Bachand, B., Overman, L.E., and Poon, D.J. (1998) *J. Am. Chem. Soc.*, **120**, 6477–6487; (b) Overman, L.E. and Poon, D.J. (1997) *Angew. Chem. Int. Ed. Engl.*, **36**, 518–521.
80 (a) Ozawa, F., Kubo, A., and Hayashi, T. (1991) *J. Am. Chem. Soc.*, **113**, 1417–1419; (b) Ozawa, F., Kubo, A., and Hayashi, T. *Tetrahedron Lett.*, **33**, 1485–1488.
81 Tschoerner, M. and Pregosin, P.S. (1999) *Organometallics*, **18**, 670–678.
82 Tietze, L.F., Thede, K., and Sannicolò, F. (1999) *Chem. Commun.*, 1811–1812.
83 Ozawa, F., Kobatake, Y., and Hayashi, T. (1993) *Tetrahedron Lett.*, **34**, 2505–2508.
84 Tietze, L.F. and Thede, K. (2000) *Synlett*, 1470–1472.
85 (a) Carreira, E.M. (1999) in *Comprehensive Asymmetric Catalysis*, Vol. 3 (eds E.N. Jacobsen, A. Pfaltz, and H. Yamamoto), Springer-Verlag, Berlin, Heidelberg, pp. 997–1065; (b) Carreira, E.M. (2000) in *Catalytic Asymmetric Synthesis*, 2nd edn (ed. I. Ojima), John Wiley & Sons, Inc., New York, pp. 513–541.
86 Sodeoka, M., Tokunoh, R., Miyazaki, F., Hagiwara, E., and Shibasaki, M. (1997) *Synlett*, 463–466.
87 Fujimura, O. (1998) *J. Am. Chem. Soc.*, **120**, 10032–10039.
88 Naodovic, M. and Yamamoto, H. (2008) *Chem. Rev.*, **108**, 3132–3148.
89 Yanagisawa, A., Nakatsuka, Y., Asakawa, K., Wadamoto, M., Kageyama, H., and Yamamoto, H. (2001) *Bull. Chem. Soc. Jpn.*, **74**, 1477–1484.
90 Yanagisawa, A., Matsumoto, Y., Nakashima, H., Asakawa, K., and Yamamoto, H. (1997) *J. Am. Chem. Soc.*, **119**, 9319–9320.
91 (a) Momiyama, N. and Yamamoto, H. (2003) *J. Am. Chem. Soc.*, **125**, 6038–6039; (b) Momiyama, N. and Yamamoto, H. (2004) *J. Am. Chem. Soc.*, **126**, 5360–5361.
92 (a) Krüger, J. and Carreira, E.M. (1998) *J. Am. Chem. Soc.*, **120**, 837–838; (b) Pagenkopf, B.L., Krüger, J., Stojanovic, A., and Carreira, E.M. (1998) *Angew. Chem. Int. Ed.*, **37**, 3124–3126.
93 Moreau, X., Bazán-Tejeda, B., and Campagne, J.-M. (2005) *J. Am. Chem. Soc.*, **127**, 7288–7289.
94 Hamashima, Y., Sasamoto, N., Hotta, D., Somei, H., Umebayashi, N., and Sodeoka, M. (2005) *Angew. Chem. Int. Ed.*, **44**, 1525–1529.
95 (a) Yanagisawa, A. (1999) in *Comprehensive Asymmetric Catalysis*, Vol. 2 (eds E.N. Jacobsen, A. Pfaltz, and H. Yamamoto), Springer-Verlag, Berlin, Heidelberg, pp. 965–979; (b) Denmark, S.E. and Fu, J. (2003) *Chem. Rev.*, **103**, 2763–2793.
96 Yanagisawa, A., Kageyama, H., Nakatsuka, Y., Asakawa, K., Matsumoto, Y., and Yamamoto, H. (1999) *Angew. Chem. Int. Ed.*, **38**, 3701–3703.
97 Wadamoto, M. and Yamamoto, H. (2005) *J. Am. Chem. Soc.*, **127**, 14556–14557.
98 Tomita, D., Wada, R., Kanai, M., and Shibasaki, M. (2005) *J. Am. Chem. Soc.*, **127**, 4138–4139.

99 Saaby, S., Fang, X., Gathergood, N., and Jørgensen, K.A. (2000) *Angew. Chem. Int. Ed.*, **39**, 4114–4116.
100 Cho, C.-W. and Krische, M.J. (2007) in *The Handbook of Homogeneous Hydrogenation*, Vol. **2** (eds J.G. de Vries and C.J. Elsevier), Wiley-VCH Verlag GmbH, Weinheim, pp. 713–741.
101 Kong, J.R. and Krische, M.J. (2006) *J. Am. Chem. Soc.*, **128**, 16040–16041.
102 Komanduri, V. and Krische, M.J. (2006) *J. Am. Chem. Soc.*, **128**, 16448–16449.
103 Kurono, N., Arai, K., Uemura, M., and Ohkuma, T. (2008) *Angew. Chem. Int. Ed.*, **47**, 6643–6646.
104 Kurono, N., Uemura, M., and Ohkuma, T. (2010) *Eur. J. Org. Chem.*, 1455–1459.
105 Hamashima, Y. and Sodeoka, M. (2007) *J. Synth. Org. Chem., Jpn.*, **65**, 1099–1107.
106 Hamashima, Y., Yagi, K., Takano, H., Tamás, L., and Sodeoka, M. (2002) *J. Am. Chem. Soc.*, **124**, 14530–14531.
107 (a) Hamashima, Y., Suzuki, T., Shimura, Y., Shimizu, T., Umebayashi, N., Tamura, T., Sasamoto, N., and Sodeoka, M. (2005) *Tetrahedron Lett.*, **46**, 1447–1450; (b) Suzuki, T., Goto, T., Hamashima, Y., and Sodeoka, M. (2007) *J. Org. Chem.*, **72**, 246–250; (c) Hamashima, Y., Suzuki, T., Takano, H., Shimura, Y., and Sodeoka, M. (2005) *J. Am. Chem. Soc.*, **127**, 10164–10165; (d) Kim, S.M., Kim, H.R., and Kim, D.Y. (2005) *Org. Lett.*, **7**, 2309–2311.
108 Kang, Y.K. and Kim, D.Y. (2006) *Tetrahedron Lett.*, **47**, 4565–4568.
109 Lee, J.H., Bang, H.T., and Kim, D.Y. (2008) *Synlett*, 1821–1824.
110 Åhman, J., Wolfe, J.P., Troutman, M.V., Palucki, M., and Buchwald, S.L. (1998) *J. Am. Chem. Soc.*, **120**, 1918–1919.
111 Spielvogel, D.J. and Buchwald, S.L. (2002) *J. Am. Chem. Soc.*, **124**, 3500–3501.
112 Evans, D.A. and Thomson, R.J. (2005) *J. Am. Chem. Soc.*, **127**, 10506–10507.
113 (a) Yamaguchi, M. (1999) in *Comprehensive Asymmetric Catalysis*, Vol. **3** (eds E.N. Jacobsen, A. Pfaltz, and H. Yamamoto), Springer-Verlag, Berlin, Heidelberg, pp. 1121–1139; (b) Kanai, M. and Shibasaki, M. (2000) in *Catalytic Asymmetric Synthesis*, 2nd edn (ed. I. Ojima), John Wiley & Sons, Inc., New York, pp. 569–592.
114 (a) Hamashima, Y., Hotta, D., and Sodeoka, M. (2002) *J. Am. Chem. Soc.*, **124**, 11240–11241; (b) Hamashima, Y., Hotta, D., Umebayashi, N., Tsuchiya, Y., Suzuki, T., and Sodeoka, M. (2005) *Adv. Synth. Catal.*, **347**, 1576–1586.
115 Evans, D.A., Thomson, R.J., and Franco, F. (2005) *J. Am. Chem. Soc.*, **127**, 10816–10817.
116 Li, K., Cheng, X., and Hii, K.K. (2004) *Eur. J. Org. Chem.*, 959–964.
117 (a) Hamashima, Y., Somei, H., Shimura, Y., Tamura, T., and Sodeoka, M. (2004) *Org. Lett.*, **6**, 1861–1864; (b) Li, K. and Hii, K.K. (2003) *Chem. Commun.*, 1132–1133.
118 (a) Hayashi, T. and Yamasaki, K. (2003) *Chem. Rev.*, **103**, 2829–2844; (b) Shintani, R. and Hayashi, T. (2007) in *New Frontiers in Asymmetric Catalysis* (eds K. Mikami and M. Lautens), John Wiley & Sons, Inc., Hoboken, New Jersey, pp. 59–100; (c) Christoffers, J., Koripelly, G., Rosiak, A., and Rössle, M. (2007) *Synthesis*, 1279–1300.
119 Hayashi, T. (2004) *Bull. Chem. Soc. Jpn.*, **77**, 13–21.
120 (a) Takaya, Y., Ogasawara, M., and Hayashi, T. (2004) *J. Am. Chem. Soc.*, **120**, 5579–5580; (b) Hayashi, T., Takahashi, M., Takaya, Y., and Ogasawara, M. (2002) *J. Am. Chem. Soc.*, **124**, 5052–5058.
121 (a) Takaya, Y., Senda, T., Kurushima, H., Ogasawara, M., and Hayashi, T. (1999) *Tetrahedron: Asymmetry*, **10**, 4047–4056; (b) Sakuma, S., Sakai, M., Itooka, R., and Miyaura, N. (2000) *J. Org. Chem.*, **65**, 5951–5955; (c) Pucheault, M., Michaut, V., Darses, S., and Genêt, J.P. (2004) *Tetrahedron Lett.*, **45**, 4729–4732; (d) Senda, T., Ogasawara, M., and Hayashi, T. (2001) *J. Org. Chem.*, **66**, 6852–6856; (e) Sakuma, S. and Miyaura, N. (2001) *J. Org. Chem.*, **66**, 8944–8946; (f) Hayashi, T., Senda, T., Takaya, Y.,

and Ogasawara, M. (1999) *J. Am. Chem. Soc.*, **121**, 11591–11592; (g) Hayashi, T., Senda, T., and Ogasawara, M. (2000) *J. Am. Chem. Soc.*, **122**, 10716–10717.

122 (a) Oi, S., Taira, A., Honma, Y., and Inoue, Y. (2003) *Org. Lett.*, **5**, 97–99; (b) Otomaru, Y. and Hayashi, T. (2004) *Tetrahedron: Asymmetry*, **15**, 2647–2651.

123 Takaya, Y., Ogasawara, M., and Hayashi, T. (1998) *Tetrahedron Lett.*, **39**, 8479–8482.

124 Oi, S., Sato, T., and Inoue, Y. (2003) *Tetrahedron Lett.*, **45**, 5051–5055.

125 Wang, S.-Y., Ji, S.-J., and Loh, T.-P. (2007) *J. Am. Chem. Soc.*, **129**, 276–277.

126 (a) Evans, D.A. and Johnson, J.S. (1999) in *Comprehensive Asymmetric Catalysis*, Vol. **3** (eds E.N. Jacobsen, A. Pfaltz, and H. Yamamoto), Springer-Verlag, Berlin, Heidelberg, pp. 1177–1235; (b) Maruoka, K. (2000) in *Catalytic Asymmetric Synthesis*, 2nd edn (ed. I. Ojima), John Wiley & Sons, Inc., New York, pp. 467–491.

127 (a) Oi, S., Kawagishi, K., and Inoue, Y. (1998) *Tetrahedron Lett.*, **39**, 6253–6256; (b) Ghosh, A.K. and Matsuda, H. (1999) *Org. Lett.*, **1**, 2157–2159.

128 (a) Ooi, T. and Maruoka, K. (1999) in *Comprehensive Asymmetric Catalysis*, Vol. **3** (eds E.N. Jacobsen, A. Pfaltz, and H. Yamamoto), Springer-Verlag, Berlin, Heidelberg, pp. 1237–1254; (b) Pellissier, H. (2009) *Tetrahedron*, **65**, 2839–2877.

129 Oi, S., Terada, E., Ohuchi, K., Kato, T., Tachibana, Y., and Inoue, Y. (1999) *J. Org. Chem.*, **64**, 8660–8667.

130 Mikami, K., Kakuno, H., and Aikawa, K. (2005) *Angew. Chem. Int. Ed.*, **44**, 7257–7260.

131 Hori, K., Kodama, H., Ohta, T., and Fukukawa, I. (1999) *J. Org. Chem.*, **64**, 5017–5023.

132 Yao, S., Johannsen, M., Hazell, R.G., and Jørgensen, K.A. (1998) *Angew. Chem. Int. Ed.*, **37**, 3121–3124.

133 Yamamoto, Y. and Yamamoto, H. (2004) *J. Am. Chem. Soc.*, **126**, 4128–4129.

134 (a) Mikami, K. and Terada, M. (1999) in *Comprehensive Asymmetric Catalysis*, Vol. **3** (eds E.N. Jacobsen, A. Pfaltz, and H. Yamamoto), Springer-Verlag, Berlin, Heidelberg, pp. 1143–1174; (b) Mikami, K. and Nakai, T. (2000) in *Catalytic Asymmetric Synthesis*, 2nd edn (ed. I. Ojima), John Wiley & Sons, Inc., New York, pp. 543–568.

135 Drury, III, W.J., Ferraris, D., Cox, C., Young, B., and Lectka, T. (1998) *J. Am. Chem. Soc.*, **120**, 11006–11007.

136 Mikami, K., Aikawa, K., Kainuma, S., Kawakami, Y., Saito, T., Sayo, N., and Kumobayashi, H. (2004) *Tetrahedron: Asymmetry*, **15**, 3885–3889.

137 Doherty, S., Knight, J.G., Smyth, C.H., Harrington, R.W., and Clegg, W. (2006) *J. Org. Chem.*, **71**, 9751–9764.

138 Hatano, M., Terada, M., and Mikami, K. (2001) *Angew. Chem. Int. Ed.*, **40**, 249–253.

139 Hatano, M. and Mikami, K. (2003) *J. Am. Chem. Soc.*, **125**, 4704–4705.

140 (a) Lei, A., He, M., Wu, S., and Zhang, X. (2002) *Angew. Chem. Int. Ed.*, **41**, 3457–3460; (b) Lei, A., Waldkirch, J.P., He, M., and Zhang, X. (2002) *Angew. Chem. Int. Ed.*, **41**, 4526–4529.

141 Jang, H.-Y., Hughes, F.W., Gong, H., Zhang, J., Brodbelt, J.S., and Krische, M.J. (2005) *J. Am. Chem. Soc.*, **131**, 6174–6175.

142 Nishimura, T., Kawamoto, T., Sasaki, K., Tsurumaki, E., and Hayashi, T. (2007) *J. Am. Chem. Soc.*, **129**, 1492–1493.

143 Larksarp, C. and Alper, H. (1997) *J. Am. Chem. Soc.*, **119**, 3709–3715.

144 Melhado, A.D., Luparia, M., and Toste, F.D. (2007) *J. Am. Chem. Soc.*, **129**, 12638–12639.

145 Wender, P.A., Haustedt, L.O., Lim, J., Love, J.A., Williams, T.J., and Yoon, J.Y. (2006) *J. Am. Chem. Soc.*, **128**, 6302–6303.

146 Tanaka, K. (2007) *Synlett*, 1977–1993.

147 Tanaka, K., Nishida, G., Wada, A., and Noguchi, K. (2004) *Angew. Chem. Int. Ed.*, **43**, 6510–6512.

148 Tanaka, K., Nishida, G., Ogino, M., Hirano, M., and Noguchi, K. (2005) *Org. Lett.*, **7**, 3119–3121.

149 Tanaka, K., Sasage, H., Toyoda, K., Noguchi, K., and Hirano, M. (2007) *J. Am. Chem. Soc.*, **129**, 1522–1523.

150 Evans, P.A., Lai, K.W., and Sawyer, J.R. (2005) *J. Am. Chem. Soc.*, **127**, 12466–12467.

151 Tanaka, K., Osaka, T., Noguchi, K., and Hirano, M. (2007) *Org. Lett.*, **9**, 1307–1310.

152 Buchwald, S.L. and Hicks, F.A. (1999) in *Comprehensive Asymmetric Catalysis*, Vol. **2** (eds E.N. Jacobsen, A. Pfaltz, and H. Yamamoto), Springer-Verlag, Berlin, Heidelberg, pp. 491–510.

153 Hiroi, K., Watanabe, T., Kawagishi, R., and Abe, I. (2000) *Tetrahedron Lett.*, **41**, 891–895.

154 Gibson, S.E., Lewis, S.E., Loch, J.A., Steed, J.W., and Tozer, M.J. (2003) *Organometallics*, **22**, 5382–5384.

155 Gibson, S.E., Kaufmann, K.A.C., Loch, J.A., Steed, J.W., and White, A.J.P. (2005) *Chem. Eur. J.*, **11**, 2566–2576.

156 Schmid, T.M. and Consiglio, G. (2004) *Tetrahedron: Asymmetry*, **15**, 2205–2208.

157 Jeong, N., Sung, B.K., and Choi, Y.K. (2000) *J. Am. Chem. Soc.*, **122**, 6771–6772.

158 Shibata, T. and Takagi, T. (2000) *J. Am. Chem. Soc.*, **122**, 9852–9853.

159 (a) Schmid, T.M. and Consiglio, G. (2004) *Chem. Commun.*, 2318–2319; (b) Kwong, F.Y., Lee, H.W., Qui, L., Lam, W.H., Li, Y.-M., Kwong, H.L., and Chan, A.S.C. (2005) *Adv. Synth. Catal.*, **347**, 1750–1754; (c) Kim, D.E., Choi, C., Kim, I.S., Jeulin, S., Ratovelomanana-Vidal, V., Genêt, J.-P., and Jeong, N. (2007) *Adv. Synth. Catal.*, **349**, 1999–2006; (d) Kwong, F.Y., Li, Y.-M., Lam, W.H., Qui, L., Lee, H.W., Yeung, C.H., Chan, K.S., and Chan, A.S.C. (2005) *Chem. Eur. J.*, **11**, 3872–3880.

160 Lautens, M., Fagnou, K., and Hiebert, S. (2003) *Acc. Chem. Res.*, **36**, 48–58.

161 Lautens, M., Chiu, P., Ma, S., and Rovis, T. (1995) *J. Am. Chem. Soc.*, **117**, 532–533.

162 (a) Lautens, M. and Rovis, T. (1997) *J. Org. Chem.*, **62**, 5246–5247; (b) Lautens, M. and Rovis, T. (1997) *J. Am. Chem. Soc.*, **119**, 11090–11091.

2
Bisphosphacycles – From DuPhos and BPE to a Diverse Set of Broadly Applied Ligands

Weicheng Zhang and Xumu Zhang

2.1
Introduction

Since Burk's pioneering work on DuPhos and BPE [1], bisphosphacycle ligands have been recognized as a versatile ligand class due to their broad applications in transition metal catalyzed asymmetric reactions. Dramatic advances in this technology have led to efficient synthesis of chiral building blocks for natural products and medicinal compounds. Starting with DuPhos and BPE, this chapter presents an overview of representative bisphosphacycle ligands and showcases their important applications in asymmetric catalysis [2].

2.2
Development of Bisphosphacycle Ligands

2.2.1
Structural Features of DuPhos and BPE

DuPhos (**1**) and BPE (**2**) are composed of a bridging scaffold (1,2-phenylene or ethylene) and two chelating phosphacycles. The original synthesis relies on a key intermediate **3** to prepare different *trans*-2,5-dialkyl substituted phospholanes (Scheme 2.1) [3]. The crystal structure of [Rh(*(S,S)*-Me-DuPhos)(nbd)]BF$_4$ shows a well-defined chiral environment around Rh, wherein a pair of diagonal quadrants are blocked by two protruding methyl groups (Figure 2.1) [4]. Hence the steric environment of **1** and **2** can be adjusted through systematic variation of the R groups. However, Burk's method fails to afford 2,5-diaryl analogues due to the base-sensitive nature of the corresponding cyclic sulfate or bismesylate. Later, based on Fiaud's synthesis and resolution of **4** [5a, b], Ph-BPE (**5**) was prepared via an alternative route [5c] (Scheme 2.2a). This approach was also adopted to explore

Privileged Chiral Ligands and Catalysts. Edited by Qi-Lin Zhou
Copyright © 2011 WILEY-VCH Verlag GmbH & Co. KGaA, Weinheim
ISBN: 978-3-527-32704-1

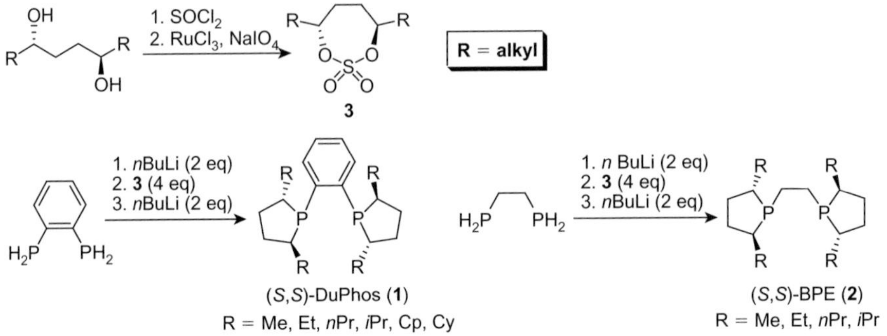

Scheme 2.1 Burk's synthesis of DuPhos and BPE.

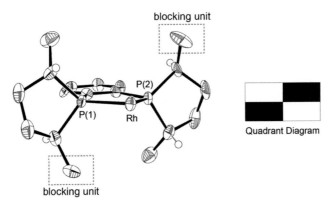

Figure 2.1 X-ray crystal structure of [Rh{(S,S)-Me-DuPhos}(nbd)]BF$_4$ with ellipsoids drawn at 50% probability (nbd, BF$_4$, and non-stereogenic hydrogen atoms are omitted for clarity) [4].

a methylene bridged BPE-analogue BPM (**6**) for asymmetric hydrogenation [5d] (Scheme 2.2b).

The prototypical DuPhos and BPE ligand series have been successfully applied in various asymmetric catalytic reactions, particularly asymmetric hydrogenation [1]. Together with other broad spectrum chiral ligands and catalysts they are considered as privileged structures [6]. A prominent feature of DuPhos is structural modularity, which not only facilitates fine-tuning of its steric and electronic properties, but also inspires the design of new ligands. Towards this goal, two fundamental strategies are backbone variation and modification of the phosphacycles. In addition, it is worth exploring dissimilar ligand structures, which is more challenging but likely to realize interesting new catalytic scopes.

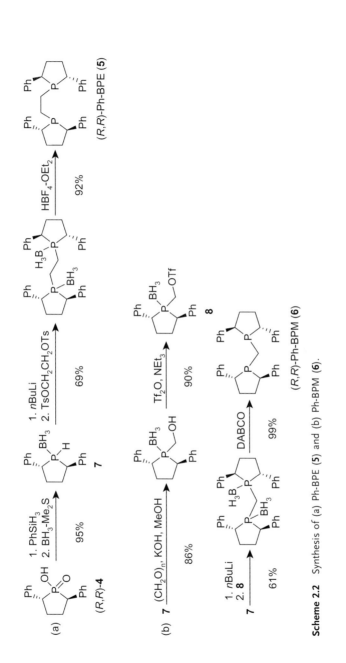

Scheme 2.2 Synthesis of (a) Ph-BPE (**5**) and (b) Ph-BPM (**6**).

2.2.2
Strategies of Ligand Design

2.2.2.1 Backbone Variation

The importance of backbone variation has been documented in the development of DuPhos [3]. The backbone mobility of BPE would generate a mixture of equilibrating conformations upon chelation to Rh, as exemplified with two limiting conformers (λ and δ, Figure 2.2). According to molecular modeling, only the former renders a desirable asymmetric environment, whereas the latter is less efficient for chiral induction. To eliminate conformational ambiguity, an aromatic backbone was introduced. The refined ligand – DuPhos – generally displays better enantiodifferentiating capability than BPE in Rh-catalyzed asymmetric hydrogenation reactions [3, 7]. However, for hindered substrates such as β,β-disubstituted α-dehydroamino acid derivatives [8a] and α-aryl enamides [8b], flexible and more sterically accessible Me-BPE offers better enantioselectivities (Table 2.1).

An alternative way to fix the mobility of BPE is attachment of new stereogenic centers to its aliphatic backbone. A matching combination of chiralities from the backbone and the phosphacycles shall reinforce enantioselectivity. An elegant example is the design of cyclopentane based ligands **9** and **10** (Figure 2.3) [9]. Their restricted cyclic backbones result in more rigid structures than BPE. Yet only the matched diastereomer (**10**) leads to improved ee in Rh-catalyzed hydrogenation of methyl α-acetamidocinnamate (MAC). In a parallel study, Burk also described the beneficial effect of backbone chirality in ligand design [10]. Insertion of one methylene unit into the BPE linker results in a more flexible ligand (**11**), which unsurprisingly showed moderate enantioselectivity (60% ee) toward MAC. Nevertheless, such a deleterious increase in conformational uncertainty is remedied by installing two new chiral centers on the backbone. Owing to enhanced rigidity, both of the modified ligands **12** and **13** gave better ees than **11**, while the stereochemically matched ligand **13** leads to more effective chiral induction.

Following the Burk route (Scheme 2.1), employing other primary diphosphines as the starting material generates new DuPhos type ligands with different sp^2 carbon linkers, including UlluPHOS (**14**) [11], Butiphane (**15**) [12], R-5-Fc (**16**, R = Me, Et) [13], Kephos (**17**) [14], MPL-SegPhos (**18**) [15], duxantphospholane (**19**) [16], and duthixantphospholane (**20**) [16] (Figure 2.4). Starting from resolved phospholanic acid **4**, Ph-Pyrazine (**21**) and Ph-Quinoxaline (**22**) were prepared via the monophospholane route (Scheme 2.2) [17]. To prepare Ph-5-Fc (**16**, R = Ph),

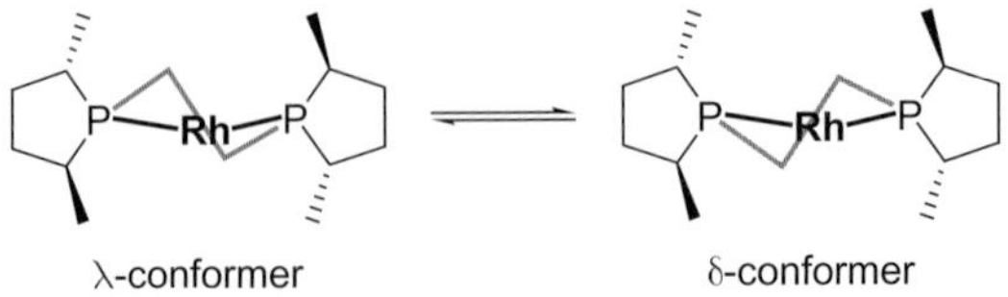

Figure 2.2 Conformational equilibration of Rh/Me-BPE catalyst precursor in solution.

Table 2.1 Ligand screening results in Rh-catalyzed asymmetric hydrogenation reactions.

Substrate	ee (%) by R-DuPhos (1)				ee (%) by R-BPE (2)				Reference
	Me-1	Et-1	nPr-1	iPr-1	Me-2	Et-2	nPr-2	iPr-2	
Ph−(COOMe)(NHAc)	98	99	>99	87	85	93	92	93	[3]
Me₂C=C(COOMe)(NHAc)	96	74	45	14	98.2	–	–	–	[8a]
Ph−C(=CH₂)NHAc	94.7	–	–	–	95.2	–	–	–	[8b]
(2-Me-C₆H₄)−C(=CH₂)NHAc	58.0	–	–	–	74.8	–	–	–	[8b]
(2-Br-C₆H₄)−C(=CH₂)NHAc	62.1	–	–	–	81.3	–	–	–	[8b]

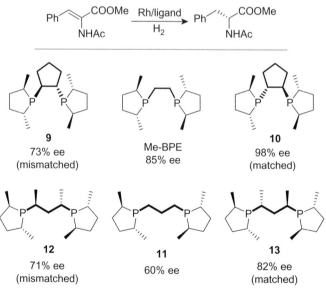

Figure 2.3 Matched/mismatched effect of BPE-analogues on asymmetric hydrogenation of methyl α-acetamidocinnamate (MAC).

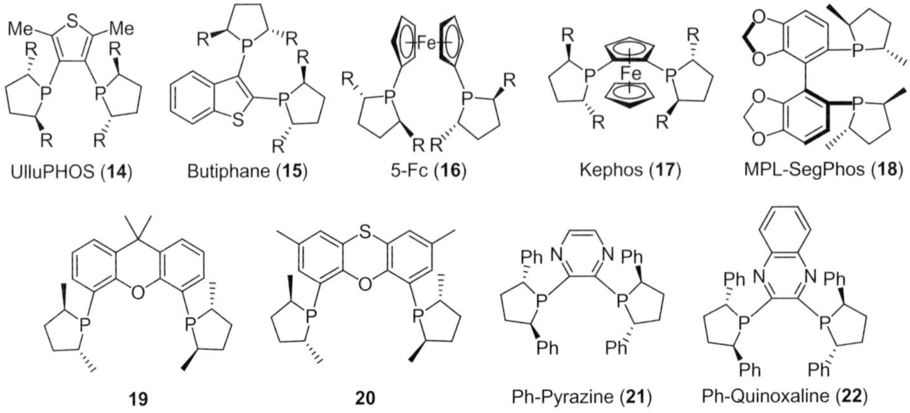

Figure 2.4 DuPhos-type ligands with various sp² carbon linkers.

a straightforward synthesis was developed involving dilithiation of ferrocene followed by electrophilic quenching with 2,5-diphenylphospholane chloride [17] (Scheme 2.3).

Practically, a major issue in asymmetric catalysis is matching the steric and electronic properties of the ligand with those of a specific substrate of interest. Thus it is desirable to have a comprehensive ligand library with varying geometric, steric, and electronic properties. An illustrative example is the catASium® ligand series (23a–q) [18]. Their preparation features late stage assembly of ligand backbone, which allows convenient access to a large set of modular bisphospholanes (Figure 2.5) [18c]. These ligands have been investigated mainly in Rh-catalyzed hydrogenation reactions. While a five-membered maleic anhydride or maleimide backbone usually gives superior enantioselectivities, no general trend could be deduced, and the solvent was identified as a critical factor for optimization.

2.2.2.2 Modification of the Phosphacycles

Modification of the chelating phosphacycles occurs in close proximity to the coordination site, thereby exerting a strong impact on the enantiodifferentiation process. Electronic permutation, which correlates to the activity of the whole organometallic catalyst, may also result from such change. Based on these considerations, numerous ligands have been created (Figure 2.6). For example, CnrPHOS (24) [19] and BPE-4 (25) [20] were reported as four-membered-ring

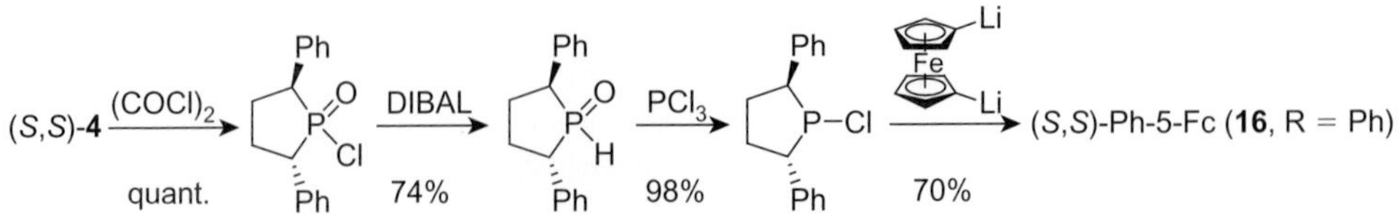

Scheme 2.3 Synthesis of Ph-5-Fc (16, R = Ph).

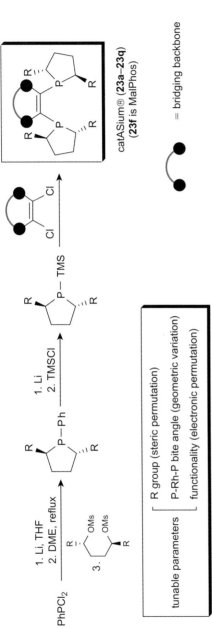
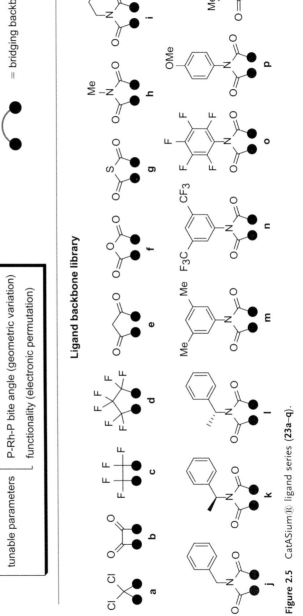

Figure 2.5 CatASium® ligand series (**23a–q**).

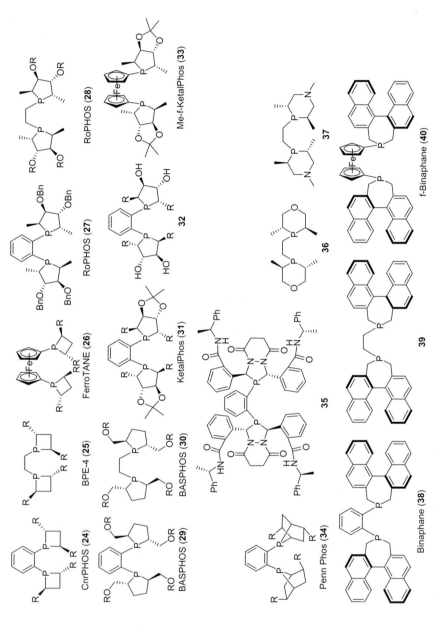

Figure 2.6 Ligands bearing modified phosphacycles.

2.2 Development of Bisphosphacycle Ligands

analogues of DuPhos (**1**) and BPE (**2**), respectively. To enhance catalyst activity as well as selectivity, Burk and coworkers designed FerroTANE (**26**), which combines a flexible and electron-rich ferrocene backbone with strained phosphetanes [21, 22]. This ligand gave excellent results in Rh-catalyzed hydrogenation of itaconates, succinimide derivatives, and (*E*)-β-dehydroamino acids [4, 21]. Analysis of the crystal structure of [Rh(Et-FerroTANE)(NBD)]BF$_4$ reveals a remarkably large P–Rh–P bite angle of 98.3°, which is in contrast to an average value of 84.5° calculated from other DuPhos type ligands [4].

Since homochiral 1,4-diol for the synthesis of phospholanes is not readily available, natural D-mannitol is a suitable replacement. Based on this chiral pool strategy, Holz and Borner developed RoPhos (**27, 28**) and BASPHOS (**29, 30**) [23], while Zhang [24] and RajanBabu [25] independently reported KetalPhos (**31**) and the related ligand **32**. Further, Zhang's group sought a 1,1′-ferrocenyl group as ligand backbone and developed Me-f-KetalPhos (**33**) [26]. Compared with structurally relevant ligands Me-Fc-5 (**16**) and Me-FerroTANE (**26**), the extra stereocenters of **33** derived from mannitol are essential to achieve excellent results in Rh-catalyzed hydrogenation of α-dehydroamino acids.

To reduce the conformational flexibility inherent in most DuPhos-type ligands, Zhang designed a fused phosphabicyclic ligand known as PennPhos (**34**) [27]. There are several distinct features of PennPhos as compared with DuPhos, namely (i) more rigid structure, (ii) more hindered steric environment, (iii) better air-stability, and (iv) easy preparation from 1,4-dialkylbenzene (Scheme 2.4). This ligand has been applied in asymmetric hydrogenation of challenging cyclic enamides [27d], cyclic enol acetates [27c], and, most importantly, unfunctionalized alkyl alkyl ketones [27b].

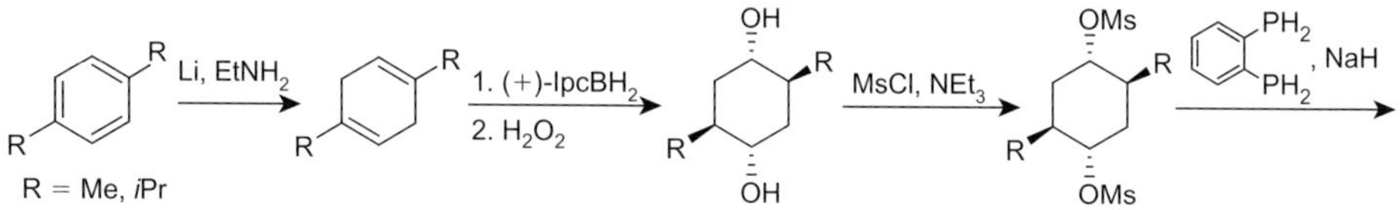

Scheme 2.4 Synthesis of PennPhos (**34**) from 1,4-dialkylbenzene.

Recently, bisazaphospholane **35** was developed by Landis as an interesting variant of DuPhos [28]. The ligand synthesis takes advantage of highly *rac* selective condensation of diphosphinobenzene with an azine promoted by a diacid chloride, which produces a tetra-acid intermediate for subsequent coupling with excess chiral amines (Scheme 2.5). Chromatographic separation of the derivatized amide diastereomers afforded enantiomerically pure **35**, which was applied in Rh-catalyzed asymmetric hydroformylation reactions.

Ligands of ring sizes larger than phospholanes are also available (Figure 2.6). In searching for six-membered-ring analogues of BPE, Helmchen and Zhang developed **36** [29] and **37** [30], respectively. In addition, Zhang designed a series of seven-membered binaphthyl-based bisphosphepines (**38–40**) [31–33]. Among them, f-Binaphane (**40**) is particularly effective for imine reduction [33].

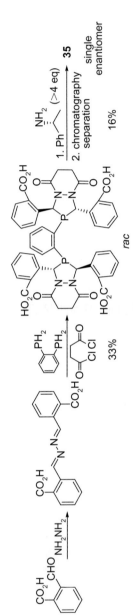

Scheme 2.5 Synthesis of bisazaphospholane **35** from simple building blocks.

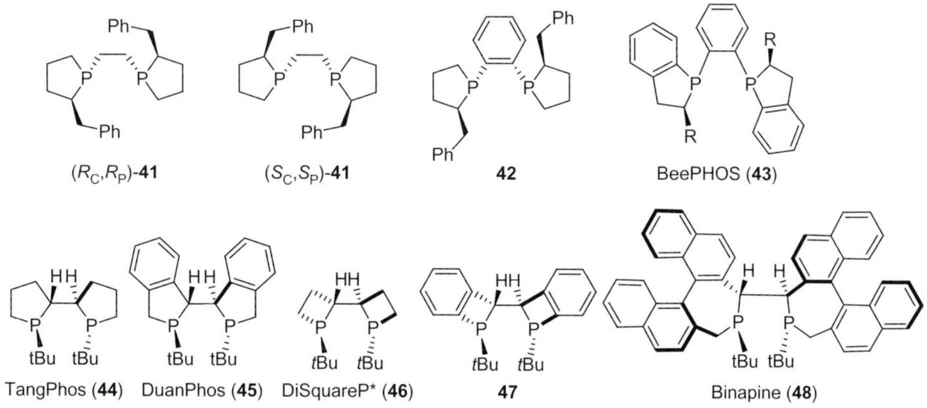

(R_C,R_P)-**41**   (S_C,S_P)-**41**   **42**   BeePHOS (**43**)

TangPhos (**44**)   DuanPhos (**45**)   DiSquareP* (**46**)   **47**   Binapine (**48**)

Figure 2.7 P-chiral bisphosphacycle ligands.

2.2.2.3 Design of New Bisphosphacycle Ligands

The excellent enantioselectivity of DuPhos is attributed to its well-defined asymmetric pattern generated by two protruding stereogenic R substituents (Figure 2.1). In contrast, the remaining two R groups stay backward in the open quadrants and as such make little contribution to chiral induction. To remove these redundant components and generate chirality at the chelating phosphorus sites for effective stereocontrol, Hoge designed ligands **41** and **42** (Figure 2.7) [34]. The two enantiomers of **41** were synthesized via different routes (Scheme 2.6a, b) [34a]. For the preparation of phenylene-bridged ligand **42** with correct stereochemistry, a one-pot process of sequential cyclization and epimerization was developed (Scheme 2.6c) [34b]. A relevant P-chiral ligand BeePHOS (**43**) was developed by Saito, which incorporates two phenyl substituents in each phosphorus donors [35]. This ligand was initially intended for Ru-catalyzed asymmetric hydrogenation reactions, although it was found to be better suited to Rh-catalyzed hydrogenation reactions.

Another noteworthy class of P-chiral ligands were created by Zhang, including TangPhos (**44**) [36], DuanPhos (**45**) [37], and Binapine (**48**) [38]. These rigid ligands are characterized by the presence of two bulky *tert*-butyl P-substituents, which serve as effective steric shields to implement a chiral environment (Figure 2.8). Structurally similar ligands DiSquareP* (**46**) and **47** were reported by Imamoto [39].

2.3 Applications of Bisphosphacycle Ligands

2.3.1 Asymmetric Hydrogenation

2.3.1.1 Asymmetric Hydrogenation of C=C Bonds

Dehydroamino Acid Derivatives The Rh/Et-DuPhos catalyst shows excellent enantioselectivities (>95% ee) in asymmetric hydrogenation of α-dehydroamino

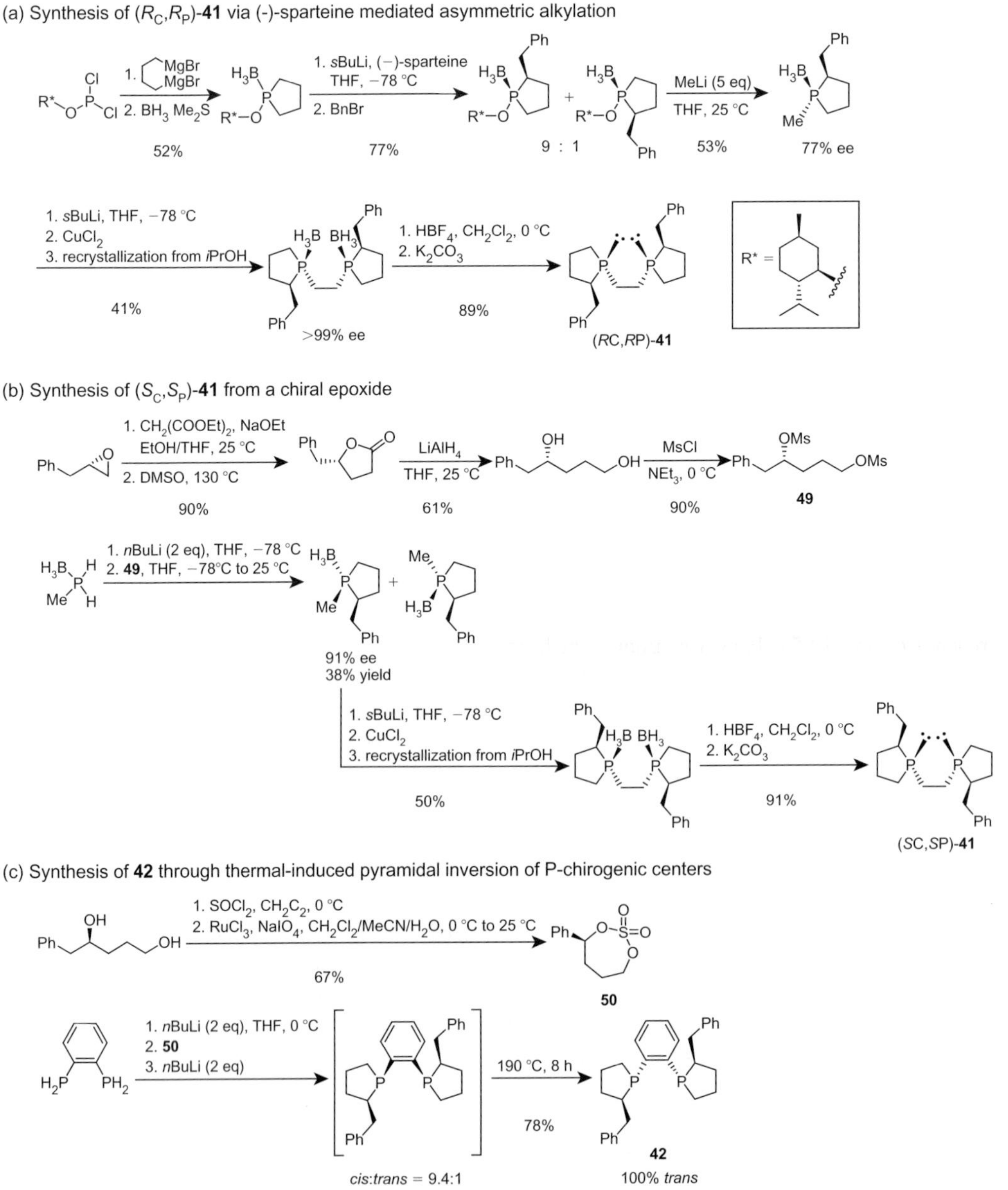

66 | *2 Bisphosphacycles – From DuPhos and BPE to a Diverse Set of Broadly Applied Ligands*

(a) Synthesis of (R_C,R_P)-**41** via (−)-sparteine mediated asymmetric alkylation

(b) Synthesis of (S_C,S_P)-**41** from a chiral epoxide

(c) Synthesis of **42** through thermal-induced pyramidal inversion of P-chirogenic centers

Scheme 2.6 Synthesis of P-chiral phospholanes (R_C,R_P)-**41**, (S_C,S_P)-**41**, and **42**. (a) Synthesis of (R_C,R_P)-**41** via (−)-sparteine-mediated asymmetric alkylation; (b) synthesis of (S_C,S_P)-**41** from a chiral epoxide; (c) synthesis of **42** through thermal-induced pyramidal inversion of P-chirogenic centers.

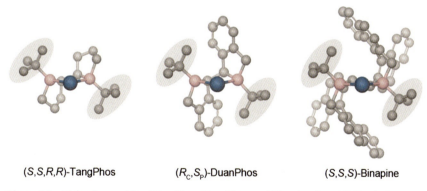

(S,S,R,R)-TangPhos   (R_c,S_p)-DuanPhos   (S,S,S)-Binapine

Figure 2.8 Molecular models of TangPhos, DuanPhos, and Binapine from MM2 calculations.

acid derivatives [3], which has been utilized during asymmetric synthesis of several natural products such as sanglifehrin A [40] and ecteinascidin 743 [41] (Figure 2.9). For more hindered β,β-disubstituted enamides, Me-BPE provides better stereoselectivity: by choosing the double bond configuration of **51**, either *threo*-**52a** or *erythro*-**52b** can be prepared in good to excellent ees [8a].

Compared with α-dehydroamino acid derivatives, the presence of configurational isomers poses a challenging task for enantioselective hydrogenation of β-(acylamino)acrylates. As shown in Table 2.2, most ligands display effective stereocontrol for *(E)*-**53**. However, upon switching to the *(Z)* isomer, considerable drop in ee occurs. An exception is Binapine (**48**), which enables better chiral induction for *(Z)*-**53**. Thus, the Rh/**48** system was applied for the preparation of chiral β-aryl-β-amino acids through asymmetric hydrogenation of the corresponding *(Z)*-enamides (Scheme 2.7). Another noteworthy class of *(Z)* substrates are *N*-aryl-β-enamines, for which TangPhos was identified as the optimal ligand to attain up to 96.3% ee (Scheme 2.8) [36f].

Simple Enamides Asymmetric hydrogenation of simple enamides is an attractive method for the synthesis of chiral amines. For example, α-aryl enamides were hydrogenated with Rh/Me-DuPhos or Rh/Me-BPE catalyst to afford the protected form of chiral α-aryl amines in up to 98.5% ee (Scheme 2.9) [8b]. Similarly, chiral β- and γ-aryl amines can be prepared via asymmetric hydrogenation (Scheme 2.10) [36d, 37d]. Enamides bearing *exo*- or *endo*-cyclic double bond have also been successfully hydrogenated using Me-DuPhos or Me-PennPhos as the ligand (Scheme 2.11) [42, 27d].

Unsaturated Carboxylic Acids and Esters Unsaturated carboxylic acids, including itaconic acid derivatives, were smoothly hydrogenated by the use of different bisphosphacycle ligands (Scheme 2.12) [21, 37c]. The products are useful pharmaceutical intermediates leading to chiral drugs such as pregabalin (Scheme 2.13) [7d] and candoxatril (Scheme 2.14) [7c].

Apart from ketone hydrogenation, asymmetric hydrogenation of enol esters is an alternative method for the synthesis of chiral alcohols. For example,

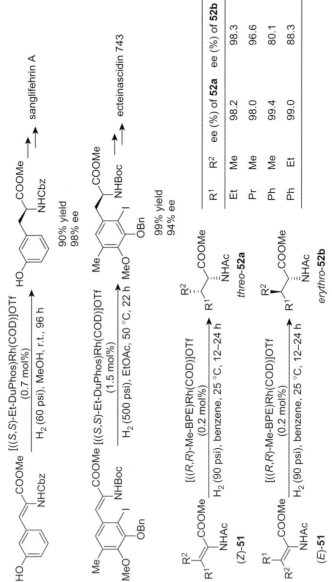

Figure 2.9 Application of DuPhos and BPE in asymmetric hydrogenation of α-dehydroamino acid derivatives.

Table 2.2 Comparative results of Rh-catalyzed asymmetric hydrogenation of *(Z)*- and *(E)*-methyl 3-acetamido-2-butenoate (**53**).

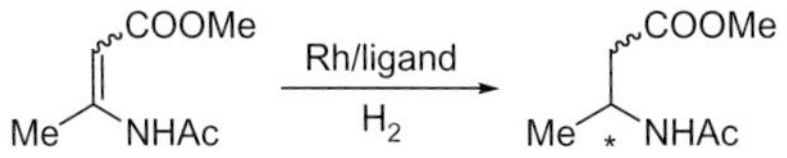

Ligand	ee (%) for (Z)-53	ee (%) for (E)-53	Ref.
Et-DuPhos (**1**, R = Et)	87	98	[4]
Et-Ferrotane (**26**, R = Et)	28	99	[4]
37	96	>99	[30]
TangPhos (**44**)	98.5	99.6	[36e]
Binapine (**48**)	99.2	32.7	[38]

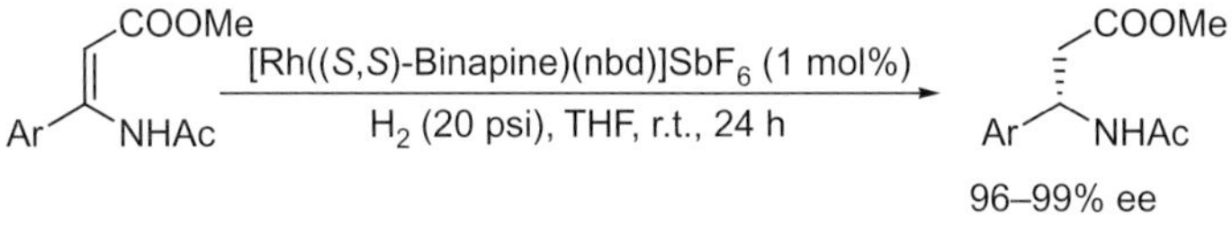

Scheme 2.7

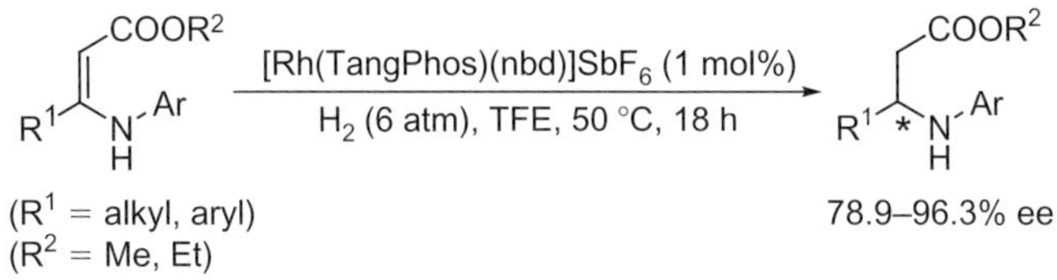

Scheme 2.8

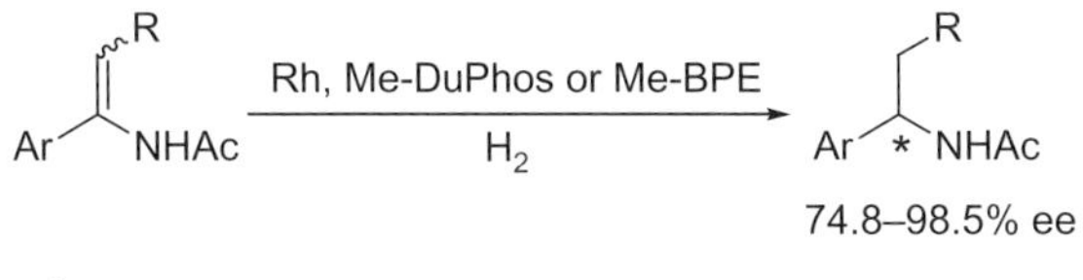

(R^1 = alkyl, aryl)
(R^2 = Me, Et)

Scheme 2.9

Rh/Et-DuPhos catalyzed hydrogenation of **54** leads to chiral α-hydroxy esters and 1,2-diols (Scheme 2.15) [43], while Rh/TangPhos allows highly enantioselective hydrogenation of α-aryl enol acetates (Scheme 2.16) [36g]. In the case of asymmetric hydrogenation of challenging cyclic enol acetates, PennPhos is the ligand of choice and up to 99.1% ee was achieved (Scheme 2.17) [27c].

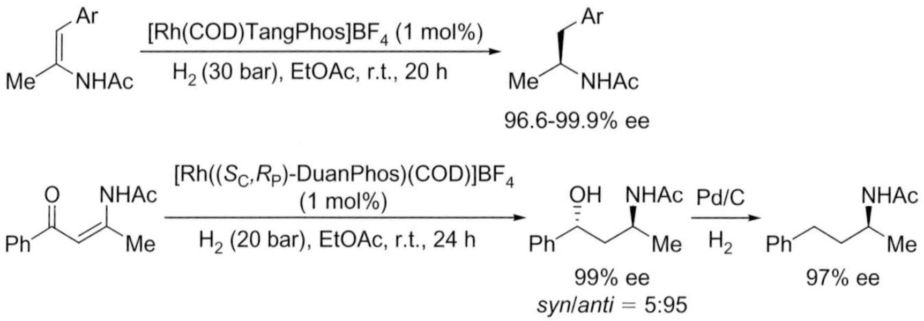

Scheme 2.10

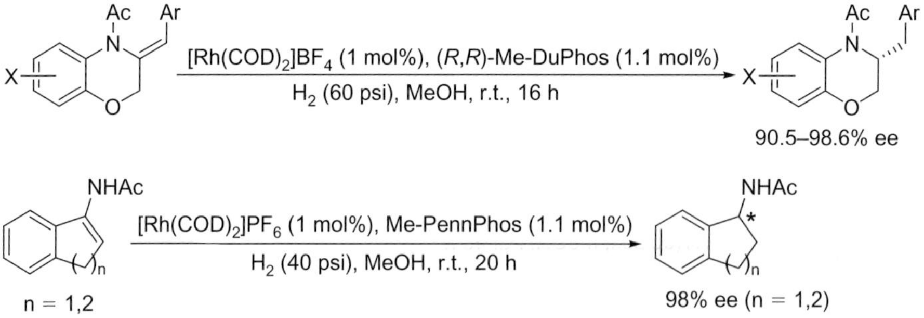

Ar/Me/NHAc alkene + [Rh(COD)TangPhos]BF$_4$ (1 mol%), H$_2$ (30 bar), EtOAc, r.t., 20 h → Ar/Me/NHAc product, 96.6–99.9% ee

PhC(O)–C(NHAc)=CHMe + [Rh((S$_C$,R$_P$)-DuanPhos)(COD)]BF$_4$ (1 mol%), H$_2$ (20 bar), EtOAc, r.t., 24 h → Ph–CH(OH)–CH(NHAc)–Me, 99% ee, syn/anti = 5:95; then Pd/C, H$_2$ → Ph–CH$_2$–CH(NHAc)–Me, 97% ee

Scheme 2.11

Benzoxazine with Ac/Ar exocyclic alkene (X substituent) + [Rh(COD)$_2$]BF$_4$ (1 mol%), (R,R)-Me-DuPhos (1.1 mol%), H$_2$ (60 psi), MeOH, r.t., 16 h → saturated product, 90.5–98.6% ee

Indene/tetrahydronaphthalene-NHAc (n = 1,2) + [Rh(COD)$_2$]PF$_6$ (1 mol%), Me-PennPhos (1.1 mol%), H$_2$ (40 psi), MeOH, r.t., 20 h → chiral product, 98% ee (n = 1,2)

Scheme 2.12

HOOC–CH=C(R)–C(O)–N(morpholine) + [Rh((S,S)-Et-Ferrotane)(COD)]BF$_4$ (0.1 mol%), H$_2$ (0.55 MPa), MeOH, 20 °C → saturated product, 92–99% ee

MeOOC–C(R)=CH–COOH + Rh, ligand, H$_2$ → MeOOC–C*H(R)–CH$_2$–COOH

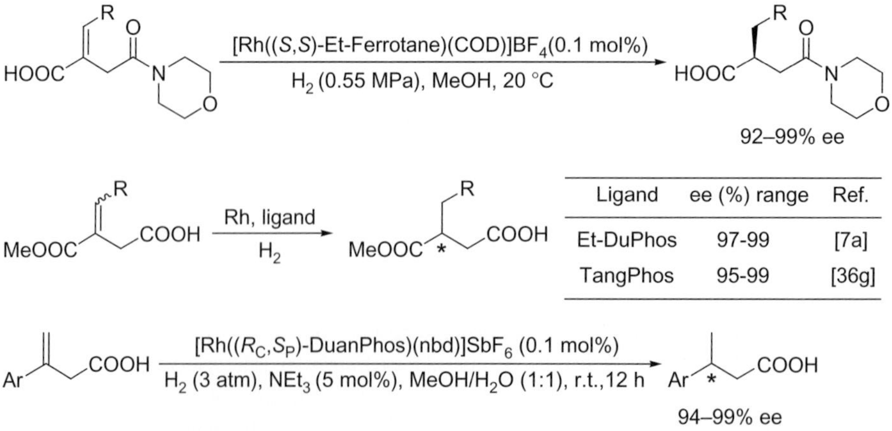

Ligand	ee (%) range	Ref.
Et-DuPhos	97–99	[7a]
TangPhos	95–99	[36g]

Ar–C(=CH$_2$)–CH$_2$–COOH + [Rh((R$_C$,S$_P$)-DuanPhos)(nbd)]SbF$_6$ (0.1 mol%), H$_2$ (3 atm), NEt$_3$ (5 mol%), MeOH/H$_2$O (1:1), r.t., 12 h → Ar–C*H(Me)–CH$_2$–COOH, 94–99% ee

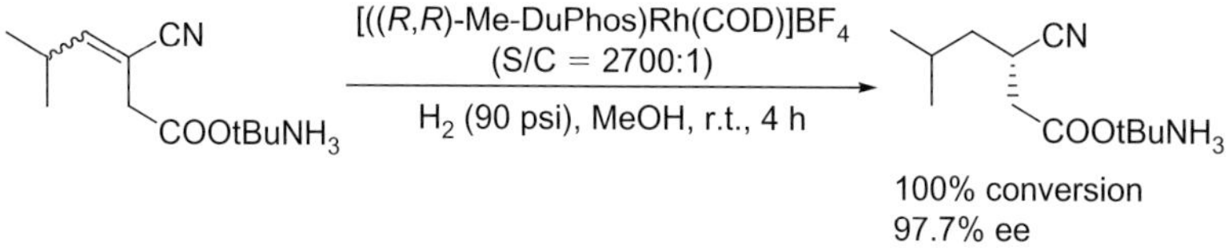

Scheme 2.13

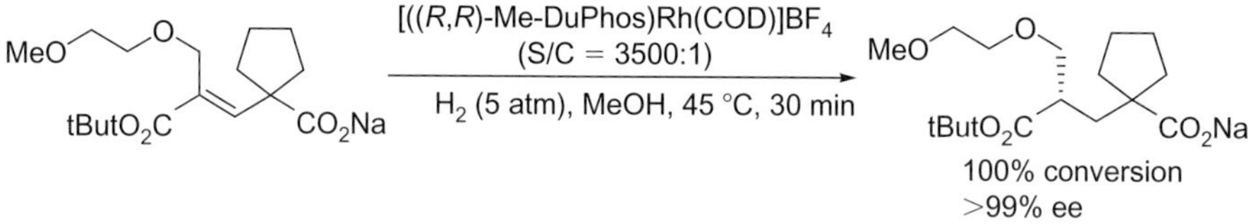

Reaction: alkene with CN and COOtBuNH₃ groups + [((R,R)-Me-DuPhos)Rh(COD)]BF₄ (S/C = 2700:1), H₂ (90 psi), MeOH, r.t., 4 h → saturated product with CN and COOtBuNH₃. 100% conversion, 97.7% ee.

Scheme 2.14

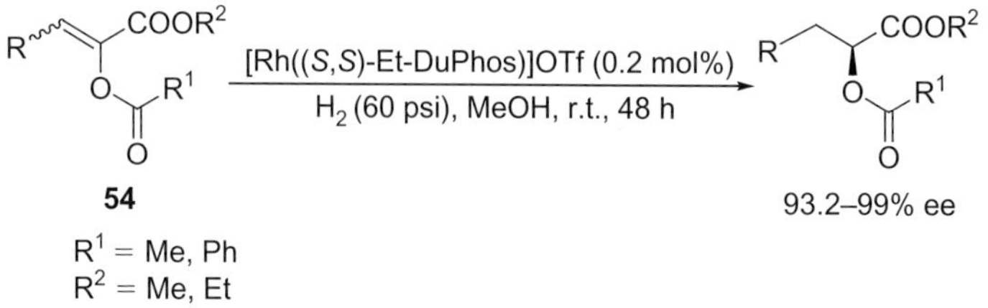

Substrate with MeO, tButO₂C, and CO₂Na groups + [((R,R)-Me-DuPhos)Rh(COD)]BF₄ (S/C = 3500:1), H₂ (5 atm), MeOH, 45 °C, 30 min → hydrogenated product. 100% conversion, >99% ee.

Scheme 2.15

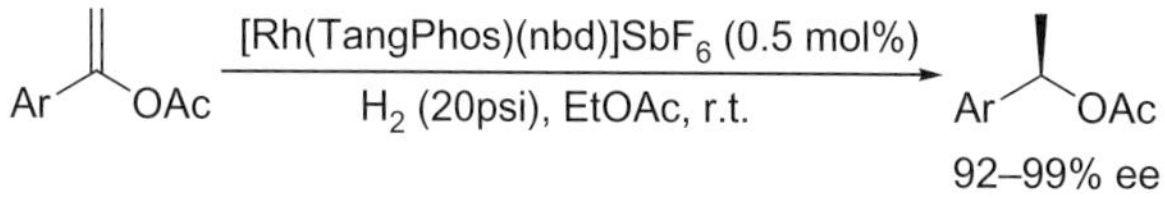

Substrate **54** (R¹ = Me, Ph; R² = Me, Et) + [Rh((S,S)-Et-DuPhos)]OTf (0.2 mol%), H₂ (60 psi), MeOH, r.t., 48 h → product with 93.2–99% ee.

Scheme 2.16

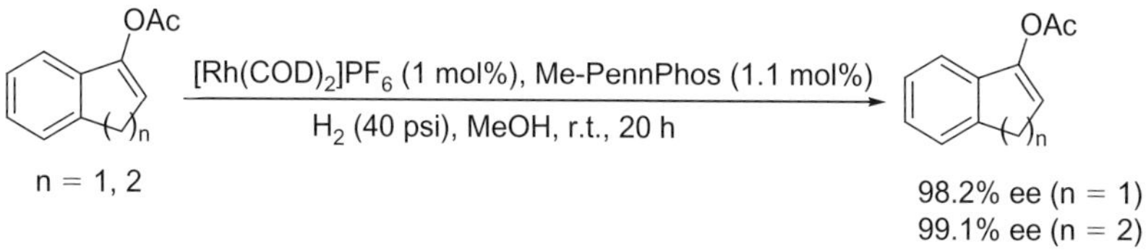

Ar–C(=CH₂)–OAc + [Rh(TangPhos)(nbd)]SbF₆ (0.5 mol%), H₂ (20 psi), EtOAc, r.t. → Ar–CH(CH₃)–OAc, 92–99% ee.

Scheme 2.17

Indene/benzofused cyclic enol acetate (n = 1, 2) + [Rh(COD)₂]PF₆ (1 mol%), Me-PennPhos (1.1 mol%), H₂ (40 psi), MeOH, r.t., 20 h → saturated OAc product. 98.2% ee (n = 1), 99.1% ee (n = 2).

2.3.1.2 Asymmetric Hydrogenation of C=O Bonds

Under low hydrogen pressure, Ru/iPr-BPE catalyst led to asymmetric hydrogenation of β-keto esters with high ees [44]. Another effective ligand for these substrates is TangPhos, which tolerates both alkyl and aryl substituents at the β-position (Scheme 2.18) [36h].

Chiral amino alcohols are widely found building blocks for pharmaceutical products. So far, various α- and β-amino ketones have been hydrogenated successfully by the use of a suitable bisphosphacycle ligand (Scheme 2.19) [37e, 37b]. This transformation is key to efficient synthesis of chiral drugs *(S)*-fluoxetine and *(S)*-duloxetine via asymmetric hydrogenation (Scheme 2.20) [37b].

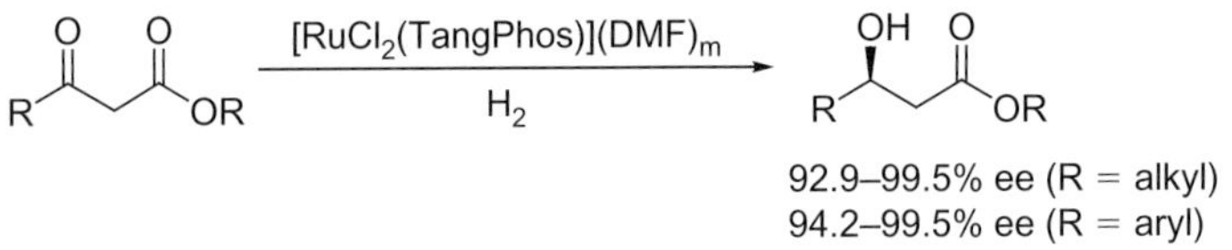

Scheme 2.18

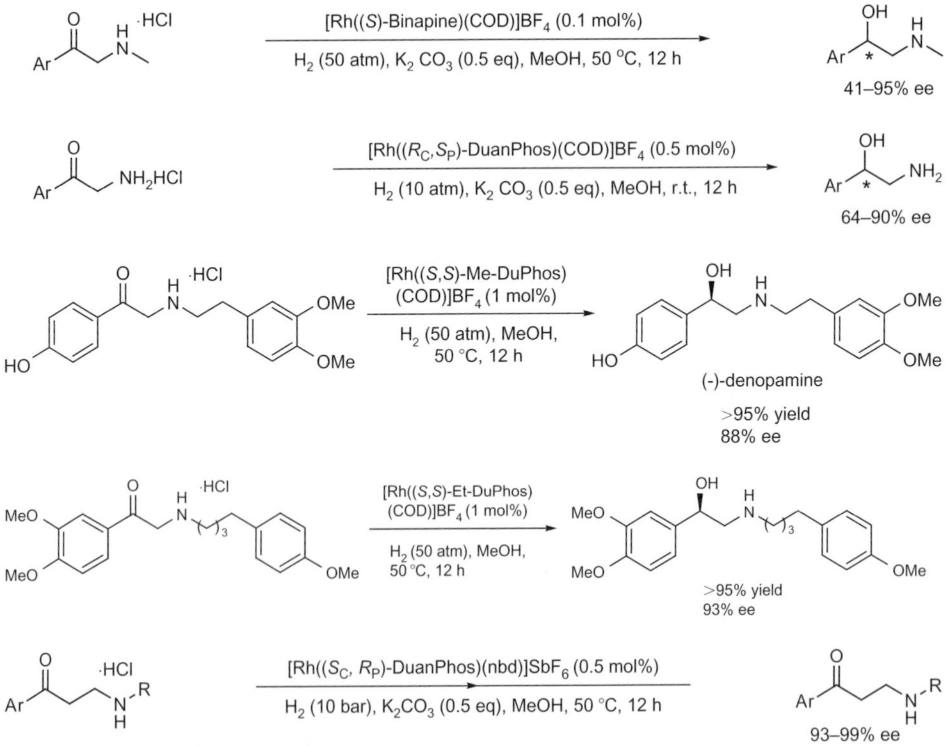

Scheme 2.19

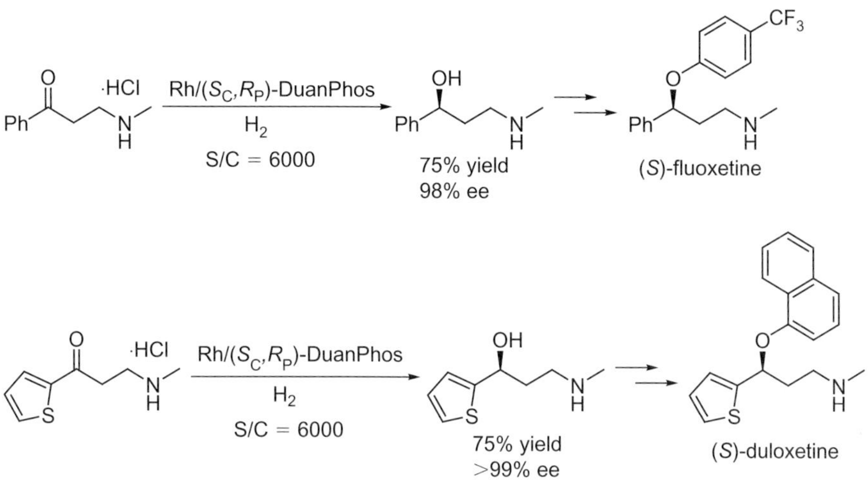

Scheme 2.20

Table 2.3 Selected results of Rh/PennPhos-catalyzed hydrogenation of alkyl alkyl ketones.

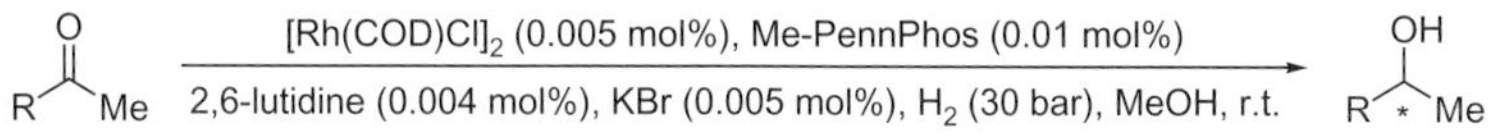

R^1	Yield (%)	ee (%)
nBu	96	75
Me$_2$CHCH$_2$	66	85
iPr	99	84
cyclohexyl	90	92
tBu	51	94

Although Noyori's Ru/BINAP/diamine system leads to highly enantioselective hydrogenation of simple aryl alkyl ketones [45], asymmetric hydrogenation of alkyl alkyl ketones remains a challenging problem. Zhang have found that in the presence of a Rh/PennPhos catalyst these substrates were hydrogenated with up to 94% ee (Table 2.3) [27b].

2.3.1.3 Asymmetric Hydrogenation of C=N Bonds

The combination of a Ru precursor, Et-DuPhos, and a chiral diamine was found to induce 94% ee in asymmetric hydrogenation of an N-phenyl imine **55** under optimized conditions (Scheme 2.21) [46]. DuanPhos and f-Binaphane give excellent ees in Ir-catalyzed asymmetric hydrogenation of imine substrates with different N-aryl groups (Scheme 2.22) [37e, 33a]. Further, f-Binaphane allows enantioselective hydrogenation of iminium salts with satisfactory ees (Scheme 2.23). For

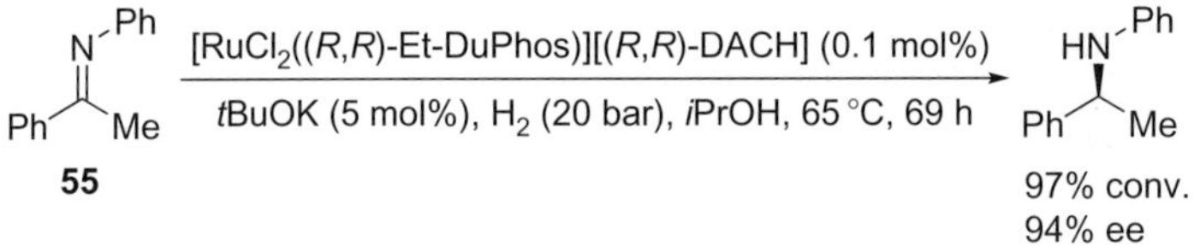

Ph
N
Ph
Me
55
[RuCl2((R,R)-Et-DuPhos)][(R,R)-DACH] (0.1 mol%)
tBuOK (5 mol%), H2 (20 bar), iPrOH, 65 °C, 69 h
HN
Ph
Ph
Me
97% conv.
94% ee

Scheme 2.21

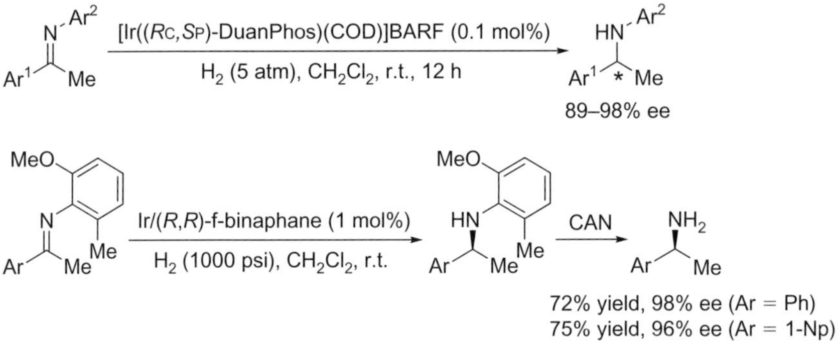

Ar2
N
Ar1
Me
[Ir((Rc,Sp)-DuanPhos)(COD)]BARF (0.1 mol%)
H2 (5 atm), CH2Cl2, r.t., 12 h
HN
Ar2
Ar1
*
Me
89–98% ee
MeO
N
Ar
Me
Me
Ir/(R,R)-f-binaphane (1 mol%)
H2 (1000 psi), CH2Cl2, r.t.
MeO
HN
Ar
Me
Me
CAN
NH2
Ar
Me
72% yield, 98% ee (Ar = Ph)
75% yield, 96% ee (Ar = 1-Np)

Scheme 2.22

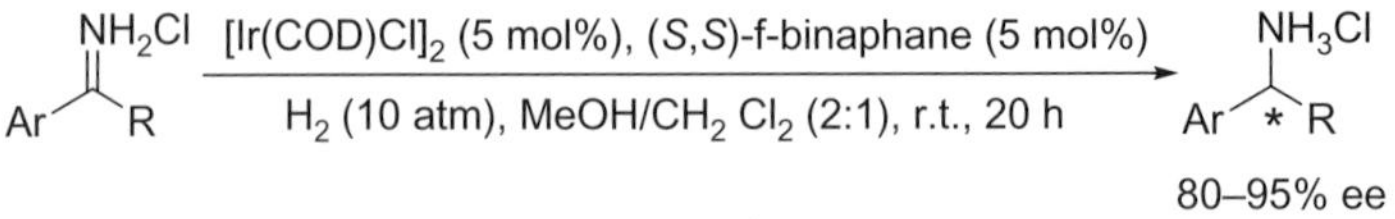

NH2Cl
Ar
R
[Ir(COD)Cl]2 (5 mol%), (S,S)-f-binaphane (5 mol%)
H2 (10 atm), MeOH/CH2 Cl2 (2:1), r.t., 20 h
NH3Cl
Ar
*
R
80–95% ee

Scheme 2.23

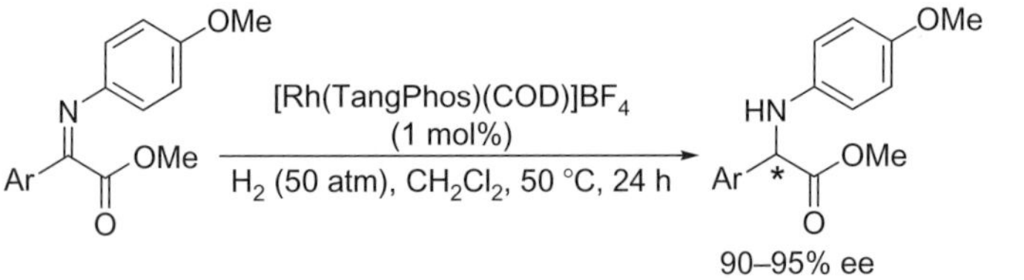

OMe
N
Ar
OMe
O
[Rh(TangPhos)(COD)]BF4
(1 mol%)
H2 (50 atm), CH2Cl2, 50 °C, 24 h
OMe
HN
Ar
*
OMe
O
90–95% ee

Scheme 2.24

asymmetric hydrogenation of *N*-aryl imino esters, a Rh/TangPhos complex was the optimal catalyst to give up to 95% ee (Scheme 2.24).

Asymmetric hydrogenation of *N*-aroylhydrazones has been accomplished by the use of a Rh/Et-DuPhos catalyst (Scheme 2.25) [47]. Palladium-catalyzed asymmetric hydrogenation is also effective for imine hydrogenation. Thus, asymmetric hydrogenation of both acyclic and cyclic *N*-tosyl imines were achieved by the use of TangPhos (Scheme 2.26) [36b] or f-Binaphane (Scheme 2.27) [33c, 33d].

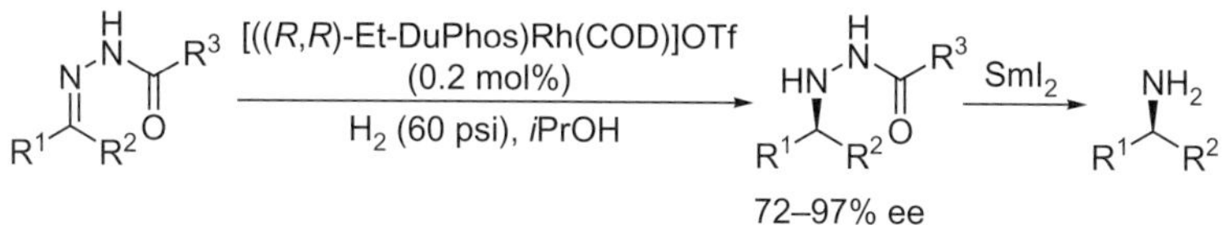

Scheme 2.25

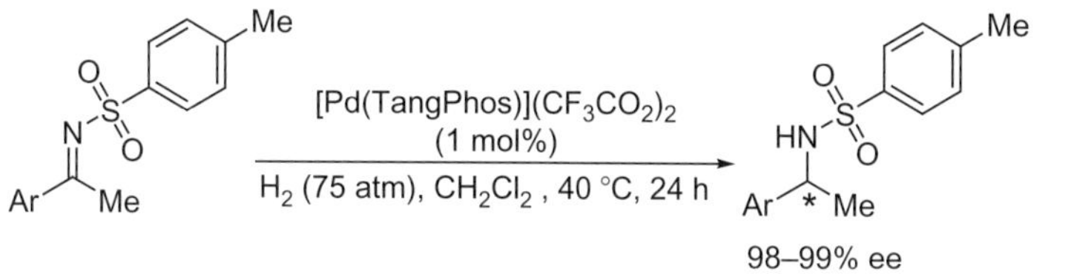

Scheme 2.26

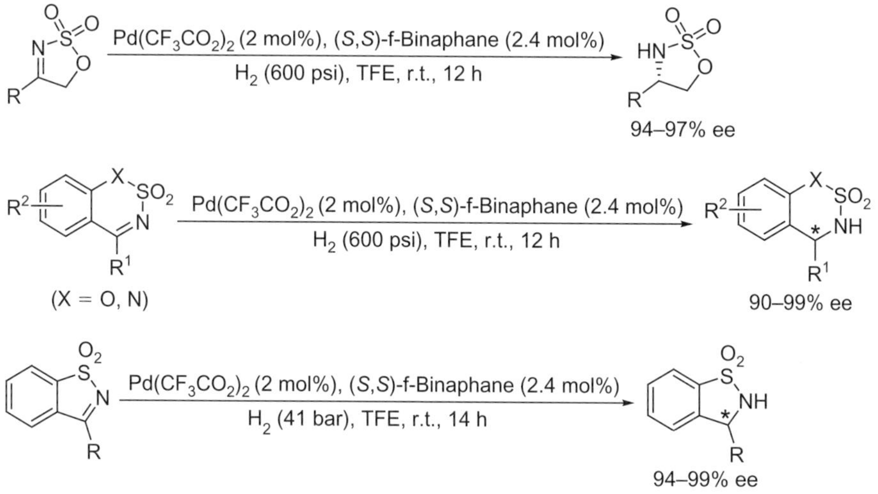

Scheme 2.27

2.3.2
Asymmetric Hydroformylation

Rhodium-catalyzed asymmetric hydroformylation (AHF) is a powerful reaction for one-carbon homologation of widely available olefins [48]. The product aldehyde is a versatile intermediate leading to various useful chiral compounds such as alcohols, acids, and amines. Recently, diazaphospholane **35** [28] and Ph-BPE (**5**) [49] were found to be the desirable ligands. Table 2.4 lists the results of AHF of three standard substrates (styrene, allyl cyanide, vinyl acetate) by the use of different bisphospholanes. Evidently, **35** and **5** display much better activities and/or

Table 2.4 Results of AHF with different Rh/bisphosphine catalysts.[a]

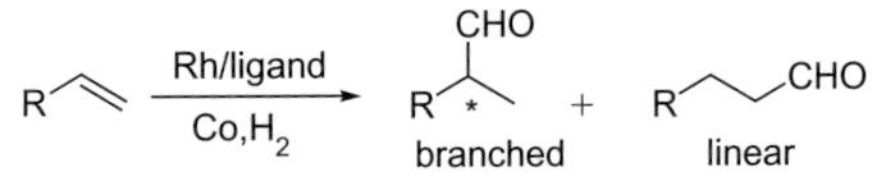

$$R\diagdown\diagdown \xrightarrow[CO, H_2]{Rh/ligand} R\underset{branched}{\overset{CHO}{\diagdown*\diagdown}} + R\diagdown\diagdown\diagdown\underset{linear}{CHO}$$

Ligand	Styrene			Allyl cyanide			Vinyl acetate		
	Conversion (%)	B/l	ee (%)	Conversion (%)	B/l	ee (%)	Conversion (%)	B/l	ee (%)
Me-DuPhos	10	15.7	44	42	6.6	32	26	176	51
Et-DuPhos	14	13.7	52	49	7.8	35	27	371	66
iPr-DuPhos	15	11.3	83	55	7.2	82	29	322	74
35	100	6.6	82	100	4.1	87	100	37	96
Me-BPE	8	14.0	43	36	5.8	37	23	97	59
Et-BPE	10	11.3	55	40	6.2	49	23	97	59
iPr-BPE	11	9.5	82	48	6.7	83	28	142	70
Ph-BPE 5	57	45.0	94	96	7.1	90	52	340	82

[a]Reaction condition: ligand/Rh = 1.2:1, 10 atm of CO/H_2 (1:1), TON = 5000, 80 °C, toluene, 3 h.

selectivities than the other ligands [49]. In a subsequent study, a series of bis(2,5-diphenylpholane) ligands, including **5**, were subjected to AHF under the same conditions to gain insight into the influence of the bridging ligand backbone [50]. The conclusion from this study is that electron deficiency at the phosphorus donors is responsible for superior catalyst activity. Further analysis of the X-ray crystal structures of several Rh/bisphospholane complexes implies that, at least for this type of ligands, a P–Rh–P bite angle of around 85° is optimal for maximum enantioselectivity.

The success of Ph-BPE in both asymmetric hydrogenation (AH) and AHF prompted evaluation of other AH ligands in AHF reactions [51]. Despite moderate activities, outstanding enantioselectivities were accomplished by TangPhos (**44**) and Binapine (**48**). The P–Rh–P bite angle of [Rh(TangPhos)(acac)] is 85.88°, which is in good agreement with the optimal value found in BPE ligand series. Moreover, TangPhos gave excellent results in AHF of bicyclic [2.2.1] olefins such as **56** at room temperature (Scheme 2.28) [52]. The product was further converted into optically pure *exo*-norbornylamine as a key precursor to a pharmaceutical candidate.

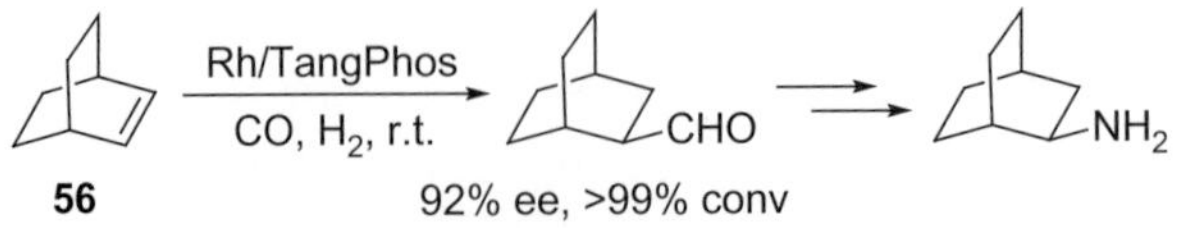

Scheme 2.28

2.3.3
Asymmetric Hydrosilylation

Asymmetric hydrosilylation is an alternative way to reduce unsaturated carbon–heteroatom bonds [53]. In 1992 Burk reported the use of DuPhos in Rh-catalyzed intramolecular hydrosilylation of α-hydroxy ketones [54]. Up to 93% ee was observed for methyl- and ethyl ketones, whereas enantioselectivity diminished in the case of phenyl ketone (Scheme 2.29).

Recently, Beller developed asymmetric hydrosilylation of ketones based on cheap and non-toxic iron catalysts [55]. Extensive screening of iron precursors, chiral ligands, hydride sources, and solvents determined $Fe(OAc)_2$/Me-DuPhos/$(EtO)_2SiMeH$/THF as the optimal combination. Under this condition, several aryl ketones were reduced with good to excellent ees (Scheme 2.30).

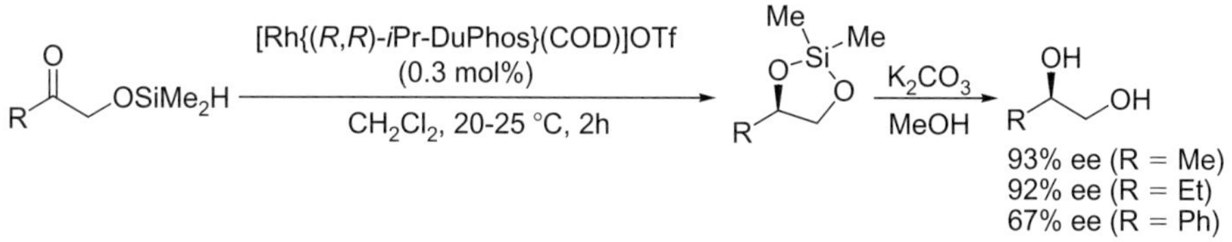

Scheme 2.29

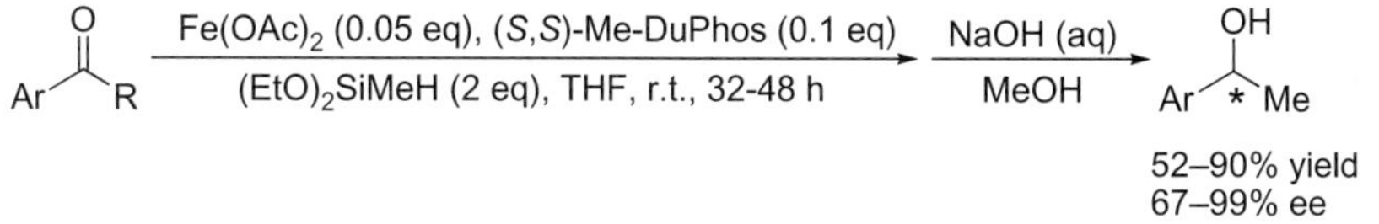

Scheme 2.30

2.3.4
Asymmetric Hydroacylation

Rhodium-catalyzed asymmetric hydroacylation is useful to prepare chiral cyclic ketones [56]. The utility of DuPhos was first reported by Bosnich for highly enantioselective cyclization of 4-substituted pent-4-enals bearing primary and secondary substituents (Scheme 2.31) [57]. Dong and coworkers further extended

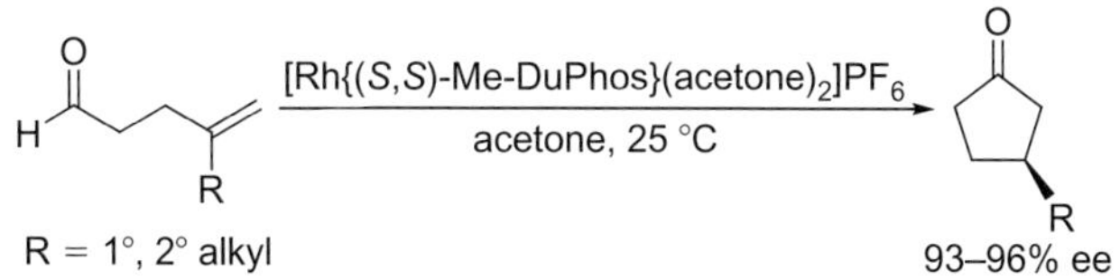

Scheme 2.31

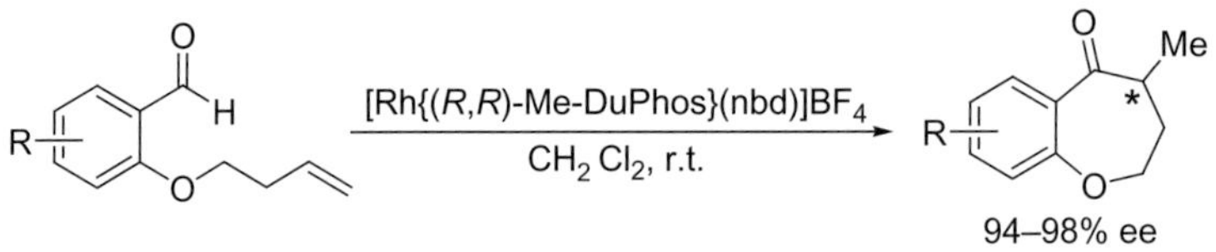

Scheme 2.32

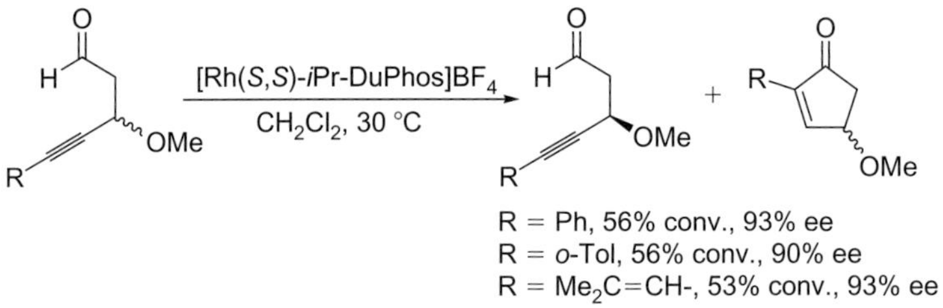

Scheme 2.33

this methodology to efficient synthesis of medium-sized rings (Scheme 2.32) [58]. In addition, kinetic resolution of 4-alkynals mediated by Rh/DuPhos catalyst was reported by Fu [59]. In this process, a prerequisite for high levels of stereoselection is the presence of a chelating methoxy group in the substrate (Scheme 2.33).

Despite the progress in intramolecular hydroacylation, the intermolecular reaction is less studied because of competitive metal-catalyzed decarbonylation. However, Willis recently showed that such a reaction is feasible given allenes as the substrate (Scheme 2.34) [60]. The authors noted the beneficial effect of a rigid aromatic aldehyde on the observed regioselectivity and enantioselectivity.

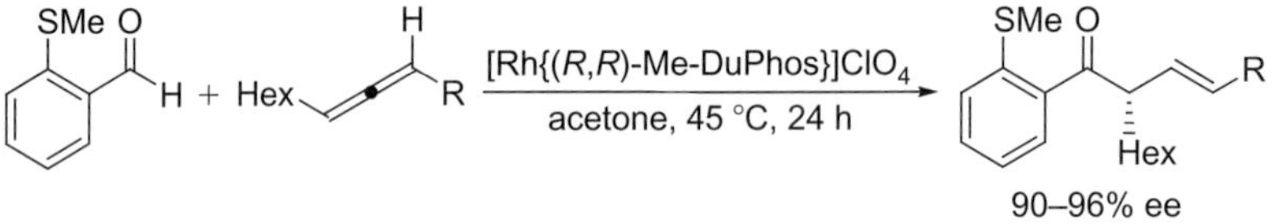

Scheme 2.34

2.3.5
Asymmetric Cycloisomerization, Cycloaddition, and Cyclization

The synthesis of chiral carbocycles via asymmetric catalysis is an intensively studied subject, wherein DuPhos as a privileged ligand has found several remarkable applications. For example, Gilbertson developed Rh/DuPhos catalyzed asymmetric [4+2] cycloisomerization of dieneynes (Scheme 2.35) [61]. In an effort to expand the substrate scope, Mikami found an intriguing synergy between chiral

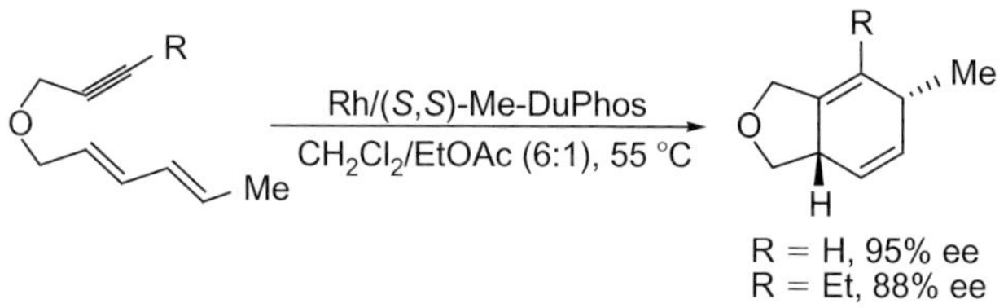

Scheme 2.35

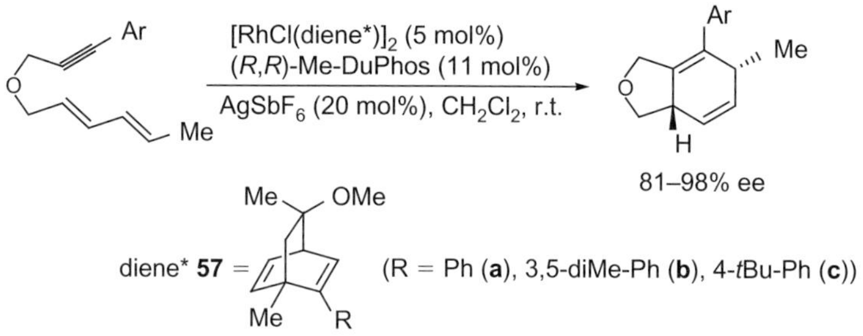

Scheme 2.36

dienes and diphosphines [62]. Namely, the use of (R,R)-Me-DuPhos and a chiral diene (**57a**) afforded the product (Ar = Ph) with 88% ee, whereas changing the ligand to (S,S)-Me-DuPhos resulted in 9% ee with the opposite chirality. Thus, employing the matched combination of DuPhos and diene, a series of aryl substituted dieneynes were transformed into the products with excellent enantioselectivities (Scheme 2.36).

In another report, Rh-catalyzed asymmetric [4 + 1] cycloaddition of vinylallenes with carbon monoxide was realized by the use of DuPhos (Scheme 2.37) [63a]. Interestingly, when the metal was changed from rhodium to platinum, reversal of enantioselectivity occurred [63b].

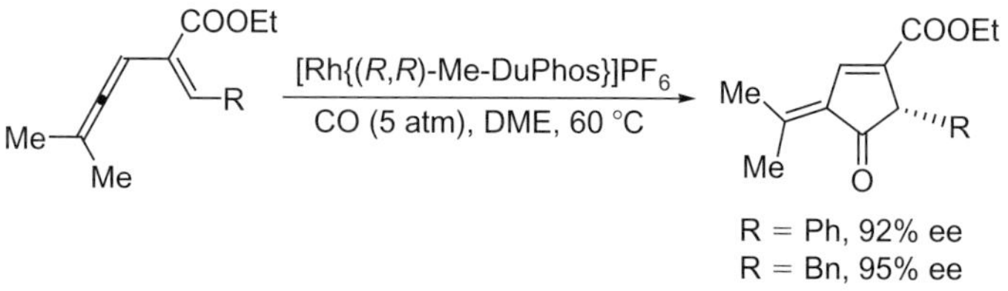

Scheme 2.37

Besides DuPhos, other bisphosphacycle ligands have been applied in asymmetric cyclization reactions. In 2007 Tanaka described the first example of the asymmetric catalytic synthesis of planar-chiral paracyclophanes, wherein Binaphane (**38**) gave best results among other tested ligands (Scheme 2.38) [64].

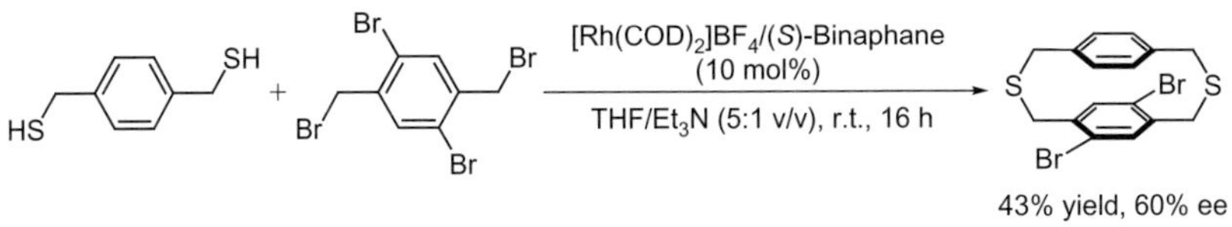

Scheme 2.38

Moreover, Toste reported Pd-catalyzed asymmetric cyclization of silyloxy-1,6-enynes [65]. In this reaction, Binaphane played an indispensable role for efficient synthesis of chiral spiro stereocenters (Table 2.5).

Equally useful in asymmetric cyclization is TangPhos (**44**). For example, Jacobsen developed a Ni/TangPhos-catalyzed asymmetric intramolecular arylcyanation reaction via C–CN bond activation, which is an efficient entry to chiral quaternary carbon stereogenic centers (Scheme 2.39) [66]. Very recently, Bergman

Table 2.5 Rh/Binaphane-catalyzed asymmetric synthesis of chiral spiro stereocenters.

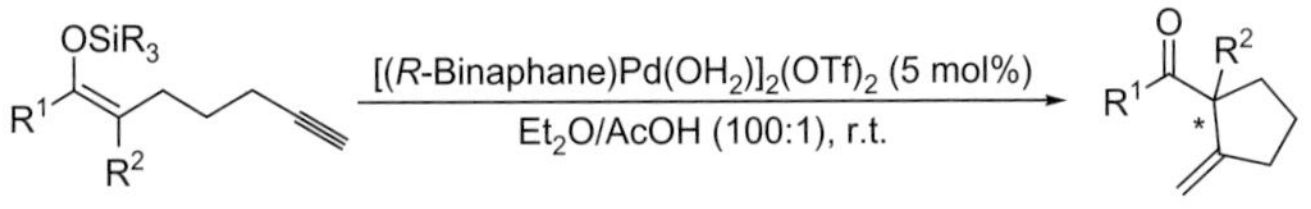

Substrate	Product	t (min)	Yield (%)	ee (%)
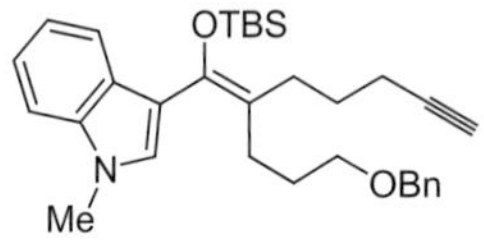 3-indolyl, OTBS, OBn alkyne	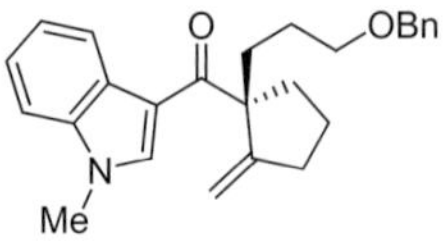 cyclopentane with OBn	30	83	89
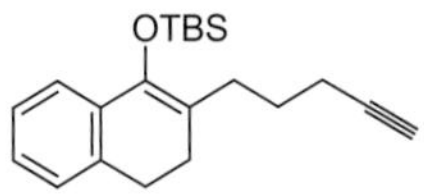 dihydronaphthalene-OTBS alkyne	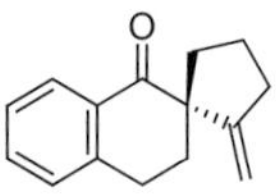 spiro ketone	30	91	87
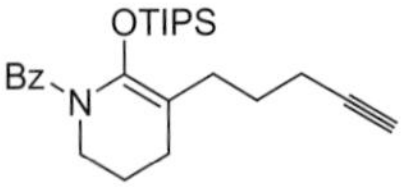 Bz-N tetrahydropyridine OTIPS alkyne	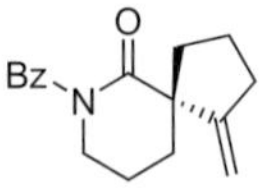 Bz-N piperidinone spiro	20	80	98
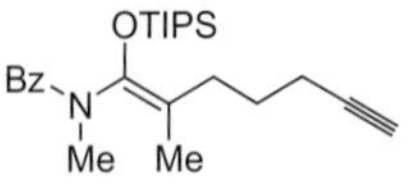 Bz-N, Me, Me, OTIPS alkyne	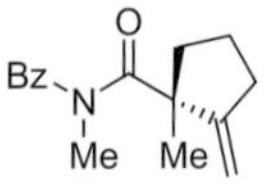 Bz-N, Me, Me spiro ketone	120	79	80
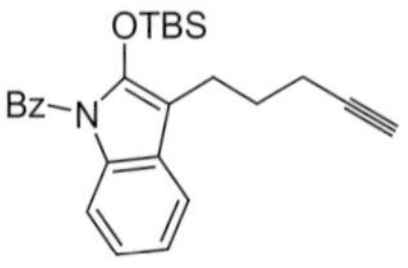 Bz-N indole OTBS alkyne	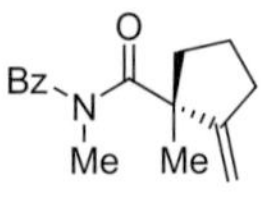 Bz-N, Me, Me spiro ketone	20	83	91

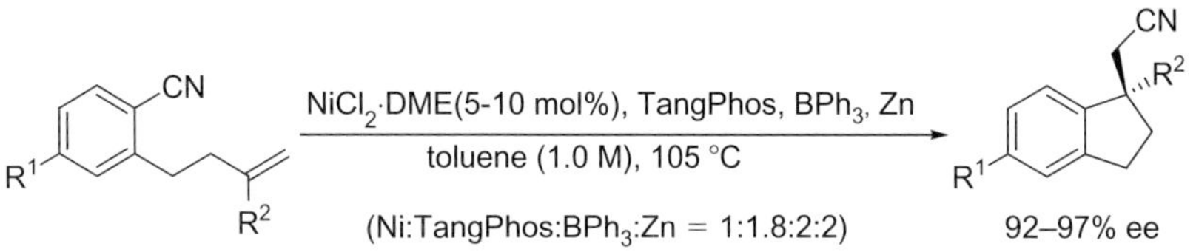

Scheme 2.39

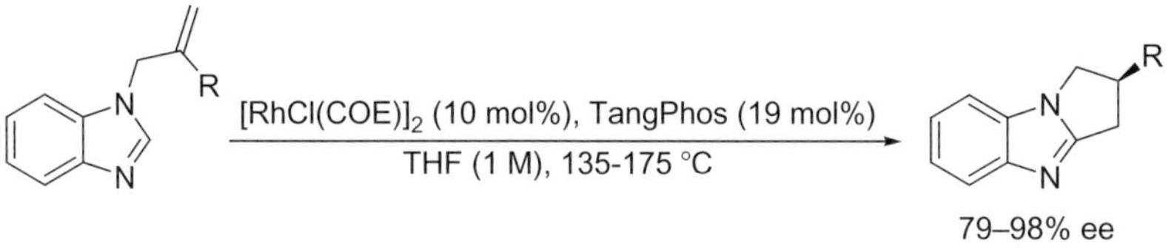

R¹—[Ar(CN)(CH₂CH₂C(=CH₂)R²)] → NiCl₂·DME (5–10 mol%), TangPhos, BPh₃, Zn, toluene (1.0 M), 105 °C, (Ni:TangPhos:BPh₃:Zn = 1:1.8:2:2) → R¹—indane(CH₂CN)(R²), 92–97% ee

Scheme 2.40

benzimidazole-N-CH₂C(=CH₂)R → [RhCl(COE)]₂ (10 mol%), TangPhos (19 mol%), THF (1 M), 135–175 °C → fused pyrrolo-benzimidazole with R, 79–98% ee

and Ellman showed that asymmetric cyclization of imidazoles via C–H bond activation can be achieved with the Rh/TangPhos catalytic system (Scheme 2.40) [67]. Notably, in both examples, excellent enantioselectivities were obtained at high temperature (>100 °C). Obviously, such uncompromised selectivity at elevated temperature originates from the conformational rigidity of TangPhos.

2.3.6
Asymmetric Phosphination

Palladium-catalyzed asymmetric cross-coupling of a secondary phosphine with an aryl halide is a promising method to prepare chiral phosphines. Glueck first found DuPhos as a suitable ligand for this reaction (Scheme 2.41), presumably by means of facile dynamic resolution of the phosphido ligand [68]. Inspired by this result, Helmchen reported asymmetric synthesis of chiral triarylphosphines with Pd/FerroTANE catalyst, which is a potential method to prepare P-chiral P,N-ligands via double stereodifferentiation [69]. Further, this approach was explored by Bergman and Toste for enantioselective arylation of silylphosphines [70]. The best aryl iodides are those bearing hindered N,N-diisopropyl amide at the *ortho*

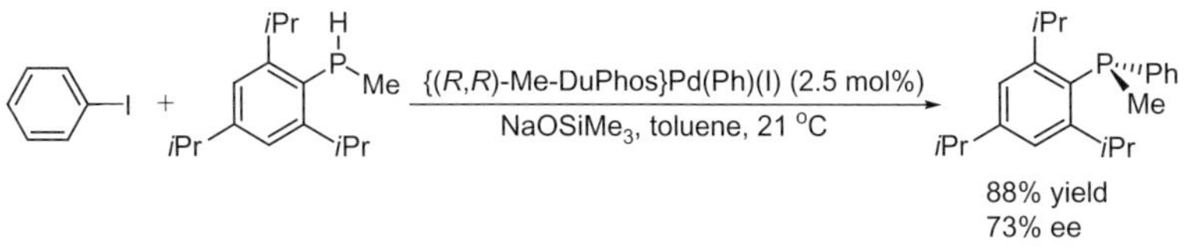

Scheme 2.41

Ph–I + (2,6-iPr₂C₆H₃)P(H)Me → {(R,R)-Me-DuPhos}Pd(Ph)(I) (2.5 mol%), NaOSiMe₃, toluene, 21 °C → (2,6-iPr₂C₆H₃)P(Ph)(Me), 88% yield, 73% ee

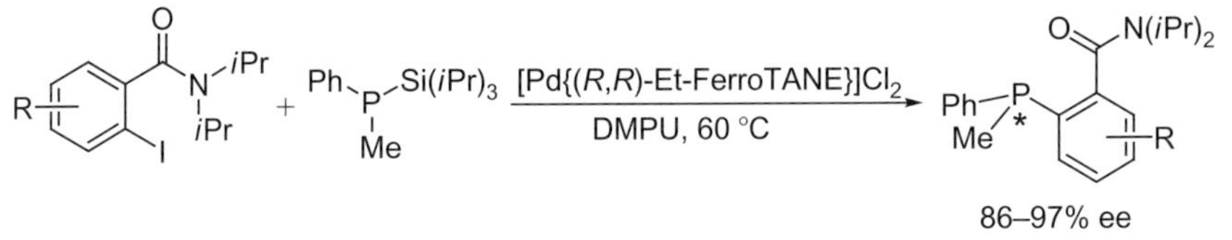

Scheme 2.42

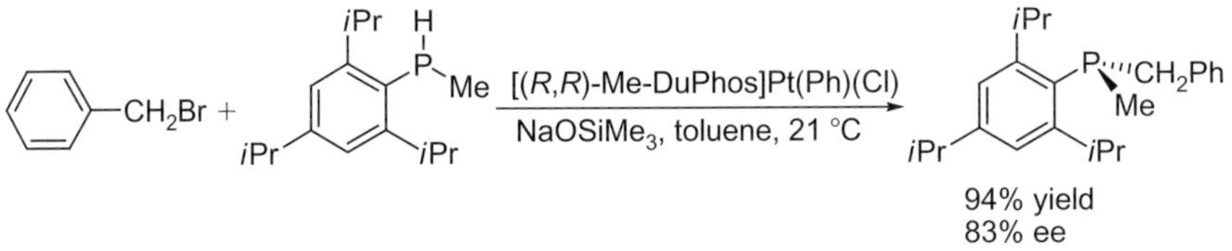

Scheme 2.43

position, for which excellent ees were attained by the use of Pd/FerroTANE catalyst (Scheme 2.42).

Apart from Pd chemistry, Glueck also developed a Pt-based catalytic system for asymmetric alkylation of secondary phosphines (Scheme 2.43) [71]. An attractive application of this methodology lies in enantioselective tandem alkylation/arylation of primary phosphines, which allows efficient synthesis of new chiral phosphines such as DuAcePhos (**58**) (Scheme 2.44) [72].

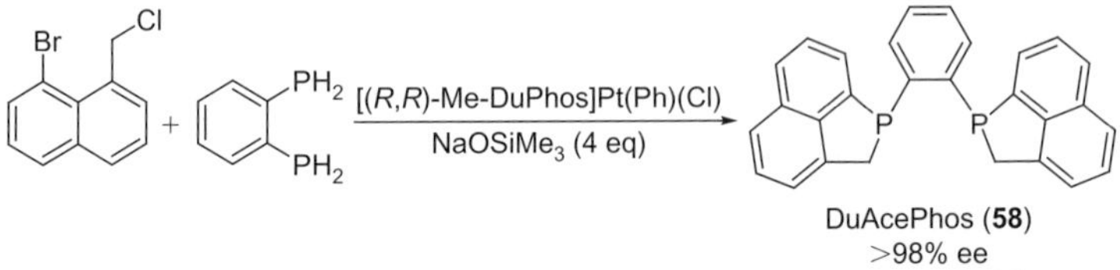

Scheme 2.44

2.3.7
Asymmetric Nucleophilic Addition to Ketones and Ketimines

The preparation of chiral tertiary alcohols and α-tertiary amines via Cu-catalyzed asymmetric nucleophilic addition to ketones and ketimines has been developed by Shibasaki and coworkers as a powerful synthetic methodology [73]. In this emerging field, DuPhos and BPE have found important applications. For example, asymmetric allylation of ketones [74] and ketimines [75] was achieved by iPr-DuPhos and cyclopentyl-DuPhos, respectively (Scheme 2.45). Despite the structural similarity of these two types of substrates, different additives were required for optimal activity and enantioselectivity. Recently, this catalytic protocol

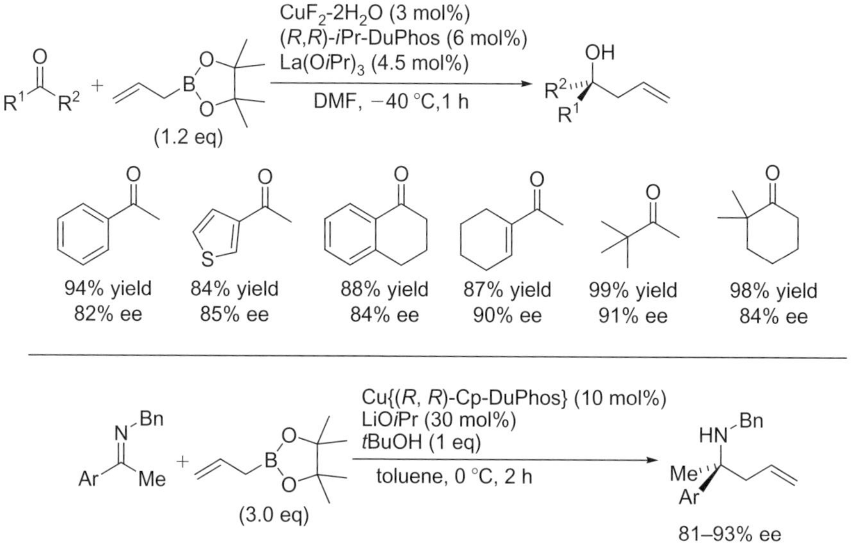

Scheme 2.45

was further expanded to an intramolecular arylation reaction during the synthesis of SM-130686 [76]. In this case, however, Ph-BPE turned out to be the ligand of choice (Scheme 2.46).

The merit of Ph-PBE is also seen in asymmetric addition of allyl cyanide to ketones (Scheme 2.47) [77] and ketimines (Scheme 2.48) [78]. Interestingly, distinct regioselectivities were observed for the two substrates: the addition to ketones involves the α-position of allyl cyanide to afford homoallylic alcohols; in contrast, the addition to ketimines occurs at the γ-position, which after double bond isomerization leads predominantly to conjugated cyanides with a *(Z)*-configuration.

In the development of Cu-catalyzed asymmetric Mannich-type reactions, Shibasaki found **59**, a special member of DuPhos with bulky 4-*trans*-tBu-substituted cyclohexyl groups, afforded high levels of stereoselection toward aliphatic ketimines (Scheme 2.49) [79]. This result is complementary to that given by axially chiral DTBM-SegPhos, whose scope is limited to aromatic substrates.

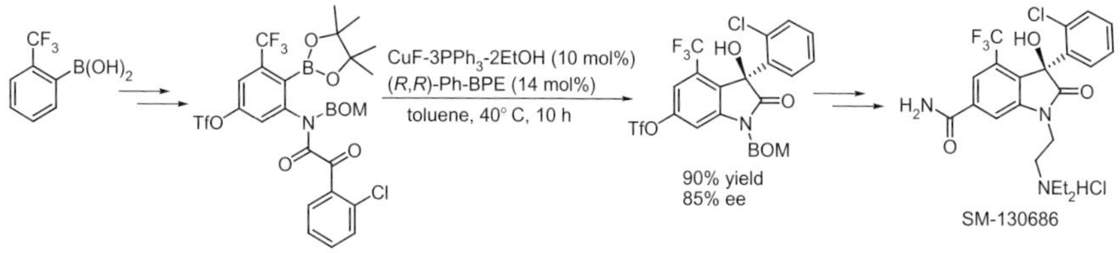

Scheme 2.46 Copper-catalyzed asymmetric intramolecular arylation for the synthesis of SM-130686.

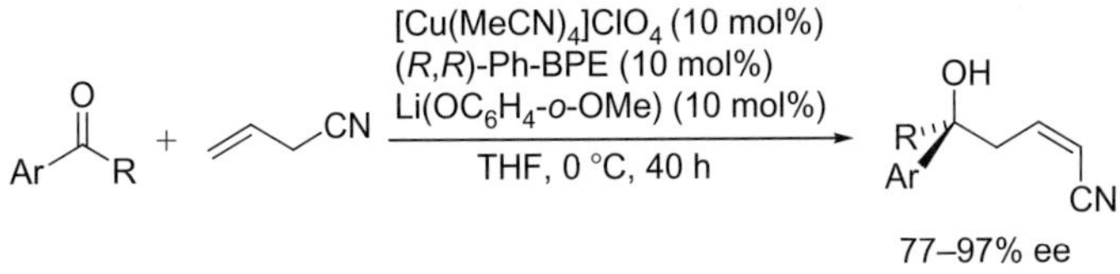

Scheme 2.47

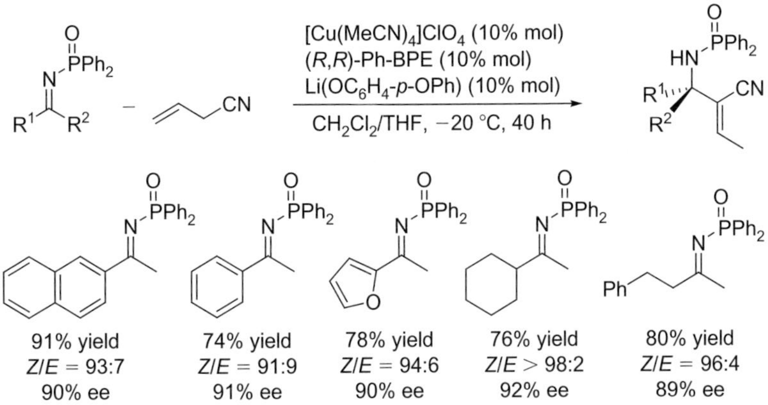

Ar–C(O)–R + allyl–CN → [Cu(MeCN)$_4$]ClO$_4$ (10 mol%), (R,R)-Ph-BPE (10 mol%), Li(OC$_6$H$_4$-o-OMe) (10 mol%), THF, 0 °C, 40 h → Ar–C(OH)(R)–CH$_2$–CH=CH–CN

77–97% ee

Scheme 2.48

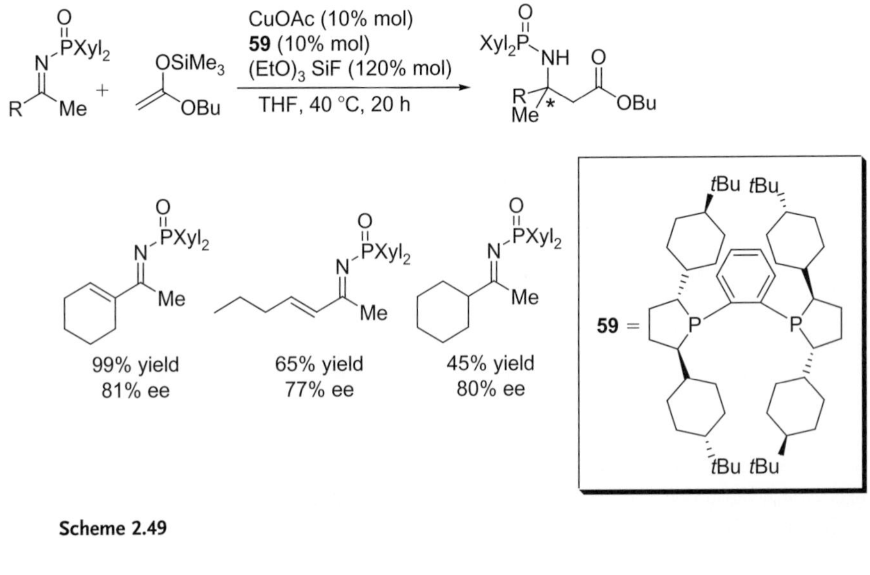

R^{1}R^{2}C=N-P(O)Ph$_2$ + allyl–CN → [Cu(MeCN)$_4$]ClO$_4$ (10% mol), (R,R)-Ph-BPE (10% mol), Li(OC$_6$H$_4$-p-OPh) (10% mol), CH$_2$Cl$_2$/THF, −20 °C, 40 h → R^1(HN-P(O)Ph$_2$)(R^2)C–CH=CH–CN

Naphthyl	Phenyl	Furyl	Cyclohexyl	PhCH$_2$CH$_2$
91% yield	74% yield	78% yield	76% yield	80% yield
Z/E = 93:7	Z/E = 91:9	Z/E = 94:6	Z/E > 98:2	Z/E = 96:4
90% ee	91% ee	90% ee	92% ee	89% ee

Scheme 2.49

R–C(Me)=N–PXyl$_2$(O) + CH$_2$=C(OSiMe$_3$)(OBu) → CuOAc (10% mol), 59 (10% mol), (EtO)$_3$SiF (120% mol), THF, 40 °C, 20 h → R–C*(Me)(NH–PXyl$_2$(O))–CH$_2$–C(O)OBu

Cyclohexenyl: 99% yield, 81% ee
Pentenyl: 65% yield, 77% ee
Cyclohexyl: 45% yield, 80% ee

59 = [bisphospholane ligand with tBu-substituted cyclohexyl groups on a phenylene backbone]

2.3.8
Asymmetric Conjugate Addition

Catalytic asymmetric conjugate addition is an important carbon–carbon bond formation reaction, for which bisphosphacycle ligands have shown valuable applications. In 2005 Minnaard reported the first Pd-catalyzed asymmetric conjugate addition of arylboronic acids to cyclic α,β-unsaturated ketones [80]. The key to this reaction is the use of $Pd(O_2CCF_3)_2$ as the metal precursor as well as Me-DuPhos as the chiral ligand (Scheme 2.50). Toward the same type of substrates, Frost developed Rh/DuPhos-catalyzed conjugate addition reactions involving 2-thienylzinc reagents (Scheme 2.51) [81].

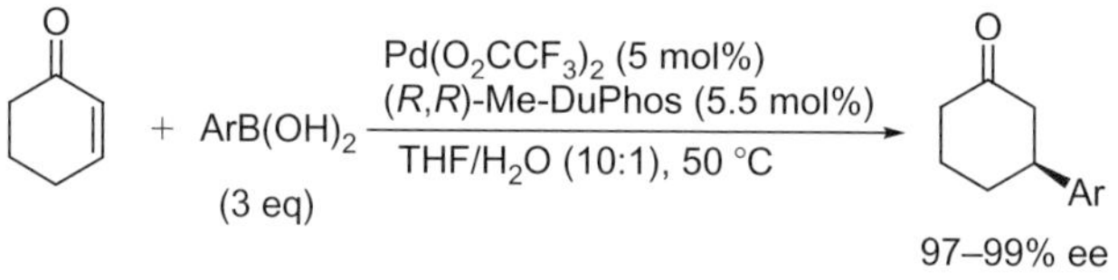

Scheme 2.50

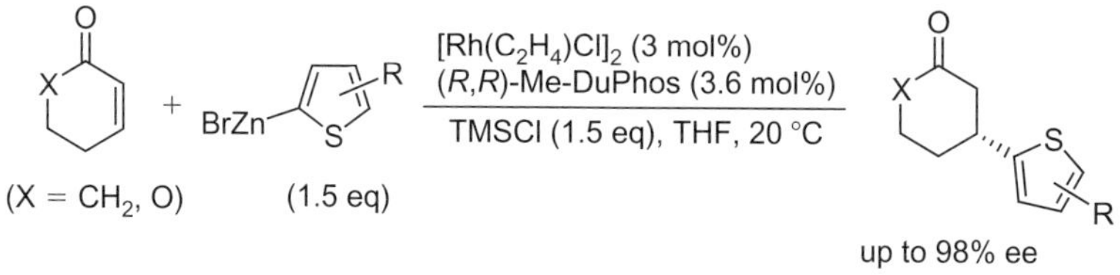

Scheme 2.51

Another remarkable example is Rh-catalyzed asymmetric conjugate addition of arylboronic acids to 4-oxobutenamides, which are electronically differentiated 1,4-unsaturated dicarbonyl compounds [82]. Ligand screening identified conformationally rigid P-chiral ligands, in particular DuanPhos (**45**), afforded excellent enantioselectivities and high regioselectivities toward various Michael donors and acceptors (Table 2.6). The desired product **61** can be further modified at both sides to generate useful chiral compounds.

2.3.9
Miscellaneous Reactions

Bisphosphacycle ligands have also been applied in other asymmetric catalytic reactions, such as Ni-catalyzed isomerization of 4,7-dihydro-1,3-dioxepines (Scheme 2.52) [83] and Pt-catalyzed Bayer–Villiger oxidation of a *meso* cyclic ketone (Scheme 2.53) [84]. Moreover, both DuPhos [85] and duthixantphospholane (**20**)

Table 2.6 Rh/DuanPhos-catalyzed asymmetric conjugate addition reactions.

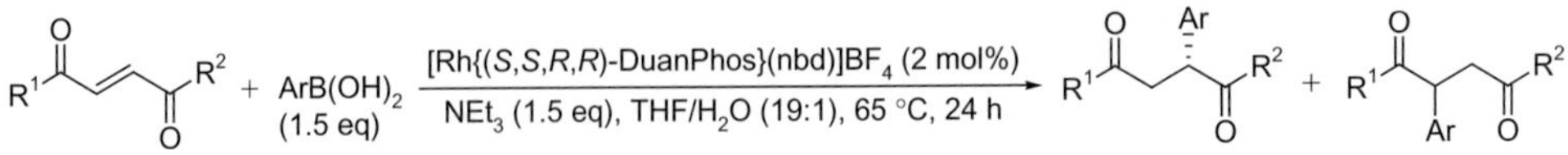

R^1-CH=CH-C(O)-R^2 (60) + ArB(OH)$_2$ (1.5 eq) → [Rh{(S,S,R,R)-DuanPhos}(nbd)]BF$_4$ (2 mol%), NEt$_3$ (1.5 eq), THF/H$_2$O (19:1), 65 °C, 24 h → R^1-CH(Ar)-CH$_2$-C(O)-R^2 (61) + R^1-CH$_2$-CH(Ar)-C(O)-R^2 (62)

R¹	R²	Ar	Product	61:62	Yield (%)	ee (%)
Ph	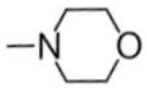 –N(morpholine)	4-MeO-Ph	61a	>99:1	93	98
Ph	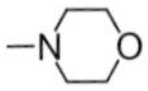 –N(morpholine)	3-Br-Ph	61b	>99:1	99	93
Ph	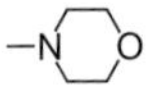 –N(morpholine)	2-Naphthyl	61c	99:1	60	98
Ph	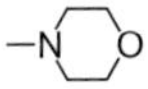 –N(morpholine)	(E)-Ph-CH=CH-	61d	>99:1	92	99
Ph	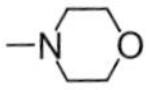 –N(morpholine)	2-Me-Ph	61e	>99:1	99	>99
Me	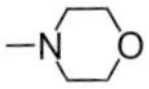 –N(morpholine)	4-MeO-Ph	61f	97:3	90	97
nPr	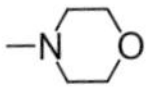 –N(morpholine)	4-MeO-Ph	61g	>95:5	91	99
iPr	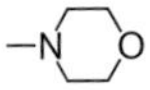 –N(morpholine)	4-MeO-Ph	61h	>95:5	82	99
tBu	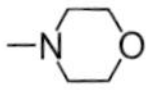 –N(morpholine)	4-MeO-Ph	61i	>95:5	78	99
Ph	–NMe$_2$	4-MeO-Ph	61j	>99:1	89	97
Me	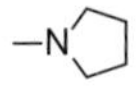 –N(pyrrolidine)	4-MeO-Ph	61k	>99:1	90	99
Me	–N(Me)(OMe)	4-MeO-Ph	61l	2:1	91	97

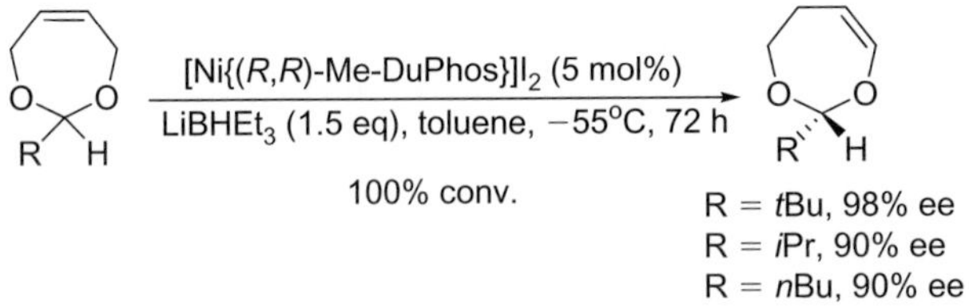

Scheme 2.52

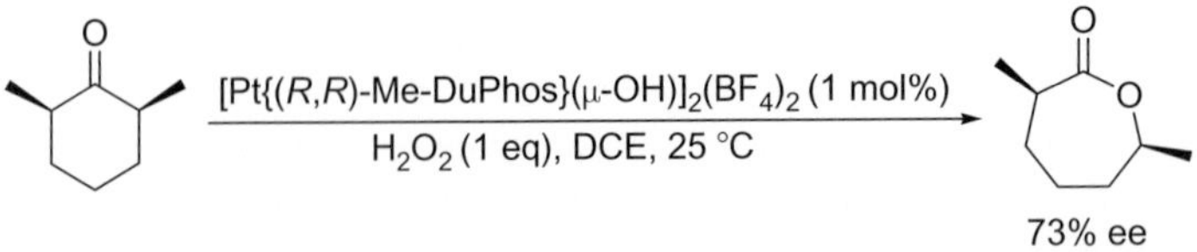

Dioxepine (R,H) → [Ni{(R,R)-Me-DuPhos}]I$_2$ (5 mol%), LiBHEt$_3$ (1.5 eq), toluene, −55 °C, 72 h, 100% conv. → dioxepane (R,H)

R = tBu, 98% ee
R = iPr, 90% ee
R = nBu, 90% ee

Scheme 2.53

2,6-dimethylcyclohexanone → [Pt{(R,R)-Me-DuPhos}(μ-OH)]$_2$(BF$_4$)$_2$ (1 mol%), H$_2$O$_2$ (1 eq), DCE, 25 °C → lactone, 73% ee

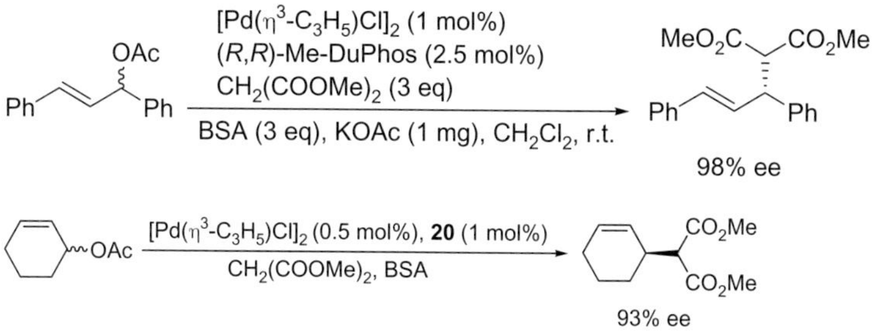

Scheme 2.54

[16] gave good results in Pd-catalyzed asymmetric allylic alkylation reactions (Scheme 2.54).

Very recently, Yun reported Cu-catalyzed asymmetric hydroboration of styrenes (Scheme 2.55) [86]. Excellent regioselectivities and enantioselectivities were attained by the use of TangPhos (**44**), the strong electron-donating property of which is believed to facilitate transmetalation while suppressing competitive β-hydride elimination.

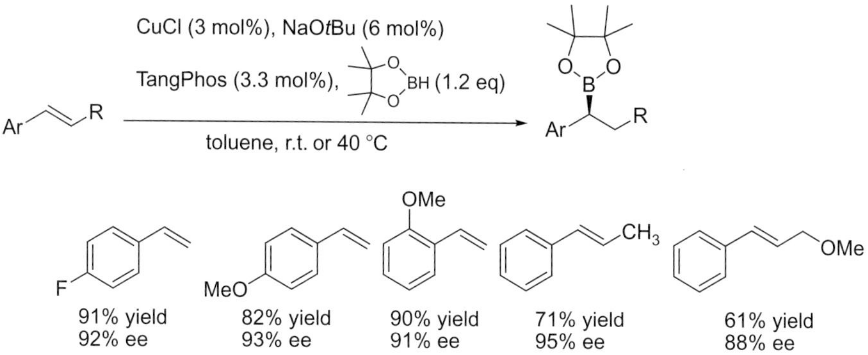

Scheme 2.55

2.4
Concluding Remarks

This chapter has summarized bisphosphacycle ligands pioneered by DuPhos and BPE with respect to their development and representative applications. It is a major achievement that these ligands have been applied in a wide range of highly enantioselective catalytic reactions. As most breakthroughs achieved in this fascinating field are associated with the discovery of new metal–ligand systems showing unprecedented scopes [87], the exploration of new ligands and catalysts has been and will continue to be the central theme.

References

1 Burk, M.J. (2000) *Acc. Chem. Res.*, **33**, 363–370.
2 Two excellent reviews on phospholane ligands have been published, see: (a) Clark, T.P. and Landis, C.R. (2004) *Tetrahedron: Asymmetry*, **15**, 2123–2137; (b) Cobley, C.J. and Moran, P.H. (2007) Enantioselective hydrogenation: phospholane ligands, in *The Handbook of Homogeneous Hydrogenation* (eds J.G. de Vries and C.J. Elsevier), Wiley-VCH Verlag, GmbH, Weinheim, Ch. 24.
3 Burk, M.J., Feaster, J.E., Nugent, W.A., and Harlow, R.L. (1993) *J. Am. Chem. Soc.*, **115**, 10125–10138.
4 You, J., Drexler, H.-J., Zhang, S., Fischer, C., and Heller, D. (2003) *Angew. Chem. Int. Ed.*, **42**, 913–916.
5 (a) Guillen, F. and Fiaud, J.-C. (1999) *Tetrahedron Lett.*, **40**, 2939–2942; (b) Guillen, F., Rivard, M., Toffano, M., Legros, J.-Y., Darran, J.-C., and Fiaud, J.-C. (2002) *Tetrahedron*, **58**, 5895–5904; (c) Pilkington, C.J. and Zanotti-Gerosa, A. (2003) *Org. Lett.*, **5**, 1273–2375; (d) Jackson, M. and Lennon, I.C. (2007) *Tetrahedron Lett.*, **48**, 1831–1834.
6 Yoon, T.P. and Jacobsen, E.N. (2003) *Science*, **299**, 1691–1693.
7 (a) Burk, M.J., Bienewald, F., Harris, M., and Zanotti-Gerosa, A. (1998) *Angew. Chem. Int. Ed.*, **37**, 1931–1933; (b) Burk, M.J., Stammers, T.A., and Straub, J.A. (1999) *Org. Lett.*, **1**, 387–390; (c) Burk, M.J., Bienewald, F., Challenger, S., Derrick, A., and Ramsden, J.A. (1999) *J. Org. Chem.*, **64**, 3290–3298; (d) Burk, M.J., de Koning, P.D., Grote, T.M., Hoekstra, M.S., Hoge, G., Jennings, R.A., Kissel, W.S., Le, T.V., Lennon, I.C., Mulhern, T.A., Ramsden, J.A., and Wade, R.A. (2003) *J. Org. Chem.*, **68**, 5731–5734.
8 (a) Burk, M.J., Gross, M.F., and Martinez, J.P. (1995) *J. Am. Chem. Soc.*, **117**, 9375–9376; (b) Burk, M.J., Wang, Y.M., and Lee, J.R. (1996) *J. Am. Chem. Soc.*, **118**, 5142–5143.
9 Fernandez, E., Gillon, A., Heslop, K., Horwood, E., Hyett, D.J., Orpen, A.G., and Pringle, P.G. (2000) *Chem. Commun.*, 1663–1664.
10 Burk, M.J., Pizzano, A., Martin, J.A., Liable-Sands, L.M., and Rheingold, A.L. (2000) *Organometallics*, **19**, 250–260.
11 Benincori, T., Pilati, T., Rizzo, S., Sannicol, F., Burk, M.J., de Ferra, L., Ullucci, E., and Piccolo, O. (2005) *J. Org. Chem.*, **70**, 5436–5441.
12 Berens, U. (2003) WO 03/031456.
13 Burk, M.J. and Gross, M.F. (1994) *Tetrahedron Lett.*, **35**, 9363–9366.
14 Lotz, M., Kesselgruber, M., Thommen, M., and Pugin, B. (2005) WO 2005/056568.
15 Shimizu, H., Ishizaki, T., Fujiwara, T., and Saito, T. (2004) *Tetrahedron Asymmetry*, **15**, 2169–2172.
16 (a) Dierkes, P., Ramdeehul, S., Barloy, L., De Cian, A., Fischer, J., Kamer, P.C.J., van Leeuwen, P.W.N.M., and Osborn, J.A. (1998) *Angew. Chem. Int. Ed.*, **37**, 3116–3118; (b) Ramdeehul, S., Dierkes, P., Aguado, R., Kamer, P.C.J., van Leeuwen, P.W.N.M., and Osborn, J.A. (1998) *Angew. Chem. Int. Ed.*, **37**, 3118–3121.
17 Fox, M.E., Jackson, M., Lennon, I.C., Klosin, J., and Abboud, K.A. (2008) *J. Org. Chem.*, **73**, 775–784.
18 (a) Holz, J., Monsees, A., Jiao, H., You, J., Komarov, I.V., Fischer, C., Drauz, K., and Borner, A. (2003) *J. Org. Chem.*, **68**, 1701–1707; (b) Almena, J., Monsees, A., Kadyrov, R., Riermeier, T.H., Gotov, B., Holz, J., and Borner, A. (2004) *Adv. Synth. Catal.*, **346**, 1263–1266; (c) Holz, J., Zayas, O., Jiao, H., Baumann, W., Spannenberg, A., Monsees, A., Riermeier, T.H., Almena, J., Kadyrov, R., and Borner, A. (2006) *Chem. Eur. J.*, **12**, 5001–5013; (d) Riermeier, T., Monsees, A., Almena Perea, J., Kadyrov, R., Gotov, B., Zeiss, W., Iris, N., Börner, A., Holz, J., and Drauz, K. (2005) WO 2005/049629 A1; (e) Enthaler, S., Erre, G., Junge, K., Addis, D., Kadyrov, R., and Beller, M. (2008) *Chem. Asian J.*, **3**, 1104–1110.
19 (a) Marinetti, A., Kruger, V., and Buzin, F.-X. (1997) *Tetrahedron Lett.*, **38**, 2947–2950; (b) Marinetti, A., Genêt, J.-P., Jus, S., Blanc, D., and

Ratovelomanana-Vidal, V. (1999) *Chem. Eur. J.*, **5**, 1160–1165.
20 (a) Marinetti, A., Jus, S., and Genet, J.-P. (1999) *Tetrahedron Lett.*, **40**, 8365–8368; (b) Marinetti, A., Jus, S., Genet, J.-P., and Ricard, L. (2001) *J. Organomet. Chem.*, **624**, 162–166.
21 Berens, U., Burk, M.J., Gerlach, A., and Hems, W. (2000) *Angew. Chem. Int. Ed.*, **39**, 1981–1984.
22 Independent of Burk's work, Marinetti and Genet also developed FerroTANE: (a) Marinetti, A., Labrue, F., and Genet, J.-P. (1999) *Synlett*, 1975–1977; (b) Marinetti, A., Labrue, F., Pons, B., Jus, S., Ricard, L., and Genet, J.-P. (2003) *Eur. J. Inorg. Chem.*, 2583–2590.
23 (a) Holz, J., Quirmbach, M., Schmidt, U., Heller, D., Stürmer, R., and Borner, A. (1998) *J. Org. Chem.*, **63**, 8031–8034; (b) Stürmer, R., Börner, A., and Holz, J. (1998) DE 19725796.
24 (a) Li, W., Zhang, Z., Xiao, D., and Zhang, X. (1999) *Tetrahedron Lett.*, **40**, 6701–6704; (b) Li, W., Zhang, Z., Xiao, D., and Zhang, X. (2000) *J. Org. Chem.*, **65**, 3489–3496.
25 (a) Yan, Y.-Y. and RajanBabu, T.V. (2000) *J. Org. Chem.*, **65**, 900–906; (b) RajanBabu, T.V., Yan, Y.-Y., and Shin, S. (2001) *J. Am. Chem. Soc.*, **123**, 10207–10213.
26 Liu, D., Li, W., and Zhang, X. (2002) *Org. Lett.*, **4**, 4471–4474.
27 (a) Chen, Z., Jiang, Q., Zhu, G., Xiao, D., Cao, P., and Zhang, X. (1997) *J. Org. Chem.*, **62**, 4521–4523; (b) Jiang, Q., Jiang, Y., Xiao, D., Cao, P., and Zhang, X. (1998) *Angew. Chem. Int. Ed.*, **37**, 1100–1103; (c) Jiang, Q., Xiao, D., Zhang, Z., Cao, P., and Zhang, X. (1999) *Angew. Chem. Int. Ed.*, **38**, 516–518; (d) Zhang, Z., Zhu, G., Jiang, Q., Xiao, D., and Zhang, X. (1999) *J. Org. Chem.*, **64**, 1774–1775.
28 (a) Clark, T.P., Landis, C.R., Freed, S.L., Klosin, J., and Abboud, K.A. (2005) *J. Am. Chem. Soc.*, **127**, 5040–5041; (b) Klosin, J. and Landis, C.R. (2007) *Acc. Chem. Res.*, **40**, 1251–1259.
29 Ostermeier, M., Prieb, J., and Helmchen, G. (2002) *Angew. Chem. Int. Ed.*, **41**, 612–614.
30 Yan, Y. and Zhang, X. (2006) *Tetrahedron Lett.*, **47**, 1567–1569.
31 Xiao, D., Zhang, Z., and Zhang, X. (1999) *Org. Lett.*, **1**, 1679–1681.
32 Chi, Y. and Zhang, X. (2002) *Tetrahedron Lett.*, **43**, 4849–4852.
33 (a) Xiao, D. and Zhang, X. (2001) *Angew. Chem. Int. Ed.*, **40**, 3425–3428; (b) Chi, Y., Zhou, Y.-G., and Zhang, X. (2003) *J. Org. Chem.*, **68**, 4120–4122; (c) Wang, Y.-Q., Yu, C.-B., Wang, D.-W., Wang, X.-B., and Zhou, Y.-G. (2008) *Org. Lett.*, **10**, 2071–2074; (d) Yu, C.-B., Wang, D.-W., and Zhou, Y.-G. (2009) *J. Org. Chem.*, **74**, 5633–5635; (e) Hou, G., Gosselin, F., Li, W., McWilliams, J.C., Sun, Y., Weisel, M., O'Shea, P.D., Chen, C.-y., Davies, I.W., and Zhang, X. (2009) *J. Am. Chem. Soc.*, **131**, 9882–9883.
34 (a) Hoge, G. (2003) *J. Am. Chem. Soc.*, **125**, 10219–10227; (b) Hoge, G. (2004) *J. Am. Chem. Soc.*, **126**, 9920–9921.
35 Shimizu, H., Saito, T., and Kumobayashi, H. (2003) *Adv. Synth. Catal.*, **345**, 185–189.
36 (a) Tang, W. and Zhang, X. (2002) *Angew. Chem. Int. Ed.*, **41**, 1612–1614; (b) Yang, Q., Shang, G., Gao, W., Deng, J., and Zhang, X. (2006) *Angew. Chem. Int. Ed.*, **45**, 3832–3835; (c) Shang, G., Yang, Q., and Zhang, X. (2006) *Angew. Chem. Int. Ed.*, **45**, 6360–6362; (d) Chen, J., Zhang, W., Geng, H., Li, W., Hou, G., Lei, A., and Zhang, X. (2009) *Angew. Chem. Int. Ed.*, **48**, 800–803; (e) Tang, W. and Zhang, X. (2002) *Org. Lett.*, **4**, 4159–4162; (f) Dai, Q., Yang, W., and Zhang, X. (2005) *Org. Lett.*, **7**, 5343–5346; (g) Tang, W., Liu, D., and Zhang, X. (2003) *Org. Lett.*, **5**, 205–208; (h) Wang, C.-J., Tao, H., and Zhang, X. (2006) *Tetrahedron Lett.*, **47**, 1901–1903.
37 (a) Liu, D. and Zhang, X. (2005) *Eur. J. Org. Chem.*, 646–649; (b) Liu, D., Gao, W., Wang, C., and Zhang, X. (2005) *Angew. Chem. Int. Ed.*, **44**, 1687–1689; (c) Sun, X., Zhou, L., Wang, C.-J., and Zhang, X. (2007) *Angew. Chem. Int. Ed.*, **46**, 2623–2626; (d) Geng, H., Zhang, W., Chen, J., Hou, G., Zhou, L., Zou, Y., Wu, W., and Zhang, X. (2009) *Angew. Chem. Int. Ed.*, **48**, 6052–6054; (e) Li, W., Hou, G., Chang, M., and Zhang, X. (2009) *Adv. Synth. Catal.*, **351**, 3123–3127.
38 Tang, W., Wang, W., Chi, Y., and Zhang, X. (2003) *Angew. Chem. Int. Ed.*, **42**, 3509–3511.

39 (a) Imamoto, T., Oohara, N., and Takahashi, H. (2004) *Synthesis*, 1353–1358; (b) Imamoto, T., Crepy, K.V.L., and Katagiri, K. (2004) *Tetrahedron: Asymmetry*, **15**, 2213–2218.
40 Nicolaou, K.C., Xu, J., Murphy, F., Barluenga, S., Baudoin, O., Wei, H.-X., Gray, D.L.F., and Ohshima, T. (1999) *Angew. Chem. Int. Ed.*, **38**, 2447–2451.
41 Endo, A., Yanagisawa, A., Abe, M., Tohma, S., Kan, T., and Fukuyama, T. (2002) *J. Am. Chem. Soc.*, **124**, 6552–6554.
42 Zhou, Y.-G., Yang, P.-Y., and Han, X.-W. (2005) *J. Org. Chem.*, **70**, 1679–1683.
43 Burk, M.J., Kalberg, C.S., and Pizzano, A. (1998) *J. Am. Chem. Soc.*, **120**, 4345–4353.
44 Burk, M.J., Harper, G.P., and Kalberg, C.S. (1995) *J. Am. Chem. Soc.*, **117**, 4423–4424.
45 Noyori, R. (2002) *Angew. Chem. Int. Ed.*, **41**, 2008–2022.
46 Cobley, C.J. and Henschke, J.P. (2003) *Adv. Synth. Catal.*, **345**, 195–201.
47 Burk, M.J. and Feaster, J.E. (1992) *J. Am. Chem. Soc.*, **114**, 6266–6267.
48 Claver, C. and van Leeuwen, P.W.N.M. (2000) *Rhodium Catalyzed Hydroformylation*, Kluwer Academic Publishers, Dordrecht.
49 Axtell, A.T., Cobley, C.J., Klosin, J., Whiteker, G.T., Zanotti-Gerosa, A., and Abboud, K.A. (2005) *Angew. Chem. Int. Ed.*, **44**, 5834–5838.
50 Axtell, A.T., Klosin, J., Whiteker, G.T., Cobley, C.J., Fox, M.E., Jackson, M., and Abboud, K.A. (2009) *Organometallics*, **28**, 2993–2999.
51 Axtell, A.T., Klosin, J., and Abboud, K.A. (2006) *Organometallics*, **25**, 5003–5007.
52 Huang, J., Bunel, E., Allgeier, A., Tedrow, J., Storz, T., Preston, J., Correll, T., Manley, D., Soukup, T., Jensen, R., Syed, R., Moniz, G., Larsen, R., Martinelli, M., and Reider, P.J. (2005) *Tetrahedron Lett.*, **46**, 7831–7834.
53 Marciniec, B. (2008) *Hydrosilylation, A Comprehensive Review on Recent Advances*, Vol. **1**, Springer, The Netherlands.
54 Burk, M.J. and Feaster, J.E. (1992) *Tetrahedron Lett.*, **33**, 2099–2102.
55 Shaikh, N.S., Enthaler, S., Junge, K., and Beller, M. (2008) *Angew. Chem. Int. Ed.*, **47**, 2497–2501.
56 Fu, G.C. (2005) Recent advances in rhodium(I)-catalyzed asymmetric olefin isomerization and hydroacylation reactions, in *Modern Rhodium-Catalyzed Organic Reactions* (ed. P.A. Evans), Wiley-VCH Verlag GmbH, Weinheim, Ch. 4.
57 Barnhart, R.W., McMorran, D.A., and Bosnich, B. (1997) *Chem. Commun.*, 589–590.
58 Coulter, M.M., Dornan, P.R., and Dong, K.V.M. (2009) *J. Am. Chem. Soc.*, **131**, 6932–6933.
59 Tanaka, K. and Fu, G.C. (2002) *J. Am. Chem. Soc.*, **124**, 10296–10297.
60 Osborne, J.D., Randell-Sly, H.E., Currie, G.S., Cowley, A.R., and Willis, M.C. (2008) *J. Am. Chem. Soc.*, **130**, 17232–17233.
61 Gilbertson, S.R., Hoge, G.S., and Genov, D.G. (1998) *J. Org. Chem.*, **63**, 10077–10080.
62 Aikawa, K., Akutagawa, S., and Mikami, K. (2006) *J. Am. Chem. Soc.*, **128**, 12648–12649.
63 (a) Murakami, M., Itami, K., and Ito, Y. (1997) *J. Am. Chem. Soc.*, **119**, 2950–2951; (b) Murakami, M., Itami, K., and Ito, Y. (1999) *J. Am. Chem. Soc.*, **121**, 4130–4135.
64 Tanaka, K., Hori, T., Osaka, T., Noguchi, K., and Hirano, M. (2007) *Org. Lett.*, **9**, 4881–4884.
65 Corkey, B.K. and Toste, F.D. (2007) *J. Am. Chem. Soc.*, **129**, 2764–2765.
66 Watson, M.P. and Jacobsen, E.N. (2008) *J. Am. Chem. Soc.*, **130**, 12594–12595.
67 Tsai, A.S., Wilson, R.M., Harada, H., Bergman, R.G., and Ellman, J.A. (2009) *Chem. Commun.*, 3910–3912.
68 (a) Moncarz, J.R., Laritcheva, N.F., and Glueck, D.S. (2002) *J. Am. Chem. Soc.*, **124**, 13356–13357; (b) Blank, N.F., Moncarz, J.R., Brunker, T.J., Scriban, C., Anderson, B.J., Amir, O., Glueck, D.S., Zakharov, L.N., Golen, J.A., Incarvito, C.D., and Rheingold, A.L. (2007) *J. Am. Chem. Soc.*, **129**, 6847–6858.
69 Korff, C. and Helmchen, G. (2004) *Chem. Commun.*, 530–531.

70 Chan, V.S., Bergman, R.G., and Toste, F.D. (2007) *J. Am. Chem. Soc.*, **129**, 15122–15123.
71 (a) Scriban, C. and Glueck, D.S. (2006) *J. Am. Chem. Soc.*, **128**, 2788–2789; (b) Scriban, C., Glueck, D.S., Golen, J.A., and Rheingold, A.L. (2007) *Organometallics*, **26**, 1788–1800.
72 Anderson, B.J., Guino-o, M.A., Glueck, D.S., Golen, J.A., DiPasquale, A.G., Liable-Sands, L.M., and Rheingold, A.L. (2008) *Org. Lett.*, **10**, 4425–4428.
73 Shibasaki, M. and Kanai, M. (2008) *Chem. Rev.*, **108**, 2853–2873.
74 Wada, R., Oisaki, K., Kanai, M., and Shibasaki, M. (2004) *J. Am. Chem. Soc.*, **126**, 8910–8911.
75 Wada, R., Shibuguchi, T., Makino, S., Oisaki, K., Kanai, M., and Shibasaki, M. (2006) *J. Am. Chem. Soc.*, **128**, 7687–7691.
76 Tomita, D., Yamatsugu, K., Kanai, M., and Shibasaki, M. (2009) *J. Am. Chem. Soc.*, **131**, 6946–6948.
77 Yazaki, R., Kumagai, N., and Shibasaki, M. (2009) *J. Am. Chem. Soc.*, **131**, 3195–3197.
78 Yazaki, R., Nitabaru, T., Kumagai, N., and Shibasaki, M. (2008) *J. Am. Chem. Soc.*, **130**, 14477–14479.
79 Suto, Y., Kanai, M., and Shibasaki, M. (2007) *J. Am. Chem. Soc.*, **129**, 500–501.
80 Gini, F., Hessen, B., and Minnaard, A.J. (2005) *Org. Lett.*, **7**, 5309–5312.
81 Le Notre, J., Allen, J.C., and Frost, C.G. (2008) *Chem. Commun.*, 3795–3797.
82 Zigterman, J.L., Woo, J.C.S., Walker, S. D., Tedrow, J.S., Borths, C.J., Bunel, E. E., and Faul, M.M. (2007) *J. Org. Chem.*, **72**, 8870–8876.
83 Frauenrath, H., Brethauer, D., Reim, S., Maurer, M., and Raabe, G. (2001) *Angew. Chem. Int. Ed.*, **40**, 177–179.
84 Paneghetti, C., Gavagnin, R., Pinna, F., and Strukul, G. (1999) *Organometallics*, **18**, 5057–5065.
85 Drago, D. and Pregosin, P.S. (2000) *J. Chem. Soc., Dalton Trans.*, 3191–3196.
86 Noh, D., Chea, H., Ju, J., and Yun, J. (2009) *Angew. Chem. Int. Ed.*, **48**, 6062–6064.
87 Trost, B.M. (2004) *Proc. Natl. Acad. Sci. U.S.A.*, **101**, 5348–5355.

3
Josiphos Ligands: From Discovery to Technical Applications

Hans-Ulrich Blaser, Benoît Pugin, Felix Spindler, Esteban Mejía, and Antonio Togni

3.1
Introduction and Background

Metal complexes bearing chiral ligands arguably constitute the most versatile catalysts for enantioselective transformations [1, 2]. An impressively large number of chiral ligands is recorded in the literature with very high enantioselectivity in various catalytic reactions. However, if one has a closer look at which ligands are really used by the synthetic organic chemist in academia, and even more so in industry, a very different picture emerges. In fact, very few chiral catalytic systems are used on a regular basis for the synthesis of target molecules [3, 4] and Jacobsen has coined the term "privileged ligands" for these selected few [5].

Several prerequisites must be fulfilled to make a chiral ligand attractive for the industrial chemist who wants to solve a particular synthetic problem. We have identified the following success factors:

1. Scope (and limitations) and specificity should be well described in the literature and the catalysts should have a high functional group tolerance (i.e., should also work with multifunctional substrates).
2. Catalyst performance (ee, TON, and TOF) must be very good – not just for model substrates but for "real world" applications.
3. Ligands should be commercially available (or their synthesis must be very simple) both for screening purposes (large selection) as well as on a larger scale (multi gram to kilogram).
4. Conditions for patent protected technology should be simple and clear.

There is little doubt that the Josiphos ligand family described in this chapter fulfils all of these criteria and, indeed, numerous synthetic as well as industrial applications using Josiphos ligands have been implemented in the last decade. Because the discovery and the early development of the Josiphos ligands has been described in some detail [6], only a brief summary is given here. In Section 3.3 we analyze the reasons why Josiphos is such a versatile ligand for various very

Privileged Chiral Ligands and Catalysts. Edited by Qi-Lin Zhou
Copyright © 2011 WILEY-VCH Verlag GmbH & Co. KGaA, Weinheim
ISBN: 978-3-527-32704-1

different reactions types. Section 3.4 presents an up-dated overview on the synthetic and industrial applications of Josiphos ligands.

3.2
Discovery and Development of the Josiphos Ligand Family

Very often, the most striking new developments in ligand synthesis are due to serendipity and the Josiphos ligands are a very good illustration of such a discovery. While studying the Au(I)-catalyzed aldol reaction originally reported by Hayashi, Sawamura, and Ito [7], Antonio Togni prepared several ferrocenyl ligands bearing different amine side chains (left-hand reaction of Scheme 3.1, X = PPh_2). A key finding was that the acetate group at the stereogenic center could be smoothly substituted by various nucleophiles with complete retention, which led to the synthesis of novel chelating ligands bearing two different PR_2 groups (right-hand reaction of Scheme 3.1, X = H) [8]. The first example (with PCy_2) was named after Josi Puleo, the technician who actually prepared it but the name is now used for the whole family of ligands.

Why did the story not stop here as in so many other cases, where only one or two ligands are actually prepared, tested, and published? In retrospect, three reasons can be identified:

1. The first catalytic tests were highly successful (e.g., the Rh-catalyzed enamide hydrogenation occurred with >90% ee).
2. It was evident that both enantiomers of a wide variety of sterically and electronically different diphosphine ligands could be prepared in two steps from Ugi amine (very high modularity; see Figure 3.1).
3. Josiphos ligands were one of the major success factors for the economical manufacture of *(S)*-Metolachlor (see below).

These of course were reasons enough to devote a lot of manpower to the synthesis and technical development of various different ligands with a broad variation of the two substituents at the two phosphorous atoms. Indeed, to date about 150 different Josiphos ligands have been prepared, with the substituents R and R' at the P atoms encompassing phenyl, substituted phenyl, heteroaryl, alkyl, and cycloalkyl (Figure 3.2). Forty derivatives are available in a ligand kit for

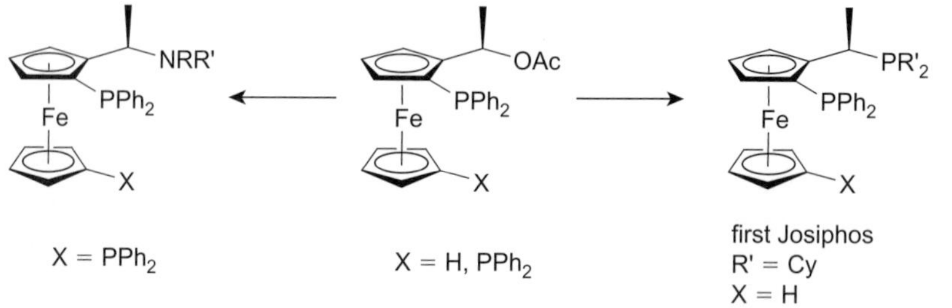

Scheme 3.1 Substitution reactions at the stereogenic center of ethyl ferrocene derivatives.

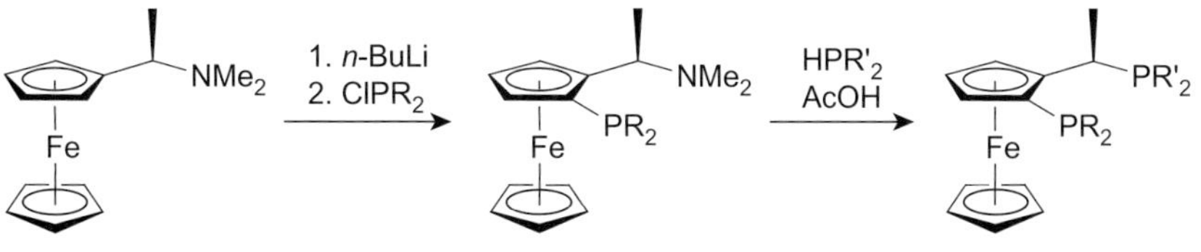

(*R*)-Ugi amine (*R,S*)-ppfa (*R,S*)-Josiphos

Name	R	R'	Important applications
SL-J001	Ph	Cy	Process for Jasmonate; hydrogenation of enamides and itaconates; hydroboration; allylic alkylations; Michael additions; PMHS reduction of C=C; addition to meso anhydrides
SL-J002	Ph	*t*-Bu	Opening of oxabicycles; Biotin and Sitagliptin process
SL-J003	Cy	Cy	Hydrogenation of phosphinylimines
SL-J004	Cy	Ph	Michael additions
SL-J005	Ph	3,5-Xyl	Metolachlor process; methoxycarboxylation
SL-J009	Cy	*t*-Bu	Hydrophosphination; cross-coupling
SL-J011	4-CF_3-Ph	*t*-Bu	Hydrogenation of β-imino acid derivates
SL-J013	4-MeO-Xyl	*t*-Bu	Dextromethorphane process

Figure 3.1 Technical synthesis and applications of important Josiphos ligands.

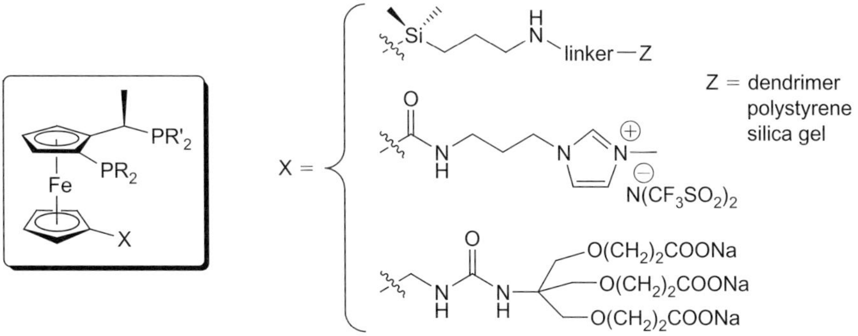

Figure 3.2 Functionalized Josiphos ligands.

screening and on a multi-kg scale for production [9]. The ligands are well soluble in most organic solvents (such as diethyl ether, THF, CH_2Cl_2, etc.), partially soluble in alcohols (such as MeOH, EtOH, etc.) but insoluble in water. They are supplied as orange or yellow solids and are air-stable for several years in the solid form in closed bottles.

Naming of the ligand: When Solvias started to market its own ligands, it was decided to use Josiphos as brand name for the whole ligand family and each ligand was given its own J-number [9]. Selected members of the family, that is, the ones referred to in the next section, are depicted with their numbering in Table 3.1. In most cases, the (R,S_{Fc}) enantiomers (numbering SL-JXXX-1) have been used, but both enantiomers of all ligands are available from Aldrich and Strem, and in technical quantities from Solvias. The individual ligands are also abbreviated as (R,S)-R_2PF-PR$'_2$, where the P-substituent at the Cp ring is in front of the F (for ferrocene) and the one at the stereogenic center after the F. For historical reasons, Ph$_2$PF-PR$'_2$ is abbreviated as PPF-PR$'_2$ Concerning the absolute configuration, the first descriptor stands for the stereogenic center, the second for the planar chirality. Up to now, only the (R,S)-family (and its enantiomers, of course) but not the (R,R) diastereomers have led to high enantioselectivities (see Figure 3.4 below).

The Josiphos ligands can be functionalized by covalent attachment to the lower Cp ring to prepare immobilized catalysts or catalysts soluble in water or ionic liquids (Figure 3.2). Here, we will not describe these catalysts or their application but refer the reader to recent reviews [6, 10]. Sufficient to say that in general these catalysts have very similar catalytic properties to those of the homogeneous analogs.

Table 3.1 Name and CAS number of selected Josiphos ligands.

R	R'	Name	CAS registry number	Name	CAS registry number
Ph	Cy	SL-J001-1	155806–35–2	SL-J001-2	162291–02–3
Ph	t-Bu	SL-J002-1	155830–69–6	SL-J002-2	277306–29–3
Cy	Cy	SL-J003-1	167416–28–6	SL-J003-2	158923–07–0
Cy	Ph	SL-J004-1	158923–09–2	SL-J004-2	162291–01–2
Ph	3,5-Xyl	SL-J005-1	184095–69–0	SL-J005-2	223121–07–1
3,5-(CF$_3$)$_2$Ph	Cy	SL-J006-1	292638–88–1	SL-J006-2	849923–15–5
4-MeO-Xyl	Cy	SL-J007-1	360048–63–1	SL-J007-2	849923–88–2
Cy	t-Bu	SL-J009-1	158923–11–6	SL-J009-2	194020–16–1
4-CF$_3$-Ph	t-Bu	SL-J011-1	246231–79–8	SL-J011-2	849924–37–4
4-MeO-Xyl	t-Bu	SL-J013-1	187733–50–2	SL-J013-2	849924–40–9
Ph	3-Tolyl	SL-J102-1	184095–68–9	SL-J102-2	Not available
Ph	2-Anisyl	SL-J106-1	155830–66–3	SL-J106-2	Not available
Ph	4-CF$_3$Ph	SL-J110-1	155830–73–2	SL-J110-2	Not available
2-Furyl	Xyl	SL-J212-1	849924–41–0	SL-J212-2	849924-42-1
Xyl	Xyl	SL-J408-1	349150–76–1	SL-J408-2	158923–10–5
t-Bu	o-Tol	SL-J505-1	849924–76–1	SL-J505-2	849924–77–2

3.3
Why Are Josiphos Ligands So Effective?

3.3.1
General Considerations

The short answer would be: We do not really understand! Obviously, an effective ligand must coordinate to the metal center and provide the necessary chiral environment to induce enantioselectivity. Furthermore, it must sufficiently stabilize the various reaction intermediates encompassing different oxidation states, coordination numbers, and geometries to allow for a high reaction rate. If one analyzes the structures of various effective ligands, several design principles can be identified that might lead to good enantiocontrol. Generally speaking, these measures create the necessary flexibility of the ligand to give high turnover rates and also impart sufficient rigidity to control stereoselectivity. According to the literature [11], a ligand is more likely to induce high enantioselectivity if it:

- Is bidentate, usually cis-chelating for more rigidity (first example: diop), the optimal chelate size strongly dependent on the nature of the backbone (e.g., miniphos forms four-, dipamp five-, bppfoh six-, diop or binap seven- and walphos eight-membered metallocycles). The trap ligands form nine-membered chelates for which the trans configuration is accessible for the first time.
- Has C_2 symmetry (first example: diop) to reduce the number of possible isomeric catalyst–substrate complexes [12]. An alternative idea is that coordinating P moieties with very different electronic and steric properties should be advantageous for polar substrates [13]. In this respect, Josiphos is one of the few ligands allowing a combination of a P with all alkyl and a P with all aryl substituents (this is rather rare).
- Has two aryl or bulky alkyl groups at the P atom to "transfer" chirality from the more remote chiral backbone (first example, diop).
- Has chiral elements as close as possible to the coordinated substrate, for example, the stereogenic P atom (first example, dipamp).

The first generation of chiral ligands all had central chirality (often from chiral pool starting materials) and carried PPh_2 groups. Later it was found that backbones with axial (e.g., binap) as well as planar chirality (e.g., ferrocenyl phosphines) and that phosphorous moieties with other P-alkyl as well as P–O or P–N bonds can be just as effective.

It appears that the ferrocene moiety is generally very well suited to serve as backbone for phosphine-containing ligands [10]. In particular, several highly effective diphosphine ligands such as BoPhoz, Walphos, or Taniaphos (Figure 3.3) have been prepared where the two P moieties are attached in some way in the 1,2-positions of one of the Cp rings. However, the catalytic profile of these four ligand families is quite different, most likely due to the difference in the geometries of the resulting metal complexes.

To understand a little more about the effect, several ligands have been prepared that have the same relationship of the two phosphine moieties (Figures 3.4

(*R,S*)-BoPhoz (*R,R*)-Walphos (*S,S*)-Taniaphos third generation R'' = Me, Et, Ph

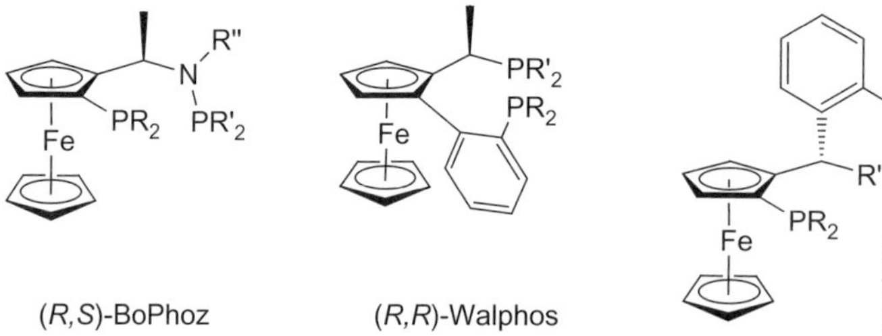

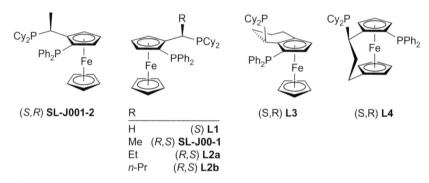

Figure 3.3 Generic structures of BoPhoz, Walphos, and Taniaphos ligands.

(*R,S*) **SL-J005-1** ee 80% (*S*)

(*R,R*) **L5** ee 60% (*R*)

(*R,R*) **L6** ee 39% (*R*)

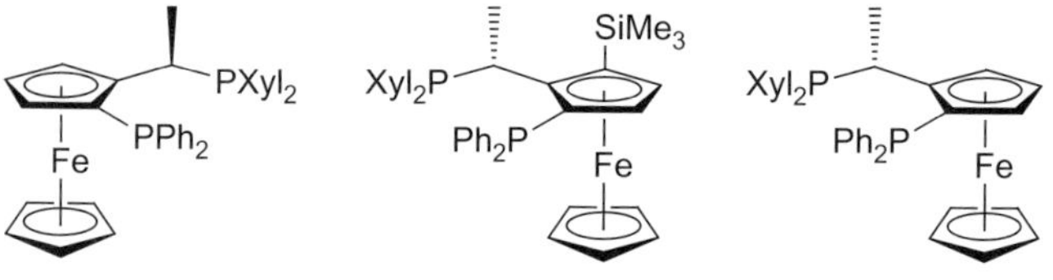

Figure 3.4 Josiphos with *(R,R)* configuration and enantioselectivity for 2-methyl-5-ethyl-aniline (MEA) imine hydrogenation (see Scheme 3.29).

(*S,R*) **SL-J001-2**

R	
H	(*S*) **L1**
Me	(*R,S*) **SL-J00-1**
Et	(*R,S*) **L2a**
n-Pr	(*R,S*) **L2b**

(S,R) **L3**

(S,R) **L4**

Figure 3.5 Structures of Josiphos-related diphosphines.

and 3.5). We have already indicated that the *(R,R)*-Josiphos are much less effective, as evidenced by the results for the hydrogenation of MEA imine (see Scheme 3.29 below). Not only is the enantioselectivity significantly lower, also the sense of induction is reversed, indicating that the central chirality dominates. Similarly, the replacement of the methyl group in the side chain by H (**L1**, [14]), by Et (**L2a**, [15]) or *n*-Pr (B. Pugin, unpublished results) as well as the restriction of the conformational flexibility in **L3** and **L4** [16, 17] lead to a considerable decrease in enantioselectivity for various transformations (Table 3.2).

Table 3.2 Comparison of catalytic performance for various Josiphos analogs.

Reaction (metal)	SL-J001-1	(S)-L1	(R,S)-L2a	(R,S)-L2b	(R,S)-L3	(R,S)-L4
Allylic alkylation (Pd)	93 (S)	62 (S)	N.a.[a]	N.a.	34 (S)	61 (R)
Hydrogenation of MAC[b] (Rh)	96 (R)	34 (S)	N.a.	71 (R)	71 (S)	35 (S)
Hydrogenation of DMI[c] (Rh)	99 (S)	93 (S)	N.a.	93 (S)	34 (S)	61 (R)
Hydrogenation of MEAI[d] (Ir)	73 (R)	N.a.	N.a.	N.a.	32 (R)	73 (R)
Hydroboration of styrene (Rh)	91 (R) (b/l 99 : 1)	N.a.	46 (R) (b/l 92/8)	46 (R) (b/l 92/8)	N.a.	N.a.

[a]N.a. = not available.
[b]MAC = methyl acetamido cinnamate.
[c]DMI = dimethyl itaconate.
[d]MEAI = MEA imine.

3.3.2
Structural Aspects of Transition Metal Complexes Containing Josiphos and Josiphos-Like Ligands

Since the Josiphos scaffold is well-suited for modular syntheses, it unsurprising that there is a relatively large number of derivatives differing by the nature of the substituents on the two phosphorus atoms. Furthermore, there are several derivatives in which the stereogenic side chain is part of a tether connecting the two cyclopentadienyl rings (ferrocenophanes, 2,1'-substitution) or forms a second, typically six-membered, ring (2,3-substitution). For many of these ligands, the X-ray crystal structure of corresponding transition-metal complexes relevant as catalyst precursors, or as intermediates in catalytic reactions, is known. Though the solid-state structure of a complex does not necessarily reflect the structure of the same species in solution, a large set of representative compounds may be used to identify and analyze possible structural trends. In particular, simple geometrical parameters, such as bond angles and torsion angles, serve to define the conformational properties of a chiral ligand coordinated to a metal center and hence to define the chiral environment of the catalytically active center.

For over 70 different compounds, all conforming to the general formulation of Figure 3.6, we have compiled the parameters collected in the Appendix [18–42] (B. Pugin, unpublished results; A. Togni et al., unpublished material). The data concern complexes containing one of the metal centers Cu(I), Ru(II), Rh(I), Pd(0), Pd(II), W(0), Re(I), Re(V), Ir(I), Ir(III), or Pt(II).

Unsurprisingly, the two metal–phosphorus distances are more dependent on the size and formal oxidation state of the metal center but are not necessarily relevant in terms of overall conformational properties of the complexes. Sufficient to say that for most of the compounds analyzed the distance M–P^1 is shorter than M–P^2, reflecting the fact that P^1 and P^2 generally bear aryl and bulky alkyl

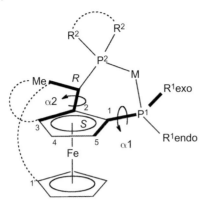

Figure 3.6 General structure and numbering scheme for complexes containing Josiphos-type ligands of *(R)-(S)* absolute configuration. Dotted lines indicate the possible presence of a tether between the corresponding atoms.

substituents, respectively. In contrast, the angle P^1–M–P^2, that is, the bite angle of the ligand, is a more important parameter. Thus, an average value of 92.7° (median value 93.7°) is observed for most cases involving either square-planar or octahedral complexes. However, significant deviations from this typical value are possible, indicating that Josiphos ligands may well adapt to geometric requirements dictated by the metal center. The lowest observed value of 84.4° is displayed by the octahedral complex [W(CO)$_4$(Josiphos)] (A. Togni *et al.*, unpublished material), whereas [CuBr(Josiphos)] [31] shows a distorted trigonal planar geometry around the Cu atom and as a consequence a relatively wide bite angle of 104.1°, the highest value of the series. This means that Josiphos ligands may adopt bite angles deviating by as much as 10° from the average value, though this seems to occur only in rare cases.

Conformational rigidity, understood as a highly preferred conformation of the common, basic scaffold in a series of derivatives, is often put forward and discussed as a qualitative feature that may or may not correlate with the typical reactivity of a system. In our case, two important issues arise. Is it possible to claim conformational rigidity of Josiphos ligands as displayed by the known complexes? In addition, is there a possible correlation with the stereoselectivity of the corresponding catalysts? To identify typical conformational properties of Josiphos ligands we have chosen the two torsion angles illustrated in Figure 3.6. The first angle, $\alpha 1$ (C^5–C^1–P^1–R^1_{exo}), describes the torsion around the C^1–P^1 bond and defines the relative orientation of the substituent R^1 in the *exo* position with respect to the ferrocene core. Thus, a large absolute value of $\alpha 1$ indicates a pseudo-axial orientation, with the P^1–R^1_{exo} bond vector approaching a parallel orientation with the main ferrocene axis. In contrast, values close to 0° mean that the same substituent is pseudo-equatorial, that is, the P^1–R^1_{exo} bond vector lies approximately in the plane of the substituted Cp ring. Analogously, the second angle, $\alpha 2$ (C^3–C^2–C^6–C^7), indicates primarily the orientation of the methyl group at the stereogenic center. Note that the sign of the two torsion angles given in the

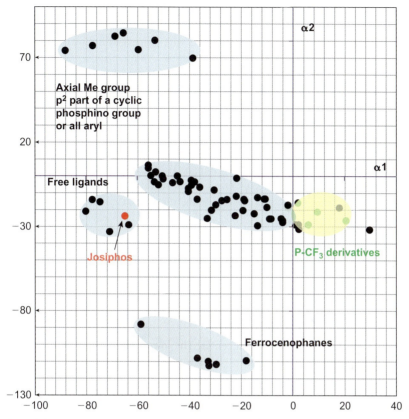

Figure 3.7 Conformational space of Josiphos-type ligands and their complexes. Each circle represents the α1,α2-combination for the corresponding compound.

Appendix are valid for the *(R)-(S)* absolute configuration of the ligands, as shown in Figure 3.6. Thus, negative values of α2 indicate that the methyl group is located in the region between the planes of the two Cp rings, that is, below the substituted Cp ring (endo position). Figure 3.7 depicts a plot of the two angles α1 and α2 and represents part of the accessible, preferred conformational space of Josiphos and its complexes.

Inspection of Figure 3.7 clearly shows that there are preferred regions of the conformational space, as indicated by ovals. Thus, the region of highest concentration of derivatives is characterized by a relatively narrow range of α2 values between about 0° and −30°, with an average of about −15°, indicating that the methyl group is situated slightly below the Cp ring (endo position). In contrast, the corresponding values for α1 vary over a wider range, from 0° to about −60°. This means that conformational rigidity is rather connected with the stereogenic side chain, with the methyl group in a preferred endo-equatorial orientation and, hence, P^2 being rather axial. Depending on the metal and its ancillary ligands, the Josiphos ligands respond and adjust their conformation by more strongly varying α1. The main consequence of altering this torsion angle is a changing orientation of the P^1 lone pair. Eventually, this means that the plane defined by the two P and the metal atoms will increase its inclination with respect to the plane of the Cp ring as α1 gradually increases from large negative towards positive values. This is illustrated by the three representative examples of Figure 3.8. Note that an even small but positive value of α1 means that both R^1 substituents will display an endo orientation, as it is the case for the complexes in Figure 3.8c and d.

Another region of the α1, α2 conformational space is characterized by large positive values of α2 between about 70° and 85°, meaning that the methyl group has an exo-axial orientation. In the corresponding compounds, α1 is still strongly varying, however, over a more negative range. This indicates a more pronounced axial orientation of R^1_{exo}. Compounds in this region are characterized by the substituents R^2 being either aryl rings or being interconnected to form an aliphatic

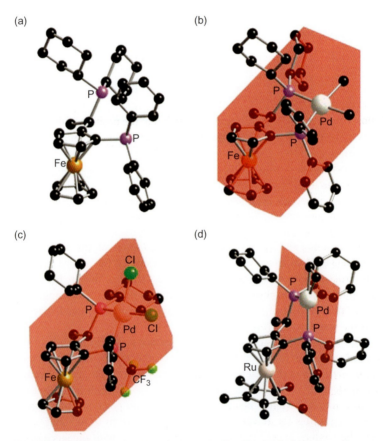

Figure 3.8 Representations of the structures of Josiphos and of corresponding complexes illustrating the effect of a progressively increasing value of α1 on the inclination of the P^1–M–P^2 plane (in red): (a) WIJJII (the original Josiphos); (b) MEDFUX; (c) N10; (d) ZUYDOM.

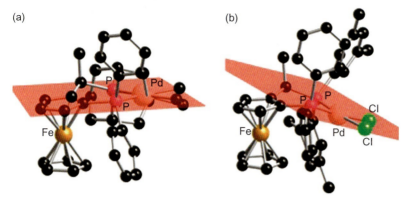

Figure 3.9 Representations of the structures of complexes, illustrating the conformational features of complexes having large, positive α2 values (P^1–M–P^2 plane in red): (a) YOJLAK; (b) XIKTAN.

ring. However, given the restricted number of examples, it is impossible to discern whether the altered combination of α1 and α2 values is solely due to the nature of the substituents on P^2. Interestingly, in these cases, the inclination of the P^1–M–P^2 plane is quite different from that observed for most complexes previously discussed. In fact, the inclination is very weak, such that the coordination plane of the metal is almost coincident with the Cp ring, and such that the position of the metal center may even be defined as endo. The two examples given in Figure 3.9 are illustrative of this conformational behavior.

An endo-axial orientation of the carbon atom corresponding to the methyl group for most compounds is obviously obtained in the case when this carbon atom is part of the tether connecting the two Cp rings (ferrocenophanes). The corresponding region of the conformational space displays α2 values between −90° and −110°. Again, also in these case, α1 seems to vary over a similarly wide range as for the other two major regions, though the number of examples is too small for a more appropriate analysis.

Notably, the region in conformational space characterizing the free ligands does not overlap with any of the regions concerning metal complexes. This is mainly due to a shift of the corresponding α1 torsion angle to more negative values than is the case in the complexes, meaning that the orientation of R^1_{exo} is almost perfectly axial. The simple conclusion to draw from this is that the preferred overall conformation of the free ligands is not retained in any of the complexes, despite the comparable arrangement of the stereogenic side chain, as indicated by the α2 values between about −15° and −30°.

A new region in the α1,α2 conformational space has been disclosed by novel derivatives in which one or both of the R^1 substituents is a CF_3 group (A. Togni et al., unpublished material). These compounds display α1 values between about 0° and +20°, while still maintaining α2 within the anticipated, slightly negative range (i.e., they shift to the right in Figure 3.7). Whether this is a consequence of the small size of the CF_3 group remains to be clarified.

One final structural feature worth mentioning because it is common to essentially all complexes analyzed is the distortion of the Cp ring that results upon coordination. The important parameter characterizing this is the distance of P^1 from the plane of the ring. Thus, P^1 is lifted by as much as 0.56 Å (Figure 3.8d) out of its ideal position.

In conclusion, Josiphos-type ligands show some unique conformational features when coordinated to catalytically active transition-metal centers. A proper, simplistically understood conformational rigidity is, however, not present. The analysis, based on the available X-ray crystal structures of a significant number of derivatives, rather shows a remarkable adaptability of the basic scaffold. However, the subtlety resides, somewhat surprisingly, in the higher flexibility of the P^1 phosphino group, whereas the stereogenic side chain displays a more constant conformation for a large majority of cases. This dual behavior probably constitutes the open secret as to why Josiphos ligands are so successful in various catalytic applications.

3.4
Catalytic Profile of the Josiphos Ligand Family

The choice of transformation and the (model) substrate first tried with a new ligand could well decide whether it will make it. Therefore, it is quite important to have a good strategy for ligand testing. Over the years, several compounds have emerged that serve as benchmarks for the comparison of the catalytic performance of new ligands and catalysts. Examples are acetamidocinnamic or itaconic acid derivatives and ethyl acetoacetate for hydrogenation, acetophenone for reduction, 1,3-diphenyl-1-acetoxypropene for allylic alkylation or styrene for various addition reactions. When assessing the results obtained for the Josiphos family described in the following sub-sections, one has to keep in mind that the quality and relevance of the results may differ widely. In some cases just a few experiments were carried out, whereas for other reactions a broad structure–activity profile was made and/or the reaction conditions were optimized.

3.4.1
Enantioselective Reductions of C=C, C=O and C=N Bonds

3.4.1.1   Enantioselective Hydrogenation of C=C Bonds

Enamides and Itaconic Acid Derivatives   Enamides and itaconates are so-called privileged substrates due to the possibility of a secondary coordination with an amide or ester carbonyl group, respectively [11]. As shown in Table 3.3, Rh/Josiphos catalysts give excellent enantioselectivities; under optimized reaction conditions and using substrates of high purity, s/c > 20 000 were achieved and the TOF at 1 bar easily reached > 300 h^{-1} even under non-optimized conditions (F. Spindler, unpublished results).

Table 3.3 Hydrogenation of enamide and itaconate model substrates.

Substrate	Best catalyst	Reaction conditions	Ee range (%)	TON [TOF (h^{-1})]^a	Reference
R-C(COOR')=CH-NHCOCH$_3$	[Rh(nbd)$_2$]BF$_4$ SL-J001	1 bar H$_2$ 25 °C	25–97	100 (330)	[8]
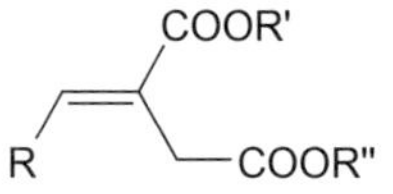 R-C(COOR')=CH-COOR"	[Rh(nbd)$_2$]BF$_4$ SL-J001	1 bar H$_2$ 25 °C	90–99.6	100 (200)	[8] (F. Spindler unpublished results)

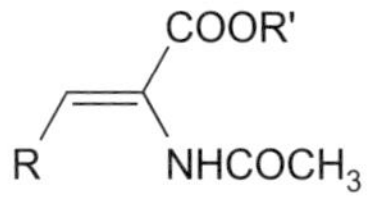

^{a}Turnover number (TON) and turnover frequency (TOF); standard experiments, values not optimized.

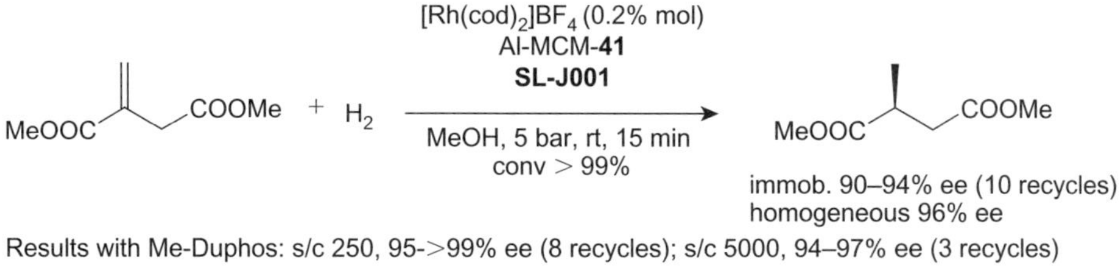

MeOOC-CH$_2$-C(=CH$_2$)-COOMe + H$_2$

[Rh(cod)$_2$]BF$_4$ (0.2% mol)
Al-MCM-41
SL-J001
────────────────→
MeOH, 5 bar, rt, 15 min
conv > 99%

MeOOC-CH$_2$-CH(CH$_3$)-COOMe

immob. 90–94% ee (10 recycles)
homogeneous 96% ee

Results with Me-Duphos: s/c 250, 95->99% ee (8 recycles); s/c 5000, 94–97% ee (3 recycles)

Scheme 3.2

Dimethyl itaconate can be hydrogenated with high efficiency using a Rh/Josiphos complex immobilized on Al-MCM-41 (Scheme 3.2) [43]. The continuous hydrogenation of dimethyl itaconate in scCO$_2$ using Rh/SL-J001 immobilized on PTA-alumina was carried out with similar s/c ratios at 160 bar, 35–55 °C with up to 83% ee [44].

2-Piperazinecarboxylic acid derivatives are interesting intermediates for the HIV protease inhibitor crixivan. A pilot process was developed by Lonza [45] for the asymmetric hydrogenation of the cyclic enamide with up to 97% ee with an s/c ratio of 1000 (Scheme 3.3). Important for good catalyst performance was the choice of the ligand (best ee with SL-J002) and the substituents at the tetrahydropyrazine. The hydrogenation of several tetra- substituted N-sulfonated

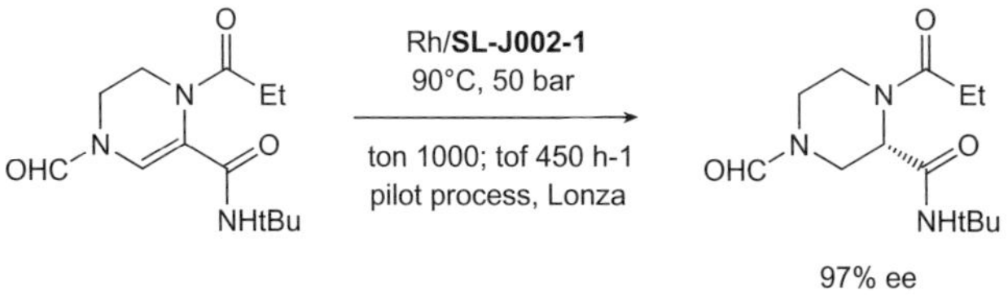

Rh/SL-J002-1
90°C, 50 bar
────────────────→
ton 1000; tof 450 h^{-1}
pilot process, Lonza

97% ee

Scheme 3.3

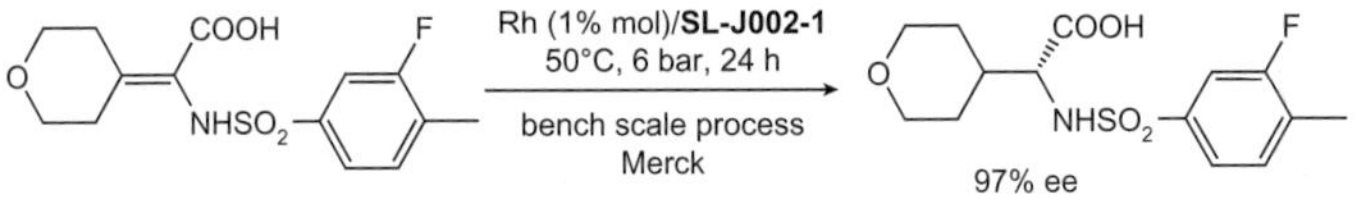

Scheme 3.4

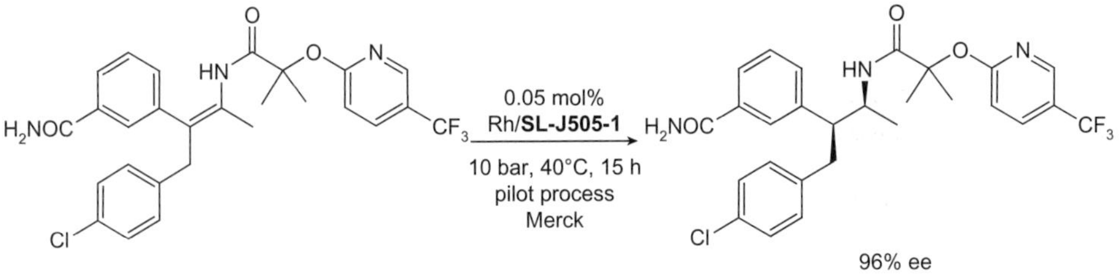

Scheme 3.5

α-dehydroamino acids was reported by Shultz et al. [46] using Ru/SL-J002-1 catalysts with up to 99 % ee, and a bench-scale process was developed for an intermediate of an anthrax lethal factor inhibitor (Scheme 3.4). A tetrasubstituted enamide was also starting material for introducing two stereogenic centers for the synthesis of taranabant, a cannabinoid-1 receptor (CB1R) inverse agonist for the treatment of obesity (Scheme 3.5). The Rh/SL-J505-1 catalyst proved to be very active, allowing full conversion within a reasonable time in 96% ee. Moreover, crystallization of the product proceeded with an upgrade in enantiomeric purity, leading to a 94% isolated yield with >99.5% ee [47].

Dehydro Amino Acid Derivatives Surprisingly, unprotected dehydro β-amino acids were shown to be good substrates for Rh/Josiphos catalysts with enantioselectivities up to 97% [48]. Not only simple derivatives (Table 3.4) but also the complex intermediate for sitagliptin, a DPP-IV inhibitor for type 2 diabetes (Scheme 3.6) was hydrogenated successfully; the latter is now produced on a

Table 3.4 Hydrogenation of enamide and itaconate model substrates..

Substrate	R	Best ligand (solvent)	Reaction conditions	Ee range (%)	Ton [TOF (h^{-1})]
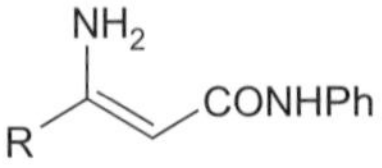 R-CH=C(NH$_2$)-CONHPh	Ph, 4-FPh, 4-MeOPh, Bn	SL-J002 (MeOH)	6 bar H$_2$ 50 °C	82–97	300 (50)
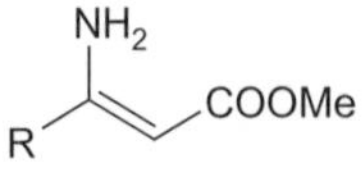 R-CH=C(NH$_2$)-COOMe	3-Py, Ph, 4-FPh, 4-MeOPh, Bn	SL-J011 (CF$_3$CH$_2$OH)	6 bar H$_2$ 50 °C	93–96	300 (15–50)

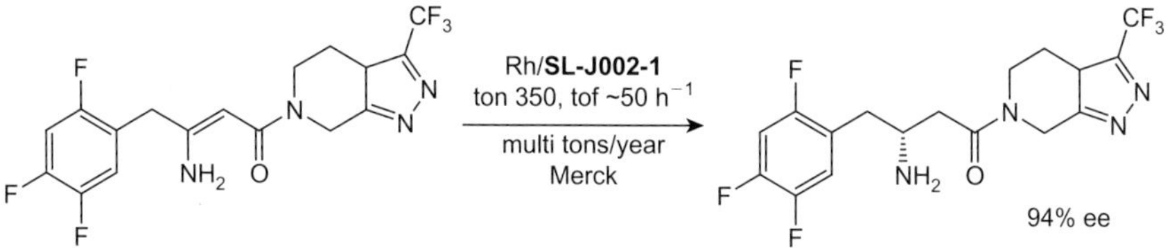

Scheme 3.6

multi-ton scale. Interestingly, deuteration experiments indicate that it is not the enamine C=C bond that is reduced but the tautomeric imine! Almost at the same time Takasago has reported that a Ru/segphos catalyst can hydrogenate unprotected dehydro β-amino esters with ees of 94–97% [49].

β-Enamido phosphonates β-Enamido phosphonates can be hydrogenated with good to high enantioselectivity using Rh/SL-J001 or SL-J002 in aprotic solvents (Scheme 3.7 [50]) with high conversion and up to 91% ee. Both *(E)* and *(Z)* enamides can give high ee, but usually with opposite absolute configuration. Similar results have been reported for Rh/SL-J001 in MeOH (Scheme 3.8 [51]).

α,β-Unsaturated Carboxylic Acids Several industrial feasibility studies have shown that Rh/Josiphos complexes are suitable catalysts for the hydrogenation of functionalized and sterically congested α,β-unsaturated carboxylic acids. A bench-scale process was developed for the Rh/SL-J001-catalyzed hydrogenation of an exocyclic C=C bond for an intermediate of a GPIIb/IIIa receptor antagonist (Scheme 3.9) with 99% ee and s/c ratios up to 2300 [48]. Rh/SL-J002 in presence of triethylamine was the catalyst of choice for the hydrogenation step of a synthesis of a

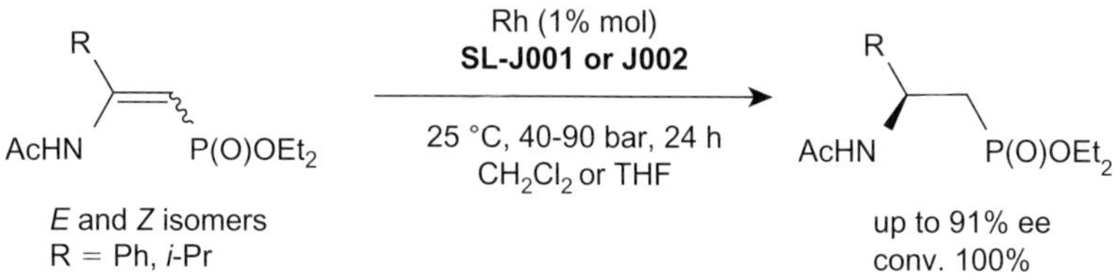

Scheme 3.7

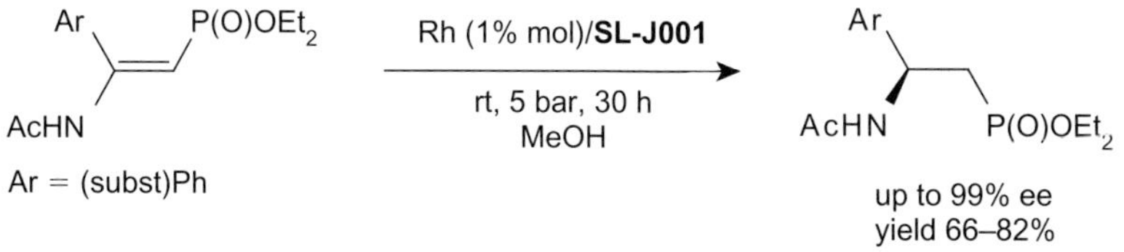

Scheme 3.8

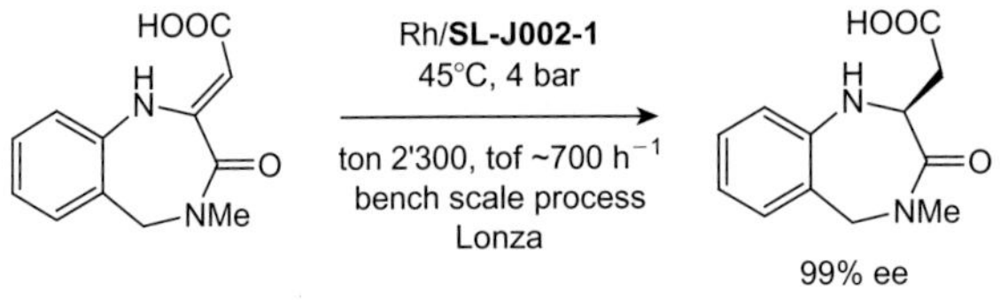

Scheme 3.9

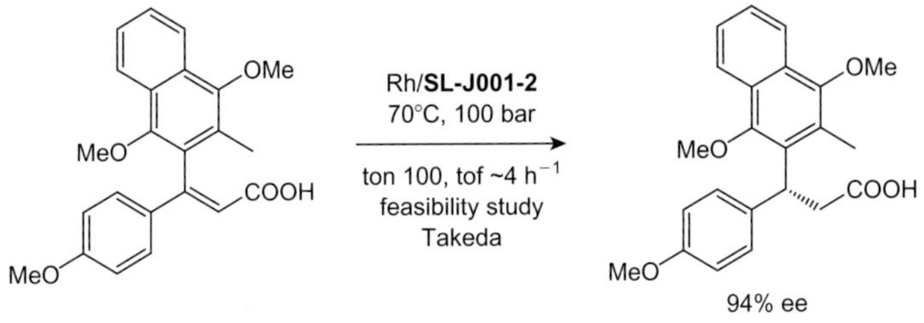

Rh/SL-J002-1
45°C, 4 bar
ton 2'300, tof ~700 h^{-1}
bench scale process
Lonza

99% ee

Scheme 3.10

Rh/SL-J001-2
70°C, 100 bar
ton 100, tof ~4 h^{-1}
feasibility study
Takeda

94% ee

neurodegenerative disease agent (Scheme 3.10), achieving up to 94% ee [52]. A feasibility study using Rh/Josiphos complexes for the hydrogenation of an unsaturated acid that was an intermediate for the synthesis of (S)-naproxen (Scheme 3.11) showed quite a striking ligand effect: at 5 bar H$_2$ and 25 °C 62% ee was obtained using catalysts prepared *in situ* from [Rh(nbd)$_2$]BF$_4$ and SL-J001. With the 4-CF$_3$Ph analog, enantioselectivity increased to 76% and with SL-J006 [R = 3,5-(CF$_3$)$_2$Ph] 86% ee were observed, which increased to 92% ee at 5 °C [11, 45]. While these results are promising, they did not quite reach those of the Ru/Binap catalysts, where up to 97% ee were reported [53].

Miscellaneous Substrates Tetra-substituted C=C bonds, especially without privileged functional groups are usually very difficult to hydrogenate with homogeneous catalysts. It is therefore noteworthy that two production processes (Schemes 3.12 and 3.13) were developed using Rh/ and Ru/Josiphos complexes, respectively.

The asymmetric hydrogenation of a tetrasubstituted C=C bond (Scheme 3.12) was the key step for a new biotin synthesis developed by Lonza [45]. While the diastereomeric excess (de) with a heterogeneous Rh/Al$_2$O$_3$ catalyst was only in the order of 60%, Rh/Josiphos complexes with alkyl groups at the side chain

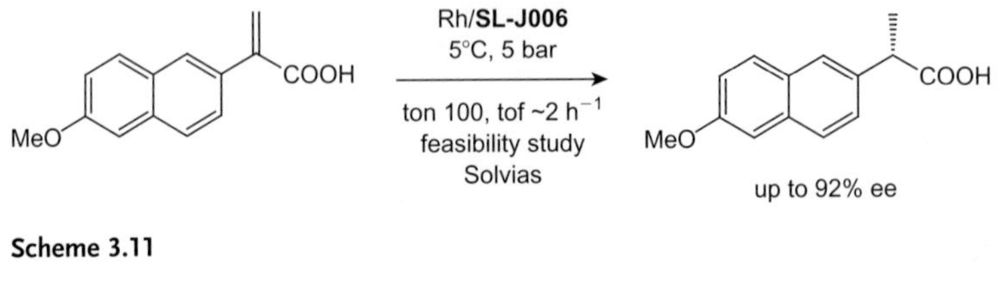

Scheme 3.11

Rh/SL-J006
5°C, 5 bar
ton 100, tof ~2 h^{-1}
feasibility study
Solvias

up to 92% ee

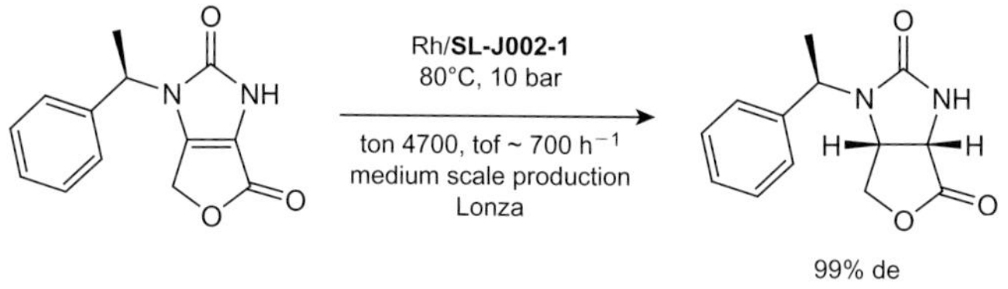

3.4 Catalytic Profile of the Josiphos Ligand Family | 109

Scheme 3.12

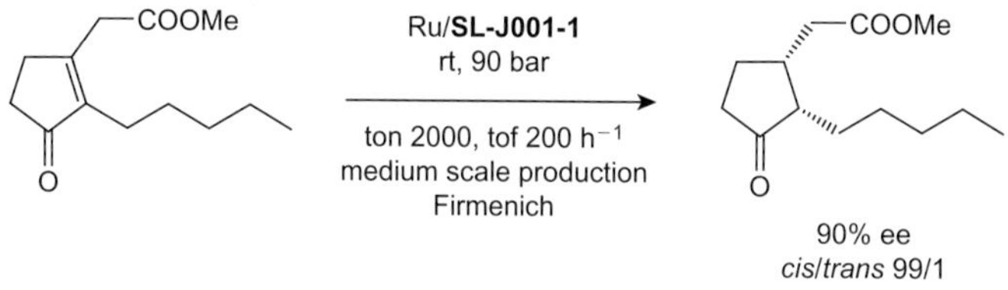

Rh/SL-J002-1
80°C, 10 bar
ton 4700, tof ~ 700 h^{-1}
medium scale production
Lonza

99% de

Scheme 3.13

Ru/SL-J001-1
rt, 90 bar
ton 2000, tof 200 h^{-1}
medium scale production
Firmenich

90% ee
cis/trans 99/1

provided high selectivities. Best results were obtained with SL-J002-1, which also was the ligand of choice for the enantioselective hydrogenation (N-Bn instead of N-phenethyl), achieving up to 90% ee. For the production process the diastereoselective variant was chosen and the process was operated on a multi ton/y scale for several years.

For the stereoselective manufacture of (+)-cis methyl dihydrojasmonate (Scheme 3.13) developed by Firmenich [54] a novel Ru precursor and a new reaction system had to be found because the classical Ru complexes and conditions for the hydrogenation of C=C bonds did not work. A broad screening of Ru catalysts (partly in collaboration with Solvias) showed that [Ru(H)(cyclooctatriene)(ligand)]BF$_4$ complexes with selected Josiphos ligands and DuPhos satisfied the prerequisites. Besides the enantioselectivity, chemo- and cis-selectivity and activity problems (tetrasubstituted C=C) were solved on a very good level.

The Rh/Josiphos-catalyzed hydrogenation of a chromanoylpyridine derivative (Scheme 3.14), an intermediate for a hair growth and antihypertensive agent, was achieved with remarkable enantioselectivity but low activity [55]. In this case, not only the enantio- but also the chemoselectivity (N−O, C≡N bonds) of the catalyst

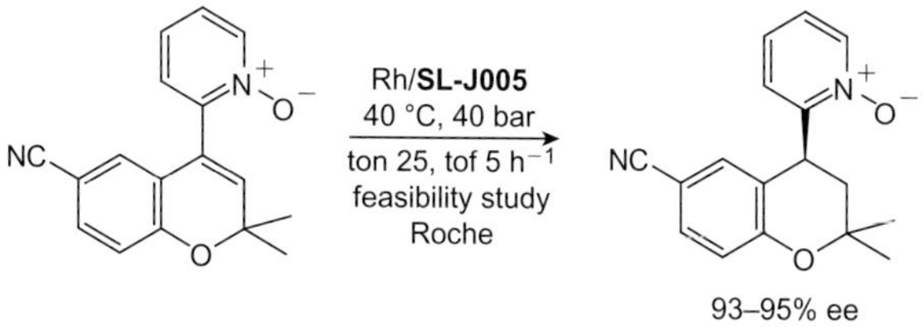

Rh/SL-J005
40 °C, 40 bar
ton 25, tof 5 h^{-1}
feasibility study
Roche

93–95% ee

Scheme 3.14

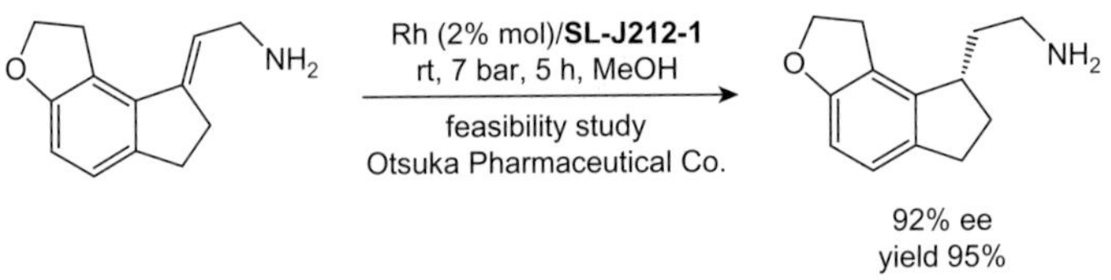

Scheme 3.15

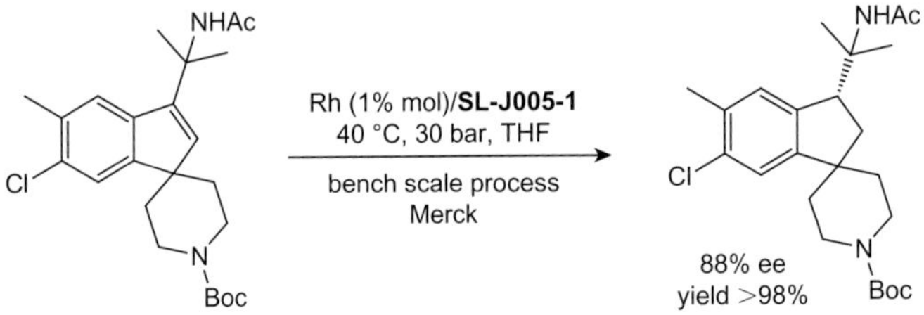

Scheme 3.16

is remarkable. The melatonin receptor antagonist ramelteon can be synthesized via the Rh/SL-J212-1-catalyzed hydrogenation of an allylamine with good enantioselectivity (Scheme 3.15 [56]). Somewhat lower ee values were reported for the hydrogenation of a cyclic allylic amide using Rh/SL-J005 to give an intermediate for an active compound in good yields (Scheme 3.16 [57]). The Rh/SL-J002-catalyzed hydrogenation of α,β-unsaturated ester phosphonates proceeds with good to very good enantioselectivities (Scheme 3.17 [58]).

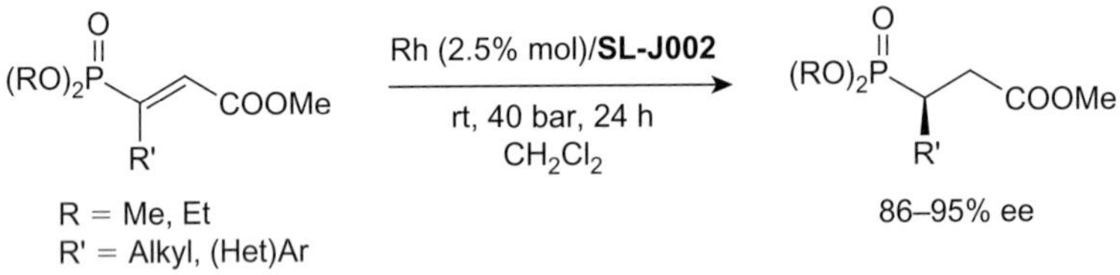

Scheme 3.17

3.4.1.2 Copper-Catalyzed Reduction of Activated C=C bonds with PMHS (Conjugate Reduction)

Alkyl Josiphos are the ligands of choice for the Cu-catalyzed reduction of activated C=C bonds with polymethylhydrosiloxane (PMHS), with very high enantioselectivities for nitro alkenes [59], α,β-unsaturated ketones [60], esters [61], and nitriles [62] (Schemes 3.18 and 19). Since both Cu complexes and PMHS are cheap and commercially available, these transformations have industrial potential, even though work-up of the polymeric reducing agent could be problematic on a large

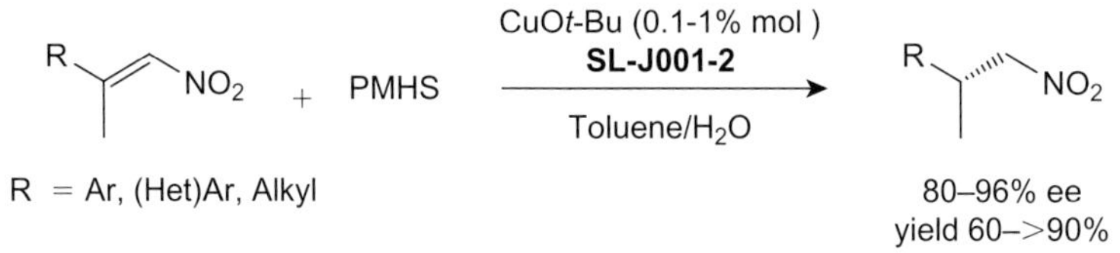

Scheme 3.18

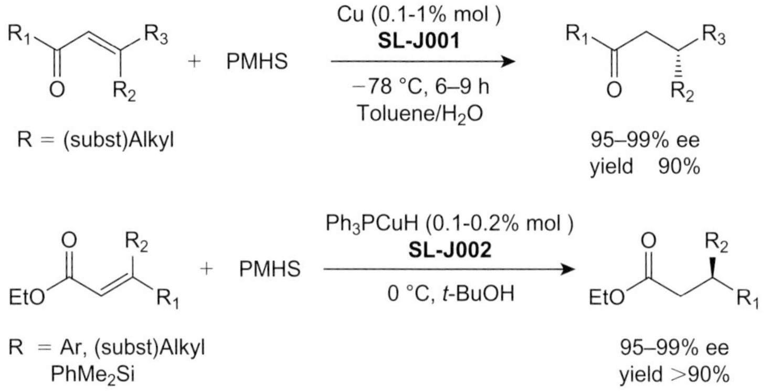

R−CH=CH−NO$_2$ + PMHS →[CuOt-Bu (0.1-1% mol) / SL-J001-2 / Toluene/H$_2$O] R−*CH(−)−CH$_2$−NO$_2$

R = Ar, (Het)Ar, Alkyl

80–96% ee
yield 60–>90%

Scheme 3.19

R$_1$−C(=O)−C(R$_2$)=CH−R$_3$ + PMHS →[Cu (0.1-1% mol) / SL-J001 / −78 °C, 6–9 h / Toluene/H$_2$O] R$_1$−C(=O)−CH(R$_2$)−CH(R$_3$)

R = (subst)Alkyl

95–99% ee
yield 90%

EtO−C(=O)−C(R$_1$)=CH−R$_2$ + PMHS →[Ph$_3$PCuH (0.1-0.2% mol) / SL-J002 / 0 °C, t-BuOH] EtO−C(=O)−CH(R$_1$)−CH(R$_2$)

R = Ar, (subst)Alkyl
PhMe$_2$Si

95–99% ee
yield >90%

scale. The reduction of nitroalkenes (Scheme 3.18) gives access to chiral nitro compounds that can further be transformed into amines, oximes, and nitriles.

Very few catalysts can enantioselectively reduce the C=C bond of enones; the Rh/SL-J002-catalyzed reaction has been applied to the synthesis of multifunctional products without reduction of the non-activated C=C bonds (Scheme 3.20 [63]).

The reduction of α,β-unsaturated nitriles (Scheme 3.21) deserves special mention since in contrast to the nitro or carbonyl group the lone pair of the nitrile function is not able to coordinate with the Cu atom. The catalyst can even differentiate between two similar aryl substituents; the position of the two aryl groups with respect to CN is important for induction and determines the absolute configuration. The reaction has been applied to the synthesis of tolderodine [64]. Finally, the methodology can also be applied to Cu-catalyzed Reduction of 2-alkenylheteroarenes with good to excellent yields and enantioselectivities (Scheme 3.22 [65]).

3.4.1.3 Enantioselective Hydrogenation of C=O Bonds

Functionalized Ketones Ruthenium complexes with atropisomeric ligands such as Binap, MeOBIPHEP, and Segphos are generally the preferred catalysts for the reduction of activated C=O double bonds [11] Screening experiments summarized in Table 3.5 (F. Spindler, unpublished results) and a recent feasibility study

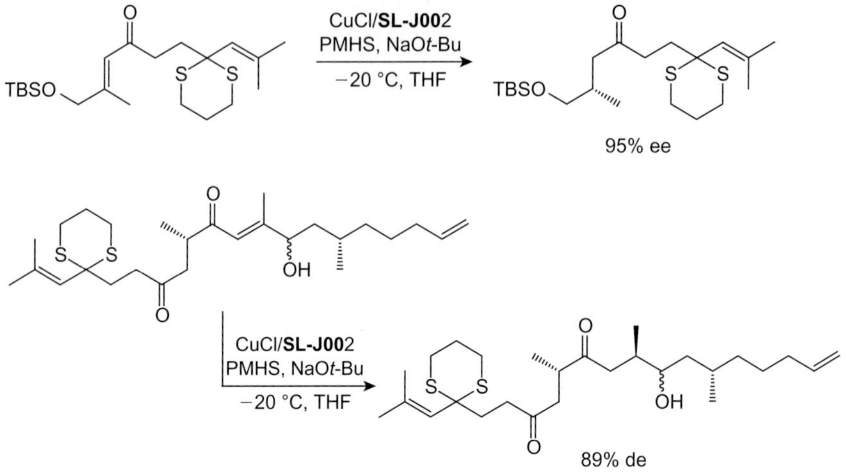

CuCl/SL-J002
PMHS, NaOt-Bu
−20 °C, THF
TBSO
95% ee
OH
89% de

Scheme 3.20

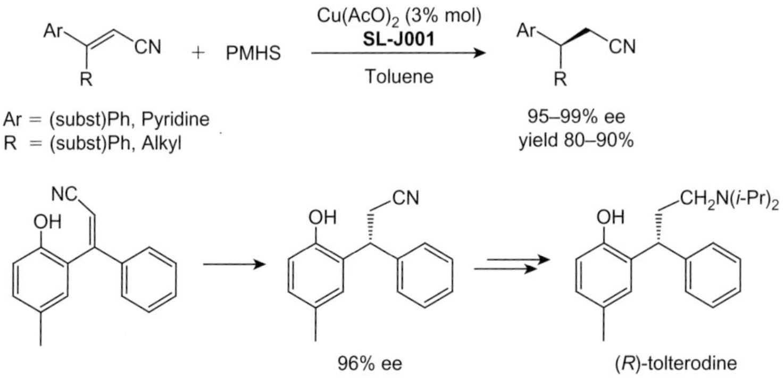

Cu(AcO)2 (3% mol)
SL-J001
Toluene
PMHS
Ar = (subst)Ph, Pyridine
R = (subst)Ph, Alkyl
95–99% ee
yield 80–90%
96% ee
(R)-tolterodine

Scheme 3.21

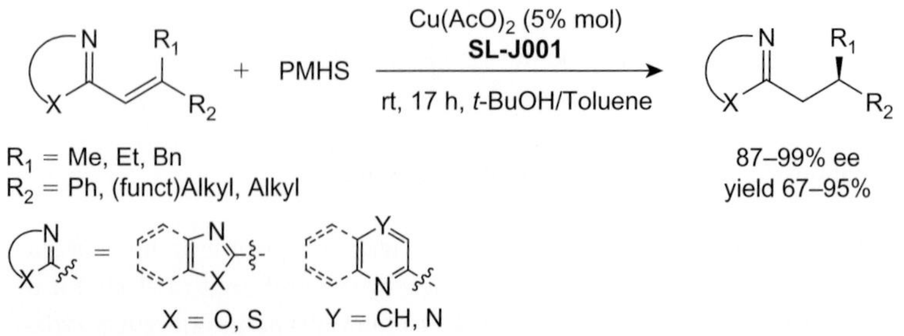

Cu(AcO)2 (5% mol)
SL-J001
rt, 17 h, t-BuOH/Toluene
PMHS
R1 = Me, Et, Bn
R2 = Ph, (funct)Alkyl, Alkyl
87–99% ee
yield 67–95%
X = O, S
Y = CH, N

Scheme 3.22

Table 3.5 Rh/Josiphos-catalyzed hydrogenation of α- and β-keto acid derivatives..

Substrate	Best catalyst	Reaction conditions	Ee^a range (%)
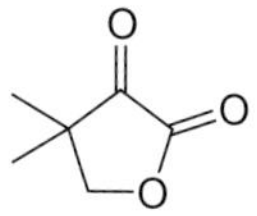 (α-keto lactone)	[Rh(nbd)(OAc)]₂ **SL-J011**	40 bar, 10 °C	Up to 95
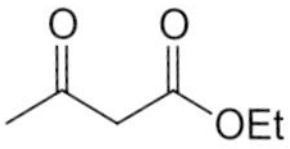 (β-ketoester OEt)	[Rh(nbd)₂]BF₄ **SL-J006**	5 bar, room temp.	84–97

^{a}Standard experiments, values not optimized.

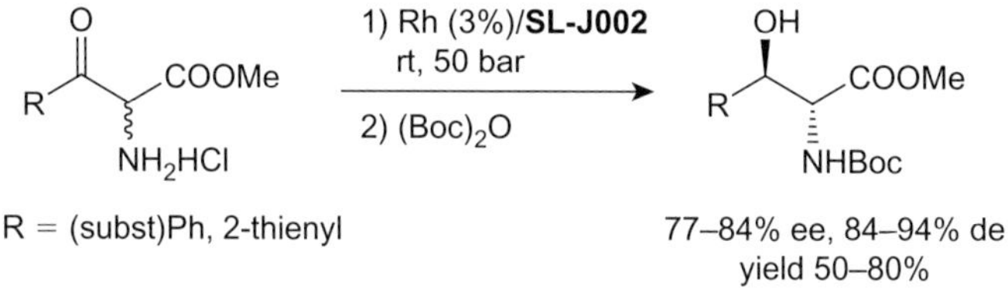

Scheme 3.23

Reaction: R-C(O)-CH(NH₂HCl)-COOMe → (1) Rh (3%)/**SL-J002**, rt, 50 bar; (2) (Boc)₂O → R-CH(OH)-CH(NHBoc)-COOMe

R = (subst)Ph, 2-thienyl

77–84% ee, 84–94% de, yield 50–80%

(Scheme 3.23) [66] have shown that for α- and β-keto acid derivatives high enantioselectivities are also possible with Rh/Josiphos catalysts but, at present, the TON and TOF are usually lower than those achieved with the best atropisomeric ligands. Of special interest is the recently published Ni/SL-J001-catalyzed hydrogenation of an α-substituted β-keto esters with good ee values for aryl ketones but low to medium enantioselectivity and very low conversion for aliphatic ketones; turnover numbers are still low (Scheme 3.24 [67]).

An Ir/SL-J001 was recently reported to be feasible for the hydrogenation of an α-amino acetophenone, an intermediate for an opioid receptor antagonist CJ-15,161 (Scheme 3.25 [68]). Even though no details were given concerning catalyst activity, this is remarkable since it the first case were an Ir complex gave better performance than Rh (up to now the catalysts of choice).

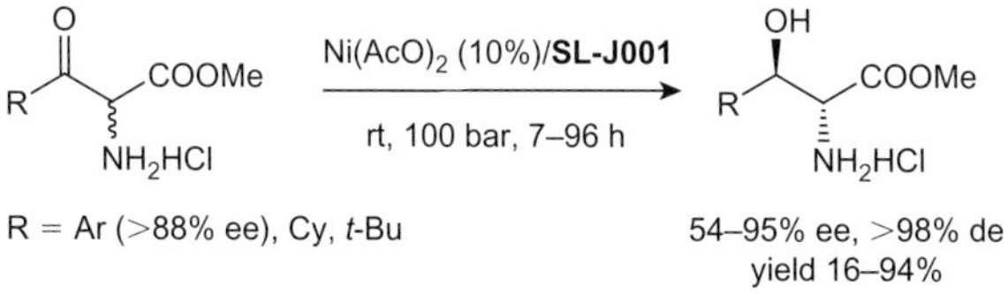

Scheme 3.24

Reaction: R-C(O)-CH(NH₂HCl)-COOMe → Ni(AcO)₂ (10%)/**SL-J001**, rt, 100 bar, 7–96 h → R-CH(OH)-CH(NH₂HCl)-COOMe

R = Ar (>88% ee), Cy, t-Bu

54–95% ee, >98% de, yield 16–94%

Non-functionalized Ketones In a series of papers Baratta and coworkers described novel Ru and Os complexes with very impressive catalytic properties for the

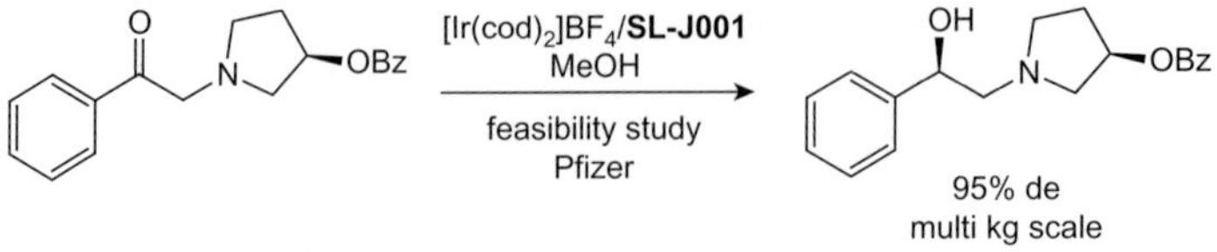

Scheme 3.25

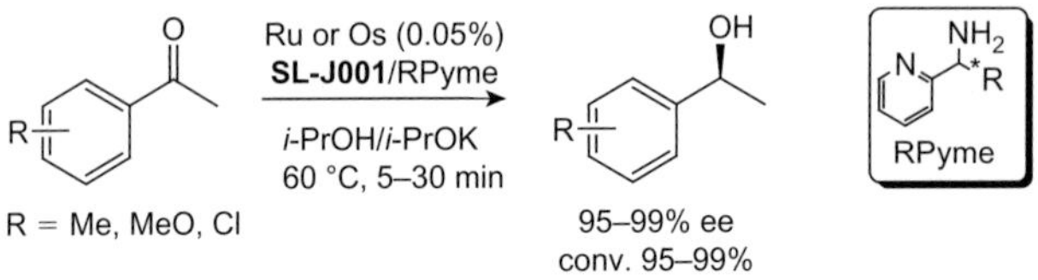

Scheme 3.26

hydrogenation and transfer hydrogenation of aryl ketones. Very high catalyst activities (TON up to 20 000, TOF >20 000 h^{-1}) were reported for the catalyzed transfer hydrogenation of acetophenone (Scheme 3.26) using Ru [37] and Os [69] complexes containing 2-(1-aminoalkyl) pyridines (RPyme) and SL-J001.

Even better performances were described by Baratta for a new class of Ru [70] and Os [71] pincer complexes (Scheme 3.27). Again best performance was obtained with SL-J001, both for the transfer hydrogenation as well as hydrogenation of aryl ketones, with enantioselectivities up to 99% and H TONs of up to 50 000.

The Rh/Josiphos-catalyzed hydrogenation of 3-oxoquinuclidine to 3-hydroxyquinuclidine (Scheme 3.28), an intermediates for M$_1$-receptor antagonists, is unprecedented but at the moment not selective enough to be of technical interest [72].

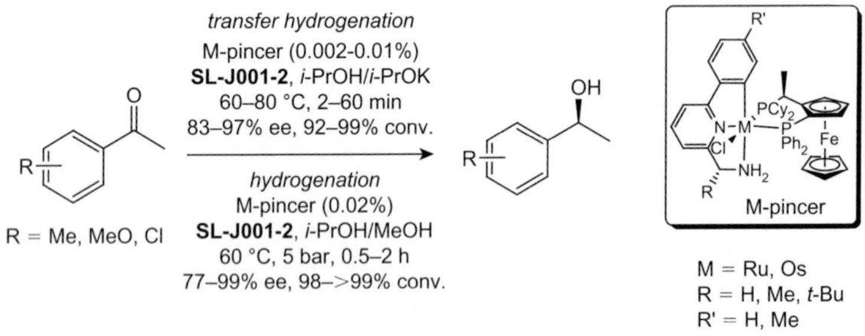

Scheme 3.27

3.4.1.4 Enantioselective Hydrogenation of C=N Bonds

Except for N-alkyl imines, Josiphos is among the most effective ligand families for the hydrogenation of C=N functions [73, 74]. For the hydrogenation of N-aryl imines with Ir/Josiphos catalysts, the s/c ratio can reach >1 000 000 (see below), but the typical range is 50–5000 (Table 3.6). The degree of enantioselectivity depends

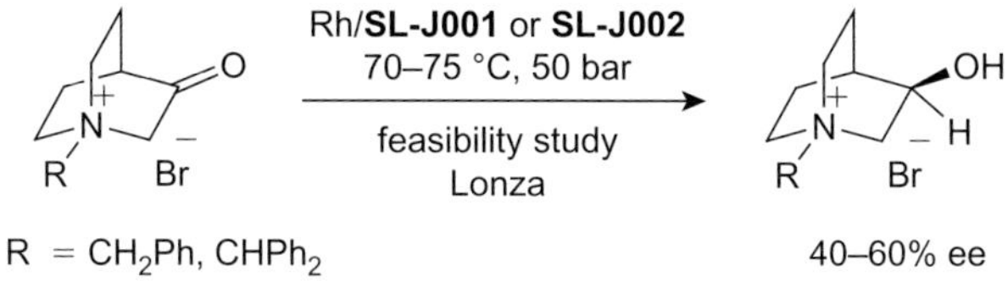

Scheme 3.28

R = CH$_2$Ph, CHPh$_2$

Conditions: Rh/**SL-J001** or **SL-J002**, 70–75 °C, 50 bar, feasibility study Lonza

40–60% ee

strongly on the substituents in 2 and 6 position of the N-aryl group [75]. The average TOF typically varies between 10 and 300 000 h^{-1}. The results for the Rh/Josiphos-catalyzed hydrogenation of phosphinylimines are also unprecedented, but the scope of the catalyst is still relatively narrow [76]. Similar enantioselectivities have also been achieved with Rh/Mandyphos catalysts (F. Spindler, unpublished results).

Production Process for (S)-Metolachlor (DUAL Magnum®) Metolachlor is the active ingredient of Dual®, one of the most important grass herbicides for use in maize and several other crops. It is an N-chloroacetylated, N-alkoxyalkylated ortho-disubstituted aniline. The commercial product was introduced onto the market in 1976 as a racemic mixture of four stereoisomers (Figure 3.10) and was produced via a Pt/C-catalyzed reductive alkylation of 2-methyl-5-ethyl-aniline (MEA) with aqueous methoxyacetone, followed by chloroacetylation [77]. In 1982 it was found that about 95% of the herbicidal activity of Metolachlor was due to the two (1'S)-diastereomers. In 1997, after years of intensive research, Dual Magnum® with a content of approximately 90% (1'S)-diastereomers and with the same

Table 3.6 Rh/Josiphos-catalyzed hydrogenation of selected C=N bonds..

Substrate	Best catalyst	Reaction conditions	Ee range (%)	TON [TOF (h^{-1})]^a	Reference
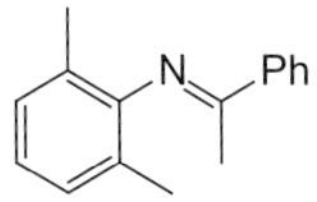 (2,6-dimethylphenyl)-N=C(Ph)(Me)	[Ir(cod)Cl]$_2$	80 bar H$_2$	67–96	200	F. Spindler and A.G. Solvias, unpublished results, [73]
	SL-J110, SL-J005 TBAI, CF$_3$COOH	25 °C		Not available	
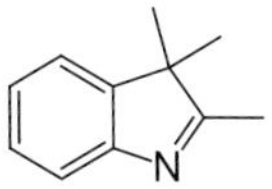 2,3,3-trimethyl-3H-indole	[Ir(cod)Cl]$_2$ SL-J408 TBAI, CH$_3$COOH	80 bar H$_2$ 15 °C	76–93	250 (56)	[73]
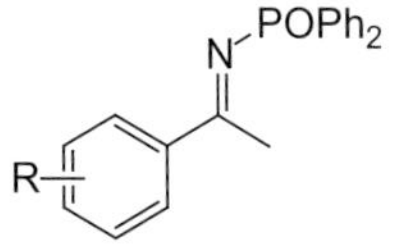 R-C$_6$H$_4$-C(Me)=N-POPh$_2$	[Rh(nbd)$_2$]BF$_4$ SL-J003	70 bar H$_2$ 60 °C	35–99	500 (500)	[74]

^{a}Standard experiments, values not optimized.

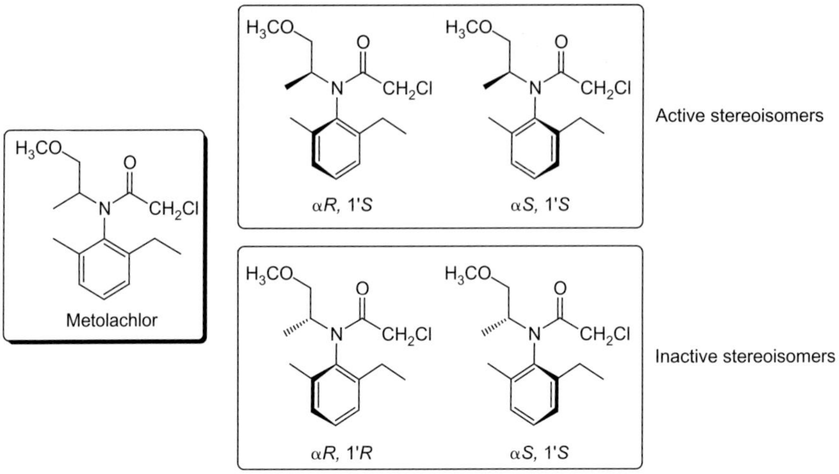

Figure 3.10 Structure of Metolachlor and its individual stereoisomers.

biological effect at about 65% of the use rate of the racemate was introduced into the market [78]. Today, this is by far the largest volume application of any homogeneous chiral catalyst.

The key step of this new synthesis is the enantioselective hydrogenation of the distilled MEA imine (Scheme 3.29). The optimized process operates at 80 bar hydrogen and 50 °C with a catalyst generated *in situ* from [Ir(cod)Cl]$_2$ and SL-J005-1 at a substrate to catalyst ratio (s/c) of >1 000 000. Complete conversion is reached within 3–4 h, the initial TOF exceeds 1 800 000 h^{-1} and enantioselectivity is approximately 80%. Key success factors of the process are the novel, very active catalyst Ir/Josiphos catalysts, the use of iodide and acid as additives, and the high purity of MEA imine.

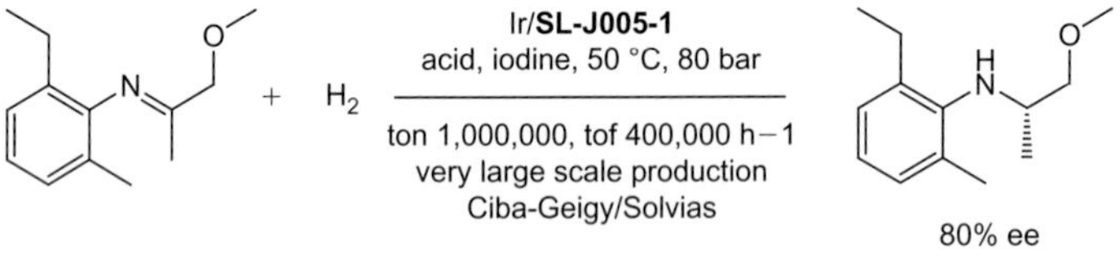

Scheme 3.29

Alternative processes such as the direct reductive amination (Scheme 3.30) [79] as well as the application of immobilized Josiphos [80] were investigated to avoid the distillation of the N-alkylated aniline. While both variants reached respectable turnover numbers of 10 000–100 000, the processes were not competitive.

Pilot Process for Dextromethorphane A pilot for the preparation of dextromethorphane, a traditional antitussive agent, in a two-phase process (toluene/water)

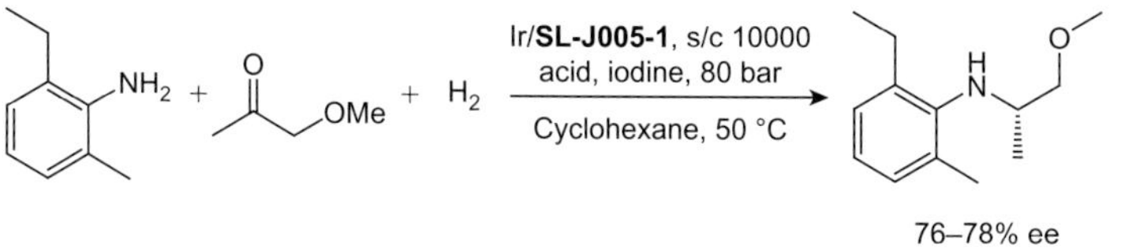

Scheme 3.30

[Scheme 3.30: Hydrogenation of 2-ethyl-6-methylaniline with methyl pyruvate using Ir/SL-J005-1, s/c 10000, acid, iodine, 80 bar, cyclohexane, 50 °C, giving product with 76–78% ee]

Scheme 3.31

[Scheme 3.31: Hydrogenation of a dihydroisoquinoline·H₃PO₄ salt with Ir/SL-J013-1, 20 °C, 70 bar, ton 1500, tof 170 h⁻¹, pilot process Lonza, giving tetrahydroisoquinoline with up to 90% ee]

was developed by Lonza [45] using an Ir/Josiphos catalyst (Scheme 3.31). Key success factors were ligand fine tuning, the choice of the phosphoric acid salt of the imine, the reaction medium, and the addition of base and iodide.

Synthesis of an HIV Integrase Inhibitor A feasibility study by Merck [81] described the hydrogenation of an exocyclic methyl imine with 90% ee as key step for the synthesis of a HIV integrase inhibitor (Scheme 3.32). Best results were obtained in 2,2,2-trifluoroethanol in presence of trifluoroacetic acid, and deuteration experiments indicated that the hydrogenation occurs via the tautomeric enamine. The reaction was carried out on a 500 g scale.

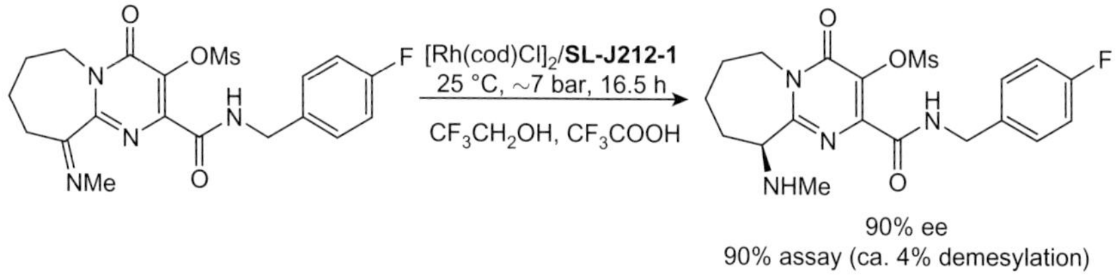

Scheme 3.32

[Scheme 3.32: Hydrogenation of exocyclic methyl imine intermediate using [Rh(cod)Cl]₂/SL-J212-1, 25 °C, ~7 bar, 16.5 h, CF₃CH₂OH, CF₃COOH, giving product with 90% ee, 90% assay (ca. 4% demesylation)]

Enantioselective Hydrogenation of Heteroarenes Very few homogeneous catalysts can catalyze the enantioselective hydrogenation of heteroarenes. While in general preferred catalysts are Ir complexes of atropisomeric ligands [73, 74] two feasibility studies demonstrated that Rh/Josiphos catalysts have potential as well. The hydrogenation of a 2-pyrazine-carboxylate (Scheme 3.33) is an alternative access to

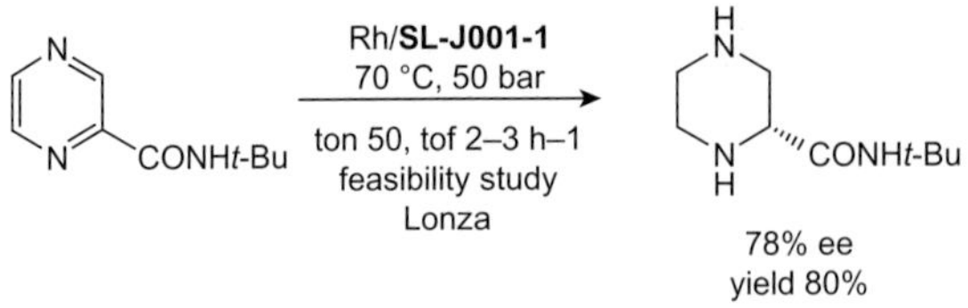

Scheme 3.33

2-piperazinecarboxylic acid derivatives, which are potential intermediates for crixivan (Scheme 3.3). In a feasibility study by Lonza [82], up to 78% ee was achieved with Rh/SL-J001-1, albeit with rather low catalyst activities.

3.4.2
Enantioselective Hydrofunctionalizations

3.4.2.1 Hydroboration

Hydroboration of olefins followed by oxidation is an interesting alternative method for preparing chiral alcohols or amines. Several studies have been published concerning the addition of B-H across C=C bonds in presence of SL-J001 as ligand. Bench marks for this transformation are Hayashi's Rh/Binap catalysts [83] and Togni's Rh/pyrazolyl ferrocenes [84] complexes with up to 99% ee. Preliminary experiments by Togni et al. [8] have shown that Rh/SL-J001 catalyzes the hydroboration of styrene using catecholborane with ee values up to 92% (at −78 °C). Crudden et al. [85] used pinacol borane, which is easier to handle than catecholborane, and reported enantioselectivities of 76–88% for various vinyl arenes with moderate to good regioselectivities and low to moderate yields (Scheme 3.34). Diazanorbornene was desymmetrized via Ir-catalyzed hydroboration followed by oxidation with H_2O_2 with moderate enantioselectivity (Scheme 3.35 [86]).

Yun et al. [87] reported the Cu-catalyzed hydroboration of α,β-unsaturated nitriles or acids with bis-pinacol borane–MeOH with good ee values (Scheme 3.36). Similar reactions have also been obtained with various α,β-unsaturated ketones, also with very good enantioselectivities (Scheme 3.37) [88]. In many cases, Cu/Mandyphos catalysts are also suitable, albeit with somewhat lower ee.

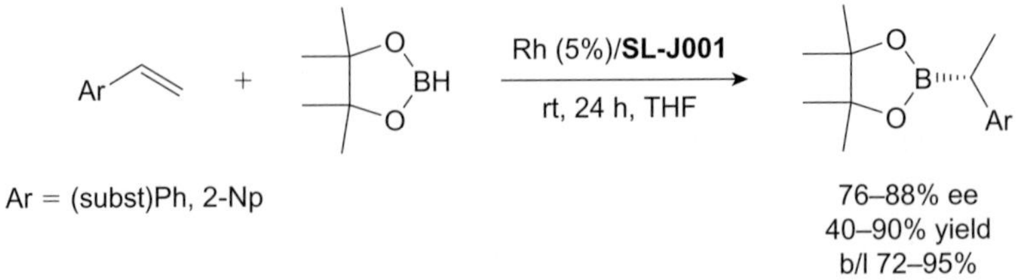

Scheme 3.34

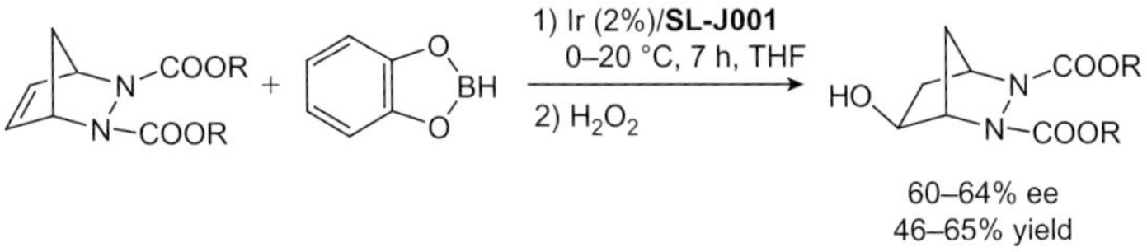

3.4 Catalytic Profile of the Josiphos Ligand Family

Scheme 3.35

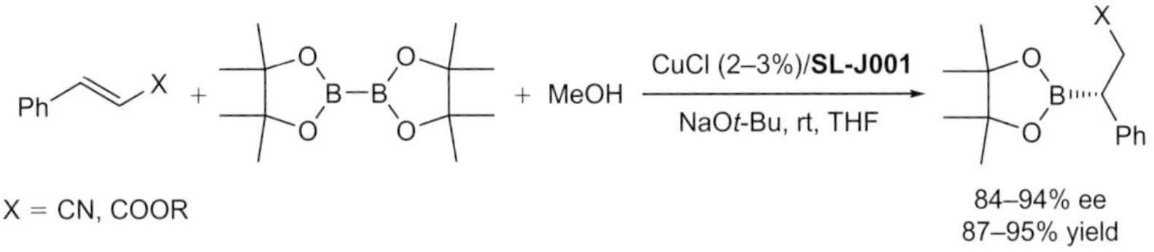

Scheme 3.36

X = CN, COOR

Scheme 3.37

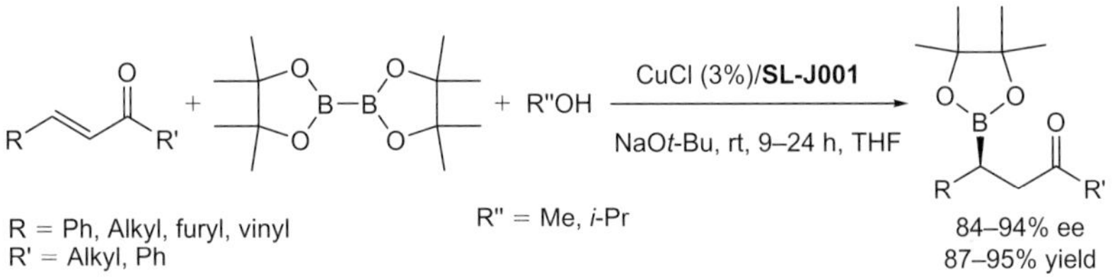

R = Ph, Alkyl, furyl, vinyl
R' = Alkyl, Ph
R" = Me, i-Pr

3.4.2.2 Hydroamination and Hydrophosphonation

The enantioselective hydroamination of alkenes would be an attractive method for the synthesis of chiral amines but very few successful examples are known. Ir/SL-J001 complexes exhibit unprecedented activity (TOF up to 3.4 h^{-1}) but moderate enantioselectivity for the model reaction of norbornene with aniline (Scheme 3.38 [89]). Up to 95% ee but lower activities were observed for Ir/Binap catalysts. For both ligands, a remarkable fluoride ion effect on both activity and enantioselectivity was found. Togni et al. [90] also described the Ir-catalyzed cyclization of N-tosyl-2-allylaniline to give indoline in moderate ee but rather low yields (Scheme 3.39).

Hydrophosphonation presents an interesting access to chiral phosphines. Xu and Han [91] have shown that Pd/Josiphos complexes are moderately active (reaction

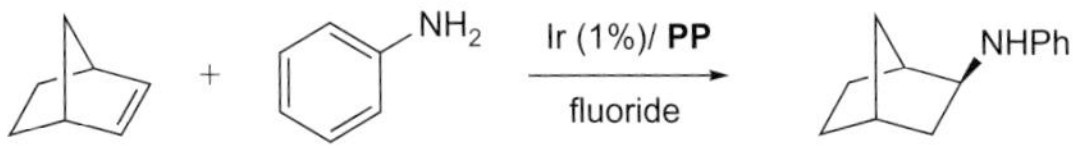

PP	
SL-J001	50% ee, tof 1.1 h^{-1}
Binap	95% ee, tof 1.1 h^{-1}

Scheme 3.38

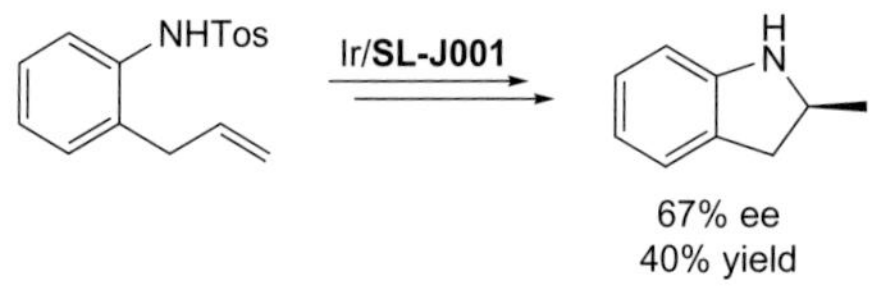

Scheme 3.39

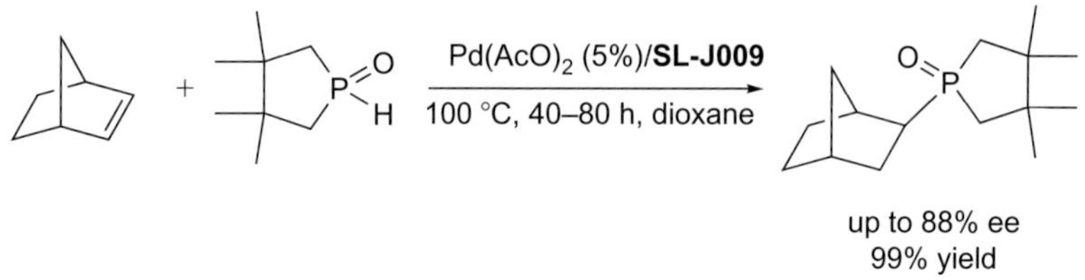

Scheme 3.40

time 40–80 h) for the addition of phosphine oxides to C=C double bonds. For norbornene (Scheme 3.40) very high yields and up to 88% ee were reported. The addition to styrene occurred with 73% ee and a branched/linear ratio of >15.

3.4.2.3 Hydrocarboxylation

While several catalysts are known for the effective hydroformylation of C=C bonds [92] only few examples for the corresponding hydrocarboxylation reaction have been reported. In a quite recent study, Claver and coworkers [93] have shown that selected Pd/Josiphos catalysts can lead to moderate to good enantioselectivities for styrene (Scheme 3.41). At the moment, yields and regioselectivities are too low for synthetic applications.

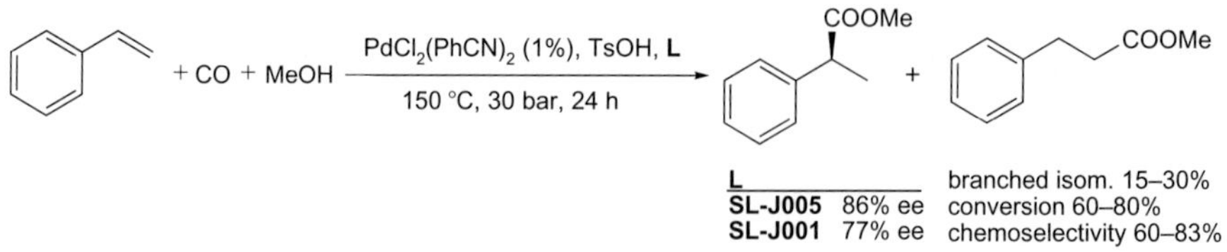

Scheme 3.41

3.4.3
Enantioselective C–C Bond Forming Reactions

3.4.3.1 Allylic Alkylation
With ee values of <93% Pd/Josiphos catalysts [8] do not quite match the best performance of catalysts developed by Trost, Pfaltz, and Helmchen for the classical reaction of allyl acetates with soft C-nucleophiles [94]. Two more recent papers

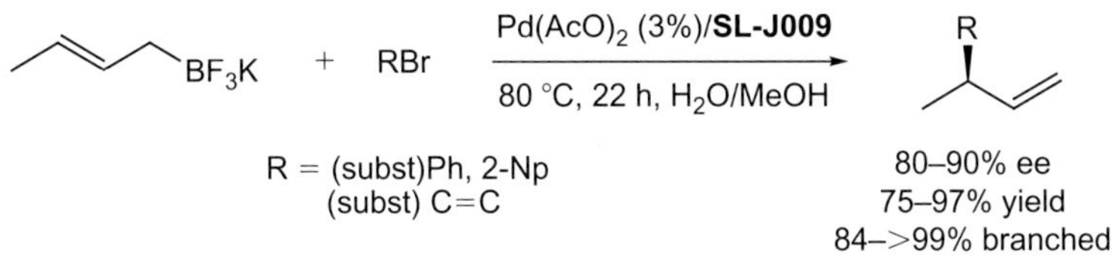

3.4 Catalytic Profile of the Josiphos Ligand Family | 121

Scheme 3.42

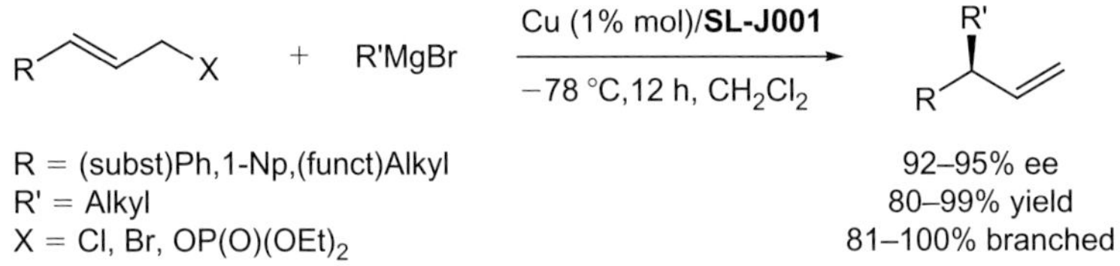

R = (subst)Ph, 2-Np
(subst) C=C

80–90% ee
75–97% yield
84–>99% branched

Scheme 3.43

R = (subst)Ph,1-Np,(funct)Alkyl
R' = Alkyl
X = Cl, Br, OP(O)(OEt)$_2$

92–95% ee
80–99% yield
81–100% branched

report interesting variations for this reaction type. Yamamoto et al. [95] described the Pd-catalyzed reaction of allyl trifluoroborates with aryl bromides with up to 90% ee and good to very high yields and regioselectivities (Scheme 3.42). The group of Feringa [96] achieved remarkable results for the Cu/Josiphos-catalyzed reaction of various allyl derivatives with Grignard reagents (Scheme 3.43). With Cu/Taniaphos complexes ee values up to 95% and branched product yield of up to 99% were obtained.

3.4.3.2 Michael Addition

Cu/Josiphos complexes have been shown by the Feringa group to be excellent catalysts for the 1,4-addition of Grignard reagents to α,β-unsaturated esters [26], thioesters [97], and cyclic enones [98]. SL-J004 is preferred for aryl substituted α,β-unsaturated esters (Scheme 3.44). SL-J001 is the ligand of choice for the addition to alkyl substituted α,β-unsaturated esters and the addition of branched Grignard reagents to cyclohexenone (Scheme 3.45) and SL-J002 for α,β-unsaturated thioesters (Scheme 3.46).

Genet and coworkers [99] have developed the Rh/Binap-catalyzed Michael addition of various organotrifluoroborates to enones. One example indicates that

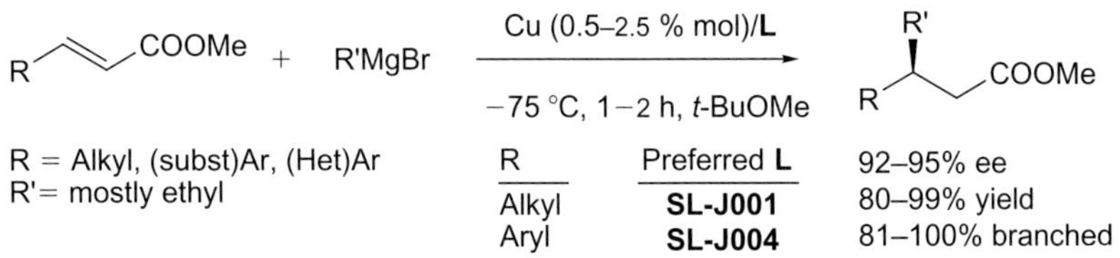

Scheme 3.44

R = Alkyl, (subst)Ar, (Het)Ar
R'= mostly ethyl

R	Preferred L
Alkyl	SL-J001
Aryl	SL-J004

92–95% ee
80–99% yield
81–100% branched

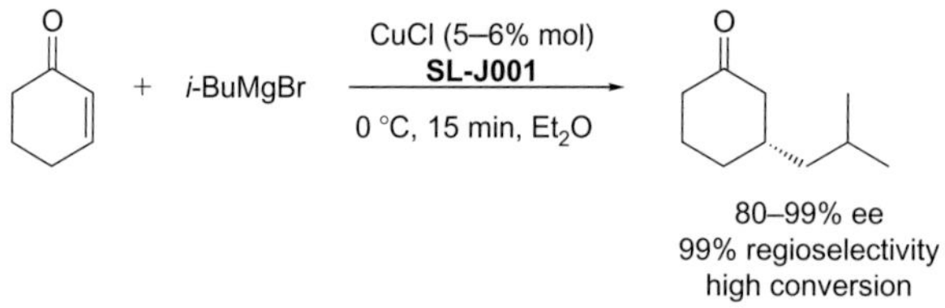

Scheme 3.45

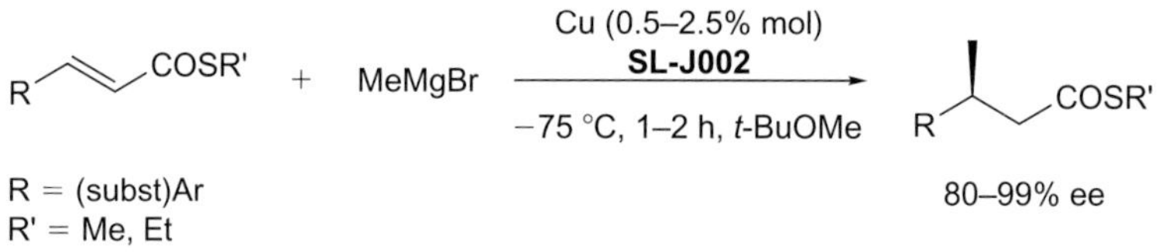

Cyclohexenone + i-BuMgBr, CuCl (5–6% mol)/SL-J001, 0 °C, 15 min, Et$_2$O → 3-isobutylcyclohexanone

80–99% ee
99% regioselectivity
high conversion

Scheme 3.46

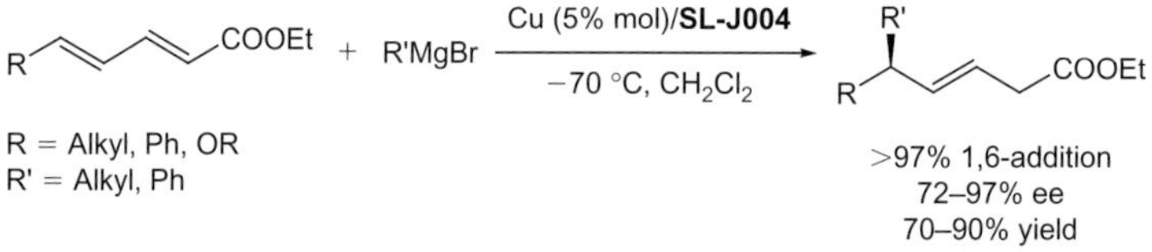

R-CH=CH-COSR' + MeMgBr, Cu (0.5–2.5% mol)/SL-J002, −75 °C, 1–2 h, t-BuOMe → R-CH(Me)-CH$_2$-COSR'

R = (subst)Ar
R' = Me, Et

80–99% ee

Scheme 3.47

R-CH=CH-CH=CH-COOEt + R'MgBr, Cu (5% mol)/SL-J004, −70 °C, CH$_2$Cl$_2$ → R-CH(R')-CH=CH-CH$_2$-COOEt

R = Alkyl, Ph, OR
R' = Alkyl, Ph

>97% 1,6-addition
72–97% ee
70–90% yield

Josiphos ligands also have a good potential for this reaction: Rh/SL-J001 achieved 99% ee for the addition of PhBF$_3$K to cyclohexenone. The Cu/SL-J004 system is also able to catalyze the addition of Grignard reagents to linear dienoates with ee values up to 97% and in satisfactory yield (Scheme 3.47 [100]).

3.4.3.3 Heck Reaction

The intramolecular Heck reaction depicted in Scheme 3.48 allows an elegant access to tetracyclic systems [101]. Of the many ligands tested, only Josiphos led to sizeable enantioselectivities; the TON and TOF are still very low.

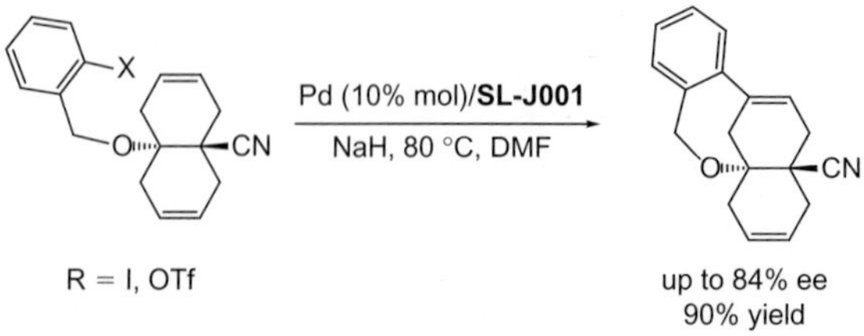

Scheme 3.48

Pd (10% mol)/SL-J001, NaH, 80 °C, DMF

R = I, OTf

up to 84% ee
90% yield

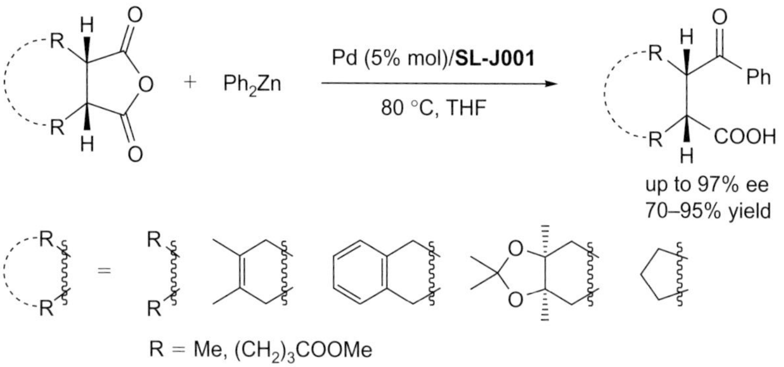

Scheme 3.49

3.4.3.4 Miscellaneous C–C Reactions

Several C–C bond forming reactions of potential synthetic interest have been described using various Josiphos complexes. Bercot and Rovis [102] developed a desymmetrization reaction of anhydrides via the Pd/Josiphos-catalyzed addition of diphenylzinc (Scheme 3.49). Best results were obtained with SL-J001 but other Josiphos ligands were also effective. The TON and TOF must be improved to make it of preparative-technical interest. Leitner et al. [103] described the Pd/Josiphos-catalyzed addition of dicarbonyl compounds to dienes (Scheme 3.50) with enantioselectivities up to 81% and good to very good chemical yields. The Rh/Josiphos-catalyzed [4 + 2]-addition of 4-alkenals with acrylic acid derivatives developed by Tanaka's group [104] gives an elegant access to substituted tetralone derivatives (Scheme 3.51)

The Cu/Josiphos-catalyzed reductive aldol cyclization developed by Lipshutz et al. [105] represents an efficient, highly enantio- and diastereoselective approach

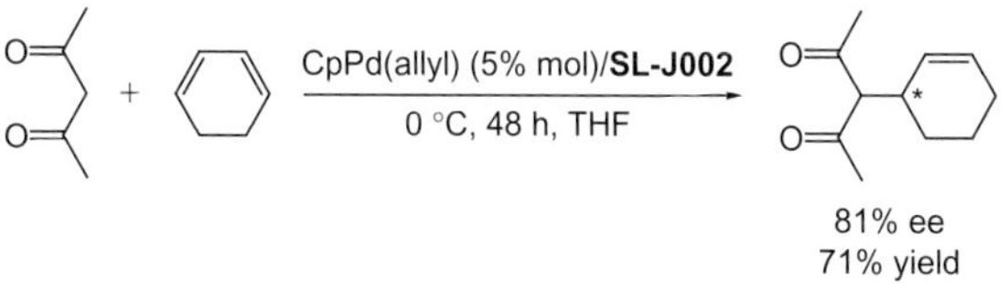

Scheme 3.50

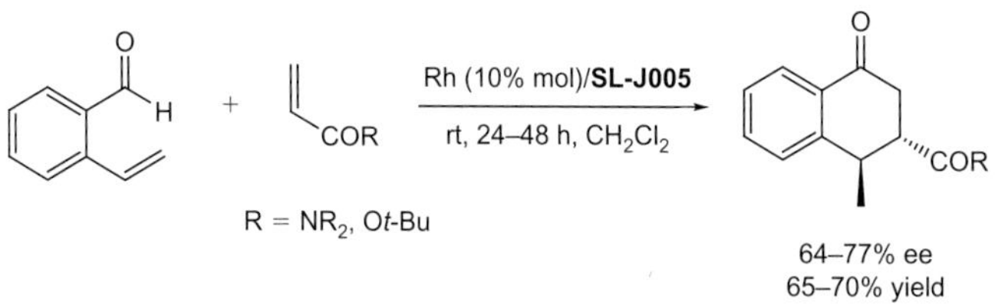

Scheme 3.51

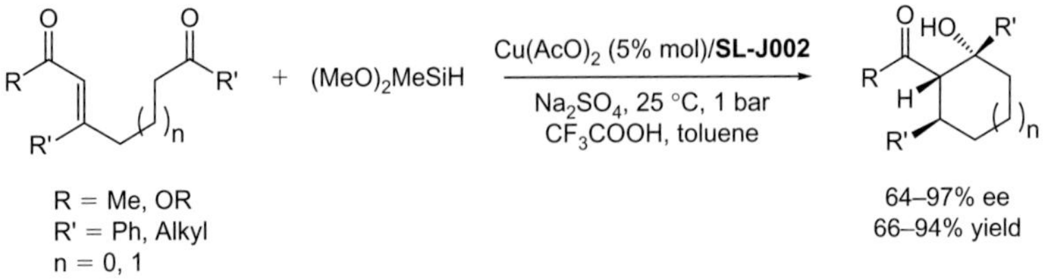

Scheme 3.52

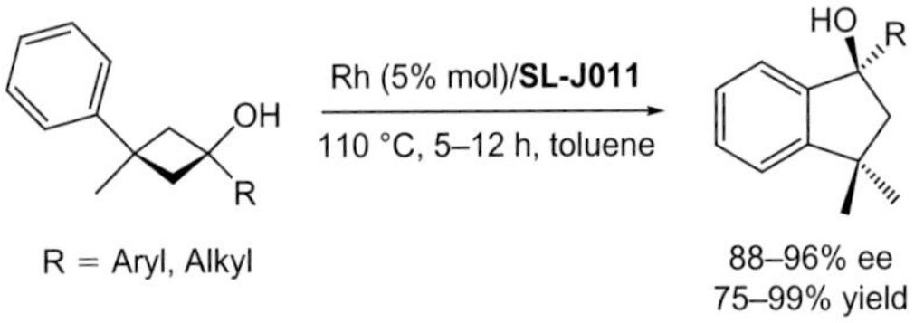

R = Me, OR
R' = Ph, Alkyl
n = 0, 1

64–97% ee
66–94% yield

Scheme 3.53

R = Aryl, Alkyl

88–96% ee
75–99% yield

to polyfunctionalized cyclic systems (Scheme 3.52). The reaction generates three contiguous stereogenic centers. The novel Rh-catalyzed rearrangement of phenyl cyclobutanols developed by Cramer and coworkers occurs with C–H activation and represents a new enantioselective synthesis of indanols (Scheme 3.53 [106]).

The Cu-catalyzed kinetic resolution of ene-epoxides via ring-opening reaction was described to occur with up to 90% ee and very good yields by the addition of RMgBr (Scheme 3.54 [107]). Among the different ligands screened, SL-J005 achieved the best performance. Important side reactions were epoxide ring opening and C=C isomerization. Bicyclic hydrazines are ring opened by Rh/SL-J002 with up to 99% ee and in good to very high yields (Scheme 3.55 [108]). Other

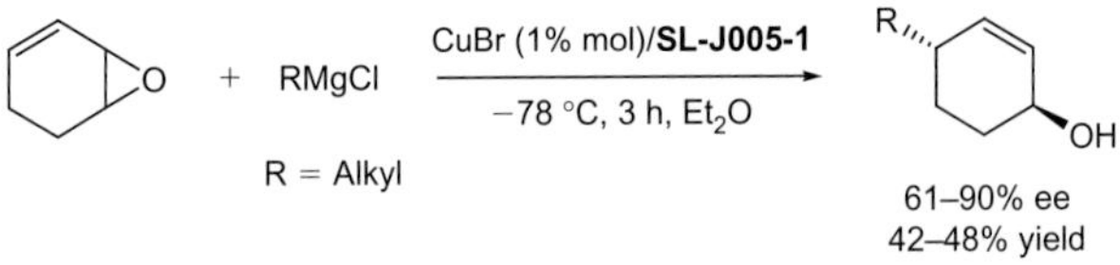

R = Alkyl

61–90% ee
42–48% yield

Scheme 3.54

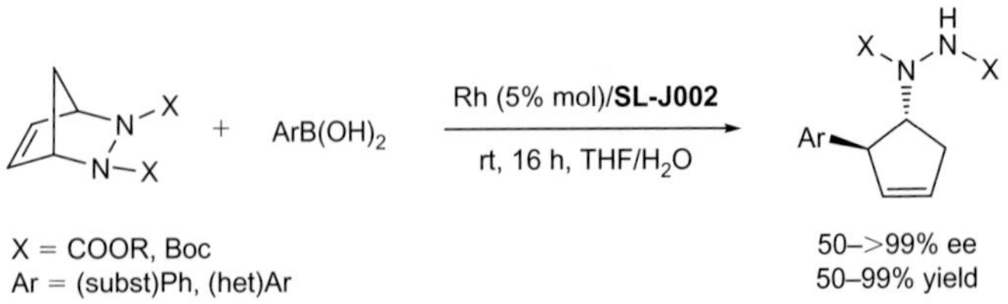

X = COOR, Boc
Ar = (subst)Ph, (het)Ar

50–>99% ee
50–99% yield

Scheme 3.55

Josiphos and Walphos complexes can also be used but SL-J002 usually gives the best results. Some reductive arylation can occur, depending on ligand type and substituent.

3.4.4
Miscellaneous Enantioselective Reactions

3.4.4.1 Isomerization of Allylamines

This isomerization reaction of diethylgeranylamine using Rh/Binap catalysts is one of the milestones of industrial homogeneous catalysis [53]. A feasibility study carried out by Firmenich [109] revealed that Rh/Josiphos catalysts (best results were obtained SL-J002) can catalyze the transformation of both the *(E)* and *(Z)* isomers with 92–97% ee and TON up to 1500 (Scheme 3.56). Interestingly, an immobilized catalyst showed similar catalytic performance to the soluble analog. While these results are not (yet) up to the performance of the Ru/Binap catalyst, they show that alternatives to this proprietary process are possible.

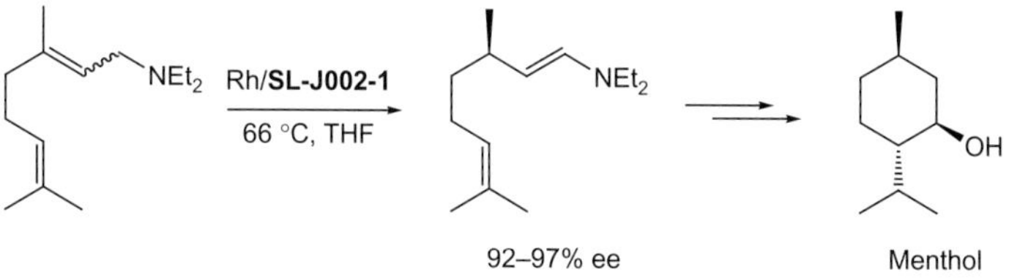

Scheme 3.56

3.4.4.2 Ring-Opening of Oxabicycles

The Rh/SL-J002-catalyzed reaction of oxabicyclic substrates with various nucleophiles developed by the Lautens [110] group gives a synthetically useful access to substituted dihydronaphthalenes [111] (Scheme 3.57) and cyclohexene derivatives (Scheme 3.58). S/c ratios were between 50 and 400 and reaction times 6 to >24 h. At 100 °C and without solvent, an s/c up to 10 000 is feasible with similar ees. The reaction has been scaled up to the kilogram scale by Solvias (D. Spielvogel and F. Naud, unpublished results). Unsymmetrical oxabicyclic alkenes can be resolved

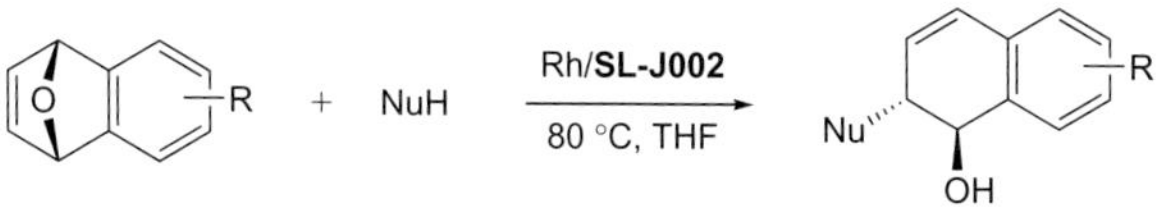

NuH	%ee
(subst)phenols	95→>99
alcohols	93→>99
PhSO$_2$NH$_2$	95

Scheme 3.57

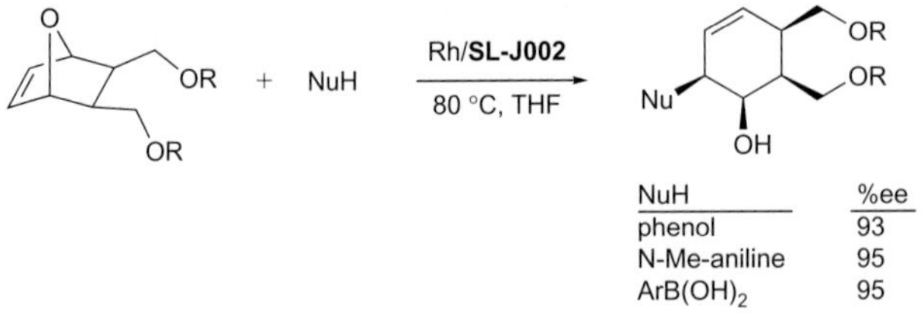

O
OR
OR
+ NuH
Rh/SL-J002
80 °C, THF
OR
OR
Nu
OH
NuH %ee
phenol 93
N-Me-aniline 95
ArB(OH)2 95

Scheme 3.58

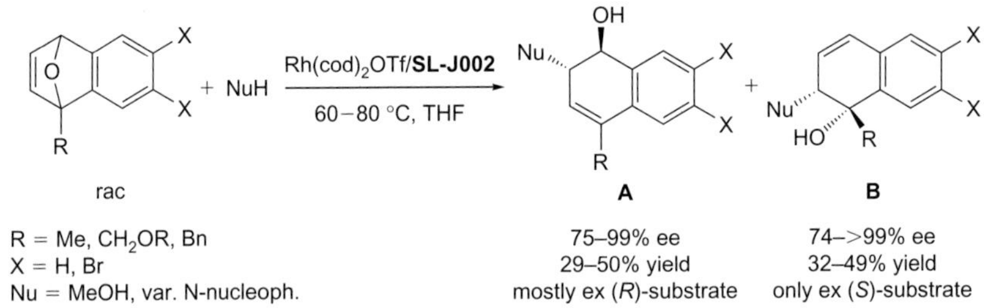

X
O
X
R
+ NuH
rac
Rh(cod)2OTf/SL-J002
60–80 °C, THF
OH
Nu
X
X
R
A
+
X
X
Nu
HO
R
B
R = Me, CH2OR, Bn
X = H, Br
Nu = MeOH, var. N-nucleoph.
75–99% ee
29–50% yield
mostly ex (R)-substrate
74–>99% ee
32–49% yield
only ex (S)-substrate

Scheme 3.59

in a Rh/Josiphos-catalyzed reagent-controlled regiodivergent reaction (Scheme 3.59 [112]).

3.4.4.3 Allylic Substitution

An intramolecular allylic amination reaction was described by Berkowitz and Maiti [113] mediated by Ni/Josiphos as well as Ni/MeOBIPHEP complexes (Scheme 3.60). Up to 82% ee was achieved with SL-J002 but yields were rather low. Better yields (85–95%) but lower ees (53–72%) were obtained with Ni/MeOBIPHEP,

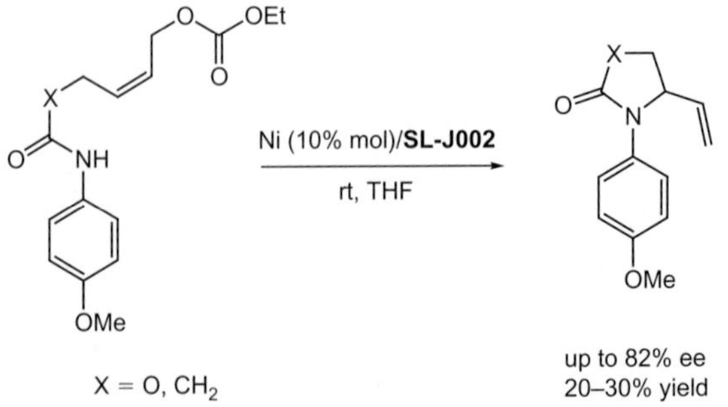

O
OEt
O
X
O
NH
OMe
X = O, CH2
Ni (10% mol)/SL-J002
rt, THF
X
O
N
OMe
up to 82% ee
20–30% yield

Scheme 3.60

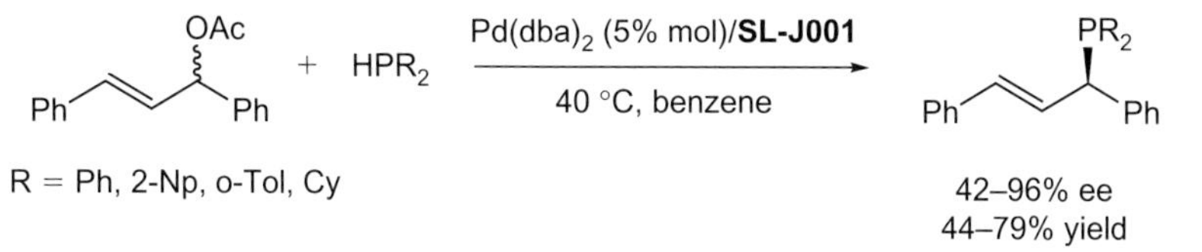

Scheme 3.61

Ph–CH=CH–CH(OAc)–Ph + HPR$_2$ →[Pd(dba)$_2$ (5% mol)/**SL-J001**][40 °C, benzene] Ph–CH=CH–CH(PR$_2$)–Ph

R = Ph, 2-Np, o-Tol, Cy

42–96% ee
44–79% yield

which at the moment is the more practical variant. Togni and coworkers [114] developed the corresponding allylic phosphination using a Pd/SL-J001 catalysts (Scheme 3.61). The new reaction type has as yet narrow scope and catalytic activities are rather low.

3.4.5
Application in Non-Enantioselective Reactions

The chiral Josiphos ligands were obviously developed for the catalysis of enantioselective reactions. Surprisingly, the Josiphos ligands with bulky alkyl phosphine groups such as SL-J002, SL-J003, and SL-J009 were shown to be ideal ligands for the Pd-catalyzed reaction of aryl chlorides and aryl and vinyl tosylates with various nucleophiles. Catalytic activity and turnover numbers (up to 20 000) were comparable or superior to the best catalysts (usually monodentate bulky and basic phosphines) reported in the literature [115]. Described were Pd/Josiphos-catalyzed carbonylation reactions [116, 117], coupling with Grignard reagents [25, 118], Negishi coupling [119], and reaction with amines [118, 119–122] and with sulfides [29, 123, 124].

3.5
Concluding Remarks

The Josiphos ligand family has proven to be highly effective in a large number of enantioselective reactions, encompassing the hydrogenation of C=C, C=O, and C=N bonds, hydrofunctionalization reactions, various C–C bond forming reactions as well as isomerization and ring-opening reactions. Furthermore, several applications in non-enantioselective reactions have been reported to proceed efficiently in the presence of Josiphos complexes. Several hydrogenation reactions are already being applied on a technical scale, including the world largest-scale process, or have been shown to be feasible for the solution of "real-world" industrial problems. In contrast, the application of these ligands for other transformations has generally been carried out using model compounds aimed at showing the viability of the catalytic system. It will be interesting to see their application to industrial problems.

Appendix: Selected Geometrical Parameters of Transition Metal Complexes Containing Josiphos-Type Ligands

Compound, CCDC code or new numbering	M	R^1	R^2	L_n, X	α1 (°)	α2 (°)	M–P^1 (Å)	M–P^2 (Å)	P^1–M–P^2 (°)	Further substituents	Reference
ASEQOE	Ir	Ph	3,5-Me$_2$Ph	HI(COD), BF$_4$	−50.66	0	2.357	2.35	90.87		[19]
ASEQUK	Ir	Ph	3,5-Me$_2$Ph	I(μ-I)$_3$, I$_3$	−43.61	−2.3	2.324	2.366	92		[19]
ASERAR	Ir	Ph	3,5-Me$_2$Ph	COD, BF$_4$	−30.34	−8.34	2.318	2.325	92.76		[19]
ASEREV	Ir	Ph	3,5-Me$_2$Ph	I(COD)	−53.62	−3.6	2.358	2.352	89.89		[19]
CAQPEQ	Pd	Ph	t-Bu	I(Ph)	2.33	−31.25	2.248	2.439	97.01		[20]
CAQSAP	Pd	Cy	t-Bu	I(Ph)	−13.77	−29.16	2.3	2.415	98.01		[20]
CAQSET	Pd	Cy	t-Bu	η^2-PhCHCHPh	−4.7	−25.47	2.314	2.344	10.27		[20]
EJAKIJ	Pd	Ph	Ph	Cl$_2$	−32.75	−109.99	2.252	2.252	94.32	μ-CH(P^2)-CH$_2$CH(Me)-	[20]
EZIGOJ	Rh	Ph	c-(CHMeCH$_2$-)$_2$	NBD, BF$_4$	−53.4	80.15	2.31	2.302	92.99		[22]
EZIGUP	Rh	Ph	c-(CHMeCH$_2$-)$_2$	same L, BF$_4$	−59.68	74.58	2.328	2.306	89.21		[22]
EZUGEL	Ru	Ph	Ph	H(Cp)	−58.84	−87.97	2.227	2.235	87.95	μ-CH(P^2)-CH$_2$CH$_2$-	[23]
EZUGIP	Ru	Ph	Ph	TFA(Cp)	−29.79	−111.67	2.342	2.305	92.03	μ-CH(P^2)-CH$_2$CH$_2$-	[23]
GAXLUM	Ir	Ph	Ph	HI(η-OH)	−54.83	0.38	2.198	2.228	91.17		[24]
GAXPAW	Ir	Ph	Ph	HCl(CH$_2$Ph)(NH$_2$)Me	−53.35	2.24	2.225	2.326	90.19		[24]
HOHFIT	Ir	Ph	3,5-Me$_2$Ph	HI(I-COD), BF$_4$	−55.95	4.53	2.355	2.354	89.83		[24]
HOHFOZ	Ir	Ph	3,5-Me$_2$Ph	I(η^3,2-COD), BF$_4$	−50.2	−1.27	2.347	2.37	88.74		[24]

Code	Metal	R1	R2	Ligand	v1	v2	v3	v4	Extra	Ref
HUXNET	Ru	Ph	Cy	Cl$_2$(py)$_2$	−55.04	0.74	2.288	2.331	88.54		[25]
IKEPUJ	Pd	Ph	t-Bu	Br(Ph)	1.89	−29.04	2.255	2.428	96.75		[26]
IKEQAQ	Pd	Cy	t-Bu	Br(Ph)	−8.28	−25.16	2.286	2.419	98.01		[26]
JAPTAW	Cu	Ph	Cy	Br	−33.11	−25.09	2.281	2.257	101.74		[27]
KAKFUY	Ru	Ph	Cy	Cl$_2$(py-2-CH$_2$NH$_2$)	−43.55	−3.21	2.281	2.314	93.24		[28]
KATYEJ	Ir	Ph	Ph	(μ-NHPh)	−77.58	77	2.192	2.209	92.46		[28]
KENJAP	Pd	Cy	t-Bu	NH$_3$(4-OMePh)	−3.92	−27.62	2.276	2.401	97.23		[30]
MEFDUX	Pd	Ph	Cy	Me$_2$	−32.39	−112.35	2.295	2.318	93.84	μ-CH(P^2)CH$_2$CH$_2$-	[31]
MEQVOU	Cu	Cy	Ph	Br	−31.55	−20.13	2.266	2.269	104.13		[32]
MEQVUA	Cu	Ph	Cy	(μ-Br)	17.83	−18.8	2.247	2.264	101.07		[33]
ODAPAK	Pd	4-OMePh	Cy	Cl(Me)	−18.66	−14.63	2.236	2.365	97.08		[33]
ODAQUF	Pd	Ph	Cy	Cl(Me)	−15.1	−22.21	2.199	2.343	97.24		[33]
ODARAM	Pd	Ph	Cy	Me$_2$	−25.47	−13.8	2.298	2.327	93.25		[33]
ODAREQ	Pd	3,5-(CF$_3$)$_2$Ph	Cy	Cl$_2$	−13.73	−12.74	2.236	2.275	96.02		[33]
ODARIU	Pd	Ph	Ph	(Me)Cl	−68.82	82.57	2.348	2.236	96.8		[33]
OKOGAW	Pd	Ph	t-Bu	I(2-OMePh)	0.87	−28.98	2.245	2.442	95.65		[18]
PEJVEG	Pd	Ph	Cy	1,3-Ph$_2$Allyl, OTf	−21.67	−1.22	2.293	2.323	96.27	c-(CH$_2$)$_3$	[18]
PEVSOZ	Ru	Ph	Cy	(Cp)Cl	−44.77	−0.25	2.281	2.311	90.23		[35]
QICBAG	Pd	Cy	t-Bu	Cl$_2$	−3.95	−26.59	2.27	2.307	96.38		[36]
QICBEK	Pt	Cy	t-Bu	Cl$_2$	−9.07	−25.53	2.263	2.287	97.33		[36]
RINROW	Ru	Ph	Cy	Cl$_2$(py-2-CH(Me)NH$_2$	−40.28	−7.22	2.285	2.314	93.4		[37]
RINRUC	Ru	Ph	Cy	Cl$_2$(py-2-CH(Ph)NH$_2$	−40.5	−9.07	2.266	2.309	91.83		[37]
WAYQAO	Pt	Ph	Ph	Cl$_2$	−37.01	−108.17	2.24	2.24	95.39	μ-CH(P^2)CH$_2$CH$_2$-	[38]
WAYQES	Pt	Ph	Cy	Cl$_2$	−17.97	−109.67	2.237	2.244	97.11	μ-CH(P^2)CH$_2$CH$_2$-	[38]

(Continued)

Compound, CCDC code or new numbering	M	R¹	R²	L$_m$, X	α1 (°)	α2 (°)	M–P¹ (Å)	M–P² (Å)	P¹–M–P² (°)	Further substituents	Reference
WIJJUU	Rh	Ph	Cy	NBD, BF$_4$	−10.53	−13.48	2.299	2.33	93.27		[39]
WIJKAB	Pd	Ph	Cy	η³-Allyl, OTf	−27.51	−14.63	2.297	2.314	95.68		[39]
WIJKEF	Pt	Ph	Cy	Cl$_2$	−19.21	−20.28	2.237	2.248	95.19		[39]
XIKTAN	Pd	Ph	3,5-Me$_2$Ph	Cl$_2$	−88.1	74.03	2.214	2.256	93.59		[40]
YOJLAK	Pd	Ph	Phobiphos	η³-Allyl, OTf	−65.57	83.88	2.282	2.329	97.78		[41]
YOJLEO	Pd	Ph	Phobiphos	η³-1,3-Ph$_2$Allyl, OTf	−38.79	69.72	2.284	2.3	92.86		[41]
ZUJPOJ	Ru	Ph	Cy	(TFA)$_2$(EtOH)$_2$	−39.22	−2.46	2.272	2.273	93.05		[42]
ZUJPUP	Ru	Ph	Cy	(η⁶-Cym)Cl, PF$_6$	−38.05	−3.68	2.368	2.389	92.56		[42]
ZUYDIG	Pd	Ph	Cy	η³-Allyl, OTf	−37	−13.79	2.296	2.313	95.37	Cp*	[43]
ZUYDOM	Pd	Ph	Cy	η³-Allyl, OTf	29.55	−31.69	2.284	2.353	94.02	RuCp*	[43]
N1	Re	Ph	Cy	Br(CO)$_3$	−52.08	−5.19	2.483	2.511	87.45		[44]
N2	Re	Ph	Cy	Cl$_2$O(OiPr)	−22.33	−23.54	2.458	2.456	92.29		[44]
N3	Re	CF$_3$	Cy	Cl$_3$O	−38.81	−4.79	2.418	2.506	92.03		[44]
N4	Re	Cy	Cy	Cl$_3$O	−29.89	−17.08	2.489	2.493	92.23		[44]
N5	Re	Ph	Cy	Cl$_3$O	−54.79	0.19	2.46	2.487	89.09		[44]
N6	Pd	Ph	Cy	η³-1,3-Ph$_2$Allyl, PF$_6$	−11.29	−13.44	2.309	2.338	97.55		[44]
N7	Pd	CF$_3$	Cy	η³-1,3-Ph$_2$Allyl, PF$_6$	−46.57	−4.07	2.289	2.361	92.66		[44]
N8	Pd	CF$_3$	Ad	Cl$_2$	6	−28.79	2.219	2.332	96.3		[44]
N9	Pd	CF$_3$	Cy	Cl$_2$	−1.5	−18.57	2.233	2.304	94.89		[44]

N10	Pd	CF$_3$/Ph-exo	Cy	Cl$_2$	9.44	−21.36	2.232	2.291	95.53	[44]
N11	Pd	Ph/CF$_3$-exo	Me/1-Ad	Cl$_2$	20.72	−26.27	2.225	2.301	93.24	[44]
N12	Pd	Ph	H/Ad	Cl$_2$	−18.9	−13.97	2.263	2.224	94.25	[44]
N13	Pt	CF$_3$	Cy	Cl$_2$	−2.02	−17.32	2.193	2.289	95.42	[44]
N14	Pt	CF$_3$/2-Np-exo	Cy	Cl$_2$	1.91	−15.82	2.21	2.267	96.24	[44]
N15	Pt	Me/Ad-exo	Cy	Cl$_2$	−36.13	−6.64	2.248	2.261	95.1	[44]
N16	Pt	Ph	OH/Cy-exo	Cl$_2$	−21.9	−11.97	2.247	2.206	95.38 3-TMS	[44]
N17	W	CF$_3$	Cy	(CO)$_4$	−10.15	−18.37	2.453	2.527	87.9	[44]
N18	W	Ph	Cy	(CO)$_4$	−55.89	6.43	2.509	2.567	84.38	[44]
WIJJII		Ph	Cy		−64.98	−23.87				[39]
ODAPOY		3,5-(CF$_3$)$_2$Ph	Cy		−74.83	−15.42				[33]
ODAPUE		3,5-(OMe)$_2$Ph	Cy		−80.25	−20.99				[33]
ODAQAL		4-CF$_3$Ph	Cy		−70.96	−33.36				[33]
ODAQIT		4-OMePh	Cy		−77.74	−14.18				[33]
ODAQOZ		2-CF$_3$Ph	Cy		−63.72	−29.11				[33]

References

1 Jacobsen, E.N., Pfaltz, A., and Yamamoto H. (eds) (1999) *Comprehensive Asymmetric Catalysis*, Springer, Berlin. (a) Jacobsen, E.N., pp. 1473–1477; (b) Blaser, H.U. and Spindler, F., pp. 247–265; (c) Hayashi, T., pp. 351–364; (d) Pfaltz, A. and Lautens, M., pp. 833–884; (e) Schmid, R. and Scalone, M., pp. 1439–1449; (f) Nozaki, K., pp. 381–413.
2 Ojima I. (ed.) (2000) *Catalytic Asymmetric Synthesis*, 2nd edn, Wiley-VCH Verlag GmbH, Weinheim.
3 Blaser, H.U., Hoge, G., Pugin, B., and Spindler, F. (2009) in *Handbook of Green Chemistry, Vol. 1 Green Catalysis* (eds P.T. Anastas and R.H. Crabtree), Wiley-VCH Verlag GmbH, pp. 153–203.
4 Blaser, H.U., Pugin, B., and Studer M. (2010) in *Encyclopedia of Catalysis*, 2nd edn (ed. I.T. Horvath), John Wiley & Sons, Inc., Hoboken, DOI/10.1002/0471227617.200025.
5 Yoon, T.P. and Jacobsen, E.N. (2003) *Science*, **299**, 1691–1693.
6 Blaser, H.U., Brieden, W., Pugin, B., Spindler, F., Studer, M., and Togni, A. (2002) *Top. Catal.*, **19**, 3–16.
7 Ito, Y., Sawamura, M., and Hayashi, T. (1986) *J. Am. Chem. Soc.*, **108**, 6405–6406.
8 Togni, A., Breutel, C., Schnyder, A., Spindler, F., Landert, H., and Tijani, A. (1994) *J. Am. Chem. Soc.*, **116**, 4062–4066.
9 For more information: Thommen, M. and Blaser, H.U. (2002) *PharmaChem*, July/August, 33–34.
10 Chen, W. and Blaser, H.U. (2008) in *Trivalent Phosphorus Compounds in Asymmetric Catalysis: Synthesis and Applications* (ed. A. Börner), Wiley-VCH Verlag GmbH, pp. 359–393.
11 For a more detailed discussion and references, see: Blaser, H.U., Malan, C., Pugin, B., Spindler, F., Steiner, H., and Studer, M. (2003) *Adv. Synth. Catal.*, **345**, 103–151.
12 For many years this was almost treated as a dogma for an overview, see: Whitesell, J.K. (1989) *Chem. Rev.*, **89**, 1581–1590.
13 For a concept discussion, see: Inoguchi, K., Sakuraba, S., and Achiwa, K. (1992) *Synlett*, 169–178.
14 Argouarch, G., Samuel, O., and Kagan, H.B. (2000) *Eur. J. Org. Chem.*, 2885–1891.
15 Yasuike, S., Kofink, C.C., Kloetzing, R.J., Gommermann, N., Tappe, K., Gavryushin, A., and Knochel, P. (2005) *Tetrahedron: Asymmetry*, **16**, 3385–3393.
16 Sturm, T., Weissensteiner, W., Spindler, F., Mereiter, K., Lopez-Agenjo, A.M., Manzano, B.R., and Jalon, F.A. (2002) *Organometallics*, **21**, 1766–1774.
17 Sturm, T., Abad, B., Weissensteiner, W., Mereiter, K., Manzano, B.R., and Jalon, F.A. (2006) *J. Mol. Catal. A: General*, **255**, 209–219.
18 Dorta, R. and Broggini, D., Stoop, R., Rüegger, H., Spindler, F., and Togni, A. (2004) *Chem. Eur. J.*, **10**, 267–278.
19 Brunker, T.J., Blank, N.F., Moncarz, J.R., Scriban, C., Anderson, B.J., Glueck, D.S., Zakharov, L.N., Golen, J.A., Sommer, R.D., Incarvito, C.D., and Rheingold A.L. (2005) *Organometallics*, **24**, 2730–2746.
20 Liptau, P., Seki, T., Kehr, G., Abele, A., Fröhlich, R., Erker, G., and Grimme, S. (2003) *Organometallics*, **22**, 2226–2232.
21 Braun, W., Calmuschi, B., Haberland, J., Hummel, W., Liese, A., Nickel, T., Stelzer, O., and Salzer, A. (2004) *Eur. J. Inorg. Chem.*, 2235–2243.
22 Cayuela, E. Jalon, F.A. Manzano, B.R. Espino, G. Weissensteiner, W. Mereiter, K. (2004) *J. Am. Chem. Soc.*, **126**, 7049–7062.
23 Dorta, R. and Togni, A. (1998) *Organometallics*, **17**, 3423–3428.
24 Leong, C.G., Akotsi, O.M., Ferguson, M.J., and Bergens, S.H. (2003) *Chem. Commun.*, 750–751;. Erratum 1779.
25 Roy, A.H. and Hartwig, J.F. (2003) *J. Am. Chem. Soc.*, **125**, 8704–8705.
26 Lopez, F., Harutyunyan, S.R., Meetsma, A., Minnaard, A.J., and

Feringa, B.L. (2005) *Angew. Chem. Int. Ed.*, **44**, 2752–2756.

27 Baratta, W., Herdtweck, E., Siega, K., Toniutti, M., and Rigo, P. (2005) *Organometallics*, **24**, 1660–1669.

28 Dorta, R. Togni, A. (2000) *Helv. Chim. Acta*, **83**, 119–127.

29 Shen, Q. and Hartwig, J.F. (2006) *J. Am. Chem. Soc.*, **128**, 10028–10029.

30 Carrion, M.C., Jalon, F.A., Lopez-Agenjo, A., Manzano, B.R., Weissensteiner, W., and Mereiter, K. (2006) *J. Organomet. Chem.*, **691**, 1369–1381.

31 Harutyunyan, S.R., Lopez, F., Browne, W.R., Correa, A., Pena, D., Badorrey, R., Meetsma, A., Minnaard, A.J., and Feringa, B.L. (2006) *J. Am. Chem. Soc.*, **128**, 9103–9118.

32 Gambs, C., Consiglio, G., and Togni, A. (2001) *Helv. Chim. Acta*, **84**, 3105–3126.

33 Moncarz, J.R., Brunker, T.J., Jewett, J.C., Orchowski, M., Glueck, D.R., Sommer, D.S., Lam, K.-C., Incarvito, C.D., Concolino, T.E., Ceccarelli, C., Zakharov, L.N., and Rheingold, A.L. (2003) *Organometallics*, **22**, 3205–3221.

34 Ng, S.Y., Fang, G., Leong, W.K., Goh, L.Y., and Garland, M.V. (2007) *Eur. J. Inorg. Chem.*, 452–462.

35 Ghent, B.L., Martinak, S.L., Sites, L.A., Golen, J.A., Rheingold, A.L., and Nataro, C. (2007) *J. Organomet. Chem.*, **692**, 2365–2374.

36 Baratta, W., Chelucci, G., Herdtweck, E., Magnolia, S., Siega, K., and Rigo, P. (2007) *Angew. Chem. Int. Ed.*, **46**, 7651–7654.

37 Sturm, T., Weissensteiner, W., Mereiter, K., Kegl, T., Jeges, G., Petolz, G., and Kollar, L. (2000) *J. Organomet. Chem.*, **595**, 93–101.

38 Togni, A., Breutel, C., Soares, M.C., Zanetti, N., Gerfin, T., Gramlich, V., Spindler, F., and Rihs, G. (1994) *Inorg. Chim. Acta*, **222**, 213–224.

39 Wang, Y., Weissensteiner, W., Spindler, F., Arion, V.B., and Mereiter, K. (2007) *Organometallics*, **26**, 3530–3540.

40 Abbenhuis, H.C.L., Burckhardt, U., Gramlich, V., Köllner, C., Pregosin, P.S., Salzmann, R., and Togni, A. (1995) *Organometallics*, **14**, 759–766.

41 Zanetti, N.C., Spindler, F., Spencer, J., Togni, A., and Rihs, G. (1996) *Organometallics*, **15**, 860–866.

42 Abbenhuis, H.C.L., Burckhardt, U., Gramlich, V., Martelletti, A., Spencer, J., Steiner, I., and Togni, A. (1996) *Organometallics*, **15**, 1614–1621.

43 Hems, W.P., McMorn, P., Riddel, S., Watson, S., Hancock, F.E., and Hutchings, G. (2005) *J. Org. Biomol. Chem.*, **3**, 1547–1550.

44 Stephenson, P., Kondor, B., Licence, P., Scovell, K., Ross, S.K., and Poliakoff, M. (2006) *Adv. Synth. Catal.*, **348**, 1605–1610.

45 McGarrity, J.F., Brieden, W., Fuchs, R., Mettler, H.-P., Schmidt, B., and Werbitzky, O. (2003) in *Large Scale Asymmetric Catalysis* (eds H.U. Blaser and E. Schmidt), Wiley-VCH Verlag GmbH, Weinheim, pp. 283–308;.

46 Shultz, C.S., Dreher, S.D., Ikemoto, N., Williams, J.M., Grabowski, E.J.J., Krska, S.W., Sun, Y., Dormer, P.G., and DiMichele, L. (2005) *Org. Lett.*, **7**, 3405–3408.

47 (a) Shultz, C.S. and Krska, S.W. (2007) *Acc. Chem. Res.*, **40**, 1320–1326; (b) Wallace, D.J., Campos, K.R., Shultz, C.S., Klapars, A., Zewge, D., Crump, B.R., Phenix, B.D., McWilliams, C., Krska, S., Sun, Y., Chen, C., and Spindler, F. (2009) *Org. Process Res. Dev.*, **13**, 84–90.

48 (a) Rouhi, M. (2004) *Chem. Eng. News*, **82** (37), 28–32; (b) Hansen, K.B., Hsiao, Y., Xu, F., Rivera, N., Clausen, A., Kubryk, M., Krska, S., Rosner, T., Simmons, B., Balsells, J., Ikemoto, N., Sun, Y., Spindler, F., Malan, C., Grabowski, E.J.J., and Armstrong, J.D., III (2009) *J. Am. Chem. Soc.*, **131**, 8798–8799.

49 Matsumura, K., Zhang, X., and Saito, T. (2004) EP 1386901 assigned to Takasago and T. Saito, personal communication.

50 Kadyrov, R., Holz, J., Schäffner, B., Zayas, O., Almena, J., and Börner, A. (2008) *Tetrahedron: Asymmetry*, **19**, 1189–1192.

51 Doherty, S., Knight, J.G., Bell, A.L., El-Menabawey, S., Vogels, C.M., Decken, A., and Westcott, S.A. (2009) *Tetrahedron: Asymmetry*, **20**, 1438–1444.

52 Ikemoto, T., Nagata, T., Yamano, M., Ito, T., Mizuno, Y., and Tomimatsu, K. (2004) *Tetrahedron Lett.*, **45**, 7757–7760.

53 Kumobayashi, H. (1996) *Recl. Trav. Chim. Pays-Bas*, **115**, 201–210.

54 Dobbs, D.A., Vanhessche, K.P.M., Brazi, E., Rautenstrauch, V., Lenoir, J.-Y., Genet, J.-P., Wiles, J., and Bergens, S.H. (2000) *Angew. Chem. Int. Ed.*, **39**, 1992–1995.

55 Broger, E., Buchecker, R., Crameri, Y., and Lukac, T. (1995) EP 749,953 assigned to Hoffmann-La Roche.

56 Yamashita, M. and Yamano, T. (2009) *Chem. Lett.*, **38**, 100–101.

57 Limanto, J., Shultz, C.S., Corner, B., Desmond, R.A., Devine, P.N., and Krska, S.W. (2008) *J. Org. Chem.*, **73**, 1639–1642.

58 Huang, Y., Berthiol, F., Stegink, B., Pollard, M.A., and Minnaard, A.J. (2009) *Adv. Synth. Catal.*, **351**, 1423–1430.

59 (a) Czekelius, C. and Carreira, E.M. (2003) *Angew. Chem. Int. Ed.*, **42**, 4793–4795; (b) Czekelius, C. and Carreira, E.M. (2004) *Org. Lett.*, **6**, 4575–4578.

60 Lipshutz, B.H. and Servesko, J.M. (2003) *Angew. Chem. Int. Ed.*, **42**, 4789–4792.

61 (a) Lipshutz, B.H., Servesko, J.M., and Taft, B.R. (2004) *J. Am. Chem. Soc.*, **126**, 8352–8353; (b) Lipshutz, B.H., Tanaka, N., Taft, B.R., and Lee, C.-T. (2006) *Org. Lett.*, **8**, 1963–1966.

62 Lee, D., Kim, D., and Yun, J. (2006) *Angew. Chem. Int. Ed.*, **45**, 2785–2788.

63 Lipshutz, B.H., Lee, C.-T., and Serversko, J.M. (2007) *Org. Lett.*, **9**, 4713–4716.

64 Yoo, K., Kim, H., and Yun, J. (2009) *J. Org. Chem.*, **74**, 4232–4235.

65 Rupnicki, L., Saxena, A., and Lam, H.W. (2009) *J. Am. Chem. Soc.*, **131**, 10386–10387.

66 Makino, K., Fijii, T., and Hamada, Y. (2006) *Tetrahedron: Asymmetry*, **17**, 481–485.

67 Hamada, Y., Koseki, Y., Fujii, T., Maeda, T., Hibino, T., and Makino, K. (2008) *Chem. Commun*, 6206–6208.

68 Andresen, B.M., Caron, S., Couturier, M., DeVries, K.M., Do, N.M., Dupont, K., Gosh, A., Girardin, M., Hawkins, J.M., Makowski, T.M., Riou, M., Sieser, J.E., Tucker, J.L., Vanderplas, B.C., and Watson, T.J.N. (2006) *Chimia*, **60**, 554–560.

69 Baratta, W., Ballico, M., Del Zotto, A., Siega, K., Magnolia, S., and Rigo, P. (2008) *Chem. Eur. J.*, **14**, 2557–2563.

70 Baratta, W., Chelucci, G., Magnolia, S., Siega, K., and Rigo, P. (2009) *Chem. Eur. J.*, **15**, 726–732.

71 Baratta, W., Ballico, M., Chelucci, G., Siega, K., and Rigo, P. (2008) *Angew. Chem. Int. Ed.*, **47**, 4362–4365.

72 Brieden, W. (1996) EP 785198 assigned to Lonza AG.

73 Spindler, F. and Blaser, H.U. (2004) in *Transition Metals for Organic Synthesis*, Vol. **2**, 2nd edn (eds C. Bolm and M. Beller), Wiley-VCH Verlag GmbH, Weinheim, pp. 113–123;.

74 Blaser, H.U. and Spindler, F. (2007) in *Handbook of Homogeneous Hydrogenation* (eds J.G. de Vries and C.J. Elsevier) Wiley-VCH Verlag GmbH, Weinheim, pp. 1193–1214;.

75 Blaser, H.U., Buser, H.P., Häusel, R., Jalett, H.P., and Spindler, F. (2001) *J. Organomet. Chem.*, **621**, 34–38.

76 Spindler, F. and Blaser, H.U. (2000) *Adv. Synth. Catal.*, **343**, 68–70.

77 (a) Bader, R.R. and Blaser, H.U. (1997) *Stud. Surf. Sci. Catal.*, **108**, 17–29; (b) Hofer, R. (2004) *Chimia*, **59**, 10–12.

78 Blaser, H.U. (2002) *Adv. Synth. Catal.*, **344**, 17–31.

79 Blaser, H.U., Buser, H.P., Jalett, H.P., Pugin, B., and Spindler, F. (1999) *Synlett*, 867–868.

80 Pugin, B., Landert, H., Spindler, F., and Blaser, H.U. (2002) *Adv. Synth. Catal.*, **344**, 974–997.

81 Zhong, Y.-L., Krska, S.W., Zhou, H., Reamer, R.A., Lee, J., Sun, Y., and Askin, D. (2009) *Org. Lett.*, **11**, 369–373.

82 Fuchs, R. (1997) EP 0803502 assigned to Lonza AG.

83 Hayashi, T. (1999) in *Comprehensive Asymmetric Catalysis* (eds E.N. Jacobsen, A. Pfaltz, and H. Yamamoto), Springer, Berlin, pp. 351–364.
84 Togni, A. (1996) *Chimia*, **50**, 86–93.
85 Crudden, C.M., Hleba, Y.B., and Chen, A.C. (2004) *J. Am. Chem. Soc.*, **126**, 9200–9201.
86 Perez Luna, A., Bonin, M., Micouin, L., and Husson, H.-P. (2002) *J. Am. Chem. Soc.*, **124**, 12098–12099.
87 (a) Mun, S., Lee, J.-E., and Yun, J. (2006) *Org. Lett.*, **8**, 4887–4890; (b) Lee, J.-E. and Yun, J. (2007) *Angew. Chem. Int. Ed.*, **47**, 145–148.
88 Sim, H.-S., Feng, X., and Yun, J. (2009) *Chem. Eur. J.*, **15**, 1939–1943.
89 Dorta, R., Egli, P., Zürcher, F., and Togni, A. (1997) *J. Am. Chem. Soc.*, **119**, 10857–10858.
90 Togni, A., Bieler, N., Burckhardt, U., Köllner, C., Pioda, G., Schneider, R., and Schnyder, A. (1999) *Pure Appl. Chem.*, **71**, 1531–1537.
91 Xu, G. and Han, L.-B. (2006) *Org. Lett.*, **8**, 2099–2102.
92 Nozaki, K. and Ojima I. (2000) in *Catalytic Asymmetric Synthesis*, 2nd edn (ed. I. Ojima), Wiley-VCH Verlag GmbH, Weinheim, pp. 429–463;.
93 Godard, C., Ruiz, A., and Claver, C. (2006) *Helv. Chim. Acta*, **89**, 1610–1622.
94 Trost, B.M. and Lee, C. (2000) in *Catalytic Asymmetric Synthesis*, 2nd edn (ed. I. Ojima), Wiley-VCH Verlag GmbH, Weinheim, pp. 593–649;.
95 Yamamoto, Y., Takada, S., and Miyaura, N. (2006) *Chem. Lett.*, **35**, 1368–1369.
96 Lopez, F., van Zijl, A.W., Minnaard, A.J., and Feringa, B.L. (2006) *Chem. Commun.*, 409–411.
97 (a) Des Mazery, R., Pullez, M., Lopez, F., Harutyunyan, S.R., Minnaard, A.J., and Feringa, B.L. (2005) *J. Am. Chem. Soc.*, **127**, 9966–9967; (b) Ruiz, B.M., Geurts, K., Fernandez-Ibanez, M.A., ter Horst, B., Minnaard, A.J., and Feringa, B.L. (2007) *Org. Lett.*, **9**, 5123–5126.
98 Feringa, B.L., Badorrey, R., Pena, D., Harutyunyan, S.R., and Minnaard, A.J. (2004) *Proc. Natl. Acad. Sci. U.S.A.*, **101**, 5834–5838.
99 Pucheault, M., Darses, S., and Genet, J.-P. (2002) *Eur. J. Org. Chem.*, 3552–3557.
100 den Hartog, T., Harutyunyan, S.R., Font, D., Minnaard, A.J., and Feringa, B.L. (2008) *Angew. Chem. Int. Ed.*, **47**, 398–401.
101 Lormann, M.E.P., Bräse, S., and Nieger, M. (2006) *J. Organomet. Chem.*, **691**, 2159–2161.
102 Bercot, E.A. and Rovis, T. (2004) *J. Am. Chem. Soc.*, **126**, 10248–10249.
103 Leitner, A., Larsens, J., Steffens, C., and Hartwig, J.F. (2004) *J. Org. Chem.*, **69**, 7552–7557.
104 Tanaka, K., Hojo, D., Shoji, T., and Hirano, M. (2007) *Org. Lett.*, **9**, 1307–1310.
105 Lipshutz, B.H., Amorelli, B., and Unger, J.B. (2008) *J. Am. Chem. Soc.*, **130**, 14378–14379.
106 Seiser, T., Roth, O.A., and Cramer, N. (2009) *Angew. Chem. Int. Ed.*, **48**, 6320–6323.
107 Millet, R. and Alexakis, A. (2007) *Synlett*, 435–438.
108 Panteleev, L.J., Menard, F., and Lautens, M. (2008) *Adv. Synth. Catal.*, **350**, 2893–2902.
109 Chapuis, C., Barthe, M., and de Saint Laumer, J.-Y. (2001) *Helv. Chim. Acta*, **84**, 230–242.
110 Reviews: (a) Lautens, M., Fagnou, K., and Hiebert, S. (2003) *Acc. Chem. Res.*, **36**, 48–58; (b) Lautens, M. and Fagnou, K. (2004) *Proc. Natl. Acad. Sci. U.S.A.*, **101**, 5455–5460.
111 (a) Lautens, M., Fagnou, K., and Taylor, M. (2000) *Org. Lett.*, **2**, 1677–1680; (b) Lautens, M., Fagnou, K., and Rovis, T. (2000) *J. Am. Chem. Soc.*, **122**, 5650–5661.
112 Webster, R., Böing, C., and Lautens, M. (2009) *J. Am. Chem. Soc.*, **131**, 444–445.
113 Berkowitz, D.B. and Maiti, G. (2004) *Org. Lett.*, **6**, 2661–1664.
114 Butti, P., Rochat, R., Sadow, A.D., and Togni, A. (2008) *Angew. Chem. Int. Ed.*, **47**, 4878–4881.

115 Littke, A.F. and Fu, G.C. (2002) *Angew. Chem. Int. Ed.*, **41**, 4176–4179.

116 (a) Mägerlein, W., Indolese, A.F., and Beller, M. (2001) *Angew. Chem. Int. Ed.*, **40**, 2865–2868; (b) Mägerlein, W., Indolese, A.F., and Beller, M. (2002) *J. Organomet. Chem.*, **641**, 30–40.

117 Cai, C., Rivera, N.R., Balsells, J., Sidler, R.S., McWilliams, J.C., Shultz, C.S., and Sun, Y. (2006) *Org. Lett.*, **8**, 5161–5164.

118 Limmert, M.E., Roy, A.J., and Hartwig, J.F. (2005) *J. Org. Chem.*, **70**, 9364–9370.

119 (a) Lindhardt, A.T. and Skrydstrup, T. (2008) *Chem. Eur. J.*, **14**, 8756–8766; (b) Lindhardt, A.T., Gøgsig, T.M., and Skrydstrup, T. (2009) *J. Org. Chem.*, **74**, 135–143.

120 (a) Shen, Q., Shekar, S., Stambuli, J., and Hartwig, J.F. (2005) *Angew. Chem. Int. Ed.*, **44**, 1371–1374; (b) Shen, Q., Ogata, T., and Hartwig, J.F. (2008) *J. Am. Chem. Soc.*, **130**, 6586–6596.

121 Shen, Q. and Hartwig, J.F. (2008) *Org. Lett.*, **10**, 4109–4112.

122 Shen, Q. and Hartwig, J.F. (2006) *J. Am. Chem. Soc.*, **128**, 2180–2181.

123 (a) Fernandez, M.A., Shen, Q., and Hartwig, J.F. (2006) *Chem Eur. J.*, **12**, 7782–7796; (b) Fernandez, M.A., Shen, Q., and Hartwig, J.F. (2009) *J. Org. Chem.*, **74**, 1663–1672.

124 Alvaro, E. and Hartwig, J.F. (2009) *J. Am. Chem. Soc.*, **131**, 7858–7868.

4
Chiral Spiro Ligands
Shou-Fei Zhu and Qi-Lin Zhou

4.1
Introduction

Molecules containing a spirocyclic framework are ubiquitous in nature. Researchers accomplished the synthesis of molecules with spiro structures in the late 1890s [1]. The word "spiro" comes from the Latin meaning spiral. As early as 1900, von Baeyer named bicyclic hydrocarbons having two rings with a conjunct carbon atom (spiro carbon atom) "spirocyclane," constructing them as shaped like pretzels [2]. In fact, due to the tetrahedron structure of the spiro carbon, the two rings of spiro compounds lie in perpendicular planes. Rotation of the two rings in bicyclic spiro compounds is restricted and as a result gives rise to an axial chirality in spiro compounds with substituents on the rings. The inherent molecular rigidity and the quaternary structure of the spiro carbon atom make racemization virtually impossible. Given the special properties of chiral spiro compounds, they essentially are ideal scaffold candidates for chiral ligands design. However, partially due to difficulty in obtaining the enantiopure form, chiral spiro molecules were not used in asymmetric synthesis until 1990s.

In 1992, Kumar *et al.* [3] applied a chiral spiro diol, *cis,cis*-(+)- or (−)-spiro[4.4]nonane-1,6-diol (**1**, Figure 4.1) as a chiral auxiliary in the reduction of ketones with lithium aluminum hydride, yielding the corresponding alcohols with high enantioselectivities (up to 90% ee). The mono-pivalate mono-acrylate esters of *cis,cis*-**1** have also been applied in the asymmetric Diels–Alder reaction with cyclopentadiene to produce *endo* bicyclo adduct in >97% de [4]. These pioneering works not only proved that the chiral spiro compounds can accomplish chiral induction in asymmetric synthesis but also provided an important chiral spiro compound with two hydroxy groups in the "bay region" for further ligand design. In 1997, Chan and Jiang developed chiral spiro diphosphinite ligand **2** (Figure 4.1), starting with the spiro diol **1**, and successfully applied it in the rhodium-catalyzed asymmetric hydrogenation of α-dehydroamino acid derivatives [5]. This seminal work gives impetus to the development of chiral spiro ligands in transition metal-catalyzed asymmetric reactions.

Privileged Chiral Ligands and Catalysts. Edited by Qi-Lin Zhou
Copyright © 2011 WILEY-VCH Verlag GmbH & Co. KGaA, Weinheim
ISBN: 978-3-527-32704-1

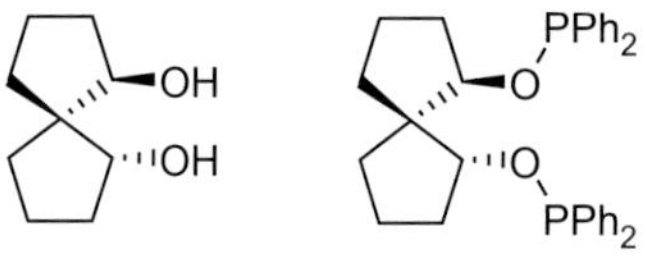

Figure 4.1 Chiral spiro diol **1** and spiro diphosphinite ligand **2**.

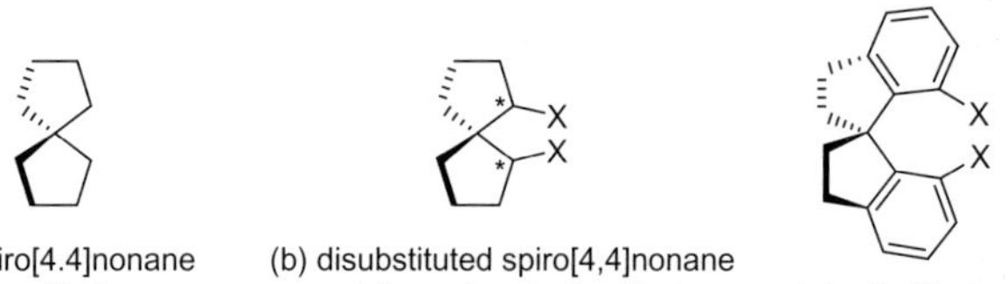

(a) spiro[4.4]nonane
achiral

(b) disubstituted spiro[4,4]nonane
axially and centrally chiral
C_2-symmetric
rigid

(c) spirobiindane
axially chiral
C_2-symmetric
highly rigid

Figure 4.2 Concept of spiro ligand design.

Chiral spiro compounds **1** and **2** are derived from spiro[4.4]nonane (Figure 4.2a). Spiro[4.4]nonane itself is achiral, and only becomes chiral after introducing groups at the 1,6-positions (also called "bay positions") (Figure 4.2b). Although it possesses a perfect C_2 symmetry and looks very simple, the ligands derived from 1,6-disubstituted spiro[4.4]nonane have one axial chirality and two central chiralities, and thus have six stereoisomers, which makes the preparation of optically pure ligands a very tough task [6]. Moreover, the introduction of suitable coordination groups in the ligands shown in Figure 4.2b is difficult because of significant steric hindrance at the "bay positions" and possible racemization of the two central chiralities in the reaction. When benzo groups are introduced into spiro[4.4]nonane to generate the 1,1′-spirobiindane structure (Figure 4.2c), the limitations lying on the spiro[4.4]nonane back-bond can be avoided. Firstly, 1,1′-spirobiindane has only one axial chirality, which significantly facilitates the preparation of optically pure ligands. Secondly, the fused benzo groups of 1,1′-spirobiindane further increase the chemical robustness and conformational rigidity of the spiro backbone and benefit the chiral inducement of the ligands. The functional groups (X) and benzo rings of the 1,1′-spirobiindane backbone (Figure 4.2c) leave positions for further structure modifications of ligands to meet different requirements of various reactions. As it combines high rigidity, perfect C_2 symmetric, simple chirality, and easy modification, chiral ligands based on the 1,1′-spirobiindane skeleton are expected to be highly efficient in enantioselective reactions.

4.2
Preparation of Chiral Spiro Ligands

Spiro ligands with 1,1′-spirobiindane were prepared conveniently from the same starting material, 1,1′-spirobiindane-7,7′-diol (SPINOL). SPINOL was first prepared by Birman et al. [7], starting with m-anisaldehyde, in six steps in 28% overall yield. All the reagents used in the synthetic procedure are commercially available. Racemic SPINOL has been successfully synthesized on a kilogram scale. The convenient resolution of racemic SPINOL through the formation of inclusion complexes with N-benzylcinchonidinium chloride has also been established [8]. Since the inclusion resolution process avoids the use of chromatography for separation, it can be easily scaled up to the kilogram scale. Starting with optically pure SPINOL, more than 100 chiral spiro ligands with the 1,1′-spirobiindane scaffold have been prepared through a single or multiple steps (Figure 4.3), including diphosphine ligand, SDPs (3) [9], diphosphinite ligand, SDPO (4) [10], diimine ligand, SIDIMs (5) [11], bis(oxazoline) ligand, SpiroBOXs (6) [12], phosphine-oxazoline ligand, SIPHOXs (7) [13] and a wide range of monodentate phosphorous ligands that include phosphine SITCP (8) [14], phosphite ShiPs (9) [15], phosphonite FuPs (10) [16], and phosphoramidite SIPHOSs (11) [17]. Some of the ligands, such as SIPHOS (11a and 11e), ShiP (9a), and SDP (3a, 3b, and 3d) are now commercially available from Aldrich and Strem Co. Efficient procedures for introducing substituents at the 4,4′-position or 6,6′-position of the spirobiindane framework, resulting in ligands 12 [18] and 13 [19], have also been developed to tune the electronic or steric properties of ligands. Since the chiral spirobiindane scaffold cannot undergo racemization, relatively harsh conditions are tolerated during the preparation of spiro ligands. For all the 1,1′-spirobiindane ligands, the steric and electronic properties are readily modified according to general procedures, which provided additional opportunities to achieve the best result in certain catalytic asymmetric transformations. Notably, all of the chiral spiro ligands with trivalent phosphorous atoms are capable of being purified through chromatography on silica gel.

Following a similar concept of spirobiindane ligands, the spirobifluorene diphosphine ligand SFDP (14) [20] and spirobitetraline monophosphoramidite ligand (15) [21] were also developed (Figure 4.4). The spirobifluorene scaffold can be considered to be a benzospirobiindane. The additional fused phenyl groups on the spiro rings further increase the rigidity of the scaffold. The spirobitetraline structure can be thought of as the result of inserting a CH_2 group into each of the five-membered spiro rings of spirobiindane, and the rigidity and skew pattern of the ligand 15 are adjusted accordingly. The spiro monophosphoramidite ligand 16, with a spirobixanthene backbone, which can be regarded as oxy-inserted spirofluorene, was prepared by Zhang et al. (Figure 4.4) [22].

Ligands with spiro scaffolds showed special steric and electronic properties, which are the origin of the outstanding chiral inducements of the ligands and catalysts in various asymmetric transformations. The relationship between structure and catalytic property of ligands and catalysts will be discussed for a particular

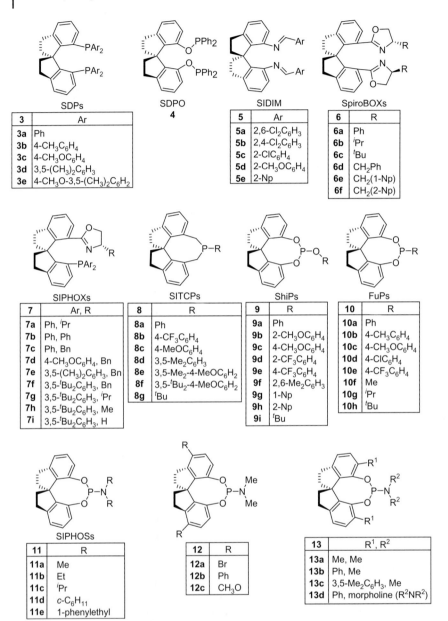

Figure 4.3 Chiral ligands with a 1,1'-spirobiindane backbone.

reaction. Here, the typical structures of spiro-diphosphine ligands SDP (**3**) and SFDP (**14**) are illustrated as representative examples of spiro ligands and catalysts. The structures of ligands **3** and **14** were revealed by X-ray crystal analysis of their Pd-complexes [PdCl$_2${(S)-**3a**}] and [PdCl$_2${(R)-**14a**}] (Figure 4.5) [9b, 20]. Both complexes exhibit a square-planar configuration, and the eight-membered

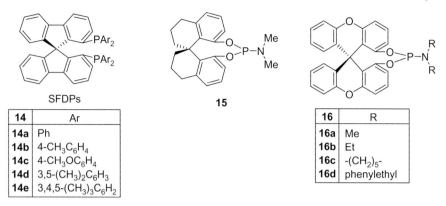

Figure 4.4 Other spiro ligands.

heterometallocyclic rings formed by chelating of ligands **3a** and **14a** to palladium are highly rigid. Both chiral SDP and SFDP ligands create a perfect C_2 symmetric chiral environment around the palladium center. The P–Pd–P bite angles of [PdCl$_2${(S)-**3a**}] and [PdCl$_2${(R)-**14a**}] are 96.0° and 96.7°, respectively, which are significant greater than that of [PdCl$_2${(R)-BINAP}] (92.7°) [23]. The bigger bite angles indicate that the rigid scaffolds effect the coordination property of ligands to metal, and thus influence the activity and selectivity of catalysts. In each complex, two P-phenyl groups are oriented axially and the other two P-phenyl groups are directed equatorially to the P–Pd–P plane. The axial P-phenyl groups lie parallel to the indane or fluorene ring. The central distances between the axial P-phenyls and the indane or fluorene ring are 3.5 and 4.2 Å in [PdCl$_2${(S)-**3a**}] and 3.5 and 3.6 Å in [PdCl$_2${(R)-**14a**}], respectively. Such short distances indicates the likely existence of π–π stacking interactions between the P-phenyl rings and the indane or fluorene of backbones.

Along with the development of spirobiindane chiral ligands, several other spiro ligands based on different frameworks have also been prepared by several research groups. The diphosphite ligands **17** [24], the bisphosphinamidite ligand SpiroNP (**18**) [25], the phosphine-oxazine ligand **19** [26], and the monodentate or bidentate phosphorous ligands **20** [27] have a 1,6-disubstituted spiro[4.4]nonane backbone. The diphosphinite ligand SpiroBIP (**21**) [28] has a 2,2′-spirobiindane-1,1′-diol structure, and the phosphine-oxazoline ligand SpinPHOX (**22**) [29] has a spiro[4.4] nona-1,6-diene framework (Figure 4.6). These chiral spiro ligands were demonstrated to be highly efficient for rhodium-catalyzed hydroformylation of styrene, rhodium-catalyzed hydrogenation of α-dehydroamino acid derivatives, iridium-catalyzed hydrogenation of imines, and palladium-catalyzed alkylative allylation reactions and other asymmetric reactions.

In the past decade, Sasai's group [30] has developed a series of spiro bis(isoxazoline) ligands (**23** and **24**), spiro bis(pyrazole) ligands (**25**), spiro bis(isoxazole) ligands (**26**), spiro bis(oxazoline) ligands (**27**), and spiro isoxazole-isoxazoline ligands (**28**) with fused spiro cyclic backbones (Figure 4.7). The palladium

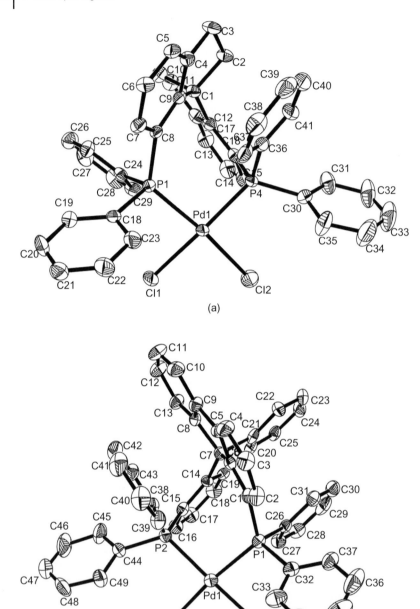

Figure 4.5 Crystal structures of [PdCl$_2${(S)-**3a**}] (a) and [PdCl$_2${(R)-**14a**}] (b).

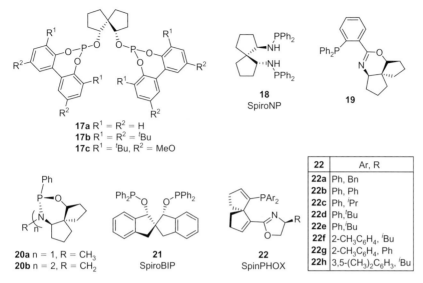

Figure 4.6 Chiral ligands based on 1,6-disubstituted spiro[4.4]nonane.

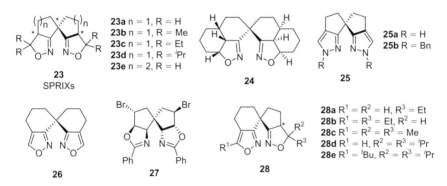

Figure 4.7 Chiral spiro dinitrogen ligands.

complexes of SPRIXs (**23**) exhibited an unprecedented activity in tandem asymmetric oxidative cyclization reactions.

Compounds with a spiro backbone have also been used as chiral organocatalysts in enantioselective transformations. For instance, Fu and Chung [31] reported that the spiro monophosphine (*R*)-**8a** can promote the enantioselective cyclization of hydroxy-2-alkynotes to chiral saturated oxygen-containing heterocycles with good to excellent enantioselectivities (63–94% ee). A hypervalent iodine(III) reagent (*R*)-**29** with a 1,1′-spirobiindane backbone afforded the highest level of asymmetric induction (up to 86% ee) in the field of iodine(III)-mediated asymmetric transformations (Figure 4.8) [32].

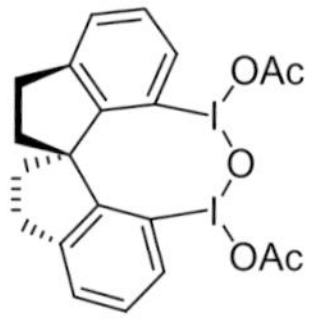

Figure 4.8 A chiral spiro organocatalyst containing hypervalent iodine(III).

4.3
Asymmetric Hydrogenation

Although the history of the development of chiral spiro ligands is very short, many chiral spiro ligands have been applied in various mechanistically unrelated reactions. In many reactions, such as hydrogenation, carbon–carbon bond formation, and carbon–heteroatom bond formation, chiral spiro ligands exhibit unique enantioselectivity and reactivity. Consequently, chiral spiro ligands have, now, become one of the "privileged" chiral ligands [33].

Transition metal-catalyzed asymmetric hydrogenation utilizing molecular hydrogen to reduce prochiral unsaturated bonds is one of the most efficient and atom-economic methods for the preparation of optically active compounds. Many chiral ligands and catalysts with high reactivity and enantioselectivity have been developed and some of them have been successfully applied in industrial production [34]. Chiral spiro ligands including bidentate phosphine ligands SDP and SFDP (**3** and **14**), phosphinite ligands (**2**, **4**, and **21**), phosphine-oxazoline ligands (**7** and **22**), and monodentate phosphorous ligands (**10**, **11**, **15**, and **16**) exhibited high activity and excellent enantioselectivity in the asymmetric hydrogenations of olefins, ketones, aldehydes, imines, and quinolines.

4.3.1
Hydrogenation of Functionalized Olefins

4.3.1.1 Hydrogenation of Enamides

Rhodium-catalyzed asymmetric hydrogenation of enamides provides one of the most efficient methods for the preparation of natural or non-natural α- or β-amino acids and optically active amines. In the past four decades, many chiral bidentate phosphorous ligands and their rhodium complexes have been developed to achieve high enantioselectivity in asymmetric hydrogenation of enamides. When spiro phosphinite ligand SpirOP (1R,5R,6R)-**2** was used in the Rh-catalyzed hydrogenation of α-dehydroamino acids and esters, the hydrogenation products, saturated α-dehydroamino acids and esters, were obtained in perfect enantioselectivities (94–99% ee) (Scheme 4.1, Reaction 1) [5]. Two analogous spiro ligands (1R,5R,6R)-**18** and (1R,2R,1′R)-**21** were also efficient for this hydrogenation reaction, affording good to excellent enantioselectivities (94– >99% ee [25] and 67–93% ee [28] respectively). The monodentate spiro phosphoramidite ligand (S)-**11a**

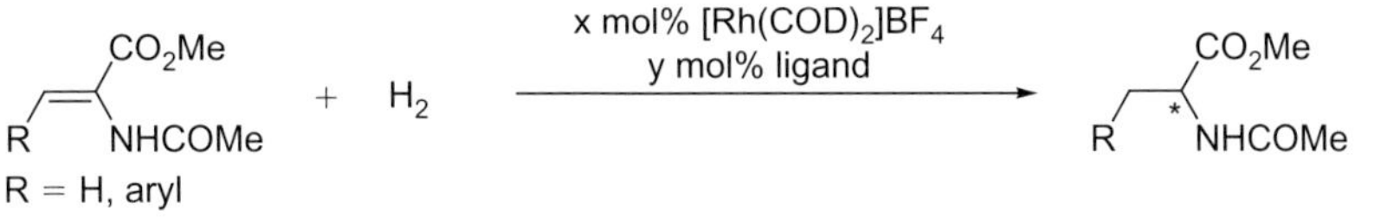

4.3 Asymmetric Hydrogenation | 145

Reaction (1): Ar/R-substituted enamide (R = H, aryl) with CO₂Me and NHCOMe groups + H₂, catalyzed by x mol% [Rh(COD)₂]BF₄ and y mol% ligand, gives the chiral α-amino ester product.

ligand	conditions	ee [%]
(1R,5R,6R)-**2**	x = y = 1, 1 atm, MeOH, 25 °C	94–99 (R)
(1R,5R,6R)-**18**	x = y = 0.2, 1 atm, acetone, rt, 1 h	94–>99 (R)
(1R,2R,1'R)-**21**	x = 1, y = 1.2, 5 atm, MeOH, rt, 24 h	67–93 (R)
(S)-**11a**	x = 1, y = 2.1, 1 atm, CH₂Cl₂, rt, 24 h	96–99.3 (S)
(S)-**10c**	x = 1, y = 2, 10 atm, toluene, rt, 6 h	97–99 (S)
(S)-**15**	x = 1, y = 2.1, 10 atm, toluene/CH₂Cl₂, rt, 3 h	92–99.3 (S)
(R)-**16a***	x = 1, y = 2.2, 1.7 atm, CH₂Cl₂, rt, 12 h	98–99.9 (S)

* [Rh(nbd)₂]BF₄ was used as precursor.

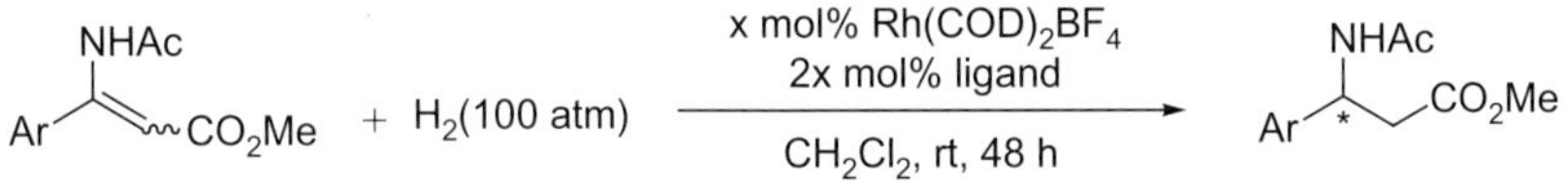

Reaction (2): Ar-substituted β-enamide with NHAc and CO₂Me + H₂ (100 atm), catalyzed by x mol% Rh(COD)₂BF₄ and 2x mol% ligand in CH₂Cl₂, rt, 48 h.

ligand	conditions	ee [%]
(R)-**11a**	x = 2	90–94 (S)
(S)-**10c**	x = 1	85–98 (R)

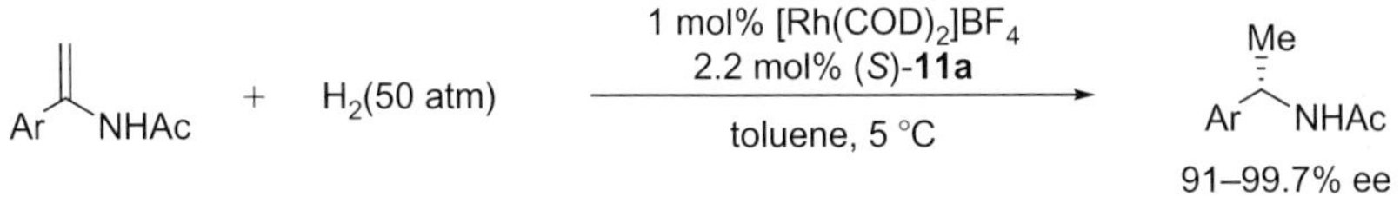

Reaction (3): Ar-NHAc enamide + H₂ (50 atm) with 1 mol% [Rh(COD)₂]BF₄, 2.2 mol% (S)-**11a** in toluene, 5 °C, gives chiral product with 91–99.7% ee.

Scheme 4.1 Rhodium-catalyzed asymmetric hydrogenation of enamides.

exhibited a comparable enantioselectivity in the rhodium-catalyzed asymmetric hydrogenation of α-dehydroamino acid derivatives (96–99.3% ee) [35]. In the design of chiral ligands applied in catalytic asymmetric reactions, most considerations focus on the steric effect. In contrast, the electronic effect of ligands has been less explored. A study of the electronic effects of spiro phosphonite ligands **10** was carried out for the hydrogenation reaction and clearly indicated that electron-rich ligands are beneficial to both enantioselectivity and reactivity [16a]. The chiral spiro monodentate phosphoramidite ligands (S)-**15**, with a spirobitetraline backbone, and (R)-**16a**, with a spirobixanthene backbone, also showed excellent enantioselectivities (92–99.3% ee [21] and 98–99.9% ee [22a], respectively) in the rhodium-catalyzed asymmetric hydrogenation of α-dehydroamino esters.

Chiral spiro phosphoramidite (R)-**11a** and phosphonite (S)-**10c** were efficient ligands in the rhodium-catalyzed hydrogenation of β-dehydroamino acid derivatives (Scheme 4.1, Reaction 2). Although a higher hydrogen pressure (100 atm) was required, both (R)-**11a** and (S)-**10c** gave high enantioselectivities (90–94% ee [35b] and 85–98% ee, [16a], respectively) in this transformation. Notably, the rhodium complexes of (R)-**11a** and (S)-**10c** can hydrogenate a (Z/E) mixture of β-(acylamino)acrylates (Z/E = 98 : 2–50 : 50) with excellent enantioselectivities. This is of practical importance because the β-(acylamino)acrylate substrates are normally prepared as a mixture of (Z)- and (E)-isomers.

In the asymmetric hydrogenation of α-arylethenyl acetamides many well-known chiral diphosphorous ligands, which are successful in the asymmetric hydrogenation of α- or β-dehydroamino acid derivatives, gave unsatisfying enantioselectivities. The spiro bis(phosphinite) ligand (1R,5R,6R)-**2** showed good enantioselectivity in the rhodium-catalyzed hydrogenation reaction of α-arylethenyl acetamides (86–90% ee) [36]. When chiral spiro phosphoramidite ligand (S)-**11a** was introduced into this reaction, the enantioselectivity was further improved to 91–99.7% ee (Scheme 4.1, Reaction 3) [37].

4.3.1.2 Hydrogenation of Enamines

In transition metal-catalyzed asymmetric hydrogenation of enamides, a substantial drawback is the requirement of an N-acyl group in the substrate. This N-acyl group is considered indispensable for the substrate to form a chelate complex with the metal of the catalyst in the transition state, giving good reactivity and enantioselectivity. The asymmetric hydrogenation of unprotected enamines is quite challenging because of the absence of a chelating N-acyl group in the substrates [38]. The phosphonite (S)-**10h**, which has a *tert*-butyl group on the P-atom, was an admirable ligand for the rhodium-catalyzed asymmetric hydrogenation of linear enamines (Scheme 4.2, Reaction 1) [16b]. In the presence of 2 mol.% I_2 and 20 mol.% acetic acid as additives, various (E)-1-(1-pyrrolidinyl)-1,2-diarylethenes were hydrogenated by the Rh/(S)-**10h** catalyst, yielding the corresponding tertiary amines with excellent enantioselectivities (up to 99.9% ee). By contrast, other ligands such as the diphosphine ligands BINAP and Josiphos, and the monophosphorus ligands MonoPhos and H-MOP, provided significantly low enantioselectivities (not exceeding 56% ee) under the same reaction conditions.

The asymmetric hydrogenation of cyclic N,N-dialkyl enamines provides a direct approach to the synthesis of optically active cyclic tertiary amines, which are essential structural units in natural products and drugs. Although the catalyst Rh/(R)-**10h** only afforded very poor enantioselectivity (20% ee) for the hydrogenation of cyclic enamines, the complex Ir/(R_a,S,S)-**11e** was found to be a highly efficient catalyst for this reaction [39]. In the presence of 5 mol.% I_2 as additive, the hydrogenation of cyclic enamines ran smoothly under ambient hydrogen pressure and afforded chiral cyclic tertiary amines with good to excellent enantioselectivities (72–97% ee) (Scheme 4.2, Reaction 2). This new reaction provided an efficient methodology for the synthesis of the natural product crispine A.

A recent study revealed that the iridium complex of (S_a,R,R)-**11e** complex was also a highly efficient catalyst for the asymmetric hydrogenation of enamines with an exocyclic double bond (Scheme 4.2, Reaction 3) [40]. In combination with iodine or potassium iodide, the catalyst Ir/(S_a,R,R)-**11e** accomplished the asymmetric hydrogenation of N-alkyl-1-alkylidenetetrahydroisoquinolines to produce N-alkyl-tetrahydroisoquinolines in high yields with up to 98% ee.

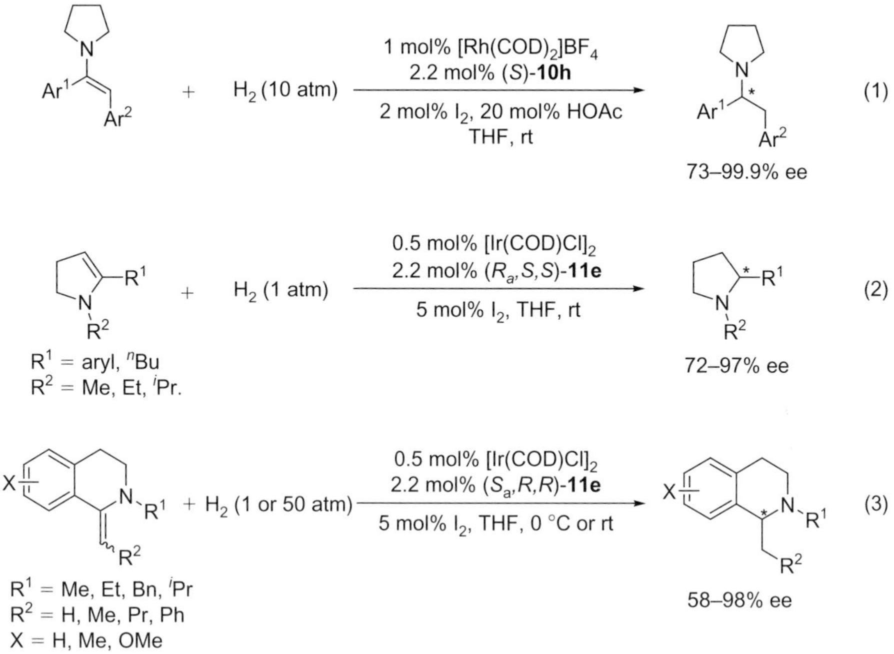

Scheme 4.2 Asymmetric hydrogenation of enamines.

4.3.1.3 Hydrogenation of α,β-Unsaturated Acids

Catalytic enantioselective hydrogenation of α,β-unsaturated carboxylic acids is a straightforward method for the synthesis of chiral carboxylic acids, which are important intermediates for the construction of biologically active compounds. In past decades, various ruthenium catalysts with chiral diphosphine ligands and rhodium catalysts with chiral phosphorous or nitrogen-containing ligands have been developed for the hydrogenation of different α,β-unsaturated carboxylic acids in high enantioselectivities [34]. However, the efficiency of the ruthenium and rhodium catalysts are highly substrate-dependent. For most reported catalysts, a high catalyst loading, drastic reaction conditions, or long reaction time is needed to achieve satisfying results. Thus, more efficient chiral catalysts are highly desirable in this useful reaction. The ruthenium diacetate complexes ligated by **14**, a highly rigid diphosphine with a large dihedral angle, were highly efficient in the asymmetric hydrogenation of tiglic acid [20]. The results illustrated in Scheme 4.3 clearly show that the larger bite angle is beneficial to high enantioselectivity [41]. The catalyst Ru/(R)-**14d** has a very high reactivity, and the S/C (the molar ratio of substrate to catalyst) of the reaction was as high as 10 000. The catalyst Ru/(R)-**14d** also exhibited high reactivity and enantioselectivity (90–97% ee) in the hydrogenation of cinnamic acid derivatives [20]. The ligand **14b** with *para*-methyl groups on the *P*-phenyl rings gave high yields (89–93%) and enantioselectivities (up to 95% ee) in the ruthenium-catalyzed hydrogenation of α-aryloxybutenoic acids [42].

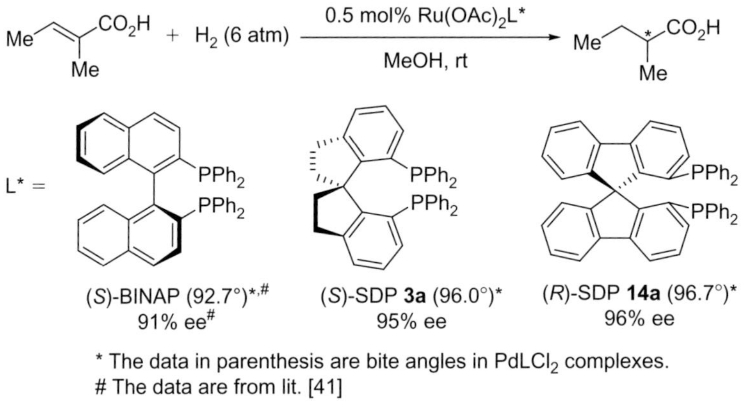
Me
CO2H
Me
+ H2 (6 atm)
0.5 mol% Ru(OAc)2L*
MeOH, rt
Me
*
CO2H
Me
L* =
PPh2
PPh2
PPh2
PPh2
PPh2
PPh2
(S)-BINAP (92.7°)*,#
91% ee#
(S)-SDP 3a (96.0°)*
95% ee
(R)-SDP 14a (96.7°)*
96% ee
* The data in parenthesis are bite angles in PdLCl2 complexes.
# The data are from lit. [41]

Scheme 4.3 Ruthenium-catalyzed asymmetric hydrogenation of tiglic acid.

Recently, a highly enantioselective iridium-catalyzed hydrogenation of α,β-unsaturated carboxylic acids was developed [13b]. Chiral cationic iridium complex **30** derived from spiro phosphine-oxazoline ligands SIPHOX (**7**) showed excellent reactivity and enantioselectivity in the hydrogenation of α-substituted cinnamic acid derivatives. The bulky aryl groups on the phosphorus atom of ligands were found to be critical for catalysts **30** to have high reactivity and enantioselectivity. For instance, the catalyst (S_a,S)-**30f**, which has 3,5-di-*tert*-butyl groups on the *P*-phenyl, exhibited extremely high reactivity (TOF, turnover frequency, 800 h^{-1}) and unprecedented enantioselectivity (94–99.2% ee) in the hydrogenation of α-substituted cinnamic acid derivatives (Scheme 4.4, Reaction 1). The iridium catalyst (S_a,S)-**30g** was also efficient for the enantioselective hydrogenation of various aliphatic α,β-unsaturated acids, affording the corresponding optically active acids with 90–99.4% ee (Scheme 4.4, Reaction 2).

Compared with the hydrogenation of α-aryl or α-alkyl substituted α,β-unsaturated acids, highly efficient catalysts for the asymmetric hydrogenation of α-aryloxy or α-alkoxy substituted α,β-unsaturated acids are limited. Although tremendous efforts have been devoted in this field, none of the present chiral catalysts gave acceptable results for the asymmetric hydrogenations of both α-aryloxy and α-alkyloxy substituted α,β-unsaturated acids. Moreover, the catalytic asymmetric hydrogenation of α-aryloxy substituted cinnamic acids has not been documented yet. The catalysts Ir/SIPHOX (**30**) showed a high efficiency in asymmetric hydrogenation of α-alkoxy and α-aryloxy substituted α,β-unsaturated acids (Scheme 4.4, Reactions 3 and 4) [43]. Under mild reaction conditions, a broad range of α-alkoxy and α-aryloxy substituted α,β-unsaturated acids were hydrogenated with exceptional enantioselectivities (up to 99.8% ee) at high reactivities (S/C up to 10 000). The key intermediates for the syntheses of the new blood pressure-lowering drug aliskiren and rhinovirus protease inhibitor rupintrivir were prepared using the asymmetric hydrogenation catalyzed by 0.01 mol.% (S_a)-**30i**. In view of extremely high activity and enantioselectivity, broad substrate scope,

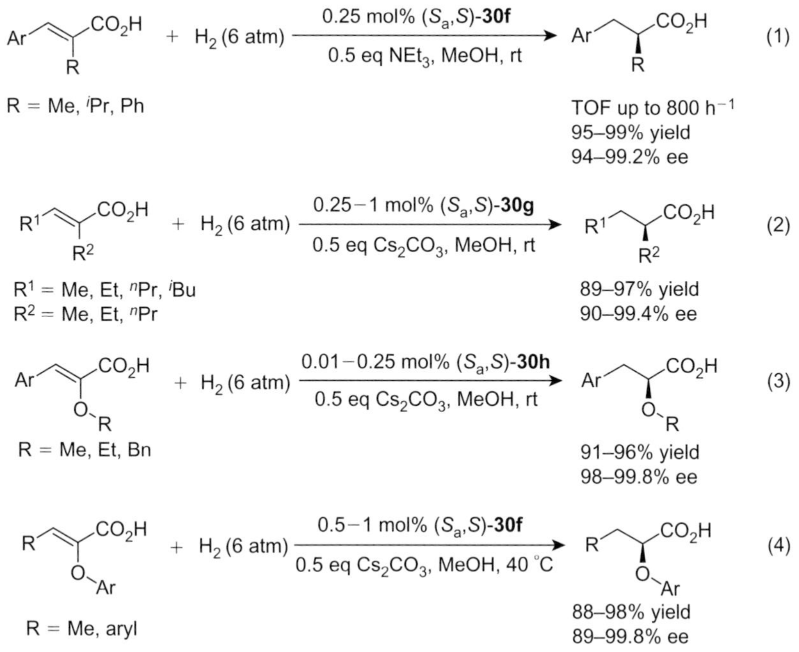

4.3 Asymmetric Hydrogenation | 149

$$\text{Ar}\underset{R}{\overset{}{=}}\text{CO}_2\text{H} + \text{H}_2\text{ (6 atm)} \xrightarrow[\text{0.5 eq NEt}_3\text{, MeOH, rt}]{0.25\text{ mol\% }(S_a,S)\text{-30f}} \text{Ar}\underset{R}{\overset{}{\text{-}}}\text{CO}_2\text{H} \quad (1)$$

R = Me, ^{i}Pr, Ph

TOF up to 800 h^{-1}
95–99% yield
94–99.2% ee

$$R^1\underset{R^2}{\overset{}{=}}\text{CO}_2\text{H} + \text{H}_2\text{ (6 atm)} \xrightarrow[\text{0.5 eq Cs}_2\text{CO}_3\text{, MeOH, rt}]{0.25\text{–}1\text{ mol\% }(S_a,S)\text{-30g}} R^1\underset{R^2}{\overset{}{\text{-}}}\text{CO}_2\text{H} \quad (2)$$

R^1 = Me, Et, ^{n}Pr, ^{i}Bu
R^2 = Me, Et, ^{n}Pr

89–97% yield
90–99.4% ee

$$\text{Ar}\underset{\text{O-R}}{\overset{}{=}}\text{CO}_2\text{H} + \text{H}_2\text{ (6 atm)} \xrightarrow[\text{0.5 eq Cs}_2\text{CO}_3\text{, MeOH, rt}]{0.01\text{–}0.25\text{ mol\% }(S_a,S)\text{-30h}} \text{Ar}\underset{\text{O-R}}{\overset{}{\text{-}}}\text{CO}_2\text{H} \quad (3)$$

R = Me, Et, Bn

91–96% yield
98–99.8% ee

$$R\underset{\text{O-Ar}}{\overset{}{=}}\text{CO}_2\text{H} + \text{H}_2\text{ (6 atm)} \xrightarrow[\text{0.5 eq Cs}_2\text{CO}_3\text{, MeOH, 40 °C}]{0.5\text{–}1\text{ mol\% }(S_a,S)\text{-30f}} R\underset{\text{O-Ar}}{\overset{}{\text{-}}}\text{CO}_2\text{H} \quad (4)$$

R = Me, aryl

88–98% yield
89–99.8% ee

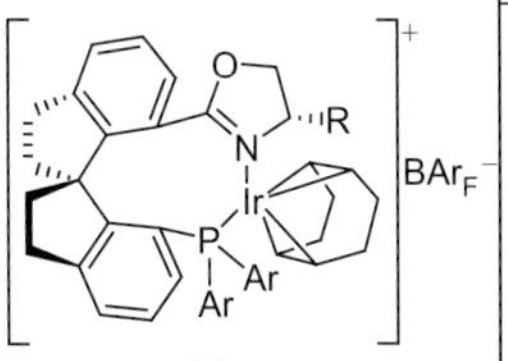

30	Ar, R
30a	C$_6$H$_5$, ^{i}Pr
30b	C$_6$H$_5$, Ph
30c	C$_6$H$_5$, Bn
30d	4-CH$_3$OC$_6$H$_4$, Bn
30e	3,5-(CH$_3$)$_2$C$_6$H$_3$, Bn
30f	3,5-^{t}Bu$_2$C$_6$H$_3$, Bn
30g	3,5-^{t}Bu$_2$C$_6$H$_3$, ^{i}Pr
30h	3,5-^{t}Bu$_2$C$_6$H$_3$, Me
30i	3,5-^{t}Bu$_2$C$_6$H$_3$, H

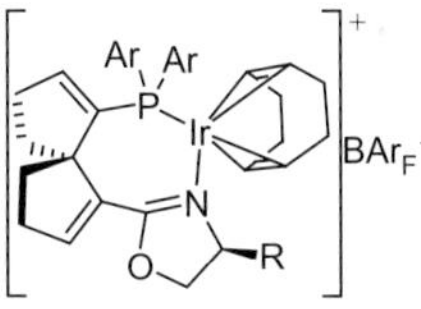

31	Ar, R
31a	Ph, Bn
31b	Ph, Ph
31c	Ph, ^{i}Pr
31d	Ph, ^{t}Bu
31e	Ph, ^{i}Bu
31f	2-CH$_3$C$_6$H$_4$, ^{i}Bu
31g	2-CH$_3$C$_6$H$_4$, Ph
31h	3,5-(CH$_3$)$_2$C$_6$H$_3$, ^{i}Bu

Scheme 4.4 Iridium-catalyzed asymmetric hydrogenation of α,β-unsaturated carboxylic acids. BAr$_F$ = tetrakis[3,5-bis(trifluoromethyl)phenyl]borate.

and mild reaction conditions, Ir/SIPHOX represents one of the most efficient catalysts for asymmetric hydrogenation of α,β-unsaturated carboxylic acids.

A single-crystal analysis of the structure of (S_a,S)-**30f** disclosed that the spiro phosphine-oxazoline ligand (S_a,S)-**7f** created a rigid and sterically hindered chiral environment ("chiral pocket") around the iridium center, which resulted in the high stability of the catalyst by preventing auto-aggregation. Moreover, the crowded "chiral pocket" of the catalyst minimized the number of possible transition states in the reaction and benefited chiral inducement. In a stereo-recognition model, a part of the spirobiindane backbone of ligand (S_a,S)-**7f** blocks one of the quadrants in front of the central iridium. One of the aryl groups on the phosphorus atom and the *tert*-butyl group on another P-phenyl hindered two other quadrants. This structure of catalyst directs the olefin double bond of α,β-unsaturated acid coordinated to iridium by its *Re* face, leading to the saturated carboxylic acid product with *S* configuration, which is consistent with the experimental results (Figure 4.9) [43].

Recently, Ding and coworkers [44] achieved the asymmetric hydrogenation of α-aryl-β-substituted acrylic acids by using iridium complexes of ligand SpinPHOX

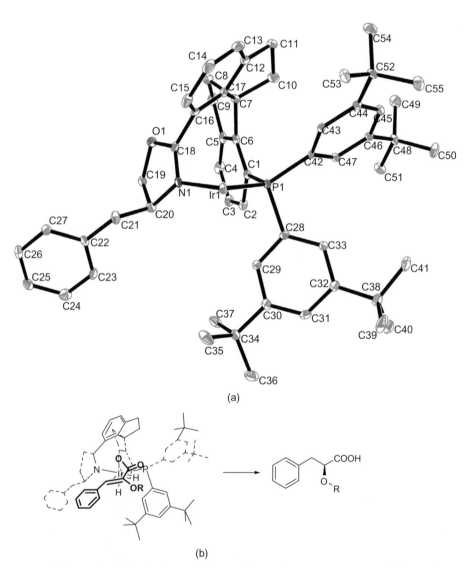

Figure 4.9 Single-crystal structure of (S_a,S)-**30f** (a) and stereo-recognition model (b).

(R_a,S)-**31g**. A series of chiral carboxylic acids have been obtained with excellent yields and up to 96% ee under optimal reaction conditions.

4.3.2
Hydrogenation of Ketones and Aldehydes

The enantioselective hydrogenation of carbonyl compounds catalyzed by well-defined transition metal complexes is an efficient synthetic protocol for producing optically active alcohols. One of the best catalysts for ketone hydrogenation is the $RuCl_2$[(diphosphine)(1,2-diamine)] complex, which was initially reported by Noyori [45]. The ruthenium complexes $RuCl_2$[(SDPs)(1,2-diamine)] (**32**) of chiral spirobiindane diphosphines SDP (**3**), were proven to be highly effective catalysts not only for the asymmetric hydrogenation of prochiral ketones, but also for the hydrogenation of racemic α-substituted ketones and aldehydes via dynamic kinetic resolution (DKR).

4.3.2.1 Hydrogenation of Simple Ketones
The catalyst (S_a,R,R)-**32d** having 3,5-dimethylphenyl groups was found to be one of the most effective catalysts in the asymmetric hydrogenation of prochiral ketones (Scheme 4.5, Reaction 1) [9a]. The catalytic activity of (S_a,R,R)-**32d** was extremely high, such that the hydrogenation of acetophenone can be carried out at a catalyst loading of 0.001 mol% (S/C = 100 000). Various simple ketones, including aromatic, heteroaromatic, and α,β-unsaturated ketones, can be hydrogenated by catalyst (S_a,R,R)-**32d**, providing the corresponding chiral secondary alcohols in excellent enantioselectivities (96–99.5% ee).

4.3.2.2 Hydrogenation of Racemic 2-Substituted Ketones via DKR
In the hydrogenation a racemic chiral ketone, if a catalyst can selectively hydrogenate one of the two enantiomers and the remaining enantiomer can be rapidly racemized under the reaction conditions, two enantiomers of ketone will be ultimately converted into chiral alcohol enantioselectively. This process is termed asymmetric hydrogenation via dynamic kinetic resolution (DKR) [46]. The $RuCl_2$[(diphosphine)(1,2-diamine)] complex-catalyzed asymmetric hydrogenation of ketones was performed under basic reaction conditions. Two enantiomers of α-substituted ketones may racemize rapidly in the presence of suitable bases, allowing DKR of the racemate by asymmetric hydrogenation. Based on this concept, several racemic α-substituted ketones have been enantioselectively hydrogenated to the corresponding alcohols [47]. By using catalysts $RuCl_2$[(SDPs)(1,2-diamine)] (**32**) several aryl-, aryloxy-, and amino-substituted ketones were hydrogenated via DKR in extremely high reactivity and enantioselectivity.

With catalyst (S_a,R,R)-**32d**, racemic α-arylcycloalkanones were hydrogenated via DKR to α-arylcycloalkanols in excellent *cis/trans* selectivity (>99 : 1) and enantioselectivity (up to 99.9% ee for the cis isomer) (Scheme 4.5, Reaction 2) [48]. Similarly, various *racemic* α-aryloxydialkyl ketones underwent the (S_a,R,R)-**32d**-catalyzed hydrogenation via DKR to produce β-aryloxy alcohols with perfect *anti/syn*

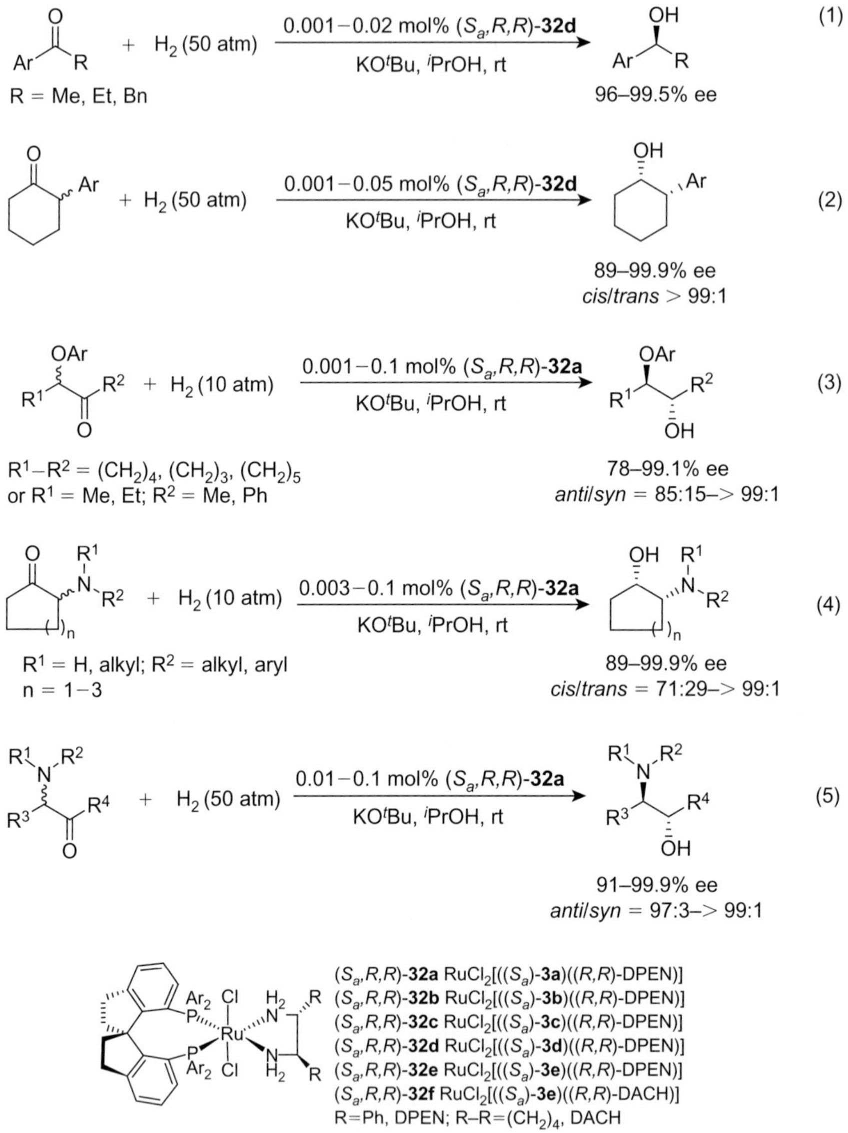

4 Chiral Spiro Ligands

$$Ar\underset{R=Me, Et, Bn}{\overset{O}{\underset{\|}{C}}}R + H_2 \text{ (50 atm)} \xrightarrow[\text{KO}^tBu, ^iPrOH, rt]{0.001-0.02 \text{ mol\% }(S_a,R,R)\text{-32d}} Ar\underset{96-99.5\% \text{ ee}}{\overset{OH}{\underset{|}{C}}}R \quad (1)$$

$$\text{cyclohexanone-Ar} + H_2 \text{ (50 atm)} \xrightarrow[\text{KO}^tBu, ^iPrOH, rt]{0.001-0.05 \text{ mol\% }(S_a,R,R)\text{-32d}} \text{cyclohexanol-Ar} \quad (2)$$

89–99.9% ee
cis/trans > 99:1

$$R^1\underset{O}{\overset{OAr}{\underset{\|}{C}}}R^2 + H_2 \text{ (10 atm)} \xrightarrow[\text{KO}^tBu, ^iPrOH, rt]{0.001-0.1 \text{ mol\% }(S_a,R,R)\text{-32a}} R^1\underset{OH}{\overset{OAr}{\underset{|}{C}}}R^2 \quad (3)$$

$R^1-R^2 = (CH_2)_4, (CH_2)_3, (CH_2)_5$
or $R^1 = Me, Et; R^2 = Me, Ph$

78–99.1% ee
anti/syn = 85:15–> 99:1

$$\text{cyclic ketone with } NR^1R^2 + H_2 \text{ (10 atm)} \xrightarrow[\text{KO}^tBu, ^iPrOH, rt]{0.003-0.1 \text{ mol\% }(S_a,R,R)\text{-32a}} \text{cyclic alcohol} \quad (4)$$

$R^1 = H$, alkyl; $R^2 =$ alkyl, aryl
n = 1–3

89–99.9% ee
cis/trans = 71:29–> 99:1

$$R^3\underset{O}{\overset{R^1\underset{|}{N}R^2}{\underset{\|}{C}}}R^4 + H_2 \text{ (50 atm)} \xrightarrow[\text{KO}^tBu, ^iPrOH, rt]{0.01-0.1 \text{ mol\% }(S_a,R,R)\text{-32a}} R^3\underset{OH}{\overset{R^1\underset{|}{N}R^2}{\underset{|}{C}}}R^4 \quad (5)$$

91–99.9% ee
anti/syn = 97:3–> 99:1

(S_a,R,R)-**32a** RuCl$_2$[((S_a)-**3a**)((R,R)-DPEN)]
(S_a,R,R)-**32b** RuCl$_2$[((S_a)-**3b**)((R,R)-DPEN)]
(S_a,R,R)-**32c** RuCl$_2$[((S_a)-**3c**)((R,R)-DPEN)]
(S_a,R,R)-**32d** RuCl$_2$[((S_a)-**3d**)((R,R)-DPEN)]
(S_a,R,R)-**32e** RuCl$_2$[((S_a)-**3e**)((R,R)-DPEN)]
(S_a,R,R)-**32f** RuCl$_2$[((S_a)-**3e**)((R,R)-DACH)]
R=Ph, DPEN; R–R=(CH$_2$)$_4$, DACH

Scheme 4.5 Ruthenium-catalyzed hydrogenation of ketones.

selectivity (>99 : 1) and enantioselectivity (up to 99.1% ee for *anti* isomer) (Scheme 4.5, Reaction 3) [49]. A new type of non-steroidal glucocorticoid modulator, 1-aryl-1-indazolyloxypropan-2-amine, has been synthesized according to this protocol.

The asymmetric hydrogenation of racemic α-amino ketones is one of the most efficient methods for synthesizing chiral amino alcohols. However, the hydrogenation of α-amino ketones via DKR has achieved very limited progress. The difficulty for this reaction presumably arose from the amino group, which coordinated to the metal center of the catalyst, resulting in a low catalytic activity. By using catalyst (S_a,R,R)-**32a**, which has a rigid and bulky spirobiindane diphosphine, highly enantioselective hydrogenation of racemic α-aminocycloalkanones via DKR was successfully realized (Scheme 4.5, Reaction 4) [50]. The enantioselectivities (up to 99.9% ee) and diastereoselectivities (*cis/trans* up to >99 : 1) are excellent, with a S/C ratio up to 30 000.

The catalyst (S_a,R,R)-**32a** was efficient enough to accomplish the highly selective hydrogenation of the conformationally flexible acyclic α-amino aliphatic ketones, which are challenging substrates to hydrogenate (Scheme 4.5, Reaction 5) [51]. In the asymmetric hydrogenation of racemic acyclic α-N-alkyl/arylamino aliphatic ketones with an unprotected α-amino group, the catalyst (S_a,R,R)-**32a** also showed high enantioselectivity and diastereoselectivity [52]. This protocol provided a practical approach to the synthesis of all four isomers of the piperidine alkaloid conhydrine.

4.3.2.3 DKR Hydrogenation of Racemic 2-Substituted Aldehydes

In the catalytic asymmetric hydrogenation of prochiral ketones, at least one new stereogenic center was generated. However, no new stereogenic center was generated in the hydrogenation of α-branched aldehydes, which makes enantiocontrol of the reaction extremely difficult. Thus, the asymmetric hydrogenation of α-branched aldehydes, which provides a potential approach towards synthetically important chiral primary alcohols, remains a challenge to chemists. The complex (S_a,R,R)-**32f** was found to be competent catalyst for the asymmetric hydrogenation of racemic α-branched aldehydes via DKR, providing practical access to chiral primary alcohols in high enantiomeric excess (up to 96% ee) (Scheme 4.6, Reaction 1) [53]. An important leukotriene receptor antagonist and lipoxygenase inhibitor BAY X 1005 were synthesized by this protocol.

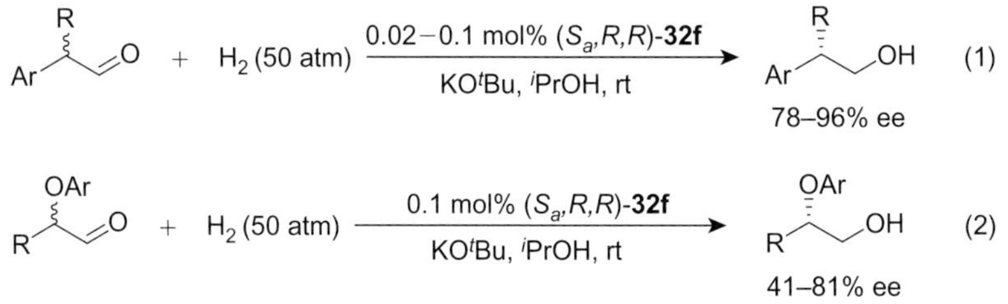

Scheme 4.6 Ruthenium-catalyzed dynamic kinetic resolution (DKR) hydrogenation of α-branched aldehydes.

By using the same catalysts, the enantioselective hydrogenation of racemic α-aryloxy aldehydes via DKR has also been accomplished with moderate to good enantioselectivity (Scheme 4.6, Reaction 2) [54].

4.3.3
Hydrogenation of Imines

The asymmetric hydrogenation of imines provides an efficient and direct route for the synthesis of chiral amines. Although great efforts have been made in recent decades, catalytic asymmetric hydrogenation of imines remains a challenge in modern synthesis. The Ir/SIPHOX complex (S_a,S)-**30e** can catalyze the hydrogenation of acyclic N-aryl ketimines under ambient pressure with excellent enantioselectivities (up to 97% ee) and full conversions (Scheme 4.7, Reaction 1) [13a]. Thanks to the high rigidity and bulk of SIPHOX, the Ir/SIPHOX catalysts efficiently prevents the formation of inactive trimers and exhibit unprecedented stability under a hydrogen atmosphere.

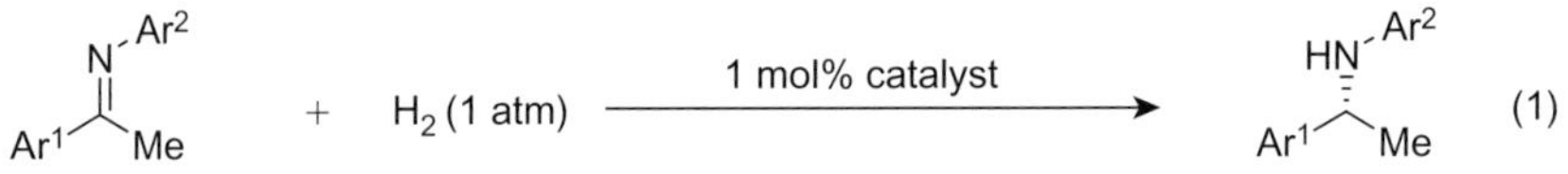

catalyst	conditions	ee [%]
(S_a,S)-**30e**	^{t}BuOMe, 10 °C, 20 h	90–97 (R)
(R_a,S)-**31c**	CH_2Cl_2, 10 °C, 8 h	88–95 (R)

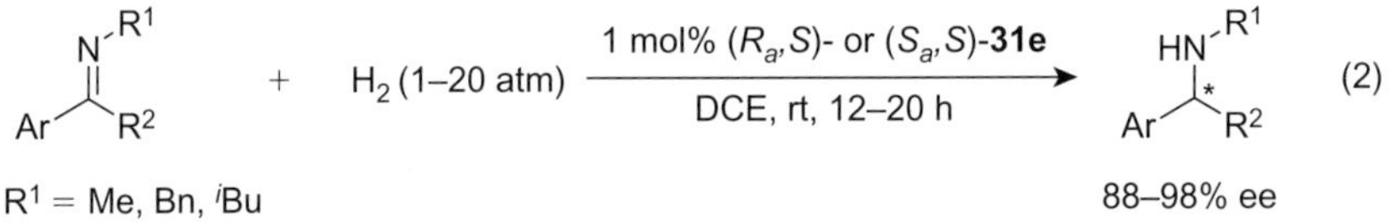

R^1 = Me, Bn, ^{i}Bu

88–98% ee

Scheme 4.7 Iridium-catalyzed asymmetric hydrogenation of ketimines.

Recently, Ding [29] communicated a highly enantioselective hydrogenation of ketimines by using Ir/SpinPHOX catalysts **31**. Under mild reaction conditions, the hydrogenation of N-aryl and N-alkyl ketimine were achieved with good to excellent enantioselectivities (up to 98% ee) (Scheme 4.7). The hydrogenation reaction has been applied successfully in the asymmetric synthesis of the chiral antidepressant drug sertraline [(+)-cis-(1S,4S)-1-methylamino-4-(3,4-dichlorophenyl)- tetralin].

4.3.4
Hydrogenation of 2-Substituted Quinolines

The enantioselective hydrogenation of heteroaromatic compounds, which is a convenient method for the preparation of enantiomerically pure heterocycles, remains a challenging task. Since Zhou and coworkers found that the iridium-diphosphine complexes were capable of catalyzing hydrogenation of α-substituted quinolines in good yields with high enantioselectivities, various efficient catalysts

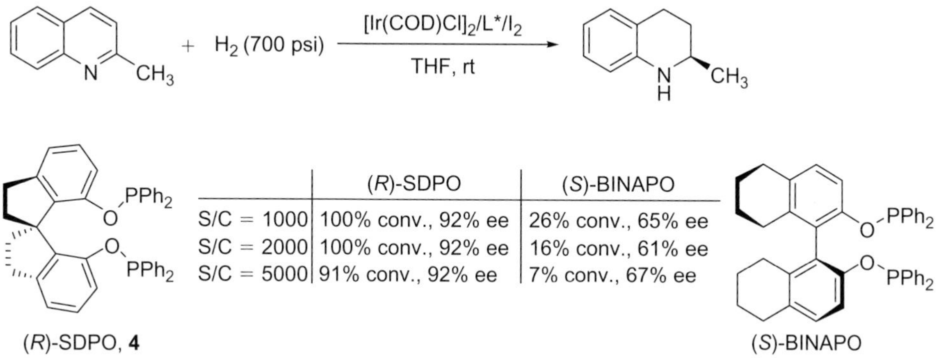

4.4 Asymmetric Carbon–Carbon Bond Forming Reaction

Scheme 4.8

	(R)-SDPO	(S)-BINAPO
S/C = 1000	100% conv., 92% ee	26% conv., 65% ee
S/C = 2000	100% conv., 92% ee	16% conv., 61% ee
S/C = 5000	91% conv., 92% ee	7% conv., 67% ee

have been developed [55]. However, all iridium-diphosphine catalysts suffered from low efficiency, as evidenced by the fact that good results could only be obtained at a low S/C ratio. Recently, the chiral spiro diphosphinite (R)-4 (SPDO) was reported to be a highly efficient ligand for iridium-catalyzed asymmetric hydrogenation of quinolines. The catalyst Ir/(R)-4 has a very high reactivity (S/C up to 5000) and enantioselectivity (up to 92% ee) (Scheme 4.8) [10]. In comparison, the ligand (S)-H$_8$-BINAPO with a binaphthyl backbone displayed significantly lower reactivity and enantioselectivity.

4.4
Asymmetric Carbon–Carbon Bond Forming Reaction

Transition metal-catalyzed asymmetric carbon–carbon bond-forming reactions are the essence of organic synthesis. Chiral spiro monodentate and bidentate ligands with phosphorous or nitrogen as coordination atoms have been successfully used in various asymmetric carbon–carbon bond forming reactions. In most investigated reactions, the spiro ligands showed superior chiral inducements than ligands with other scaffolds.

4.4.1
Rhodium-Catalyzed Arylation of Carbonyl Compounds and Imines

The enantioselective addition of aryl organometallic reagents to aldehydes, ketones, and imines remains an extremely attractive topic because the products, chiral alcohols or amines, are important intermediates for the synthesis of biologically and pharmaceutically active compounds [56]. Although significant progress has been achieved in this field, the requirement for the use of toxic, moisture sensitive, or non-commercially available organometallic reagents for high enantioselectivity without doubt limits their application. The use of arylboronic acids, which are less toxic, stable towards air and moisture, and tolerant towards various functional groups, may greatly advance the applicability of this important transformation. However, the application of arylboronic acids in the catalytic asymmetric addition reactions achieved only limited progress despite considerable efforts. Miyaura and coworkers initially reported the enantioselective rhodium-catalyzed addition of phenylboronic acid to naphthaldehyde, giving naphthylphenylmethanol in 78% yield with 41% ee [57]. The asymmetric addition of aryl boronic acids to imines is rare [58]. Moreover, the catalytic enantioselective addition of arylboronic acids to ketones to produce chiral tertiary alcohols was even more difficult [59].

The chiral spiro phosphite (S)-**9h** was found to be an efficient ligand for the asymmetric rhodium-catalyzed addition of arylboronic acids to aldehydes, providing diarylmethanols in excellent yields (88–98%) with up to 87% ee (Scheme 4.9, Reaction 1) [15c]. The related phosphite ligand with binaphthyl backbone was less efficient (59% yield, 49% ee) than the spiro phosphite ligands **9**, indicating that rigid spiro phosphite ligands with a large dihedral angle benefit the formation of an effective catalyst with high activity and enantioselectivity.

The Rh/ShiPs catalysts also exhibited high enantioselectivity for the addition of arylboronic acids to N-tosylarylimines (Scheme 4.9, Reaction 2) [60]. The reaction proceeded in aqueous toluene to give diarylmethylamines in good yields (56–85%) with up to 96% ee. This result expands the scope of application of arylboronic acids as aryl sources and provides a practical entry to the synthesis of optically active diarylmethylamines.

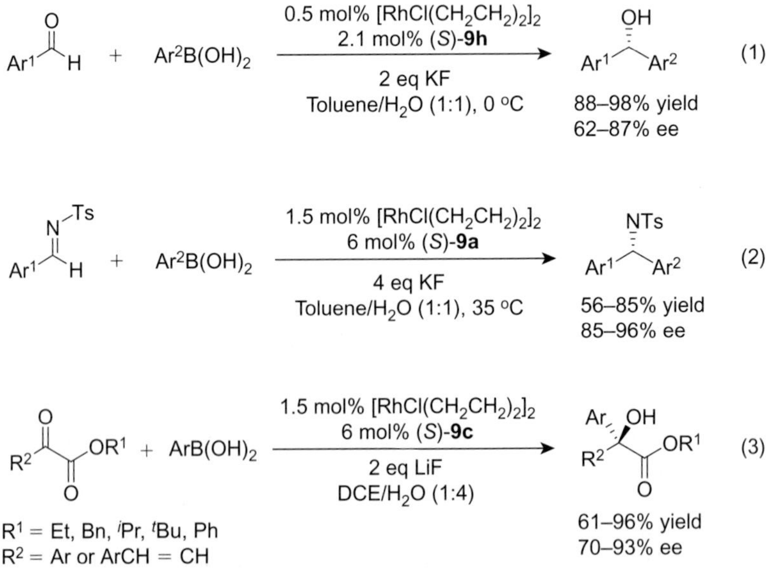

Scheme 4.9 Rhodium-catalyzed asymmetric arylation reactions.

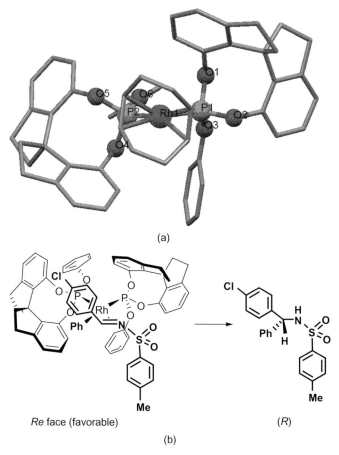

Figure 4.10 Single-crystal structure of Rh/(S)-**9a** (a) and stereo-recognition model (b).

An X-ray analysis of the crystal structure of Rh/(S)-**9a** showed that two coordinating spiro phosphite ligands (S)-**9a** created an effective asymmetric environment around the rhodium (Figure 4.10). Two phenyl groups of the phosphite ligands blocked the back-side of the complex, and both arylboronic acid and, consequently, N-tosylarylimines could not access and coordinate with rhodium from the back-side. It can be seen from the model (Figure 4.10b) that the transfer of the phenyl group to the Re face of N-tosylarylimines to generate the (R)-product is much more favorable, which is consistent with the experimental results.

By using catalyst Rh/(S)-**9c**, the enantioselective addition of arylboronic acids to α-ketoesters was realized, yielding chiral tertiary α-hydroxyesters, which are desirable chiral intermediates for the synthesis of antagonists of muscarinic receptors, in good yields and high enantiomeric excesses (up to 93% ee) (Scheme 4.9, Reaction 3) [61].

4.4.2
Palladium-Catalyzed Umpolung Allylation of Aldehydes

Compared with well-established electrophilic π-allylpalladium chemistry, the catalytic asymmetric reaction via umpolung of π-allylpalladium has received very limited exploration [62]. In 2004, Zanoni et al. [63] reported the palladium-catalyzed asymmetric umpolung allylation of aldehydes with allylic acetate in the presence of Et$_2$Zn as reductive agent and obtained moderate enantioselectivities (up to 70% ee). Recently, the palladium-catalyzed asymmetric umpolung allylation of aldehydes by directly using stable and readily available allylic alcohols as allylic reagents was realized. The chiral homoallylic alcohol products were obtained in good enantioselectivity (up to 83% ee) by using chiral spiro monophosphine ligand (R)-**8a** (Scheme 4.10) [14a]. The fact that the reaction with the simplest allylic alcohol, propan-2-en-1-ol, could give a good enantioselectivity (80% ee) showed that a chiral palladium intermediate must be involved in the chiral discriminating step.

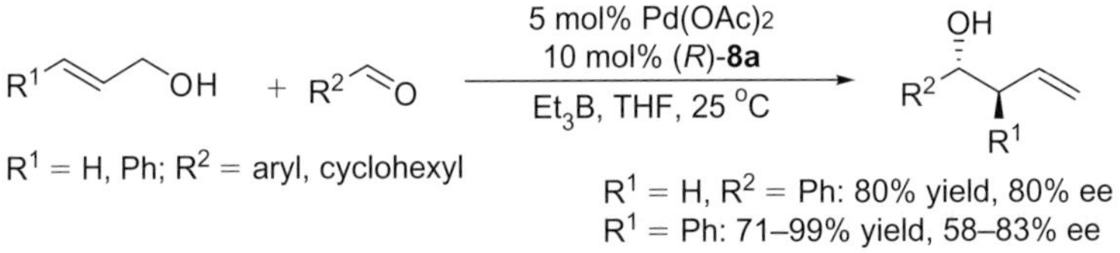

R^1 = H, Ph; R^2 = aryl, cyclohexyl

5 mol% Pd(OAc)$_2$
10 mol% (R)-**8a**
Et$_3$B, THF, 25 °C

R^1 = H, R^2 = Ph: 80% yield, 80% ee
R^1 = Ph: 71–99% yield, 58–83% ee

Scheme 4.10

4.4.3
Copper-Catalyzed Conjugate Addition Reaction

The transition metal-catalyzed conjugate addition of organometallic reagents to α,β-unsaturated carbonyl compounds is one of the most useful processes for carbon–carbon bond formation in organic synthesis [64]. Sasai et al. [65] tested spiro bis(isoxazoline) ligand (**23**) in the copper-catalyzed conjugate addition reaction of diisopropylzinc with cyclohexenone and obtained modest enantioselectivity (49% ee). When the spiro phosphoramidite ligands (R_a,R,R)-**11e** was introduced into the copper-catalyzed conjugate addition of Et$_2$Zn to cyclic enones high yield (88%) and excellent enantioselectivity (98% ee) were obtained (Scheme 4.11) [17]. The analog ligand (S_a,S,S)-**16d** with a spirobixanthene framework afforded the same level of enantioselectivity (Scheme 4.11) [22b].

4.4.4
Copper-Catalyzed Ring-Opening Reaction with Grignard Reagents

The enantioselective ring-opening reaction of meso-oxabicyclic alkenes with organometallic reagents is a powerful method to construct cyclic compounds with multiple chiral centers while simultaneously generating a new C–C bond [66]. This

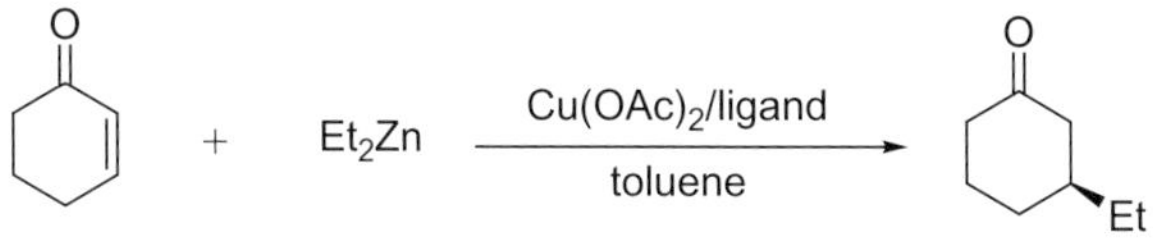

Scheme 4.11

ligand	catalyst loading	T [°C]	yield [%]	ee [%]
(R_a,R,R)-**11e**	1 mol%	0	88	98
(S_a,S,S)-**16d**	3 mol%	−20	92	99

asymmetric reaction has been accomplished successfully with various organometallic reagents, such as organozinc, organoaluminium, and organoboron in the presence of palladium, nickel, rhodium, copper, and zirconium catalysts. However, Grignard reagents, the most accessible organometallic reagent, have seldom been used as nucleophiles in asymmetric ring-opening reactions. There is only one example of ring-opening of meso-oxabicyclic alkenes with Grignard reagents, that is, the reaction of 2,4-dimethyl-3-(benzyloxy)-8-oxabicyclo-6-octene and ethylmagnesium bromide catalyzed by chiral zirconium catalyst to give the ring-opening product in 27% yield with 48% ee [67]. The spiro ligand (S_a,S,S)-**11e** showed a high enantioselectivity (up to 88% ee) in the copper-catalyzed ring-opening reaction of meso-oxabicyclic alkenes with Grignard reagents [68]. Further study found that the copper complexes of electron-rich chiral phosphine ligands (R)-**8e** with a 3,5-dimethyl-4-methoxyphenyl moiety is a more efficient catalyst for this reaction (Scheme 4.12) [14b]. This catalyst has a broad substrate generality, excellent enantioselectivity (80–99.6% ee) and *anti*-selectivity (>99 : 1). The turnover number (TON) of the reaction reached 9000, which represents one of the highest TONs reported in an asymmetric carbon–carbon bond formation reaction employing Grignard reagents.

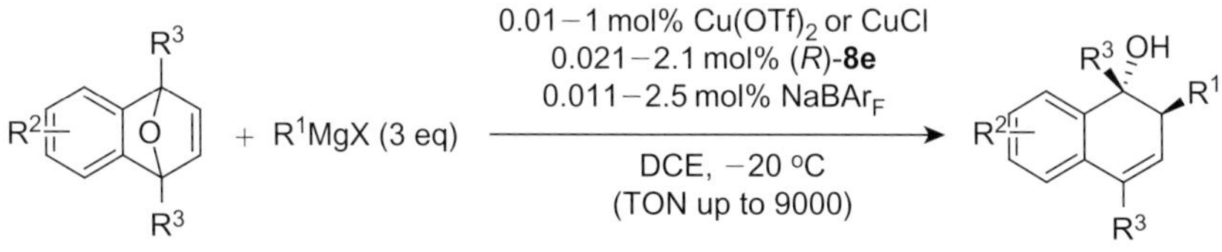

R^1 = Me, Et, ^{i}Pr, ^{n}Bu, ^{i}Bu, ^{t}Bu, Bn
R^2 = H, Me, OMe, Br
R^3 = H, Me

71–95% yield
trans/cis > 99:1
80–99.6% ee

Scheme 4.12

4.4.5
Nickel-Catalyzed Three-Component Coupling Reaction

Nickel-catalyzed multicomponent coupling reactions are an efficient protocol for carbon–carbon bond formation and have drawn increasing attention [69].

Although great progress has been achieved in this field, successful examples of high enantioselective nickel-catalyzed multicomponent coupling reactions were quite few [70]. The nickel-catalyzed asymmetric intermolecular reductive coupling of 1,3-dienes and aldehydes with Et_2Zn as a reducing reagent was realized by using 6,6′-substituted phosphoramidite ligand (R)-**13d** (Scheme 4.13, Reaction 1) [19]. The coupling products, chiral bishomoallylic alcohols, were produced in high yields (78–99%) with excellent diastereoselectivity (*anti/syn* = 98 : 2–99 : 1) and enantioselectivity (up to 96% ee).

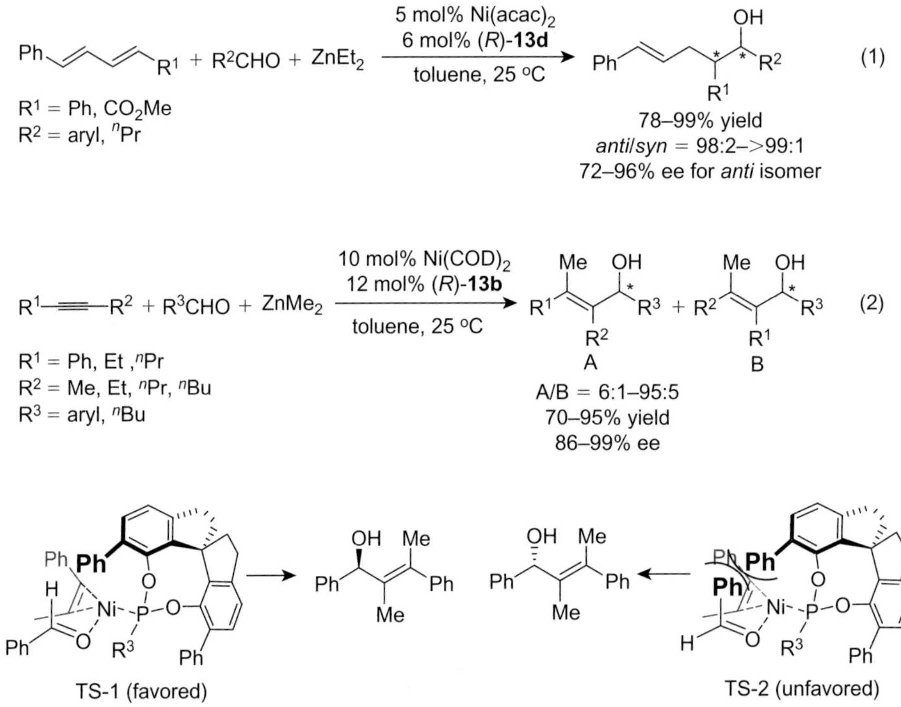

Scheme 4.13 Nickel-catalyzed asymmetric three-component coupling reactions.

The alkylative coupling reaction includes a transfer of an alkyl group instead of a hydrogen atom from organometallic reagents to the alkyne or diene and, therefore, is a more atom-efficient process than its reductive counterpart. The alkylative coupling reaction produces allylic alcohols with a tetrasubstituted olefin, which are hard to assess but are useful building blocks for bioactive compounds. The nickel complex of 6,6′-disubstituted spiro phosphoramidite ligand (R)-**13b** was demonstrated to be a efficient catalyst for the asymmetric alkylative coupling reaction of alkyne, aldehyde, and Me_2Zn (Scheme 4.13, Reaction 2) [71]. Various chiral allylic alcohols with tetrasubstituted olefin moiety were produced in high yields (70–95%), high regioselectivity (6 : 1–95 : 5), and excellent enantioselectivity (up to 99% ee). The groups on the 6,6′-position of spiro phosphoramidite ligand (R)-**13b** are crucial for obtaining high enantioselectivity. It can be seen from the proposed transition state model, the phenyl groups on the 6,6′-position of ligand directed the nickel-activated alkyne to approach aldehyde from its *Si* face (model TS-1), leading to the formation of allylic alcohols with *R* configuration (Scheme 4.13).

4.4.6
Nickel-Catalyzed Hydrovinylation Reaction

Nickel-catalyzed asymmetric hydrovinylation of vinylarenes is an important carbon–carbon bond-forming reaction, and impressive progress has been achieved in this field [72]. However, most investigations focused on the hydrovinylation of vinylarenes, yielding chiral 3-arylbut-1-enes with a chiral tertiary carbon center. By using spiro phosphoramidite ligand (S_a,R,R)-**11e**, the asymmetric hydrovinylation of α-alkyl vinylarenes was accomplished under ambient pressure to afford the hydrovinylation products bearing a chiral all-carbon quaternary center. The reaction exhibited high yield (76–96%), good chemoselectivity (80–89%), and good to excellent enantioselectivity (70–99% ee) (Scheme 4.14) [73]. This reaction provided a new methodology for the enantioselective construction of a chiral all-carbon quaternary center from very simple feedstock materials. The hydrovinylation products of this reaction are potentially useful intermediates for the synthesis of versatile optically active molecules such as chiral carboxylic acids, aldehydes, and alcohols.

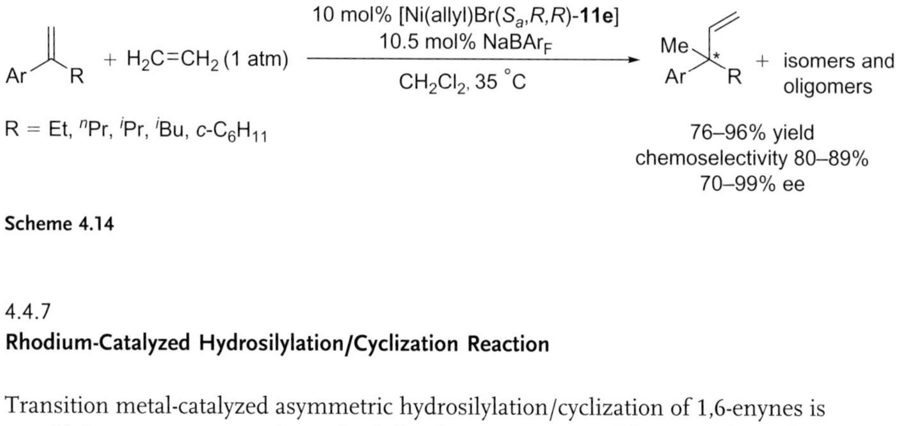

$$\text{Ar}\overset{\text{O}}{\underset{\text{R}}{\|}} + H_2C=CH_2 \,(1\text{ atm}) \xrightarrow[\text{CH}_2\text{Cl}_2,\,35\,°\text{C}]{\substack{10\text{ mol\% [Ni(allyl)Br}(S_a,R,R)\text{-}\mathbf{11e}] \\ 10.5\text{ mol\% NaBAr}_F}} \overset{\text{Me}}{\underset{\text{Ar}\quad\text{R}}{\diagdown\!\!\diagup^*}} + \text{ isomers and oligomers}$$

R = Et, ^{n}Pr, ^{i}Pr, ^{i}Bu, c-C$_6$H$_{11}$

76–96% yield
chemoselectivity 80–89%
70–99% ee

Scheme 4.14

4.4.7
Rhodium-Catalyzed Hydrosilylation/Cyclization Reaction

Transition metal-catalyzed asymmetric hydrosilylation/cyclization of 1,6-enynes is an efficient, atom-economic method for the preparation of functionalized five-membered carbocycles and heterocycles [74]. In this tandem reaction not only a cyclic carbon–carbon bond but also a carbon–silica bond was formed. The rhodium complex of spiro diphosphine (*R*)-**3a** was proven to be an efficient catalyst for this reaction and afforded unprecedented high enantioselectivity (89–99% ee) for a wide range of substrates (Scheme 4.15, Reaction 1) [75]. This highly enantioselective reaction provides a facile access to optically active silyl-alkylidene cyclopentane and pyrrolidine derivatives, along with other functionalized carbocyclic

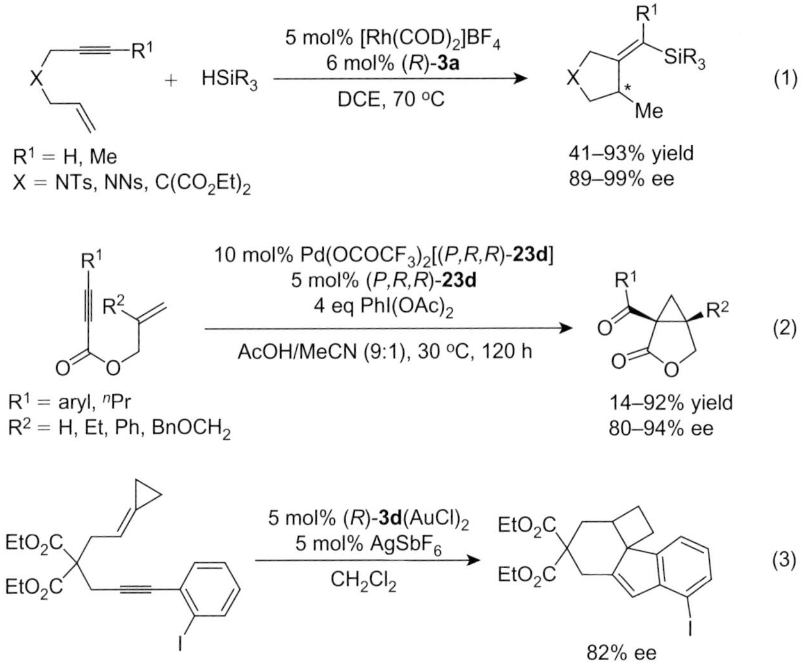

Scheme 4.15 Asymmetric cyclization reactions.

and heterocyclic compounds after subsequent appropriate transformations. Other commercially available chelating diphosphine ligands such as BINAP, SYNPHOS, JOSIPHOS, and Me-DuPHOS gave lower enantioselectivity under the same reaction conditions, which indicates that the rigid spiro backbone of SDP ligands is the key factor for achieving high enantioselectivity in the silyl-cyclization of 1,6-enynes.

4.4.8
Palladium-Catalyzed Asymmetric Oxidative Cyclization

Compared to the impressive development of enantioselective reactions through the Pd^0/Pd^{II} catalytic cycle, only minimal attention has been paid to the enantioselective reactions involving the Pd^{II}/Pd^{IV} catalytic cycle [76]. Recently, Sasai and coworkers [77] developed an oxidative cyclization of enynes using spiro bis(isoxazoline) ligand (P,R,R)-**23d** and hypervalent iodine reagent as oxidant, yielding bicyclic lactone products with up to 94% ee (Scheme 4.15, Reaction 2). This is an excellent example of asymmetric reactions that proceed through a Pd^{II}/Pd^{IV} catalytic cycle. Notably, no enantioselectivity was observed in the reactions with other known chiral ligands such as (R)-BINAP, (−)-sparteine, (S,S)-t-Bu-BOX, and (S,S)-i-Pr-BOXAX. The high stability of ligand **23d** under such oxidative and acidic conditions proved to be crucial for this asymmetric Pd^{II}/Pd^{IV} catalysis.

4.4.9
Gold-Catalyzed Ring Expanding Cycloisomerization

Toste et al. [78] tested chiral spiro diphosphine ligands **3** in a gold-catalyzed ring-expanding cycloisomerization reaction (Scheme 4.15, Reaction 3). They found that the reaction catalyzed by (R)-**3d**(AuCl)$_2$/AgSbF$_6$ afforded multicyclic product with 82% ee. However, other diphosphine ligands gave the same product with significantly lower enantioselectivities.

4.5
Asymmetric Carbon–Heteroatom Bond Forming Reaction

4.5.1
Palladium-Catalyzed Hydrosilylation

The asymmetric palladium-catalyzed hydrosilylation of olefins with trichlorosilane is a highly potent approach to chiral silanes [79]. Despite the palladium-catalyzed hydrosilylation of alkenes being a facile method for the synthesis of chiral alcohols upon oxidation, only a few chiral ligands, mostly monophosphorus ligands, have been applied in this reaction. The spiro phosphoramidite ligand (R_a,R,R)-**11e** was found to be very efficient for the palladium-catalyzed asymmetric hydrosilylation of styrene derivatives, providing 1-aryl-1-silyalkanes in high yields (75–99%) with excellent enantioselectivity (up to 99.1% ee) (Scheme 4.16) [80].

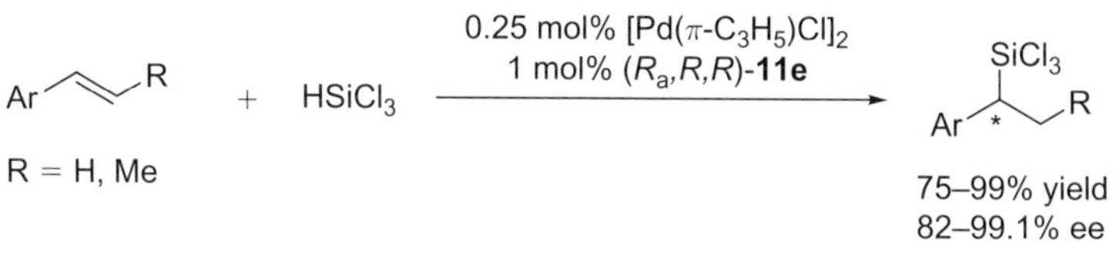

Scheme 4.16

4.5.2
Palladium-Catalyzed Wacker-Type Oxidative Cyclization Reaction

The chiral spiro bis(isoxazole) ligand (M,S,S)-**23d** and hybrid isoxazole-isoxazoline ligands (R_a,S)-**28d** and (R_a,S)-**28e** showed high enantioselectivities in the palladium-catalyzed Wacker-type cyclization of alkenyl alcohols. The tandem cyclization of an alkenyl alcohol via an oxy-palladation gave a unique bicyclic ether compound with up to 97% ee (Scheme 4.17) [81]. The rigid spiro[4.4]nonane skeleton of the bis(isoxazole) ligand (M,S,S)-**23d** and its analogs was regarded as a crucial fact in preventing β-elimination and thus making the tandem reaction possible.

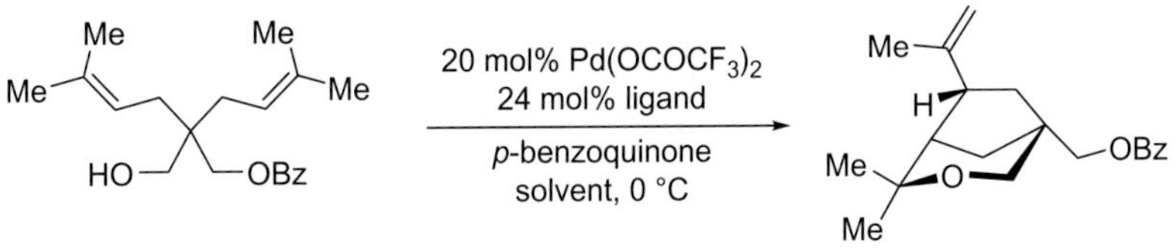

Scheme 4.17

ligand	solvent	yield [%]	ee [%]
(M,S,S)-23d	CH$_2$Cl$_2$	68	95
(R$_a$,S)-28d	CH$_2$Cl$_2$/MeOH (1:1)	59	97
(R$_a$,S)-28e	CH$_2$Cl$_2$/MeOH (1:1)	74	95

4.5.3 Copper-Catalyzed Carbene Insertion into X–H Bonds

The catalytic insertion of α-diazocarbonyl compounds into X–H (X = C, N, O, S) bonds is a very powerful organic transformation [82]. Remarkable advances have been made in the development of methodology for catalytic asymmetric diazo insertion into C–H bonds, but only limited success has been achieved for asymmetric diazo insertions into heteroatom–hydrogen bonds.

The asymmetric N–H insertion of α-diazo carbonyl compounds provided an efficient method for preparing chiral α-amino ketones, α-amino esters, and nitrogen-containing heterocycles. The concept of metal-carbenoid insertions into N–H bonds has been known for more than five decades, but only two catalytic asymmetric versions have been documented, and both of them gave low enantioselectivity (<50% ee) [83]. By using copper complexes of chiral spiro bis(oxazoline) ligand (S$_a$,S,S)-6a as catalyst, highly enantioselective catalytic insertion of α-diazoesters into N–H bonds was realized in high yields (78–96%) and high enantioselectivities (85–98% ee) (Scheme 4.18, Reaction 1) [12b]. Bis(oxazoline) ligands with other scaffolds, however, gave significantly lower enantioselectivity, which showed that the chiral spirobiindane structure of ligands is essential for obtaining optimum enantioselectivity in this reaction.

The catalyst Cu/(S$_a$,S,S)-6a was also efficient for the insertion into O–H bonds of phenols (Scheme 4.18, Reaction 2) [84]. Under similar conditions as those for N–H insertion reactions, a wide range of phenol derivatives underwent an O–H bond insertion reaction with α-diazopropionates, providing α-aryloxypropionates with excellent enantioselectivities (95–99.6% ee).

The asymmetric insertion of α-diazoesters into the O–H bond of water provides an extremely simple approach for the synthesis of chiral α-hydroxyesters in an efficient and atom-economical way. Since most metal catalysts were unstable in the presence of water, the diazo insertion with water remains a great challenge [85]. When the catalyst Cu/(S$_a$,S,S)-6a was employed the asymmetric insertion of α-diazo-α-arylacetates with water was accomplished in high enantioselectivity (up

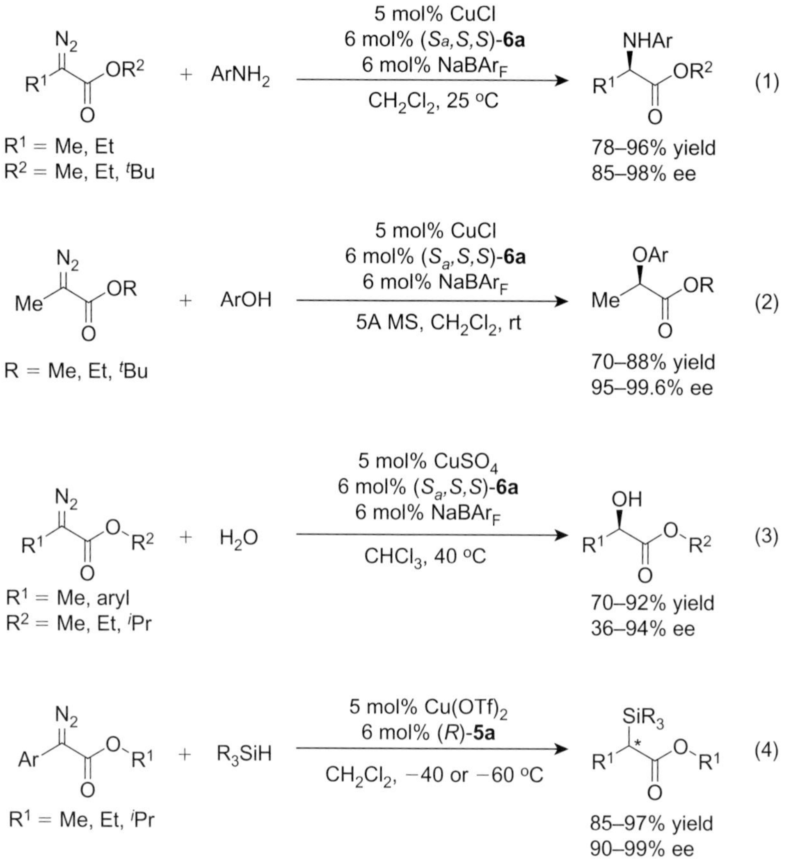

4.5 Asymmetric Carbon–Heteroatom Bond Forming Reaction

$$\underset{\substack{R^1 = Me, Et \\ R^2 = Me, Et, {}^tBu}}{\overset{N_2}{\underset{\|}{R^1}}\underset{O}{\overset{}{\bigvee}}OR^2} + ArNH_2 \xrightarrow[\text{CH}_2\text{Cl}_2,\ 25\ ^\circ\text{C}]{\substack{5\ \text{mol}\%\ \text{CuCl} \\ 6\ \text{mol}\%\ (S_a,S,S)\text{-}\mathbf{6a} \\ 6\ \text{mol}\%\ \text{NaBAr}_F}} \underset{\substack{78-96\%\ \text{yield} \\ 85-98\%\ \text{ee}}}{R^1\underset{O}{\overset{NHAr}{\bigvee}}OR^2} \quad (1)$$

$$\underset{R = Me, Et, {}^tBu}{\overset{N_2}{\underset{\|}{Me}}\underset{O}{\overset{}{\bigvee}}OR} + ArOH \xrightarrow[\text{5A MS, CH}_2\text{Cl}_2,\ \text{rt}]{\substack{5\ \text{mol}\%\ \text{CuCl} \\ 6\ \text{mol}\%\ (S_a,S,S)\text{-}\mathbf{6a} \\ 6\ \text{mol}\%\ \text{NaBAr}_F}} \underset{\substack{70-88\%\ \text{yield} \\ 95-99.6\%\ \text{ee}}}{Me\underset{O}{\overset{OAr}{\bigvee}}OR} \quad (2)$$

$$\underset{\substack{R^1 = Me, aryl \\ R^2 = Me, Et, {}^iPr}}{\overset{N_2}{\underset{\|}{R^1}}\underset{O}{\overset{}{\bigvee}}O^{R^2}} + H_2O \xrightarrow[\text{CHCl}_3,\ 40\ ^\circ\text{C}]{\substack{5\ \text{mol}\%\ \text{CuSO}_4 \\ 6\ \text{mol}\%\ (S_a,S,S)\text{-}\mathbf{6a} \\ 6\ \text{mol}\%\ \text{NaBAr}_F}} \underset{\substack{70-92\%\ \text{yield} \\ 36-94\%\ \text{ee}}}{R^1\underset{O}{\overset{OH}{\bigvee}}O^{R^2}} \quad (3)$$

$$\underset{R^1 = Me, Et, {}^iPr}{\overset{N_2}{\underset{\|}{Ar}}\underset{O}{\overset{}{\bigvee}}O^{R^1}} + R_3SiH \xrightarrow[\text{CH}_2\text{Cl}_2,\ -40\ \text{or}\ -60\ ^\circ\text{C}]{\substack{5\ \text{mol}\%\ \text{Cu(OTf)}_2 \\ 6\ \text{mol}\%\ (R)\text{-}\mathbf{5a}}} \underset{\substack{85-97\%\ \text{yield} \\ 90-99\%\ \text{ee}}}{R^1\underset{O}{\overset{SiR_3}{\overset{*}{\bigvee}}}O^{R^1}} \quad (4)$$

Scheme 4.18 Copper-catalyzed asymmetric carbene insertion reactions.

to 94% ee) (Scheme 4.18, Reaction 3) [86]. This reaction represents one of only a few catalytic asymmetric procedures using water as a reactant.

Transition metal-catalyzed carbene insertion into a Si–H bond provides a direct and efficient method for the synthesis of silica-containing compounds. A few highly enantioselective Si–H insertion reactions have been developed and all of them used chiral dirhodium catalysts [82]. Although copper catalysts had been applied to Si–H bond insertion reactions before rhodium catalysts, the copper-catalyzed asymmetric Si–H bond insertion afforded only modest enantioselectivity (not higher than 85% ee). When chiral spiro diimine ligand (R)-**5a** was applied in a copper-catalyzed asymmetric insertion of α-diazo-α-arylacetates with silanes, the Si–H insertion products were obtained in high yields (85–97%) and excellent enantioselectivities (90–99% ee) (Scheme 4.18, Reaction 4) [11]. The results represent the highest level of enantiocontrol so far in catalytic asymmetric Si–H bond insertion reactions.

4.5.4
Allene-Based Allylic Cyclization Reactions

Transition metal-catalyzed asymmetric allylic substitution of allylic acetate or carbonates is a well-established process for the highly enantioselective formation of carbon–carbon or carbon–heteroatom bonds [87]. Several examples of the asymmetric allylation based on the carbopalladation of allenes have also been reported, but only moderate enantioselectivity has been achieved. Most known ligands do not work well for the allene-based enantioselective allylation, probably due to the *syn/anti*-complexity caused by the introduction of the substituent at the 2-position of the π-allylic moiety. Ma and coworkers [12c] found that the spiro bis(oxazoline) (R_a,S,S)-**6e**, which contains α-naphthylmethyl groups on the oxazoline rings, is a suitable ligand for the enantioselective cyclization of 3,4-allenyl-hydrazines with organic halides, affording the optically active 3-substituted pyrazolidines in good yields (71–90%) with 92–95% ee (Scheme 4.19, Reaction 1). These results are superior to those obtained with analog ligand (R,R)-Bn-BOX (51–77% yield, 75–84% ee) [88]. The β-naphthylmethyl-substituted spiro bis(oxazoline) ligand (R_a,S,S)-**6f** was also successfully applied to the palladium-catalyzed enantioselective cyclization reaction of simple allenes with *o*-aminoiodobenzenes, yielding 3-alkylideneindolines in good yields (52–87%) with excellent enantiomeric excesses (94–98% ee) (Scheme 4.19, Reaction 2) [12d]. Again, the spiro ligand (R_a,S,S)-**6f** exhibited significantly higher enantioselectivities than those obtained with other chiral ligands in this reaction [89].

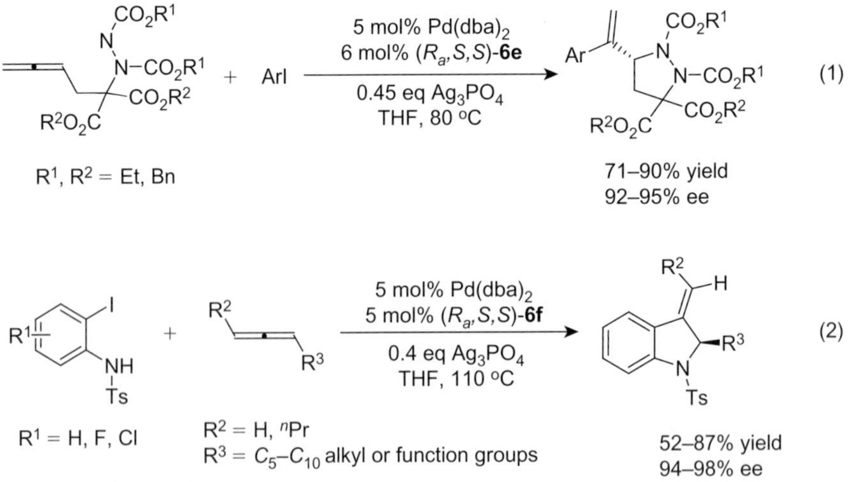

Scheme 4.19 Palladium-catalyzed allylic cyclization reactions.

4.6
Conclusion

This chapter has highlighted the design and updated applications of chiral spiro ligands in transition metal-catalyzed asymmetric reactions. Although the history for the application of chiral spiro ligands in asymmetric catalysis is short, many chiral spiro ligands have been synthesized and applied in various seemingly unrelated reactions, including hydrogenation of prochiral unsaturated bonds (carbon–carbon double bonds, carbonyl, imines, etc.) and carbon–carbon bond-forming and carbon–heteroatom bond-forming reactions. In most of these reactions, spiro ligands exhibit a remarkable capability for enantiocontrol. It can be expected that the spiro ligands will have a broad application in asymmetric catalysis in the future.

The features that make spiro ligands "privileged" may involve extreme chemical robustness, high conformational rigidity, perfect C_2 symmetric, and easy modification. The exciting results obtained with spiro ligands will stimulate future efforts to understand the features that account for the broad applicability of ideal chiral ligands, which are of significance for designing new chiral ligands and catalysts.

References

1 Krapcho, A.P. (1974) *Synthesis*, 383–419.
2 von Baeyer, A. (1900) *Ber. Dtsch. Chem. Ges.*, **33**, 3771–3775.
3 Srivastava, N., Mital, A., and Kumar, A. (1992) *J. Chem. Soc., Chem. Commun.*, 493–494.
4 Nieman, J.A. and Keay, B.A. (1996) *Tetrahedron: Asymmetry*, **7**, 3521–3526.
5 Chan, A.S.C., Hu, W.-H., Pai, C.-C., Lau, C.-P., Jiang, Y.-Z., Mi, A.-Q., Yan, M., Sun, J., Lou, R.-L., and Deng, J.-G. (1997) *J. Am. Chem. Soc.*, **119**, 9570–9571.
6 (a) Cram, D.J. and Steinberg, H. (1954) *J. Am. Chem. Soc.*, **76**, 2753–2757; (b) Nieman, J.A., Parvez, M., and Keay, B.A. (1993) *Tetrahedron: Asymmetry*, **4**, 1973–1976; (c) Chan, A.S.C., Lin, C.-C., Sun, J., Hu, W.-H., Li, Z., Pan, W.-D., Mi, A.-Q., Jiang, Y.-Z., Huang, T.-M., Yang, T.-K., Chen, J.-H., Wang, Y., and Lee, G.-H. (1995) *Tetrahedron: Asymmetry*, **6**, 2953–2959.
7 Birman, V.B., Rheingold, A.L., and Lam, K.-C. (1999) *Tetrahedron: Asymmetry*, **10**, 125–131.
8 Zhang, J.-H., Liao, J., Cui, X., Yu, K.-B., Zhu, J., Deng, J.-G., Zhu, S.-F., Wang, L.-X., Zhou, Q.-L., Chung, L.W., and Ye, T. (2002) *Tetrahedron: Asymmetry*, **13**, 1363–1366.
9 (a) Xie, J.-H., Wang, L.-X., Fu, Y., Zhu, S.-F., Fan, B.-M., Duan, H.-F., and Zhou, Q.-L. (2003) *J. Am. Chem. Soc.*, **125**, 4404–4405; (b) Xie, J.-H., Duan, H.-F., Fan, B.-M., Cheng, X., Wang, L.-X., and Zhou, Q.-L. (2004) *Adv. Synth. Catal.*, **346**, 625–632.
10 Tang, W.-J., Zhu, S.-F., Xu, L.-J., Zhou, Q.-L., Fan, Q.-H., Zhou, H.-F., Lam, K., and Chan, A.S.C. (2007) *Chem. Commun.*, 613–615.
11 Zhang, Y.-Z., Zhu, S.-F., Wang, L.-X., and Zhou, Q.-L. (2008) *Angew. Chem. Int. Ed.*, **47**, 8496–8498.
12 (a) Liu, B., Zhu, S.-F., Wang, L.-X., and Zhou, Q.-L. (2006) *Tetrahedron: Asymmetry*, **17**, 634–641; (b) Liu, B., Zhu, S.-F., Zhang, W., Chen, C., and Zhou, Q.-L. (2007) *J. Am. Chem. Soc.*, **129**, 5834–5835; (c) Shu, W. and Ma, S.-M. (2009) *Chem. Commun.*,

6198–6200; (d) Shu, W., Yu, Q., and Ma, S.-M. (2009) *Adv. Synth. Catal.*, **351**, 2807–2810.

13 (a) Zhu, S.-F., Xie, J.-B., Zhang, Y.-Z., Li, S., and Zhou, Q.-L. (2006) *J. Am. Chem. Soc.*, **128**, 12886–12891; (b) Li, S., Zhu, S.-F., Zhang, C.-M., Song, S., and Zhou, Q.-L. (2008) *J. Am. Chem. Soc.*, **130**, 8584–8585.

14 (a) Zhu, S.-F., Yang, Y., Wang, L.-X., Liu, B., and Zhou, Q.-L. (2005) *Org. Lett.*, **7**, 2333–2335; (b) Zhang, W., Zhu, S.-F., Qiao, X.-C., and Zhou, Q.-L. (2008) *Chem. Asian J.*, **3**, 2105–2111.

15 (a) Shi, W.-J., Wang, L.-X., Fu, Y., Zhu, S.-F., and Zhou, Q.-L. (2003) *Tetrahedron: Asymmetry*, **14**, 3867–3872; (b) Shi, W.-J., Xie, J.-H., and Zhou, Q.-L. (2005) *Tetrahedron: Asymmetry*, **16**, 705–710; (c) Duan, H.-F., Xie, J.-H., Shi, W.-J., Zhang, Q., and Zhou, Q.-L. (2006) *Org. Lett.*, **8**, 1479–1481.

16 (a) Fu, Y., Hou, G.-H., Xie, J.-H., Xing, L., Wang, L.-X., and Zhou, Q.-L. (2004) *J. Org. Chem.*, **69**, 8157–8160; (b) Hou, G.-H., Xie, J.-H., Wang, L.-X., and Zhou, Q.-L. (2006) *J. Am. Chem. Soc.*, **128**, 11774–11775.

17 Zhou, H., Wang, W.-H., Fu, Y., Xie, J.-H., Shi, W.-J., Wang, L.-X., and Zhou, Q.-L. (2003) *J. Org. Chem.*, **68**, 1582–1584.

18 Zhu, S.-F., Fu, Y., Xie, J.-H., Liu, B., Xing, L., and Zhou, Q.-L. (2003) *Tetrahedron: Asymmetry*, **14**, 3219–3224.

19 Yang, Y., Zhu, S.-F., Duan, H.-F., Zhou, C.-Y., Wang, L.-X., and Zhou, Q.-L. (2007) *J. Am. Chem. Soc.*, **129**, 2248–2249.

20 Cheng, X., Zhang, Q., Xie, J.-H., Wang, L.-X., and Zhou, Q.-L. (2005) *Angew. Chem. Int. Ed.*, **44**, 1118–1121.

21 Huo, X.-H., Xie, J.-H., Wang, Q.-S., and Zhou, Q.-L. (2007) *Adv. Synth. Catal.*, **349**, 2477–2484.

22 (a) Wu, S.-L., Zhang, W.-C., Zhang, Z.-G., and Zhang, X.-M. (2004) *Org. Lett.*, **6**, 3565–3567; (b) Zhang, W.-C., Wang, C.-J., Gao, W.-Z., and Zhang, X.-M. (2005) *Tetrahedron Lett.*, **46**, 6087–6090.

23 Ozawa, F., Kubo, A., Matsumoto, Y., Hayashi, T., Nishioka, E., Yanagi, K., and Moriguchi, K. (1993) *Organometallics*, **12**, 4188–4196.

24 Jiang, Y.-Z., Xue, S., Li, Z., Deng, J.-G., Mi, A.-Q., and Chan, A.S.C. (1998) *Tetrahedron: Asymmetry*, **9**, 3185–3189.

25 Lin, C.W., Lin, C.-C., Lam, L.F.-L., Au-Yeung, T.T.-L., and Chan, A.S.C. (2004) *Tetrahedron Lett.*, **45**, 7379–7381.

26 Lait, S.M., Parvez, M., and Keay, B.A. (2004) *Tetrahedron: Asymmetry*, **15**, 155–158.

27 Benoit, W.L., Parvez, M., and Keay, B.A. (2009) *Tetrahedron: Asymmetry*, **20**, 69–77.

28 Guo, Z.-Q., Guan, X.-Y., and Chen, Z.-Y. (2006) *Tetrahedron: Asymmetry*, **17**, 468–473.

29 Han, Z.-B., Wang, Z., Zhang, X.-M., and Ding, K.-L. (2009) *Angew. Chem. Int. Ed.*, **48**, 5345–5349.

30 For an award account, see: Bajracharya, G.B., Arai, M.A., Koranne, P.S., Suzuki, T., Takizawa, S., and Sasai, H. (2009) *Bull. Chem. Soc. Jpn.*, **82**, 285–302.

31 Chung, Y.K. and Fu, G.C. (2009) *Angew. Chem. Int. Ed.*, **48**, 2225–2227.

32 Dohi, T., Maruyama, A., Takenaga, N., Senami, K., Minamitsuji, Y., Fujioka, H., Caemmerer, S.B., and Kita, Y. (2008) *Angew. Chem. Int. Ed.*, **47**, 3787–3790.

33 (a) For an account of asymmetric catalysis with chiral diphosphine and monodentate phosphorus ligands on a spiro scaffold, see: Xie, J.-H. and Zhou, Q.-L. (2008) *Acc. Chem. Res.*, **41**, 581–593. (b) For a recent review on chiral spiro ligands, see: Ding, K.L., Han, Z.-B., and Wang, Z. (2009) *Chem. Asian J.*, **4**, 32–41.

34 (a) Jacobsen, E.N., Pfaltz, A., and Yamamoto, H. (eds.) (1999) *Comprehensive Asymmetric Catalysis*, Springer, Berlin; (b) de Vries, J.G. and Elsevier, C.J. (eds.) (2007) *Handbook of Homogeneous Hydrogenation*, Wiley-VCH Verlag GmbH, Weinheim; (c) Tang, W.-J. and Zhang, X.-M. (2003) *Chem. Rev.*, **103**, 3029–3069.

35 (a) Fu, Y., Xie, J.-H., Hu, A.-G., Zhou, H., Wang, L.-X., and Zhou, Q.-L. (2002) *Chem. Commun.*, 480–481; (b) Fu, Y., Guo, X.-X., Zhu, S.-F., Hu, A.-G., Xie, J.-H., and Zhou, Q.-L. (2004) *J. Org. Chem.*, **69**, 4648–4655.

36 Hu, W.-H., Yan, M., Chan, A.S.C., Jiang, Y.-Z., and Mi, A.-Q. (1999) *Tetrahedron Lett.*, **40**, 973–976.
37 Hu, A.-G., Fu, Y., Xie, J.-H., Zhou, H., Wang, L.-X. and Zhou, Q.-L. (2002) *Angew. Chem. Int. Ed.*, **41**, 2348–2350.
38 (a) Lee, N.E. and Buchwald, S.L. (1994) *J. Am. Chem. Soc.*, **116**, 5985–5986; (b) Tararov, V.I., Kadyrov, R., Riermeier, T. H., Holz, J., and Börner, A. (2000) *Tetrahedron Lett.*, **41**, 2351–2355.
39 Hou, G.-H., Xie, J.-H., Yan, P.-C., and Zhou, Q.-L. (2009) *J. Am. Chem. Soc.*, **131**, 1366–1367.
40 Yan, P.-C., Xie, J.-H., Hou, G.-H., Wang, L.-X., and Zhou, Q.-L. (2009) *Adv. Synth. Catal.*, **351**, 3174–3176.
41 Ohta, T., Takaya, H., Kitamura, M., Nagai, K., and Noyori, R. (1987) *J. Org. Chem.*, **52**, 3176–3178.
42 Cheng, X., Xie, J.-H., Li, S., and Zhou, Q.-L. (2006) *Adv. Synth. Catal.*, **348**, 1271–1276.
43 Li, S., Zhu, S.-F., Xie, J.-H., Song, S., Zhang, C.-M., and Zhou, Q.-L. (2010) *J. Am. Chem. Soc.*, **132**, 1172–1179.
44 Zhang, Y., Han, Z.-B., Li, F.-Y., Ding, K.-L., and Zhang, A. (2010) *Chem. Commun.*, 156–158.
45 For a review, see: Noyori, R. and Ohkuma, T. (2001) *Angew. Chem. Int. Ed.*, **40**, 40–73.
46 For reviews on dynamic kinetic resolution, see: (a) Noyori, R., Tokunaga, M., and Kitamura, M. (1995) *Bull. Chem. Soc. Jpn.*, **68**, 36–56; (b) Vedejs, E. and Jure, M. (2005) *Angew. Chem. Int. Ed.*, **44**, 3974–4001; (c) Pellissier, H. (2008) *Tetrahedron*, **64**, 1563–1601.
47 (a) Ohkuma, T., Ishii, D., Takeno, H., and Noyori, R. (2000) *J. Am. Chem. Soc.*, **122**, 6510–6511; (b) Scalone, M. and Waldmeier, P. (2003) *Org. Process Res. Dev.*, **7**, 418–425; (c) Ohkuma, T., Li, J., and Noyori, R. (2004) *Synlett*, 1383–1386.
48 Xie, J.-H., Liu, S., Huo, X.-H., Cheng, X., Duan, H.-F., Fan, B.-M., Wang, L.-X., and Zhou, Q.-L. (2005) *J. Org. Chem.*, **70**, 2967–2973.
49 Bai, W.-J., Xie, J.-H., Li, Y.-L., Liu, S., and Zhou, Q.-L. (2010) *Adv. Synth. Catal.*, **352**, 81–84.
50 Liu, S., Xie, J.-H., Wang, L.-X., and Zhou, Q.-L. (2007) *Angew. Chem. Int. Ed.*, **46**, 7506–7508.
51 Xie, J.-H., Liu, S., Kong, W.-L., Bai, W.-J., Wang, X.-C., Wang, L.-X., and Zhou, Q.-L. (2009) *J. Am. Chem. Soc.*, **131**, 4222–4223.
52 Liu, S., Xie, J.-H., Li, W., Kong, W.-L., Wang, L.-X., and Zhou, Q.-L. (2009) *Org. Lett.*, **11**, 4994–4997.
53 Xie, J.-H., Zhou, Z.-T., Kong, W.-L., and Zhou, Q.-L. (2007) *J. Am. Chem. Soc.*, **129**, 1868–1869.
54 Zhou, Z.-T., Xie, J.-H., and Zhou, Q.-L. (2009) *Adv. Synth. Catal.*, **351**, 363–366.
55 Zhou, Y.-G. (2007) *Acc. Chem. Res.*, **40**, 1357–1366.
56 Fagnou, K. and Lautens, M. (2003) *Chem. Rev.*, **103**, 169–196.
57 Sakai, M., Ueda, M., and Miyaura, N. (1998) *Angew. Chem. Int. Ed.*, **37**, 3279–3281.
58 For the diastereoselective addition of arylboronic acids to chiral sulfinimines, see: (a) Weix, D.J., Shi, Y.-L., and Ellman, J.A. (2005) *J. Am. Chem. Soc.*, **127**, 1092–1093; (b) Bolshan, Y. and Batey, R.A. (2005) *Org. Lett.*, **7**, 1481–1484.
59 Riant, O. and Hannedouche, J. (2007) *Org. Biomol. Chem.*, **5**, 873–888.
60 Duan, H.-F., Jia, Y.-X., Wang, L.-X., and Zhou, Q.-L. (2006) *Org. Lett.*, **8**, 2567–2569.
61 Duan, H.-F., Xie, J.-H., Qiao, X.-C., Wang, L.-X., and Zhou, Q.-L. (2008) *Angew. Chem. Int. Ed.*, **47**, 4351–4353.
62 Zanoni, G., Pontiroli, A., Marchetti, A., and Vidari, G. (2007) *Eur. J. Org. Chem.*, 3599–3611.
63 Zanoni, G., Gladiali, S., Marchetti, A., Picoinini, P., Tredic, I., and Vidari, G. (2004) *Angew. Chem. Int. Ed.*, **43**, 846–849.
64 Sibi, M.P. and Manyem, S. (2000) *Tetrahedron*, **56**, 8033–8061.
65 Arai, M.A., Arai, T., and Sasai, H. (1999) *Org. Lett.*, **1**, 1795–1797.
66 Lautens, M., Fagnou, K., and Hiebert, S. (2003) *Acc. Chem. Res.*, **36**, 48–58.
67 Millward, D.B., Sammis, G., and Waymouth, R.M. (2000) *J. Org. Chem.*, **65**, 3902–3909.

68 Zhang, W., Wang, L.-X., Shi, W.-J., and Zhou, Q.-L. (2005) *J. Org. Chem.*, **70**, 3734–3736.

69 (a) Ikeda, S. (2003) *Angew. Chem. Int. Ed.*, **42**, 5120–5122; (b) Montgomery, J. (2004) *Angew. Chem. Int. Ed.*, **43**, 3890–3908; (c) Moslin, R.M., Miller-Moslin, K., and Jamison, T.F. (2007) *Chem. Commun.*, 4441–4449.

70 (a) Ikeda, S. Cui, D.-M., and Sato, Y. (1999) *J. Am. Chem. Soc.*, **121**, 4712–4713; (b) Sato, Y., Saito, N., and Mori, M. (2000) *J. Am. Chem. Soc.*, **122**, 2371–2372; (c) Miller, K.M., Huang, W.-S., and Jamison, T.F. (2003) *J. Am. Chem. Soc.*, **125**, 3442–3443; (d) Chaulagain, M.R., Sormunen, G.J., and Montgomery, J. (2007) *J. Am. Chem. Soc.*, **129**, 9568–9569.

71 Yang, Y., Zhu, S.-F., Zhou, C.-Y., and Zhou, Q.-L. (2008) *J. Am. Chem. Soc.*, **130**, 14052–14053.

72 RajanBabu, T.V. (2003) *Chem. Rev.*, **103**, 2845–2860.

73 Shi, W.-J., Zhang, Q., Xie, J.-H., Zhu, S.-F., Hou, G.-H., and Zhou, Q.-L. (2006) *J. Am. Chem. Soc.*, **128**, 2780–2781.

74 Chakrapani, H., Liu, C., and Widenhoefer, R.A. (2003) *Org. Lett.*, **5**, 157–159.

75 Fan, B.-M., Xie, J.-H., Li, S., Wang, L.-X., and Zhou, Q.-L. (2007) *Angew. Chem. Int. Ed.*, **46**, 1275–1277.

76 (a) Tietze, L.F., Ila, H., and Bell, H.P. (2004) *Chem. Rev.*, **104**, 3453–3516; (b) Trost, B.M., Machacek, M.R., and Aponick, A. (2006) *Acc. Chem. Res.*, **39**, 747–760.

77 Tsujihara, T., Takenaka, K., Onitsuka, K., Hatanaka, M., and Sasai, H. (2009) *J. Am. Chem. Soc.*, **131**, 3452–3453.

78 Sethofer, S.G., Staben, S.T., Hung, O.Y., and Toste, F.D. (2008) *Org. Lett.*, **10**, 4315–4318.

79 Hayashi, T. (2000) *Acc. Chem. Res.*, **33**, 354–362.

80 Guo, X.-X., Xie, J.-H., Hou, G.-H., Shi, W.-J., Wang, L.-X., and Zhou, Q.-L. (2004) *Tetrahedron: Asymmetry*, **15**, 2231–2234.

81 (a) Arai, M.A., Kuraishi, M., Arai, T., and Sasai, H. (2001) *J. Am. Chem. Soc.*, **123**, 2907–2908; (b) Koranne, P.S., Tsujihara, T., Arai, M.A., Bajracharya, G.B., Suzuki, T., Onitsuka, K., and Sasai, H. (2007) *Tetrahedron: Asymmetry*, **18**, 919–923.

82 Doyle, M.P., McKervey, M.A., and Ye, T. (1998) *Modern Catalytic Methods for Organic Synthesis with Diazocompounds*, John Wiley & Sons, Inc., New York.

83 (a) García, C.F., McKervey, M.A., and Ye, T. (1996) *Chem. Commun.*, 1465–1466; (b) Bachmann, S., Fielenbach, D., and Jørgensen, K.A. (2004) *Org. Biomol. Chem.*, **2**, 3044–3049.

84 Chao, C., Zhu, S.-F., Liu, B., Wang, L.-X., and Zhou, Q.-L. (2007) *J. Am. Chem. Soc.*, **129**, 12616–12617.

85 Maier, T.C. and Fu, G.C. (2006) *J. Am. Chem. Soc.*, **128**, 4594–4595.

86 Zhu, S.-F., Chen, C., Cai, Y., and Zhou, Q.-L. (2008) *Angew. Chem. Int. Ed.*, **47**, 932–934.

87 Lu, Z. and Ma, S.-M. (2008) *Angew. Chem. Int. Ed.*, **47**, 258–297.

88 Shu, W., Yang, Q., Jia, G.-C., and Ma, S.-M. (2008) *Tetrahedron*, **49**, 11159–11166.

89 Zenner, J.M. and Larock, R.C. (1999) *J. Org. Chem.*, **64**, 7312–7322.

5
Chiral Bisoxazoline Ligands

Levi M. Stanley and Mukund P. Sibi

5.1
Introduction

Chiral bisoxazolines (Box) comprise a class of ligands that figure prominently in metal-catalyzed asymmetric synthesis [1–7]. The basic design of these ligands includes two oxazoline moieties connected by a spacer unit. The parent bisoxazoline ligands contain a single carbon spacer unit that is generally substituted with two identical substituents (Figure 5.1). Ligands of this type (**1–5**) represent the most prevalent structural motif for the bisoxazolines. In general, this chapter will focus on the utility of bisoxazoline ligands containing this structural motif. However, many bisoxazoline ligands containing different linker units are now well established in metal-catalyzed asymmetric synthesis. Figure 5.2 presents a selection of bisoxazoline scaffolds (**6–18**) containing alternative linker units. Among the ligands, the pyridine bisoxazoline (Pybox) scaffold **6** is most common [8]. The Pybox ligands are intentionally omitted from this chapter in the interest of space. Bisoxazoline ligands, such as the DBFOX ligands **7**, containing an alternative linker units will, however, be discussed in cases where these scaffolds complement or improve upon the utility of the parent bisoxazoline ligands.

This chapter is organized into three major sections on enantioselective carbon–carbon bond formation, enantioselective carbon–heteroatom bond formation, and enantioselective cycloaddition processes that are catalyzed by metal-bisoxazoline complexes. Each of these sections is further divided by reaction type. The content of the chapter is designed to allow the reader to identify appropriate metal-bisoxazoline complexes that lead to efficient and enantioselective transformations for reaction types in which these catalysts are prominent, and is not meant to provide a comprehensive list of all metal-bisoxazoline complexes known to catalyze a given process. However, a brief discussion of the genesis of the bisoxazoline ligands and a short analysis of some of the prominent structural features of common metal-bisoxazoline complexes are presented prior to the discussion of the application of bisoxazoline ligands in asymmetric catalysis.

Privileged Chiral Ligands and Catalysts. Edited by Qi-Lin Zhou
Copyright © 2011 WILEY-VCH Verlag GmbH & Co. KGaA, Weinheim
ISBN: 978-3-527-32704-1

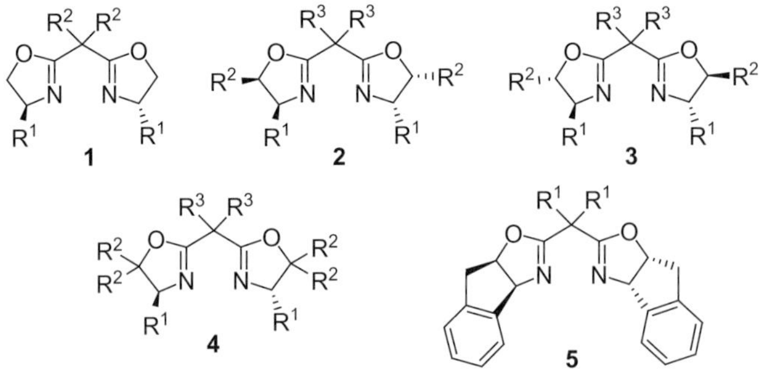

5 Chiral Bisoxazoline Ligands

Figure 5.1 Bisoxazoline structures.

Primary efforts from four research groups in the late 1980s through the early 1990s initiated the intense study of bisoxazoline ligands that has followed over the past two decades and, ultimately, has led this class of chiral ligands to be considered among the privileged ligand architectures. In 1988 Pfaltz and coworkers reported on the synthesis of chiral semicorrin ligands and the utility of these ligands in enantioselective cyclopropanation reactions [9, 10]. Although the semicorrin ligands lack bisoxazoline moieties and thus cannot be considered as part of the bisoxazoline family, the architecture of the semicorrins is analogous to the bisoxazoline ligands and certainly served as inspiration to the genesis of the bisoxazoline

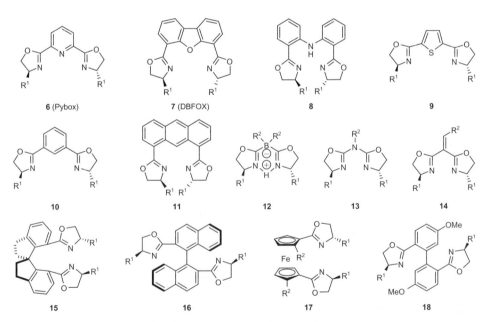

Figure 5.2 Bisoxazoline structures containing alternative linker units.

ligand scaffold. In 1990 Masamune and coworkers reported the first syntheses of chiral bisoxazoline ligands. These ligands contained an unsubstituted methylene spacer unit (see structure **1**, $R^2 = H$ and structure **3**, $R^3 = H$ in Figure 5.1) [11]. Furthermore, Masamune demonstrated for the first time the potential utility of chiral bisoxazoline ligands in asymmetric catalysis and showed that these ligands led to improved enantioselectivities compared to the semicorrin ligands in enantioselective cyclopropanation reactions. Early work from the Masamune group also showed that various bisoxazoline ligands could be readily synthesized and that access to these ligands containing different substitution patterns could be used to improve the stereoselectivity and reactivity of a catalyst in a chosen reaction. For example, Lowenthal and Masamune in 1991 demonstrated that a bisoxazoline ligand based on structure **2** ($R^1, R^2 = Ph, R^3 = H$) in Figure 5.1 could be used to expand the scope of highly enantioselective cyclopropanation reactions to encompass trisubstituted and unsymmetrical *cis*-1,2-disubstituted alkenes [12].

The groups of Evans and Corey independently established the foundation for the utility of bisoxazoline ligands that began their path to be considered among the privileged ligand architectures. In 1991 in back-to-back communications in the *Journal of the American Chemical Society* the groups of Evans [13] and Corey [14] reported the synthesis and utility of bisoxazoline ligands of type **1** containing a fully substituted spacer unit ($R^2 = Me$). Bisoxazoline scaffolds of this type have since become the most heavily studied in the literature on the application of bisoxazoline ligands in asymmetric catalysis. In hindsight, the versatility that led to the emergence of the bisoxazoline scaffold could be envisioned based on the results of early studies by the groups of Evans and Corey. Evans showed that enantioselectivities of greater than 99% and catalyst turnover of greater than 1000 were possible in cyclopropanations catalyzed by a Cu(I) complex of the ligand that came to be known as *t*-Bu-Box (structure **1**, $R^1 = t$-Bu, $R^2 = Me$). In the following communication Corey reported an enantioselective Diels–Alder cycloaddition between cyclopentadiene and oxazolidinone acrylate in the presence of an Fe(III) complex of the ligand that came to be known as Ph-Box. This cycloaddition occurred to form the corresponding cycloadduct with up to 99 : 1 *endo* : *exo* selectivity and up to 86% enantiomeric excess. The activity and selectivity of these complexes of different bisoxazoline ligands with metals salts in seemingly very different reaction types led many other groups to study and ultimately expand the chemistry of metal-bisoxazoline catalysts to ligands, metal precursors, and reactions that would have been difficult to imagine in 1991.

Corey and coworkers recognized in their communication on the enantioselective Diels–Alder reaction catalyzed by Fe(III)-Ph-Box that reasonable models of a reactant (in this case the oxazolidinone acrylate dienophile) bound to a metal-bisoxazoline complex could be generated based on knowledge of the metal geometry to rationalize the stereochemical outcomes of processes catalyzed by these complexes. This fundamental realization led numerous groups to seek information on the structure of metal-bisoxazoline complexes through X-ray analysis, NMR spectroscopy, and computational methods. Perhaps the most insightful pieces of information, at least in the early years of metal-bisoxazoline catalysis,

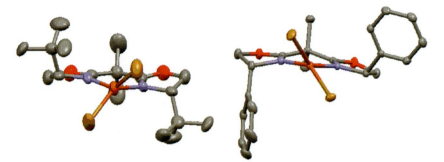

Figure 5.3 Molecular structures of [Cu((S)-t-Bu-Box)Br$_2$] and [Cu((R)-Ph-Box)Br$_2$].

were X-ray crystal structures of complexes prepared from various bisoxazoline ligands and inorganic metal salts. These X-ray crystal structures have proven invaluable over the years by providing a representation of the geometry of the metal center, the orientation of the chiral bisoxazoline ligand, the steric bias inherent to the bisoxazoline ligand, and the locations of counter-anions that can be displaced or reorganized upon binding of a substrate molecule to generate the activated intermediate that reacts to form the product. Given the importance of these X-ray crystal structures to the area of metal-bisoxazoline catalysis, a selection of structures of common complexes prepared from metal salts and bisoxazoline ligands are illustrated and discussed in the following paragraphs.

Among the various metal salts that have been used to form crystalline metal-bisoxazoline complexes suitable for X-ray analysis, the Cu(II) complexes of this type are the most prevalent and well studied [1]. In general, Cu(II) complexes of bisoxazoline ligands possess a distorted square-planar geometry (Figure 5.3) [15, 16]. However, distorted square-pyramidal structures have been solved for Cu(II) complexes containing triflate counterions and two water molecules as an ancillary ligands (Figure 5.4). Copper(II) halide complexes of t-Bu-Box and Ph-Box are

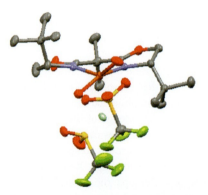

Figure 5.4 Molecular structure of [Cu((S)-t-Bu-Box)(H$_2$O)$_2$OTf]OTf.

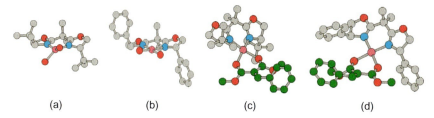

Figure 5.5 Molecular structures of (a) [Cu((S)-t-Bu-Box)(H$_2$O)$_2$](SbF$_6$)$_2$, (b) [Cu((S)-Ph-Box)(H$_2$O)$_2$](SbF$_6$)$_2$, (c) [Cu((S)-t-Bu-Box)(PhCH=C(CO$_2$Me)$_2$)](SbF$_6$)$_2$, and (d) [Cu((S)-Ph-Box)(PhCH=C(CO$_2$Me)$_2$)](SbF$_6$)$_2$ (SbF$_6$ counteranions omitted for clarity).

distorted square planar complexes with large distortions of the halide anions away from the substituent at the 4-position of the oxazoline substructures and out of the N–C–N plane. In contrast, the hydrated complexes of Cu(SbF$_6$)$_2$ with t-Bu-Box and Ph-Box have prominent differences from each other (Figure 5.5) [15, 17–20]. The Cu(SbF$_6$)$_2$/(S)-t-Bu-Box complex [Cu((S)-t-Bu-box)(H$_2$O)$_2$](SbF$_6$)$_2$ has a distorted square-planar geometry around the copper center with large distortion of the water molecules away from the t-butyl substituents of the ligand. The Cu(SbF$_6$)$_2$/(S)-Ph-Box complex [Cu((S)-Ph-Box)(H$_2$O)$_2$](SbF$_6$)$_2$ is also a square-planar structure, but the distortion of the water molecules is relatively small and is instead toward the phenyl substituents of the ligand. This degree and sense of distortion is consistent in Cu(SbF$_6$)$_2$ complexes of (S)-t-Bu-Box or (S)-Ph-Box with benzylidene dimethyl malonate [21, 22]. The different distortions in the complexes with each ligand lead to distinct modes of coordination of the benzylidene malonate and an opposite sense of asymmetric induction by Cu(II) complexes of (S)-t-Bu-Box or (S)-Ph-Box.

X-Ray crystal structures have also been solved for metal-bisoxazoline complexes that are important in chiral Lewis acid catalysis beyond Cu(II) complexes of (S)-t-Bu-Box or (S)-Ph-Box (Figure 5.6). Copper(II) complexes of aminoindanol-derived bisoxazoline complexes are distorted square-planar complexes with the distortion of the anions away from the indane groups [16, 23]. Zinc(II) halide [16] and Ni(II)

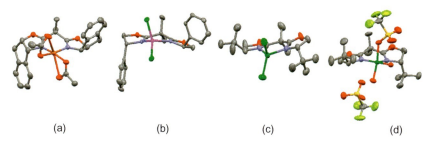

Figure 5.6 Molecular structures of (a) [Cu(R,S)-indane-Box(OAc)$_2$], (b) [Zn((R)-Ph-Box)Cl$_2$], (c) [Ni((S)-t-Bu-Box)Cl$_2$], and (d) [Ni((S)-t-Bu-Box)(H$_2$O)$_2$OTf]OTf.

5.2
Enantioselective Carbon–Carbon Bond Formation

5.2.1
Addition of Carbon Nucleophiles to C=O and C=N Bonds

5.2.1.1 Aldol Reactions

Chiral Lewis acids prepared from bisoxazoline ligands and various metal salts are common catalysts for enantioselective aldol reactions. Evans and coworkers reported aldol additions of enolsilanes to pyruvate esters catalyzed by t-Bu-Box/Cu(OTf)$_2$ (Scheme 5.1) [25, 26]. The additions of both silylketene acetals and ketone-derived enolsilanes to methyl pyruvate occur with high yields and excellent enantioselectivities. The stereoinduction results from the formation of a square-planar Cu(II)box intermediate with the pyruvate carbonyl oxygens forming

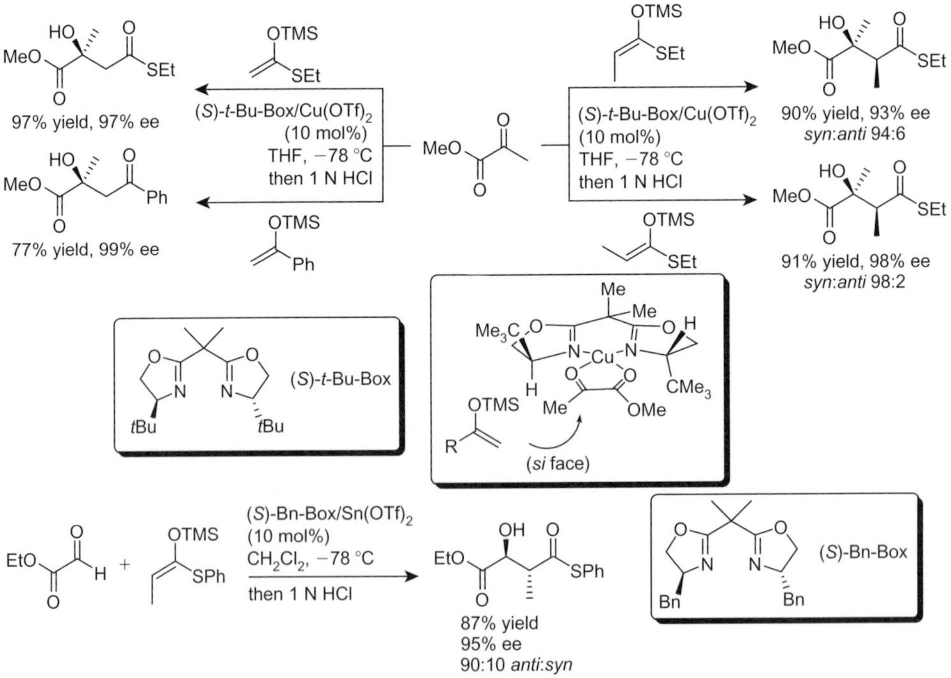

Scheme 5.1 Enantioselective aldol reactions using bisoxazoline ligands.

a bidentate chelate to the Cu(II) center. Furthermore, the reactions of both *(Z)*- and *(E)*-propionate derived silylketene acetals occur in a stereoconvergent manner to form the *syn* aldol products in high yields and enantioselectivities. The stereoconvergent addition of *(Z)*- and *(E)*-propionate derived silylketene acetals to pyruvate esters precludes access to both *syn* and *anti* aldol adducts based on the geometry of the nucleophile. However, Evans and coworkers showed that a Bn-Box/Sn(OTf)$_2$ complex catalyzes the *anti* aldol addition of silylketene acetals to ethyl glyoxylate [27].

Although Lewis acidic complexes containing chiral bisoxazoline ligands are effective catalysts for additions of enolsilanes to aldehydes or ketones capable of bidentate coordination, the analogous reactions of single-point binding alkyl or aryl aldehydes were less selective. This problem has been successfully addressed by using nucleophiles, instead of electrophiles, capable of bidentate coordination to the chiral Lewis acid. Evans and coworkers found that the *t*-Bu-Box/Ni(OTf)$_2$-catalyzed aldol additions of *N*-propionylthiazolidinethione to aromatic and aliphatic aldehydes form *syn* aldol adducts in good yields with excellent selectivities (Scheme 5.2) [24].

Scheme 5.2 syn-Propionate aldol reactions.

The metal complexes of bisoxazoline ligands that catalyze and substrates that undergo Mukaiyama–aldol reactions are not limited to those discussed in the previous paragraph. Entries 1 and 2 in Scheme 5.3 illustrate additional combinations of copper(II) salts and bisoxazoline ligands that catalyze Mukaiyama–aldol reactions of additional substrates, such as pyridine *N*-oxide aldehydes (entry 1) and dienolsilanes (entry 2) [28, 29]. Entries 3 and 4 show the utility of copper(II)-bisoxazoline as catalysts of asymmetric aldol reactions in or on water [30, 31]. In addition, several groups have developed supported copper complexes of bisoxazoline or aza(bisoxazoline) ligands for Mukaiyama–aldol reactions [32–35]. Specific supports include laponite, polystyrene, polysiloxane derivatives, silica, and an imidazolium-modified silica.

5.2.1.2 Mannich-Type Reactions

Metal complexes of bisoxazoline ligands have also found utility as catalysts for Mannich-type reactions of enolates and enolate equivalents with imine derivatives. Several reports detailing the application of Cu(II)/Box complexes as catalysts for Mannich reactions have appeared in the literature (Scheme 5.4). Jørgensen and coworkers established Ph-Box/Cu(OTf)$_2$ as an efficient and highly selective catalyst for additions of pyruvate esters to an *N*-tosyl-α-imino ester [36] and *t*-Bu-Box/Cu

Entry	Nucleophile	Acceptor	Ligand	Lewis Acid	Solvent	Product	Yield (%)	syn/anti	ee (%)
1	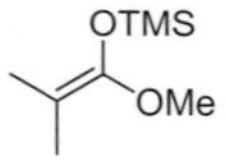 OTMS / OMe	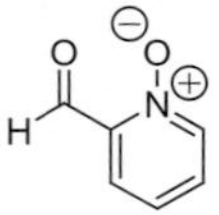 pyridine N-oxide aldehyde	(S)-t-Bu-Box	Cu(OTf)$_2$	CH$_2$Cl$_2$	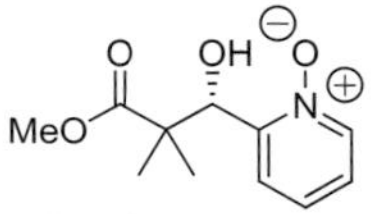	79	—	93
2	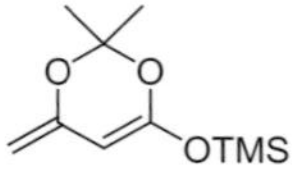 dioxinone OTMS	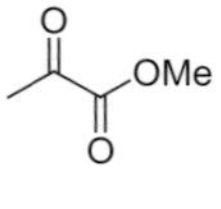 methyl pyruvate	19	CuCl$_2$	THF	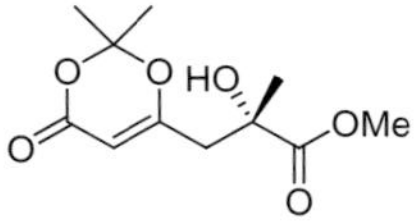	81	—	94
3	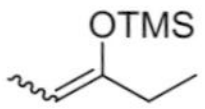 OTMS (E/Z)	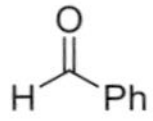 PhCHO	(S)-i-Pr-Box	Cu(OSOC$_{12}$H$_{25}$)$_2$	H$_2$O	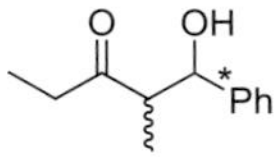	72	2.8:1	69 (syn)
4	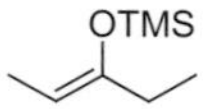 OTMS (E/Z)	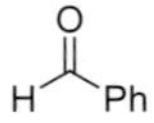 PhCHO	(S)-i-Pr-Box	Cu(OTf)$_2$	EtOH/H$_2$O	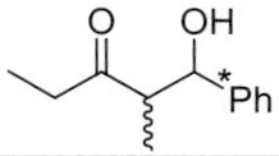	81	3.5:1	81 (syn)

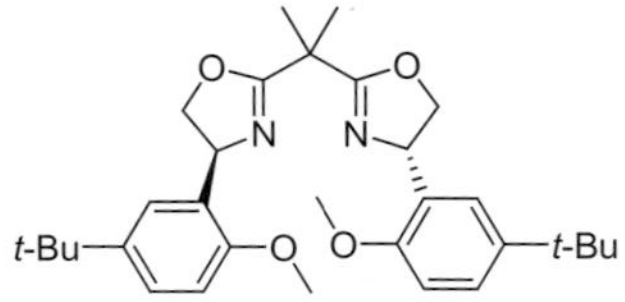

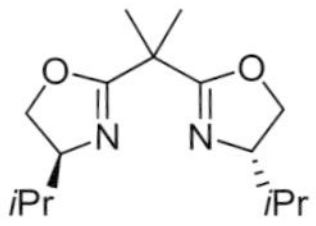

(S)-i-Pr-Box

19

Scheme 5.3 Examples of enantioselective aldol reactions using various substrates and ligands.

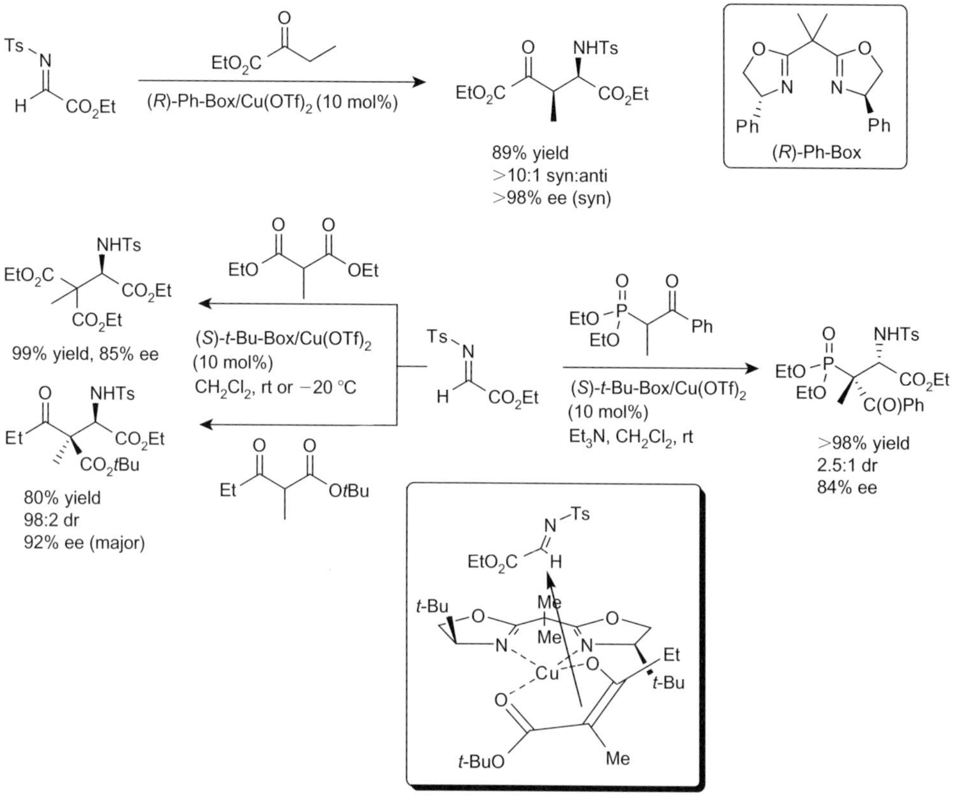

Scheme 5.4 Mannich reactions using N-tosylglyoxylimine.

(OTf)$_2$ as a suitable catalyst for the addition of malonates, β-keto esters, and β-keto phosphonates to an N-tosyl-α-imino ester [37, 38]. The absolute and relative stereochemistry is proposed to arise from bidentate coordination of the malonate (shown in Scheme 5.4), β-keto ester, or β-keto phosphonate nucleophile to form a distorted square planar copper(II) complex. Approach of the Re-face of the activated nucleophile to the N-tosyl-α-imino ester oriented to avoid interaction between the tosyl group and the t-Bu group of the bisoxazoline ligands generates the observed products with (R,R) stereochemistry.

5.2.1.3 Nitroaldol (Henry) Reactions

The nitroaldol or Henry reaction involves the addition of a nitroalkane to an aldehyde or ketone and constitutes an attractive approach to β-nitro alcohol moieties [39, 40]. Metal complexes of bisoxazoline ligands are among the most important catalysts for enantioselective versions of this reaction, and copper(II) complexes of bisoxazoline ligands have been prominent catalysts in the development of the asymmetric Henry reaction. A Cu(OTf)$_2$/t-Bu-Box complex was the first catalyst containing a bisoxazoline ligand to be reported for the enantioselective Henry reaction. Jørgensen and coworkers showed that the addition of nitromethane to an α-keto ester, such as ethyl pyruvate, occurs to form the β-nitro α-hydroxy ester product in high yield and enantioselectivity [41, 42]. Later the Jørgensen group reported a related asymmetric Henry reaction using silyl nitronates as nucleophiles instead of nitroalkanes [43]. Du, Xu, and coworkers also studied the addition of nitromethane to α-keto esters. They showed that a Cu(OTf)$_2$ complex of the tridentate bisoxazoline **20** promotes the addition of nitromethane to ethyl pyruvate with modest enantioselectivity [44], while a complex prepared from Et$_2$Zn and tridentate bisoxazoline **21** promotes the same reaction with good enantioselectivity to form the product with the opposite absolute stereochemistry (Scheme 5.5) [45].

To this point in the development of enantioselective Henry reactions, Lewis acid-bisoxazoline complexes had not been shown to be stereoselective catalysts for reactions of simple aldehyde substrates that are incapable of two-point binding to the chiral Lewis acid. In 2003 Evans *et al.* reported a complex of Cu(OAc)$_2$ and aminoindanol-derived bisoxazoline **22** as a catalyst for the addition of nitromethane to various aromatic and aliphatic aldehydes [23]. These reactions occur to form the nitroaldol products in good to high yield with excellent enantioselectivity. One proposal accounting for the observed absolute stereochemistry involves simultaneous binding of the nucleophile perpendicular to the ligand plane and the electrophile to an equatorial site in the ligand plane as shown in Scheme 5.6. Furthermore, this complex has been applied in catalyst-controlled, diastereoselective Henry reactions that are the key steps in concise syntheses of protected (2S,4R)- and (2S,4S)-4-hydroxyornithine [46]. In subsequent years many Cu(II) complexes of ligands containing bisoxazoline units have been studied as catalysts

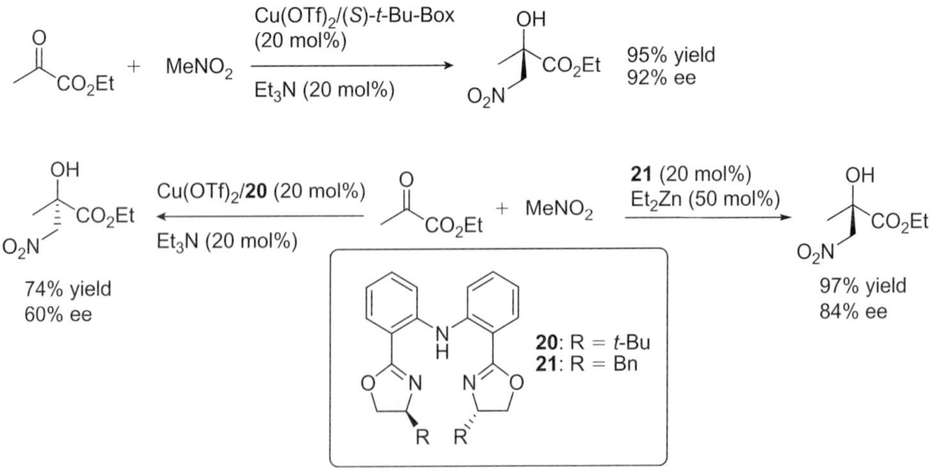

Scheme 5.5 Nitroaldol reactions with keto esters.

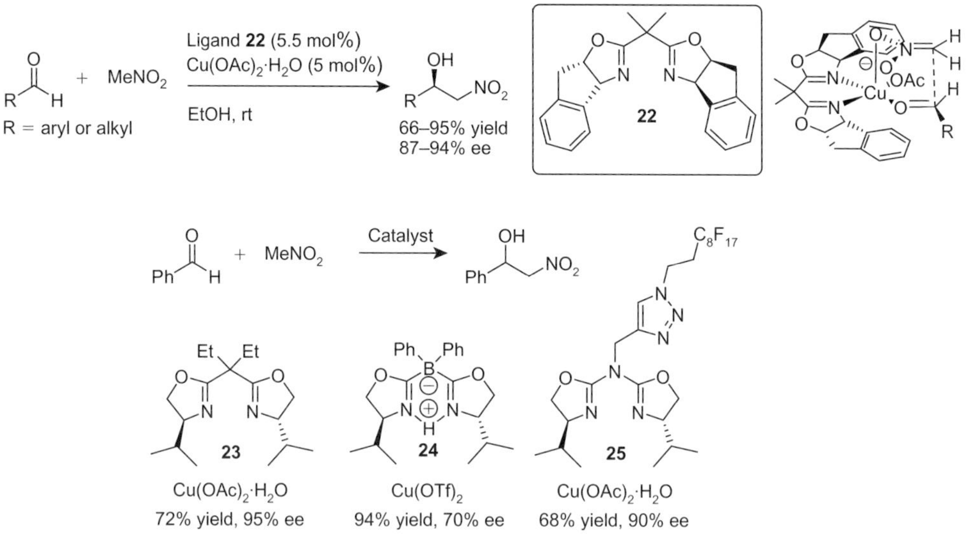

Scheme 5.6 Nitroaldol reactions with aldehydes using various bisoxazolines.

for the enantioselective Henry reaction. These ligands include additional parent bisoxazoline ligands [47, 48], such as **23**; borabox ligands [49], such as **24**; and fluorous-tagged aza(bisoxazoline) ligands [50], such as **25**. Scheme 5.6 summarizes illustrative results from these studies.

5.2.1.4 Nitro-Mannich (Aza-Henry) Reactions

The utility of copper-bisoxazoline complexes as catalysts extends beyond their application in nitroaldol reactions to include the nitro-Mannich or aza-Henry reaction. These reactions encompass additions of silyl nitronates or nitroalkanes to imine derivatives (Scheme 5.7). Jørgensen and coworkers reported Cu(SbF$_6$)$_2$/**26** as catalyst for diastereo- and enantioselective addition of silyl nitronates, such as **27**, to α-imino esters [51] and Cu(OTf)$_2$/Ph-Box as catalyst for diastereo- and enantioselective addition of nitroalkanes to an α-imino ester [52]. The groups of

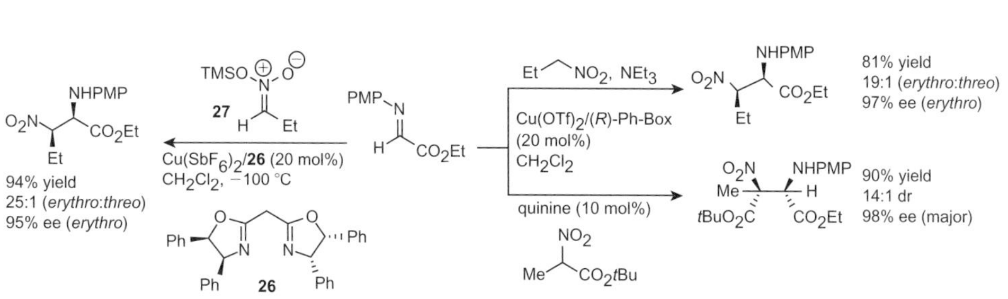

Scheme 5.7 Aza-Henry reactions using bisoxazolines.

Hyeon and Kim subsequently developed an immobilized derivative of the Ph-Box ligand that was grafted onto mesoporous silica gel and upon complexation of Cu(OTf)$_2$ performed with nearly identical selectivity as the homogeneous catalyst [53]. Knudsen and Jørgensen also developed a chiral molecular recognition approach for the aza-Henry reaction of 2-nitropropanoic acid esters and an α-imino ester. They used a Cu(OTf)$_2$/Ph-Box catalyst to activate and control the facial selectivity of the α-imino ester electrophile, while the chiral base, quinine, activates and controls the approach of the 2-nitropropanoic acid ester [54].

5.2.1.5 Addition of Activated Carbon Nucleophiles to Carbonyl Electrophiles

The enantioselective addition of alkyl nucleophiles, such as alkyllithiums and dialkylzincs, to electrophiles containing carbonyl groups is now well established to occur in the presence of chiral bisoxazoline ligands. These reactions generally fall into one of two classes. The first class encompasses the chiral bisoxazoline-mediated addition of α-heteroatom-substituted benzyllithium or α-heteroatom-substituted allyllithium reagents to carbonyl electrophiles, such as carbon dioxide or aldehydes. The second class includes diethylzinc additions to aldehydes that occur in the presence of bisoxazoline ligands that contain O–H or N–H functional groups.

The groups of Nakai and Toru have studied extensively the bisoxazoline-mediated addition of α-heteroatom-substituted benzyllithium or α-heteroatom-substituted allyllithium reagents to carbonyl electrophiles. These reaction encompass additions of α-alkoxybenzyllithium [55–57], α-thiobenzyllithium [58–60], α-thioallyllithium [61], α-lithiated dithioacetal [62], and α-lithiated trifluoromethyl sulfones reagents [63] to various electrophiles, including carbon dioxide, aldehydes, and ketones. Typically, ligands based on the *i*-Pr-Box, *t*-Bu-Box, and Ph-Box scaffolds provide the best asymmetric induction. However, the appropriate choice of ligand is often dependent on the nucleophile and the electrophile. Scheme 5.8 illustrates a subset of the nucleophiles and carbonyl electrophiles that have been employed in these bisoxazoline-mediated addition reactions. The additions of trifluoromethyl sulfones to aldehydes are particularly noteworthy since a substoichiometric quantity (10–30 mol%) of the bisoxazoline ligand is required to achieve excellent diastereo- and enantioselectivity (Scheme 5.9).

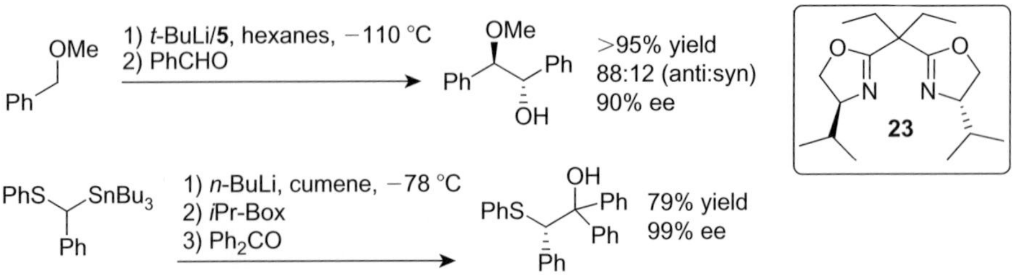

Scheme 5.8 Enantioselective reactions of α-heteroatom stabilized nucleophiles.

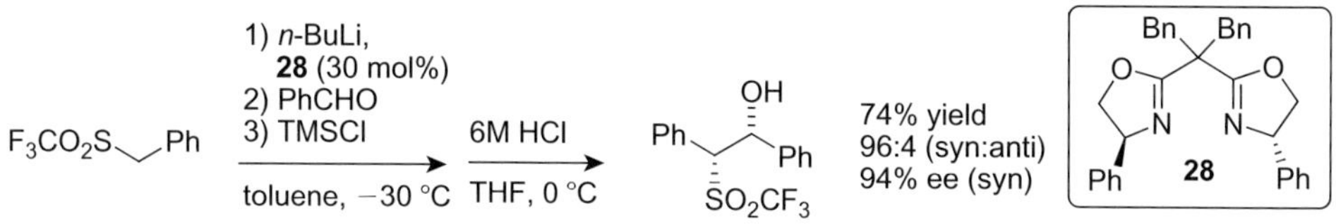

Scheme 5.9 Enantioselective addition of a trifluoromethylsulfone to benzaldehyde.

5.2.1.6 Addition of Activated Carbon Nucleophiles to Imines

Bisoxazoline ligands also promote the addition of organolithium reagents to imine electrophiles. However, the scope of these reactions differ in that bisoxazoline-mediated additions of unfunctionalized organolithium and Grignard reagents to imines occur with high enantioselectivity, while the analogous additions to carbonyl compounds, such as aldehydes and ketones, are less selective. Denmark and coworkers reported the first bisoxazoline-mediated, enantioselective additions of organolithium reagents to imines [64–66]. They showed that the additions of unfunctionalized organolithium reagents, such as methyllithium, to imines derived from aromatic or aliphatic aldehydes occur with high enantioselectivity when conducted in the presence of bisoxazoline ligands based on the *t*-Bu-Box scaffold. The Denmark group also detailed the effect of the bisoxazoline ligand structure on enantioselectivity and the scope of organolithium and imine reagents that are suitable substrates in these addition reactions. This strategy has been utilized in the first asymmetric synthesis of *(R)*-desmethyl-sibutramine **(32)** [67]. In this synthesis the *t*-Bu-Box-mediated addition of *iso*-butyllithium to imine **31** generated desmethyl-sibutramine **(32)** with 40% ee, which was enriched to >99% ee by a single crystallization (Scheme 5.10).

Although bisoxazoline-mediated additions of organolithium and Grignard reagents to imines are more common, Lewis acidic copper(II) and magnesium(II) complexes of bisoxazoline ligands also promote asymmetric additions of carbon radicals to glyoxylic oxime ethers and hydrazones, respectively. Naito and coworkers showed that isopropyl radical addition to an oxime ether of methyl

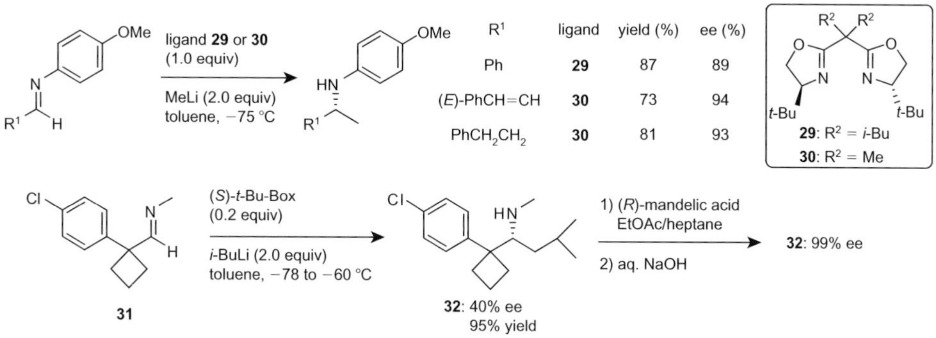

Scheme 5.10 Enantioselective addition of organolithiums to imines.

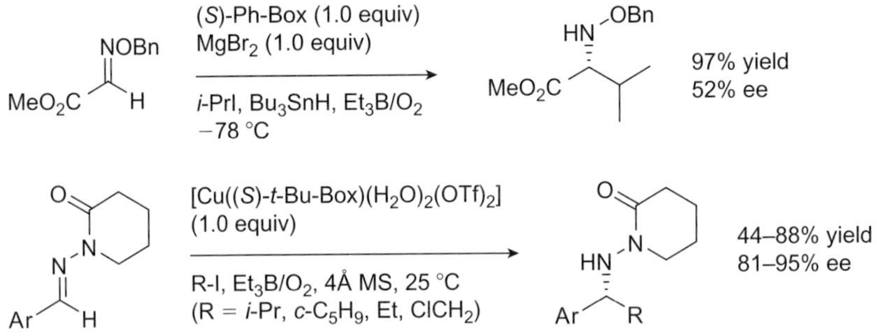

Scheme 5.11 Enantioselective radical addition to oxime ethers and hydrazones.

glyoxylate occurs with modest selectivity when the reaction is conducted in the presence of one equivalent of MgBr$_2$ and (S)-Ph-Box (Scheme 5.11) [68]. Friestad and coworkers later reported the additions of primary and secondary alkyl radicals to N-acyl hydrazones derived from aromatic aldehydes [69]. These reactions are promoted by a complex of Cu(OTf)$_2$ and (S)-t-Bu-Box and form addition products in moderate to good yields with good to excellent enantioselectivities. However, loadings of the chiral Lewis acid below 50 mol% lead to a decrease in the enantioselectivity, presumably as a result of uncatalyzed background reaction.

5.2.1.7 Ene Reactions

In 1998 Evans and coworkers reported that copper(II) complexes of t-Bu-Box and (S)-Ph-Box catalyze highly enantioselective glyoxylate-ene reactions [70]. Notably, [Cu((S)-t-Bu-Box)(H$_2$O)$_2$](SbF$_6$)$_2$ (**33**) and [Cu((S)-Ph-Box)](OTf)$_2$ (**34**) are both efficient catalysts for reactions of various olefins with ethyl glyoxylate and generate products with opposite absolute stereochemistry. Reactions of isobutylene with ethyl glyoxylate are illustrative (Scheme 5.12). The reaction of isobutylene with ethyl glyoxylate occurs in the presence of 1 mol% [Cu((S)-t-Bu-Box)(H$_2$O)$_2$] (SbF$_6$)$_2$ and forms (S)-**35** in 83% yield and 96% ee. In contrast, (R)-**35** is generated in 92% yield and 92% ee when the same reaction is conducted in the presence of 10 mol% [Cu((S)-Ph-Box)](OTf)$_2$. The related catalyst [Cu(S)-t-Bu-Box)](SbF$_6$)$_2$ (**36**) is also an efficient catalyst for additions of olefins to pyruvate esters [71]. For example, the reaction of isobutylene with methyl pyruvate occurs in the presence of 10 mol% [Cu(S)-t-Bu-Box)](SbF$_6$)$_2$ to form (S)-**37** in 76% yield and 98% ee. The observed absolute stereochemistry and enantioselection is consistent with bidentate coordination of the pyruvate ester to a distorted square planar Cu(II)-t-Bu-Box complex that effectively shields the Re face of the pyruvate electrophile.

5.2.1.8 Friedel–Crafts Reactions of Aromatic Compounds with C=O and C=N Bonds

Chiral Lewis acids prepared from bisoxazoline ligands and Cu(OTf)$_2$ are catalysts for enantioselective Friedel–Crafts alkylations of electron-rich aromatic compounds with ethyl glyoxylate, ethyl trifluoropyruvate, and N-sulfonyl imines [72].

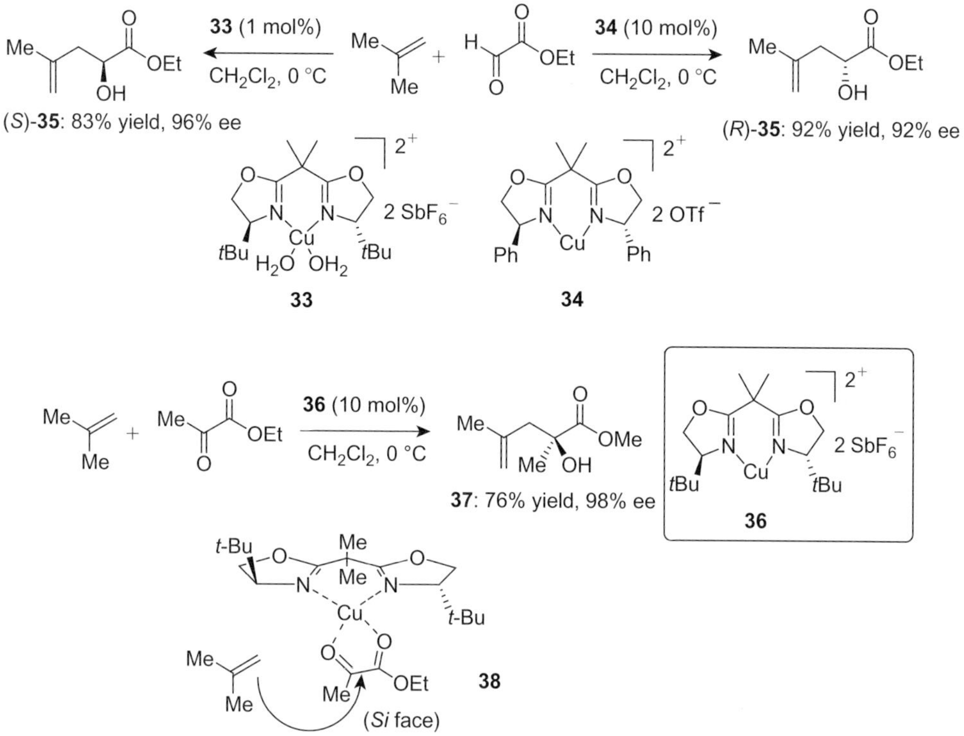

Scheme 5.12 Enantioselective glyoxylate-ene reactions.

In 2000 Jørgensen and coworkers reported enantioselective Friedel–Crafts alkylations of N,N-dimethylanilines with ethyl glyoxylate catalyzed by a complex formed from (S)-t-Bu-Box and Cu(OTf)$_2$ [73]. For example, the alkylation of N,N-dimethylaniline with ethyl glyoxylate occurs in 82% yield with 94% ee when the reaction is conducted in THF (Scheme 5.13). The alkylations of electron-rich aromatic compounds are, however, not limited to N,N-dimethylanilines. Jørgensen et al. also showed that N-methylindoline, N-methyltetrahydroquinoline, and julolidine undergo enantioselective alkylation with ethyl glyoxylate (78–93% ee).

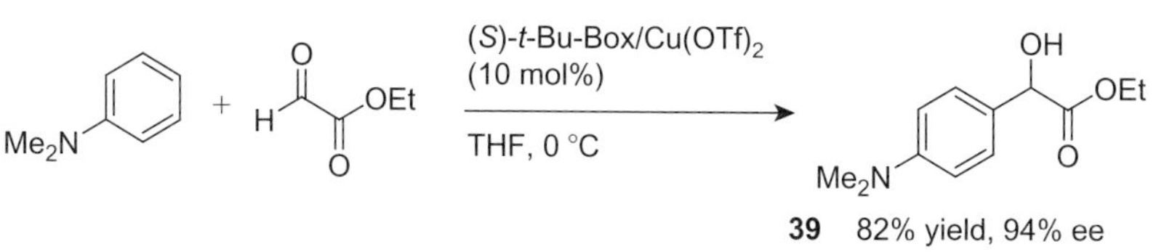

Scheme 5.13 Enantioselective Friedel–Crafts alkylation of ethyl glyoxylate.

5.2.2
1,4-Addition of Carbon Nucleophiles to α,β-Unsaturated Acceptors

Chiral bisoxazoline-Lewis acid complexes are now well established as catalysts in enantioselective 1,4-additions of carbon nucleophiles to α,β-unsaturated acceptors [74, 75]. These reactions encompass a wide range of carbon nucleophiles that includes, but is not limited to, silylketene acetals, enol silanes, 2-trialkylsiloxyfurans, nitroalkanes, 1,3-dicarbonyl compounds, indoles, and carbon-centered radicals. Common α,β-unsaturated acceptors in enantioselective 1,4-additions of carbon nucleophiles include arylidene malonates, imides, nitroalkenes, and α′-hydroxyl ketones. Copper(II) complexes of chiral bisoxazoline ligands are, in general, the most common catalysts for enantioselective 1,4-additions of carbon nucleophiles to α,β-unsaturated acceptors, especially in reactions of arylidene malonate and oxazolidinone imide acceptors. However, magnesium(II), zinc(II), and nickel(II) complexes of bisoxazoline ligands are also effective catalysts in enantioselective 1,4-additions of carbon nucleophiles.

The Mukaiyama–Michael addition of a silylketene acetal or an enol silane to α,β-unsaturated carbonyl compounds is one of the most studied 1,4-addition reactions catalyzed by Lewis acidic chiral bisoxazoline complexes. Evans and coworkers have conducted extensive studies on Cu(II)-bisoxazoline-catalyzed Mukaiyama–Michael reactions involving additions of silylketene thioacetals to arylidene malonates [21, 22] and α,β-unsaturated oxazolidinone imides [20, 76]. Their studies revealed that [Cu((S)-t-Bu-Box)](SbF$_6$)$_2$ is an efficient and highly enantioselective catalyst for both substrate classes. For example, the addition of silylketene thioacetal **41** (R = t-Bu) to β-phenyl arylidene malonate **40** and the addition of silylketene thioacetal **41** to α,β-unsaturated oxazolidinone imide **43** proceed to form products in high yields and enantioselectivities (Scheme 5.14). As is common for a substrate capable of bidentate coordination to a Cu(II)-box complex, the stereochemical outcomes of reactions of both arylidene malonates and α,β-unsaturated oxazolidinone imide are best represented by a distorted square planar substrate–chiral Lewis acid complex. However, the X-ray crystal structure of the [Cu((S)-t-Bu-Box)](SbF$_6$)$_2$ complex with arylidene malonate **40** provides evidence that the malonate ligand is significantly distorted from planarity, allowing the t-Bu group of the ligand to effectively shield the *Re* face of the β-substituted arylidene malonate. Notably, the addition of prochiral silylketene thioacetal **45** to imide **43** occurs with excellent diastereoselectivity. Sibi and Chen further expanded the scope of Cu(II)-bisoxazoline-catalyzed Mukaiyama–Michael reactions to include β-enamidomalonates as α,β-unsaturated acceptors [77]. In contrast to the high selectivities observed in reactions of silylketene thioacetals with β-substituted arylidene malonates catalyzed by [Cu((S)-t-Bu-Box)](SbF$_6$)$_2$, the corresponding additions of silylketene thioacetals to β-enamidomalonates proceed with only modest enantioselectivity. However, a complex prepared from Cu(OTf)$_2$ and an aminoindanol-derived bisoxazoline ligand leads to good enantioselectivities in reactions of silylketene thioacetals with β-enamidomalonates.

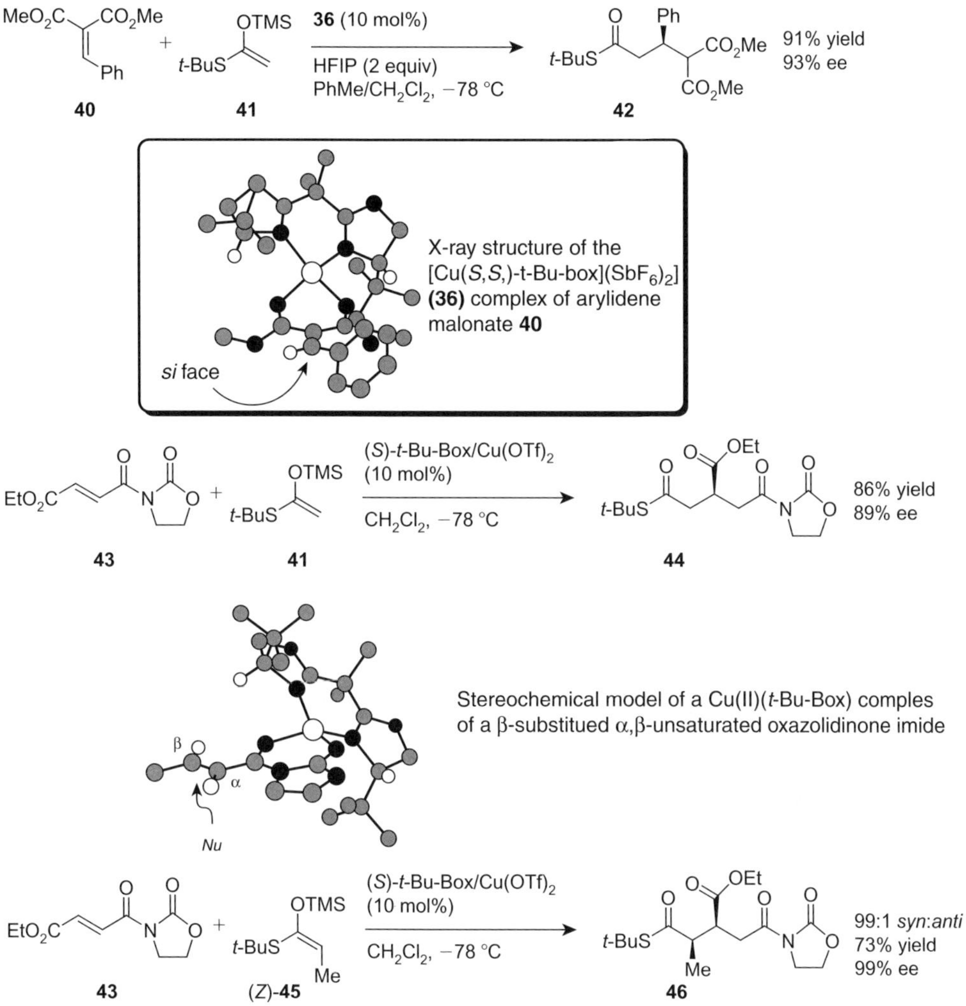

Scheme 5.14 Enantioselective Mukaiyama–Michael reactions.

Enantioselective Michael additions catalyzed by metal complexes of chiral bisoxazoline ligands are not limited to pre-activated nucleophiles. Direct additions 1,3-dicarbonyl compounds, malononitriles, Schiff bases of glycine esters, nitroalkanes, allylsilanes, diorganozinc reagents, N,N-dialkylhydrazones, and electron-rich aromatic compounds to α,β-unsaturated carbonyl compounds and nitroalkenes occur with this class of catalyst. A selection of examples follows to illustrate the types of 1,4-additions of carbon nucleophiles that are catalyzed by metal complexes of chiral bisoxazoline ligands. In 1999 Ji and Barnes developed a method for the enantioselective addition of β-ketoesters and dialkyl malonates to

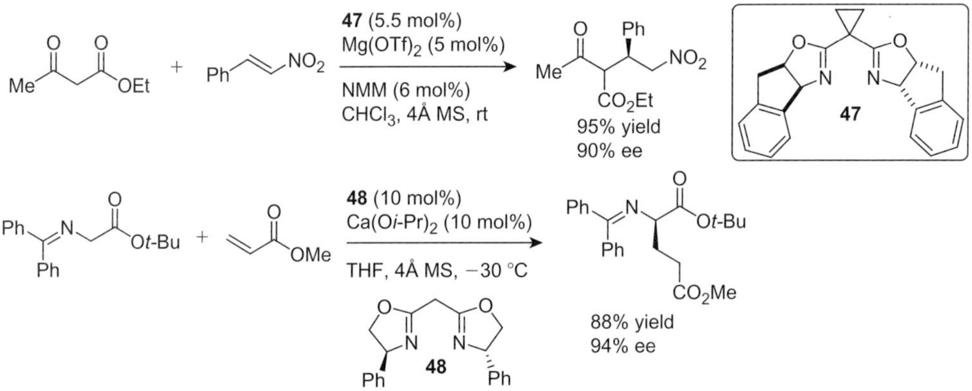

Scheme 5.15 Conjugate addition of nucleophiles to α,β-unsaturated acceptors.

nitroalkenes (Scheme 5.15) [78]. They showed that the addition of ethyl acetoacetate, as well as various additional 1,3-dicarbonyl compounds, to nitrostyrene occurs in high yield and enantioselectivity when conducted in the presence of a complex prepared from Mg(OTf)$_2$ and the aminoindanol-derived bisoxazoline ligand **47**. In the following years many other combinations of chiral bisoxazoline ligands and metal salts have been shown to catalyze enantioselective additions of 1,3-dicarbonyl compounds and malononitriles to α,β-unsaturated acceptors. For example, t-Bu-Box/Cu(OTf)$_2$ catalyzes additions of 4-hydroxycoumarins to α,β-unsaturated α-ketoesters [79], and hydrated complexes of DBFOX-Ph/Ni(ClO$_4$)$_2$ catalyze additions of dimedone, β-hydroxy lactones, and malononitriles to α,β-unsaturated 3,5-dimethylpyrazole amides [80–82]. More recently, Kobayashi and coworkers demonstrated that a calcium complex of the Ph-Box derivative **48** is an effective Brønsted base catalyst for enantioselective additions of glycine Schiff bases to acrylates.

Bisoxazoline ligands also figure prominently in the development of enantioselective Michael additions of electron-rich aromatic compounds to α,β-unsaturated acceptors. These reactions include additions of indoles, furans, electron-rich benzenes, and pyrroles to α,β-unsaturated carbonyl compounds and nitroalkenes, and the catalysts of these reactions are most commonly copper(II) and zinc(II) complexes of chiral bisoxazoline ligands. In 1999 Jørgensen and coworkers have reported enantioselective, Cu(OTf)$_2$/(S)-t-Bu-Box-catalyzed additions of indoles, 2-methylfuran, and 1,3-dimethoxybenzene to α,β-unsaturated α-ketoesters [83]. For example, the addition of 5-methoxyindole to 4-phenyl-2-oxo-butenoate forms the corresponding alkylation product in 95% yield and >99% ee (Scheme 5.16). Palomo and coworkers showed that Cu(OTf)$_2$/t-Bu-Box is an excellent catalyst for addition of indoles and pyrroles to α′-hydroxyenones [84]. With this catalyst the addition of N-methylpyrrole to α′-hydroxyenone **49** forms addition product **50** in 86% yield with 92% ee. Sibi et al. also reported Zn(NTf$_2$)$_2$/DBFOX-Ph as a catalyst

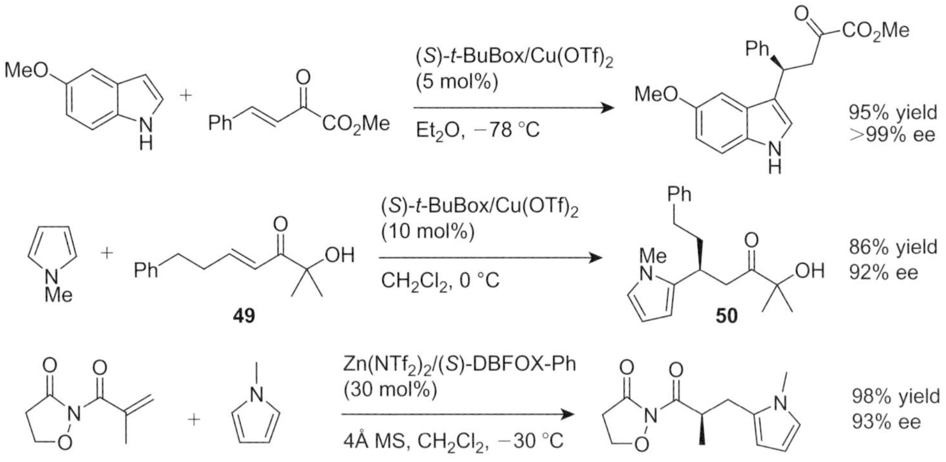

Scheme 5.16 Michael additions of electron-rich aromatic compounds to α,β-unsaturated acceptors.

for tandem 1,4-addition/enantioselective protonation reactions of N-alkyl pyrroles with α-substituted α,β-unsaturated isoxazolidinone imides [85].

The 1,4-additions additions of carbon nucleophiles to α,β-unsaturated acceptors described to this point in the chapter involve reactions that proceed through ionic pathways. However, catalysis by metal complexes of chiral bisoxazoline ligands was also central to the development of enantioselective radical conjugate additions [86]. In 1996 Sibi and Porter reported the first examples of highly enantioselective, catalytic conjugate additions of alkyl radicals to α,β-unsaturated oxazolidinone imides [87]. In these reactions the combinations of MgI_2 with t-Bu-Box and $Zn(OTf)_2$ with Ph-Box were found to provide the highest levels of enantioselectivity (Scheme 5.17). Subsequently, Sibi and Ji showed that the combination of MgI_2 and bisoxazoline **47** derived from aminoindanol provides even higher levels of enantioselectivity at much lower catalyst loading [88]. The same chiral Lewis acid also provides excellent levels of asymmetric induction in enantioselective tandem radical reactions. For example, the MgI_2/**47**-catalyzed addition of t-butyl radical to oxazolidinone cinnamate followed by trapping of the resulting radical with tri(butyl)allylstannane occurs in 84% yield with 99 : 1 diastereoselectivity and 97% ee [89]. The utility of Mg(II)/**47**-catalyzed radical conjugate additions is further demonstrated by the addition of t-butyl radical to α,β-disubstituted α,β-unsaturated imide **51** [90]. This reaction occurs with 99 : 1 *anti* : *syn* selectivity and 90% ee (*anti*) and provides an alternative approach to *anti*-propionate aldol-like products. The absolute and relative stereochemical outcomes of this reaction are consistent with an octahedral MgI_2-**47** complex in which approach of the radical to the top face opposite the aryl group of the ligand is favored. Subsequent hydrogen atom transfer is controlled by the newly formed β-stereocenter leading to approach of the hydrogen atom opposite the radical R_1 group.

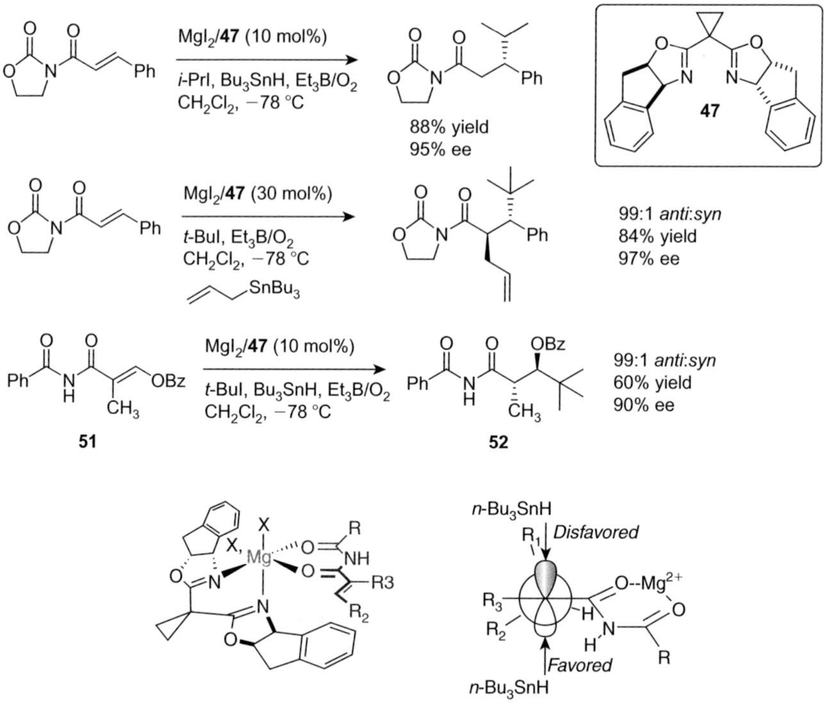

Scheme 5.17 Enantioselective conjugate radical additions.

5.2.3
Reactions of Radicals Alpha to Carbonyls

The application of Lewis acidic metal complexes of chiral bisoxazoline ligands to enantioselective radical conjugate addition reactions is a well established strategy to set stereocenters at the β-position of carbonyl compounds (Section 5.2.2). However, the utility of these complexes is not limited to radical conjugate additions. In 1995 Porter and coworkers reported the $Zn(OTf)_2/(S)$-Ph-Box-mediated addition of alkyl radicals, such as t-butyl radical, to oxazolidinone acrylate followed by enantioselective trapping of the resulting α-radical with allyltributylstannane (Scheme 5.18) [91]. The ability of chiral Lewis acids based on bisoxazoline ligands to mediate enantioselective functionalizations of radicals alpha to carbonyls led to the development of enantioselective atom-transfer radical cyclizations promoted by $Mg(ClO_4)_2/t$-Bu-Box [92, 93], group transfer radical cyclizations promoted by $Mg(ClO_4)_2/t$-Bu-Box [94], and cascade radical addition-cyclization-trapping sequences promoted by a combination of $Zn(OTf)_2$ and aminoindanol-derived bisoxazoline 47 [95, 96]. Furthermore, Sibi and coworkers demonstrated that radical conjugate addition to α-substituted acrylate derivatives followed by enantioselective hydrogen atom transfer is an effective strategy to generate N-acyl-α-amino acid esters [97],

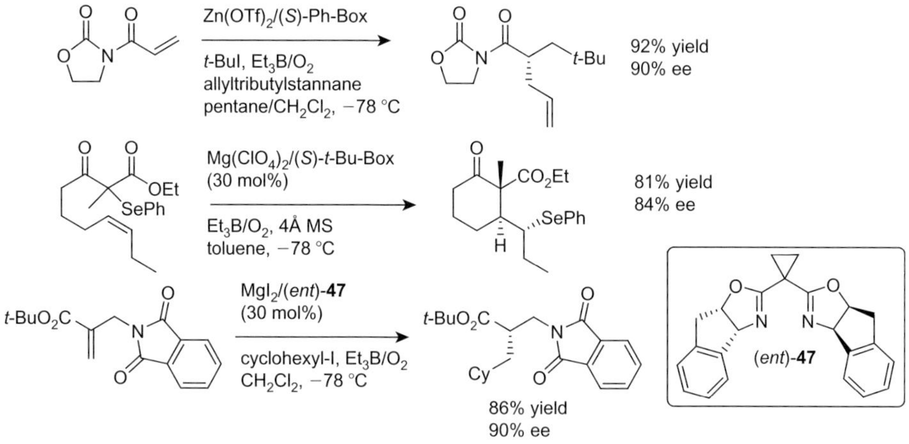

Scheme 5.18 Enantioselective tandem radical reactions.

and β²-amino acid derivatives [98], when these reactions are conducted in the presence of Mg(II) salts and the aminoindanol-derived bisoxazoline *ent*-**47** [99].

5.2.4
Cyclization Reactions

Various stereoselective, C–C bond-forming cyclization and cycloisomerization reactions are known to occur in the presence of metal complexes of bisoxazoline ligands [100–108]. For example, Aggarwal and Belfield reported that Cu(SbF$_6$)$_2$/ (*S*)-*t*-Bu-Box promotes the enantioselective Nazarov cyclization of divinyl ketones bearing α-amide groups (Scheme 5.19) [100]. The asymmetric induction is consistent with a highly distorted square planar copper(II) complex that is similar to geometry at copper observed for the crystal structure of [Cu((*S*)-*t*-Bu-Box)](SbF$_6$)$_2$ complex with β-phenyl arylidene dimethyl malonate (see Scheme 5.14). Cyclization reactions mediated by metal-bisoxazoline complexes are, however, not limited to electrocyclization reactions. In fact, palladium(II) complexes of bisoxazoline ligands catalyze enantioselective cyclizations of 4′-acetoxy-2′-butenyl-2-alkynoates and other related processes [103, 104].

5.2.5
Rearrangement Reactions

Chiral bisoxazoline ligands are stereo-directing elements in various rearrangement processes. These processes include [2,3]- and [1,2]-Wittig rearrangements, Claisen rearrangements, and [2,3]-sigmatropic rearrangements of sulfur ylides. Nakai *et al.* used *i*-Pr-Box derivative **23** to mediate asymmetric lithiation of *(E)*-crotyl propargylic ether **53**, which undergoes a [2,3]-Wittig rearrangement to generate

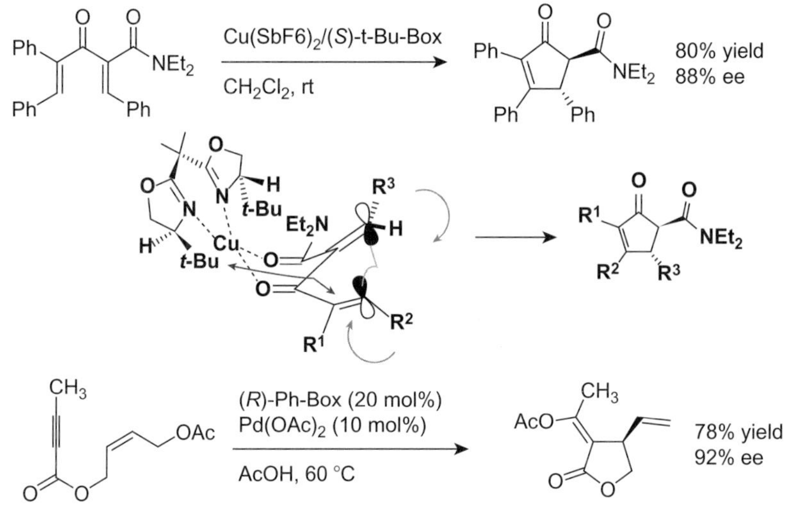

Scheme 5.19 Enantioselective cyclization reactions.

propargylic alcohol **54** in 89% ee (Scheme 5.20) [109, 110]. The utility of chiral bisoxazoline ligands in enantioselective rearrangement processes extends beyond asymmetric lithiation followed by rearrangement. Hiersemann and coworkers found that Cu(OTf)$_2$/(S)-Ph-Box catalyzes enantioselective Claisen rearrangements of 2-alkoxycarbonyl-substituted allyl vinyl ethers that form enantioenriched β-ketoesters with high diastereo- and enantioselectivity (Scheme 5.20) and also showed that [Cu((S)-t-Bu-box)(H$_2$O)$_2$](SbF$_6$)$_2$ provides even higher enantioselectivity in Claisen rearrangements of similar substrates that lack substitution at the terminal allyl moiety [111–113].

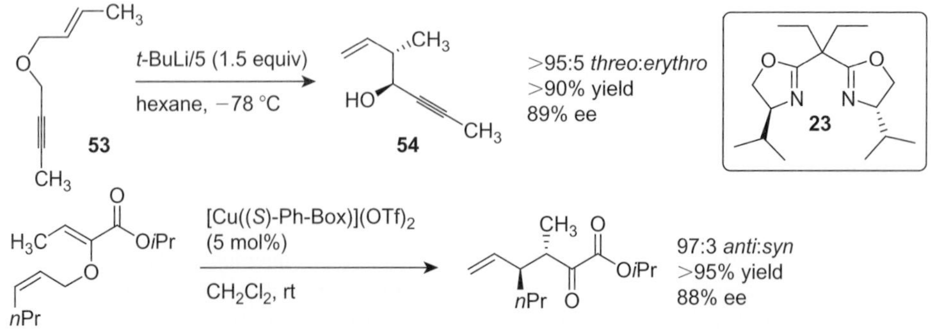

Scheme 5.20 Enantioselective Claisen rearrangements.

5.3
Enantioselective Carbon–Heteroatom Bond Formation

5.3.1
1,4-Addition of Heteroatom Nucleophiles to α,β-Unsaturated Acceptors

5.3.1.1 1,4-Addition of Nitrogen Nucleophiles

Conjugate addition of nitrogen nucleophiles to α,β-unsaturated acceptors using chiral Lewis acids derived from bisoxazolines has been investigated in some detail by several groups since these reactions provide direct access to β-amino acid derivatives. Sibi and coworkers have studied 1,4-additions of hydroxylamine derivatives to α,β-unsaturated acceptors [114–117]. The first of these reports detailed the addition of O-benzylhydroxylamine to 3,5-dimethylpyrazole amides using a chiral Lewis acid prepared from MgBr$_2$ and aminoindanol-derived bisoxazoline 47 (Scheme 5.21) [117]. 3,5-Dimethylpyrazole was the optimal achiral template for these reactions, and the additions to β-alkyl substituted substrates occurred in high yields and selectivities. However, the addition to a substrate containing a β-phenyl substituent was not efficient and gave the product in low yield with moderate selectivity. Jørgensen and coworkers reported a protocol for the addition of modestly nucleophilic anilines to β-substituted oxazolidinone enoates using Ni(ClO$_4$)$_2$/Ph-DBFOX as the catalyst [118]. Reactions of N-methylanilines with electron-donating substituents on the aryl ring resulted in the highest yields and selectivities of the conjugate addition products.

Conjugate additions to α,β-disubstituted acceptors are often difficult because these substrates generally exhibit low reactivity and a lack of rotamer control. Sibi and coworkers reported the use of NH-imides as achiral templates in the conjugate addition of N-benzylhydroxylamine to α,β-disubstituted substrates using Mg(NTf$_2$)$_2$/47 as the chiral Lewis acid catalyst (Scheme 5.22) [119]. Various imide substituents were evaluated in this study, with i-propyl and cyclohexyl (R^3) groups being optimal. The reaction shows excellent substrate scope and proceeds with high yield, diastereoselectivity, and enantioselectivity. The product isoxazolidinones are readily converted into the parent β-amino acids by hydrogenation.

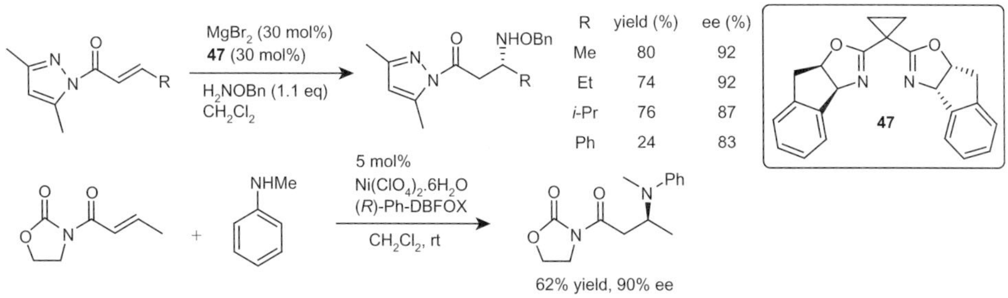

Scheme 5.21 Enantioselective conjugate amine additions.

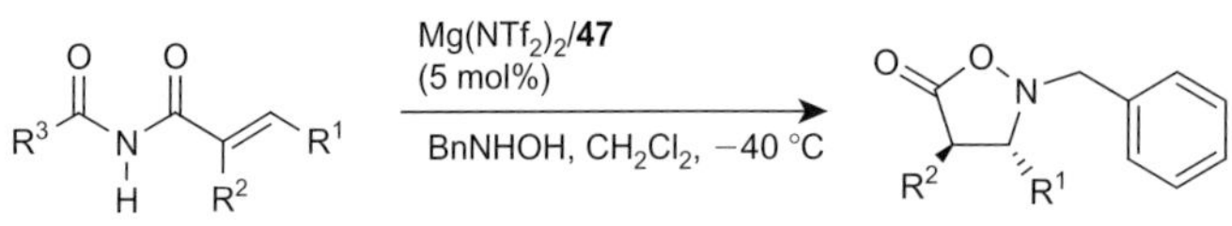

Scheme 5.22 Conjugate amine additions: establishment of two chiral centers.

R¹	R²	R³	yield (%)	ee (%)
Me	Me	i-Pr	95	96
Me	Me	c-Hexyl	66	90
Me	Ph	i-Pr	90	95
Et	Et	i-Pr	72	96
Ph	Me	i-Pr	38	95

Palomo and coworkers demonstrated that α′-hydroxyenones are excellent substrates in the conjugate additions of carbamates catalyzed by Cu(II)-bisoxazoline complex **55** (Scheme 5.23) [120]. This method is noteworthy because it uses the less nucleophilic carbamates as nucleophiles. The method has good substrate scope and provides the addition products in moderate to good yields and high enantioselectivity. The products can be readily converted into β-amino acids by an oxidative cleavage reaction.

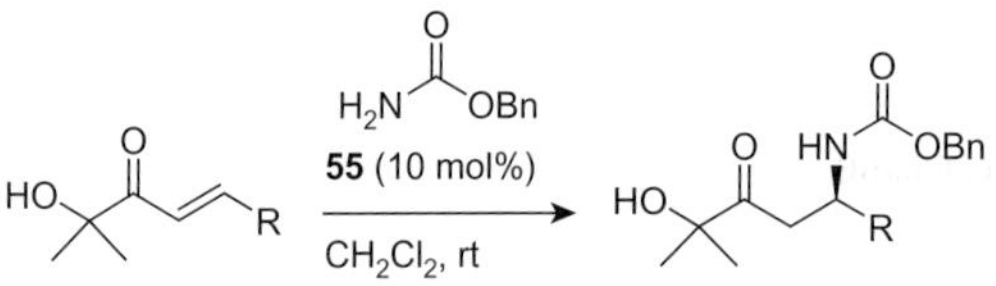

R	yield (%)	ee (%)
Et	83	96
c-C$_6$H$_{11}$	57	95
i-Pr	53	98
t-Bu	65	94

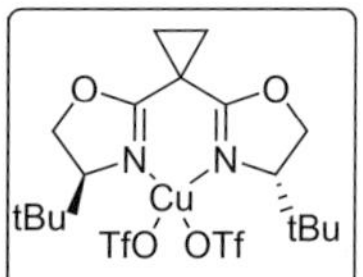

Scheme 5.23 Conjugate addition of carbamates.

Sibi and coworkers also reported the enantioselective conjugate addition of hydrazines to β-substituted imides using 30 mol% Mg(ClO$_4$)$_2$/*ent*-**47** as the chiral Lewis acid catalyst (Scheme 5.24) [121]. The products from this reaction, 5-substituted-pyrazolidin-3-ones, are formed with good chemo- and enantioselectivity. The isomer containing N1 substitution is formed as the major product with >98 : 2 selectivity. The absolute stereochemistry of the pyrazolidinone products is

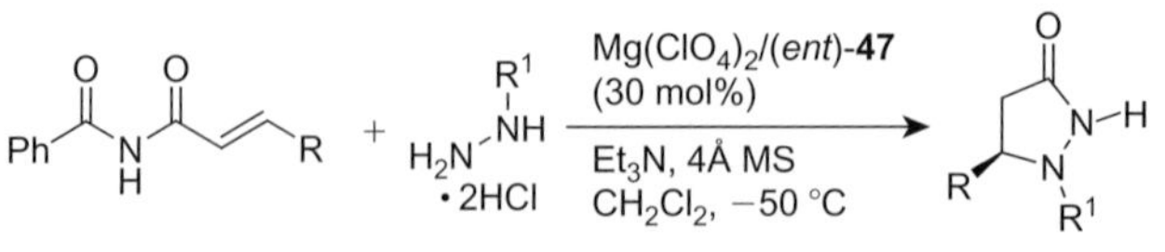

R	R¹	yield (%)	ee (%)
Me	Bn	92	84
Et	Bn	93	80
CH$_2$OBn	Bn	70	95
Me	i-Pr	79	87

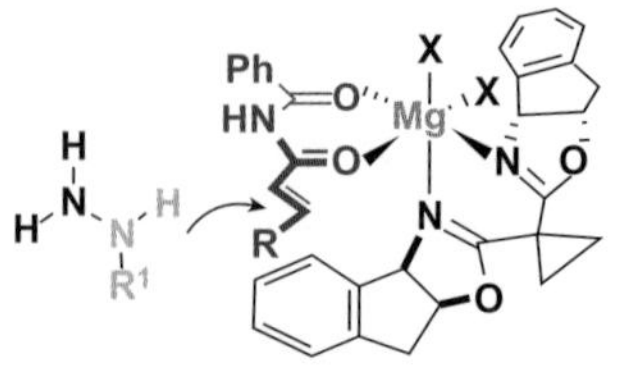

Scheme 5.24 Conjugate hydrazine additions.

5.3.1.2 1,4-Addition of Sulfur and Oxygen Nucleophiles

Kanemasa and coworkers reported the first example of enantioselective conjugate thiol addition using a catalyst prepared from Ph-DBFOX and Ni(ClO$_4$)$_2$ [122]. The additions of various thiophenols to oxazolidinone crotonate were evaluated. The products were generally obtained in high yield with excellent enantioselectivity. Notably, the addition of the bulky mesitylthiol to oxazolidinone crotonate occurred with very high enantioselectivity. There is only one report on the use of less nucleophilic oxygen nucleophiles in conjugate additions catalyzed by metal-bisoxazoline complexes. Jørgensen and coworkers reported the conjugate addition of an electron-rich phenol to an α,β-unsaturated keto ester using a catalyst prepared from Mg(OTf)$_2$ and the 1-naphthyl-substituted bisoxazoline ligand 56 (Scheme 5.25) [123]. Conjugate addition followed by an intramolecular Friedel–Crafts cyclization furnished the benzopyran product in good yield and selectivity.

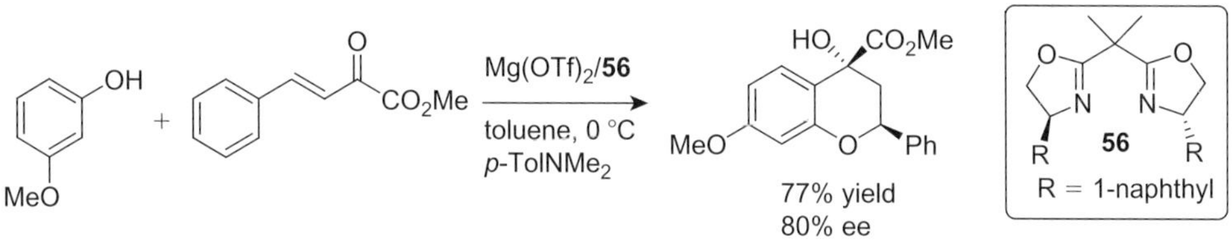

Scheme 5.25 Conjugate addition of an oxygen nucleophile.

5.3.1.3 1,4-Addition of Boron Nucleophiles

The enantioselective addition of boron reagents to α,β-unsaturated carbonyl compounds has traditionally been catalyzed by metal complexes of phosphine, NHC, diene, or P,N ligands. However, Nishiyama and coworkers recently reported the use of a chiral rhodium-bis(oxazolinyl)phenyl acetate complex (57) in β-boration of α,β-unsaturated carbonyl compounds [124]. β-Boration using bis(pinacolato)diboron as pro-nucleophile, sodium t-butoxide as base, and 1 mol% 57 as catalyst occurred with high yield and selectivity (Scheme 5.26).

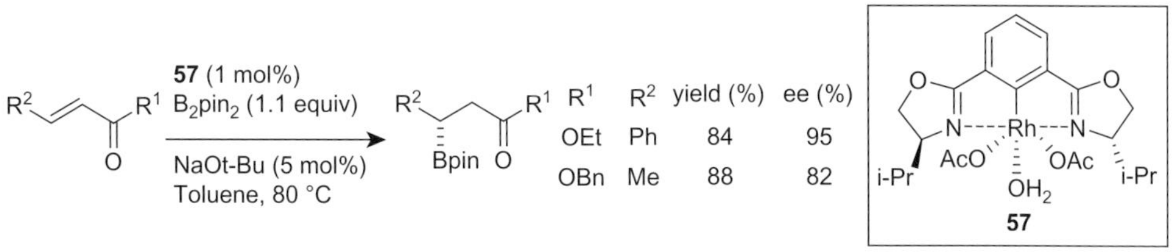

R^1	R^2	yield (%)	ee (%)
OEt	Ph	84	95
OBn	Me	88	82

Scheme 5.26 β-Boration of α,β-unsaturated carbonyl compounds.

5.3.2
Allylic Functionalization Reactions

Bisoxazoline ligands played an important role in the early development of enantioselective methodologies for the functionalization of allylic C–H bonds. In a very interesting study, Clark and coworkers reported the allylic amination of both acyclic and cyclic olefins using a Cu(I) salt and Ph-Box as a ligand (Scheme 5.27) [125]. These reactions proceed by a radical pathway and form the amination products in modest yield and selectivity. Depending on the nature of the aryl group of the carbamate, either the allylic amination or the oxidation product could be obtained selectively. The research groups of Andrus and Pfaltz independently described the first examples of asymmetric allylic oxidation using Cu(I)/bisoxazoline catalysts [126–128]. This transformation is known as the Kharasch–Sosnovsky reaction and it proceeds via a radical intermediate. After the initial work, Andrus reported significant improvements to enantioselectivity for reactions with cyclic olefins [129].

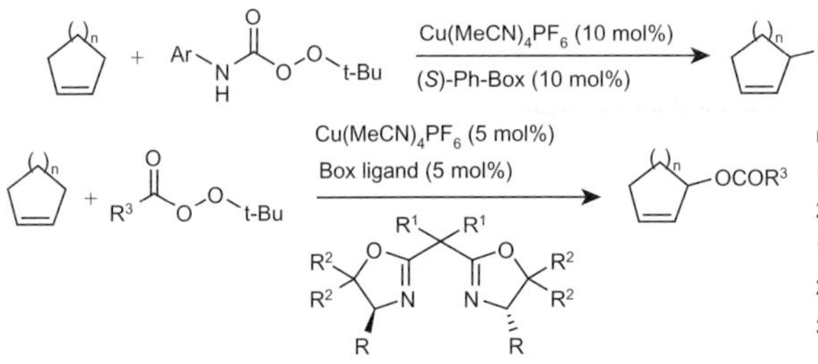

n	yield (%)	ee (%)
1	30	46
2	44	51

n	R	R¹	R²	R³	yield (%)	ee (%)
1	Ph	Me	Me	Ph	49	81
2	t-Bu	Me	H	Ph	43	80
1	Ph	Et	H	4-NO$_2$Ph	41	99
2	Ph	Me	H	4-NO$_2$Ph	44	96
3	i-Pr	Me	H	4-NO$_2$Ph	14	99

Scheme 5.27 Enantioselective allylic oxygenations.

5.3.3
α-Heteroatom Functionalization of Carbonyl Compounds

The α-functionalization of carbonyl compounds with heteroatom electrophiles encompasses a broad range of synthetic transformations [130]. These transformations primarily involve amination, oxygenation, and halogenation processes. Metal-bisoxazolines are established catalysts for each of these reaction types and the following sections summarize the developments in this area.

5.3.3.1 Amination
Neutral or ionic nucleophilic addition of carbonyl compounds to diazo esters results in the formation of hydrazides that can be readily converted into amines. Evans and coworkers reported the first example of enolsilyl ether and silylketene acetal addition to an N-acyloxazolidinone substituted diazoester (Scheme 5.28) [131]. These reactions are catalyzed by Cu(OTf)$_2$/t-Bu-Box, and the aminated products are formed in high yields with outstanding levels of enantioselectivity. Taking advantage of the acidity of α-hydrogens in 2-ketoesters, Jørgensen and

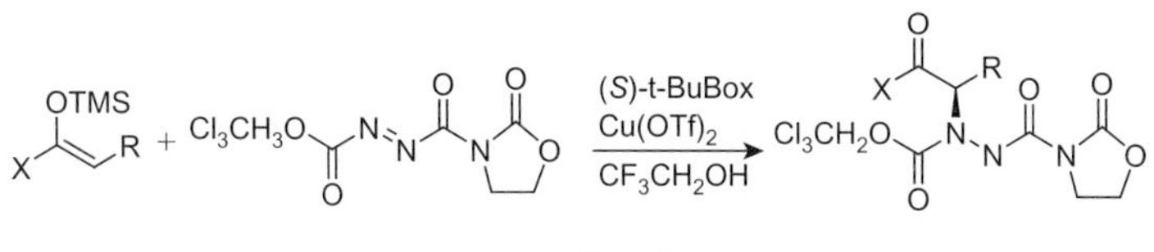

Scheme 5.28 α-Amination of carbonyl compounds.

X	R	yield (%)	ee (%)
Ph	Me	95	99
Ph	i-Pr	86	99
t-BuS	Me	85	96
Pyrrole	Me	96	99

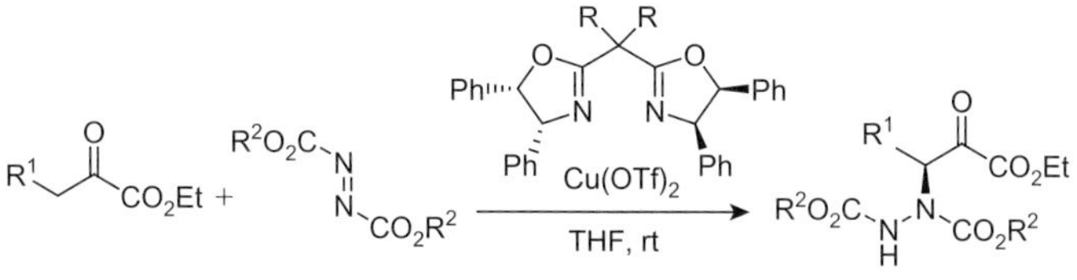

R^1	R^2	R	yield (%)	ee (%)
Bn	Bn	H	57	89
Bn	Bn	Me	58	88
Me	Bn	Me	44	92
i-Pr	Bn	H	60	95

coworkers developed a highly selective amination protocol catalyzed by complexes of $Cu(OTf)_2$ and diPh-Box derivatives [132]. The reactions gave modest yields of the amination products, but the enantioselectivities were generally high.

5.3.3.2 Oxygenation

Metal-bisoxazoline complexes are also catalysts for enolate oxidations. Shibata and coworkers developed a catalytic enantioselective method for the α-oxidation of carbonyl compounds (Scheme 5.29). Using a combination of $Zn(OAc)_2$ and DBFOX-Ph, this group showed that oxindoles undergo oxidation in high yield and enantioselectivity using oxaziridine **58** as the stoichiometric oxidant [133]. They also demonstrated that the α-hydroxylation of unsymmetrical malonates occurs with high enantioselectivity in the presence of the Ni(II)/DBFOX-Ph catalyst system [134].

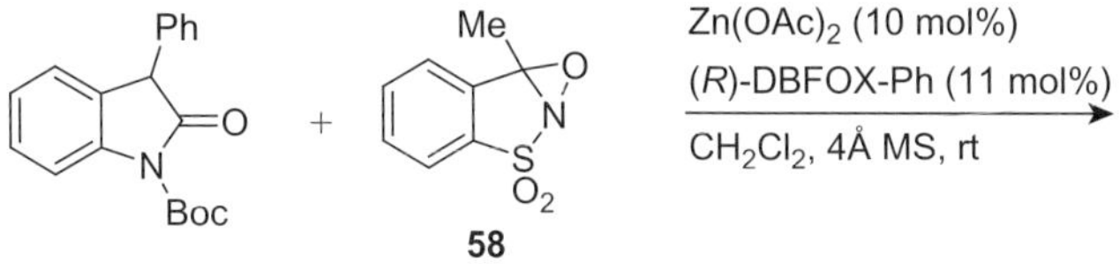

Scheme 5.29 Enantioselective enolate oxidation.

5.3.3.3 Halogenation

Fluorinated organic compounds are important intermediates in medicinal chemistry and thus the development of enantioselective methods for their synthesis is significant. Cahard and coworkers reported an efficient method for fluorination of cyclic β-ketoesters using N-fluorobenzenesulfonimide (NFSI) as the electrophile [135]. Shibata demonstrated that Ni(II) and Zn(II) complexes of DBFOX-Ph catalyze α-fluorination of carbonyl compounds with high enantioselectivities. Enantioselective α-fluorination of β-ketoesters and oxindoles occurs in the presence of Ni(II)/DBFOX-Ph [136]. Furthermore, this group developed highly

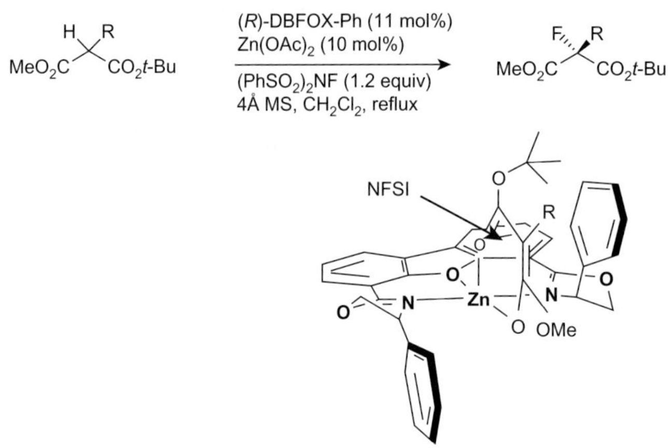

Scheme 5.30

R	yield (%)	ee (%)
Bn	90	98
Et	94	96
Ph	95	99

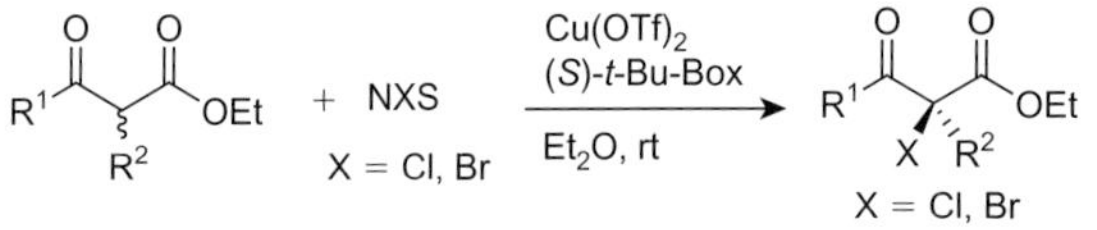

R^1	R^2	X = Cl yield (%)/ee (%)	X = Br yield (%)/ee (%)
Me	Me	98/77	98/80
Et	Me	96/57	77/57
Ph	Me	98/53	95/41

Scheme 5.30 Enantioselective enolate halogenations.

enantioselective α-fluorinations of α-substituted unsymmetrical malonates using Zn(OAc)$_2$/DBFOX-Ph as the catalyst (Scheme 5.30) [137]. Octahedral complexes ligated by a water molecule are proposed to account for the asymmetric induction observed in the halogenations catalyzed by Ni(II)/DBFOX-Ph and Zn(II)/DBFOX-Ph (shown in Scheme 5.30). N-Halosuccinimides are excellent reagents for electrophilic halogenation. Jørgensen and coworkers showed that β-ketoesters can be halogenated in high yields using Cu(OTf)$_2$/t-Bu-Box as the catalyst [138]. Both chlorinated and brominated products could be obtained with modest to good enantioselectivities.

5.3.4
X–H Insertion Reactions (X = O, N, S)

Insertion of a metal-carbenoid into a heteroatom–H bond provides a direct method for the synthesis of amines, alcohols, ethers, and sulfides. Zhou and coworkers developed novel spirobisoxazolines, such as **59**, and used these ligands in various enantioselective transformations, including insertion reactions [139]. For example, reactions of anilines with α-diazoesters in the presence of CuCl/**59** as catalyst formed α-amino esters in high yields and enantioselectivities (Scheme 5.31). Zhou and coworkers also demonstrated that the insertion of α-diazocarbonyl compounds into O–H bonds can be carried out using phenols as substrates in the presence of a Cu(I)/spirobisoxazoline **59** complex [140]. These reactions formed ethers in good to high yields and enantioselectivities. Reactions with alkyl substituents on the

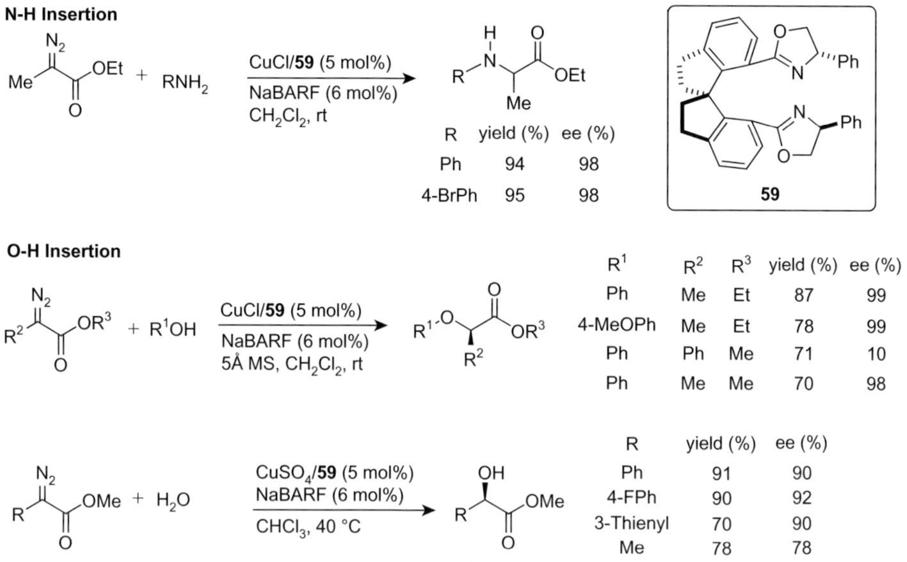

5.3 Enantioselective Carbon–Heteroatom Bond Formation

N-H Insertion

Me-C(=N$_2$)-C(=O)-OEt + RNH$_2$ →[CuCl/59 (5 mol%), NaBARF (6 mol%), CH$_2$Cl$_2$, rt] R-NH-CH(Me)-C(=O)-OEt

R	yield (%)	ee (%)
Ph	94	98
4-BrPh	95	98

Ligand **59**: spirobisoxazoline with Ph substituents

O-H Insertion

R^2-C(=N$_2$)-C(=O)-OR^3 + R^{1}OH →[CuCl/59 (5 mol%), NaBARF (6 mol%), 5Å MS, CH$_2$Cl$_2$, rt] R^1-O-CH(R^2)-C(=O)-OR^3

R^1	R^2	R^3	yield (%)	ee (%)
Ph	Me	Et	87	99
4-MeOPh	Me	Et	78	99
Ph	Ph	Me	71	10
Ph	Me	Me	70	98

R-C(=N$_2$)-C(=O)-OMe + H$_2$O →[CuSO$_4$/59 (5 mol%), NaBARF (6 mol%), CHCl$_3$, 40 °C] R-CH(OH)-C(=O)-OMe

R	yield (%)	ee (%)
Ph	91	90
4-FPh	90	92
3-Thienyl	70	90
Me	78	78

Scheme 5.31 Enantioselective insertion reactions into heteroatom–H bonds.

α-carbon (R^2) of the diazocarbonyl compounds gave higher selectivity than those with aryl substituents. Another O–H insertion of note is the direct formation of alcohols by insertion into an O–H bond of water [141]. This reaction utilizes CuSO$_4$/spirobisoxazoline **59** as the catalyst, and the product alcohols are formed in excellent yields and selectivities. Zhou and coworkers have also extended heteroatom–H insertion reactions to encompass thiols as substrates. The insertions of α-diazoesters into benzylthiol occur in the presence of CuCl/spirobisoxazoline **59** [142]. The product sulfides are formed in good yields and enantioselectivities.

5.3.5 Cyclization Reactions

5.3.5.1 Carbonylative Cyclization

Palladium(II)-catalyzed asymmetric cyclization–carbonylation has been investigated by Kato and coworkers [143]. Reactions of 2-alkyl-2-proprgyl-1,3-cyclohexanediones in the presence carbon monoxide, an alcohol, and Pd(TFA)$_2$/bisoxazoline as the catalyst gave the cyclized product in good yield and selectivity (Scheme 5.32). Of the different box ligands examined, the 3,4-dimethoxyphenyl-Box ligand **60** gave the best results.

5.3.5.2 Wacker-Type Cyclizations

Palladium(II)-catalyzed Wacker-type cyclizations [144, 145] are intermolecular variants of one of the most important industrial reactions, the Wacker reaction. Zhang and coworkers reported the Wacker-type cyclization of a 2-allylphenol that utilizes novel biphenyl-derived bisoxazolines as ligands (Scheme 5.33) [145]. These

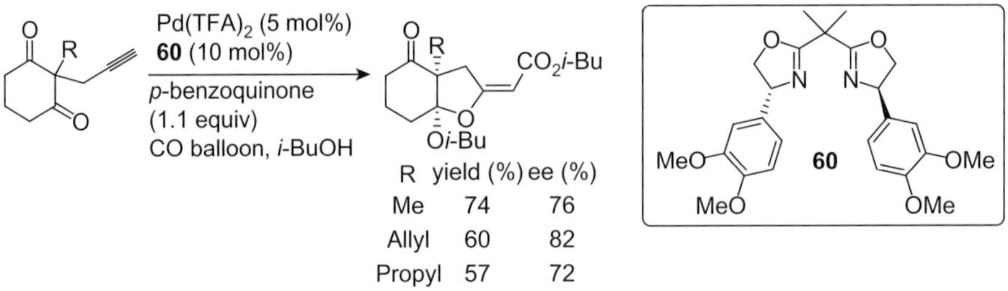

Scheme 5.32 Palladium(II)-catalyzed asymmetric cyclization–carbonylation.

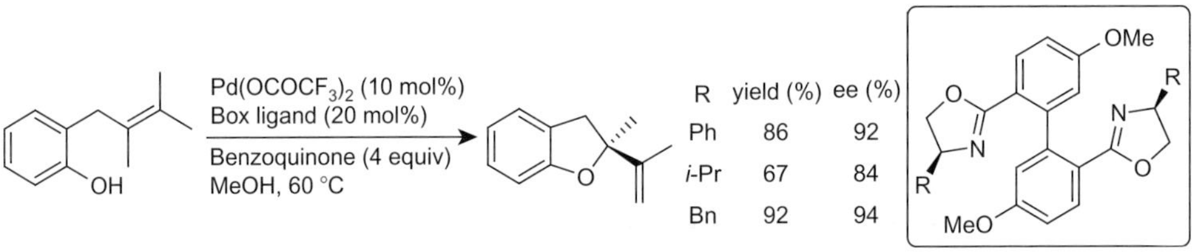

R	yield (%)	ee (%)
Me	74	76
Allyl	60	82
Propyl	57	72

Scheme 5.33 Palladium(II)-catalyzed Wacker-type cyclizations.

R	yield (%)	ee (%)
Ph	86	92
i-Pr	67	84
Bn	92	94

ligands have free rotation around the biphenyl axis and form single diastereomer metal complexes. The ligands, thus, avoid any match–mismatch issues associated with bisoxazoline ligands containing a fixed chiral axis. In this reaction benzoquinone functions as the stoichiometric oxidant, and the cyclized products are formed in good yields with high selectivity.

5.3.5.3 Hydroamination

Conjugate additions of amine nucleophiles to electron-deficient acceptors catalyzed by metal-bisoxazoline complexes are well-explored processes (Section 5.3.1). In contrast, amine or lithium amide addition to unactivated double bonds is less explored with metal-bisoxazoline catalysts. Tomioka and coworkers reported a bisoxazoline-mediated, enantioselective intramolecular hydroamination via lithium amides (Scheme 5.34) [146]. The reaction proceeds in high yield and >90% ee. Various bisoxazoline ligands were investigated, and ligand **61**, which contains an additional chiral center in the 4-substituents of the oxazoline groups, gave the best results.

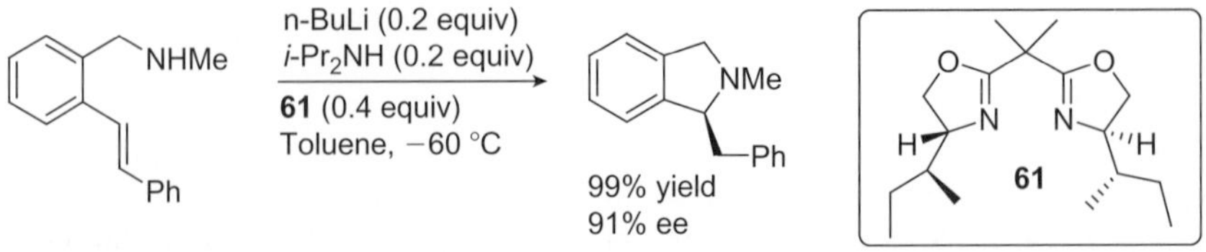

Scheme 5.34 Enantioselective hydroamination reaction.

5.3.6
Kinetic Resolution and Desymmetrization Reactions

5.3.6.1 Kinetic Resolution

Kinetic resolutions are well suited for obtaining both enantiomers of chiral materials with high enantiopurity, and several reports on the utility of bisoxazoline ligands in these transformations have emerged over the past decade. Two examples that illustrate this utility are presented in Scheme 5.35. Matsumura and coworkers showed that Cu(II)/Ph-Box effects highly selective kinetic resolution of 1,2-diols [147]. The resolution of hydrobenzoin proceeds with spectacular levels of selectivity that are among the highest ever reported for kinetic resolutions. Reiser and coworkers also developed polymer-supported azabox ligands for the kinetic resolution of diols (s factor >700) [148, 149]. Pfaltz and coworkers also investigated the utility of boron-bridged bisoxazolines (borabox) in kinetic resolutions [150]. The resolution of a hydroxytetrahydroquinoline using 1 mol% $CuCl_2$/**62** as catalyst gave the alcohol and the benzoylated products in high enantiomeric excess. In related studies, Onomura and coworkers showed Cu(II)/Ph-Box to be effective as a catalyst for kinetic resolutions of 3- and 2-hydroxyalkanamides [151, 152].

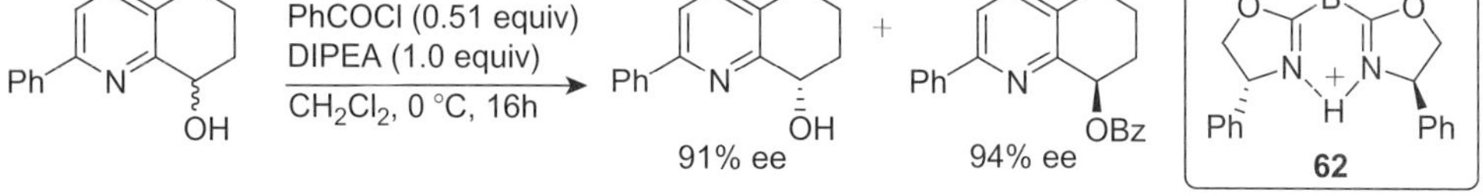

Scheme 5.35 Kinetic resolution of alcohols using bisoxazolines.

5.3.6.2 Desymmetrization

A novel method for the synthesis of enantioenriched *tert*-alkylamines has been reported by Kang and coworkers based on the desymmetrization of 2,2-disubstituted-1,3-propanediols in the presence of benzoyl chloride as the acylating agent and $CuCl_2$/(4R,5S)-diPh-Box complex **63** as catalyst (Scheme 5.36) [153]. The reaction shows good substrate scope and provides the protected diol products in high yield and enantioselectivity.

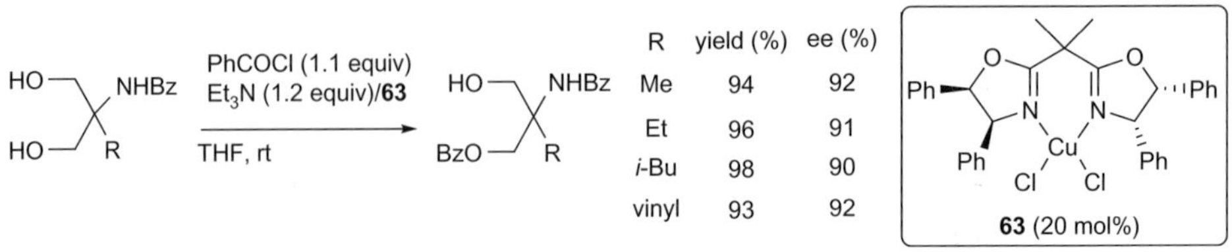

Scheme 5.36

Enantioselective desymmetrization of diols.

R	yield (%)	ee (%)
Me	94	92
Et	96	91
i-Bu	98	90
vinyl	93	92

5.4
Enantioselective Cycloaddition Reactions

5.4.1
Carbo-Diels–Alder Cycloadditions

Enantioselective Diels–Alder reactions are among the most well-studied asymmetric processes in organic synthesis [154]. In fact, the enantioselective Diels–Alder reaction is now often used as a benchmark to test the utility of new chiral ligands. The widespread application of chiral bisoxazoline ligands is responsible, at least in part, for the classification of the Diels–Alder reaction as a "benchmark" reaction. Corey and coworkers reported the first bisoxazoline-Lewis acid catalysts for enantioselective Diels–Alder cycloadditions [14]. They found that Fe(III) and Mg(II) complexes of (S)-Ph-Box catalyze the cycloaddition of oxazolidinone acrylate and cyclopentadiene with excellent diastereoselectivity and up to 91% ee (see entries 1 and 2, Scheme 5.37). These initial reports began what is now two decades of steady investigation into enantioselective Diels–Alder reactions that are catalyzed by chiral bisoxazoline-based Lewis acids. Over these two decades many combinations of chiral bisoxazoline ligands and metal salts have been used to promote enantioselective Diels–Alder cycloadditions. Furthermore, the methodology developed over this time has been applied to the synthesis of various natural products. Space does not allow a comprehensive survey of catalysts based on bisoxazoline ligands and applications of these catalysts in enantioselective Diels–Alder reactions. Thus, this section will outline combinations of chiral bisoxazoline ligands and metal salts that are common in highly enantioselective Diels–Alder reactions and will highlight additional examples that helped to move the field forward.

A list of common chiral bisoxazoline ligand/metal salt combinations that catalyze highly enantioselective Diels–Alder cycloadditions is presented in Scheme 5.37. In addition, a selection of substrates capable of two-point coordination to the chiral Lewis acid is included to highlight the types of functional group that are typically required to achieve high enantioselectivity. After Corey's initial report on the utility of Fe(III) and Mg(II) complexes of Ph-Box in enantioselective Diels–Alder reactions of oxazolidinone acrylate and cyclopentadiene [14, 155], Evans reported that a Zn(II) complex of this ligand catalyzed the cycloaddition with similar levels of diastereo- and enantioselectivity (entry 3) [156]. However, the most

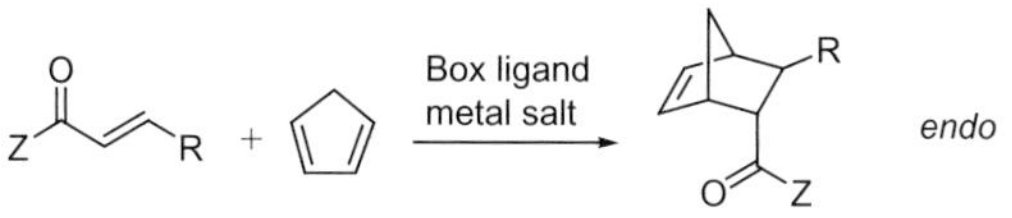

Entry	Z	R	Box ligand	metal salt	endo:exo	endo ee (%)
1	Z1	H	(S,S)-Ph-Box	FeI$_3$	99:1	−86
2		H	(S,S)-Ph-Box	Mg(SbF$_6$)$_2$	98:2	−91
3		H	(S,S)-Ph-Box	Zn(SbF$_6$)$_2$	98:2	−92
4		H	(S,S)-t-Bu-Box	Cu(OTf)$_2$	98:2	>98
5		H	(S,S)-t-Bu-Box	Cu(SbF$_6$)$_2$	96:4	>98
6		Me	(S,S)-t-Bu-Box	Cu(OTf)$_2$	96:4	97
7		Me	(S,S)-t-Bu-Box	Cu(SbF$_6$)$_2$	85:15	99
8		Ph	(S,S)-t-Bu-Box	Cu(OTf)$_2$	92:8	97
9		Ph	(S,S)-t-Bu-Box	Cu(SbF$_6$)$_2$	88:12	98
10		H	22 R = Me	Cu(OTf)$_2$	98:2	83
11		H	47	Cu(OTf)$_2$	98:2	96
12		H	64	Cu(OTf)$_2$	97:3	92
13		H	65	Cu(OTf)$_2$	97:3	90
14		H	66	Cu(OTf)$_2$	96:4	83
15		H	67 R = H	Cu(ClO$_4$)$_2$·6H$_2$O	>99:1	98
16		Me	67 R = H	Cu(ClO$_4$)$_2$·6H$_2$O	95:5	99
17		H	(R,R)-DBFOX/Ph	Ni(ClO$_4$)$_2$·6H$_2$O	97:3	>99
18		Me	(R,R)-DBFOX/Ph	Ni(ClO$_4$)$_2$·6H$_2$O	92:8	94
19		H	(R,S)-diPh-Box	Mg(ClO$_4$)$_2$	94:6	−94
20		Me	(R,S)-diPh-Box	Mg(OTf)$_2$	84:16	−91
21		H	(R,R)-2-Np-Box	Mg(OTf)$_2$	89:11	−94
22		Me	(R,R)-2-Np-Box	Mg(OTf)$_2$	80:20	−92
23	Z2	H	(R,R)-Ph-Box	Mg(ClO$_4$)$_2$	97:3	88
24	Z3	H	(S,S)-t-Bu-Box	Cu(OTf)$_2$	>99:1	>99
25		Ph	(S,S)-t-Bu-Box	Cu(SbF$_6$)$_2$	94:6	>99
26	Z4	Ph	(S,S)-Ph-Box	Cu(OTf)$_2$	97:3	96
27	Z5	Me	(S,S)-i-Pr-Box	Cu(OTf)$_2$	93:7	95
28		Me	(S,S)-i-Pr-Box	Pd(ClO$_4$)$_2$	95:5	99

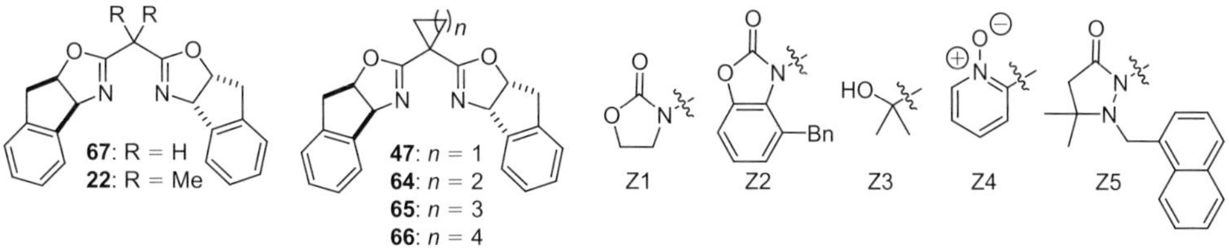

Scheme 5.37 Metal-bisoxazoline complexes in enantioselective Diels–Alder cycloadditions.

widely studied and utilized chiral Lewis acids are combinations of Cu(II) salts and the t-Bu-Box ligand. Evans showed that complexes of Cu(OTf)$_2$ and Cu(SbF$_6$)$_2$ with t-BuBox both catalyze highly enantioselective Diels–Alder reactions (entries 5–9) [15, 157–159]. In general, Cu(OTf)$_2$/t-Bu-Box leads to higher diastereoselectivities, while Cu(SbF$_6$)$_2$/t-BuBox is a more active catalyst and leads to slightly higher enantioselectivities. An alternative to the t-Bu-Box ligand that provides high levels of enantioselectivity in Cu(II)-catalyzed Diels–Alder reactions is the aminoindanol-derived Box ligand 47. A group from Merck and Co., led by Davies, synthesized a series of aminoindanol-derived Box ligands with dimethyl (**22**), cyclopropyl (**47**), cyclobutyl (**64**), cyclopentyl (**65**), and cyclohexyl (**66**) substitution at the carbon bridging the oxazoline moieties (entries 10–14) [160]. They showed that the cyclopropyl ligand **47** possessed the optimal bite angle and led to the highest enantioselectivity in Cu(II)-catalyzed Diels–Alder reactions with this series of ligands. Although this ligand scaffold, even without substitution at the bridging carbon (entries 15 and 16) [161], provides similar selectivity to *t*-Bu-Box, Cu(II)-*t*-Bu-Box complexes remain the most prominent catalysts for enantioselective Diels–Alder reactions of this type. However, the cyclopropyl-substituted Box ligand **47** has been central to the development of many additional enantioselective addition and cycloaddition reactions that are discussed in other parts of this chapter. Additional combinations of Box ligands and metals also provide high enantioselectivities in reactions of α,β-unsaturated oxazolidinone imides with cyclopentadiene. For example, Ni(II)/DBFOX-Ph [162–164], Mg(II)/(*R*,*S*)-diPh-Box [165], and Mg(II)/2-Np-Box [166] complexes all catalyze these cycloadditions in >90% ee (entries 17–22).

Highly enantioselective Diels–Alder cycloadditions catalyzed by metal-bisoxazoline catalysts are not limited to reactions of α,β-unsaturated oxazolidinone imides. Enantioselective Diels–Alder cycloadditions of α,β-unsaturated 1,3-benzoxazol-2-(3*H*)-one imides occur in the presence of a Mg(II)/Ph-Box complex (entry 23) [167], and analogous reactions of α,β-unsaturated α'-hydroxy enone [168], 2-alkenoyl pyridine *N*-oxides [169], and α,β-unsaturated pyrazolidinone imides [170, 171] occur in the presence of Cu(II) complexes of Box ligands (entries 24–27). In addition, the Diels–Alder cycloadditions of α,β-unsaturated pyrazolidinone imides also occur with high enantioselectivity in the presence of Pd(ClO$_4$)$_2$/*i*-Pr-Box (entry 28) [170]. Scheme 5.37 presents a representative cross section of bisoxazoline ligands that lead to highly enantioselective Diels–Alder reactions, but it should be noted that silica-supported [172–175] Box ligands and charge-transfer complexes [176] of Box ligands have also been studied in Diels–Alder reactions as methods to recycle the metal-Box catalysts.

Metal-bisoxazoline complexes catalyze Diels–Alder reactions of dienes and dienophiles in addition to those shown in Scheme 5.37. Copper(II) complexes of Ph-Box promote Diels–Alder cycloadditions of cyclopentadiene with 2-methylene-1,3-dicarbonyl compounds [177, 178] and α-thioacrylates [179, 180]. Copper(II) complexes of *t*-Bu-Box catalyze cycloadditions of various dienes with α-methylene lactams [181, 182]. In addition, a catalyst prepared from Cu(OTf)$_2$ and aminoindanol-derived Box ligand **47** catalyzes a rare catalytic kinetic resolution based on an enantioselective Diels–Alder reaction of α,β-unsaturated pyrazolidinone imides (Scheme 5.38) [183].

5.4.2
Hetero-Diels–Alder Cycloadditions

Applications of metal-bisoxazoline complexes as catalysts of enantioselective hetero-Diels–Alder (HDA) reactions began to appear soon after these complexes

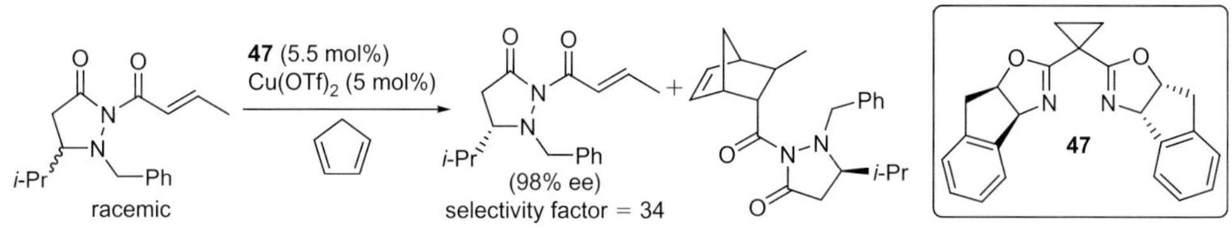

Scheme 5.38 Kinetic resolution by catalytic, asymmetric Diels–Alder cycloaddition.

had been established as highly enantioselective catalysts for Diels–Alder reactions of the type described in the previous section. Jørgensen and coworkers led the development of enantioselective hetero-Diels–Alder reaction catalyzed by metal-chiral bisoxazoline complexes. Their early studies on HDA reactions between unactivated dienes and alkyl glyoxylates demonstrated the potential of Cu(OTf)$_2$/t-Bu-Box, MgI$_2$/t-Bu-Box, and Zn(OTf)$_2$/Ph-Box complexes to provide HDA products with moderate to high enantioselectivity [184–187]. These catalysts proved most effective in HDA reactions involving cyclic dienes that were less prone to competing ene reactions. Several leading reports in the field are presented below.

In 1998 Jørgensen and coworkers reported highly enantioselective HDA reactions of ketones with electron-rich alkenes [188]. The Cu(OTf)$_2$/t-Bu-Box-catalyzed cycloaddition of Danishefsky's diene and ethyl pyruvate occurred with excellent enantioselectivity and in high yield (Scheme 5.39). In addition, Jørgensen and Evans independently showed that Cu(II) complexes of t-Bu-Box promote highly enantioselective HDA reactions of electron-rich alkenes with β,γ-unsaturated α-ketoesters [189] or α,β-unsaturated acyl phosphonates [19, 190], respectively. For example, the reaction of β,γ-unsaturated α-ketoester **68** or α,β-unsaturated acyl phosphonate **69** with dihydrofuran in the presence of Cu(OTf)$_2$/t-Bu-Box forms the corresponding dihydropyran product in high yield and enantioselectivity. The Cu(OTf)$_2$/t-Bu-Box catalyst system is also an efficient and selective catalyst of enantioselective aza-Diels–Alder reactions. This catalyst promotes the reaction of an electron-rich 2-azadiene with oxazolidinone crotonate in 96% yield and 94% ee. More recently, Esquivias, Arrayás, and Carretero showed that highly enantioselective aza-Diels–Alder reactions between 1-azadienes and electron-rich alkenes are catalyzed by a Ni(II)/DBFOX-Ph complex [191].

5.4.3
Cyclopropanations

The literature on enantioselective cyclopropanation reactions catalyzed by metal-chiral bisoxazoline complexes is extensive. In fact, over 70 studies and applications of cyclopropanation catalyzed by metal complexes of chiral bisoxazoline ligands have been published in the last 20 years. Of these studies nearly all rely on copper complexes of bisoxazoline ligands to provide asymmetric induction. Thus, the primary focus of this section will necessarily be Cu/bisoxazoline-catalyzed cyclopropanation reactions.

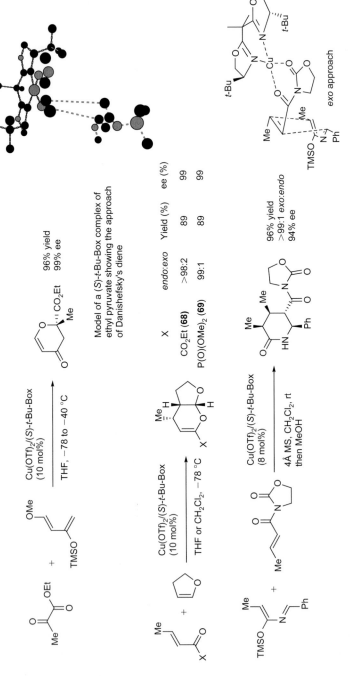

Scheme 5.39 Enantioselective hetero-Diels–Alder reactions catalyzed by metal-bisoxazoline complexes.

The benchmark reaction in this area is the cyclopropanation of styrene with a diazoacetate. Although the scope of enantioselective cyclopropanation reactions extends far beyond the copper-catalyzed cyclopropanation of styrene, this reaction illustrates the types of bisoxazoline ligands that provide the highest selectivities (Scheme 5.40). Perhaps not surprisingly, the *t*-Bu-Box ligand is often the bisoxazoline of choice for copper-catalyzed cyclopropanation reactions. For example, Evans showed in 1991 that the combination of CuOTf and *t*-Bu-Box promotes the highly enantioselective cyclopropanation of styrene with ethyl diazoacetate (entry 1) [13]. This group also found that the process could be made highly *trans*-selective if a bulky diazoacetate is employed as the carbene precursor (entry 2). Many additional copper complexes of bisoxazoline ligands (**70–75**), most of which are based on the *t*-Bu-Box framework, catalyze this process with moderate to high enantioselectivity and trans : cis selectivity (entries 3–10) [192–199]. The cyclopropyl-bridged bisoxazoline **73** and boron-bridged bisoxazoline **74** are of particular note because they lead, in certain cases, to improved trans : cis ratios relative to the parent *t*-Bu-Box ligand (compare entry 7 with entry 1 and entry 9 with entry 2). The diastereoselectivity of the cyclopropanation can also be reversed based on the

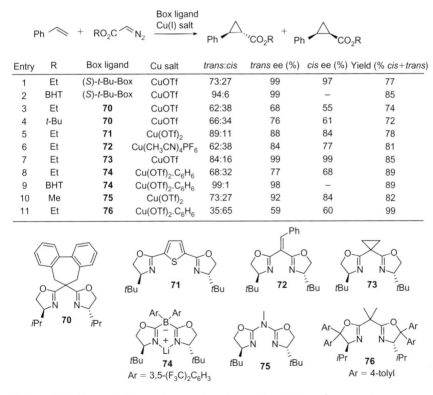

Entry	R	Box ligand	Cu salt	trans:cis	trans ee (%)	cis ee (%)	Yield (% cis+trans)
1	Et	(S)-*t*-Bu-Box	CuOTf	73:27	99	97	77
2	BHT	(S)-*t*-Bu-Box	CuOTf	94:6	99	–	85
3	Et	70	CuOTf	62:38	68	55	74
4	t-Bu	70	CuOTf	66:34	76	61	72
5	Et	71	Cu(OTf)$_2$	89:11	88	84	78
6	Et	72	Cu(CH$_3$CN)$_4$PF$_6$	62:38	84	77	81
7	Et	73	CuOTf	84:16	99	99	85
8	Et	74	Cu(OTf)$_2$·C$_6$H$_6$	68:32	77	68	89
9	BHT	74	Cu(OTf)$_2$·C$_6$H$_6$	99:1	98	–	89
10	Me	75	Cu(OTf)$_2$	73:27	92	84	82
11	Et	76	Cu(OTf)$_2$·C$_6$H$_6$	35:65	59	60	99

Scheme 5.40 Copper(I)-bisoxazoline complexes in enantioselective cyclopropanation reactions.

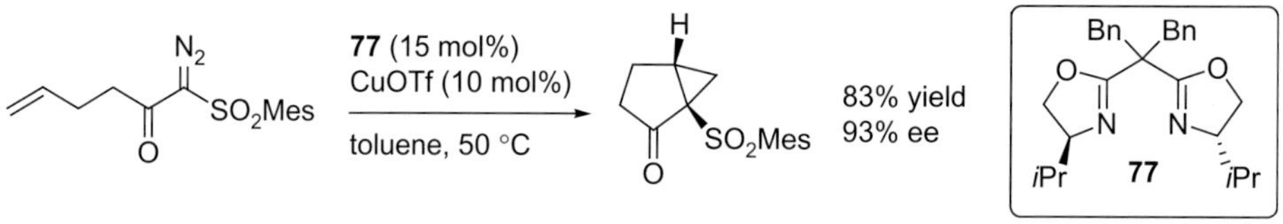

Scheme 5.41 Enantioselective intramolecular cyclopropanation.

structure of the bisoxazoline ligand. Gibson and coworkers demonstrated that the 5,5-diaryl-*i*-Pr-Box derivative **76** facilitated the cis-selective cyclopropanation of styrene with ethyl diazoacetate (entry 11) [200].

The copper/bisoxazoline-catalyzed enantioselective cyclopropanation of alkenes with diazo compounds now encompasses many classes of each substrate. The types of alkenes that can be employed in these reactions include styrenes, enol ethers, tri-substituted alkenes, cinnamate esters, dienes, pyrroles, and furans. Diazo compounds that have been used in these reactions include diazoacetates, (trimethylsilyl)diazomethane, α-nitro-α-diazo carbonyl compounds, diazosulfonate esters, and α-diazophosphonates. An extensive body of literature exists on enantioselective intramolecular cyclopropanation that are catalyzed by copper/bisoxazoline complexes. Doyle and coworkers, in particular, have conducted extensive studies that define and expand the scope of intramolecular cyclopropanation processes [201–206]. One notable example in this subset of cyclopropanation reactions is the enantioselective intramolecular cyclopropanation of α-diazo-β-keto sulfones catalyzed by a complex prepared from CuOTf and *i*-Pr-Box derivative **77** [207]. This reaction occurred to form the cyclopropanation product in 83% yield with 93% ee (Scheme 5.41).

5.4.4
Aziridination

Copper complexes of chiral bisoxazoline ligands are known to catalyze two distinct enantioselective aziridination processes. The first process, the aziridination of an alkene with a nitrene precursor, is more developed with copper/chiral bisoxazoline catalysts. The second process, the synthesis of an aziridine from an imine and a carbene precursor, is less selective with these copper-bisoxazoline catalyst systems. In 1993, Evans and coworkers reported enantioselective aziridination of cinnamate esters and styrenes with PhI=NTs as the nitrene precursors [208]. The aziridinations of cinnamate esters occurred with highest enantioselectivity in the presence of CuOTf/Ph-Box, while the reactions of styrenes are best conducted with CuOTf/*t*-BuBox (Scheme 5.42). In subsequent years the CuOTf/*t*-Bu-Box catalyst was found to promote the aziridination of styrenes with nitrene precursors generated *in situ* from PhI=O and a sulfonamide with similar enantioselectivity [209]. Hutchings *et al.* demonstrated that Ph-Box-modified Cu^{2+}-exchanged zeolite Y is also a highly enantioselective catalyst for aziridinations of styrenes with PhI=NNs [210]. Enantioselective aziridinations of additional alkene substrates

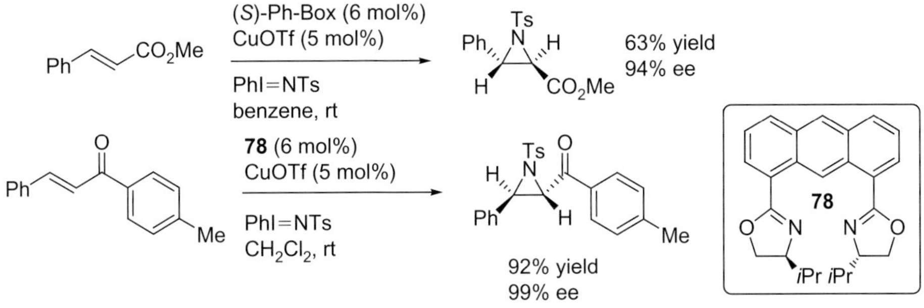

Scheme 5.42 Enantioselective aziridinations using bisoxazolines.

that are catalyzed by copper bisoxazoline complexes have been reported and include enol acetates and enol ethers [211], chalcones [212], and α,β,γ,δ-unsaturated ketones [213]. Notably, the reaction of a chalcone with PhI=NTs occurs in the presence of CuOTf and 1,8-bisoxazolinylanthracene-Box **78** to form the aziridine product in 92% yield and 99% ee [212].

In contrast to aziridinations of alkenes with nitrene precursors, the copper/bisoxazoline-catalyzed synthesis of aziridines from imines and diazo compounds is generally less efficient and less enantioselective [214–216]. For example, the Cu(ClO$_4$)$_2$/Ph-Box-catalyzed reaction of the N-tosyl α-imino ester of ethyl glyoxylate with trimethylsilyldiazomethane forms the aziridine product in 28% yield with 2 : 1 cis : trans selectivity, 40% cis ee, and 63% trans ee [215].

5.4.5
1,3-Dipolar Cycloadditions

Chiral Lewis acids prepared from combinations of bisoxazoline ligands and metal salts have been integral to the development of highly enantioselective 1,3-dipolar cycloadditions reactions [217, 218]. Scheme 5.43 presents a summary of the Lewis acid/bisoxazoline ligand combinations that have been studied in enantioselective 1,3-dipolar cycloadditions of nitrones with various α,β-unsaturated carbonyl compounds. Jørgensen et al. first showed that MgI$_2$/Ph-Box catalyzes *endo* and enantioselective cycloadditions of nitrones with α,β-unsaturated oxazolidinone imides (entry 1) [219, 220]. Subsequently, Kanemasa and coworkers reported improved diastereo- and enantioselectivity in *endo*-selective cycloadditions of nitrones with α,β-unsaturated oxazolidinone catalyzed by a Ni(II)/DBFOX-Ph complex (entry 2) [221]. Notably, this catalyst also promotes enantioselective cycloadditions of nitrones with α,β-unsaturated aldehydes that are not capable of two-point coordination to the Lewis acid [222, 223]. Copper(II) complexes of Box ligands are also enantioselective catalysts of nitrone cycloadditions. For example, chiral Lewis acids prepared from Cu(OTf)$_2$ and *t*-Bu-Box or (4R,5S)-diPh-Box promote cycloadditions of nitrones with α′-hydroxy enones [224] and α′-phosphoric enones [225], respectively, with nearly perfect *endo* selectivity and high enantioselectivity (entries 3 and 4). *Exo* selective nitrone cycloadditions are also known to

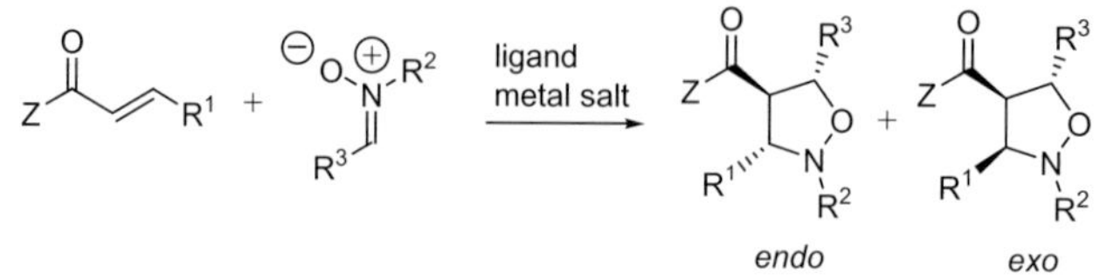

Entry	Z	R^1	R^2	R^3	ligand	metal salt	yield (%)	endo:exo	endo ee (%)	exo ee (%)
1	Z1	Me	Ph	Ph	(R)-Ph-Box	MgI$_2$	81	84:16	75	–
2	Z1	Me	Me	Ph	(R)-DBFOX-Ph	Ni(ClO$_4$)$_2$	63	99:1	99	–
3	Z3	H	Ph	Ph	(S)-t-Bu-Box	Cu(OTf)$_2$	85	>98:2	94	–
4	Z6	Me	Bn	Ph	(4R,5S)-diPh-Box	Cu(OTf)$_2$	90	>99:1	91	–
5	Z7	CH$_3$	Me	Ph	47	Cu(OTf)$_2$	94	4:96	–	98
6	Z1	H	Ph	Ph	(4R,5R)-diPh-Box	Ni(ClO$_4$)$_2$	99	10:90	75	99

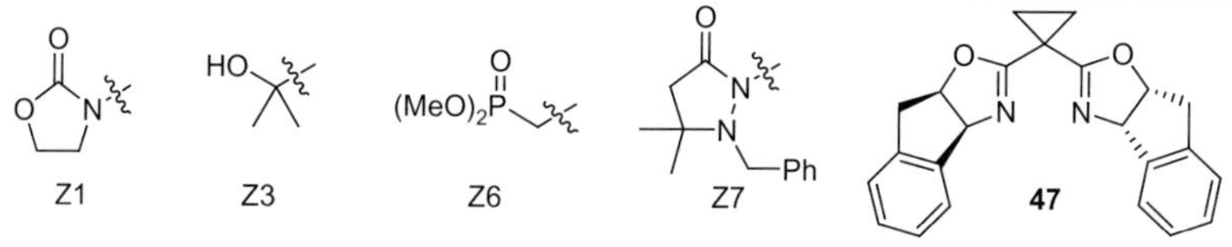

Scheme 5.43 Enantioselective nitrone cycloadditions using bisoxazolines.

occur in the presence of metal-bisoxazoline catalysts. Sibi and coworkers showed that Cu(OTf)$_2$ and aminoindanol-derived Box 47 form a chiral Lewis acid that catalyzes *exo* and enantioselective cycloadditions of nitrones with α,β-unsaturated pyrazolidinone imides (entry 5) [226], while Desimoni et al. reported a Ni(II)/(4R,5S)-diPh-Box complex as a catalyst for *exo* selective nitrone cycloadditions with oxazolidinone acrylate (entry 6) [227].

The cycloadditions presented in Scheme 5.43 involve reactions of nitrones with electron-deficient dipolarophiles. However, enantioselective nitrone cycloadditions catalyzed by metal-bisoxazoline complexes are not limited to this class of substrates. Jørgensen et al. reported Cu(OTf)$_2$/t-Bu-Box as a catalyst for enantioselective cycloadditions between nitrones derived from glyoxylate esters and electron-rich alkenes [228]. Miura reported the first catalytic enantioselective Kinugasa reaction of phenyl acetylene with a nitrone to form a β-lactam in the presence of catalytic amounts of CuI, iPr-Box, and (−)-sparteine [229]. In addition, Sibi and coworkers showed that Ni(ClO$_4$)$_2$/DBFOX-Ph promotes enantioselective additions of nitrones to activated cyclopropanes (Scheme 5.44) [230].

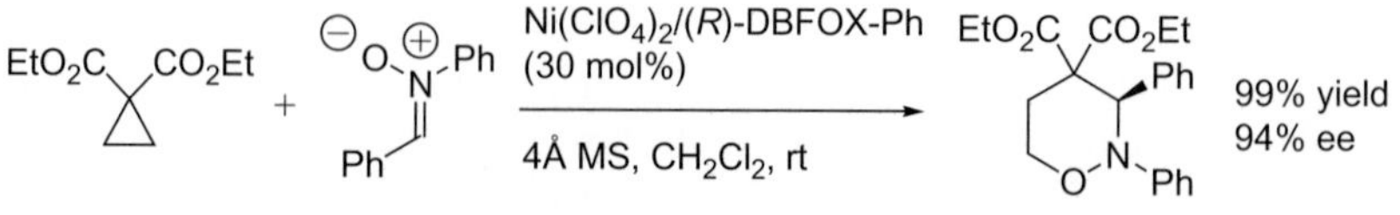

Scheme 5.44 Enantioselective [3 + 3] nitrone cycloadditions.

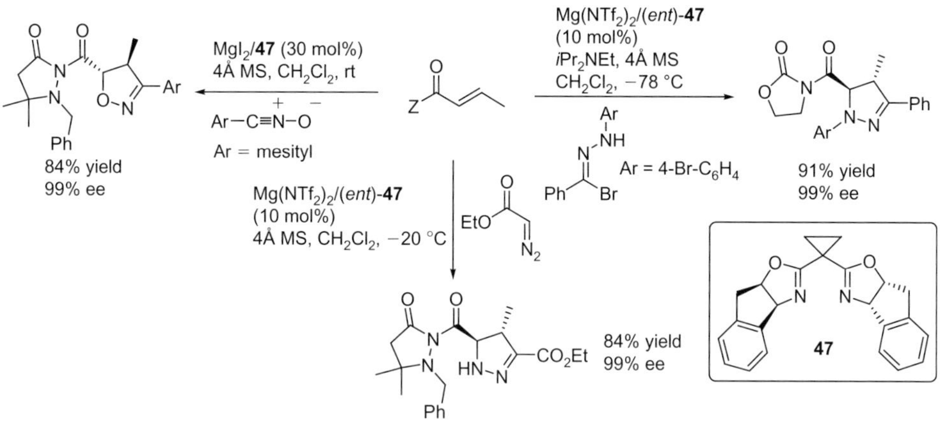

Scheme 5.45 Enantioselective dipolar cycloadditions.

Metal-bisoxazoline complexes also promote enantioselective cycloadditions of azomethine ylides [231], azomethine imines [232], diazoalkane [233], diazoacetates [234], nitrile oxides [235, 236], and nitrile imines [237, 238] with electron-deficient dipolarophiles. $Zn(OTf)_2/t$-Bu-Box catalyzes enantioselective cycloadditions of azomethine ylides derived from glycine alkyl esters with acrylate esters [231], while the combination of $Cu(OTf)_2$ and aminoindanol-derived bisoxazoline ent-**47** promotes enantioselective cycloadditions of azomethine imines with α,β-unsaturated pyrazolidinone imides [232]. Kanemasa and Kanai also reported Zn(II), Ni(II), and Mg(II) complexes of DBFOX-Ph as catalysts in highly enantioselective 1,3-dipolar cycloadditions of trimethylsilyldiazomethane with β-substituted α,β-unsaturated oxazolidinone imides [233]. However, complexes prepared from Mg(II) salts and aminoindanol-derived bisoxazoline **47** are probably the most versatile metal-bisoxazoline chiral Lewis acids for enantioselective 1,3-dipolar cycloadditions. The chiral Lewis acids prepared from Mg(II) salts and ligand **47** catalyze enantioselective 1,3-dipolar cycloadditions of nitrile oxides with α,β-unsaturated pyrazolidinone [235] and N–H imides [236], nitrile imines with α,β-unsaturated oxazolidinone [237] and pyrazolidinone imides [238], and diazoacetates with α,β-unsaturated pyrazolidinone imides (Scheme 5.45) [232]. In addition, the combination of MgI_2 and bisoxazoline ent-**47** promotes enantioselective ring expansions of a methylene cyclopropane in the presence of N-tosylimines to form methylenepyrrolidines as formal 1,3-dipolar cycloaddition adducts [239].

5.4.6
Additional Cycloaddition Reactions

Metal-chiral bisoxazoline complexes catalyze cycloaddition reactions beyond the Diels–Alder cycloadditions, cyclopropanations, aziridinations, and 1,3-dipolar cycloadditions discussed to this point of Section 5.4. In 2001 Evans and Janey reported Cu(II)/*t*-Bu-Box complexes as catalysts for enantioselective [2 + 2] cycloadditions of silyl ketenes with glyoxylate and pyruvate esters that form β-lactones with high enantioselectivity [240]. Furthermore, Huang and Hsung showed that Cu(OTf)$_2$/(4R,5S)-diPh-Box promotes enantioselective [4 + 3] cycloadditions of allenamide-derived, nitrogen-stabilized oxyallyl cations with furans [241]. Epoxidations reactions are also known to occur in the presence of metal-bisoxazoline catalysts [242, 243]. However, the enantioselectivity of these processes is generally low, and these catalysts do not compare favorably with metal catalysts for epoxidation reactions that are based on other ligand scaffolds.

5.5
Conclusions

Bisoxazoline ligands have played an important role in the development of various enantioselective transformations, and in a broader context have helped to shape the ways that chemists approach asymmetric catalysis. The popularity of bisoxazoline ligands is due, at least in part, to their ease of preparation from readily available amino alcohols. In addition to the array of structures accessible from various amino alcohols, the development of bisoxazolines with additional spacer units lends further diversity and utility to this powerful class of ligands. In general, the box ligands discussed in this chapter form well-organized bidentate complexes with various metal salts that allow for predictable stereochemical outcomes in reactions catalyzed by metal-bisoxazoline complexes. The past success in the development of highly enantioselective transformations using bisoxazoline ligands to control stereoselectivity portends an equally bright future. In the coming years, it is highly likely that bisoxazoline ligands will find further utility as the key to controlling the stereochemistry in many new reactions; these ligands will also undoubtedly play a prominent role in improving the enantioselectivity of existing processes. Thus, the bisoxazoline ligands are indeed a well-deserving member of the privileged ligand architectures and are likely to remain so for the foreseeable future.

References

1 Desimoni, G., Faita, G., and Jørgensen, K.A. (2006) *Chem. Rev.*, **106**, 3561–3651.
2 Ghosh, A.K., Packiarajan, M., and Cappiello, J. (1998) *Tetrahedron: Asymmetry*, **9**, 1–45.
3 Rechavi, D. and Lemaire, M. (2002) *Chem. Rev.*, **102**, 3467–3494.
4 Fraile, J.M., Garcia, J.I., Herrerias, C.I., Mayoral, J.A., Pires, E., and Salvatella, L. (2009) *Catal. Today*, **140**, 44–50.
5 Fraile, J.M., Garcia, J.I., and Mayoral, J.A. (2008) *Coord. Chem. Rev.*, **252**, 624–646.
6 Hargaden, G.C. and Guiry, P.J. (2009) *Chem. Rev.*, **109**, 2505–2550.
7 Rasappan, R., Laventine, D., and Reiser, O. (2008) *Coord. Chem. Rev.*, **252**, 702–714.
8 Desimoni, G., Faita, G., and Quadrelli, P. (2003) *Chem. Rev.*, **103**, 3119–3154.
9 Fritschi, H., Leutenegger, U., and Pfaltz, A. (1988) *Helv. Chim. Acta*, **71**, 1553–1565.

10 Fritschi, H., Leutenegger, U., Siegmann, K., Pfaltz, A., Keller, W., and Kratky, C. (1988) *Helv. Chim. Acta*, **71**, 1541–1552.

11 Lowenthal, R.E., Abiko, A., and Masamune, S. (1990) *Tetrahedron Lett.*, **31**, 6005–6008.

12 Lowenthal, R.E. and Masamune, S. (1991) *Tetrahedron Lett.*, **32**, 7373–7376.

13 Evans, D.A., Woerpel, K.A., Hinman, M.M., and Faul, M.M. (1991) *J. Am. Chem. Soc.*, **113**, 726–727.

14 Corey, E.J., Imai, N., and Zhang, H.Y. (1991) *J. Am. Chem. Soc.*, **113**, 728–729.

15 Evans, D.A., Miller, S.J., Lectka, T., and von Matt, P. (1999) *J. Am. Chem. Soc.*, **121**, 7559–7573.

16 Thorhauge, J., Roberson, M., Hazell, R.G., and Jørgensen, K.A. (2002) *Chem. Eur. J.*, **8**, 1888–1898.

17 Evans, D.A., Johnson, J.S., Burgey, C.S., and Campos, K.R. (1999) *Tetrahedron Lett.*, **40**, 2879–2882.

18 Evans, D.A., Peterson, G.S., Johnson, J.S., Barnes, D.M., Campos, K.R., and Woerpel, K.A. (1998) *J. Org. Chem.*, **63**, 4541–4544.

19 Evans, D.A., Johnson, J.S., and Olhava, E.J. (2000) *J. Am. Chem. Soc.*, **122**, 1635–1649.

20 Evans, D.A., Scheidt, K.A., Johnston, J.N., and Willis, M.C. (2001) *J. Am. Chem. Soc.*, **123**, 4480–4491.

21 Evans, D.A., Rovis, T., Kozlowski, M.C., and Tedrow, J.S. (1999) *J. Am. Chem. Soc.*, **121**, 1994–1995.

22 Evans, D.A., Rovis, T., Kozlowski, M.C., Downey, C.W., and Tedrow, J.S. (2000) *J. Am. Chem. Soc.*, **122**, 9134–9142.

23 Evans, D.A., Seidel, D., Rueping, M., Lam, H.W., Shaw, J.T., and Downey, C.W. (2003) *J. Am. Chem. Soc.*, **125**, 12692–12693.

24 Evans, D.A., Downey, C.W., and Hubbs, J.L. (2003) *J. Am. Chem. Soc.*, **125**, 8706–8707.

25 Evans, D.A., Kozlowski, M.C., Burgey, C.S., and MacMillan, D.W.C. (1997) *J. Am. Chem. Soc.*, **119**, 7893–7894.

26 Evans, D.A., Burgey, C.S., Kozlowski, M.C., and Tregay, S.W. (1999) *J. Am. Chem. Soc.*, **121**, 686–699.

27 Evans, D.A., MacMillan, D.W.C., and Campos, K.R. (1997) *J. Am. Chem. Soc.*, **119**, 10859–10860.

28 Landa, A., Minkkila, A., Blay, G., and Jørgensen, K.A. (2006) *Chem. Eur. J.*, **12**, 3472–3483.

29 Le, J.C.D. and Pagenkopf, B.L. (2004) *Org. Lett.*, **6**, 4097–4099.

30 Kobayashi, S., Mori, Y., Nagayama, S., and Manabe, K. (1999) *Green Chem.*, **1**, 175–177.

31 Kobayashi, S., Nagayama, S., and Busujima, T. (1999) *Tetrahedron*, **55**, 8739–8746.

32 Fraile, J.M., Perez, I., and Mayoral, J.A. (2007) *J. Catal.*, **252**, 303–311.

33 Fabra, M.J., Fraile, J.M., Herrerias, C.I., Lahoz, F.J., Mayoral, J.A., and Perez, I. (2008) *Chem. Commun.*, 5402–5404.

34 Doherty, S., Goodrich, P., Hardacre, C., Parvulescu, V., and Paun, C. (2008) *Adv. Synth. Catal.*, **350**, 295–302.

35 Mandoli, A., Lessi, M., Pini, D., Evangelisti, C., and Salvadoria, P. (2008) *Adv. Synth. Catal.*, **350**, 375–379.

36 Juhl, K., Gathergood, N., and Jørgensen, K.A. (2001) *Angew. Chem. Int. Ed.*, **40**, 2995–2997.

37 Marigo, M., Kjaersgaard, A., Juhl, K., Gathergood, N., and Jørgensen, K.A. (2003) *Chem. Eur. J.*, **9**, 2359–2367.

38 Kjaersgaard, A. and Jørgensen, K.A. (2005) *Org. Biomol% Chem.*, **3**, 804–808.

39 Palomo, C., Oiarbide, M., and Mielgo, A. (2004) *Angew. Chem. Int. Ed.*, **43**, 5442–5444.

40 Palomo, C., Oiarbide, M., and Laso, A. (2007) *Eur. J. Org. Chem.*, 2561–2574.

41 Christensen, C., Juhl, K., and Jørgensen, K.A. (2001) *Chem. Commun.*, 2222–2223.

42 Christensen, C., Juhl, K., Hazell, R.G., and Jørgensen, K.A. (2002) *J. Org. Chem.*, **67**, 4875–4881.

43 Risgaard, T., Gothelf, K.V., and Jørgensen, K.A. (2003) *Org. Biomol% Chem.*, **1**, 153–156.

44 Lu, S.F., Du, D.M., Zhang, S.W., and Xu, J.X. (2004) *Tetrahedron: Asymmetry*, **15**, 3433–3441.

45 Du, D.M., Lu, S.F., Fang, T., and Xu, J.X. (2005) *J. Org. Chem.*, **70**, 3712–3715.

46 Paintner, F.F., Allmendinger, L., Bauschke, G., and Klemann, P. (2005) *Org. Lett.*, **7**, 1423–1426.

47 Ginotra, S.K. and Singh, V.K. (2007) *Org. Biomol% Chem.*, **5**, 3932–3937.

48 Yuryev, R. and Liese, A. (2009) *Synlett*, 2589–2592.

49 Toussaint, A. and Pfaltz, A. (2008) *Eur. J. Org. Chem.*, 4591–4597.

50 Rasappan, R., Olbrich, T., and Reiser, O. (2009) *Adv. Synth. Catal.*, **351**, 1961–1967.

51 Knudsen, K.R., Risgaard, T., Nishiwaki, N., Gothelf, K.V., and Jørgensen, K.A. (2001) *J. Am. Chem. Soc.*, **123**, 5843–5844.

52 Nishiwaki, N., Knudsen, K.R., Gothelf, K.V., and Jørgensen, K.A. (2001) *Angew. Chem. Int. Ed.*, **40**, 2992–2995.

53 Lee, A., Kim, W., Lee, J., Hyeon, T., and Kim, B.M. (2004) *Tetrahedron: Asymmetry*, **15**, 2595–2598.

54 Knudsen, K.R. and Jørgensen, K.A. (2005) *Org. Biomol% Chem.*, **3**, 1362–1364.

55 Komine, N., Wang, L.F., Tomooka, K., and Nakai, T. (1999) *Tetrahedron Lett.*, **40**, 6809–6812.

56 Tomooka, K., Wang, L.F., Komine, N., and Nakai, T. (1999) *Tetrahedron Lett.*, **40**, 6813–6816.

57 Tomooka, K., Wang, L.F., Okazaki, F., and Nakai, T. (2000) *Tetrahedron Lett.*, **41**, 6121–6125.

58 Nakamura, S., Nakagawa, R., Watanabe, Y., and Toru, T. (2000) *Angew. Chem. Int. Ed.*, **39**, 353–355.

59 Nakamura, S., Nakagawa, R., Watanabe, Y., and Toru, T. (2000) *J. Am. Chem. Soc.*, **122**, 11340–11347.

60 Nakamura, S., Furutani, A., and Toru, T. (2002) *Eur. J. Org. Chem.*, 1690–1695.

61 Nakamura, S., Kato, T., Nishimura, H., and Toru, T. (2004) *Chirality*, **16**, 86–89.

62 Nakamura, S., Ito, Y., Wang, L., and Toru, T. (2004) *J. Org. Chem.*, **69**, 1581–1589.

63 Nakamura, S., Hirata, N., Kita, T., Yamada, R., Nakane, D., Shibata, N., and Toru, T. (2007) *Angew. Chem. Int. Ed.*, **46**, 7648–7650.

64 Denmark, S.E., Nakajima, N., and Nicaise, O.J.C. (1994) *J. Am. Chem. Soc.*, **116**, 8797–8798.

65 Denmark, S.E. and Stiff, C.M. (2000) *J. Org. Chem.*, **65**, 5875–5878.

66 Denmark, S.E., Nakajima, N., Stiff, C.M., Nicaise, O.J.C., and Kranz, M. (2008) *Adv. Synth. Catal.*, **350**, 1023–1045.

67 Krishnamurthy, D., Han, Z.X., Wald, S.A., and Senanayake, C.H. (2002) *Tetrahedron Lett.*, **43**, 2331–2333.

68 Miyabe, H., Ushiro, C., Ueda, M., Yamakawa, K., and Naito, T. (2000) *J. Org. Chem.*, **65**, 176–185.

69 Friestad, G.K., Shen, Y.H., and Ruggles, E.L. (2003) *Angew. Chem. Int. Ed.*, **42**, 5061–5063.

70 Evans, D.A., Burgey, C.S., Paras, N.A., Vojkovsky, T., and Tregay, S.W. (1998) *J. Am. Chem. Soc.*, **120**, 5824–5825.

71 Evans, D.A., Tregay, S.W., Burgey, C.S., Paras, N.A., and Vojkovsky, T. (2000) *J. Am. Chem. Soc.*, **122**, 7936–7943.

72 Poulsen, T.B. and Jørgensen, K.A. (2008) *Chem. Rev.*, **108**, 2903–2915.

73 Gathergood, N., Zhuang, W., and Jørgensen, K.A. (2000) *J. Am. Chem. Soc.*, **122**, 12517–12522.

74 Christoffers, J., Koripelly, G., Rosiak, A., and Rossle, M. (2007) *Synthesis*, 1279–1300.

75 Kanemasa, S., Hasegawa, M., and Ono, F. (2007) *Chem. Record*, **7**, 137–149.

76 Evans, D.A., Willis, M.C., and Johnston, J.N. (1999) *Org. Lett.*, **1**, 865–868.

77 Sibi, M.P. and Chen, J.X. (2002) *Org. Lett.*, **4**, 2933–2936.

78 Ji, J.G., Barnes, D.M., Zhang, J., King, S.A., Wittenberger, S.J., and Morton, H.E. (1999) *J. Am. Chem. Soc.*, **121**, 10215–10216.

79 Halland, N., Velgaard, T., and Jørgensen, K.A. (2003) *J. Org. Chem.*, **68**, 5067–5074.

80 Itoh, K., Hasegawa, M., Tanaka, J., and Kanemasa, S. (2005) *Org. Lett.*, **7**, 979–981.

81 Itoh, K., Oderaotoshi, Y., and Kanemasa, S. (2003) *Tetrahedron: Asymmetry*, **14**, 635–639.

82 Yanagita, H., Kodama, K., and Kanemasa, S. (2006) *Tetrahedron Lett.*, **47**, 9353–9357.

83 Jensen, K.B., Thorhauge, J., Hazell, R.G., and Jørgensen, K.A. (2001) *Angew. Chem. Int. Ed.*, **40**, 160–163.

84 Palomo, C., Oiarbide, M., Kardak, B.G., Garcia, J.M., and Linden, A. (2005) *J. Am. Chem. Soc.*, **127**, 4154–4155.

85 Sibi, M.P., Coulomb, J., and Stanley, L.M. (2008) *Angew. Chem. Int. Ed.*, **47**, 9913–9915.

86 Sibi, M.P., Manyem, S., and Zimmerman, J. (2003) *Chem. Rev.*, **103**, 3263–3296.

87 Sibi, M.P., Ji, J.G., Wu, J.H., Gurtler, S., and Porter, N.A. (1996) *J. Am. Chem. Soc.*, **118**, 9200–9201.

88 Sibi, M.P. and Ji, J.G. (1997) *J. Org. Chem.*, **62**, 3800–3801.

89 Sibi, M.P. and Chen, J.X. (2001) *J. Am. Chem. Soc.*, **123**, 9472–9473.

90 Sibi, M.P., Petrovic, G., and Zimmerman, J. (2005) *J. Am. Chem. Soc.*, **127**, 2390–2391.

91 Wu, J.H., Radinov, R., and Porter, N.A. (1995) *J. Am. Chem. Soc.*, **117**, 11029–11030.

92 Yang, D., Gu, S., Yan, Y.L., Zhu, N.Y., and Cheung, K.K. (2001) *J. Am. Chem. Soc.*, **123**, 8612–8613.

93 Yang, D., Gu, S., Yan, Y.L., Zhao, H.W., and Zhu, N.Y. (2002) *Angew. Chem. Int. Ed.*, **41**, 3014–3017.

94 Yang, D., Zheng, B.F., Gao, Q., Gu, S., and Zhu, N.Y. (2006) *Angew. Chem. Int. Ed.*, **45**, 255–258.

95 Miyabe, H., Asada, R., Toyoda, A., and Takemoto, Y. (2006) *Angew. Chem. Int. Ed.*, **45**, 5863–5866.

96 Miyabe, H., Toyoda, A., and Takemoto, Y. (2007) *Synlett*, 1885–1888.

97 Sibi, M.P., Asano, Y., and Sausker, J.B. (2001) *Angew. Chem. Int. Ed.*, **40**, 1293–1296.

98 Sibi, M.P. and Patil, K. (2004) *Angew. Chem. Int. Ed.*, **43**, 1235–1238.

99 Sibi, M.P. and Sausker, J.B. (2002) *J. Am. Chem. Soc.*, **124**, 984–991.

100 Aggarwal, V.K. and Belfield, A.J. (2003) *Org. Lett.*, **5**, 5075–5078.

101 Huang, J. and Frontier, A.J. (2007) *J. Am. Chem. Soc.*, **129**, 8060–8061.

102 Takacs, J.M., Weidner, J.J., and Takacs, B.E. (1993) *Tetrahedron Lett.*, **34**, 6219–6222.

103 Zhang, Q.H. and Lu, X.Y. (2000) *J. Am. Chem. Soc.*, **122**, 7604–7605.

104 Zhang, Q.H., Lu, X.Y., and Han, X.L. (2001) *J. Org. Chem.*, **66**, 7676–7684.

105 Muthiah, C., Arai, M.A., Shinohara, T., Arai, T., Takizawa, S., and Sasai, H. (2003) *Tetrahedron Lett.*, **44**, 5201–5204.

106 Zhang, Q.G., Xu, W., and Lu, X.Y. (2005) *J. Org. Chem.*, **70**, 1505–1507.

107 Ukaji, Y., Miyamoto, M., Mikuni, M., Takeuchi, S., and Inomata, K. (1996) *Bull. Chem. Soc. Jpn.*, **69**, 735–742.

108 Kato, K., Motodate, S., Mochida, T., Kobayashi, T., and Akita, H. (2009) *Angew. Chem. Int. Ed.*, **48**, 3326–3328.

109 Tomooka, K., Komine, N., and Nakai, T. (1998) *Tetrahedron Lett.*, **39**, 5513–5516.

110 Tomooka, K., Komine, N., and Nakai, T. (2000) *Chirality*, **12**, 505–509.

111 Abraham, L., Czerwonka, R., and Hiersemann, M. (2001) *Angew. Chem. Int. Ed.*, **40**, 4700–4703.

112 Abraham, L., Korner, M., and Hiersemann, M. (2004) *Tetrahedron Lett.*, **45**, 3647–3650.

113 Abraham, L., Korner, M., Schwab, P., and Hiersemann, M. (2004) *Adv. Synth. Catal.*, **346**, 1281–1294.

114 Sibi, M.P., Gorikunti, U., and Liu, M. (2002) *Tetrahedron*, **58**, 8357–8363.

115 Sibi, M.P. and Liu, M. (2000) *Org. Lett.*, **2**, 3393–3396.

116 Sibi, M.P. and Liu, M. (2001) *Org. Lett.*, **3**, 4181–4184.

117 Sibi, M.P., Shay, J.J., Liu, M., and Jasperse, C.P. (1998) *J. Am. Chem. Soc.*, **120**, 6615–6616.

118 Zhuang, W., Hazell, R.G., and Jørgensen, K.A. (2001) *Chem. Commun.*, 1240–1241.

119 Sibi, M.P., Prabagaran, N., Ghorpade, S.G., and Jasperse, C.P. (2003) *J. Am. Chem. Soc.*, **125**, 11796–11797.

120 Palomo, C., Oiarbide, M., Halder, R., Kelso, M., Gomez-Bengoa, E., and Garcia, J.M. (2004) *J. Am. Chem. Soc.*, **126**, 9188–9189.

121 Sibi, M.P. and Soeta, T. (2007) *J. Am. Chem. Soc.*, **129**, 4522–4523.

122 Kanemasa, S., Oderaotoshi, Y., and Wada, E. (1999) *J. Am. Chem. Soc.*, **121**, 8675–8676.

123 van Lingen, H.L., Zhuang, W., Hansen, T., Rutjes, F.P.J.T., and

Jørgensen, K.A. (2003) *Org. Biomol% Chem.*, **1**, 1953–1958.
124 Shiomi, T., Adachi, T., Toribatake, K., Zhou, L., and Nishiyama, H. (2009) *Chem. Commun.*, 5987–5989.
125 Clark, J.S. and Roche, C. (2005) *Chem. Commun.*, 5175–5177.
126 Andrus, M.B., Argade, A.B., Chen, X., and Pamment, M.G. (1995) *Tetrahedron Lett.*, **36**, 2945–2948.
127 Andrus, M.B. and Chen, X. (1997) *Tetrahedron*, **53**, 16229–16240.
128 Gokhale, A.S., Minidis, A.B.E., and Pfaltz, A. (1995) *Tetrahedron Lett.*, **36**, 1831–1834.
129 Andrus, M.B. and Zhou, Z.N. (2002) *J. Am. Chem. Soc.*, **124**, 8806–8807.
130 Janey, J.M. (2005) *Angew. Chem. Int. Ed.*, **44**, 4292–4300.
131 Evans, D.A. and Johnson, D.S. (1999) *Org. Lett.*, **1**, 595–598.
132 Juhl, K. and Jørgensen, K.A. (2002) *J. Am. Chem. Soc.*, **124**, 2420–2421.
133 Ishimaru, T., Shibata, N., Nagai, J., Nakamura, S., Toru, T., and Kanemasa, S. (2006) *J. Am. Chem. Soc.*, **128**, 16488–16489.
134 Reddy, D.S., Shibata, N., Nagai, J., Nakamura, S., and Toru, T. (2009) *Angew. Chem. Int. Ed.*, **48**, 803–806.
135 Ma, J.A. and Cahard, D. (2004) *Tetrahedron: Asymmetry*, **15**, 1007–1011.
136 Shibata, N., Kohno, J., Takai, K., Ishimaru, T., Nakamura, S., Toru, T., and Kanemasa, S. (2005) *Angew. Chem. Int. Ed.*, **44**, 4204–4207.
137 Reddy, D.S., Shibata, N., Nagai, J., Nakamura, S., Toru, T., and Kanemasa, S. (2008) *Angew. Chem. Int. Ed.*, **47**, 164–168.
138 Marigo, M., Kumaragurubaran, N., and Jørgensen, K.A. (2004) *Chem. Eur. J.*, **10**, 2133–2137.
139 Liu, B., Zhu, S.F., Zhang, W., Chen, C., and Zhou, Q.L. (2007) *J. Am. Chem. Soc.*, **129**, 5834–5835.
140 Chen, C., Zhu, S.F., Liu, B., Wang, L.X., and Zhou, Q.L. (2007) *J. Am. Chem. Soc.*, **129**, 12616–12617.
141 Zhu, S.F., Chen, C., Cai, Y., and Zhou, Q.L. (2008) *Angew. Chem. Int. Ed.*, **47**, 932–934.
142 Zhang, Y.Z., Zhu, S.F., Cai, Y., Mao, H.X., and Zhou, Q.L. (2009) *Chem. Commun.*, 5362–5364.
143 Kusakabe, T., Kato, K., Takaishi, S., Yamamura, S., Mochida, T., Akita, H., Peganova, T.A., Vologdin, N.V., and Gusev, O.V. (2008) *Tetrahedron*, **64**, 319–327.
144 Kawamura, Y., Kawano, Y., Matsuda, T., Ishitobi, Y., and Hosokawa, T. (2009) *J. Org. Chem.*, **74**, 3048–3053.
145 Wang, F.J., Zhang, Y.H., Yang, G.Q., and Zhang, W.B. (2007) *Tetrahedron Lett.*, **48**, 4179–4182.
146 Ogata, T., Ujihara, A., Tsuchida, S., Shimizu, T., Kaneshige, A., and Tomioka, K. (2007) *Tetrahedron Lett.*, **48**, 6648–6650.
147 Matsumura, Y., Maki, T., Murakami, S., and Onomura, O. (2003) *J. Am. Chem. Soc.*, **125**, 2052–2053.
148 Gissibl, A., Finn, M.G., and Reiser, O. (2005) *Org. Lett.*, **7**, 2325–2328.
149 Schatz, A., Scarel, A., Zangrando, E., Mosca, L., Carfagna, C., Gissibl, A., Milani, B., and Reiser, O. (2006) *Organometallics*, **25**, 4065–4068.
150 Mazet, C., Roseblade, S., Kohler, V., and Pfaltz, A. (2006) *Org. Lett.*, **8**, 1879–1882.
151 Demizu, Y., Kubo, Y., Matsumura, Y., and Onomura, O. (2008) *Synlett*, 433–437.
152 Onomura, O., Mitsuda, M., Nguyen, M.T.T., and Demizu, Y. (2007) *Tetrahedron Lett.*, **48**, 9080–9084.
153 Hong, M.S., Kim, T.W., Jung, B., and Kang, S.H. (2008) *Chem. Eur. J.*, **14**, 3290–3296.
154 Reymond, S. and Cossy, J. (2008) *Chem. Rev.*, **108**, 5359–5406.
155 Corey, E.J. and Ishihara, K. (1992) *Tetrahedron Lett.*, **33**, 6807–6810.
156 Evans, D.A., Kozlowski, M.C., and Tedrow, J.S. (1996) *Tetrahedron Lett.*, **37**, 7481–7484.
157 Evans, D.A., Barnes, D.M., Johnson, J.S., Lectka, R., von Matt, P., Miller, S.J., Murry, J.A., Norcross, R.D., Shaughnessy, E.A., and Campos, K.R. (1999) *J. Am. Chem. Soc.*, **121**, 7582–7594.

158 Evans, D.A., Miller, S.J., and Lectka, T. (1993) *J. Am. Chem. Soc.*, **115**, 6460–6461.
159 Evans, D.A., Murry, J.A., Vonmatt, P., Norcross, R.D., and Miller, S.J. (1995) *Angew. Chem. Int. Ed.*, **34**, 798–800.
160 Davies, I.W., Gerena, L., Castonguay, L., Senanayake, C.H., Larsen, R.D., Verhoeven, T.R., and Reider, P.J. (1996) *Chem. Commun.*, 1753–1754.
161 Ghosh, A.K., Cho, H., and Cappiello, J. (1998) *Tetrahedron: Asymmetry*, **9**, 3687–3691.
162 Kanemasa, S., Oderaotoshi, Y., Sakaguchi, S., Yamamoto, H., Tanaka, J., Wada, E., and Curran, D.P. (1998) *J. Am. Chem. Soc.*, **120**, 3074–3088.
163 Kanemasa, S., Oderaotoshi, Y., Tanaka, J., and Wada, E. (1998) *Tetrahedron Lett.*, **39**, 7521–7524.
164 Kanemasa, S., Oderaotoshi, Y., Yamamoto, H., Tanaka, J., Wada, E., and Curran, D.P. (1997) *J. Org. Chem.*, **62**, 6454–6455.
165 Carbone, P., Desimoni, G., Faita, G., Filippone, S., and Righetti, P. (1998) *Tetrahedron*, **54**, 6099–6110.
166 Crosignani, S., Desimoni, G., Faita, G., and Righetti, P. (1998) *Tetrahedron*, **54**, 15721–15730.
167 Quaranta, L., Corminboeuf, O., and Renaud, P. (2002) *Org. Lett.*, **4**, 39–42.
168 Palomo, C., Oiarbide, M., Garcia, J.M., Gonzalez, A., and Arceo, E. (2003) *J. Am. Chem. Soc.*, **125**, 13942–13943.
169 Barroso, S., Blay, G., and Pedro, J.R. (2007) *Org. Lett.*, **9**, 1983–1986.
170 Sibi, M.P., Stanley, L.M., Nie, X.P., Venkatraman, L., Liu, M., and Jasperse, C.P. (2007) *J. Am. Chem. Soc.*, **129**, 395–405.
171 Sibi, M.P., Venkatraman, L., Liu, M., and Jasperse, C.P. (2001) *J. Am. Chem. Soc.*, **123**, 8444–8445.
172 Rechavi, D. and Lemaire, M. (2001) *Org. Lett.*, **3**, 2493–2496.
173 Rechavi, D. and Lemaire, M. (2002) *J. Mol. Catal. A: Chem.*, **182**, 239–247.
174 Tada, M., Tanaka, S., and Iwasawa, Y. (2005) *Chem. Lett.*, 1362–1363.
175 Tanaka, S., Tada, M., and Iwasawa, Y. (2007) *J. Catal.*, **245**, 173–183.
176 Chollet, G., Rodriguez, F., and Schulz, E. (2006) *Org. Lett.*, **8**, 539–542.
177 Honda, Y., Date, T., Hiramatsu, H., and Yamauchi, M. (1997) *Chem. Commun.*, 1411–1412.
178 Yamauchi, M., Aoki, T., Li, M.Z., and Honda, Y. (2001) *Tetrahedron: Asymmetry*, **12**, 3113–3118.
179 Aggarwal, V.K., Anderson, E.S., Jones, D.E., Obierey, K.B., and Giles, R. (1998) *Chem. Commun.*, 1985–1986.
180 Aggarwal, V.K., Jones, D.E., and Martin-Castro, A.M. (2000) *Eur. J. Org. Chem.*, 2939–2945.
181 Ishihara, J., Horie, M., Shimada, Y., Tojo, S., and Murai, A. (2002) *Synlett*, 403–406.
182 Brimble, M.A., Crimmins, D., and Trzoss, M. (2005) *Arkivoc*, 39–52.
183 Sibi, M.P., Kawashima, K., and Stanley, L.M. (2009) *Org. Lett.*, **11**, 3894–3897.
184 Johannsen, M. and Jørgensen, K.A. (1995) *J. Org. Chem.*, **60**, 5757–5762.
185 Johannsen, M. and Jørgensen, K.A. (1996) *Tetrahedron*, **52**, 7321–7328.
186 Johannsen, M., Yao, S.L., Graven, A., and Jørgensen, K.A. (1998) *Pure Appl. Chem.*, **70**, 1117–1122.
187 Yao, S., Johannsen, M., and Jørgensen, K.A. (1997) *J. Chem. Soc., Perkin Trans. 1*, 2345–2350.
188 Yao, S.L., Johannsen, M., Audrain, H., Hazell, R.G., and Jørgensen, K.A. (1998) *J. Am. Chem. Soc.*, **120**, 8599–8605.
189 Thorhauge, J., Johannsen, M., and Jørgensen, K.A. (1998) *Angew. Chem. Int. Ed.*, **37**, 2404–2406.
190 Evans, D.A. and Johnson, J.S. (1998) *J. Am. Chem. Soc.*, **120**, 4895–4896.
191 Esquivias, J., Arrayás, R.G., and Carretero, J.C. (2007) *J. Am. Chem. Soc.*, **129**, 1480–1481.
192 Du, D.M., Fu, B., and Hua, W.T. (2003) *Tetrahedron*, **59**, 1933–1938.
193 Gao, M.Z., Wang, B., Kong, D.Y., Zingaro, R.A., Clearfield, A., and Xu, Z.L. (2005) *Synth. Commun.*, **35**, 2665–2673.
194 Burke, A.J., Carreiro, E.D., Chercheja, S., Moura, N.M.M., Ramalho, J.P.P., Rodrigues, A.I., and dos Santos, C.I.M. (2007) *J. Organomet. Chem.*, **692**, 4863–4874.
195 Carreiro, E.D., Chercheja, S., Moura, N.M.M., Gertrudes, C.S.C., and Burke,

A.J. (2006) *Inorg. Chem. Commun.*, **9**, 823–826.
196 Carreiro, E.P., Burke, A.J., Ramalho, J.P.P., and Rodrigues, A.I. (2009) *Tetrahedron: Asymmetry*, **20**, 1272–1278.
197 Itagaki, M. and Yamamoto, Y. (2006) *Tetrahedron Lett.*, **47**, 523–525.
198 Mazet, C., Kohler, V., and Pfaltz, A. (2005) *Angew. Chem. Int. Ed.*, **44**, 4888–4891.
199 Glos, M. and Reiser, O. (2000) *Org. Lett.*, **2**, 2045–2048.
200 Alexander, K., Cook, S., and Gibson, C.L. (2000) *Tetrahedron Lett.*, **41**, 7135–7138.
201 Doyle, M.P. and Hu, W.H. (2000) *J. Org. Chem.*, **65**, 8839–8847.
202 Doyle, M.P. and Hu, W.H. (2000) *Tetrahedron Lett.*, **41**, 6265–6269.
203 Doyle, M.P., Hu, W.H., Chapman, B., Marnett, A.B., Peterson, C.S., Vitale, J.P., and Stanley, S.A. (2000) *J. Am. Chem. Soc.*, **122**, 5718–5728.
204 Doyle, M.P., Peterson, C.S., and Parker, D.L. (1996) *Angew. Chem. Int. Ed.*, **35**, 1334–1336.
205 Doyle, M.P., Peterson, C.S., Zhou, Q.L., and Nishiyama, H. (1997) *Chem. Commun.*, 211–212.
206 Doyle, M.P. and Phillips, I.M. (2001) *Tetrahedron Lett.*, **42**, 3155–3158.
207 Honma, M., Sawada, T., Fujisawa, Y., Utsugi, M., Watanabe, H., Umino, A., Matsumura, T., Hagihara, T., Takano, M., and Nakada, M. (2003) *J. Am. Chem. Soc.*, **125**, 2860–2861.
208 Evans, D.A., Faul, M.M., Bilodeau, M.T., Anderson, B.A., and Barnes, D.M. (1993) *J. Am. Chem. Soc.*, **115**, 5328–5329.
209 Dauban, P., Saniere, L., Tarrade, A., and Dodd, R.H. (2001) *J. Am. Chem. Soc.*, **123**, 7707–7708.
210 Ryan, D., McMorn, P., Bethell, D., and Hutchings, G. (2004) *Org. Biomol. Chem.*, **2**, 3566–3572.
211 Adam, W., Roschmann, K.J., and Saha-Moller, C.R. (2000) *Eur. J. Org. Chem.*, 557–561.
212 Xu, J.X., Ma, L.G., and Jiao, P. (2004) *Chem. Commun.*, 1616–1617.
213 Ma, L., Du, D.M., and Xu, J.X. (2006) *Chirality*, **18**, 575–580.
214 Hansen, K.B., Finney, N.S., and Jacobsen, E.N. (1995) *Angew. Chem. Int. Ed. Engl.*, **34**, 676–678.
215 Juhl, K., Hazell, R.G., and Jørgensen, K.A. (1999) *J. Chem. Soc., Perkin Trans. 1*, 2293–2297.
216 Rasmussen, K.G. and Jørgensen, K.A. (1997) *J. Chem. Soc., Perkin Trans. 1*, 1287–1292.
217 Gothelf, K.V. and Jørgensen, K.A. (1998) *Chem. Rev.*, **98**, 863–910.
218 Kanemasa, S. (2002) *Synlett*, 1371–1387.
219 Gothelf, K.V., Hazell, R.G., and Jørgensen, K.A. (1996) *J. Org. Chem.*, **61**, 346–355.
220 Gothelf, K.V., Hazell, R.G., and Jørgensen, K.A. (1998) *J. Org. Chem.*, **63**, 5483–5488.
221 Kanemasa, S., Oderaotoshi, Y., Tanaka, J., and Wada, E. (1998) *J. Am. Chem. Soc.*, **120**, 12355–12356.
222 Shirahase, M., Kanemasa, S., and Hasegawa, M. (2004) *Tetrahedron Lett.*, **45**, 4061–4063.
223 Shirahase, M., Kanemasa, S., and Oderaotoshi, Y. (2004) *Org. Lett.*, **6**, 675–678.
224 Palomo, C., Oiarbide, M., Arceo, E., Garcia, J.M., Lopez, R., Gonzalez, A., and Linden, A. (2005) *Angew. Chem. Int. Ed.*, **44**, 6187–6190.
225 Lim, K.C., Hong, Y.T., and Kim, S. (2008) *Adv. Synth. Catal.*, **350**, 380–384.
226 Sibi, M.P., Ma, Z.H., and Jasperse, C.P. (2004) *J. Am. Chem. Soc.*, **126**, 718–719.
227 Desimoni, G., Faita, G., Mella, M., and Boiocchi, M. (2005) *Eur. J. Org. Chem.*, 1020–1027.
228 Jensen, K.B., Hazell, R.G., and Jørgensen, K.A. (1999) *J. Org. Chem.*, **64**, 2353–2360.
229 Miura, M., Enna, M., Okuro, K., and Nomura, M. (1995) *J. Org. Chem.*, **60**, 4999–5004.
230 Sibi, M.P., Ma, Z.H., and Jasperse, C.P. (2005) *J. Am. Chem. Soc.*, **127**, 5764–5765.
231 Gothelf, A.S., Gothelf, K.V., Hazell, R.G., and Jørgensen, K.A. (2002) *Angew. Chem. Int. Ed.*, **41**, 4236–4238.

232 Sibi, M.P., Rane, D., Stanley, L.M., and Soeta, T. (2008) *Org. Lett.*, **10**, 2971–2974.
233 Kanemasa, S. and Kanai, T. (2000) *J. Am. Chem. Soc.*, **122**, 10710–10711.
234 Sibi, M.P., Stanley, L.M., and Soeta, T. (2007) *Org. Lett.*, **9**, 1553–1556.
235 Sibi, M.P., Itoh, K., and Jasperse, C.P. (2004) *J. Am. Chem. Soc.*, **126**, 5366–5367.
236 Sibi, M.P., Ma, Z.H., Itoh, K., Prabagaran, N., and Jasperse, C.P. (2005) *Org. Lett.*, **7**, 2349–2352.
237 Sibi, M.P., Stanley, L.M., and Jasperse, C.P. (2005) *J. Am. Chem. Soc.*, **127**, 8276–8277.
238 Sibi, M.P., Stanley, L.M., and Soeta, T. (2006) *Adv. Synth. Catal.*, **348**, 2371–2375.
239 Taillier, C. and Lautens, M. (2007) *Org. Lett.*, **9**, 591–593.
240 Evans, D.A. and Janey, J.M. (2001) *Org. Lett.*, **3**, 2125–2128.
241 Huang, J. and Hsung, R.P. (2005) *J. Am. Chem. Soc.*, **127**, 50–51.
242 Aggarwal, V.K., Bell, L., Coogan, M.P., and Jubault, P. (1998) *J. Chem. Soc., Perkin Trans. 1*, 2037–2042.
243 Neves, P., Gago, S., Pereira, C.C.L., Figueiredo, S., Lemos, A., Lopes, A.D., Goncalves, I.S., Pillinger, M., Silva, C.M., and Valente, A.A. (2009) *Catal. Lett.*, **132**, 94–103.

6
PHOX Ligands
Cory C. Bausch and Andreas Pfaltz

6.1
Introduction

Phosphinooxazolines (PHOX ligands) were introduced in 1993 in three independent publications from the laboratories of Helmchen, Pfaltz, and Williams [1–3]. Although originally designed as ligands for asymmetric palladium-catalyzed allylic substitutions, they were subsequently applied to many other metal-catalyzed reactions and have established themselves as a widely used class of privileged ligands.

For a long time C_2-symmetric ligands have been dominant in asymmetric catalysis. C_2-symmetry has the advantage that it reduces the number of possible catalyst–substrate arrangements and, consequently, the number of reaction pathways and transition states by a factor of two. This is of particular advantage in mechanistic and structural studies and facilitates an analysis of catalyst–substrate interactions that may be responsible for enantioselection. Although C_2-symmetric ligands have proven useful for a wide range of applications, there are reactions for which a mechanistic analysis suggests that non-symmetrical ligands with two electronically different coordinating groups should allow more effective enantiocontrol than C_2-symmetric ligands. Transition metal-catalyzed allylic substitution via symmetric allyl complexes is a good example (Figure 6.1).

In such reactions, the regioselectivity of nucleophilic attack determines the ratio of the two enantiomeric products. If the metal center is coordinated to two electronically different groups X and Y, the two allylic termini become electronically non-equivalent and thus are expected to display different reactivity. Electronic differentiation of this type has been demonstrated by Faller in stoichiometric nucleophilic additions to allyl–molybdenum complexes [4]. Thus, it can be argued that chiral ligands possessing two different coordinating atoms should allow more effective regiocontrol than C_2-symmetric ligands. The high enantioselectivities induced by C_2-symmetric bisoxazolines (BOX ligands) in Pd-catalyzed allylic alkylations [2] led to the idea that replacement of one of the oxazoline rings by a phosphine group might result in even more efficient ligands. Indeed, as discussed in Section 6.3, PHOX ligands were found to induce excellent enantioselectivities in Pd-catalyzed allylic alkylations with 1,3-diphenylallyl acetate and other

Privileged Chiral Ligands and Catalysts. Edited by Qi-Lin Zhou
Copyright © 2011 WILEY-VCH Verlag GmbH & Co. KGaA, Weinheim
ISBN: 978-3-527-32704-1

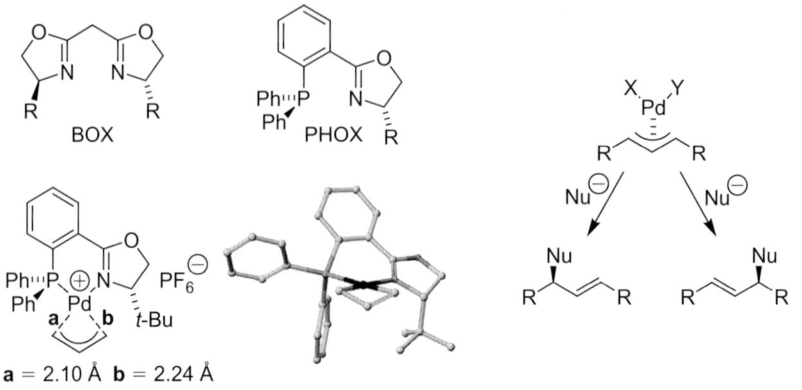

Figure 6.1 Pd(PHOX)(allyl) complex.

symmetrically substituted allyl substrates. Subsequent mechanistic studies showed that nucleophilic attack at the allylic terminus trans to the P atom is strongly favored. The different electronic effects of the P and the N atoms are also reflected in the distances between the Pd atom and the allylic termini determined by X-ray analysis, with a short Pd–C bond trans to the N atom and a long Pd–C bond trans to the P atom. This electronic differentiation of the allylic termini in combination with steric interactions of the allyl system with the chiral ligand is responsible for the observed enantioselectivities.

The promising results obtained in allylic substitutions inspired many further studies of PHOX ligands, which revealed their considerable potential for asymmetric catalysis. The modular structure and easy accessibility of phosphinooxazolines made it possible to optimize the ligand structure for a wide variety of reactions catalyzed by a range of transition metals. In addition to the applications discussed in this chapter, the success with PHOX-based catalysts stimulated the development of many other useful chiral P,N ligands such as phosphinite and phosphite analogs or pyridine-phosphinites [5–9].

6.2
Synthesis of PHOX Ligands

There are several synthetically simple, straightforward routes to this ligand family from commercially available starting materials (Scheme 6.1) [1, 3, 10–12]. The substitution pattern can be easily varied on the oxazoline ring and the P atom for reaction optimization. Various enantiomerically pure chiral amino alcohols, which serve as precursors for the oxazoline ring, are commercially available or easily prepared from α-amino acids. Nucleophilic and electrophilic phosphorus sources

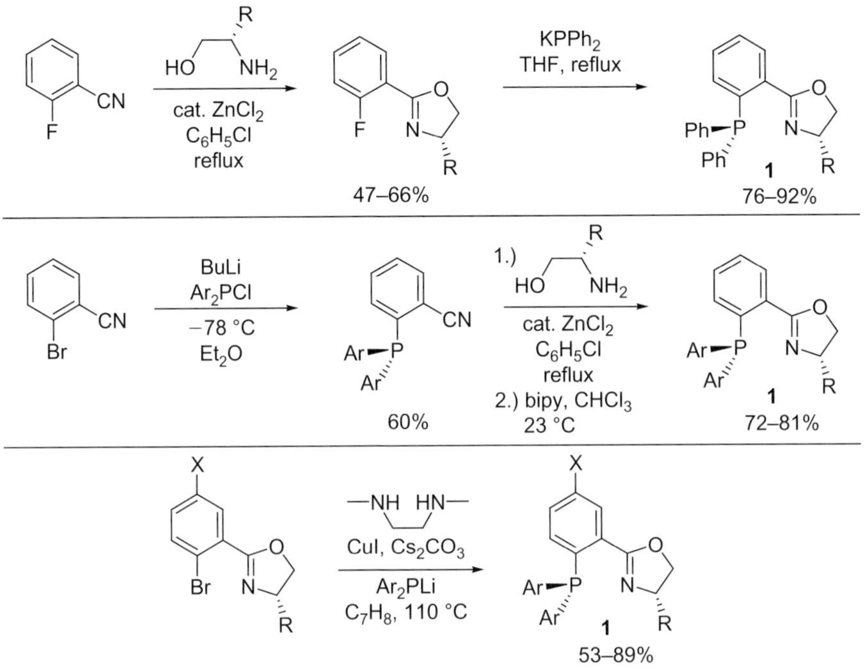

Scheme 6.1 Syntheses of PHOX ligands.

have been employed for the C–P bond-forming step. Chiral phosphines have also been employed to add further selectivity in some cases.

The concept and molecular connectivity of the PHOX ligand is unique and the concept of using two non-equivalent binding sites has been the springboard for ligand design within and beyond this family of ligands, giving a nearly infinite number of possibilities (Figure 6.2). The backbone of the ligand, the sterics, and the electronics can be manipulated to apply to a breadth of reactions [6]. In the context of this chapter, simple variants of the 2-(2-phosphinoaryl)oxazolines and their applications will be the focus.

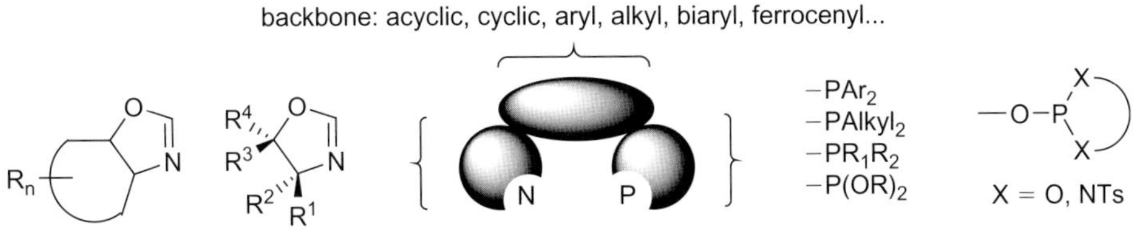

Figure 6.2 Potential variables on P,N-ligands.

6.3
Nucleophilic Allylic Substitution

6.3.1
Palladium-Catalyzed Allylic Substitution

The initial application of PHOX ligands was reported for palladium-catalyzed allylic substitution of dimethyl malonate nucleophiles [13]. In the three seminal reports [1–3], each employed 2-(o-diphenylphosphinophenyl)oxazolines **1a–e** (Figure 6.3) with varying substituents on the oxazoline ring as the ligand. The Pd-PHOX complexes showed remarkable rate enhancement and increase in selectivity compared to other catalysts, as well as little solvent dependence.

The reaction proceeds efficiently with 1 mol% palladium loading and 1.2 equivalents of ligand relative to palladium at room temperature in no more than a few hours. Three equivalents of dimethyl malonate were employed with 3.0 equivalents of N,O-bis(trimethylsilyl)acetamide (BSA) [14] and catalytic amounts of KOAc [15] in CH_2Cl_2 or THF. Table 6.1 shows examples of the conversion of racemic acetates **2** into enantioenriched malonates **3**.

The Pd-η^3-allyl intermediate with the PHOX ligand produces two electronically non-equivalent electrophilic sites even with identical terminal substituents on the allyl moiety (Section 6.1). This electronic differentiation in combination with steric effects provides nearly complete enantiocontrol. Characterization of the intermediates by ^{1}H NMR spectroscopy showed a 9 : 1 cis : trans ratio with respect of the center hydrogen on the π–allyl species to the substituent on the oxazoline ring (Scheme 6.2) [16].

The enantiomeric ratio of the allylation product is far superior to the ratio of **4a** : **4b**, so there is an inherent difference in reactivity between the two. The two allyl intermediates **4a** and **4b** are in rapid equilibrium through π–σ–π isomerization whereas nucleophilic attack, which is the enantioselectivity-determining step, is slow. According to the Curtin–Hammett principle, it is the free energy difference between the transition states of the two pathways leading to **4c** and **4d** that determines the enantiomeric ratio of the product. From the high enantioselectivity observed in this reaction, it must be concluded that intermediate **4a** is substantially more reactive than **4b**.

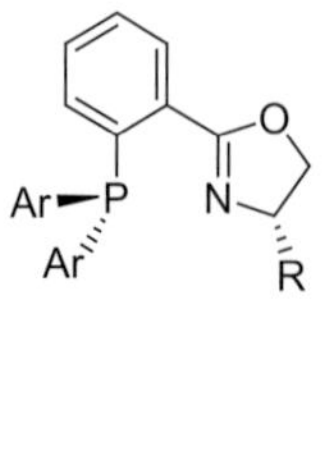

1a: R = Me, Ar = Ph
1b: R = Bn, Ar = Ph
1c: R = i-Pr, Ar = Ph
1d: R = Ph, Ar = Ph
1e: R = t-Bu, Ar = Ph
1f: R = i-Pr, Ar = 1-naphthyl
1g: R = i-Pr, Ar = 2-biphenylyl
1h: R = Ph, Ar = 2-biphenylyl
1i: R = i-Pr, Ar = 4-$CF_3C_6H_4$
1j: R = i-Pr, Ar = 3,5-$(CF_3)_2C_6H_3$

Figure 6.3 PHOX ligands.

Table 6.1 Pd-PHOX-catalyzed allylic substitution to form dimethyl malonates.

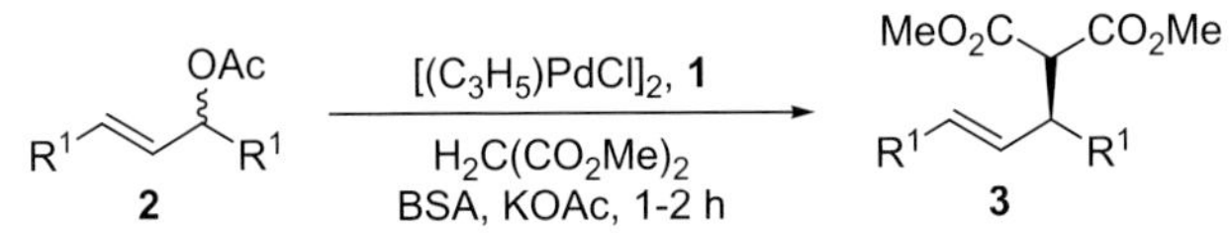

Entry	Solvent	Ligand	Substrate/R^1	Yield (%)	Ee (%)
1	CH_2Cl_2	1a	(2b)/i-Pr	91	93
2	CH_2Cl_2	1b	(2b)/i-Pr	92	92
3	CH_2Cl_2	1c	(2b)/i-Pr	88	94
4	CH_2Cl_2	1d	(2b)/i-Pr	93	89
5	CH_2Cl_2	1e	(2b)/i-Pr	88	96
6	CH_2Cl_2	1a	(2a)/Me	95	56
7	CH_2Cl_2	1e	(2a)/Me	96	69
8	CH_2Cl_2	1a	(2c)/Ph	98	89
9	CH_2Cl_2	1c	(2c)/Ph	98	98
10	THF	1b	(2c)/Ph	96	92
11	THF	1c	(2c)/Ph	92	94
12	THF	1e	(2c)/Ph	99	90

The three-dimensional structures of the intermediates, which were elucidated by extensive NMR analyses, provide a rationale for the observed enantioselectivity. Assuming electronically favored nucleophilic attack trans to the P atom, **4a** is

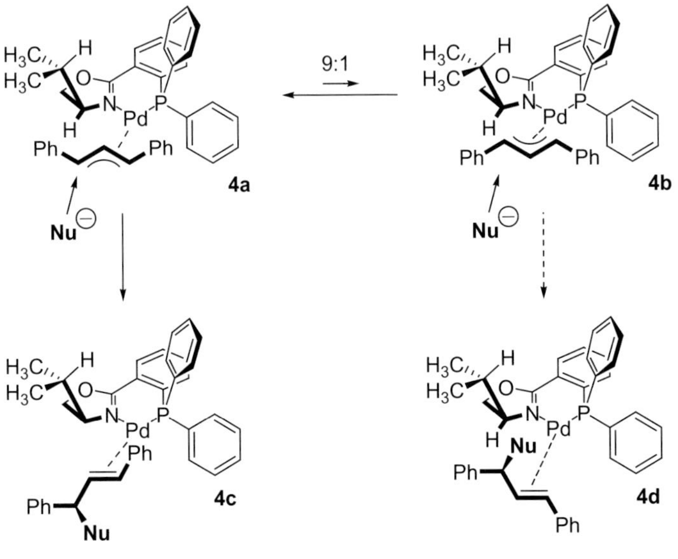

Scheme 6.2 Formation of catalytic intermediates and their structures.

converted into the Pd-olefin complex **4c** that is less sterically congested than complex **4d**. As similar steric interactions are already present in the transition states, the reaction leading to **4c** is faster than nucleophilic addition leading to **4d**. The observed absolute configuration of the product is consistent with this model.

The ratio of the cis/trans Pd-π–allyl intermediates as well as the enantioselectivity is related to the size of the substituents on the π–allyl system. The bulkier the *syn* substituents are on the allyl-termini (*syn* relative to the central C–H), the greater the cis/trans ratio and the enantioselectivity. Crystal structures of π–allyl complexes show that the substituent on the oxazoline ring and the axially oriented *P*-phenyl group are both quite remote from the allyl ligand. Therefore, the interaction between the equatorial *P*-phenyl group and the adjacent allyl substituent seems to be the primary force influencing the cis/trans ratio and the regioselectivity of nucleophilic attack. Consistent with this analysis, the cyclic π–allyl system derived from cyclohexenyl acetate gives the lowest cis/trans ratio of 64 : 36 and racemic product, while the corresponding numbers for the 1,3-dimethyl-π–allyl complex are 4 : 1 and 56% ee with **1c** as the ligand.

The discovery of this highly efficient catalytic reaction continued with the exploration of other substrate classes for allylic substitution reactions. PHOX ligands can be applied to non-symmetrical π–allyl complexes with good regio- and enantioselectivity (Table 6.2). Williams and coworkers reported an efficient allylic substitution reaction to afford diphenylallyl malonates **7** [3, 17, 18]. Both regioisomers **5** and **6** are converted into **7** in excellent yield and good enantioselectivity. Owing to the steric bulk of the geminal phenyl groups on one side of the allyl group, the substitution occurs exclusively on the less hindered side to yield the chiral product.

Table 6.2 Nucleophilic allylic substitution via unsymmetrical 1,1-diphenylallyl complexes.

Entry	Substrate, R	Solvent	Temperature (°C)	Time (h)	Ee (%)	Yield (%)
1	5a, Me	THF	Reflux	5	80	91
2	5a, Me	THF	20	24	95	95
3	5b, Ph	THF	Reflux	5	62	95
4	5b, Ph	DMF	80	5	68	96
5	6, Ph	THF	20	24	95	97
6	6, Ph	THF	Reflux	5	64	98
7	6, Ph	DMF	80	5	65	92
8	5c, 1-naphthyl	THF	30	36	>95	94

6.3 Nucleophilic Allylic Substitution

Table 6.3 Nucleophilic allylic substitution to cyclic π-allyl electrophiles.

Entry	n	X	Ligand	Solvent	Counter ion	Time (h)	Yield (%)	Ee (%)
1	2	Ac	1c	THF	K^+	72	30	0
2	2	Ac	1f	THF	K^+	18	83	32
3	2	Ac	1f	Dioxane	Na^+	1	81	45
4	2	Ac	1f	Dioxane	Li^+	24	76	47
5	2	CO_2Me	1g	Dioxane	Li^+	2 h	99	50
6	1	Ac	ent-1h	Dioxane	Li^+	1	81	62
7	3	Ac	1g	Dioxane	Li^+	17	84	83
8	4	Ac	1f	Dioxane	Li^+	48	26	36
9	1	Ac	1k	DMF	Na^+	2	73	95
10	2	Ac	1k	DMF	Na^+	3	62	93

More challenging substrates are cyclic allylic acetates or carbonates. Helmchen and coworkers reported this process with modest selectivity, obtaining enantiopure compounds through subsequent synthetic steps [19, 20]. The reactivity and enantioselectivity were not so much dependent on the size of the ring, but were more influenced by the leaving group and counter ion of the nucleophile. The ligand also had to be reevaluated as the (o-diphenylphosphino)phenyl oxazolines **1a–e** were extremely poor ligands for cyclic acetates, giving nearly racemic products. Apparently, the diphenylphosphine moiety does not provide sufficient steric bulk given the very narrow cyclic Pd-π–allyl. Alternative diarylphosphines **1f–h** (Figure 6.3) and **1k** (Table 6.3) were employed with more success. Table 6.3 shows the scope of this reaction.

Alternative C-based nucleophiles can also be utilized for allylic substitution. Rieck and Helmchen reported an allylic substitution with nitromethane [21]. Nitromethane can often lead to multiple alkylations and side products but, under these reaction conditions, mono-substitution with the Pd-π–allyl electrophile is observed. This simple 1-carbon nucleophile also provides a nitro group for further functionalization. Scheme 6.3 shows a representative example.

Complementary to the efforts of making stereogenic centers by C–C bond formation, C–N bond forming events have also received a great deal of attention and allylic aminations with nitrogen nucleophiles have been explored [22]. Various N-based nucleophiles can be used for allylic substitution, though some require a

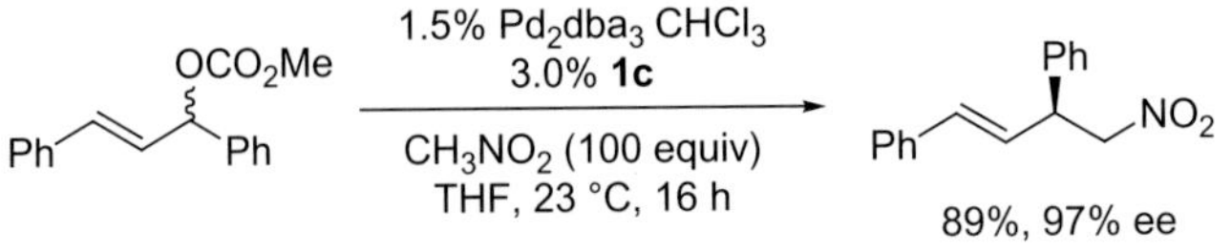

Scheme 6.3 Allylic substitution with nitromethane.

better leaving group for optimal yield and enantioselectivity (Table 6.4). The results of the allylic aminations are obtained with 2–3 mol% Pd and 2.5–3.0 mol% ligand. The reaction works well with all PHOX ligands, while **1e** usually gives slightly superior enantioselectivity. Enantioenriched allylic amines are frequently targeted functional handles due to their utility for future synthetic steps. Williams and coworkers used the allylic amines for subsequent synthesis of unnatural amino acid derivates in an efficient process [23].

As shown in Table 6.4, allylic amination works for various N-based nucleophiles from the rac-1,3-diphenyl-2-propenyl acetates and carbonates, but it also has been applied to other allyl substrates. Table 6.5 shows a compilation of alkyl substituents. In some cases with the dialkyl substitution, a more reactive leaving group is necessary as well, and so carbonates and phosphates are employed. Yields and enantioselectivities are both diminished slightly with the 1,3-dialkyl relative to the 1,3-diphenyl substituents.

Table 6.4 Nucleophilic allylic aminations.

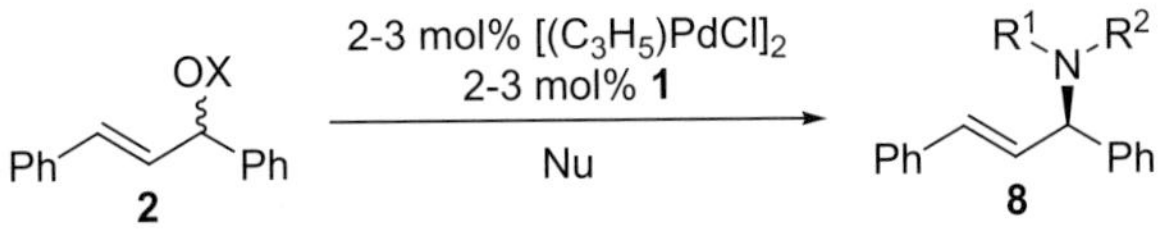

Entry	Nucleophile	Ligand	X	Temperature (°C)	Time (h)	Yield (%)	Ee (%)
1	NaNHTs	1d	Ac	50	24	97	95
2	NaNHTs	1e	Ac	50	48	96	97
3	PhC(O)NHNH$_2$/NaH	1d	Ac	50	48	63	95
4	PhC(O)NHNH$_2$/NaH	1e	Ac	50	96	95	97
5	NaN(Boc)$_2$[a]	1d	Ac	50	96	98	86
6	NaN(Boc)$_2$[a]	1e	Ac	50	96	90	67
7	PhCH$_2$NH$_2$	1d	Ac	23	96	97	73
8	PhCH$_2$NH$_2$	1e	Ac	23	96	87	89
9	PhCH$_2$NH$_2$	1b	CO$_2$Me	40	6	78	74
10	PhCH$_2$NH$_2$	1c	CO$_2$Me	40	0.5	97	87
11	PhCH$_2$NH$_2$	1d	CO$_2$Me	40	1	98	94
12	PhCH$_2$NH$_2$	1e	CO$_2$Me	40	2	93	88

[a]Boc = tert-butyl carbamate.

6.3 Nucleophilic Allylic Substitution

Table 6.5 Nucleophilic allylic aminations of 1,3-dialkyl electrophiles.

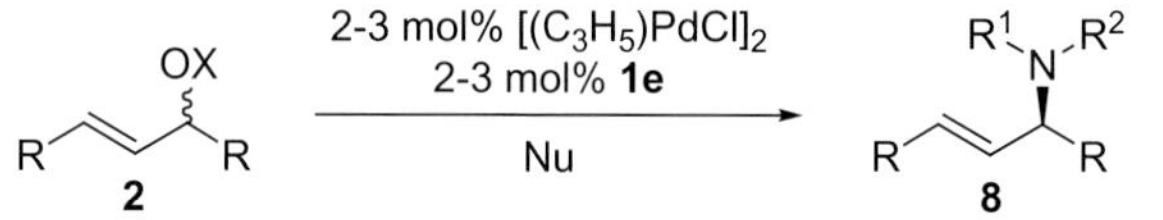

$$\underset{2}{R\diagdown\diagup\diagdown_{R}^{OX}} \xrightarrow[2-3\ mol\%\ \mathbf{1e}]{2-3\ mol\%\ [(C_3H_5)PdCl]_2} \underset{8}{R\diagdown\diagup\diagdown_{R}^{R^1\diagdown_N\diagup R^2}}$$

Entry	Nucleophile	R	X	Temperature (°C)	Time (h)	Yield (%)	Ee (%)
a1	NaNHTs	Me	Ac	23	91	61	66
2	NaNHTs	Pr	Ac	50	90	90	66
3	NaNHTs	i-Pr	CO$_2$Et	60	112	39	88
4	NaNHTs	i-Pr	PO(OEt)$_2$	50	72	55	88
5	PhC(O)NHNH$_2$/NaH	Me	Ac	50	59	52	66
6	PhC(O)NHNH$_2$/NaH	Pr	Ac	50	90	56	72
7	PhC(O)NHNH$_2$/NaH	i-Pr	CO$_2$Et	60	112	17	84
8	PhC(O)NHNH$_2$/NaH	i-Pr	PO(OEt)$_2$	50	112	20	86

In summary, Pd-PHOX complexes are efficient catalysts for allylic substitution reactions with an application range that is largely complementary to that of the diphosphine catalysts developed by Trost [6].

6.3.2
Tungsten- and Iridium-Catalyzed Allylic Substitution

Palladium-catalyzed allylic substitutions of 1- or 3-monosubstituted allyl substrates usually provide mainly the linear achiral product. However, by replacing the PHOX ligand with an analogous ligand with an electron-poor phosphorus substituent such as a phosphate group, it is possible to shift the regioselectivity toward the chiral branched product [24, 25]. Alternatively, Pd can be replaced by W or Ir, which are known to favor substitution at the substituted allyl terminus.

Lloyd-Jones and Pfaltz prepared a series of tungsten-PHOX complexes (Figure 6.4) and found that **9c** catalyzes the reaction of linear allyl esters to give mainly the chiral products (Table 6.6). Because the W catalyst was less reactive than the analogous Pd complex, the more reactive diethyl phosphates were used as substrates instead of acetates or carbonates [26, 27]. The best results were obtained with ligand **9c** (R = i-Pr), which induced high enantioselectivities and moderate to good regioselectivities between 74 : 26 and 95 : 5.

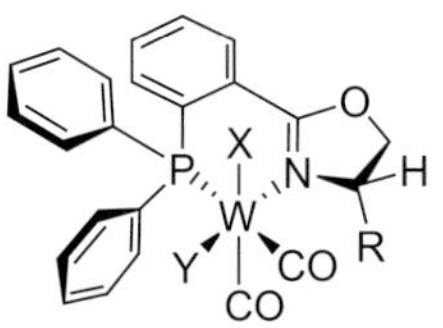

9a: X = Cl, Y = allyl
9b: X = Br, Y = 1-phenylallyl
9c: X = CH$_3$CN, Y = CO
9d: X, Y = CO

Figure 6.4 Tungsten complexes for allylic substitution.

Table 6.6 Tungsten-PHOX catalyzed nucleophilic allylic substitution.

Entry	R	Temperature (°C)	Time (h)	Yield (11a + 11b)	11a : 11b	Ee (%) of 11a
1	Ph	−10	71	89	74 : 26	96
2	4-MeC$_6$H$_4$	25	136	86	80 : 20	94
3	4-ClC$_6$H$_4$	25	136	94	78 : 22	88
4	4-PhC$_6$H$_4$	25	136	98	79 : 21	91
5	1-Naphthyl	25	136	93	95 : 5	86

Table 6.7 Iridium-PHOX catalyzed nucleophilic allylic substitution.

Entry	R	Ligand	Yield (%)	11a : 11b	Ee (%) of 11a
1	Ph	1c	61	92 : 8	30
2	Ph	1i	99	95 : 5	91
3	Ph	1j	95	89 : 11	84
4	4-(OMe)Ph	1c	89	99 : 1	72
5	4-(OMe)Ph	1i	98	99 : 1	95
6	4-(OMe)Ph	1j	71	93 : 7	62

Janssen and Helmchen have reported analogous reactions with iridium-PHOX complexes, generally giving better regioselectivity, but with slightly diminished enantioselectivity (Table 6.7) [28].

6.3.3
Allylic Substitution in Total Synthesis

A Pd-PHOX catalyzed allylic substitution has been utilized in a key step of a total synthesis. Ready and coworkers reported the total synthesis of (+)- and (−)-

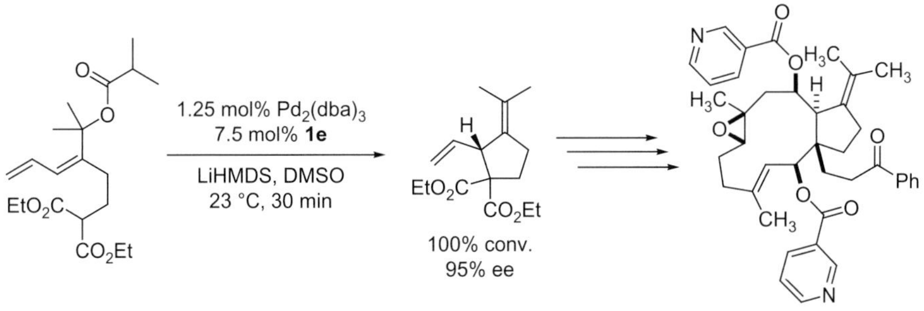

Scheme 6.4 Intramolecular allylic substitution in the total synthesis of nigellamine A_2.

nigellamine A_2 with the intramolecular allylic substitution of a malonate derivative leading to a cyclopentane intermediate with excellent yield and enantioselectivity [29]. Scheme 6.4 shows this transformation and the final product.

6.4
Decarboxylative Tsuji Allylations

6.4.1
Method Development

The Tsuji allylation, in its several forms, was reported in the 1980s [30–33]. The reactant, most well known as an allyl enol carbonate, forms a Pd-π–allyl complex and the remaining free carbonate decarboxylates with a subsequent nucleophilic attack of the free enolate on the Pd-allyl species (Scheme 6.5). This reaction is an exceptional method of regioselective enolate addition of a ketone species with multiple acidic sites. It was not until 20 years later that an enantioselective form of this reaction was addressed [34].

More recently, the laboratories of Stoltz [35] and Trost [36] have reported highly efficient enantioselective variants of this reaction. Stoltz and coworkers found optimal results with the PHOX ligands for the enantioselective Pd-catalyzed formation of quaternary centers by allylation of **12a** and analogs thereof (Scheme 6.6).

Excellent yields were achieved with several chiral ligands, but the best enantioselectivity was achieved with ligand **1e**. This reaction proceeds with complete

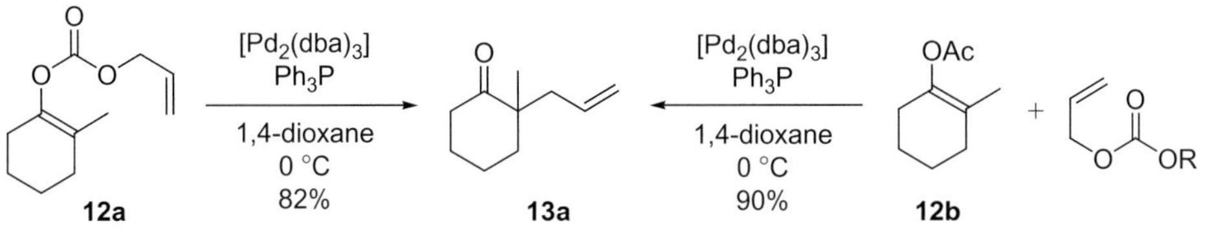

Scheme 6.5 Tsuji allylation.

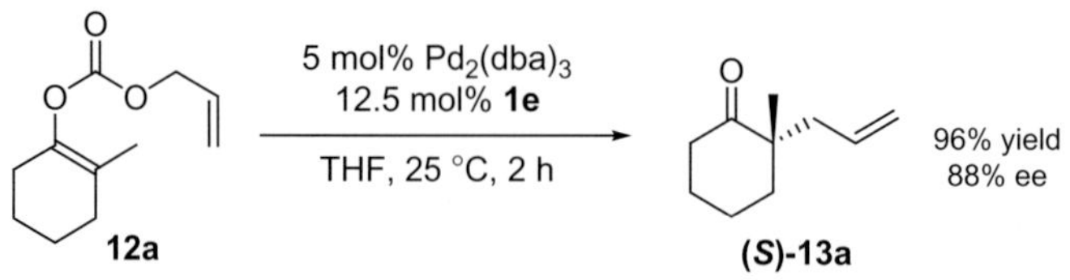

Scheme 6.6 Asymmetric Tsuji allylation.

regioselectivity, providing only allylation at the position of the pre-formed enolate. Various enantioenriched α-quaternary cycloalkanones are accessible in this way from the parent allyl enol carbonate, with optimal conditions using 2.5 mol% Pd and 6.25 mol% ligand (Scheme 6.7). Although the ee values are usually around 90%, the enantiopurity can be increased with a single recrystallization of the corresponding semicarbazone. The enantiomeric excess of *(S)*-13a was improved from 88% to 96% and further to 98% by a second recrystallization.

The same products can be achieved from the reaction of the trimethylsilyl-protected enolate and diallyl carbonate. This process, for various substrates, gives the same level of yields and enantioselectivity (Scheme 6.8).

This reaction was extended to β-ketoesters [37], more specifically racemic, quaternary β-ketoesters. In this powerful enantioconvergent process racemic starting material goes through an achiral intermediate, resulting in a highly

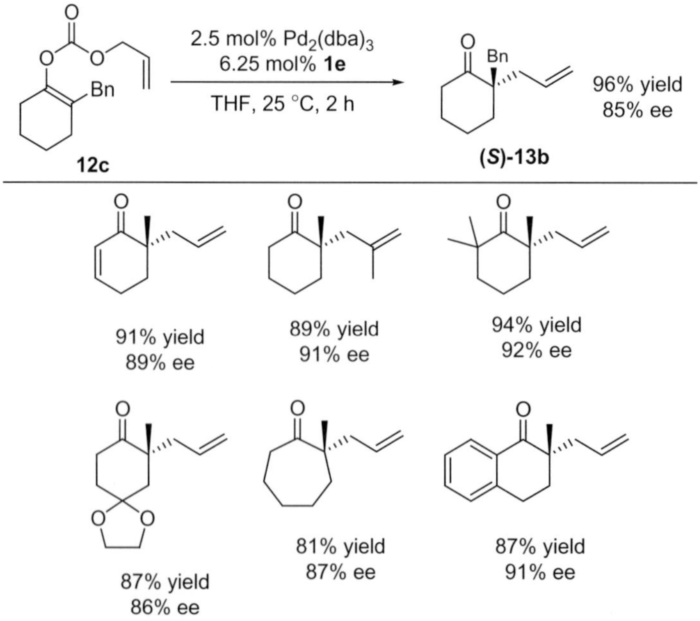

Scheme 6.7 Asymmetric Tsuji allylation of allyl enol carbonates.

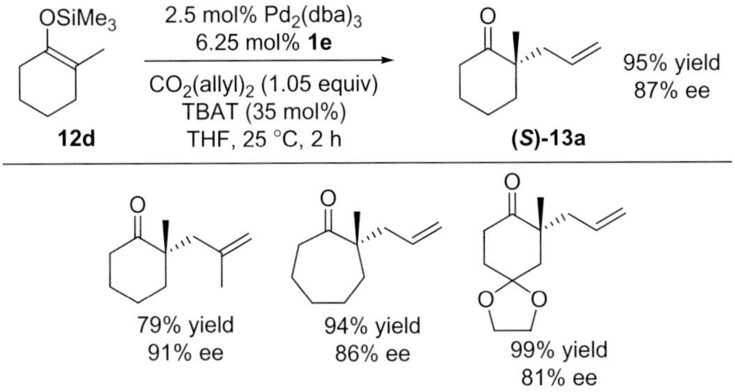

Scheme 6.8 Asymmetric Tsuji allylation of trimethylsilyl-protected enolates.

enantioenriched product (Scheme 6.9) [38, 39]. The reaction works equally well for various R groups as well as other substrate variations, giving quaternary enantioenriched products in a complementary fashion to those presented in Scheme 6.7.

With both complementary methods established, Stoltz and coworkers combined them in an elegant fashion to form two α-quaternary stereocenters in a cyclic ketone (Scheme 6.10). Both methods have been further elaborated to several more complex natural product syntheses that will be discussed below.

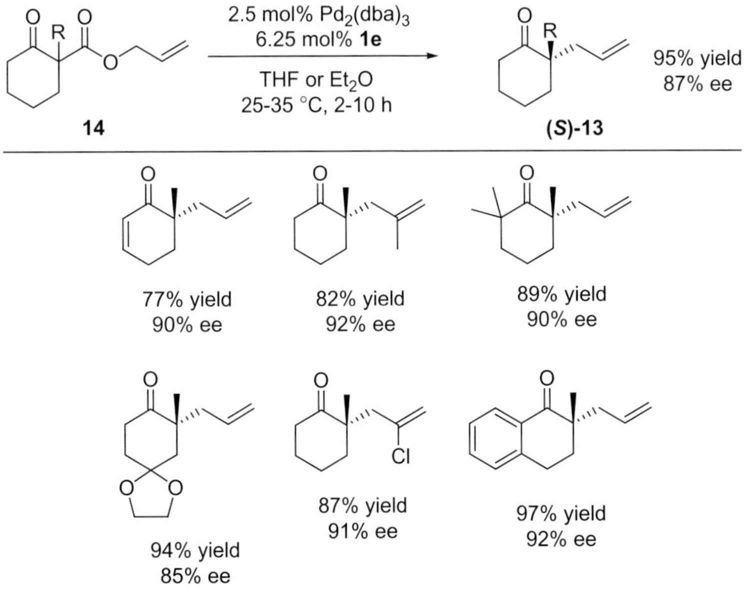

Scheme 6.9 Asymmetric Tsuji allylation of β-ketoesters.

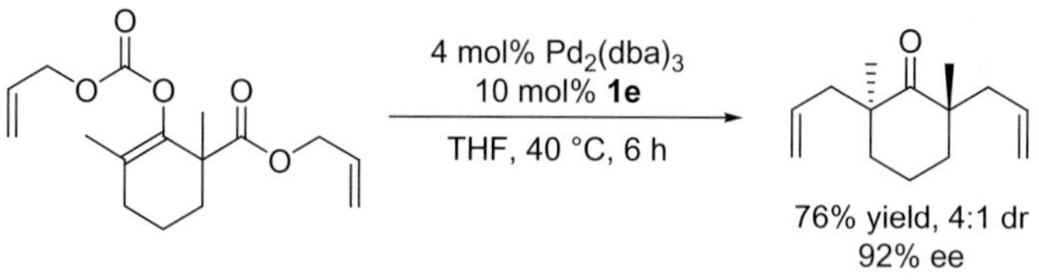

Scheme 6.10 Double asymmetric Tsuji allylation.

6.4.2
Application to Fluorinated Derivatives

Shortly after the seminal work by Stoltz and coworkers, the groups of Paquin and Nakamura reported similar processes of decarboxylative allylation to form fluorinated quaternary centers [40, 41]. Both groups examined the same ligand class, with PHOX **1e** being the optimal ligand.

Paquin and coworkers described the allylation of α-fluoro-trimethylsilyl enolates with ethyl allyl carbonate. This reaction proceeds under nearly identical conditions to those reported by Stoltz, and also in comparable yields and enantioselectivities. They found the optimal conditions to be [Pd(C$_3$H$_5$)Cl]$_2$ in toluene at 40 °C for 14–18 h. Acyclic fluorinated silyl enol ethers gave low enantioselectivities and all examples reported had only a single acidic site next to the C=O function. Scheme 6.11 shows an example of this method as well as other products achieved.

Nakamura and coworkers approached the same family of products through the decarboxylation of α-fluoro-β-ketoesters. This method gives high yields and enantioselectivities for cyclic ketone derivatives but only modest enantioselectivities for acyclic ketone derivatives. The reported conditions are Pd$_2$(dba)$_3$ as the precatalyst in THF at room temperature for 4–10 h (Scheme 6.12).

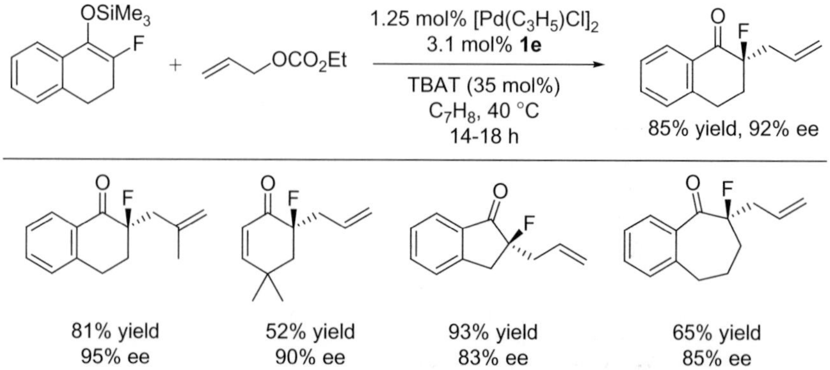

Scheme 6.11 Asymmetric Tsuji allylation of fluorinated derivatives.

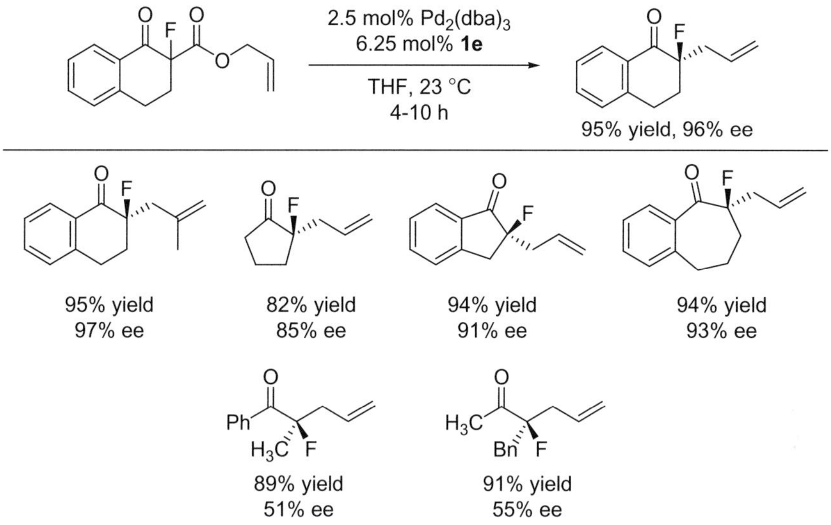

Scheme 6.12 Asymmetric Tsuji allylation of fluorinated β-ketoesters.

6.4.3
Applications in Total Synthesis

The Stoltz laboratory has demonstrated the potential of Pd-PHOX based Tsuji allylations in several total syntheses. The allylation can be used for the synthesis of intermediate (+)-**15**, a structural motif that can be converted into several natural products (Scheme 6.13) [42]. They reported the synthesis of (+)-dichroanone in a 4% overall yield over eleven steps.

In line with the allylation via the silyl enolates, Stoltz and coworkers reported the same method for substituted dioxanone enol ethers [43]. Table 6.8 shows the scope

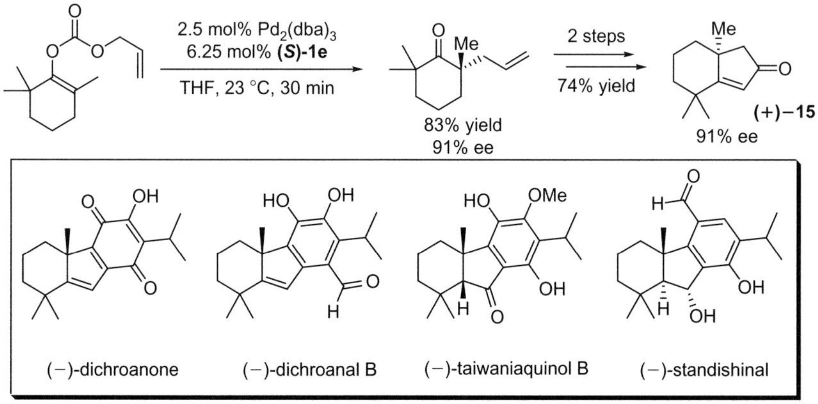

Scheme 6.13 Applications in total synthesis.

Table 6.8 Scope of allylation of dioxanone enol ethers.

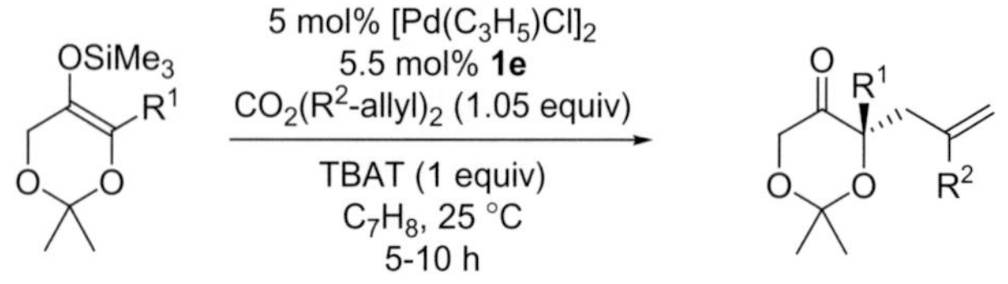

Entry	R¹	R²	Yield (%)	Ee (%)
1	Me	H	86	87
2	PhCH$_2$	H	85	86
3	Me	Me	59	89
4	Me	Ph	7	94
5	Allyl	Me	93	88
6	1-Butenyl	H	83	92

TBAT = tetrabutylammonium difluorotriphenylsilicate.

of this reaction. Enantioenriched dioxanone product can easily be converted into the α-hydroxy methyl ester or further functionalized to (−)-quinic acid, which is a key intermediate in the synthesis of (−)-dragmacidin F [44, 45].

Stoltz *et al.* also reported a double catalytic enantioselective allylation (Scheme 6.10) and a similar concept was applied to a concise total synthesis of (−)-cyanthiwigin F [46]. The crucial step in the synthesis is the double alkylation of di-β-ketoester **16** as a 1 : 1 mixture of racemic/meso diastereomers to give the desired product **17** as a 4.4 : 1 diastereomeric mixture. After further synthetic steps, the first enantioselective synthesis of (−)-cyanthiwigin F was completed (Scheme 6.14).

Another approach to natural products was recently reported by Stoltz and coworkers based on a common chiral quaternary center containing product from a

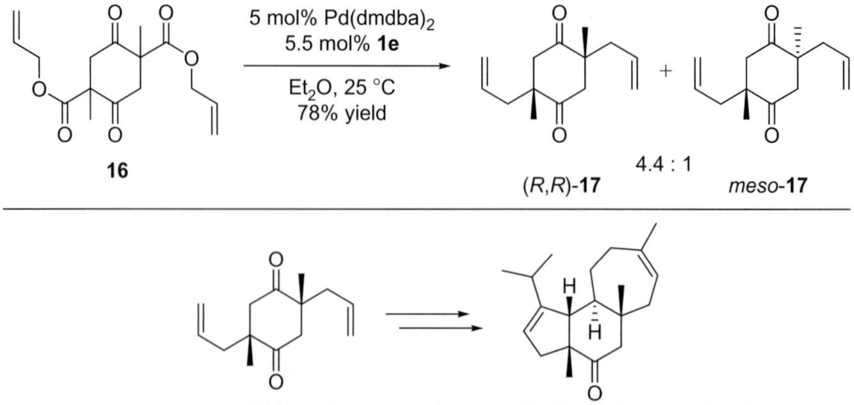

Scheme 6.14 Application in the total synthesis of (−)-cyanthiwigin F.

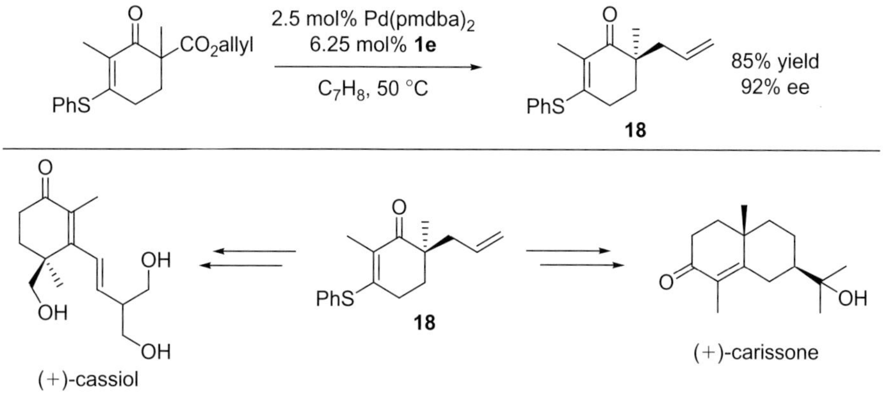

Scheme 6.15 Application in the total synthesis of (+)-cassiol and (+)-carissone.

substituted masked 1,3-dione (Scheme 6.15) [47, 48]. This common intermediate, 18, can be further manipulated to two distinct natural products, (+)-cassiol and (+)-carissone, in five and seven steps, respectively.

6.5
Heck Reaction

6.5.1
Intermolecular Heck Reaction

The Heck reaction has been an effective means of C–C bond forming reactions for many years. The coupling of an unsaturated triflate or halide with an unsubstituted double bond replaces a C–H bond with a new C–C bond in the parent alkene. The addition to cyclic alkenes provides chiral products and has been reported in an enantioselective fashion (Scheme 6.16) [49–56].

The Heck reaction proceeds through an oxidative addition of the triflate or halide and then migratory insertion into the alkene. The catalyst is regenerated after a β-hydride elimination, also replacing the element of unsaturation in the substrate. Typically, with acyclic alkenes, the substrate can rotate and the β-hydride elimination occurs with the hydrogen atom at the same position as where the new substituent resides, giving a more substituted alkene. If this reaction is applied to a cyclic alkene or a substrate where the H atom on the carbon center of initial attack

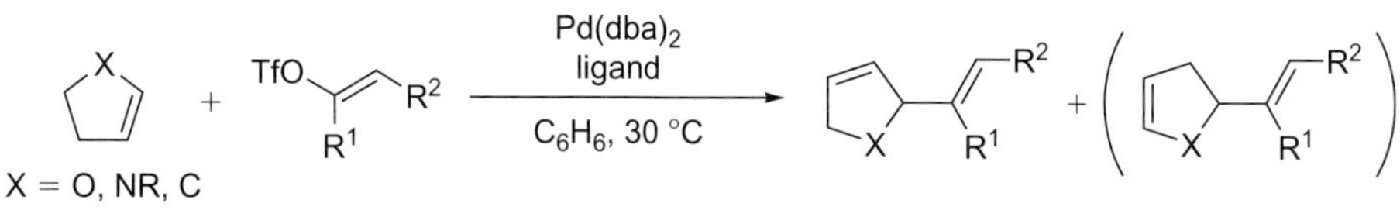

X = O, NR, C

Scheme 6.16 Asymmetric Heck reaction.

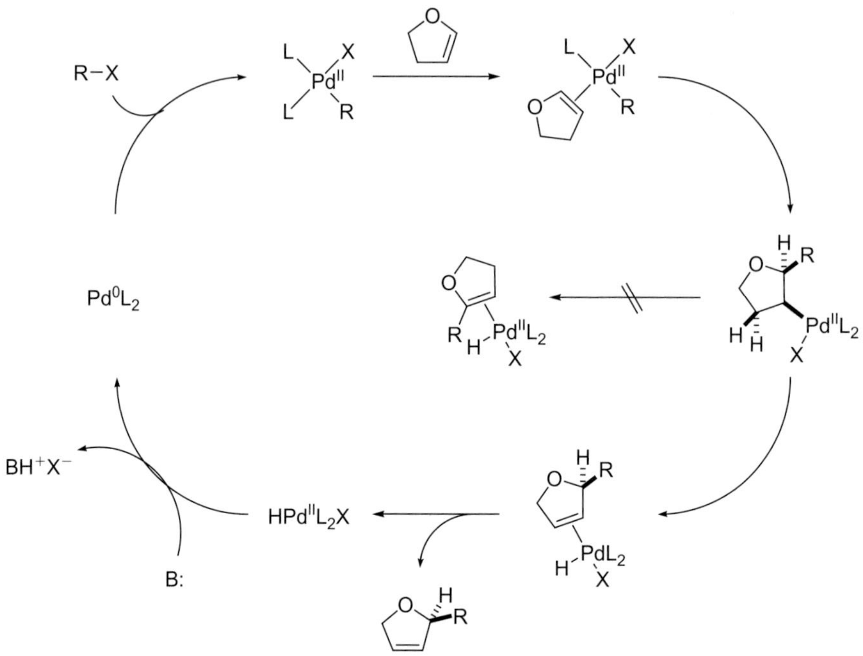

Scheme 6.17 Catalytic cycle of the Heck reaction.

is on the opposite face to the carbon-bound palladium complex, β-hydride elimination will occur in the other direction on the alkyl chain, yielding a product with a newly formed stereogenic center. Both the migratory insertion and the β-hydride elimination occur on the same face of the substrate. Scheme 6.17 shows the catalytic cycle of this Heck reaction.

The enantioselective Heck reaction with 2,3-dihydrofuran has been established with Pd(BINAP) as catalyst [57]. In this case the reaction is accompanied by double bond migration leading to the more stable 2,3-dihydrofuran with the C=C bond next to the oxygen atom. PHOX ligands have been applied to this process with great success, giving high enantioselectivities and almost exclusively the 2,5-dihydrofuran (Table 6.9) [58, 59]. This method works well for various vinyl and aryl triflates coupled with dihydrofurans, cyclopentene, and dihydropyrans (Scheme 6.18).

6.5.2
Intramolecular Heck Reaction

There are few examples of Pd-PHOX catalyzed intramolecular enantioselective Heck reactions, though this strategy is a valuable tool shown by the numerous uses in total syntheses [60]. Hallberg and coworkers have reported arylation of cyclic enamides to form spirocycles [61]. This reaction works well with Pd-PHOX (Scheme 6.19), whereas Pd-BINAP is not effective. The reaction proceeds as well with the analogous aryl iodide, but with diminished reactivity and yield.

Table 6.9 Optimization of base for the asymmetric Heck reaction.

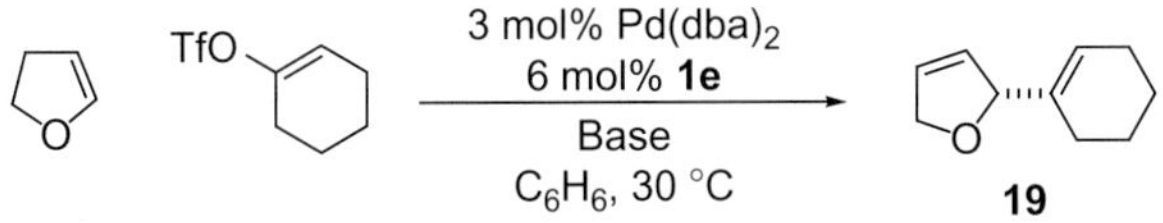
TfO
3 mol% Pd(dba)2
6 mol% 1e
Base
C6H6, 30 °C
O

19

Entry	Base	Yield (%)	Ee (%)
1	1,8-Bis(dimethylamino)naphthalene	95	98
2	2,2,6,6-Tetramethylpiperidine	95	99
3	Triethylamine	78	>99
4	*N,N*-Diisopropylamine	92	>99
5	*N,N*-Diisopropylethylamine	98	99
6	Sodium carbonate	34	98
7	Sodium acetate	50	98

dba = dibenzylideneacetone.

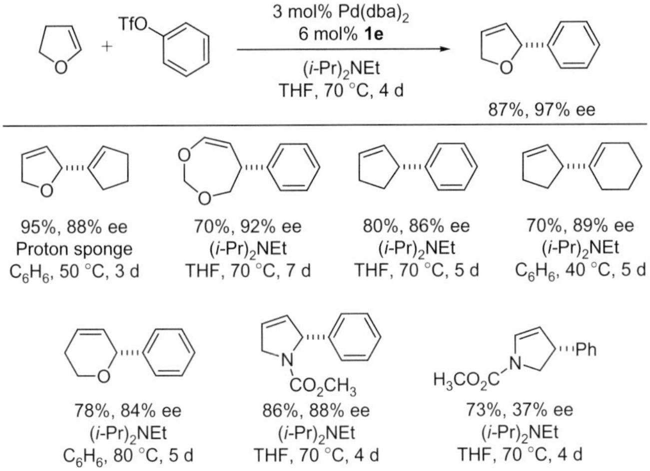
TfO
3 mol% Pd(dba)2
6 mol% 1e
(i-Pr)2NEt
THF, 70 °C, 4 d
87%, 97% ee
95%, 88% ee
Proton sponge
C6H6, 50 °C, 3 d
70%, 92% ee
(i-Pr)2NEt
THF, 70 °C, 7 d
80%, 86% ee
(i-Pr)2NEt
THF, 70 °C, 5 d
70%, 89% ee
(i-Pr)2NEt
C6H6, 40 °C, 5 d
CO2CH3
H3CO2C
Ph
78%, 84% ee
(i-Pr)2NEt
C6H6, 80 °C, 5 d
86%, 88% ee
(i-Pr)2NEt
THF, 70 °C, 4 d
73%, 37% ee
(i-Pr)2NEt
THF, 70 °C, 4 d

Scheme 6.18 Scope of the asymmetric Heck reaction.

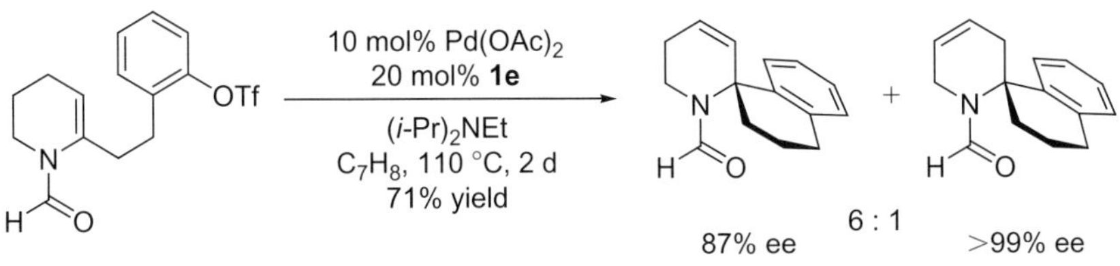
OTf
10 mol% Pd(OAc)2
20 mol% 1e
(i-Pr)2NEt
C7H8, 110 °C, 2 d
71% yield
87% ee
6 : 1
>99% ee

Scheme 6.19 Intramolecular asymmetric Heck reaction.

6.6
Hydrogenation

6.6.1
Hydrogenation of Imines

One of the most recognized uses of PHOX ligands is for iridium-catalyzed hydrogenation. Crabtree developed a robust, reactive (pyridine)(phosphine) iridium catalyst for the hydrogenation of unfunctionalized double bonds [62]. The Ir-PHOX complex is an analog in that the PHOX ligand serves as a bidentate ligand where the phosphino group and the oxazoline ring replace the monodentate phosphine and the pyridine ligands. An Ir(I) complex with a PHOX ligand is obtained from the mixture of [Ir(COD)Cl]$_2$ and the PHOX ligand and subsequent anion exchange with the more desirable PF$_6$ or BAr$_F$ (tetrakis[3,5-bis(trifluoromethyl)phenyl]borate anions. These catalysts, with the appropriate counterions, are generally air and moisture stable. The first generation of Ir-PHOX complexes has led to an immense number of chiral P,N-ligands for asymmetric hydrogenation [5, 7, 8, 63]. Figure 6.5 shows a general list of Ir-PHOX ligands used for asymmetric hydrogenation.

The first successful application of Ir-PHOX catalysts was found for the asymmetric hydrogenation of imines (Scheme 6.20) [64]. N-Phenyl acetophenone imine was reduced with 86% ee at room temperature and with 89% ee at 5 °C. Ir-PHOX catalysts proved to be highly active; 0.1 mol% catalyst was sufficient for complete conversion. The scope of Ir-PHOX catalysts for this reaction was reinvestigated recently [65]. Using complex **20e** or phosphinooxazoline complexes with aliphatic ligand backbones, up to 96% ee was obtained for N-aryl alkyl ketone imines at −20–0 °C and 5–50 bar hydrogen pressure. Cyclic imines, in contrast, gave poor results. Recently, Ir-PHOX catalysts have been successfully applied for the asymmetric hydrogenation of enamines as an alternative route to chiral amines [66].

6.6.2
Hydrogenation of Trisubstituted Olefins

Ir-PHOX catalysts were subsequently evaluated for the hydrogenation of unfunctionalized olefins [67]. With trisubstituted styrene derivatives as substrates

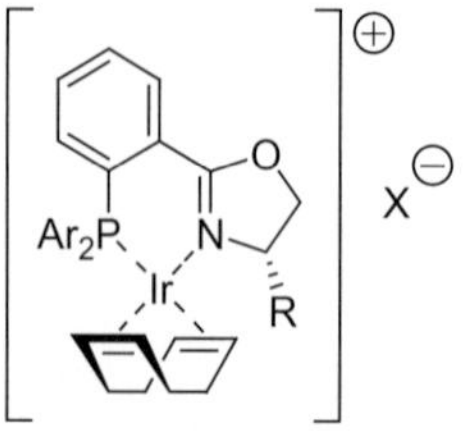

20a: R = i-Pr, Ar = Ph, X = PF$_6$
20b: R = t-Bu, Ar = Ph, X = PF$_6$
20c: R = i-Pr, Ar = o-Tol, X = PF$_6$
20d: R = t-Bu, Ar = o-Tol, X = PF$_6$
20e: R = i-Pr, Ar = Ph, X = BAr$_F$
20f: R = t-Bu, Ar = Ph, X = BAr$_F$
20g: R = i-Pr, Ar = o-Tol, X = BAr$_F$
20h: R = t-Bu, Ar = o-Tol, X = BAr$_F$
20i: R = CH$_2$t-Bu, Ar = Ph, X = BAr$_F$

Figure 6.5 Ir-PHOX catalysts.

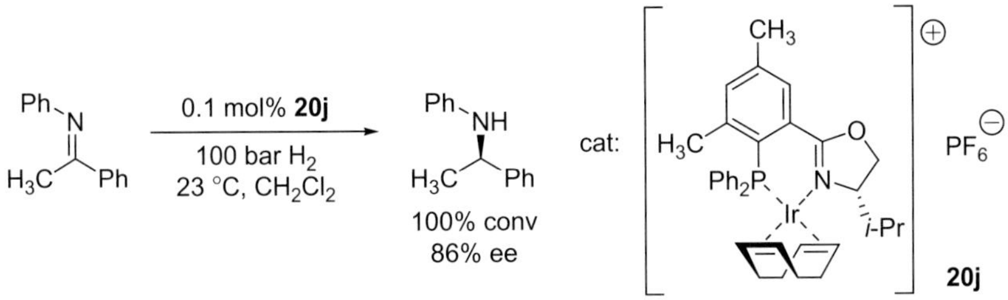

Scheme 6.20 Ir-PHOX catalyzed hydrogenation of an imine.

they showed high reactivity and excellent enantioselectivity. Table 6.10 shows a range of catalysts that have been applied and the results. Figure 6.6 shows further the substrate scope of the reaction.

Overall, complex **20h** was the best catalyst with these substrates, although enantioselectivities are catalyst-dependent to some degree. To achieve optimal enantioselectivity, multiple catalysts were tested and the best case is included with each substrate. Among various anions tested, BAr$_F$ or similar bulky, extremely weakly coordinating anions gave the best results. The BAr$_F$ salts showed very high reactivity and in addition proved to be tolerant of moisture, negating the need for rigorously anhydrous conditions. In kinetic studies, the cationic iridium complexes with BAr$_F$ and PF$_6$ both showed a high turnover frequency (TOF) of 7200 h^{-1}, but the PF$_6$ complex quickly decreased in activity whereas the BAr$_F$ salt continued as a competent catalyst until completion [68, 69]. The dependence on the hydrogen pressure was found to be weak, as 5 to 100 bar pressure gave complete conversion after 2 h with almost constant ee values, with the exception of terminal olefins, which reacted with much higher enantioselectivities at 1 bar.

Table 6.10 Screen of Ir-PHOX catalysts for asymmetric hydrogenation.

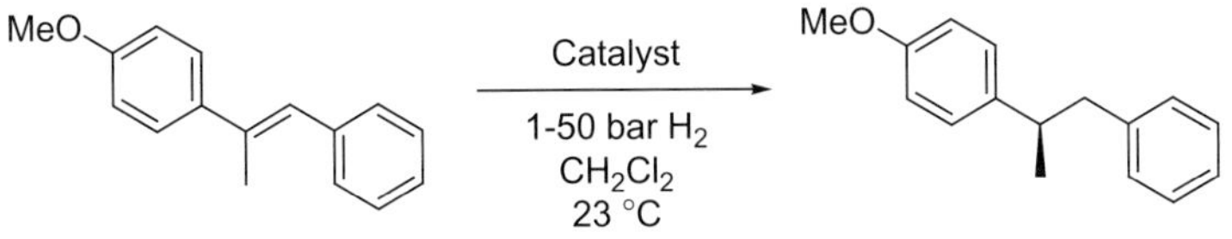

Entry	Catalyst (mol%)	Conversion (%)	Ee (%)
1	20a (4)	78	75
2	20b (4)	98	90
3	20c (4)	>99	91
4	20d (4)	57	97
5	20e (0.3)	>99	70
6	20f (0.3)	>99	98

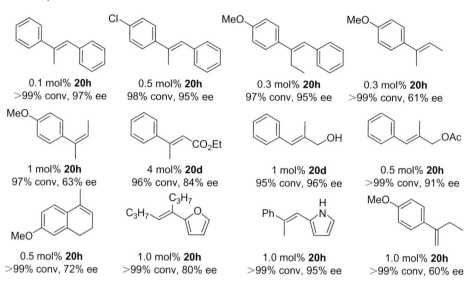

Figure 6.6 Scope of Ir-PHOX catalyzed hydrogenation.

Subsequently, many variants of PHOX ligands have been developed with different aryl or alkyl groups at the P atom or alkyl instead of aromatic backbones. A particularly useful ligand class is the NeoPHOX derivatives, which are readily accessible and in many cases give much higher enantioselectivities than the original PHOX ligands (Figure 6.8) [70].

Replacement of phosphine by phosphinite groups also led to highly versatile ligands (Figure 6.7). Especially, the ThrePHOX complexes, which are readily synthesized from threonine and have become commercially available, proved to be exceptionally efficient catalysts, which outperformed analogous PHOX catalysts in most cases. Biaryl phosphite analogs of ThreoPHOX ligands also form highly selective Ir catalysts, giving ee values between 92% and >99% ee for the standard substrates shown in Figure 6.8 [71]. SimplePHOX analogs with bis(amino)phosphine groups derived from chiral diamines have proven to be efficient ligands [72, 73].

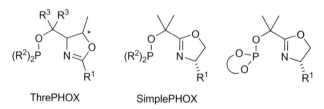

Figure 6.7 Phosphinite- and phosphite-oxazoline ligands.

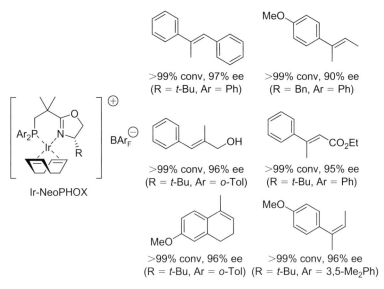

Figure 6.8 Application of Ir-NeoPHOX complexes.

6.6.3
Hydrogenation of Tetrasubstituted Olefins

PHOX ligands in general give unsatisfactory results for the hydrogenation of tetrasubstituted olefins. However, the methylene-bridged phosphinooxazolines, which form five-membered chelate complexes, performed much better and, for the first time, allowed highly enantioselective hydrogenation of this difficult substrate class (Scheme 6.21) [74]. An exception is the tricyclic olefin **21d** which gave high ee and high conversion at low catalyst loadings with PHOX complex **20e**.

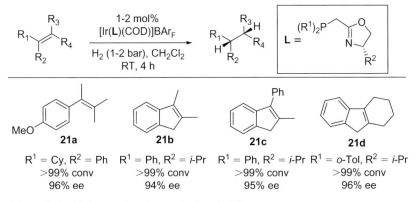

Scheme 6.21 Hydrogenation of tetrasubstituted olefins.

6.6.4
Hydrogenation of Vinyl Phosphonates

Ir-PHOX catalysts were successfully applied in the enantioselective hydrogenation of α,β-unsaturated phosphonates [75]. This work was focused on the synthesis of an enantiopure phosphorus analog of naproxen. High enantioselectivity was achieved under mild conditions and the results are summarized in Table 6.11.

6.6.5
Hydrogenation of α,β-Unsaturated Ketones

Subsequent to the application of Ir-PHOX complexes for the hydrogenation of tri- and tetrasubstituted alkenes, the laboratories of Bolm and Hou independently reported the enantioselective iridium-catalyzed hydrogenation of α,β-unsaturated ketones [76, 77]. The optimal catalyst in both cases was PHOX ligand **20f**, though Bolm chose toluene as the solvent and Hou chose dichloromethane. Both solvents were effective and gave comparable enantioselectivities. This reaction is effective for a range of aryl and alkyl substituents at the α and β positions on the C=C bond as well as at the carbonyl group. Table 6.12 shows the scope of the hydrogenation of acyclic ketones.

This catalyst also worked well for cyclic ketones, with some examples shown in Scheme 6.22. In all cases, yields and enantioselectivities were near or above 90%. Both toluene and dichloromethane were effective, with the exception of substrate **22d**, which gave low conversion and enantioselectivity in dichloromethane.

6.6.6
Hydrogenation of Ketones

While Ir-PHOX catalysts do not catalyze the hydrogenation of ketones, analogous ruthenium complexes prepared *in situ* from $Ru(PPh_3)_3Cl_2$ were found to be highly

Table 6.11 Scope of hydrogenation of vinyl phosphonates.

Entry	Ar	Time (h)	Conversion (%)	Ee (%)
1	Ph	6	100	94
2	4-PhC$_6$H$_4$	6	95	94
3	1-Naphthyl	24	93	92
4	2-Naphthyl	6	96	93
5	2-(6-MeO-Naphthyl)	115	78	95

6.6 Hydrogenation

Table 6.12 Scope of hydrogenation of α,β-unsaturated ketones.

Entry	R¹	R²	R³	Yield (%)	Ee (%)
1	Ph	Me	Me	91	98
2	Ph	Me	Ph	96	99
3	Ph	Pr	Me	88	99
4	Ph	Ph	Me	94	98
5	Et	Me	Ph	89	87
6	Ph	Ph	Ph	93	99
7[a]	H	Bn	Ph	84	86
8	2-MeOPh	Me	Me	90	99
9	3-NO2Ph	Me	Me	91	98
10	4-MeOPh	Me	Me	97	98
11	4-ClPh	Me	Me	92	99
12	4-NO2Ph	Me	Me	88	99

[a]Reaction carried out at 10 bar.

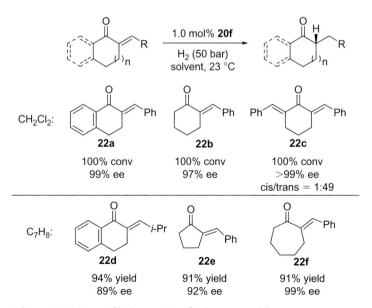

Scheme 6.22 Scope of hydrogenation of α,β-unsaturated ketones.

active catalysts for the reduction of aryl alkyl ketones with dihydrogen [78]. With PHOX ligand **1c**, 92% ee was obtained in the hydrogenation of phenyl ethyl ketone. However, clearly the best results with up to 99% ee and 50 000 turnovers were achieved with PHOX analogs with a ferrocenyl instead of a phenyl bridge. A pilot process was developed for the hydrogenation of 3,5-bistrifluoromethyl acetophenone, demonstrating the potential of these catalysts for industrial applications.

6.6.7
Transfer Hydrogenation of Ketones

In the same vein, PHOX ligands have been used in another reductive process, namely, the asymmetric transfer hydrogenation of ketones. Using a ruthenium-PHOX complex prepared from $Ru(PPh_3)_3Cl_2$ Langer and Helmchen achieved high enantioselectivities and high turnover numbers in isopropanol as solvent and reducing agent (Scheme 6.23) [79]. Preformation of the chiral metal–ligand complex gave optimal results and the reactions proceeded nearly to completion in most cases. Even higher enantioselectivities were obtained with phosphinooxazolines with a ferrocene or ruthenocene bridge replacing the phenyl ring of the PHOX ligand [78, 80–83].

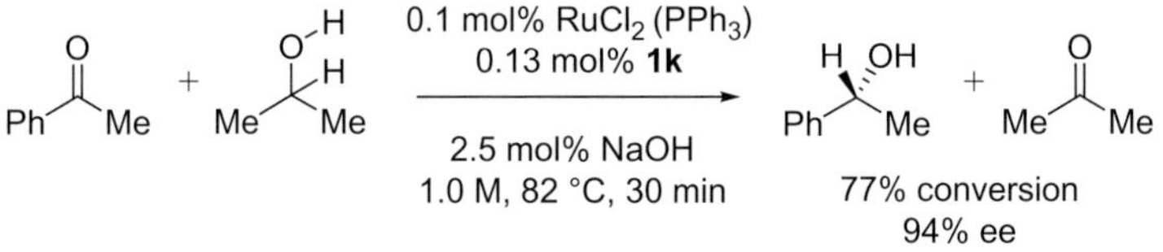

Scheme 6.23 Ru-PHOX catalyzed asymmetric transfer hydrogenation of ketones.

6.7
Cycloadditions

6.7.1
[3 + 2] Cycloadditions

Silver(I)-PHOX complexes were found to catalyze [3 + 2] cycloadditions of azomethine ylides with vinyl esters to form chiral pyrrolidines under mild conditions [84]. The intermolecular reaction shown in Scheme 6.24 proceeded in good yield, excellent diastereoselectivity, and moderate enantioselectivity. Optimization of the ligand structure for this reaction showed that introduction of geminal substituents at C5 of the oxazoline ring resulted in improved enantioselectivity, with PHOX derivative **1l** providing optimal results. While the ee values of product **23** were distinctly lower than those reported for the best literature known catalysts, much better enantioselectivities were obtained for analogous intramolecular cycloadditions. As shown in Scheme 6.24, polycyclic structures containing a chiral

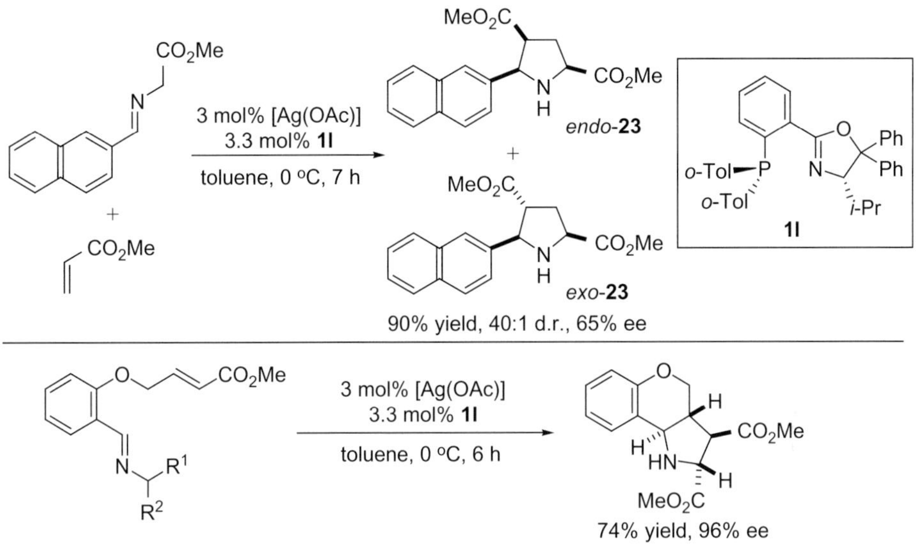

Scheme 6.24 Ag(I)-PHOX catalyzed [3 + 2] cycloaddition.

pyrrolidine ring with multiple stereocenters are accessible in this way with excellent enantio- and diastereoselectivity.

6.7.2
Diels–Alder Reactions

PHOX ligands have also been applied for enantiocontrol of Cu(II)- and Pd(II)-catalyzed Diels–Alder reactions. Sagasser and Helmchen reported the copper-catalyzed enantioselective Diels–Alder reaction of cyclic dienes and α,β-unsaturated N-acylamide dienophiles [85]. The reaction follows the precedent of Evans and coworkers' reports of Cu(II) complexes with bisoxazolines [86]. The reaction was attempted with various Cu-PHOX derivatives, varying the counterion, the P-aryl groups, and the substitution pattern of the oxazoline ring. Nitroethane and dichloromethane both were effective solvents. As shown in Scheme 6.25,

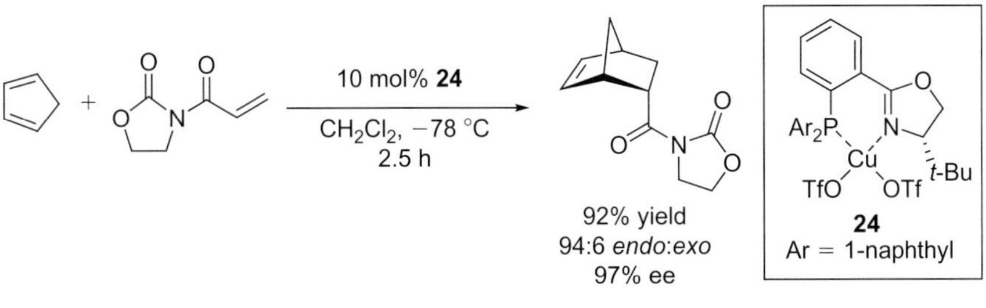

Scheme 6.25 Cu-PHOX catalyzed Diels–Alder reaction.

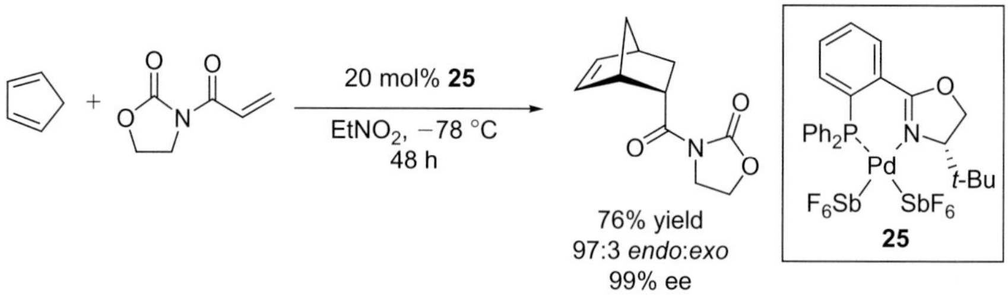

Scheme 6.26 Pd-PHOX catalyzed Diels–Alder reaction.

good reactivity and excellent enantioselectivity with a high *endo/exo* ratio was obtained.

Hiroi and Watanabe subsequently reported the same transformation with a Pd-PHOX catalyst (Scheme 6.26) [87]. This reaction is noteworthy for its exceptional enantio- and diastereoselectivity, albeit with lower yields and higher catalyst loading of a more costly metal. Examples of Ru(II)-PHOX and Os(II)-PHOX catalysts have been reported as well, but with relatively poor enantioselectivity [88].

6.8
Miscellaneous Reactions

6.8.1
Hydrosilylations

Not long after the initial reports of allylic substitution, Helmchen and Williams reported enantioselective Rh-PHOX catalyzed hydrosilylations of acetophenone derivatives with moderate enantioselectivities [89, 90]. Many ligands under various conditions were applied, and selected examples are shown in Table 6.13. More recently, Froelander and Moberg have studied analogous hydrosilylations with Ir- and Rh-PHOX catalysts [91]. They found that replacement of the alkyl substituent at C4 of the oxazoline ring with a hydroxyalkyl group led to improved enantioselectivity in the Rh-catalyzed process in the presence of Ag(I) tetrafluoroborate. With a hydroxy(phenyl)methyl-substituted PHOX ligand they achieved up to 95% ee in the hydrosilylation of acetophenone.

6.8.2
Pauson–Khand Reaction

Ir-PHOX complexes, which were previously used for asymmetric hydrogenation, were found to be active catalysts for intramolecular asymmetric Pauson–Khand reactions of enynes (Scheme 6.27) [92]. Under the reaction conditions, a dicarbonyl complex is formed, which was characterized by X-ray analysis. Relatively high catalyst loadings (2–9 mol%) are required for this process, but good yields and

Table 6.13 Screen of Rh-PHOX catalysts for hydrosilylation.

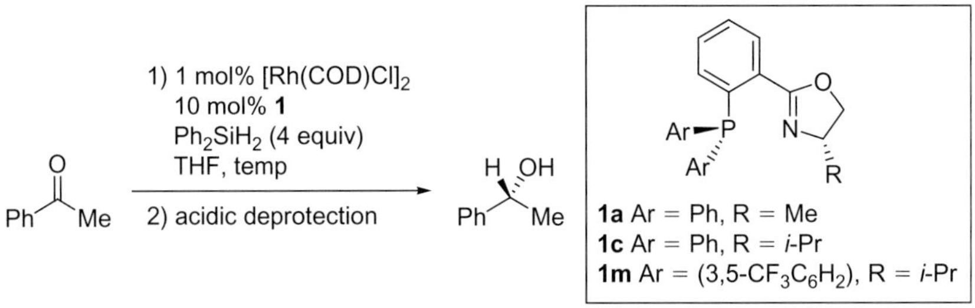

Entry	Ligand	Temperature (°C)	Yield (%)	Ee (%)
1	1c	−78	86	82
2	1a	23	92	44
3	1c	23	97	73
4	1c	10	94	81
5	1m	0	98	84

enantioselectivities were achieved for monosubstituted terminal alkenes, whereas low reactivity was observed for disubstituted terminal alkenes. The CO pressure has a strong influence on the yield and enantioselectivity. In the reaction shown in Scheme 6.27, lowering the pressure to 1.4 bar resulted in an ee increase to 97%, but the yield decreased to 51%. Catalyst loading merely influenced the yield. While at 2 mol% the yield was still high (88%), it dropped to 59% with 1 mol% catalyst. The enantioselectivities obtained in this reaction compare well with those reported for Ir-*tol*BINAP catalysts [93]. The only other application of PHOX ligands for a Pauson–Khand reaction, using stoichiometric amounts of a cobalt-PHOX complex, has been reported by Moyano and coworkers [94].

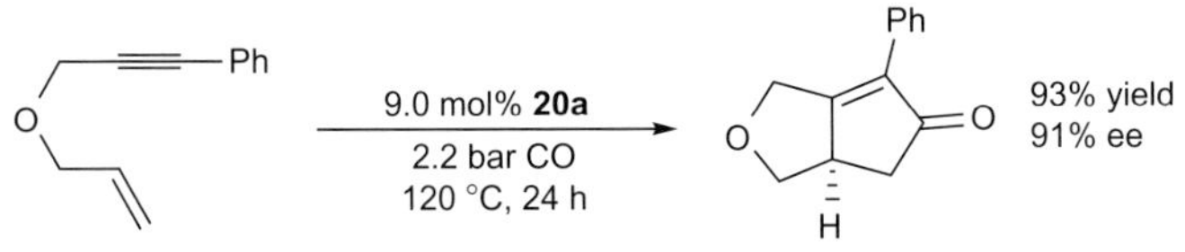

Scheme 6.27 Ir-PHOX catalyzed Pauson–Khand reaction.

6.8.3
Decarboxylative Protonation

An extension of the decarboxylative Tsuji allylation is the decarboxylative protonation developed by Stoltz to form tertiary centers adjacent to a carbonyl group

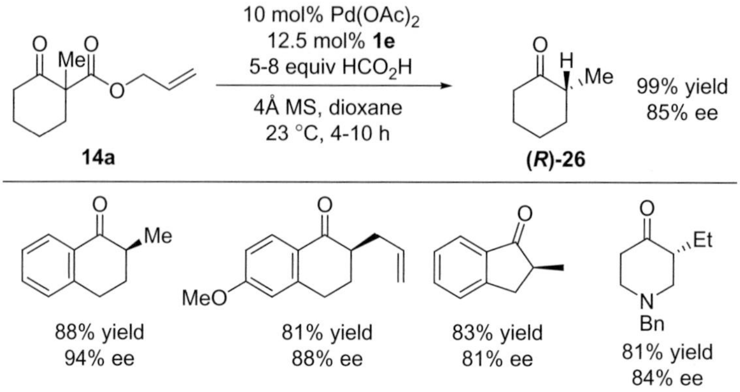

Scheme 6.28 Scope of decarboxylative protonation with formic acid.

[95]. These reactions generally work well for various substrates, with **1e** as the best ligand (Scheme 6.28).

Subsequently, other palladium catalysts and acids were evaluated to improve the yield, enantioselectivity, and substrate scope [96]. Meldrum's acid and the corresponding C-methylated derivative proved to be the optimal proton sources in terms of reactivity and selectivity. Under the optimized conditions shown in Scheme 6.29 the reactions were conducted on two different scales, both with identical concentration, providing slightly diminished selectivity on the larger scale. The results from the larger scale reactions are shown in Scheme 6.29.

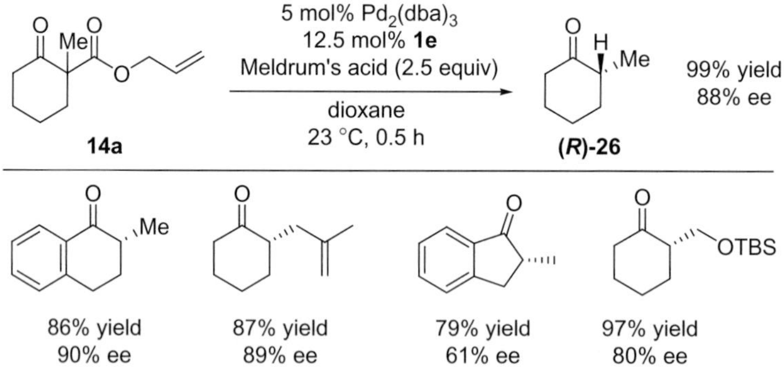

Scheme 6.29 Scope of decarboxylative allylation with Meldrum's acid.

6.8.4
Sigmatropic Rearrangements

PHOX complexes have also been applied to Pd-catalyzed enantioselective sigmatropic reactions. Hayashi and coworkers reported an asymmetric aza-Claisen

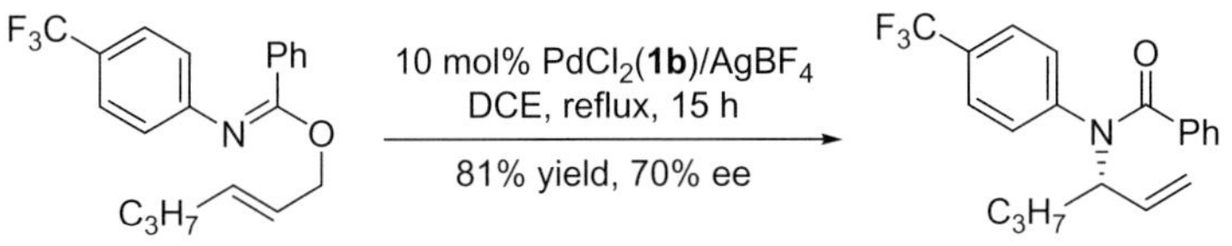

Scheme 6.30 Pd-PHOX catalyzed [3,3] sigmatropic rearrangement.

rearrangement of allyl imidates [97]. The catalyst was formed from PdCl$_2$ and PHOX ligand mixed with 1 equivalent of silver tetrafluoroborate. The reaction proceeds with modest yields and enantioselectivity (Scheme 6.30).

Another [3,3] sigmatropic rearrangement, using allyl vinyl ethers, has been reported by Linton and Kozlowski (Scheme 6.31) [98]. The dearomatization in this reaction usually requires pressing conditions but proceeds under mild conditions with palladium catalysis. The reaction requires high catalyst loading, but goes to completion in a short reaction time. A deuterium labeling experiment supported a [3,3] sigmatropic mechanism, ruling out an alternative [1,3] rearrangement via a Pd-(π–allyl) intermediate.

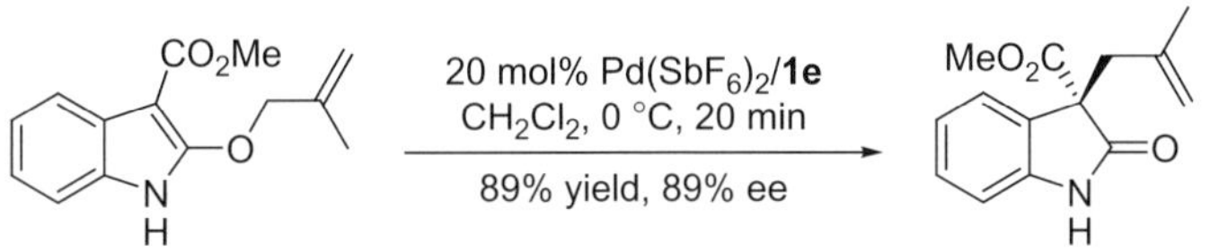

Scheme 6.31 Another Pd-PHOX catalyzed [3,3] sigmatropic rearrangement.

6.8.5
Desymmetrization Reactions

Lautens and coworkers reported a Pd-PHOX catalyzed ring opening of aza and oxabicyclic alkenes with dimethylzinc to produce highly enantioenriched allylic amines and alcohols (Scheme 6.32) [99, 100]. The ring opening of oxabenzonorbornadiene was found to be very substrate dependent and, in some cases, a ferrocene-based analog of the PHOX ligand gave superior results. The same desymmetrization reaction was also conducted with azabenzonorbornadienes as well as [3.2.1] oxabicycles. These two reactions required refluxing dichloroethane, but still gave good to excellent yields and high enantioselectivity.

Rovis and coworkers have reported several nickel- and rhodium-catalyzed desymmetrization reactions with dimethylzinc and racemic catalysts as well as some enantioselective Rh-PHOX catalyzed variants [101]. As shown in Scheme 6.33, *meso*-3,5-dimethylglutaric anhydride was converted into enantioenriched *syn*-deoxypolypropionates with a broad range of dialkylzinc reagents, including functionalized organometallic compounds transmetallated from Grignard reagents

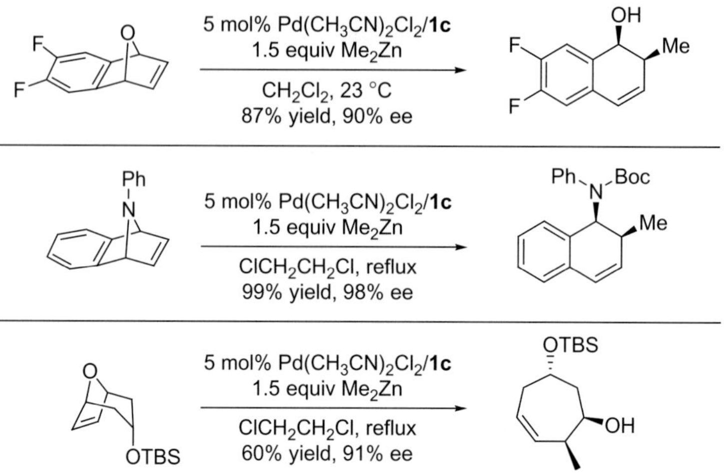

Scheme 6.32 Pd-PHOX catalyzed desymmetrization reactions.

[102]. Deoxypolypropionate structures of this type are found in several natural products and, thus, this methodology has application to total synthesis [103].

Another desymmetrization reaction, through an asymmetric Ni-catalyzed cross-coupling reaction, has been described by Hayashi and coworkers [104]. This one-step transformation from dinaphthothiophene provides axially chiral 1,1′-binaphthyls in high enantiomeric excess (Scheme 6.34).

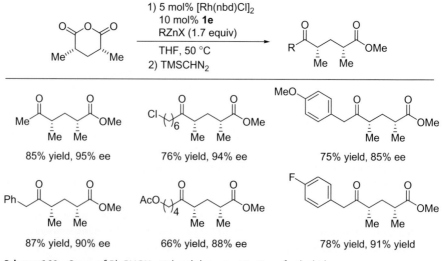

Scheme 6.33 Scope of Rh-PHOX catalyzed desymmetrization of anhydrides.

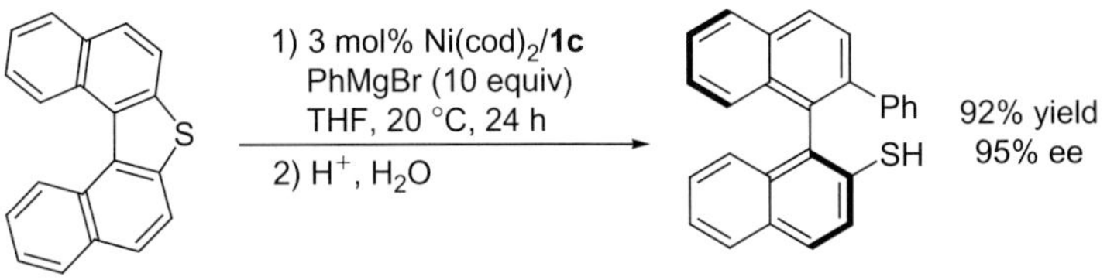

Scheme 6.34 Ni-PHOX catalyzed desymmetrization.

6.8.6
Asymmetric Arylations

Lastly, PHOX ligands have been applied for asymmetric palladium-catalyzed α-arylation of aldehydes. Buchwald and coworkers reported an intramolecular variant to produce cyclic quaternary stereocenters adjacent to a formyl group [105]. PHOX ligand **27** proved to be optimal in this case, providing good yields and high enantioselectivities for various substituents adjacent to the aldehyde as well as on the aryl ring (Scheme 6.35).

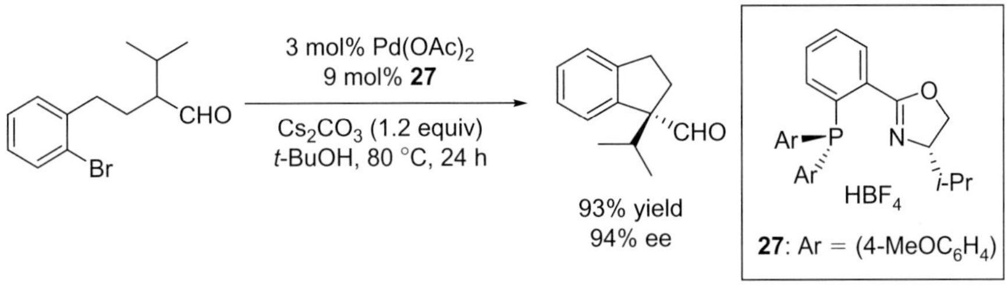

Scheme 6.35 Pd-PHOX catalyzed α-arylation.

6.9
Conclusion

As the utility of PHOX ligands has been described throughout this chapter, some reactions have been thoroughly investigated while the potential of others has yet to be realized. This ligand class has been used for a wide variety of chemical transformations, showing its potential as the need for asymmetric catalysis continues to grow. Moreover, the success with PHOX ligands has inspired the development of many other useful chiral ligands such as pyridine-phosphinites that have greatly enhanced the application range of asymmetric hydrogenation [5, 7, 8].

The design of PHOX ligands was based on the concept of electronic discrimination of two adjacent coordination sites by means of a chiral ligand with two different coordinating atoms. The superior performance of PHOX ligands in many reactions, compared to C_2-symmetric N,N or P,P ligands, suggests that this concept should be of general value for designing new ligands for asymmetric catalysis.

References

1 Sprinz, J. and Helmchen, G. (1993) *Tetrahedron Lett.*, **34**, 1769–1772.
2 von Matt, P. and Pfaltz, A. (1993) *Angew. Chem. Int. Ed.*, **32**, 566–568.
3 Dawson, G.J., Frost, C.G., Williams, J.M.J., and Coote, S.J. (1993) *Tetrahedron Lett.*, **34**, 3149–3150.
4 Faller, J.W., Chao, K.H., and Murray, H.H. (1984) *Organometallics*, **3**, 1231–1240.
5 Cui, X. and Burgess, K. (2005) *Chem. Rev.*, **105**, 3272–3296.
6 Hargaden, G.C. and Guiry, P.J. (2009) *Chem. Rev.*, **109**, 2505–2550.
7 Källström, K., Munslow, I., and Andersson, P.G. (2006) *Chem. Eur. J.*, **12**, 3194–3200.
8 Roseblade, S.J. and Pfaltz, A. (2007) *Acc. Chem. Res.*, **40**, 1402–1411.
9 Church, T.L. and Andersson, P.G. (2008) *Coord. Chem. Rev.*, **252**, 513–531.
10 Koch, G., Lloyd-Jones, G.C., Loiseleur, O., Prétôt, R., Pfaltz, A., Schaffner, S., Schnider, P., and von Matt, P. (1995) *Chim. Pays-Bas.*, **114**, 206–210.
11 Peer, M., de Jong, J.C., Kiefer, M., Langer, T., Rieck, H., Schell, H., Sennhenn, P., Sprinz, J., Steinhagen, H., Wiese, B., and Helmchen, G. (1996) *Tetrahedron*, **52**, 7547–7583.
12 Tani, K., Behenna, D.C., McFadden, R.M., and Stoltz, B.M. (2007) *Org. Lett.*, **9**, 2529–2531.
13 Trost, B.M. and Van Vranken, D.L. (1996) *Chem. Rev.*, **96**, 395–422.
14 Trost, B.M. and Murphy, D.J. (1985) *Organometallics*, **4**, 1143–1145.
15 Leutenegger, U., Umbricht, G., Fahrni, C., von Matt, P., and Pfaltz, A. (1992) *Tetrahedron*, **48**, 2143–2159.
16 Steinhagen, H., Reggelin, M., and Helmchen, G. (1997) *Angew. Chem. Int. Ed.*, **36**, 2108–2110.
17 Dawson, G.J., Williams, J.M.J., and Coote, S.J. (1995) *Tetrahedron: Asymmetry*, **6**, 2535–2546.
18 Bower, J.F. and Williams, J.M.J. (1996) *Synlett*, 685–686.
19 Sennhenn, P., Gabler, B., and Helmchen, G. (1994) *Tetrahedron Lett.*, **35**, 8595–8598.
20 Kudis, S. and Helmchen, G. (1998) *Angew. Chem. Int. Ed.*, **37**, 3047–3050.
21 Rieck, H. and Helmchen, G. (1996) *Angew. Chem. Int. Ed.*, **34**, 2687–2689.
22 von Matt, P., Loiseleur, O., Koch, G., Pfaltz, A., Lefeber, C., Feucht, T., and Helmchen, G. (1994) *Tetrahedron: Asymmetry*, **5**, 573–584.
23 Bower, J.F., Jumnah, R., Williams, A.C., and Williams, J.M.J. (1997) *J. Chem. Soc., Perkin Trans. 1*, 1411–1420.
24 Pàmies, O., Dieguez, M., and Claver, C. (2005) *J. Am. Chem. Soc.*, **127**, 3646–3647.
25 Prétôt, R. and Pfaltz, A. (1998) *Angew. Chem. Int. Ed.*, **37**, 323–325.
26 Lloyd-Jones, G.C. and Pfaltz, A. (1995) *Angew. Chem. Int. Ed.*, **34**, 462–464.
27 Prétôt, R., Lloyd-Jones, G.C., and Pfaltz, A. (1998) *Pure Appl. Chem.*, **70**, 1035–1040.
28 Janssen, J.P. and Helmchen, G. (1997) *Tetrahedron Lett.*, **38**, 8025–8026.
29 Bian, J., Van Wingerden, M., and Ready, J.M. (2006) *J. Am. Chem. Soc.*, **128**, 7428–7429.
30 Shimizu, I., Yamada, T., and Tsuji, J. (1980) *Tetrahedron Lett.*, **21**, 3199–3202.
31 Tsuji, J., Minami, I., and Shimizu, I. (1983) *Chem. Lett.*, 1325–1326.
32 Tsuji, J., Minami, I., and Shimizu, I. (1983) *Tetrahedron Lett.*, **24**, 4713–4714.
33 Tsuji, J., Minami, I., and Shimizu, I. (1983) *Tetrahedron Lett.*, **24**, 1793–1796.
34 Mohr, J.T. and Stoltz, B.M. (2007) *Chem. Asian J.*, **2**, 1476–1491.
35 Behenna, D.C. and Stoltz, B.M. (2004) *J. Am. Chem. Soc.*, **126**, 15044–15045.
36 Trost, B.M. and Xu, J. (2005) *J. Am. Chem. Soc.*, **127**, 2846–2847.
37 Meisels, G.G. (1980) *J. Am. Chem. Soc.*, **102**, 6380–6381.
38 Mohr, J.T., Behenna, D.C., Harned, A.M., and Stoltz, B.M. (2005) *Angew. Chem. Int. Ed.*, **44**, 6924–6927.
39 Mohr, J.T., Ebner, D.C., and Stoltz, B.M. (2007) *Org. Biomol. Chem.*, **5**, 3571–3576.

40 Belanger, E., Cantin, K., Messe, O., Tremblay, M., and Paquin, J.-F. (2007) *J. Am. Chem. Soc.*, **129**, 1034–1035.
41 Nakamura, M., Hajra, A., Endo, K., and Nakamura, E. (2005) *Angew. Chem. Int. Ed.*, **44**, 7248–7251.
42 McFadden, R.M. and Stoltz, B.M. (2006) *J. Am. Chem. Soc.*, **128**, 7738–7739.
43 Seto, M., Roizen, J.L., and Stoltz, B.M. (2008) *Angew. Chem. Int. Ed.*, **47**, 6873–6876.
44 Garg, N.K., Caspi, D.D., and Stoltz, B.M. (2004) *J. Am. Chem. Soc.*, **126**, 9552–9553.
45 Garg, N.K., Caspi, D.D., and Stoltz, B.M. (2005) *J. Am. Chem. Soc.*, **127**, 5970–5978.
46 Enquist, J.A., Jr. and Stoltz, B.M. (2008) *Nature*, **453**, 1228–1231.
47 Levine, S.R., Krout, M.R., and Stoltz, B.M. (2009) *Org. Lett.*, **11**, 289–292.
48 Petrova, K.V., Mohr, J.T., and Stoltz, B.M. (2009) *Org. Lett.*, **11**, 293–295.
49 Cabri, W. and Candiani, I. (1995) *Acc. Chem. Res.*, **28**, 2–7.
50 Crisp, G.T. (1998) *Chem. Soc. Rev.*, **27**, 427–436.
51 Guiry, P.J., Hennessy, A.J., and Cahill, J.P. (1997) *Top. Catal.*, **4**, 311–326.
52 Loiseleur, O., Hayashi, M., Keenan, M., Schmees, N., and Pfaltz, A. (1999) *J. Organomet. Chem.*, **576**, 16–22.
53 Ozawa, F., Kubo, A., and Hayashi, T. (1991) *J. Am. Chem. Soc.*, **113**, 1417–1419.
54 Shibasaki, M., Boden, C.D.J., and Kojima, A. (1997) *Tetrahedron*, **53**, 7371–7395.
55 Shibasaki, M., Vogl, E.M., and Ohshima, T. (2004) *Adv. Synth. Catal.*, **346**, 1533–1552.
56 Tietze, L.F., Ila, H., and Bell, H.P. (2004) *Chem. Rev.*, **104**, 3453–3516.
57 Hayashi, T., Kubo, A., and Ozawa, F. (1992) *Pure Appl. Chem.*, **64**, 421–427.
58 Loiseleur, O., Meier, P., and Pfaltz, A. (1996) *Angew. Chem. Int. Ed.*, **35**, 200–202.
59 Loiseleur, O., Hayashi, M., Schmees, N., and Pfaltz, A. (1997) *Synthesis*, 1338–1345.
60 Dounay, A.B. and Overman, L.E. (2003) *Chem. Rev.*, **103**, 2945–2963.
61 Ripa, L. and Hallberg, A. (1997) *J. Org. Chem.*, **62**, 595–602.
62 Crabtree, R. (1979) *Acc. Chem. Res.*, **12**, 331–337.
63 Pfaltz, A., Blankenstein, J., Hilgraf, R., Hormann, E., McIntyre, S., Menges, F., Schönleber, M., Smidt, S.P., Wüstenberg, B., and Zimmermann, N. (2003) *Adv. Synth. Catal.*, **345**, 33–44.
64 Schnider, P., Koch, G., Prétôt, R., Wang, G., Bohnen, F.M., Krüger, C., and Pfaltz, A. (1997) *Chem. Eur. J.*, **3**, 887–892.
65 Baeza, A. and Pfaltz, A. (2010) *Chem. Eur. J.*, **16**, 2036–2039.
66 Baeza, A. and Pfaltz, A. (2009) *Chem. Eur. J.*, **15**, 2266–2269.
67 Lightfoot, A., Schnider, P., and Pfaltz, A. (1998) *Angew. Chem. Int. Ed.*, **37**, 2897–2899.
68 Blackmond, D.G., Lightfoot, A., Pfaltz, A., Rosner, T., Schnider, P., and Zimmermann, N. (2000) *Chirality*, **12**, 442–449.
69 Smidt, S.P., Zimmermann, N., Studer, M., and Pfaltz, A. (2004) *Chem. Eur. J.*, **10**, 4685–4693.
70 Schrems, M.G. and Pfaltz, A. (2009) *Chem. Commun.*, 6210–6212.
71 Dieguez, M., Mazuela, J., Pàmies, O., Verendel, J.J., and Andersson, P.G. (2008) *Chem. Commun.*, 3888–3890.
72 Hilgraf, R. and Pfaltz, A. (2005) *Adv. Synth. Catal.*, **347**, 61–77.
73 Schönleber, M., Hilgraf, R., and Pfaltz, A. (2008) *Adv. Synth. Catal.*, **350**, 2033–2038.
74 Schrems, M.G., Neumann, E., and Pfaltz, A. (2007) *Angew. Chem. Int. Ed.*, **46**, 8274–8276.
75 Goulioukina, N.S., Dolgina, T.y.M., Bondarenko, G.N., Beletskaya, I.P., Ilyin, M.M., Davankov, V.A., and Pfaltz, A. (2003) *Tetrahedron: Asymmetry*, **14**, 1397–1401.
76 Lu, S.-M. and Bolm, C. (2008) *Angew. Chem. Int. Ed.*, **47**, 8920–8923.
77 Lu, W.-J., Chen, Y.-W., and Hou, X.-L. (2008) *Angew. Chem. Int. Ed.*, **47**, 10133–10136.
78 Naud, F.C., Malan, C., Spindler, F., Rüggeberg, C., Schmidt, A.T., and

Blaser, H.U. (2006) *Adv. Synth. Catal.*, **348**, 47–50.

79 Langer, T. and Helmchen, G. (1996) *Tetrahedron Lett.*, **37**, 1381–1384.

80 Arikawa, Y., Ueoka, M., Matoba, K., Nishibayashi, Y., Hidai, M., and Uemura, S. (1999) *J. Organomet. Chem.*, **572**, 163–168.

81 Liu, D., Xie, F., Zhao, X., and Zhang, W. (2008) *Tetrahedron*, **64**, 3561–3566.

82 Nishibayashi, Y., Takei, I., Uemura, S., and Hidai, M. (1999) *Organometallics*, **18**, 2291–2293.

83 Sammakia, T. and Strangeland, E.L. (1997) *J. Org. Chem.*, **62**, 6104–6105.

84 Stohler, R., Wahl, F., and Pfaltz, A. (2005) *Synthesis*, 1431–1436.

85 Sagasser, I. and Helmchen, G. (1998) *Tetrahedron Lett.*, **39**, 261–264.

86 Johnson, J.S. and Evans, D.A. (2000) *Acc. Chem. Res.*, **33**, 325–335.

87 Hiroi, K. and Watanabe, K. (2002) *Tetrahedron: Asymmetry*, **13**, 1841–1843.

88 Carmona, D., Vega, C., Garcia, N., Lahoz, F.J., Elipe, S., Oro, L.A., Lamata, M.P., Viguri, F., and Borao, R. (2006) *Organometallics*, **25**, 1592–1606.

89 Langer, T., Janssen, J., and Helmchen, G. (1996) *Tetrahedron: Asymmetry*, **7**, 1599–1602.

90 Newman, L.M., Williams, J.M.J., McCague, R., and Potter, G.A. (1996) *Tetrahedron: Asymmetry*, **7**, 1597–1598.

91 Froelander, A. and Moberg, C. (2007) *Org. Lett.*, **9**, 1371–1374.

92 Lu, Z.-L., Neumann, E., and Pfaltz, A. (2007) *Eur. J. Org. Chem.*, 4189–4192.

93 Shibata, T., Toshida, N., Yamasaki, M., Maekawa, S., and Takagi, K. (2005) *Tetrahedron*, **61**, 9974–9979.

94 Castro, J., Moyano, A., Pericas, M.A., Riera, A., Alvarez-Larena, A., and Piniella, J.F. (2000) *J. Am. Chem. Soc.*, **122**, 7944–7952.

95 Mohr, J.T., Nishimata, T., Behenna, D.C., and Stoltz, B.M. (2006) *J. Am. Chem. Soc.*, **128**, 11348–11349.

96 Marinescu, S.C., Nishimata, T., Mohr, J.T., and Stoltz, B.M. (2008) *Org. Lett.*, **10**, 1039–1042.

97 Uozumi, Y., Kato, K., and Hayashi, T. (1998) *Tetrahedron: Asymmetry*, **9**, 1065–1072.

98 Linton, E.C. and Kozlowski, M.C. (2008) *J. Am. Chem. Soc.*, **130**, 16162–16163.

99 Lautens, M., Renaud, J.-L., and Hiebert, S. (2000) *J. Am. Chem. Soc.*, **122**, 1804–1805.

100 Lautens, M., Hiebert, S., and Renaud, J.-L. (2000) *Org. Lett.*, **2**, 1971–1973.

101 Bercot, E.A. and Rovis, T. (2002) *J. Am. Chem. Soc.*, **124**, 174–175.

102 Cook, M.J. and Rovis, T. (2007) *J. Am. Chem. Soc.*, **129**, 9302–9303.

103 Cook, M.J. and Rovis, T. (2009) *Synthesis*, 335–338.

104 Shimada, T., Cho, Y.-H., and Hayashi, T. (2002) *J. Am. Chem. Soc.*, **124**, 13396–13397.

105 Garcìa-Fortanet, J. and Buchwald, S.L. (2008) *Angew. Chem. Int. Ed.*, **47**, 8108–8111.

7
Chiral Salen Complexes
Wen-Zhen Zhang and Xiao-Bing Lu

7.1
Introduction

Asymmetric catalysis has provided the most powerful tool to access highly enantiomerically pure compounds in organic synthesis. Numerous chiral metal complexes have been established as effective catalysts for various highly enantioselective reactions [1]. Among them, chiral salen complexes have attracted considerable attention due to their easily tunable stereochemical properties, including coordination geometries and conformations, which readily create effective chiral environments around the metal centers, resulting in exquisite selectivity in asymmetric catalysis [2].

Chiral salen complexes were not applied to asymmetric catalysis until the 1990s, although synthetic salen complexes had long been studied. In 1990, Jacobsen and Katsuki independently reported that chiral salen-Mn(III) complexes were effective chiral catalysts for enantioselective epoxidation of unfunctionalized olefins [3, 4]. These catalyst systems overcome the limitation of the Sharpless asymmetric epoxidation, and more importantly opened up the broad application of chiral salen complexes in asymmetric catalysis, especially in enantioselective ring-opening of epoxides, conjugate addition, hetero-Diels–Alder reaction, cyclopropanation, and other reactions. Chiral salen complexes have, by far, been established as privileged chiral catalysts and proven to be enantioselective over a wide range of mechanistically unrelated reactions [5]. This chapter presents synthetic strategy, structural features of chiral salen complexes, and their applications in various catalytic asymmetric reactions.

7.2
Synthesis of Chiral Salen Complexes

The abbreviation "salen" was originally used to denote N,N'-bis(salicylidine)-ethylenediamine derivatives formed by condensation of salicylaldehydes and 1,2-

Privileged Chiral Ligands and Catalysts. Edited by Qi-Lin Zhou
Copyright © 2011 WILEY-VCH Verlag GmbH & Co. KGaA, Weinheim
ISBN: 978-3-527-32704-1

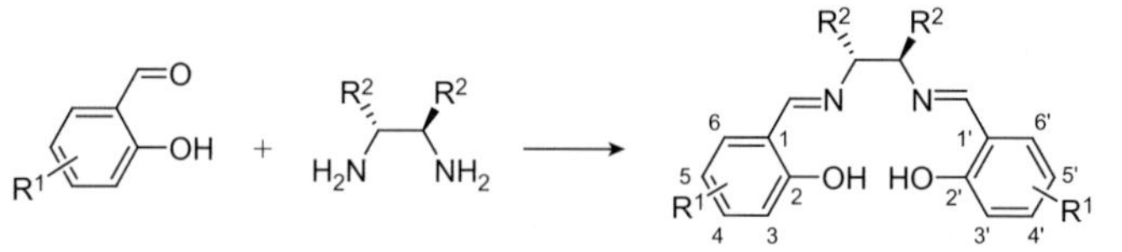

Scheme 7.1 Synthesis of chiral salen ligands.

ethylenediamine. Nowadays the denotation of "salen" has been extended to the diimine compounds obtained from condensation of salicylaldehyde derivatives with diamines. Because various chiral diamines and salicylaldehyde derivatives are commercially available or can be obtained by simple and well-established synthesis procedure, a large quantity of chiral salen ligands can be readily prepared (Scheme 7.1). The most frequently used diamines are chiral 1,2-cyclohexadiamine and 1,2-diphenylethylene-1,2-diamine, while salicylaldehyde derivatives generally contain bulky and/or chiral substituents at the position ortho to the phenolic hydroxyl group. Because hydrolysis of the imine group is liable to occur under acid conditions, chromatography of chiral salen ligands on silica should be avoided. Usually, recrystallization in a suitable solvent is a better way to purify chiral salen ligands. Coordination with metal ions significantly improves the stability of imine groups of salen ligands.

Although generating organometallic complexes *in situ* can usually circumvent multiple purifications, it is highly desirable to employ pure chiral salen complexes as catalysts straightforwardly. Metallation of chiral salen ligands with various metal precursors such as metal alkoxides, metal acetates, metal halides, or metal alkyls complexes affords the corresponding chiral salen-metal complexes (Scheme 7.2). Notably, salen-metal complexes bearing alkoxide groups are sensitive to moisture and μ-oxo complexes can be formed in the presence of adventitious water. To avoid the formation of μ-oxo salen-metal complexes, a two-step reaction, which involves firstly deprotonation of chiral salen ligand with KH or NaH and then reaction of the resulting metal phenolate compounds with metal chloride–tetrahydrofuran adducts, offers an effective way to synthesize chiral salen complexes on a large scale.

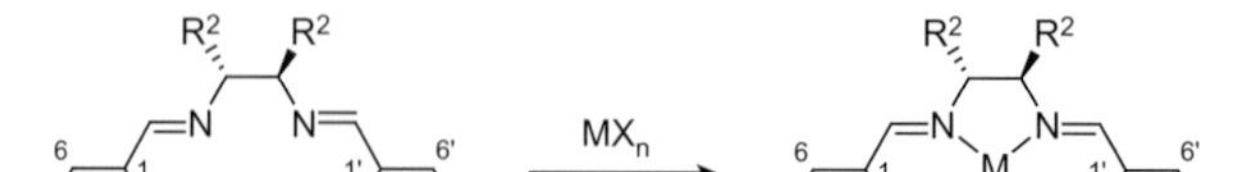

Scheme 7.2 Preparation of chiral salen-metal complexes.

7.3
Structural Properties of Chiral Salen Complexes

The structural properties of chiral salen complexes exert essential influences on their asymmetric catalytic performances. Sometimes a slight structural modification dramatically alters asymmetric induction. Most chiral salen-metal complexes adopt octahedral configuration. Dianionic tetradentate salen ligands bind metal ions securely through two nitrogen and two oxygen atoms. Two anionic or neutral ancillary ligands occupy another two coordination sites. When a non-coordinating counter anion such as PF_6^- and ClO_4^- exists in the complex, coordination solvent or water binds instead to the metal center. According to the positions of two ancillary ligands, three different configurations can potentially be adopted for octahedral chiral salen-metal complexes: *trans* (two apical positions), *cis*-β (one apical and one equatorial positions), and *cis*-α (two equatorial positions) (Figure 7.1).

When two ancillary ligands are monodentate, salen-metal complexes usually adopt a *trans*-configuration because the *trans*-isomer is more stable than the *cis*-isomer. In asymmetric ring-opening of epoxides and hetero-Diels–Alder reaction, the nucleophilic substrates substitute the ancillary ligand in a chiral *trans*-salen-metal complex and are activated by the electrophilic metal ion. In asymmetric epoxidation and cyclopropanation, the ancillary ligand in chiral *trans*-salen-metal complexes is firstly substituted with iodosylbenzene, diazo, or other nucleophilic compound, then an active oxene- or carbene-metal complex is formed and transfer of oxygen- or carbon atom to various substrates occurs (Scheme 7.3). Because chiral salen ligands have created effective asymmetric environments around the metal centers of the above *trans*-salen-metal complexes, these substrates undergo corresponding reactions in an enantioselective manner.

Owing to the flexibility of sp^3 carbon in the ethylenediamine unit, several conformations might be adopted for *trans*-salen-metal complexes. Stepped (or folded) and bowl (or umbrella) are two typical conformations (Figure 7.2). The conformation of the five-membered chelated ring, which consists of an ethylenediamine unit and the metal ion, determines the conformation of *trans*-salen-metal complex. The *trans*-salen-metal complex with a half-chair five-membered chelated ring generally adopts the stepped conformation and a complex with an envelope five-membered

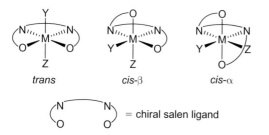

Figure 7.1 Three configurations for chiral salen-metal complexes.

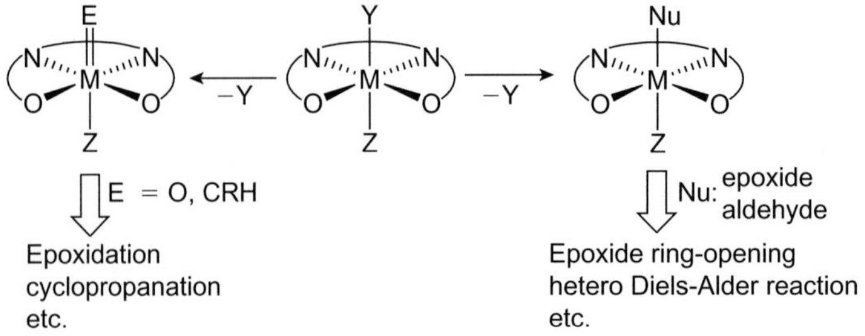

Scheme 7.3 Substitution reactions of *trans*-salen-metal complexes.

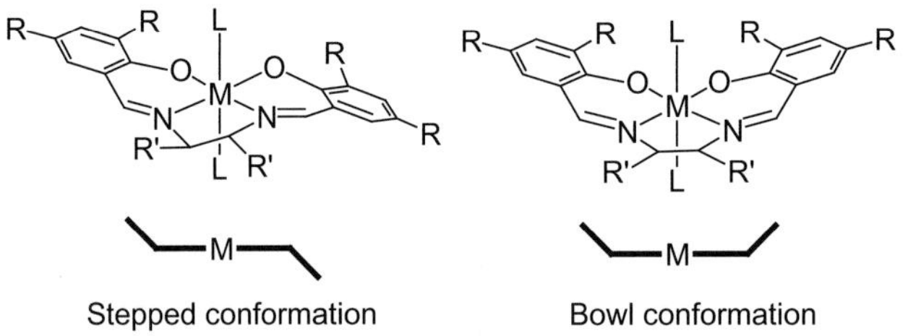

Figure 7.2 Two conformations of *trans*-salen-metal complexes.

chelated ring takes the bowl conformation. Nevertheless, chiral *trans*-salen-metal complexes generally adopt a nonplanar stepped conformation because the half-chair conformation of a five-membered chelated ring can minimize steric repulsion between two substituents in the chiral diamine unit. Furthermore, the degree of folding has a promising effect on asymmetric induction of chiral *trans*-salen-metal complexes. In general, the degree of folding is determined by the metal ion and its oxidation state, the absence or presence of apical ligand, and steric and the electronic nature of chiral salen ligand.

When chiral *trans*-salen-metal complexes adopt a nonplanar stepped conformation, there is an equilibrium between two potential conformers: diaxial and diequatorial. Generally, diequatorial conformer is more favored than diaxial conformer due to steric repulsion between the substituents at the chiral diamine unit and apical ligands (Scheme 7.4).

To obtain high enantioselectivity in asymmetric reactions catalyzed by chiral salen-metal complexes, a wise choice of metal center and delicate control of ligand conformation are of great importance. The whole conformation of a chiral salen-metal complex mainly depends on the conformation of that five-membered chelating ring, which is determined directly by the chiral diamine unit. In addition, the nature of substituents in the chiral salen ligand and of the apical ligands have significant effects on the conformation of a chiral salen-metal complex. Taking the well-developed asymmetric epoxidation catalyzed by chiral salen-Mn complexes as an example (Figure 7.3) [6], the active species oxo salen-Mn(V) complex derived from the reaction of chiral salen-Mn(III) complex with

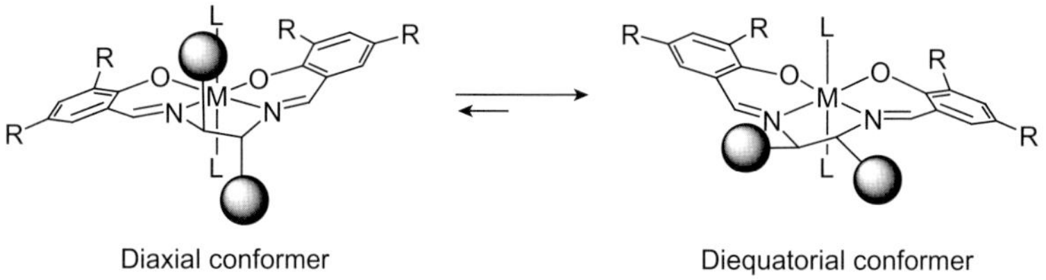

Scheme 7.4 Equilibrium between diaxial and diequatorial conformers.

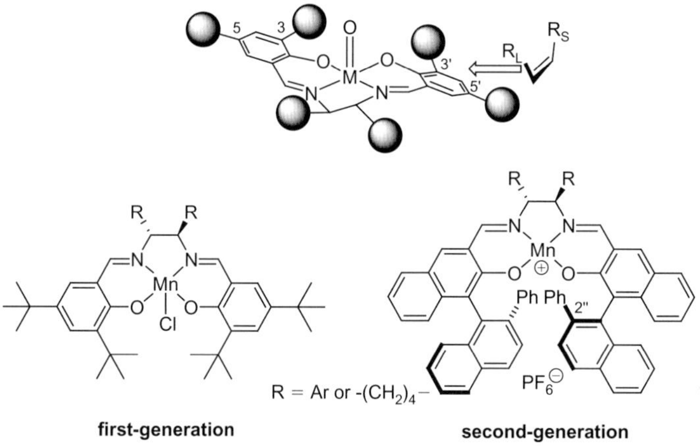

Figure 7.3 Chiral salen-Mn catalysts for asymmetric epoxidation.

iodosylbenzene is thought to adopt a nonplanar stepped conformation; the approach trajectory and the orientation of olefin are restricted. Olefins approach the reactive oxygen species over the downwardly bent benzene ring of the chiral salen ligand, directing the bulkier substituent R_L away from the 3′ position substituent to minimize repulsive interaction. The presence of a bulky 5-substituent can effectively block undesired olefins approaching over the upwardly bent benzene ring. Accordingly, the presence of bulky 3- and 3′-, 5- and 5′-substituents are quite necessary for chiral salen-Mn complexes to exhibit high enantioselectivity. Indeed, some chiral salen-Mn complexes bearing chiral 1,2-cyclohexadiamine or 1,2-diphenylethylene-1,2-diamine backbones and 3-, 3′-, 5- and 5′-t-butyl groups have proven to be highly enantioselective catalysts for asymmetric epoxidation of various olefins and have been called first-generation chiral salen-Mn catalysts. The second-generation chiral salen-Mn catalysts, which bear two different chiral elements in their ligands, show remarkably enhanced enantioselectivity because their characteristic 3- and 3′-naphthyl substituents can restrict the orientation of olefins more efficiently. Apart from the above advantage, one 2″-phenyl group can interact synergistically or conversely with an apical ligand coordinated to metal ion and

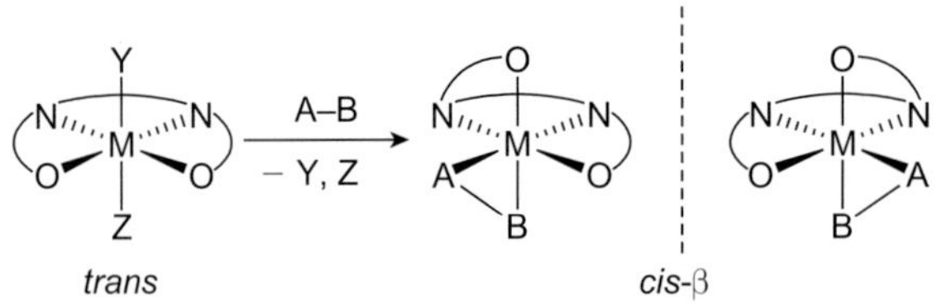

Scheme 7.5 Synthesis of cis-β-Salen-metal complexes.

thus enhance or reduce the folding degree of chiral salen-metal complex. Therefore, the conformations of chiral salen-metal complexes can be tuned by choosing the appropriate apical ligands.

As mentioned above, chiral salen-metal complexes with a *trans* configuration can function as an efficient catalyst for monodentate substrate or reagent. Conversely, when the ancillary ligands are bidentate, chiral salen-metal complexes usually adopt a *cis*-β configuration. The reactions of *trans*-salen-metal complexes with bidentate substrates or reagents usually afford *cis*-β-salen-metal complexes (Scheme 7.5). It is generally known that chiral salen-metal complexes bearing Zr, Hf, Ru, and some second or third-row transition metals are prone to adopt a *cis*-β configuration. Being different from *trans* configuration, the *cis*-β configuration itself possesses chiral coordination environment. As a result, chiral salen-metal complexes with a *cis*-β configuration might create a unique chiral environment for some catalytic asymmetric reactions.

7.4
Asymmetric Reactions Catalyzed by Chiral Salen Complexes

The malleability inherent in salen ligands has enabled access to metal centers with tunable stereochemical and electronic properties, and thus has led to their extensive use in a wide variety of enantioselective reactions. The adjustable properties, by modifying the salicylidene rings, easily create an appropriate chiral environment around the metal center, resulting in excellent enantioselectivity in asymmetric catalysis [2].

7.4.1
Asymmetric Epoxidation

Chiral epoxides are very important intermediates for the preparation of biologically active compounds and other fine chemicals. Sharpless asymmetric epoxidation was established as an effective tool to access enantiomerically pure epoxides and represented the first major breakthrough in catalytic asymmetric epoxidation. However, the applicable reaction substrates are restricted to alkenes containing a pendant functional group. In 1990, Jacobsen and Katsuki independently reported

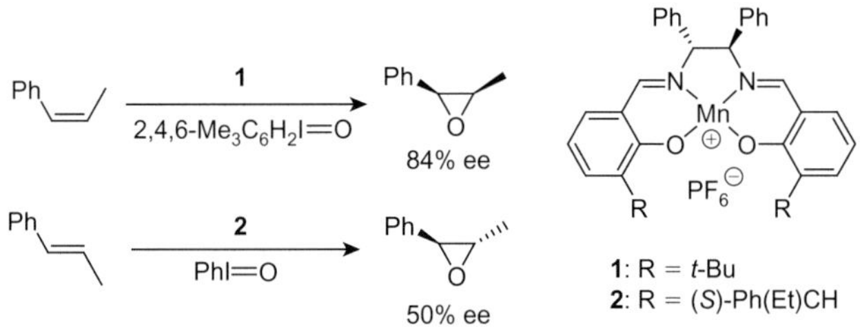

Scheme 7.6 Pioneering asymmetric epoxidation catalyzed by chiral salen-Mn(III) complexes.

that chiral salen-Mn(III) complexes can be employed as effective catalysts for enantioselective epoxidation of unfunctionalized olefins. This appeared to be the second major breakthrough in catalytic asymmetric epoxidation, and also opened up the exciting research area of chiral salen complexes as asymmetric catalysts.

Jacobsen initially reported that chiral salen-Mn(III) complex **1** bearing chiral 1,2-diphenylethylene-1,2-diamine unit and *t*-butyl groups at the position ortho to the phenolic hydroxyl group catalyzed the epoxidation of *cis*-alkenes with high levels of enantioselectivity (Scheme 7.6) [3]. Katsuki disclosed that chiral salen-Mn(III) complex **2** bearing the same diamine backbone and chiral bulky 3- and 3'-substituents gave 50% ee for the asymmetric epoxidation of *(E)*-β-methylstyrene in the presence of 2-methylimdazole (Scheme 7.6) [4], which represented the highest level of enantioselectivity for metal-catalyzed epoxidations of unfunctionalized *(E)*-alkenes at that time. Iodosylarenes were chosen as oxidants in both reactions, but sodium hypochlorite was later established as a better oxidant because it is relatively inexpensive and its reduced product is environmentally benign [7].

Subsequently, chiral salen-Mn(III) complex **3** (now known as Jacobsen's catalyst) [8] bearing chiral 1,2-cyclohexadiamine unit and four *t*-butyl substitutes proved to be an extremely effective catalyst for the asymmetric epoxidation of a wide range of unfunctionalized *cis*-alkenes, including conjugated *(Z)*-enynes (affording predominantly *trans*-epoxides), acyclic conjugated polyenes, cyclic 1,3-dienes, cinnamate esters, and conjugated trisubstituted alkenes (Figures 7.4 and 7.5) [9, 10]. Jacobsen's catalyst now is commercially available or can be readily synthesized on a large scale. Moreover, it exhibits remarkable stability and can be stored in air for a long period without significant decomposition. Indeed, asymmetric epoxidation catalyzed by Jacobsen's catalyst has been widely applied to the total synthesis of many natural products and other biologically active compounds.

Further tuning of Jacobsen's catalyst through introducing bulkier and more electron-donating groups at the 5-, 5'-positions gave chiral salen-Mn(III) complexes **4** and **5** (Figure 7.4). Compared with catalyst **3**, catalysts **4** and **5** display enhanced enantioselectivity in the asymmetric epoxidation of some cyclic dienes and conjugated tetrasubstituted alkenes (Figure 7.5) [10]. Notably, up to 89% ee can be achieved in the asymmetric epoxidation of a terminal alkene styrene using catalyst **5** and *m*-CPBA as oxidant [11].

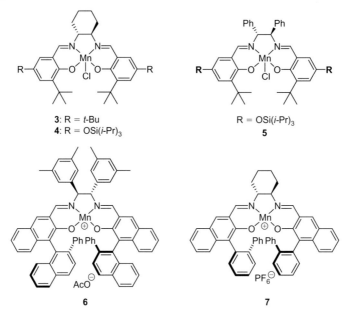

Figure 7.4 Representative chiral salen-Mn catalysts for asymmetric epoxidation.

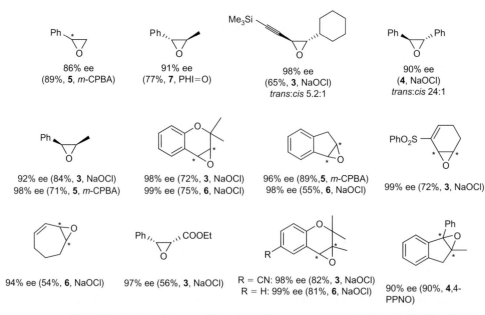

Figure 7.5 Representative epoxides achieved by asymmetric epoxidation catalyzed by chiral salen-Mn complexes.

Katsuki introduced the second-generation chiral salen-Mn catalysts bearing two different chiral elements (central chirality at diamine backbone and axial chirality at 3-, 3'- substituents) in salen ligands (Figure 7.4). Representative complex **6** catalyzed asymmetric epoxidation of *cis*-conjugated alkenes, including *cis*-enynes, cyclic 1,3-dienes, and conjugated trisubstituted alkenes, with higher enantioselectivity compared with Jacobsen's catalyst (Figure 7.5) [12]. Even though second-generation chiral salen-Mn catalysts exhibit higher enantioselectivity towards the asymmetric epoxidation and have been applied to the synthesis of several biologically active compounds, their availability limits their applications.

Asymmetric epoxidation of *trans*-alkenes using Jacobsen's catalyst often gives moderate enantioselectivity. However, modified second-generation chiral salen-Mn complex **7** has proven to be a highly enantioselective catalyst for asymmetric epoxidation of *trans*-β-substituted styrenes (Figure 7.5) [13]. The deeply-folded stepped conformation of the complex **7** is thought to account for its high enantioselectivity toward *trans*-alkenes. In addition, asymmetric epoxidation of *trans*-alkenes can be carried out using chiral salen Cr complexes with satisfactory enantioselectivity, although the substrate scope is very limited.

Elucidation of the definitive mechanism of epoxidations catalyzed by salen-metal complexes remains an important issue. With some experimental evidence in hand, researchers have proposed a reasonable mechanism. Salen-Mn(III) complex is thought to first react with oxidant, such as PhIO or NaClO, to form a reactive oxygen species salen-Mn(V)=O, and then "side-on" approach of alkene to reactive oxygen species and transfer of oxygen atom from salen-Mn(V)=O species to alkene occur, and the alkene is oxidized to epoxide and salen-Mn(V)=O is reduced to salen-Mn(III) complex.

There might be many trajectories for alkene to approach the reactive oxygen species salen-Mn(V)=O, but in asymmetric epoxidation catalyzed by chiral salen-Mn(III) complex the approach trajectory is restricted. Taking Jacobsen's catalyst as an example (Figure 7.6), the formed reactive oxygen species salen-Mn(V)=O

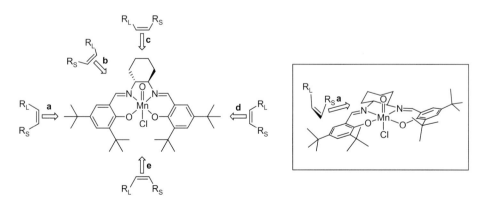

Figure 7.6 Trajectories of alkene approaching the reactive oxygen species salen-Mn(V)=O.

adopts a nonplanar stepped conformation. The approaches e and d are obviously disfavored due to steric hindrance of bulky *t*-butyl groups. Approach c can not explain exceptionally high ees observed with *cis*-alkenes or the effects of chirality at the 3,3′-positions in second-generation chiral salen-Mn catalysts. Finally, alkene is expected to approach the reactive oxygen species over the downwardly bent benzene ring of the chiral salen ligand (approach a), directing the bulkier substituent R_L away from the 3-*t*-butyl group to minimize repulsive interaction [6]. Apparently, the nonplanar stepped conformation of the reactive oxygen species salen-Mn (V)=O, which depends on the natures of chiral diamine unit and the substituents in the ligand and of the apical ligands, plays an important role in determining the exact trajectory.

7.4.2
Asymmetric Ring-Opening of Epoxides

Given that the mechanism of ring opening of epoxide involves activation of epoxide by Lewis acidic metal ion, the ground-state structure of epoxide bound to metal complex is apparently similar to the transition-state of epoxidation of alkene. As mentioned above, chiral salen ligands can create an effective asymmetric environment for oxo transfer of a metal complex in asymmetric epoxidation. Likewise, a chiral salen ligand might create an effective asymmetric environment for nucleophilic ring-opening of the activated epoxide (Figure 7.7). Based on this understanding, Jacobsen and coworkers investigated the use of chiral salen chromium(III) and cobalt(III) complexes as catalysts for asymmetric ring-opening of epoxides by various nucleophiles [14]. In addition, the availability of *meso*-epoxides and racemic epoxides renders the asymmetric ring-opening of epoxide a powerful tool to access synthetically useful compounds in enantioenriched form.

7.4.2.1 Desymmetrization of Meso-Epoxides
Desymmetrization of *meso*-epoxides by an asymmetric ring-opening reaction offers an effective strategy to generate two contiguous chiral centers from achiral starting materials. Jacobsen initially disclosed that chiral salen-chromium(III) complex **8** displayed high reactivity and enantioselectivity towards ring-opening of various five- or six-membered-ring *meso*-epoxides by TMSN$_3$ (Scheme 7.7) [15]. Another impressive advantage of this catalytic system is that asymmetric

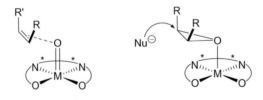

Epoxidation of Alkene   **Activation of Epoxide**

Figure 7.7 Different functions displayed by similar chiral salen-metal catalysts.

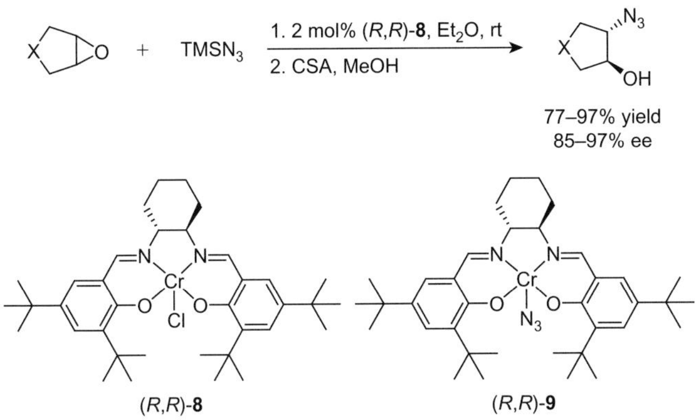

Scheme 7.7 Desymmetrization of *meso*-epoxides by asymmetric ring-opening reaction using TMSN$_3$.

ring-opening of *meso*-epoxides by TMSN$_3$ can be undertaken under solvent-free conditions; the chiral salen chromium(III) complex can be readily recovered by simple vacuum transfer distillation of product and can be recycled repeatedly without loss of activity. The steric environment around an epoxide group has an effect on the reactivity and enantioselectivity of the catalyst **8**; asymmetric ring-opening of cycloheptene oxide was carried out with low enantioselectivity, and cyclooctene oxide cannot undergo ring-opening reaction [16].

The active catalyst in the ring-opening reaction is actually chiral salen chromium(III) azide complex **9** generated after the first catalytic cycle, which suggests that the catalyst functions as an azide-delivery agent. X-Ray crystallographic and IR spectroscopic analysis of complex **9** imply that the active species is (salen)Cr(N$_3$)(epoxide) and the epoxide is activated by the chiral salen chromium(III) azide complex in epoxide ring-opening. A kinetics study revealed a second-order dependence on the azide complex **9**. This confirms that the catalyst plays a dual role in epoxide ring-opening reactions, functioning both as a Lewis acid for activating the epoxide and an azide-delivery agent [17]. Notably, the active reagent is HN$_3$ generated by hydrolysis of TMSN$_3$ since the reaction does not proceed in the absence of a catalytic amount of water.

Therefore, high enantioselectivity in asymmetric ring-opening of epoxides can be ascribed to the cooperative interaction of a chiral electrophile and a chiral nucleophile bearing the same chiral salen ligand framework. Based on such mechanistic insight, tethered dimeric chiral salen chromium(III) catalysts were constructed. Owing to the enhanced cooperative interaction between two catalyst units, the dimeric catalyst **10** (Figure 7.8), which can adopt a "head-to-tail" arrangement of two salen-Cr(III) units, exhibits dramatically enhanced reactivity and similar enantioselectivity compared with the monomeric catalyst **9** [18].

Besides nitrogen nucleophiles such as TMSN$_3$, oxygen nucleophiles such as alcohols can also be applied to the desymmetrization of *meso*-epoxides catalyzed by chiral salen-metal complexes. Jacobsen disclosed that chiral salen-Co(III) complex

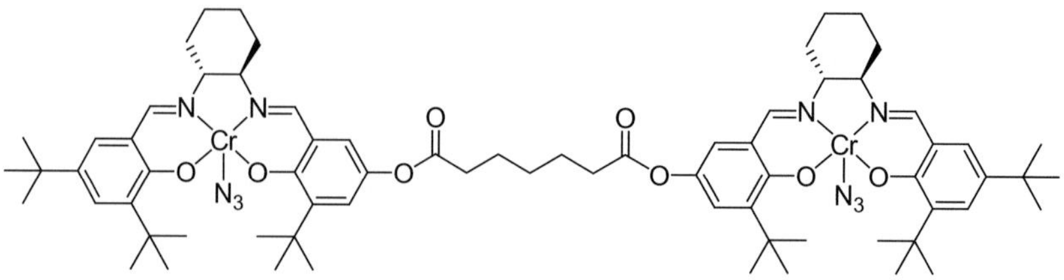

Figure 7.8 Representative tethered dimeric chiral salen chromium(III) catalyst.

11 catalyzed the intramolecular asymmetric ring-opening of *meso*-epoxy diols with excellent enantioselectivity, providing access to chiral cyclic ethers in good yields (Scheme 7.8). Complex **11** was also applied to catalyze the asymmetric Payne rearrangement of *meso*-epoxy diol and the 1,2-anhydrothreitol product was obtained in high ee [19].

More recently, Jacobsen discovered that the complex **12** catalyzed enantioselective intramolecular ring-opening of oxetanes bearing tethered oxygen nucleophiles, affording a wide variety of tetrahydrofurans or dihydrobenzofurans in good yield and high enantioselectivity (Scheme 7.9) [20].

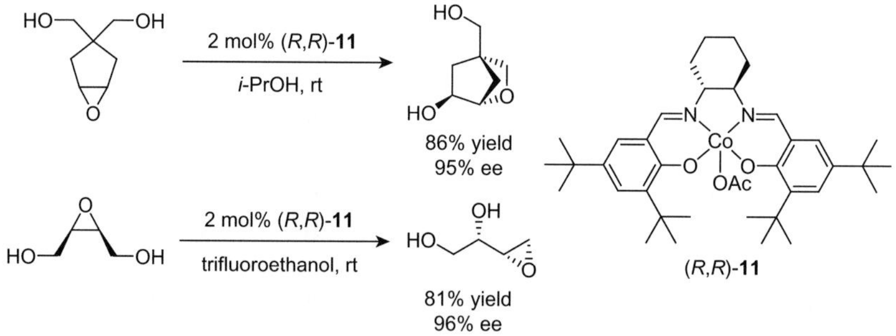

Scheme 7.8 Desymmetrization of *meso*-epoxy diols catalyzed by the complex **11**.

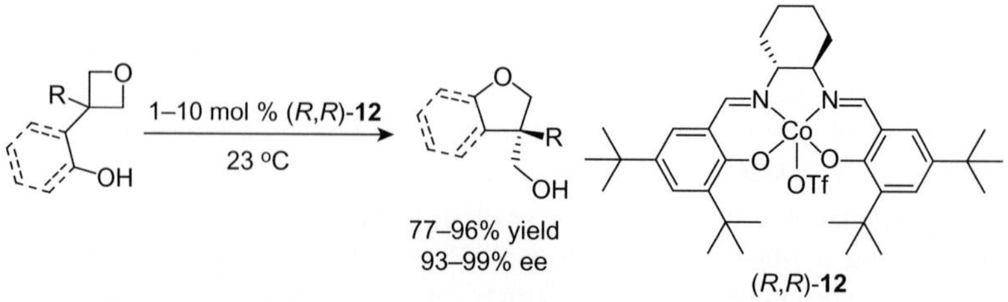

Scheme 7.9 Enantioselective intramolecular ring-opening of oxetanes.

7.4.2.2 Kinetic Resolution of Racemic Epoxides

Although significant advances have been made in the development of chiral catalysts for asymmetric epoxidation of di- or tri-substituted alkenes in high enantioselectivity, no useful method exists for the direct enantioselective synthesis of terminal epoxides. Because racemic terminal epoxides are inexpensive and commercially available, the development of kinetic resolution of racemic epoxides was highly desired for practical access to terminal epoxides in highly enantioenriched form.

Pioneering studies by Jacobsen revealed that the catalytic system in asymmetric ring-opening of *meso*-epoxides by $TMSN_3$ can be applied to the kinetic resolution of racemic terminal epoxides. Chiral salen chromium(III) complex **9** catalyzed the kinetic resolution of various racemic terminal epoxides with extraordinarily high selectivity (k_{rel} ranging from 45 to 280) (Scheme 7.10). The resulting 1-azido-2-siloxy products can be isolated in excellent enantioselectivity and can be easily converted into synthetically useful 1,2-amino alcohols [21]. The main drawbacks are that $TMSN_3$ is very expensive and must be handled with appropriate precautions due to its thermal sensibility, so it can not be established as a practical approach to preparing enantioenriched terminal epoxides.

The pursuit of ideal nucleophiles as the resolving agent led to the promising discovery of hydrolytic kinetic resolution in which water served as the resolving agent. In 1997, Jacobsen disclosed that low loading of chiral salen cobalt(III) complex **11** catalyzed hydrolytic kinetic resolution of propylene oxide with extraordinary high selectivity (k_{rel} = about 500), affording unreacted propylene oxide and 1,2-diols in extremely high ee, both are valuable chiral building block. Chiral salen cobalt(III) complex **11** can be recycled repeatedly without loss of reactivity and enantioselectivity (Scheme 7.11) [22].

Various terminal epoxides can undergo hydrolytic kinetic resolution irrespective of the electronic and steric natures of the substituents. Both the recovered terminal epoxide and 1,2-diol in highly enantioenriched form can be obtained by controlling the amount of water in hydrolytic kinetic resolution. Use of a slight excess of water (0.55 equiv relative to racemic epoxides) allowed the unreacted terminal epoxide to be isolated in >99% ee. In cases where 1,2-diol in high enantioselectivity is required, 0.45 equiv of water relative to racemic epoxide should be employed [23]. In addition, the asymmetric epoxidation–hydrolytic kinetic resolution sequence

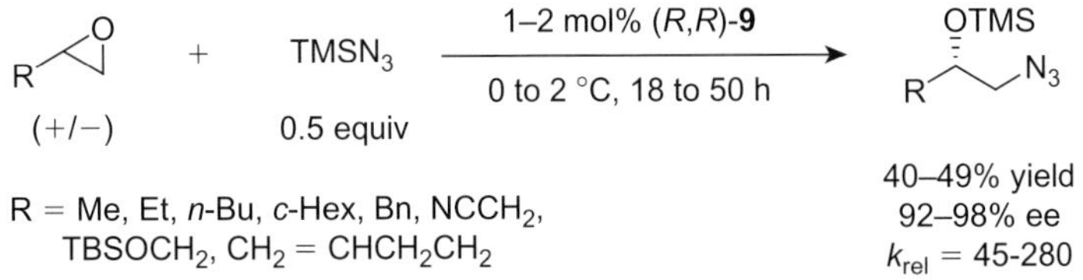

Scheme 7.10 Catalytic kinetic resolution of racemic terminal epoxides with $TMSN_3$.

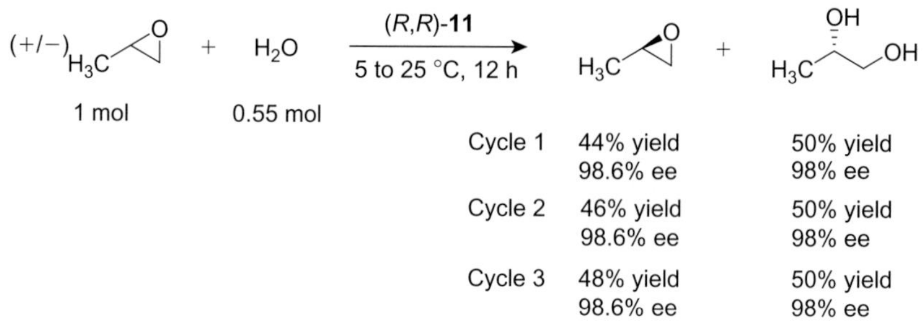

Scheme 7.11 Hydrolytic kinetic resolution of propylene oxide.

Cycle 1	44% yield 98.6% ee	50% yield 98% ee
Cycle 2	46% yield 98.6% ee	50% yield 98% ee
Cycle 3	48% yield 98.6% ee	50% yield 98% ee

can afford effectively some valuable terminal epoxides in highly enantioenriched form and satisfactory yield [24].

Kinetic studies on the hydrolytic kinetic resolution reaction revealed a second-order dependence on chiral salen cobalt(III) catalyst, which is consistent with the observation made in the asymmetric ring-opening reaction catalyzed by chiral salen chromium(III) catalyst. Therefore, the cooperative bimetallic mechanism analogous to other asymmetric ring-opening reactions involving simultaneous activation of nucleophile (water) and Lewis acid activation of epoxides is also proposed for the hydrolytic kinetic resolution (Figure 7.9). Notably, the counterion X has a profound effect on the reactivity of catalyst (salen)CoX with regard to the generation rate of (salen)Co(OH) and binding property of (salen)Co(epoxide) [25]. More recently, combined NMR and quantum chemical studies indicated that exceptionally high enantioselectivity in hydrolytic kinetic resolution can be ascribed to the situation when two catalyst units adopt "head-to-tail" arrangement: an attack of the activated OH-nucleophilic complex on the "matched" epoxide complex is sterically favored compared to an attack on the mismatched epoxide complex (Figure 7.10) [26].

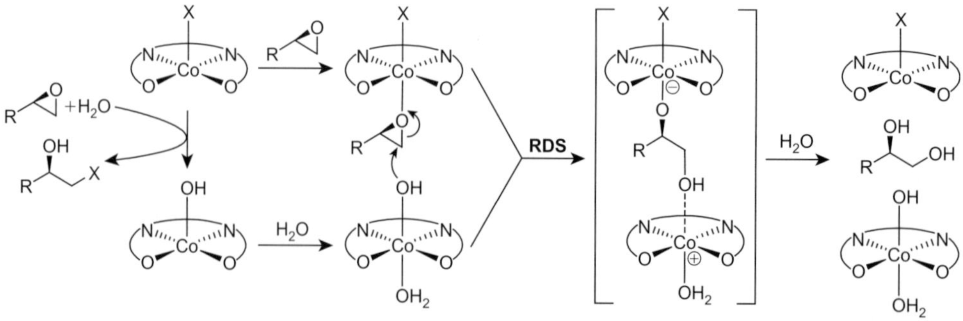

Figure 7.9 Mechanism of hydrolytic kinetic resolution catalyzed by chiral salen cobalt(III) complex.

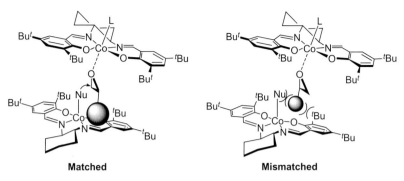

Figure 7.10 Matched and mismatched transition complexes of the attack.

7.4.2.3 Enantioselective Addition of Carbon Dioxide to Propylene Oxide

Transformations of carbon dioxide into valuable products have attracted much attention due to its potential use as an attractive, cheap, and safe C1 building block and because of its main contribution to global warming [27]. One of the most promising methodologies in this area is the synthesis of five-membered cyclic carbonates via the coupling of CO_2 and epoxides [28].

Based on the possible mechanism of epoxide ring-opening by electrophilic Lewis acids and nucleophilic co-catalyst, Lu et al. designed a binary catalyst system of a chiral salen cobalt(III) complex as the electrophile in conjunction with an quaternary ammonium halide as the nucleophile for the coupling of CO_2 and racemic epoxides to synthesize enantioenriched propylene carbonate [29]. In this system, the chiral electrophile was proposed to selectively complex one enantiomer of racemic epoxides; subsequently, the attack of nucleophile or activated CO_2 to the coordinated epoxide leads to enantioselective ring-opening of epoxides and then further forms chiral cyclic carbonates via intramolecular cyclic elimination (Figure 7.11). A chiral salen cobalt(III) complex with a sterically bulk counterion, such as *p*-toluenesulfonate, is essential with regard to high enantioselectivity of this reaction. In addition, the anion of the quaternary ammonium salts has a significant effect on the enantiomeric purity of the resulting cyclic carbonate. Consequently, chiral salen cobalt(III) complex **13** in combination with *n*-Bu$_4$NCl has proven to be relatively highly enantioselective catalytic system for the coupling of CO_2 and racemic epoxides under solvent-free conditions, and propylene carbonate with 70% ee ($k_{rel} = 9$) could be obtained in 40% yield at 0 °C (Scheme 7.12).

Based on extensive catalyst screening, Berkessel et al. have established the binary system of chiral salen cobalt(III) complex **14** as the electrophile in conjunction with bis(triphenylphosphoranylidene)ammonium fluoride (PPNF) as a highly enantioselective catalytic system for the coupling of CO_2 and racemic epoxides at atmospheric pressure – propylene carbonate with 83% ee ($k_{rel} = 19$) was obtained in 40% yield at −40 °C (Scheme 7.13) [30a]. Recently, Jing et al. reported a chiral polymer of BINOL salen cobalt(III) complexes for use in catalyzing the

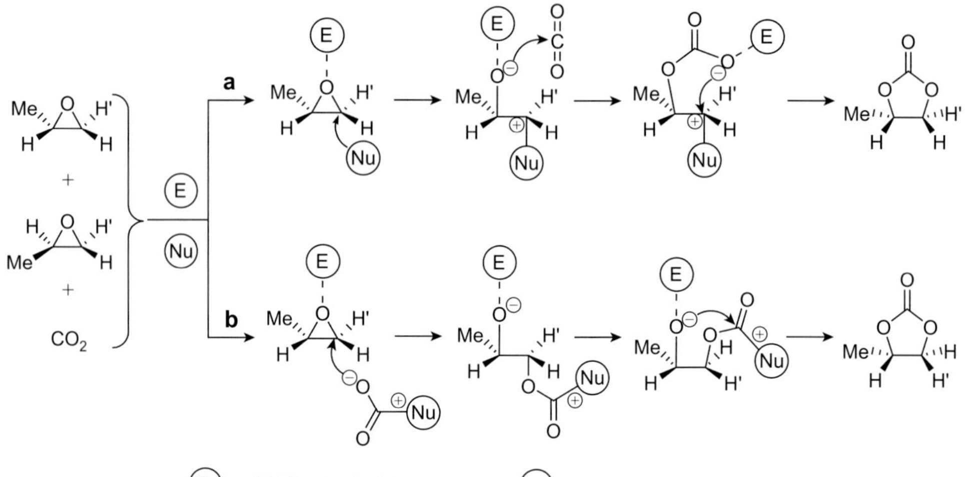

Figure 7.11 Possible mechanism for the coupling reaction of CO_2 and propylene oxide.

E = (R,R) salenCo(III) complex   Nu = nucleophilic cocatalyst

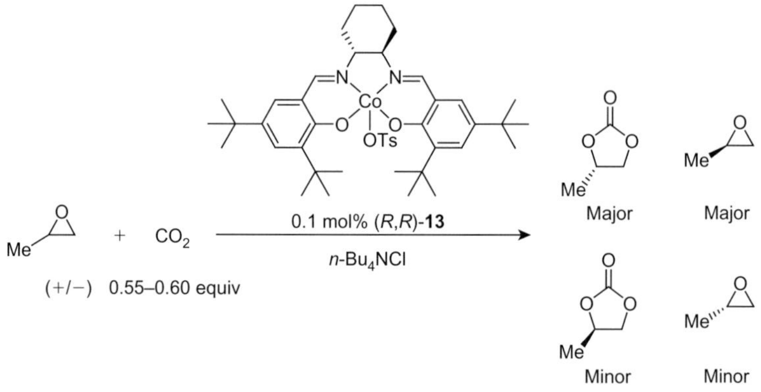

Scheme 7.12 Coupling reaction of CO_2 and propylene oxide catalyzed by **13**/n-Bu$_4$NCl system.

asymmetric addition of CO_2 and racemic propylene oxide in comparable enantioselectivity to that of the much-studied Jacobsen's catalyst. Notably, the catalyst is easily recovered and reused more than ten times without loss of activity and enantioselectivity [30b].

7.4.2.4 Asymmetric Alternating Copolymerization of Racemic Epoxides and Carbon Dioxide

Another promising reaction in the transformation of CO_2 is the alternating copolymerization of CO_2 and epoxides to synthesize polycarbonates [31].

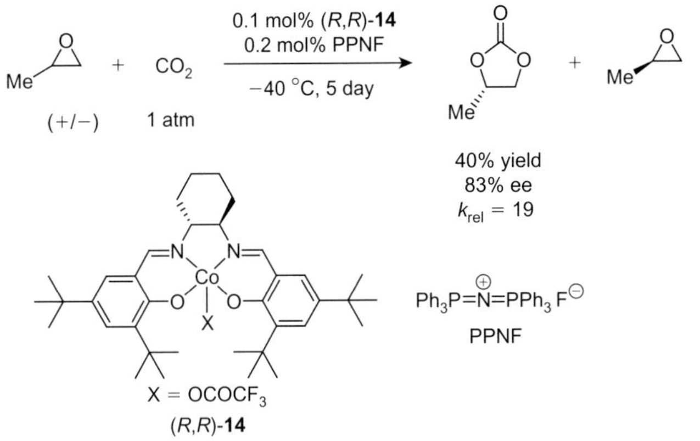

Scheme 7.13 Coupling reaction of CO_2 and propylene oxide catalyzed by 14/PPNF system.

These polycarbonates not only exhibit interesting material properties but also have additional environmental advantages resulting from their biodegradability.

Coates et al. first reported that the chiral salen cobalt(III) complex **11** catalyzed the copolymerization of CO_2 and epoxides with high selectivity for polycarbonate formation at ambient temperature and 5.5 MPa CO_2 pressure, but the complex is inactive at enhanced temperatures or low CO_2 pressures [32]. Lu and coworkers discovered a highly active, binary catalyst system for the alternating copolymerization of CO_2 with racemic aliphatic epoxides under extremely mild temperatures and pressures. In the binary system, a bulky chiral salen cobalt complex with an axial group of poor leaving ability such as complex **14** functioned as the electrophile, while a bulky ionic organic ammonium salt or a sterically hindered strong organic base with poor leaving ability and low coordination ability, such as n-Bu$_4$NCl or 7-methyl-1,5,7-triazabicyclo[4.4.0]dec-5-ene (MTBD), served as the nucleophile. Polycarbonates with more than 99% carbonate linkages were selectively formed in relatively high enantioselectivity and excellent regioselectivity (>95% head-to-tail connectivity) (Scheme 7.14) [33]. The binary catalyst system of chiral salen-Co(III) complex **15** and PPNCl also exhibited higher enantioselectivity for the alternating copolymerization of CO_2 with racemic propylene epoxide at $-20\ °C$ and a k_{rel} of 9.7 could be obtained (Scheme 7.14) [34].

7.4.2.5 Enantioselective Homopolymerization of Epoxides

Recently, Coates et al. reported a highly active catalyst **16**, in which two chiral salen cobalt complexes were tethered by a chiral binaphthol linker, for enantioselective polymerization of racemic epoxides to prepare stereoregular polyethers and enantioenriched epoxides [35]. (R,R,S,R,R)-**16**/[PPN][OAc] catalyzed the polymerization of racemic propylene oxide with extremely high activity (34% conversion after 15 min at 0 °C) and enantioselectivity ($k_{rel} = 370$) (Scheme 7.15).

The extraordinarily high activity and enantioselectivity of the **16**/[PPN][OAc] catalytic system allowed its use in the preparation of enantioenriched epoxides by

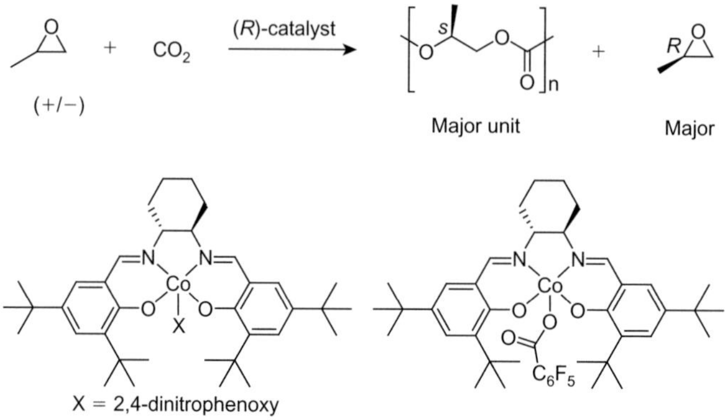

Scheme 7.14 Asymmetric alternating copolymerization of racemic epoxides and carbon dioxide.

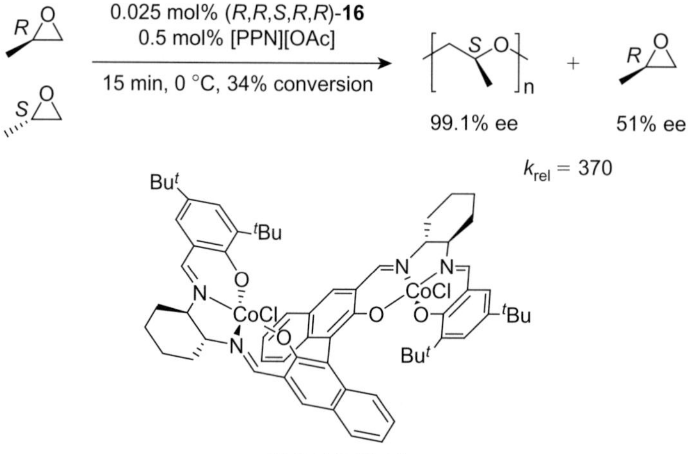

Scheme 7.15 Enantioselective polymerization of racemic propylene oxide catalyzed by (R,R,S,R,R)-**16**/[PPN][OAc].

kinetic resolution of racemic epoxides. Aliphatic epoxides were successfully resolved (>99% ee) in excellent yield.

7.4.3
Asymmetric Cyclopropanation

Optically active cyclopropane moieties are frequently found as basic structural elements in various natural products and biologically active compounds. Chiral salen-metal complexes can be used as carbene transfer agent in asymmetric

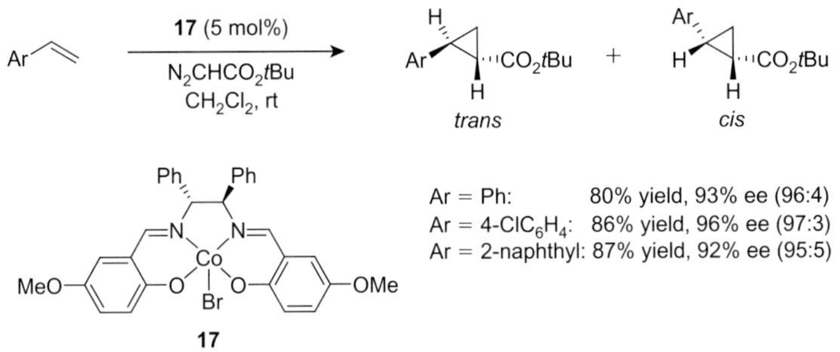

Scheme 7.16 Chiral salen-cobalt(III) catalyst for asymmetric cyclopropanation.

Ar = Ph: 80% yield, 93% ee (96:4)
Ar = 4-ClC$_6$H$_4$: 86% yield, 96% ee (97:3)
Ar = 2-naphthyl: 87% yield, 92% ee (95:5)

cyclopropanation by reacting with diazo compounds such as α-diazo ester to afford the corresponding salen-metal carbene complexes [36].

Katsuki reported chiral salen cobalt(III) complex **17** catalyzed the asymmetric cyclopropanation of styrene derivatives with high *trans*- and enantioselectivity (Scheme 7.16) [37]. The apical bromo ligand and electron-donating 5- and 5'-methoxy groups were thought to reduce the reactivity of cobalt(V) carbenoid species and thus enhanced stereoselectivity. It is considered that the reactive cobalt (V) carbenoid species adopted a nonplanar stepped conformation analogous to the reactive oxygen species salen-Mn(V)=O in asymmetric epoxidation. The ester group on the carbene atom located above the downwardly bent benzene ring of the chiral salen ligand (Figure 7.12) and the styrene approached the reactive cobalt(V) carbenoid species along Co−O bond axis with an orientation perpendicular to the Co−C bond, and thus rotated toward the open space near 3'-carbon to avoid steric repulsion; therefore, *trans*-cyclopropane product was predominantly formed [6].

Based on the mechanistic insight of asymmetric cyclopropanation, *cis*-selective cyclopropanation could be achieved using chiral salen-metal catalysts by tuning the electronic and steric properties of the substituents on the ligand, the valency of the metal ion, or the apical ligand to alter the approach trajectory and rotation direction of the alkene. Katsuki *et al.* discovered that chiral salen Ru(NO) complex **18** exhibited remarkable *cis*- and enantioselectivity for the asymmetric cyclopropanation of styrene derivatives under photoirradiation (Scheme 7.17) [38]. They proposed that the alkene approached the reactive ruthenium carbenoid species along the Ru−O bond axis and rotation direction was regulated by chiral binaphthyl units to form *cis*-cyclopropane product predominantly. In addition, chiral

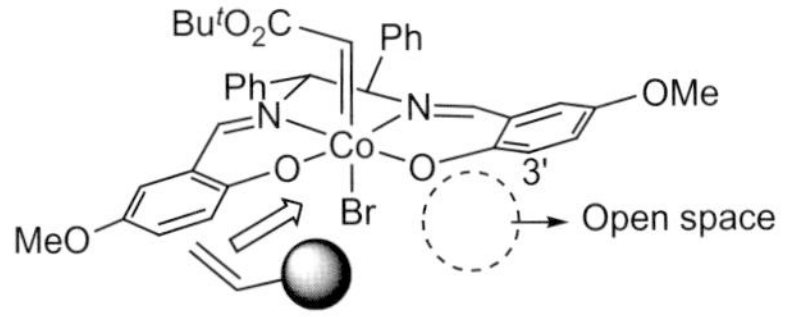

Figure 7.12 Stepped conformation of a cobalt(V) carbenoid species.

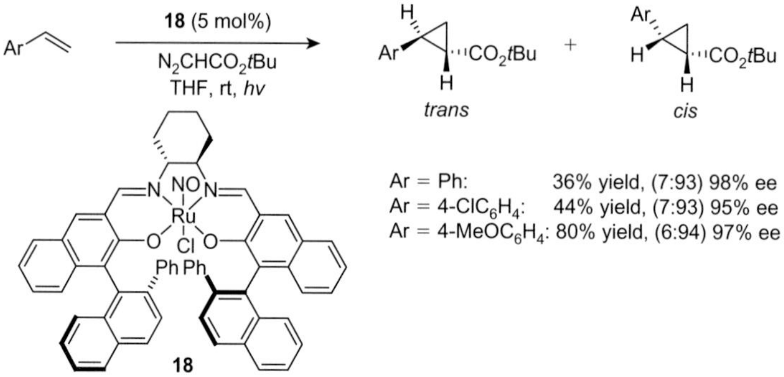

Scheme 7.17 Chiral salen-Ru(NO) catalyst for asymmetric cyclopropanation.

salen cobalt(II) complex **19** was reported as a highly *cis*- and enantioselective catalyst for the asymmetric cyclopropanation of styrene derivatives in the presence of a catalytic amount of *N*-methylimidazole (Scheme 7.18) [39]. In this catalytic system, alkene approaches the reactive cobalt carbenoid species along Co–N bond axis and thus *cis*-cyclopropane product was predominantly formed.

Recently, a stable chiral salen iridium(III) complex (**20**) with a tolyl group as apical ligand was successfully synthesized by Katsuki *et al.* and proved to be an excellent *cis*- and enantioselective catalyst for the asymmetric cyclopropanation of not only conjugated alkenes but also non-conjugated terminal alkenes, which are inactive substrates for other chiral salen-metal complexes (Scheme 7.19) [40]. The chiral salen tolyliridium carbenoid intermediate was considered to adopt a non-planar stepped conformation. The alkene was also considered to approach the carbenoid carbon along the Ir–N bond axis from the open space near the downward naphthalene ring side.

Che *et al.* disclosed chiral *cis*-β-salen ruthenium complex **21** catalyzed the asymmetric intramolecular cyclopropanation of *trans*-allylic diazoacetates under

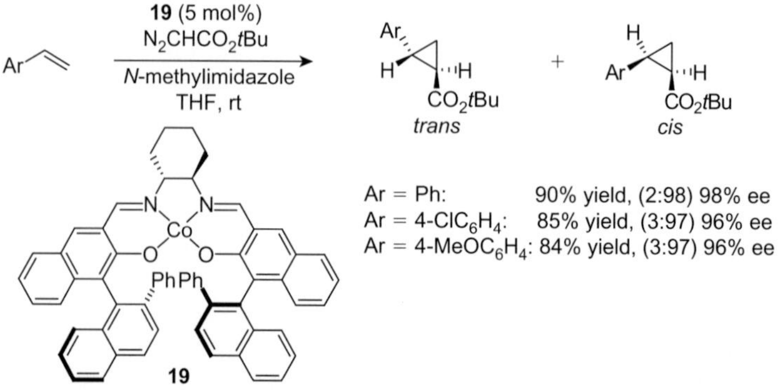

Scheme 7.18 Chiral salen-cobalt(II) catalyst for asymmetric cyclopropanation.

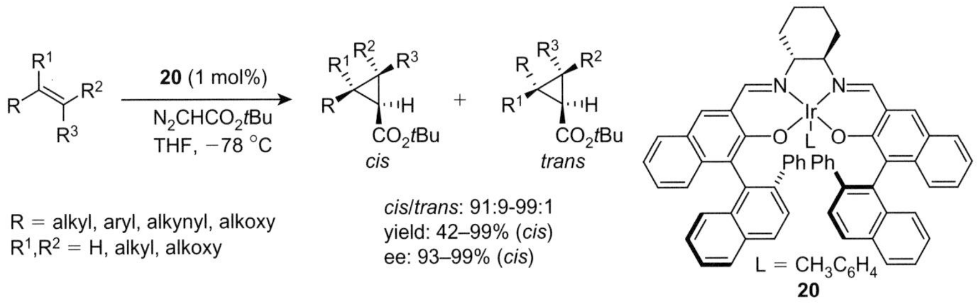

Scheme 7.19 Chiral salen-iridium(III) catalyst for asymmetric cyclopropanation.

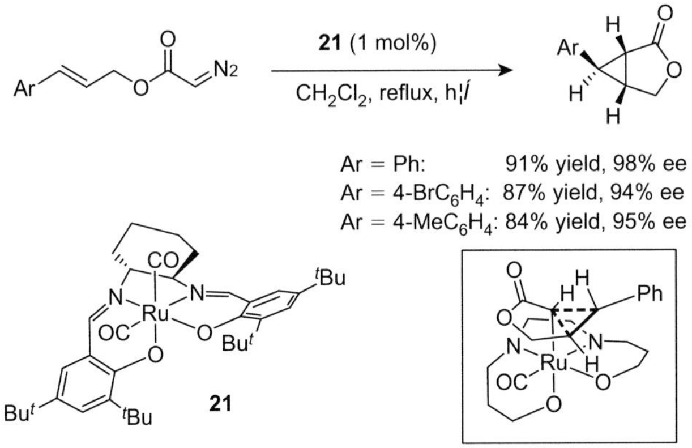

Scheme 7.20 Asymmetric intramolecular cyclopropanation catalyzed by chiral cis-β-salen-ruthenium complex **21**.

light irradiation in high enantioselectivity and good yield (Scheme 7.20) [41]. DFT calculations of the model reaction revealed that, among the ruthenium carbene intermediates possibly involved in the asymmetric cyclopropanation, the cis-β- species (Scheme 7.20) is more stable than its *trans* isomer. Therefore, intramolecular cyclopropanation from the cis-β-salen ruthenium carbene intermediate is the most favorable pathway.

7.4.4
Asymmetric Conjugate Addition Reaction

The asymmetric conjugate addition of various nucleophiles to α,β-unsaturated carbonyl compounds represents one of the most attractive approaches to access chiral building blocks for biologically active compounds. Chiral salen aluminum complexes have proven to be highly enantioselective catalysts for conjugate addition of weakly acidic nitrogen, carbon, and oxygen nucleophiles to α,β-unsaturated carbonyl compounds.

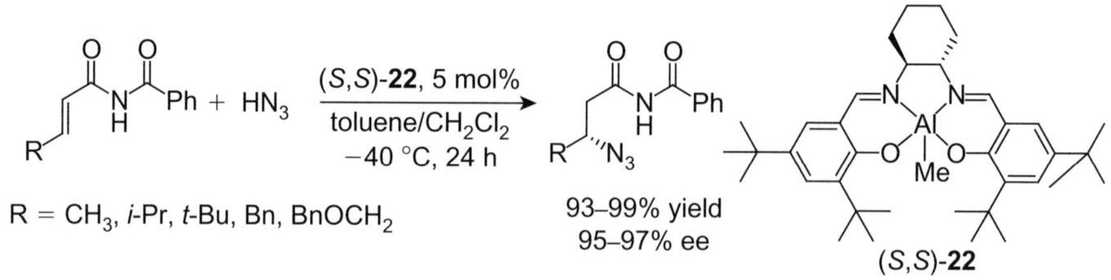

Scheme 7.21 Conjugate addition of hydrazoic acid to α,β-unsaturated imides.

In 1999, Jacobsen *et al.* disclosed that chiral salen aluminum complex **22** catalyzed the conjugate addition of hydrazoic acid (HN$_3$) to β-alkyl substituted α,β-unsaturated imides with a high level of enantioselectivity (Scheme 7.21) [42]. The steric properties of β-alkyl substituents promote the reactivity rather than the enantioselectivity. The conjugate addition of hydrazoic acid later proved successful for α,β-unsaturated ketones substrates [43]. The resulting products β-azido imides and ketones can be easily converted into β-amino acid derivatives and suitably protected β-amino ketones, respectively, both of which are versatile synthetic intermediates. Kinetic studies revealed a first-order dependence on catalyst **22**. This suggests that the high enantioselectivity in this catalytic system might be ascribed to the asymmetric induction through simple one-point binding of carbonyl compounds to catalyst **22**.

Chiral salen aluminum complex **23** is also a highly enantioselective catalyst for the conjugate addition of hydrogen cyanide (HCN) to β-alkyl substituted α,β-unsaturated imides [44]. HCN must be generated *in situ* from TMSCN and alcohols such as 2-propanol, since the reaction did not proceed when HCN alone was used. High enantioselectivities were achieved regardless of the steric properties of β-alkyl substituents. This reaction provides a powerful tool to access α-substituted-β-amino acids and β-substituted-γ-aminobutyric acids. For instance, the anticonvulsant drug pregabalin can be readily synthesized in high yield using this methodology (Scheme 7.22).

Conjugate addition of bulky β-alkyl substituted imides catalyzed by complex **23** proceeded quite slowly, and so elevated temperatures or high catalyst loadings are prerequisite for effective conversion. For β-aryl, vinyl, and alkynyl substituted α,β-unsaturated imides, catalyst **23** showed no reactivity even under harsher conditions. A mechanistic study and kinetic data revealed that the reaction rate displayed a second-order dependence on **23**. This suggests that the reaction proceeds via a cooperative bimetallic mechanism for nucleophile (cyanide) and electrophile (imides) activation. Based on this mechanistic insight, chiral lanthanide complexes were introduced to activate cyanide more effectively. The combination of chiral salen μ-oxo aluminum complex **24** and chiral erbium complex **25** (Figure 7.13) exhibited remarkably enhanced reactivity and slightly increased enantioselectivity towards conjugate addition of HCN to β-alkyl substituted α,β-unsaturated imides compared with catalyst **23** alone [45].

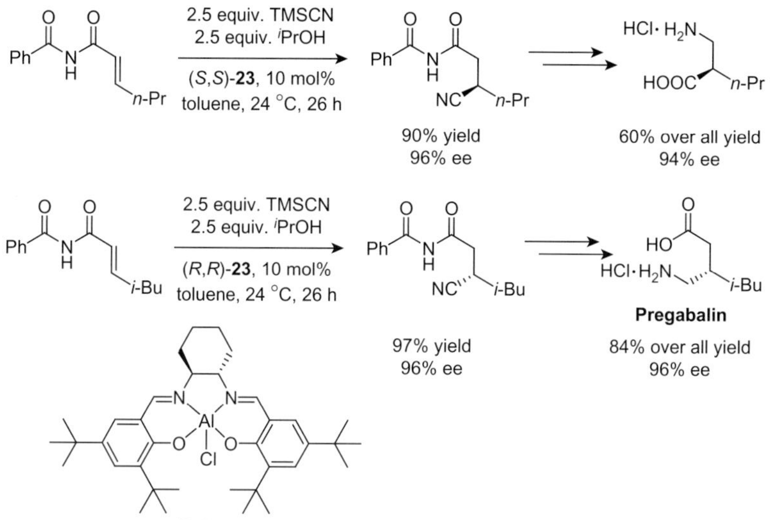

7.4 Asymmetric Reactions Catalyzed by Chiral Salen Complexes | 279

Scheme 7.22 Synthesis of α-substituted-β-amino acids and β-substituted-γ-aminobutyric acids.

Analogous to the strategy used to enforce cooperative interaction in asymmetric the ring-opening reaction of epoxides catalyzed by chiral salen chromium complexes, tethered dimeric chiral salen aluminum complexes were also applied to this reaction. Dimeric complex **26** (Figure 7.14) exhibited enhanced activity and comparable enantioselectivity for conjugate addition of HCN to β-alkyl substituted α,β-unsaturated imides compared with the monomeric catalyst **23** [46]. Moreover, substrates such as β-aryl and vinyl substituted α,β-unsaturated imides, which are unreactive in the systems of the monomeric catalyst **23** or combination of **24** and **25** as catalyst, underwent conjugate addition of HCN smoothly catalyzed by complex **26**, affording the corresponding products in high enantioselectivity and acceptable yield.

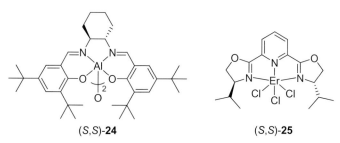

Figure 7.13 Chiral salen aluminum and chiral lanthanide complexes that have been used in combination.

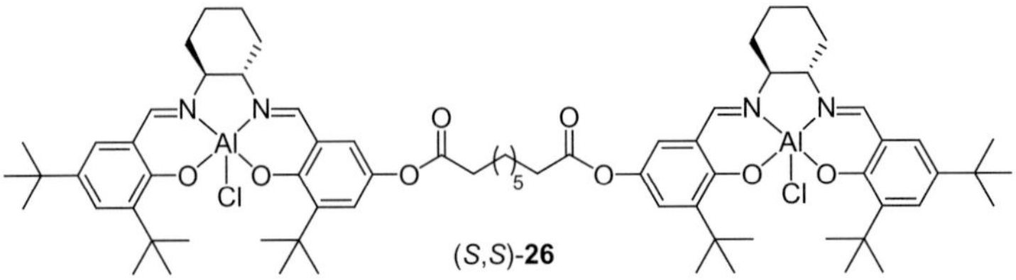

Figure 7.14 Representative tethered dimeric chiral salen aluminum catalyst.

The conjugate addition of carbon nucleophiles to α,β-unsaturated carbonyl compounds is a useful approach to C–C bond construction. Chiral salen μ-oxo aluminum complex **24** has been established as a highly enantioselective catalyst for conjugate additions of electron-deficient di- and tri-substituted nitriles to acyclic β-alkyl- and aryl-substituted α,β-unsaturated imides [47]. The resulting products can be used as versatile building block for organic synthesis. Asymmetric conjugate addition of ethyl (N-benzylamino)cyanoacetate to β-silyl substituted α,β-unsaturated imides afforded β-silyl ester; subsequent cyclization gave γ-lactam in high enantioselectivity and good yield. This provides an ideal intermediate for the concise total synthesis of the proteasome inhibitor lactacystin (Scheme 7.23) [48].

Highly enantioselective conjugate addition of oxygen nucleophiles such as salicylaldoxime to β-alkyl substituted α,β-unsaturated imides can also be carried out using chiral salen μ-oxo aluminum complex **24** as catalyst [49]. Hydrogenolysis of the crude addition products afforded enantiomerically pure β-hydroxy carboxylic acid derivatives in excellent yield. The catalytic sequence is tolerant of ester, acetal, and silyl ether functional groups (Scheme 7.24), despite its rather low activity for

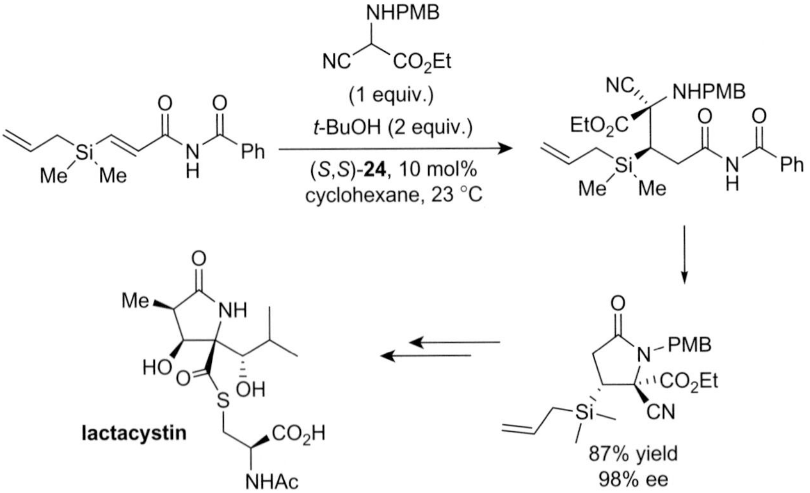

Scheme 7.23 Enantioselective conjugate addition of nitrile to α,β-unsaturated imides.

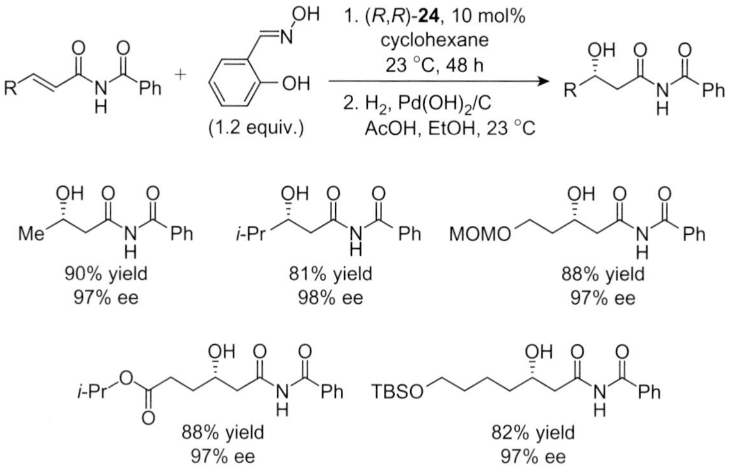

Scheme 7.24 Enantioselective conjugate addition of salicylaldoxime to α,β-unsaturated imides.

α,β-unsaturated imides bearing unsaturated or bulky β-substituents. Enantioselective conjugate addition of nitrogen nucleophiles such as weakly acidic N—H-containing heterocycles to α,β-unsaturated imides and ketones also proved successful using chiral salen μ-oxo aluminum catalyst **24** [50].

7.4.5
Asymmetric Diels–Alder Reaction

The Diels–Alder reaction has been applied extensively to construct six-membered ring compounds in modern organic synthesis. In 1998, chiral salen chromium complex **27** bearing a BF_4 counterion was identified as a highly enantioselective catalyst for the hetero-Diels–Alder reaction between 1-methoxy-3-[(trimethylsilyl) oxy]-1,3-butadiene (Danishefsky's diene) and aldehydes (Scheme 7.25) [51]. The resulting products, dihydropyranones, are synthetically important compounds. A mechanistic study excluded the possibility of a stepwise Mukaiyama aldol mechanism and proved that the hetero-Diels–Alder reaction catalyzed by chiral salen chromium complex proceeded via a concerted [4 + 2] cycloaddition pathway. The second-generation chiral cationic salen-manganese and chromium complexes also proved to be effective catalysts for asymmetric hetero-Diels–Alder reactions [52].

Jacobsen et al. disclosed that the diastereoselectivity of the hetero-Diels–Alder reaction between 1-methoxy-3-[(trimethylsilyl)oxy]-1,3-butadiene and chiral aldehydes can be elaborately controlled by appropriate choice of chiral salen chromium complexes (Scheme 7.26) [53]. Thus, a single stereoisomer of dihydropyranone products can be obtained through catalyst-controlled doubly diastereoselective reactions.

In 2000, Rawal et al. discovered that chiral salen chromium complex **28** bearing a SbF_6 counterion showed high enantioselectivity and endo-selectivity in

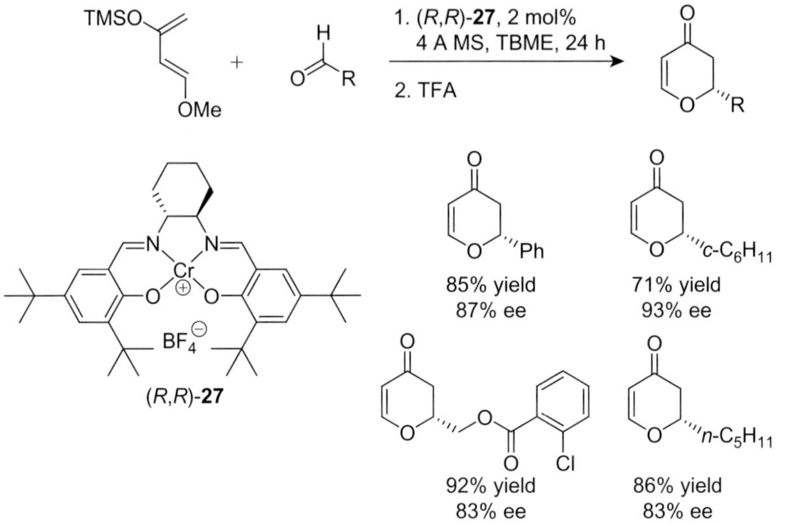

Scheme 7.25 Asymmetric hetero-Diels–Alder reaction catalyzed by a chiral salen chromium complex.

catalyzing Diels–Alder reactions of 1-amino-3-siloxy-1,3-butadienes with α-alkyl substituted acroleins, affording functionalized cyclohexene derivatives (Scheme 7.27) [54]. The steric properties of substituents on the amino group promoted the enantioselectivity. This synthetic strategy for constructing chiral six-membered carbon rings has been successfully applied to the concise synthesis of a family of *Aspidosperma* alkaloids [55].

Subsequently, Rawal and coworkers disclosed that the substrates were not limited to 1-amino-3-siloxy-1,3-butadienes as a wide range of substituted 1-amino-1,3-butadienes could also undergo Diels–Alder reactions with α-alkyl substituted acroleins using chiral salen chromium catalyst **28**, affording cycloadducts with almost complete endo-selectivity in good yield and high enantioselectivity (Scheme 7.28) [56].

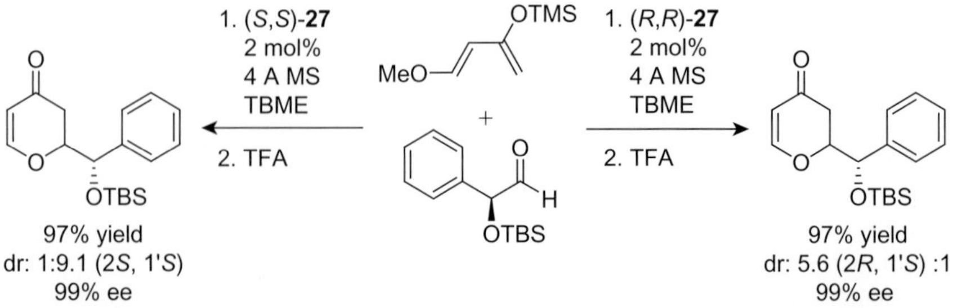

Scheme 7.26 Catalyst-controlled doubly diastereoselective hetero-Diels–Alder reactions.

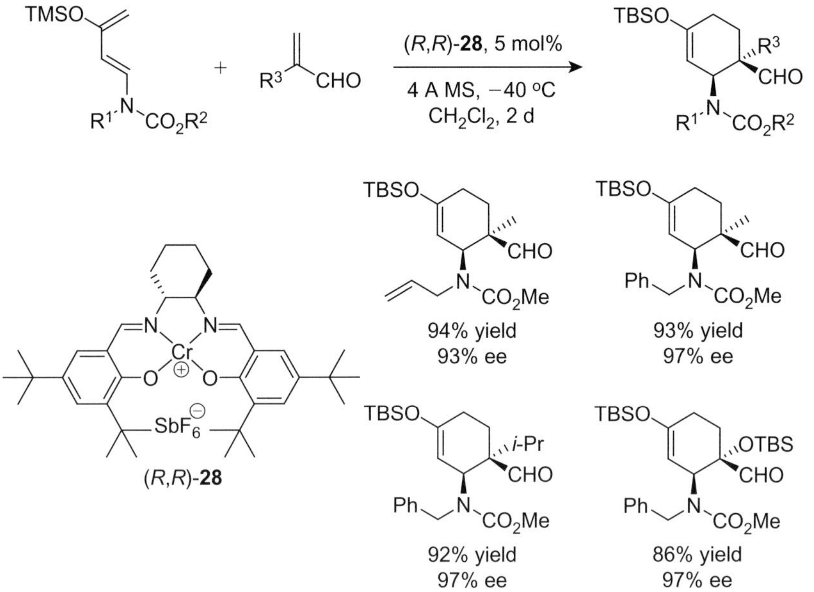

Scheme 7.27 Asymmetric Diels–Alder reactions catalyzed by a chiral salen chromium complex.

In the pursuit of more active catalysts for Diels–Alder reactions, chiral salen cobalt complex **29** was found to be more active for the reaction of 1-amino-1,3-butadiene and methacrolein (Figure 7.15). Of importance, the X-ray crystal structure of a complex formed by the coordination of two benzaldehydes to the complex **29** revealed the activation of the carbonyl group by non-perpendicular coordination to the catalyst. Based on this structural information, chiral salen cobalt complex **30** (Figure 7.15), in which 3- and 3′-silyl substituents were

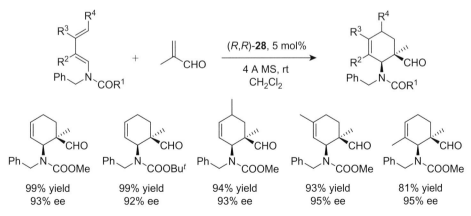

Scheme 7.28 Diels–Alder reactions of substituted 1-amino-1,3-butadienes with methacrolein.

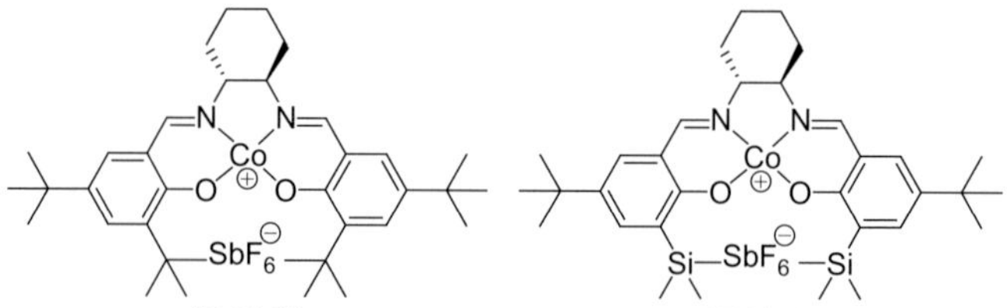

Figure 7.15 Chiral salen cobalt catalysts for Diels–Alder reactions.

introduced to force the two aromatic rings out of a near-parallel arrangement, was prepared for this reaction. Catalyst **30** exhibited dramatically enhanced reactivity and slightly increased enantioselectivity in comparison with catalyst **29** [57].

7.4.6
Asymmetric Cyanohydrin Synthesis

Cyanohydrin synthesis by asymmetric addition of cyanide to carbonyl compounds is of considerable significance because cyanohydrins are highly versatile synthetic intermediates and can be readily converted into a wide range of functional groups [58]. Initially developed chiral catalysts for this transformation suffer from high catalyst loading (10–100 mol.%), low enantioselectivity, and the need to perform at low temperatures. The breakthrough occurred in 1999, when chiral salen titanium complex **31** was established as a highly enantioselective catalyst for the trimethylsilylcyanation of various aldehydes at ambient temperature with only 0.1 mol.% catalyst loading. A trace amount of water was crucial for this catalytic system because under rigorously anhydrous conditions complex **31** is inactive in trimethylsilylcyanation. Further investigation revealed that the complex **31** with a *trans* configuration was converted into chiral salen di-μ-oxo titanium complex **32** (Figure 7.16) with *cis*-β configuration, upon treatment with 1 equivalent of water. Therefore, *in situ* generated complex **32** is probably the catalytically active species in the reaction using catalyst **31**. Indeed, chiral salen di-μ-oxo titanium complex **32** exhibited very high activity and high enantioselectivity in catalyzing trimethylsilylcyanation of various aldehydes at ambient temperature (Table 7.1), and only 0.1 mol.% catalyst loading gave complete conversion in much short times (<1 h) [59].

Based on X-ray crystallographic, NMR and IR spectroscopic studies, in combination with kinetic investigations, a possible mechanism was proposed (Figure 7.17) [60]. Di-μ-oxo complex **32** reacts with aldehyde and trimethylsilyl cyanide to afford mononuclear complexes A and B, respectively. The combination of complexes A and B generates catalytically active species *cis*-β-μ-oxo titanium complex C, in which both cyanide and aldehyde are simultaneously activated. Then cyanide is transferred intramolecularly to the aldehyde in an enantioselective manner and μ-oxo titanium complex D bonding the cyanohydrin group is formed. Silylation of complex D generates the silylated cyanohydrin product and bis-cyanide complex E, which is the rate-determining step. The reaction of complex E with aldehyde regenerates catalytically active species C.

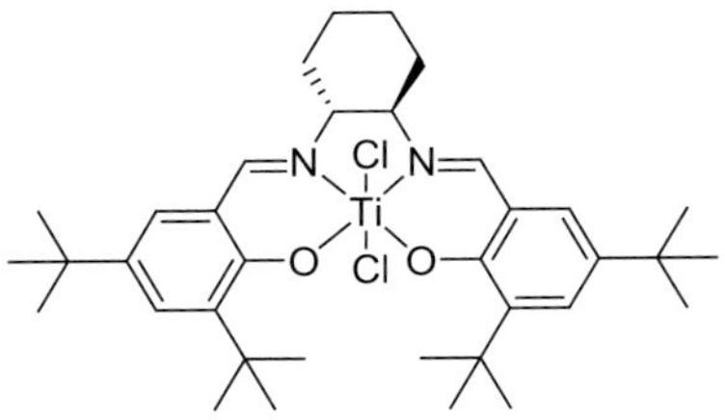

Figure 7.16 Chiral salen-metal catalysts for asymmetric synthesis of cyanohydrins.

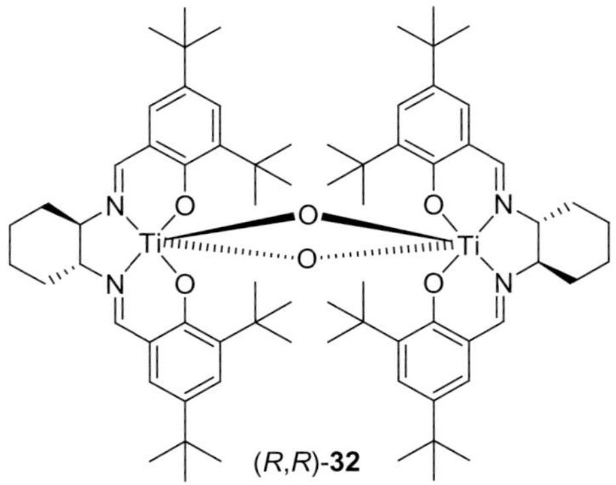

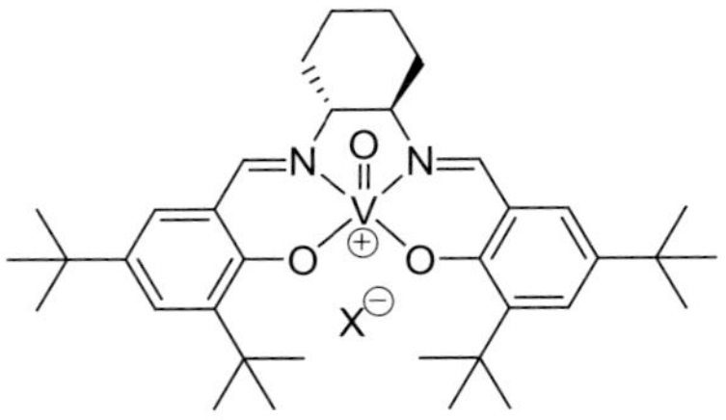

Table 7.1 Trimethylsilylcyanation of aldehydes catalyzed by chiral salen titanium catalyst **32**

$$\text{RCHO} + \text{TMSCN} \xrightarrow[\text{CH}_2\text{Cl}_2,\ \text{rt},\ 1\ \text{h}]{(R,R)\text{-}\mathbf{32},\ 0.1\ \text{mol\%}} \text{R-CH(OTMS)(CN)}$$

100% conversion

R	Ee (%)	R	Ee (%)
Ph	86	2,4-(MeO)$_2$C$_6$H$_3$	88
2-MeC$_6$H$_4$	76	3,4-(MeO)$_2$C$_6$H$_3$	85
3-MeC$_6$H$_4$	90	3,5-(MeO)$_2$C$_6$H$_3$	90
4-MeC$_6$H$_4$	87	4-CF$_3$C$_6$H$_4$	86
2-MeOC$_6$H$_4$	88	4-NO$_2$C$_6$H$_4$	50
3-MeOC$_6$H$_4$	92	Me$_3$C	66
4-MeOC$_6$H$_4$	84	CH$_3$CH$_2$	52

Subsequently, chiral salen vanadium complex **33** (Figure 7.16) was developed as an excellent catalyst for trimethylsilylcyanation of aldehyde at ambient temperature [61]. Although its activity is relatively low, catalyst **33** displayed high enantioselectivity (73–95% ee) in the reaction of aromatic aldehydes. The counterions of chiral salen vanadium complexes have a significant effect on catalyst activity. Chiral salen-vanadium complex **34** bearing a chloride counterion has proven to exhibit high activity, comparable with chiral salen titanium catalyst **32**, as well as excellent enantioselectivity comparable with the catalyst **33** [62]. In addition, introduction of appropriate co-catalysts or additives can effectively enhance the reactivity and enantioselectivity of chiral salen-metal catalysts in asymmetric cyanohydrin synthesis [63].

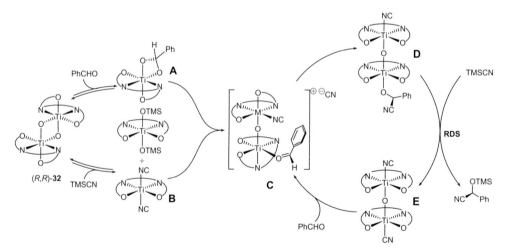

Figure 7.17 Mechanism of trimethylsilylcyanation of aldehyde catalyzed by catalyst **32**.

7.4.7
Miscellaneous Reactions

Cozzi et al. disclosed that chiral salen chromium complex **8** is an excellent catalyst for the enantioselective addition of Me$_2$Zn to various aldehydes such as aliphatic, aromatic, α,β-unsaturated, and heterocyclic aldehydes (Scheme 7.29) [64]. The reaction of more reactive Et$_2$Zn with benzaldehyde also proceeded easily using the same catalytic system, but afforded the product with low enantiomeric excess. Additionally, chiral salen-metal complexes can serve as highly effective catalysts for enantioselective addition of terminal alkynes to ketones [65] and the enantioselective Reformatsky reaction with ketones [66].

Jacobsen et al. discovered that chiral salen chromium complex **8** could effectively catalyze asymmetric α-alkylations of tetrasubstituted tin enolates of five-, six-, and seven-membered ring ketones using alkyl halides as electrophiles, affording enantioenriched products bearing α-carbonyl all-carbon-substituted quaternary stereocenters (Scheme 7.30) [67]. Asymmetric alkylation of acyclic tetrasubstituted tin enolates also proved successful using a similar catalytic system [68].

Wang reported that in the presence of a catalytic amount of Na$_2$CO$_3$, chiral salen chromium complex **8** catalyzed the asymmetric intramolecular nucleophilic addition of tertiary enamides to a carbonyl group to afford functionalized γ-lactams bearing a hydroxylated quaternary carbon center in excellent yield with high enantioselectivity (Scheme 7.31) [69]. Substrate with electron-donating substituents in the enamine double bond and/or electron-withdrawing substituents in carbonyl group displayed high reactivity, which is consistent with the nucleophilic attack of enamide at the carbonyl moiety.

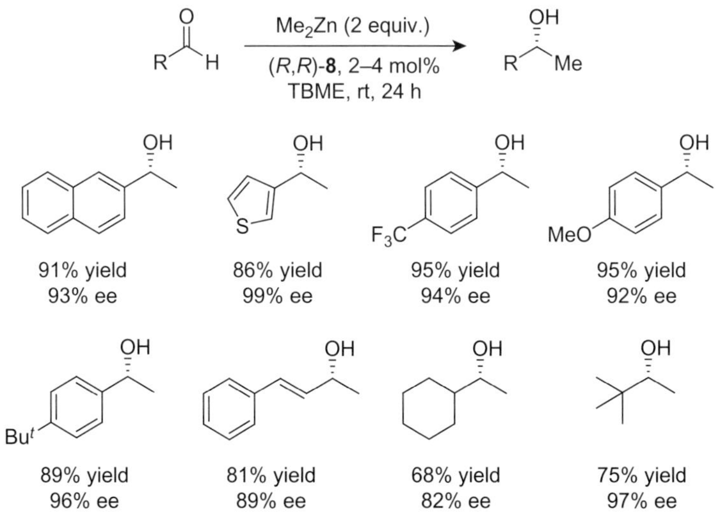

Scheme 7.29 Enantioselective addition of Me$_2$Zn to aldehydes catalyzed by chiral salen chromium complex **8**.

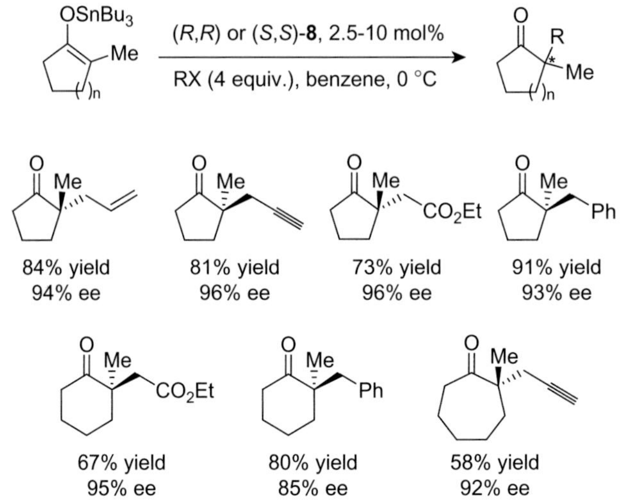

Scheme 7.30 Asymmetric α-alkylations of tetrasubstituted tin enolates with alkyl halides.

Wang and Zhu discovered that chiral salen aluminum complex **23** (shown in Scheme 7.23) catalyzed the Passerini three-component reaction, which involved the condensation of various non-chelating aldehydes, carboxylic acids, and isocyanides, with good to excellent enantioselectivity and in acceptable yield (Scheme 7.32) [70]. The Passerini-type three-component reaction of aldehydes, isocyanides, and hydrazoic acid also proceeded smoothly using chiral salen aluminum catalyst **22** (shown in Scheme 7.21) with high enantioselectivity (Scheme 7.32) [71]. The catalytic system showed extremely broad substrate applicability. Mechanistic studies indicated that chiral salen aluminum complex **22** played a role in activating aldehydes, rather than functioning as an azide deliverer.

Besides the above-mentioned reactions, chiral salen-metal complexes also exhibited promising activity and enantioselectivity in asymmetric catalysis of radical additions to cyclic ketones [72], acylvinyl additions to aldehydes [73], Baeyer–Villiger oxidation [74], aziridination [75], and sulfoxidation [76].

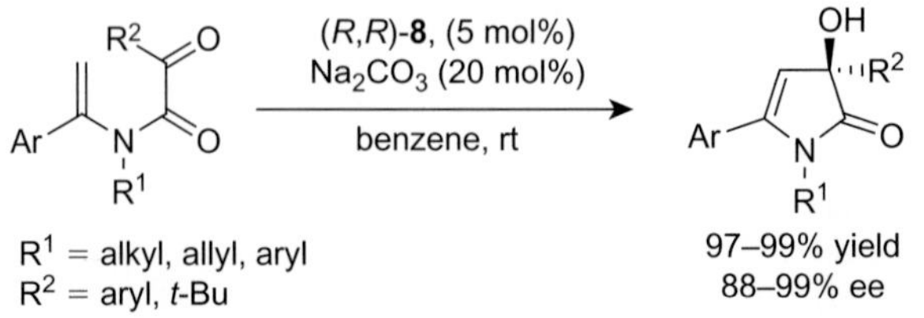

Scheme 7.31 Enantioselective intramolecular addition of tertiary enamides to ketones.

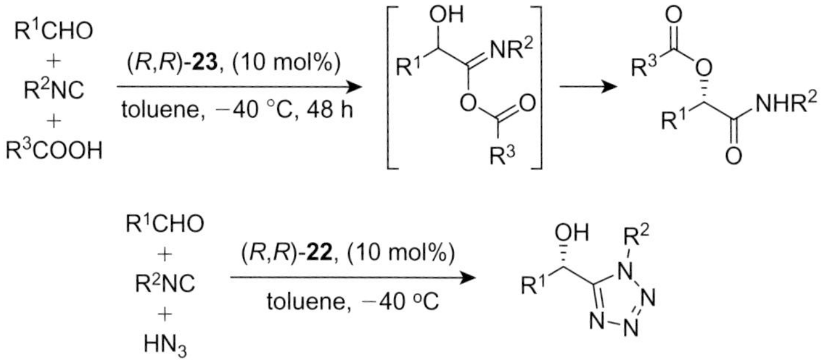

Scheme 7.32 Enantioselective Passerini-type reactions catalyzed by chiral salen aluminum complexes.

7.5
Conclusion and Outlook

Chiral salen complexes, with easily tunable stereochemical and electronic properties, have been developed as one of the most important chiral catalysts and have had a tremendous impact on asymmetric synthesis in the past two decades. With regard to the function in asymmetric catalysis, according to present understanding, chiral salen complexes serve as oxo, carbene-transfer agent or electrophile to activate epoxides and carbonyl compounds, as well as both electrophile (Lewis acidic metal ion) and nucleophile (nucleophilic counterion) to undergo cooperative homobimetallic catalysis.

Although the wide range of reactions outlined above have been studied extensively, it can be expected that more new enantioselective reactions catalyzed by well-known chiral salen complexes will be developed in the near future. Moreover, elaborate structural modifications of chiral salen complexes have afforded, and will provide further opportunity for the development of, new chiral catalysts for various asymmetric reactions. Representative examples are (i) extension of chiral salen complexes to nonplanar chiral salalen (half-reduced salen) and salan (a saturated version of Schiff-base salen ligand) complexes (Figure 7.18), which exhibited unprecedented enantioselectivity in some reactions [77]; (ii) combination of chiral scaffolds of salen with other privileged ligands such as BINAP and proline endows resultant complexes with unique asymmetric catalytic performance [35, 78];

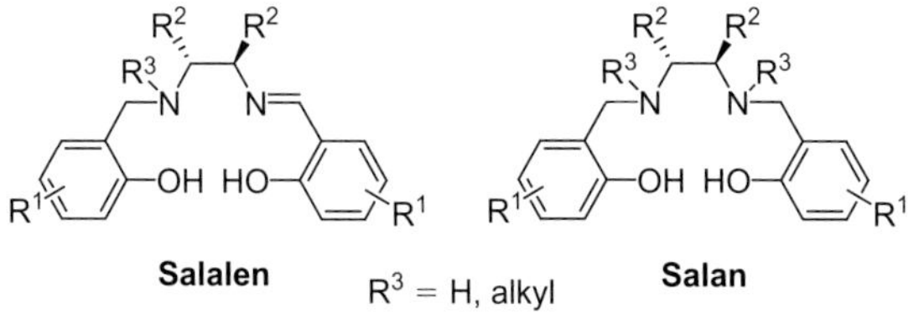

Figure 7.18 Chiral salalen and salan ligands.

(iii) introduction of nucleophiles anchored on the chiral salen ligand framework enforces intramolecular cooperative interaction [79].

Computational methods have been used to study the mechanism of some asymmetric reactions catalyzed by chiral salen complexes. It is anticipated that this approach will play an increasingly important role in understanding the features that account for the unique asymmetric catalytic performance of chiral salen complexes, and thus aid in developing new privileged catalysts.

References

1 Jacobsen, E.N., Pfaltz, A., and Yamamoto, H. (eds) (1999) *Comprehensive Asymmetric Catalysis*, Springer-Verlag, Berlin, Germany.
2 (a) Atwood, D.A. and Harvey, M.J. (2001) *Chem. Rev.*, **101**, 37–52; (b) Katsuki, T. (2003) *Synlett*, 281–297; (c) Cozzi, P.G. (2004) *Chem. Soc. Rev.*, **33**, 410–421; (d) Katsuki, T. (2004) *Chem. Soc. Rev.*, **33**, 437–444; (e) Darensbourg, D.J., Mackiewicz, R.M., Phelps, A.L., and Billodeaux, D.R. (2004) *Acc. Chem. Res.*, **37**, 836–844; (f) Venkataramanan, N.S., Kuppuraj, G., and Rajagopal, S. (2005) *Coord. Chem. Rev.*, **249**, 1249–1268; (g) McGarrigle, E.M. and Gilheany, D.G. (2005) *Chem. Rev.*, **105**, 1563–1602; (h) Baleizao, C. and Garcia, H. (2006) *Chem. Rev.*, **106**, 3987–4043; (i) Matsumoto, K., Saito, B., and Katsuki, T. (2007) *Chem. Commun.*, 3619–3627; (j) Darensbourg, D.J. (2007) *Chem. Rev.*, **107**, 2388–2410.
3 Zhang, W., Loebach, J.L., Wilson, S.R., and Jacobsen, E.N. (1990) *J. Am. Chem. Soc.*, **112**, 2801–2803.
4 Irie, R., Noda, K., Ito, Y., Matsumoto, N., and Katsuki, T. (1990) *Tetrahedron Lett.*, **31**, 7345–7348.
5 Yoon, T.P. and Jacobsen, E.N. (2003) *Science*, **299**, 1691–1693.
6 Katsuki, T. (2002) *Adv. Synth. Catal.*, **344**, 131–147.
7 Zhang, W. and Jacobsen, E.N. (1991) *J. Org. Chem.*, **56**, 2296–2298.
8 (a) Jacobsen, E.N., Zhang, W., Muci, A. R., Ecker, J.R., and Deng, L. (1991) *J. Am. Chem. Soc.*, **113**, 7063–7064; (b) Lee, N.H., Muci, A.R., and Jacobsen, E.N. (1991) *Tetrahedron Lett.*, **32**, 5055–5058.
9 (a) Lee, N.H. and Jacobsen, E.N. (1991) *Tetrahedron Lett.*, **32**, 6533–6536; (b) Chang, S., Lee, N.H., and Jacobsen, E.N. (1993) *J. Org. Chem.*, **58**, 6939–6941; (c) Hentemann, M.F. and Fuchs, P.L. (1997) *Tetrahedron Lett.*, **38**, 5615–5618; (d) Deng, L. and Jacobsen, E.N. (1992) *J. Org. Chem.*, **57**, 4320–4323; (e) Jacobsen, E.N., Deng, L., Furukawa, Y., and Martinez, L.E. (1994) *Tetrahedron*, **50**, 4323–4334.
10 (a) Chang, S., Heid, R.M., and Jacobsen, E.N. (1994) *Tetrahedron Lett.*, **35**, 669–672; (b) Brandes, B.D. and Jacobsen, E.N. (1994) *J. Org. Chem.*, **59**, 4378–4380; (c) Brandes, B.D. and Jacobsen, E.N. (1995) *Tetrahedron Lett.*, **36**, 5123–5126.
11 (a) Palucki, M., Pospisil, P.J., Zhang, W., and Jacobsen, E.N. (1994) *J. Am. Chem. Soc.*, **116**, 9333–9334; (b) Palucki, M., McCormick, G.J., and Jacobsen, E.N. (1995) *Tetrahedron Lett.*, **36**, 5457–5460.
12 (a) Sasaki, H., Irie, R., Hamada, T., Suzuki, K., and Katsuki, T. (1994) *Tetrahedron*, **50**, 11827–11838; (b) Fukuda, T., Irie, R., and Katsuki, T. (1995) *Synlett*, 197–198; (c) Fukuda, T. and Katsuki, T. (1996) *Tetrahedron Lett.*, **37**, 4389–4392.
13 Nishikori, H., Ohta, C., and Katsuki, T. (2000) *Synlett*, 1557–1560.
14 Jacobsen, E.N. (2000) *Acc. Chem. Res.*, **33**, 421–431.
15 Martinez, L.E., Leighton, J.L., Carsten, D.H., and Jacobsen, E.N. (1995) *J. Am. Chem. Soc.*, **117**, 5897–5898.
16 (a) Leighton, J.L. and Jacobsen, E.N. (1996) *J. Org. Chem.*, **61**, 389–390; (b) Martinez, L.E., Nugent, W.A., and

Jacobsen, E.N. (1996) *J. Org. Chem.*, **61**, 7963–7966; (c) Wu, M.H. and Jacobsen, E.N. (1997) *Tetrahedron Lett.*, **38**, 1693–1696; (d) Schaus, S.E., Larrow, J.F., and Jacobsen, E.N. (1997) *J. Org. Chem.*, **62**, 4197–4199.

17 Hansen, K.B., Leighton, J.L., and Jacobsen, E.N. (1996) *J. Am. Chem. Soc.*, **118**, 10924–10925.

18 Konsler, R.G., Karl, J., and Jacobsen, E.N. (1998) *J. Am. Chem. Soc.*, **120**, 10780–10781.

19 Wu, M.H., Hansen, K.B., and Jacobsen, E.N. (1999) *Angew. Chem. Int. Ed.*, **38**, 2012–2014.

20 Loy, R.N. and Jacobsen, E.N. (2009) *J. Am. Chem. Soc.*, **131**, 2786–2787.

21 Larrow, J.F., Schaus, S.E., and Jacobsen, E.N. (1996) *J. Am. Chem. Soc.*, **118**, 7420–7421.

22 Tokunaga, M., Larrow, J.F., Kakiuchi, F., and Jacobsen, E.N. (1997) *Science*, **277**, 936–938.

23 Schaus, S.E., Brandes, B.D., Larrow, J.F., Tokunaga, M., Hansen, K.B., Gould, A.E., Furrow, M.E., and Jacobsen, E.N. (2002) *J. Am. Chem. Soc.*, **124**, 1307–1315.

24 Brandes, B.D. and Jacobsen, E.N. (1997) *Tetrahedron Asymmetry*, **8**, 3927–3933.

25 Nielsen, L.P.C., Stevenson, C.P., Blackmond, D.G., and Jacobsen, E.N. (2004) *J. Am. Chem. Soc.*, **126**, 1360–1362.

26 Kemper, S., Hrobarik, P., Kaupp, M., and Schlorer, N.E. (2009) *J. Am. Chem. Soc.*, **131**, 4172–4173.

27 Sakakura, T., Choi, J.-C., and Yasuda, H. (2007) *Chem. Rev.*, **107**, 2365–2387.

28 Darensbourg, D.J. and Holtcamp, M.W. (1996) *Coord. Chem. Rev.*, **153**, 155–174.

29 Lu, X.B., Liang, B., Zhang, Y.J., Tian, Y.Z., Wang, Y.M., Bai, C.X., Wang, H., and Zhang, R. (2004) *J. Am. Chem. Soc.*, **126**, 3732–3733.

30 (a) Berkessel, A. and Brandenburg, M. (2006) *Org. Lett.*, **8**, 4401–4401; (b) Yan, P. and Jing, H. (2009) *Adv. Synth. Catal.*, **351**, 1325–1332.

31 Coates, G.W. and Moore, D.R. (2004) *Angew. Chem. Int. Ed.*, **43**, 6618–6639.

32 Qin, Z., Thomas, C.M., Lee, S., and Coates, G.W. (2003) *Angew. Chem. Int. Ed.*, **42**, 5484–5487.

33 (a) Lu, X.B. and Wang, Y. (2004) *Angew. Chem. Int. Ed.*, **43**, 3574–3577; (b) Lu, X.B., Shi, L., Wang, Y.M., Zhang, R., Zhang, Y.J., Peng, X.J., Zhang, Z.C., and Li, B. (2006) *J. Am. Chem. Soc.*, **128**, 1664–1674.

34 (a) Cohen, C.T., Chu, T., and Coates, G.W. (2005) *J. Am. Chem. Soc.*, **127**, 10869–10878; (b) Cohen, C.T. and Coates, G.W. (2006) *J. Polym. Sci., Part A: Polym. Chem.*, **44**, 5182–5191.

35 Hirahata, W., Thomas, R.M., Lobkovsky, E.B., and Coates, G.W. (2008) *J. Am. Chem. Soc.*, **130**, 17658–17659.

36 Lebel, H., Marcoux, J.F., Molinaro, C., and Charette, A.B. (2003) *Chem. Rev.*, **103**, 977–1050.

37 Fukuda, T. and Katsuki, T. (1997) *Tetrahedron*, **53**, 7201–7208.

38 Uchida, T., Irie, R., and Katsuki, T. (2000) *Tetrahedron*, **56**, 3501–3509.

39 Niimi, T., Uchida, T., Irie, R., and Katsuki, T. (2001) *Adv. Synth. Catal.*, **343**, 79–88.

40 (a) Kanchiku, S., Suematsu, H., Matsumoto, K., Uchida, T., and Katsuki, T. (2007) *Angew. Chem. Int. Ed.*, **46**, 3889–3891; (b) Suematsu, H., Kanchiku, S., Uchida, T., and Katsuki, T. (2008) *J. Am. Chem. Soc.*, **130**, 10327–10337.

41 Xu, Z.J., Fang, R., Zhao, C.Y., Huang, J.S., Li, G.Y., Zhu, N.Y., and Che, C.M. (2009) *J. Am. Chem. Soc.*, **131**, 4405–4417.

42 Myers, J.K. and Jacobsen, E.N. (1999) *J. Am. Chem. Soc.*, **121**, 8959–8960.

43 Taylor, M.S., Zalatan, D.N., Lerchner, A.M., and Jacobsen, E.N. (2005) *J. Am. Chem. Soc.*, **127**, 1313–1317.

44 Sammis, G.M. and Jacobsen, E.N. (2003) *J. Am. Chem. Soc.*, **125**, 4442–4443.

45 Sammis, G.M., Danjo, H., and Jacobsen, E.N. (2004) *J. Am. Chem. Soc.*, **126**, 9928–9929.

46 Mazet, C. and Jacobsen, E.N. (2008) *Angew. Chem. Int. Ed.*, **47**, 1762–1765.

47 Taylor, M.S. and Jacobsen, E.N. (2003) *J. Am. Chem. Soc.*, **125**, 11204–11205.

48 Balskus, E.P. and Jacobsen, E.N. (2006) *J. Am. Chem. Soc.*, **128**, 6810–6812.

49 Vanderwal, C.D. and Jacobsen, E.N. (2004) *J. Am. Chem. Soc.*, **126**, 14724–14725.
50 Gandelman, M. and Jacobsen, E.N. (2005) *Angew. Chem. Int. Ed.*, **44**, 2393–2397.
51 Schaus, S.E., Brånalt, J., and Jacobsen, E.N. (1998) *J. Org. Chem.*, **63**, 403–405.
52 Aikawa, K., Irie, R., and Katsuki, T. (2001) *Tetrahedron*, **57**, 845–851.
53 Joly, G.D. and Jacobsen, E.N. (2002) *Org. Lett.*, **4**, 1795–1798.
54 Huang, Y., Iwama, T., and Rawal, V.H. (2000) *J. Am. Chem. Soc.*, **122**, 7843–7844.
55 Kozmin, S.A., Iwama, T., Huang, Y., and Rawal, V.H. (2002) *J. Am. Chem. Soc.*, **124**, 4628–4641.
56 Huang, Y., Iwama, T., and Rawal, V.H. (2002) *Org. Lett.*, **4**, 1163–1166.
57 Huang, Y., Iwama, T., and Rawal, V.H. (2002) *J. Am. Chem. Soc.*, **124**, 5950–5951.
58 (a) Brunel, J.-M. and Holmes, I.P. (2004) *Angew. Chem. Int. Ed.*, **43**, 2752–2778; (b) North, M., Usanov, D.L., and Young, C. (2008) *Chem. Rev.*, **108**, 5146–5226.
59 (a) Tararov, V.I., Hibbs, D.E., Hursthouse, M.B., Ikonnikov, N.S., Malik, K.M.A., North, M., Orizu, C., and Belokon, Y.N. (1998) *Chem. Commun.*, 387–388; (b) Belokon, Y.N., Caveda-Cepas, S., Green, B., Ikonnikov, N.S., Khrustalev, V.N., Larichev, V.S., Moscalenko, M.A., North, M., Orizu, C., Tararov, V.I., Tasinazzo, M., Timofeeva, G.I., and Yashkina, L.V. (1999) *J. Am. Chem. Soc.*, **121**, 3968–3973.
60 Belokon, Y.N., Green, B., Ikonnikov, N.S., Larichev, V.S., Lokshin, B.V., Moscalenko, M.A., North, M., Orizu, C., Peregudov, A.S., and Timofeeva, G.I. (2000) *Eur. J. Org. Chem.*, 2655–2661.
61 (a) Belokon, Y.N., North, M., and Parsons, T. (2000) *Org. Lett.*, **2**, 1617–1619; (b) Belokon, Y.N., Green, B., Ikonnikov, N.S., North, M., Parsons, T., and Tararov, V.I. (2001) *Tetrahedron*, **57**, 771–779; (c) Belokon, Y.N., Carta, P., Gutnov, A.V., Maleev, V.I., Moscalenko, M.A., Yashkina, L.V., Ikonnikov, N.S., Voskoboev, N.V., Khrustalev, V.N., and North, M. (2002) *Helv. Chim. Acta*, **85**, 3301–3312.
62 Belokon, Y.N., Maleev, V.I., North, M., and Usanov, D.L. (2006) *Chem. Commun.*, 4614–4616.
63 (a) Kim, S.S. and Song, D.H. (2005) *Eur. J. Org. Chem.*, 1777–1780; (b) Khan, N.H., Agrawal, S., Kureshy, R.I., Abdi, S.H.R., Prathap, K.J., and Jasra, R.V. (2008) *Eur. J. Org. Chem.*, 4511–4515; (c) Alaaeddine, A., Roisnel, T., Thomas, C.M., and Carpentier, J.-F. (2008) *Adv. Synth. Catal.*, **350**, 731–740.
64 Cozzi, P.G. and Kotrusz, P. (2006) *J. Am. Chem. Soc.*, **128**, 4940–4941.
65 Cozzi, P.G. (2003) *Angew. Chem. Int. Ed.*, **42**, 2895–2898.
66 Cozzi, P.G. (2006) *Angew. Chem. Int. Ed.*, **45**, 2951–2954.
67 Doyle, A.G. and Jacobsen, E.N. (2005) *J. Am. Chem. Soc.*, **127**, 62–63.
68 Doyle, A.G. and Jacobsen, E.N. (2007) *Angew. Chem. Int. Ed.*, **46**, 3701–3705.
69 Yang, L., Wang, D.-X., Huang, Z.-T., and Wang, M.-X. (2009) *J. Am. Chem. Soc.*, **131**, 10390–10391.
70 Wang, S.-X., Wang, M.-X., Wang, D.-X., and Zhu, J. (2008) *Angew. Chem. Int. Ed.*, **47**, 388–391.
71 Yue, T., Wang, M.-X., Wang, D.-X., and Zhu, J. (2008) *Angew. Chem. Int. Ed.*, **47**, 9454–9457.
72 (a) Sibi, M.P. and Zimmerman, J. (2006) *J. Am. Chem. Soc.*, **128**, 13346–13347; (b) Sibi, M.P. and Nad, S. (2007) *Angew. Chem. Int. Ed.*, **46**, 9231–9234.
73 Reynolds, T.E. and Scheidt, K.A. (2007) *Angew. Chem. Int. Ed.*, **46**, 7806–7809.
74 (a) Uchida, T. and Katsuki, T. (2001) *Tetrahedron Lett.*, **42**, 6911–6914; (b) Bianchini, G., Cavarzan, A., Scarso, A., and Strukul, G. (2009) *Green Chem.*, **11**, 1517–1520.
75 (a) Li, Z., Conser, K.R., and Jacobsen, E.N. (1993) *J. Am. Chem. Soc.*, **115**, 5326–5327; (b) Minakata, S., Ando, T., Nishimura, M., Ryu, I., and Komatsu, M. (1998) *Angew. Chem. Int. Ed.*, **37**, 3392–3394.
76 Saito, B. and Katsuki, T. (2001) *Tetrahedron Lett.*, **42**, 3873–3876.
77 (a) Yamaguchi, T., Matsumoto, K., Saito, B., and Katsuki, T. (2007) *Angew. Chem. Int. Ed.*, **46**, 4729–4731; (b) Egami, H. and Katsuki, T. (2007) *J. Am. Chem. Soc.*, **129**, 8940–8941; (c) Saito, B.,

Egami, H., and Katsuki, T. (2007) *J. Am. Chem. Soc.*, **129**, 1978–1986; (d) Egami, H. and Katsuki, T. (2009) *J. Am. Chem. Soc.*, **131**, 6082–6083.

78 Matsumoto, K., Oguma, T., and Katsuki, T. (2009) *Angew. Chem. Int. Ed.*, **48**, 7432–7435.

79 (a) DiMauro, E.F. and Kozlowski, M.C. (2002) *J. Am. Chem. Soc.*, **124**, 12668–12669; (b) Noh, E.K., Na, S.J., Sujith, S., Kim, S.W., and Lee, B.Y. (2007) *J. Am. Chem. Soc.*, **129**, 8082–8083; (c) Handa, S., Gnanadesikan, V., Matsunaga, S., and Shibasaki, M. (2007) *J. Am. Chem. Soc.*, **129**, 4900–4901; (d) Sujith, S., Min, K.K., Seong, J.E., Na, S.J., and Lee, B.Y. (2008) *Angew. Chem. Int. Ed.*, **47**, 7306–7309; (e) Ren, W.-M., Liu, Z.-W., Wen, Y.-Q., Zhang, R., and Lu, X.-B. (2009) *J. Am. Chem. Soc.*, **131**, 11509–11518.

8
BINOL
Masakatsu Shibasaki and Shigeki Matsunaga

8.1
Introduction

Since the pioneering work of Cram and coworkers, molecules with an axially chiral 1,1′-binaphthyl unit have been utilized to realize high chiral recognition ability [1]. 1,1′-Binaphthyl-2,2′-diol (BINOL, Figure 8.1) is one of the most famous axially chiral C_2-symmetric molecules. The racemization barrier of BINOL is high under neutral conditions, and enantiomerically pure BINOL is resistant towards racemization under various experimental conditions for organic synthesis, even at high temperature [2]. BINOL and its derivatives have attracted particular interests as chiral "privileged" ligands for oxophilic Lewis acidic metals, such as group IV metals, group XIII metals, rare earth metals, alkaline earth metals, alkali metals, and others. In asymmetric catalysis, the stereoselectivity of the targeted enantioselective reactions depends on the chiral environment of catalysts. Both steric and electronic properties of the chiral ligands drastically affect the chiral environment by influencing the properties of metal centers. Thus, BINOL is ideal as a chiral ligand because it can be readily modified to fit the targeted enantioselective reactions. Various short and efficient methods have been reported for the modification of BINOL at the 3,3′-positions and 6,6′-positions to tune its steric and/or electronic properties. Methods for modification at other positions, such as the 4,4′- and 7,7′-positions, are also known, but rather long steps are required. In addition, modification of the dihedral angle of BINOL is also effective in optimizing the chiral environment. Partially hydrogenated BINOL, such as H_8-BINOL and H_4-BINOL (Figure 8.1), as well as ligands with a biphenyl backbone instead of a binaphthyl backbone are available for that purpose. This chapter introduces selected representative examples of chiral metal catalysts using BINOL and its derivatives to show the characteristic aspects of BINOL as a privileged ligand. For comprehensive reviews on BINOL and its derivatives, see several review articles [3–5]. The utility of BINOL as a scaffold for organocatalysts is also described.

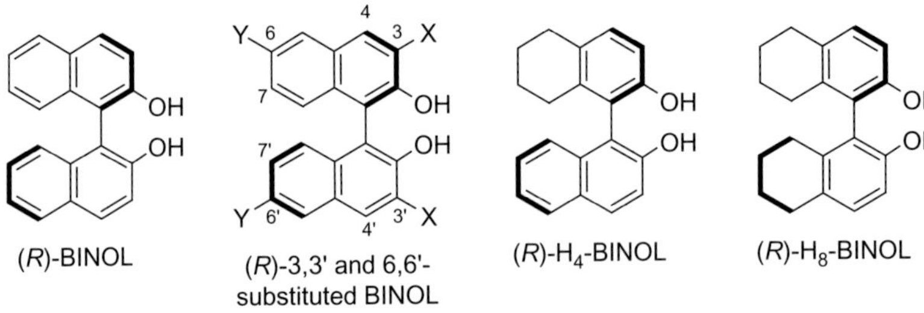

Figure 8.1 Structures of BINOL and BINOL derivatives.

8.2
Applications in Reduction and Oxidation

The potential of BINOL as a chiral ligand in metal-mediated reactions was first recognized in 1979 by Noyori in the enantioselective reduction of ketones and aldehydes (Scheme 8.1) [6]. Although a stoichiometric amount of BINOL was required to prepare the chiral reducing reagent, BINAL-H, by mixing with LiAlH$_4$ in 1 : 1 ratio, high levels of enantio-induction were achieved. A related Me$_3$Al/BINOL reagent was also utilized for a highly enantioselective Meerwein–Ponndorf–Verley reduction of N-diphenylphosphinoyl imines [7]. Several catalytic asymmetric methods with metal-BINOL catalysts were also developed using boranes [8] or silanes [9] as reducing agents, but enantioselectivities were less satisfactory than for catalysts using other chiral ligands.

BINOL and its derivatives have been intensively utilized in the catalytic asymmetric epoxidation of electron-deficient alkenes and gave synthetically versatile

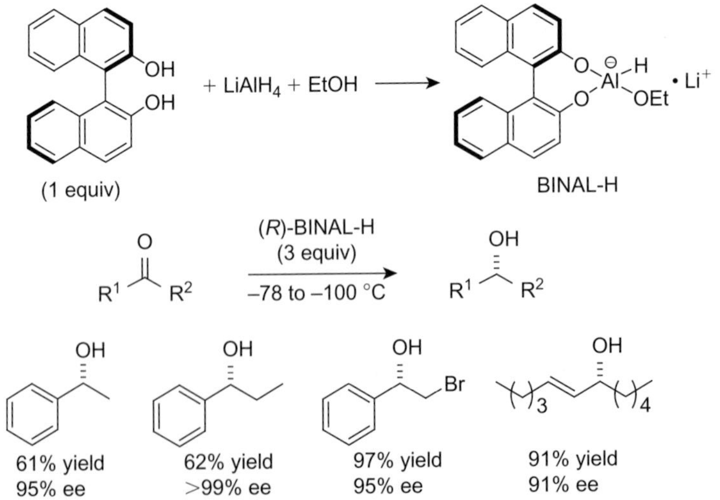

Scheme 8.1 Enantioselective reduction of ketones by BINAL-H.

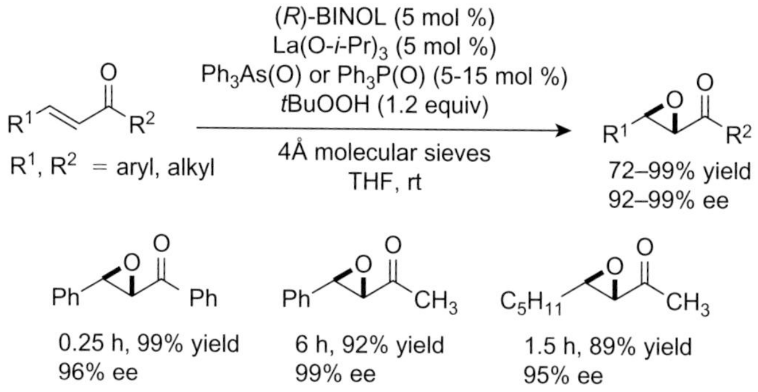

Scheme 8.2 Catalytic asymmetric epoxidation of enones with La(O-iPr)$_3$/BINOL catalyst.

epoxides. Although the first Yb(O-iPr)$_3$/3-hydroxymethyl-BINOL system realized a high enantioselectivity for enones [10], the low reaction rate was problematic. The addition of an achiral Lewis base additive, Ph$_3$P=O or Ph$_3$As=O, was effective in improving reactivity, enantioselectivity, and substrate generality (Scheme 8.2) [11]. A La(O-iPr)$_3$/BINOL/Ph$_3$P=O or Ph$_3$As=O system promoted the epoxidation of broad range of enones using *tert*-butyl hydroperoxide (TBHP) in up to 99% yield and 99% ee. Notably, less explosive cumene hydroperoxide (CMHP) was also applicable, and thus the system can be used in large-scale synthesis.

A rare earth metal alkoxide/BINOL with an achiral Lewis base system was expanded to the catalytic asymmetric epoxidation of various electron-deficient

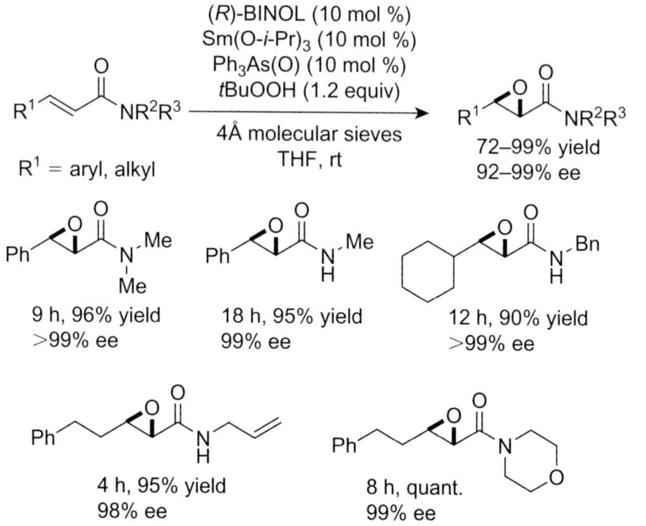

Scheme 8.3 Catalytic asymmetric epoxidation of α,β-unsaturated amides with Sm(O-iPr)$_3$/BINOL catalyst.

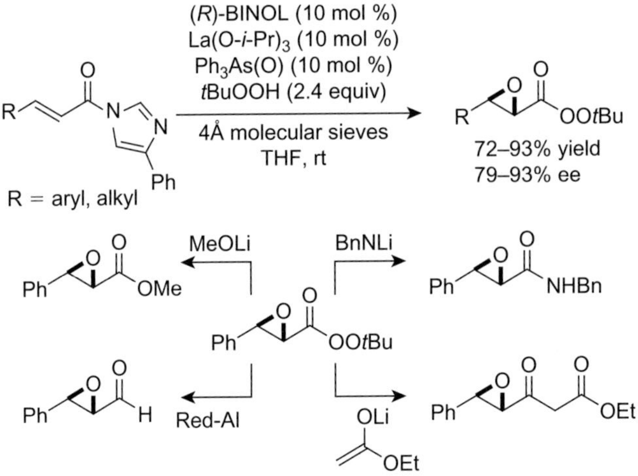

Scheme 8.4 Catalytic asymmetric epoxidation of α,β-unsaturated N-acylimidazoles with La(O-iPr)$_3$/BINOL catalyst.

alkenes, such as α,β-unsaturated amides (Scheme 8.3) [12], α,β-unsaturated N-acylimidazoles (Scheme 8.4) [13], α,β-unsaturated N-acylpyrroles (Scheme 8.5) [14], α,β-unsaturated anilides [15], α,β-unsaturated esters (Scheme 8.6) [16], and α,β-unsaturated phosphine oxides (Scheme 8.7) [17]. To achieve maximum catalytic activity and enantioselectivity, both rare earth metal alkoxides and BINOL derivatives were optimized, depending on the properties of substrates, as summarized in Schemes 8.2–8.7. Not only the rare earth metal sources, but also the dihedral

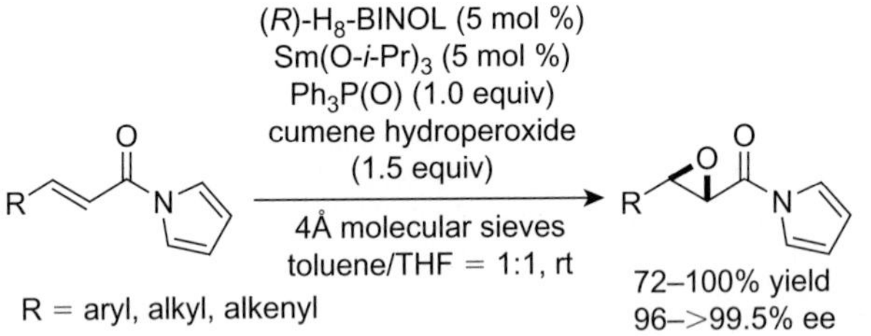

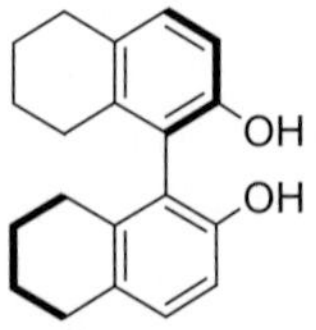

R	time	% yield	% ee
Ph	10 min	98	>99.5
cyclohexyl	10 min	90	>99.5
H$_2$C=CH(CH$_2$)$_8$–	10 min	95	99

Scheme 8.5 Catalytic asymmetric epoxidation of α,β-unsaturated N-acylpyrroles with La(O-iPr)$_3$/BINOL catalyst.

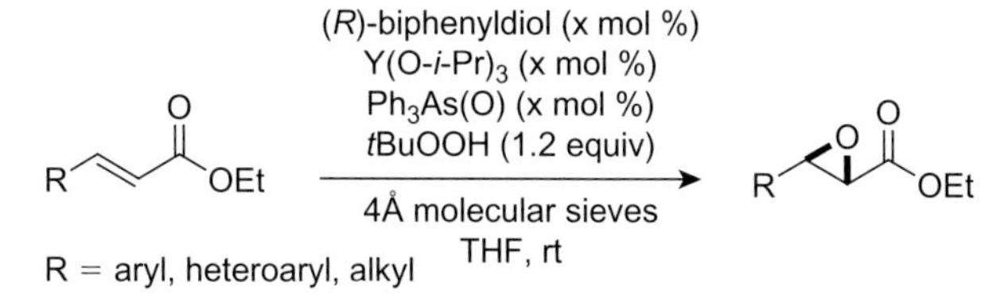

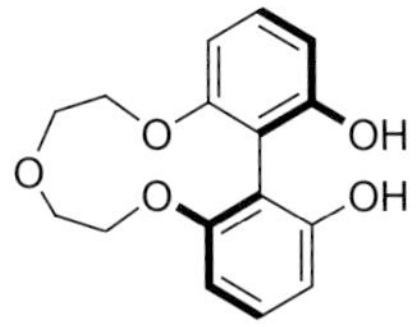

Scheme 8.6 Catalytic asymmetric epoxidation of α,β-unsaturated esters with Y(O-iPr)₃/biphenyldiol catalyst.

R	cat (x mol %)	time (h)	% yield	% ee
Ph	2	36	89	99
3-furyl	5	27	78	92
3-thienyl	3	24	93	98
3-pyridyl	3	24	97	93
PMBOCH₂CH₂	10	66	81	96

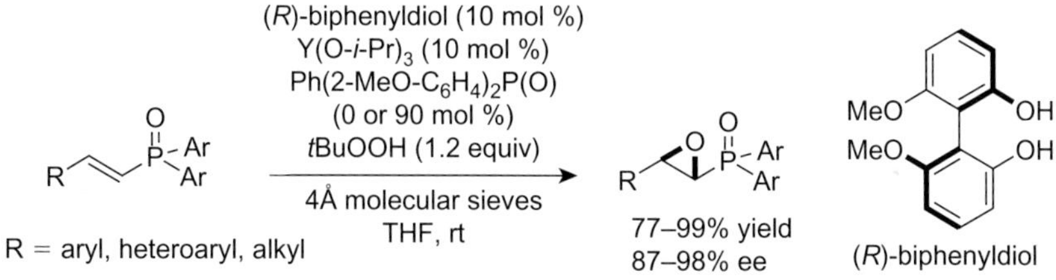

Scheme 8.7 Catalytic asymmetric epoxidation of α,β-unsaturated phosphine oxides with Y(O-iPr)₃/biphenyldiol catalyst.

angle of ligands was important in these reactions. For example, Sm(O-iPr)₃/H₈-BINOL/Ph₃P=O gave the best results for α,β-unsaturated N-acylpyrroles, while Y(O-iPr)₃/biphenyldiols gave the best reactivity and selectivity for α,β-unsaturated esters and α,β-unsaturated phosphine oxides. For other cases, BINOL gave the best results. In the epoxidation of α,β-unsaturated N-acylimidazoles, products were obtained as α,β-epoxy-peroxycarboxylic acid *t*-butyl ester, which was successfully converted into α,β-epoxy ester, aldehyde, amide, and keto esters (Scheme 8.4).

Bimetallic chiral oxovanadium(IV) complexes based on a BINOL scaffold are suitable for highly enantioselective oxidative coupling reactions of naphthols for the synthesis of BINOL derivatives (Scheme 8.8) [18, 19]. The reaction proceeded under O₂ atmosphere (1 atm), affording 7,7′-alkoxy-BINOL in up to 98% ee. Because modification at the 7,7′-positions starting from chiral BINOL is difficult, the present method is synthetically useful for providing chiral 7,7′-substituted BINOL derivatives.

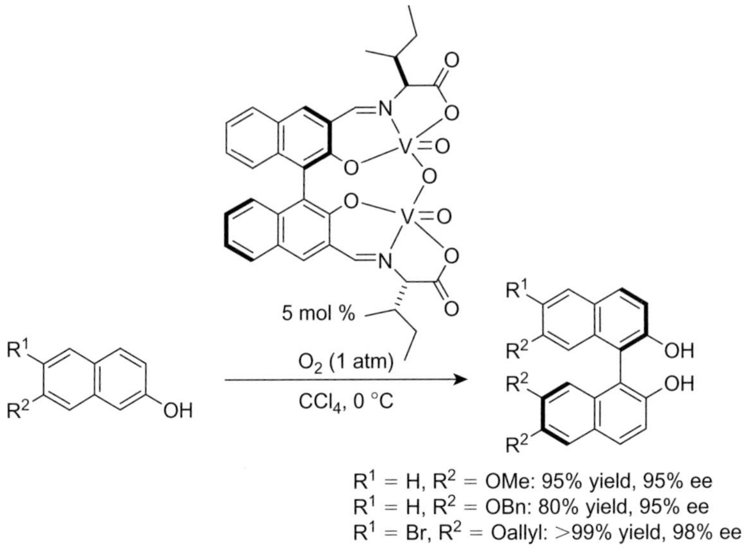

Scheme 8.8 Vanadium-catalyzed asymmetric oxidative coupling reaction of naphthols.

R^1 = H, R^2 = OMe: 95% yield, 95% ee
R^1 = H, R^2 = OBn: 80% yield, 95% ee
R^1 = Br, R^2 = Oallyl: >99% yield, 98% ee

8.3
Metal/BINOL Chiral Lewis Acid Catalysts in Asymmetric C–C Bond Forming Reactions

Since 1980s, the utility of BINOL derivatives as chiral ligands for Lewis acidic metals has been investigated intensively. The axially chiral backbone of BINOL is flexible enough to make complex with metals with various ionic radius. Chiral Lewis acid catalysts, such as B, Zn, Al, Ti, Zr, Sn, In, Yb, and many others, based on BINOL derivatives enabled a broad range of enantioselective C–C bond-forming reactions. This section introduces selected representative examples to show the utility of BINOL privileged ligands in Lewis acid catalysis. For comprehensive overviews on Lewis acidic BINOL catalysts, see other books and reviews [20].

8.3.1
Group IV Metal/BINOL Lewis Acid Catalysts

In early studies on asymmetric ene reactions [21], Al-BINOL and Zn-BINOL systems were utilized by Yamamoto and coworkers during 1980s. Drastic improvements in catalytic activity and stereoselectivity of ene reactions were achieved by Ti-BINOL catalysts. Since 1989, Mikami and Nakai have studied Ti-BINOL-catalyzed ene reactions in detail. A TiCl$_2$(O-iPr)$_2$/BINOL = 1 : 1 complex promoted asymmetric the ene reaction to glyoxylate esters. Excellent enantioselectivities of up to 98% ee were achieved using as little as 1 mol% catalyst (Scheme 8.9) [22]. A strong, positive, nonlinear effect was observed between optical purity of BINOL and product. Thus, the ene adduct was obtained in greater than

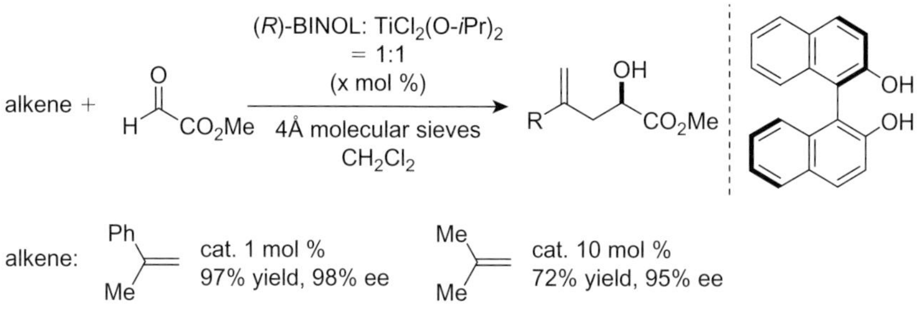

8.3 Metal/BINOL Chiral Lewis Acid Catalysts in Asymmetric C–C Bond Forming Reactions | 301

Scheme 8.9 Ti/BINOL-catalyzed asymmetric ene reaction.

90% ee even by using BINOL with only 33% ee. The utility of Ti/BINOL complexes was expanded to allylation [23], Mukaiyama aldol-type reaction through silatropic ene mechanism [24], Diels–Alder reaction [25], hetero-Diels–Alder reaction [26], 1,3-dipolar cycloaddition [27], organozinc addition to aldehydes and ketones [28], and others. Depending on the targeted reactions, molar ratio of Ti/BINOL, additives, and Ti-source were modified to achieve the best enantioselectivity and reactivity. Keck reported the importance of the preparation method of Ti(O-iPr)$_4$ + BINOL system in the asymmetric allylation of aldehydes (Scheme 8.10) [23] and the Mukaiyama aldol reaction. Mikami and coworkers also investigated in detail the effects of trace amounts of water provided by hydrated molecular sieves on the structure and catalytic activity of Ti-catalysts prepared from Ti(O-iPr)$_4$ and BINOL. Detailed studies on a Ti(O-iPr)$_2$(BINOLato) complex, including X-ray crystallographic analysis of assembled μ-oxo-Ti-species, were reported [29]. In terms of the catalytic activity of Ti-BINOL complexes, the introduction of electron-withdrawing groups at 6,6′-positions significantly increased the Lewis acidity of the Ti-catalyst. In a Diels–Alder reaction, a Ti/6,6′-Br-BINOL catalyst showed much greater catalyst activity than the Ti/BINOL catalyst (Scheme 8.11) [25]. On the basis of reports on assembled μ-oxo-Ti-species [29], Ding and coworkers

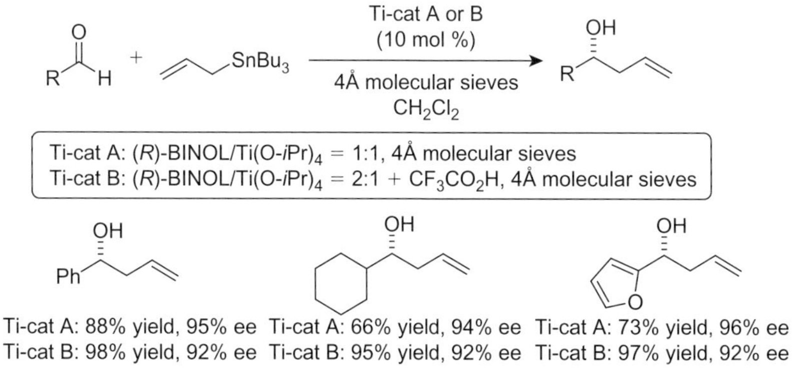

Scheme 8.10 Ti/BINOL-catalyzed asymmetric allylation.

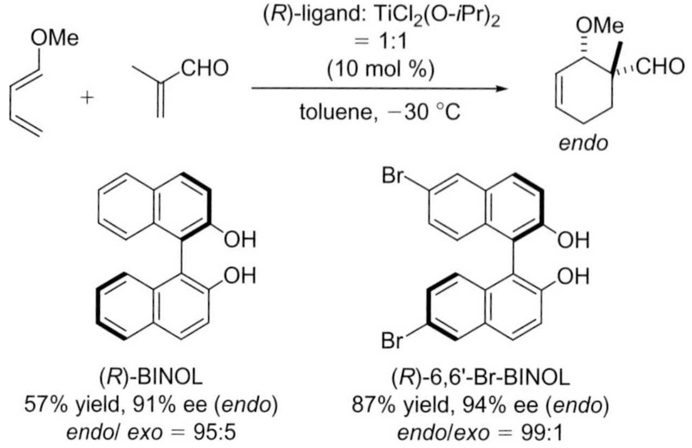

Scheme 8.11 Ti/BINOL-catalyzed asymmetric Diels–Alder reaction.

found an extremely reactive Ti-catalyst for the hetero-Diels–Alder reaction through high-throughput screening. A mixture of Ti(O-iPr)$_4$/H$_8$-BINOL/H$_4$-BINOL (1 : 1 : 1) and Ti(O-iPr)$_4$/H$_8$-BINOL (1 : 2) gave the best results, and the reaction of Danishefsky's diene with aldehydes proceeded with 0.1–0.005 mol% catalyst loading under solvent-free conditions, giving products in up to 99.8% ee (Scheme 8.12) [26]. Maruoka and coworkers reported the utility of μ-oxo-bis-Ti catalyst prepared from TiCl(O-iPr)$_3$, Ag$_2$O, and BINOL. The Ti-catalyst efficiently promoted enantioselective 1,3-dipolar cycloadditions of nitrones with acrolein (Scheme 8.13) as well as that of diazoacetate and α-substituted acrolein [27].

The utility of another group IV metal/BINOL complex, Zr/BINOL, has been intensively studied by Kobayashi and coworkers [30–35]. In 1997, Kobayashi reported the first catalytic asymmetric Mukaiyama Mannich-type reaction. A Zr(O-tBu)$_4$/BINOL (1 : 2) catalyst promoted the desired Mannich-type reaction of silyl enolates, but yield and enantioselectivity were unsatisfactory. The introduction of

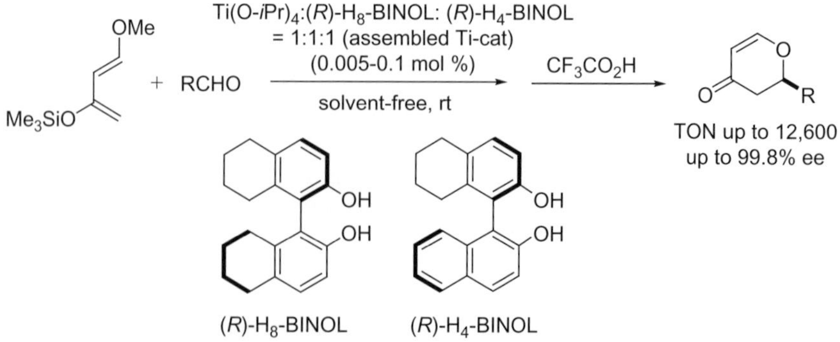

Scheme 8.12 Titanium-catalyzed asymmetric hetero-Diels–Alder reaction using a mixed-ligand system.

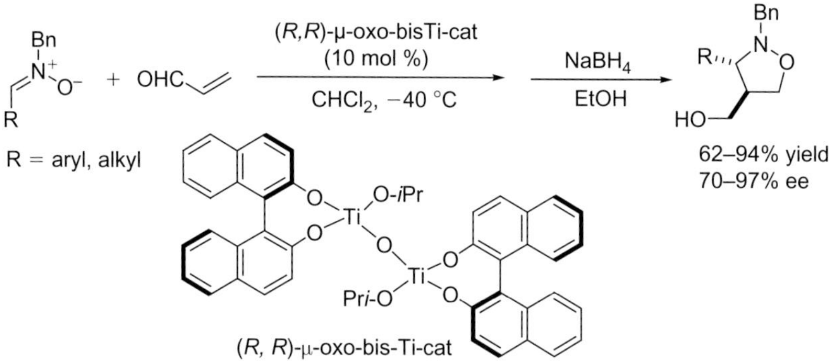

Scheme 8.13 μ-Oxo-bis-Ti catalyst for asymmetric 1,3-dipolar cycloadditions of nitrones.

electron-withdrawing groups at the 6,6′-positions of BINOL was important to increase the Lewis acidity of the Zr metal center, and to achieve high catalytic activity as well as enantioselectivity. A Zr(O-*t*Bu)$_4$/6,6′-Br-BINOL (1 : 2) catalyst (Zr-cat A) gave products in good to excellent enantioselectivity (Scheme 8.14) [30]. Catalyst turnover number was further improved by using 6,6′-CF$_3$-BINOL as a ligand (Zr-cat B). Kobayashi and coworkers systematically studied the effects of substituents at the 3,3′- and 6,6′-positions to modify the steric and electronic properties of Zr-catalysts, and reported a series of Zr-catalysts for Mannich-type reactions for the synthesis of β-amino alcohols [31], three-component Strecker reaction [32], *anti*-selective Mukaiyama aldol reaction [33], allylation of imines [34],

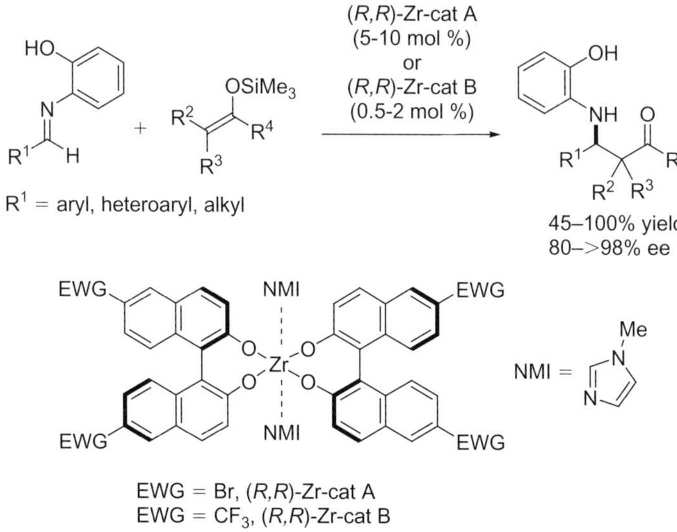

Scheme 8.14 Zirconium-catalyzed asymmetric Mannich-type reaction.

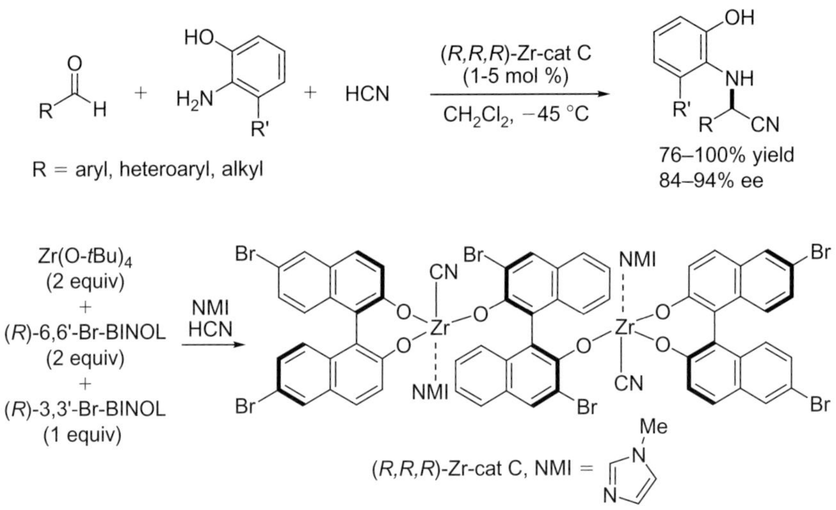

Scheme 8.15 Zirconium-catalyzed asymmetric three-component Mannich reaction.

hetero-Diels–Alder reaction [35], and others. In a Strecker-type reaction, a mixed-ligand system, prepared from Zr(O-tBu)$_4$/6,6′-Br-BINOL/3,3′-Br-BINOL (2 : 2 : 1) (Zr-cat C), proved to be the best, giving Strecker products in high yield and enantioselectivity using HCN as a cyanide source (Scheme 8.15) [32]. For the anti-selective Mukaiyama aldol reaction and allylation of imines, a sterically hindered Zr-catalyst having substituents at the 3,3′-positions gave high selectivity. In the aldol reaction, 3,3′,6,6′-I-BINOL gave the best catalytic activity and enantioselectivity. The positive effects of alcohol additive was also reported for the formation of dimeric Zr-catalyst (Zr-cat D, Scheme 8.16) [33].

8.3.2
Group XIII Metal/BINOL Lewis Acid Catalysts

The utility of group XIII metal/BINOL Lewis acid catalysts has been studied intensively by Yamamoto and coworkers since the 1980s. The importance of steric factors around the Lewis acid metal center for high stereoselectivity was recognized in early studies on Lewis acid catalysis. Sterically hindered Al-6,6′-Ar$_3$Si-BINOL catalyst (Ar = Ph or 3,5-xylyl) was reported by Yamamoto as the first catalytic asymmetric hetero-Diels–Alder reaction (Scheme 8.17), the first intermolecular ene-reaction, and others [36]. Notable catalytic activity and stereoselectivity in group XIII metal/BINOL systems was achieved by the same group in the chemistry of Brønsted acid-assisted chiral Lewis acid (BLA) catalysis [37]. By introducing phenolic substituents at the 3,3′-position, the Lewis acidic metal center is assumed to interact with the neighboring Brønsted acid moiety. Thus, the ability of chiral recognition around the Lewis acidic metal center in the transition state is enhanced through the combination of intramolecular hydrogen-bonding

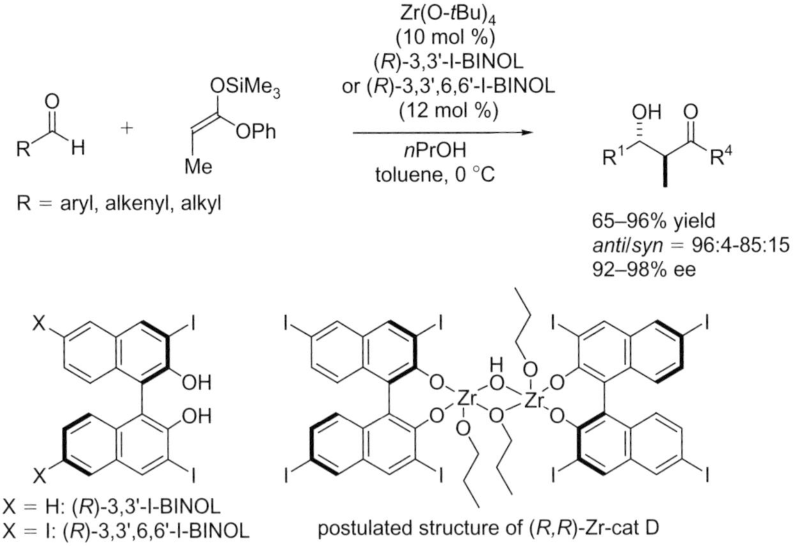

8.3 Metal/BINOL Chiral Lewis Acid Catalysts in Asymmetric C–C Bond Forming Reactions

Scheme 8.16 Zirconium-catalyzed asymmetric *anti*-selective Mukaiyama aldol reaction.

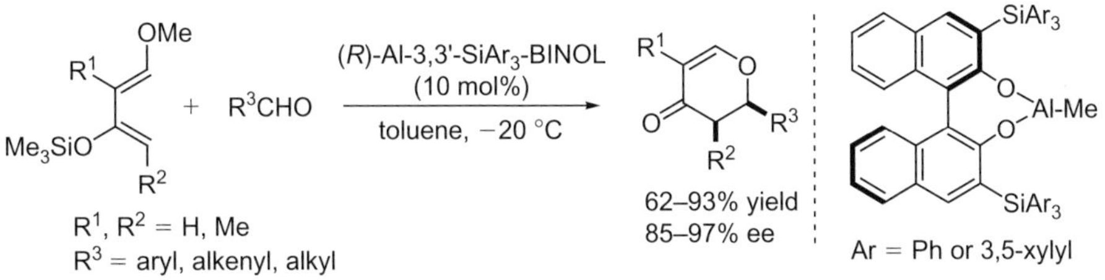

Scheme 8.17 Al/3,3′-SiAr$_3$-BINOL-catalyzed asymmetric hetero-Diels–Alder reaction.

and attractive π–π donor–acceptor interaction. A catalyst prepared from B(OMe)$_3$ and 3,3′-substituted ligand showed excellent yield, *exo/endo* selectivity, and enantioselectivity up to >99% ee in Diels–Alder reactions (BLA-cat A, Scheme 8.18) [38]. Interestingly, BLA-cat B in Scheme 8.18 resulted in complete enantio-switching. Both *(R)*-BLA-cat A and *(R)*-BLA-cat B have a *(R)*-BINOL unit. However, the absolute stereo-preference in the Diels–Alder reaction was opposite, suggesting that the 3,5-bis(trifluoromethyl)phenyl group greatly affects the asymmetric induction of BLAs. The concept of Brønsted acid-assisted chiral Lewis acid catalysis was further expanded by fine tuning the structures of phenolic substituents at the 3,3′-positions as well as by changing metal sources, depending on the targeted reaction. In similar fashion, Lewis acid-assisted Brønsted acid catalysis was also developed using Sn and BINOL [37].

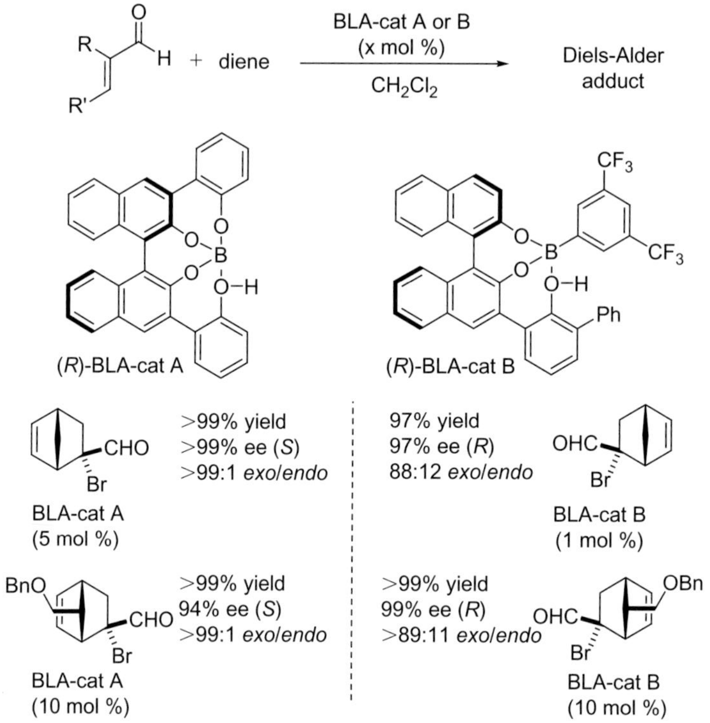

Scheme 8.18 Brønsted acid-assisted chiral Lewis acid catalysis of a highly enantioselective Diels–Alder reaction.

The utility of In-BINOL chiral Lewis acid to induce high enantioselectivity was reported by several groups. Loh and coworkers reported a novel In-BINOL catalyst prepared from InCl$_3$, BINOL, and allyltributylstannane for asymmetric Diels–Alder reaction. Although high catalyst loading (20 mol%) was required, excellent enantioselectivity up to 98% ee was achieved (Scheme 8.19) [39]. Shibasaki

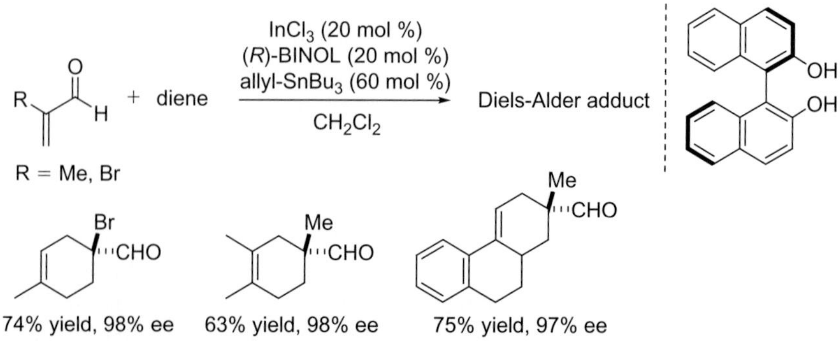

Scheme 8.19 In/BINOL-catalyzed asymmetric Diels–Alder reaction.

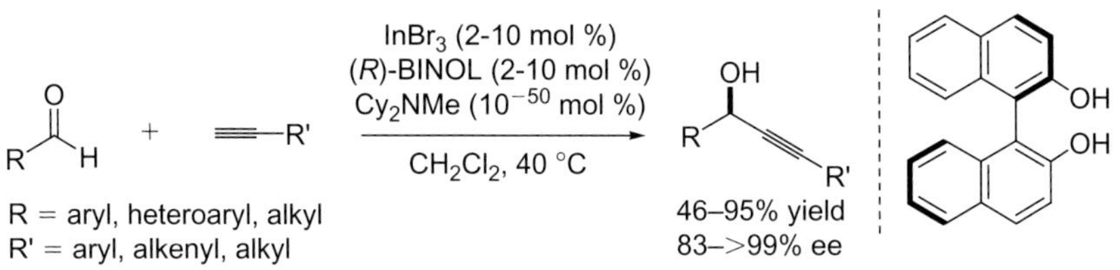

Scheme 8.20 In/BINOL-catalyzed direct asymmetric addition of terminal alkynes to aldehydes.

reported catalytic asymmetric direct alkynylation of aldehydes with a InBr$_3$/BINOL/amine base system. The system is synthetically useful, because the reaction proceeded directly from terminal alkynes. A broad range of aryl, heteroaryl, and alkyl aldehydes were applicable and high enantioselectivity was achieved (Scheme 8.20) [40].

8.3.3
Rare Earth Metal/BINOL Lewis Acid Catalysts

As a part of studies on rare earth metal triflates as Lewis acid catalysts [41], Kobayashi and coworkers utilized BINOLs as chiral ligands. A Yb(OTf)$_3$/BINOL catalyst was prepared in the presence of amines for asymmetric Diels–Alder reaction, aza-Diels–Alder reaction, and 1,3-dipolar reaction [42]. For high enantioselectivity, the addition of amine base was essential to form chiral Yb/BINOL/amine ternary complexes. In the Yb-catalyzed 1,3-dipolar cycloadditions, chiral amine was used as an additive to enhance the stereoselectivity of Yb/BINOL catalyst (Scheme 8.21) [42]. For rare earth metal triflate-catalyzed reactions, however, other chiral privileged ligands, such as pybox derivatives, often afford much superior enantioselectivity [43].

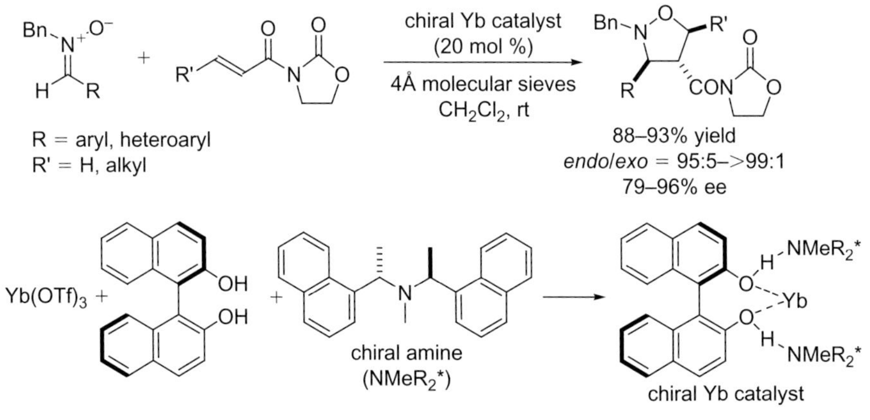

Scheme 8.21 Yb(OTf)$_3$/BINOL-catalyzed 1,3-dipolar cycloadditions.

8.4
Acid/Base Bifunctional Metal/BINOL Catalysts

8.4.1
Rare Earth Metal/Alkali Metal/BINOL Catalysts

Section 8.3 described the utility of metal/BINOL chiral Lewis acid catalysts. Since the 1990s, the utility of BINOL has been further expanded as a ligand for bimetallic acid/base bifunctional catalysts [44]. In the 1990s Shibasaki and coworkers reported a series of heterobimetallic rare earth metal/alkali metal/BINOL catalysts (abbreviated as REMB, RE: rare earth, M: alkali metal, B: BINOL) (Figure 8.2) [45]. LLB catalyst (RE = La, M = Li), a representative REMB catalyst, has been prepared from La(O-iPr)$_3$, BuLi, and BINOL in a ratio of 1 : 3 : 3. The structures of REMB complexes were determined by X-ray crystallographic analysis, elemental analysis, and mass spectroscopic analysis. A series of REMB catalysts, such as LLB, LSB, LPB, SmSB, PrPB, and YbPB, were prepared from various rare earth metal and alkali metals. Two different metal centers in REMB complexes work cooperatively as Lewis acid and Brønsted base to promote various reactions, such as nitro aldol reaction [46], direct aldol reaction [47], Michael reaction of β-keto esters and nitromethane [48], 1,4-addition of thiols [49], hydrophosphonylation of imines and aldehydes [50], cyanation of aldehydes [51], *tert*-nitroaldol resolution [52], and nitro-Mannich reaction (Figure 8.3) [53]. In the proposed reaction mechanism, the rare earth metal center is speculated to work as a Lewis acid to activate electrophiles, while the alkali metal binaphthoxide moiety functions as a Brønsted base to generate nucleophile *in situ*. Mechanistic studies by Walsh and coworkers on solid state and solution state structures of REMB catalysts revealed that the La metal center in the LLB catalyst can indeed accept seventh and eighth substrates [54], supporting the Shibasaki's proposal for La as a Lewis acid. Depending on the targeted reactions, the best chiral environment was flexibly

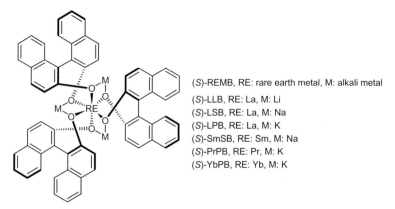

(S)-REMB, RE: rare earth metal, M: alkali metal
(S)-LLB, RE: La, M: Li
(S)-LSB, RE: La, M: Na
(S)-LPB, RE: La, M: K
(S)-SmSB, RE: Sm, M: Na
(S)-PrPB, RE: Pr, M: K
(S)-YbPB, RE: Yb, M: K

Figure 8.2 Structures of REMB complexes.

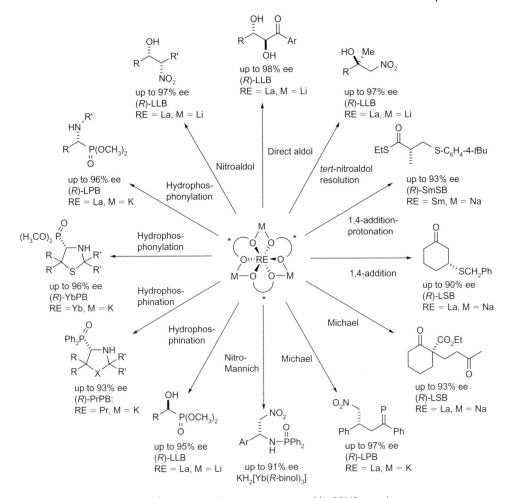

Figure 8.3 Representative catalytic asymmetric reactions promoted by REMB complexes.

constructed by suitably selecting metal combinations. In contrast to other rare earth metal/alkali metal/BINOL (1 : 3 : 3) catalysts, Sc/Li/BINOL complex in a ratio of 1 : 1 : 2 gave the best results in the Strecker reaction of ketoimines [55]. Among various applications, the most prominent achievement with REMB catalysts is the development of intermolecular direct catalytic asymmetric aldol reactions. By utilizing the Lewis acid/Brønsted base bifunctional property of a LLB catalyst, aldol adducts were obtained directly from ketones and aldehydes via a simple proton transfer process (Scheme 8.22). Thus, the process can be superior to the Mukaiyama-type reaction in terms of atom economy and step economy. In the direct aldol reaction, the addition of a catalytic amount of achiral base (KOH) had beneficial effects on accelerating the reaction, while keeping good enantioselectivity [47]. The process was further expanded to direct aldol/Tishchenko sequential

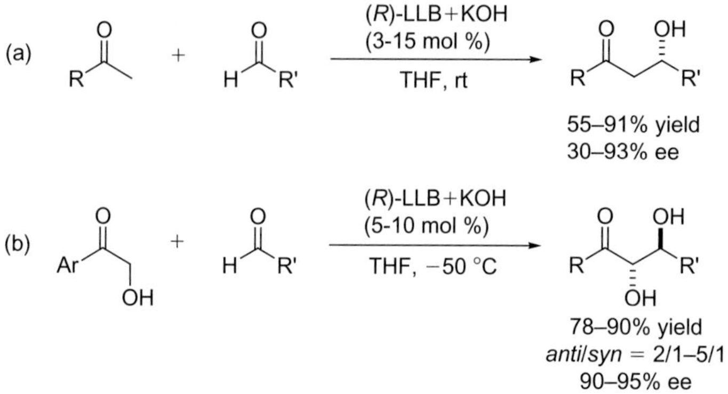

Scheme 8.22 LLB-catalyzed intermolecular asymmetric direct aldol reactions.

reactions, in which a LLB-type catalyst prepared from La(OTf)$_3$/BuLi/BINOL (1 : 5.6 : 3) was utilized to generate the LLB as a combined system with LiOTf *in situ* (Scheme 8.23) [56].

More recently, the utility of REMB catalysts was further expanded by using them as Lewis acid/Lewis acid cooperative catalysis. YLB (RE = Y, M = Li) promoted the catalytic asymmetric 1,4-addition of α-alkoxylamines to enones and α,β-unsaturated N-acylpyrroles. Dual control of both nucleophile (α-alkoxylamines) and electrophile (enone) with two Lewis acidic metal centers, Y and Li, is proposed to realize high enantioselectivity (Scheme 8.24) [57]. A similar Lewis acid/Lewis acid cooperative mechanism of REMB catalyst was utilized for the catalytic asymmetric Corey–Chaykovsky cyclopropanation of enones and α,β-unsaturated N-acylpyrroles (Scheme 8.25) [58], Corey–Chaykovsky epoxidation of ketones for the synthesis of 2,2′-disubstituted terminal epoxides (Scheme 8.26) [59], and ring-expansion of the epoxides for the synthesis of 2,2′-disubstituted oxetanes (Scheme 8.27) [60]. In those reactions, the La metal center is speculated to activate

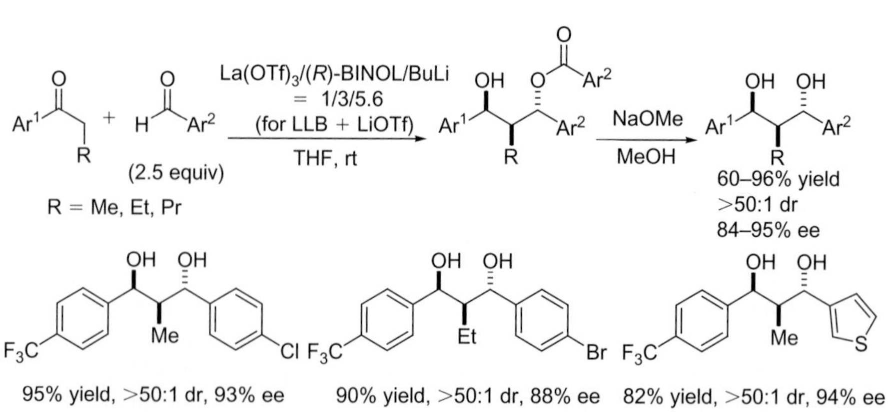

Scheme 8.23 LLB-catalyzed asymmetric direct aldol/Tishchenko sequential reactions.

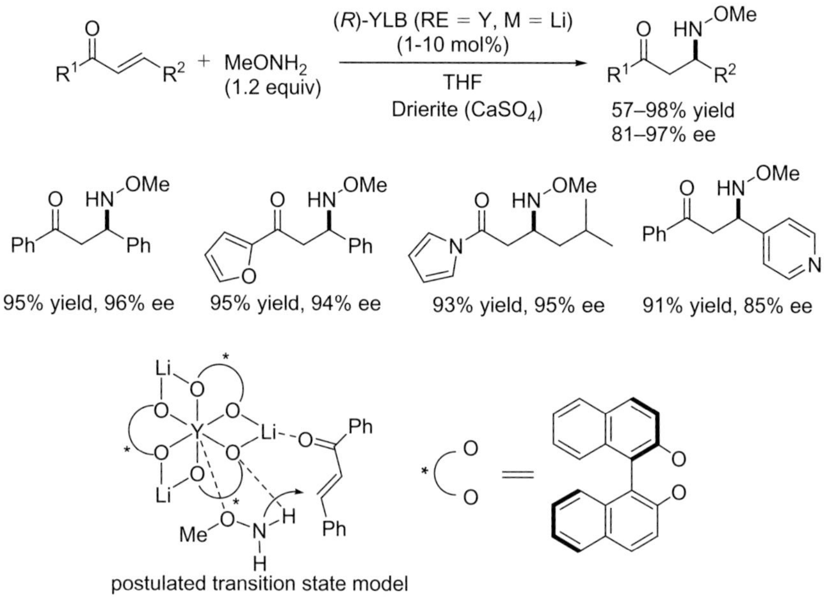

8.4 Acid/Base Bifunctional Metal/BINOL Catalysts | 311

Scheme 8.24 YLB-catalyzed asymmetric 1,4-addition of α-alkoxylamines.

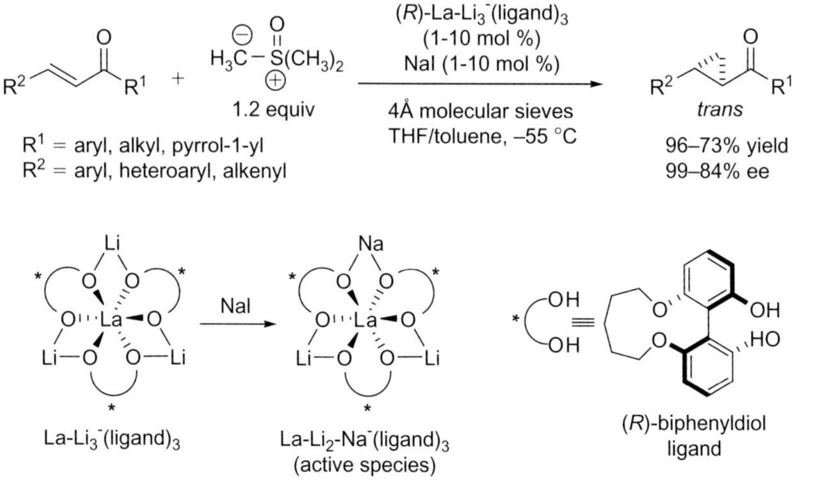

Scheme 8.25 Catalytic asymmetric Corey–Chaykovsky cyclopropanation.

electrophiles as a Lewis acid, while the Li metal center is assumed to interact with ylide and control its orientation. For the Corey–Chaykovsky cyclopropanation, a biphenyldiol ligand gave better selectivity than BINOL. The addition of NaI was also important to realize the best enantioselectivity. NaI is required to exchange one of the Li metals in an LLB-type complex with Na to construct the optimum

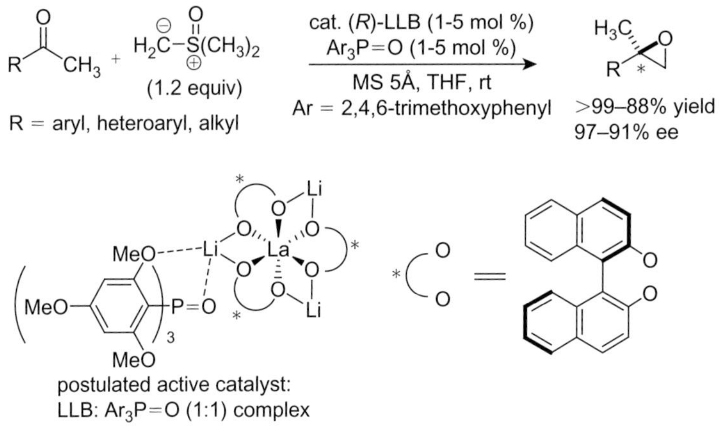

Scheme 8.26 Catalytic asymmetric Corey–Chaykovsky epoxidation of ketones.

chiral environment for cyclopropanation [58]. For the Corey–Chaykovsky epoxidation of ketones, achiral phosphine oxide played a key role in achieving excellent enantioselectivity. NMR studies suggested that LLB + phosphine oxide (1 : 1) would be the active species [59]. In the one-pot sequential process for oxetanes, chiral amplification was observed to afford 2,2′-disubstituted oxetanes in 99–>99.5% ee, while intermediate epoxides were obtained in 93–97% ee. Mechanistic studies indicated that the LLB catalyst was suitable for kinetic resolution of 2,2′-disubstituted terminal epoxides to cause chiral amplification in the sequential process [60].

8.4.2
Group XIII Metal/Alkali Metal/BINOL Catalysts

The concept of acid/base bifunctional catalysis based on BINOL scaffold is not restricted to rare earth metal complexes. Shibasaki and coworkers expanded the concept to group XIII metal/alkali metal/BINOL catalysts. Al/Li/(BINOL)$_2$ catalyst (abbreviated as ALB, A: Al, L: Li, B: BINOL) was prepared from LiAlH$_4$ and BINOL

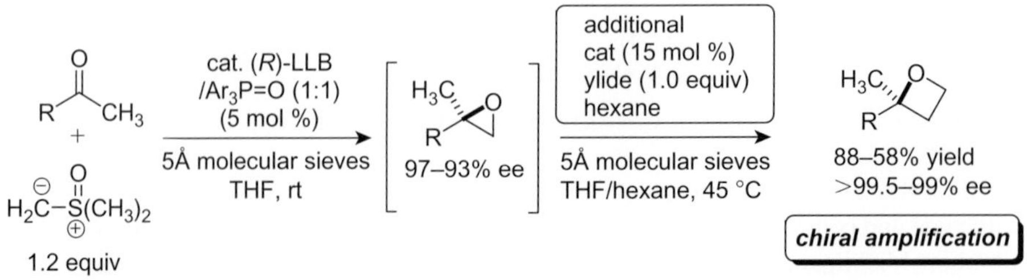

Scheme 8.27 One-pot sequential epoxidation/ring expansion process for the synthesis of chiral oxetanes.

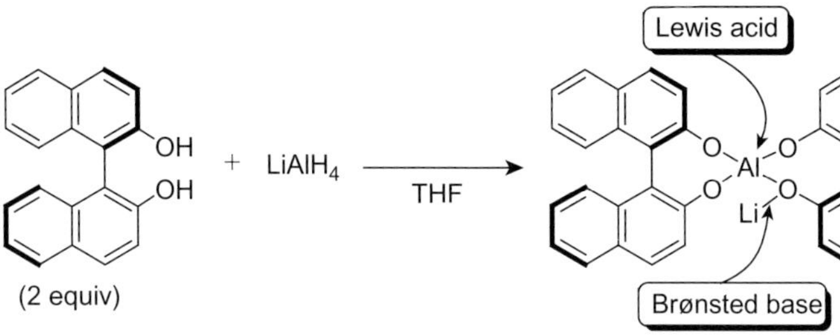

Scheme 8.28 Preparation of ALB catalyst from LiAlH₄ and BINOL.

in 1 : 2 ratio (Scheme 8.28) [61]. The ALB catalyst also works as a Lewis acid/ Brønsted base bifunctional catalyst, and showed excellent enantioselectivity for catalytic asymmetric Michael reaction of malonates to cyclic enones (Scheme 8.29a). Unfortunately, however, the reaction rate of ALB alone was unsatisfactory, possibly due to the modest Brønsted basicity of ALB. The reactivity was drastically improved by performing the reaction in the presence of a catalytic amount of achiral base (KO-*t*Bu) to accelerate the rate-limiting deprotonation step [62]. Under the optimized reaction conditions, the catalyst loading was successfully reduced to as little as 0.1 mol%, and the reaction was completed within 24 h (Scheme 8.29b). The reaction of cyclohexanone and dimethyl malonate was also performed in greater than kilogram scale using 0.1 mol% ALB under highly concentrated conditions at room temperature. The product was obtained in 99% ee and 91% yield without column chromatography (Figure 8.4) [63]. For the ALB-catalyzed Michael reaction, a similar Lewis acid/Brønsted base bifunctional mechanism under cooperative function of Al and Li metals was proposed.

ALB was also suitable for the three-component Michael–aldol reaction. In the Michael–aldol reaction, NaO-*t*-Bu as achiral base gave the best results, giving product in 84% yield and 92% ee. Scheme 8.30 shows a plausible reaction

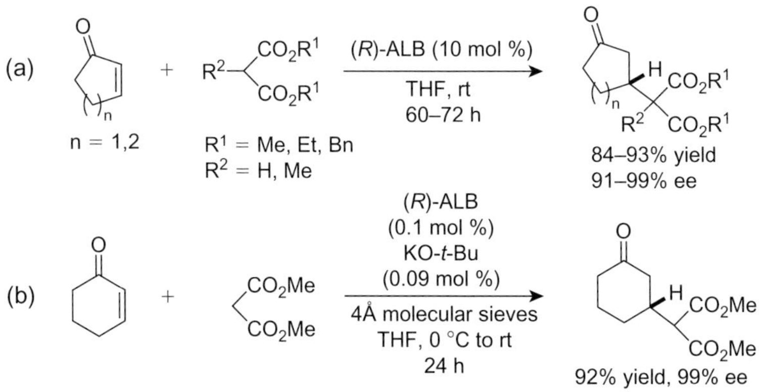

Scheme 8.29 ALB-catalyzed asymmetric Michael reactions of malonates.

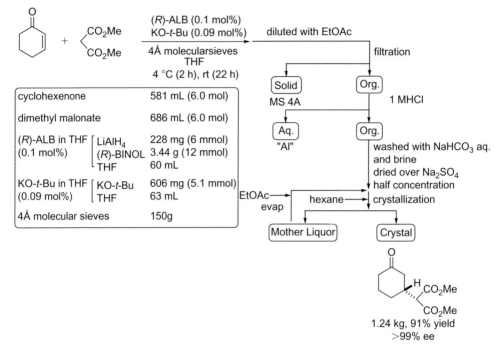

Figure 8.4 Catalytic asymmetric Michael reaction promoted by ALB on a kilogram scale.

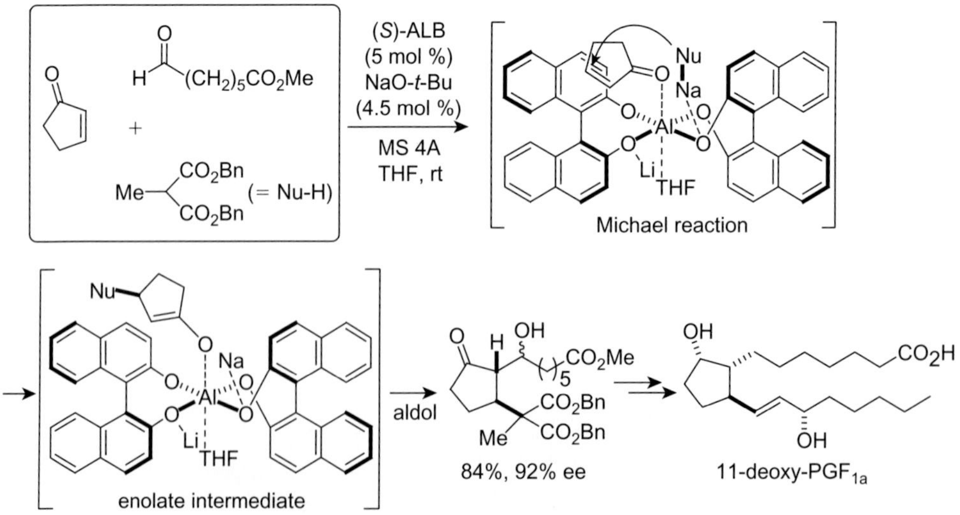

Scheme 8.30 Catalytic asymmetric three-component Michael–aldol reaction.

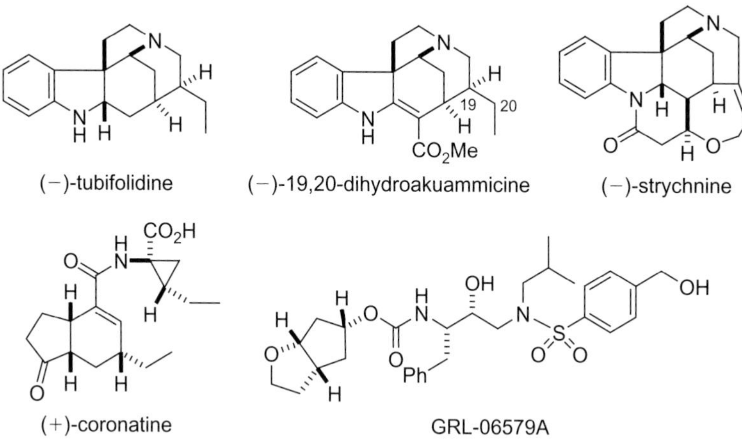

Figure 8.5 Selected examples of biologically active compounds synthesized using ALB.

mechanism of the Michael–aldol reaction as well as its application to the synthesis of 11-deoxy-PGF$_{1\alpha}$ [64]. ALB-catalyzed Michael reactions were also applied for catalytic asymmetric synthesis of various biologically active compounds, such as (−)-tubifolidine, (−)-19,20-dihydroakuammicine, (−)-strychnine, (+)-coronatine, HIV protease inhibitor GRL-06579A, and others (Figure 8.5) [62, 65]. The utility of ALB catalyst was further expanded to catalytic asymmetric 1,4-addition of Horner–Wadsworth–Emmons reagents, Michael reaction of nitroacetate, catalyst-controlled Michael reaction to racemic β-TBS-oxy-cyclopentenone, direct Mannich reaction, nitro-Mannich-type reaction, hydrophosphinylation of aldehydes, and hydrophosphonylation of aldehydes [66]. In addition to ALB, related group XIII metal/alkali metal catalysts such as Ga/Li/(BINOL)$_2$ [67], Ga/Na/(BINOL)$_2$ [68], Ga/Na/(H$_8$-BINOL)$_2$ [69], and In/K/(BINOL)$_2$ [68] complexes (Figure 8.6) were reported as Lewis acid/Brønsted base bifunctional catalysts. These catalysts were

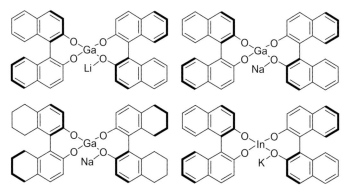

Figure 8.6 Structures of other group XIII metal/alkali metal/BINOL bifunctional catalysts.

applied for catalytic asymmetric Michael reaction of malonates and ring-opening desymmetrization of *meso*-epoxides with thiols and phenols.

8.4.3
Other Metal/BINOL Complexes as Acid/Base Bifunctional Catalysts

Of the several BINOL catalysts described in previous sections, two or three BINOLs were often incorporated in the active catalysts. To facilitate the incorporation of two BINOL units and stabilize the actives species, linked-BINOLs (Figure 8.7) were independently developed by Shibasaki and Kobayashi [70]. By linking two BINOL units through covalent bond at the 3-position, the linked-BINOLs can effectively work as a tetradentate ligand to form stable metal complexes. In addition, the introduction of heteroatom in the linker part further expanded the utility of linked-BINOLs. A La/linked-BINOL complex prepared from La(O-*i*Pr)$_3$ and O-linked-BINOL [71] in a ratio of 1 : 1 showed much superior stability than the parent La-BINOL complex, and was applied for the catalytic asymmetric Michael reaction of malonates (Scheme 8.31) [72]. For the Michael reaction with β-keto esters, the use of NMe-linked-BINOL instead of O-linked-BINOL was important to achieve high enantioselectivity, suggesting that the heteroatom in the linker affected the structure of La-linked-BINOL complexes (Scheme 8.32) [73]. A La-linked-BINOL complex was also effective in the catalytic asymmetric aza-Morita–Baylis–Hillman reaction of methyl acrylate. The utility of linked-BINOLs was further expanded as a trinuclear Zn complex [74–76]. As

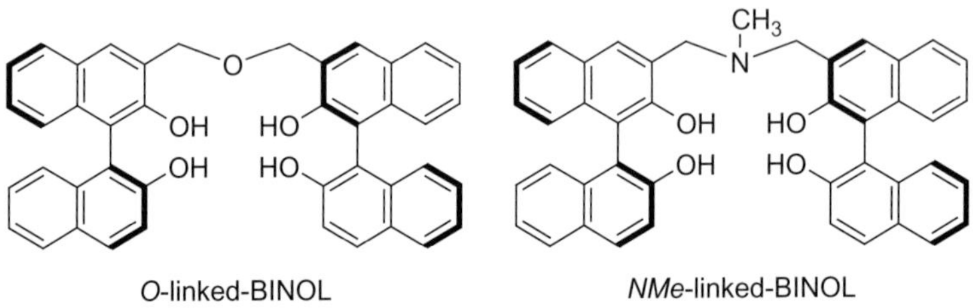

Figure 8.7 Structures of linked-BINOLs.

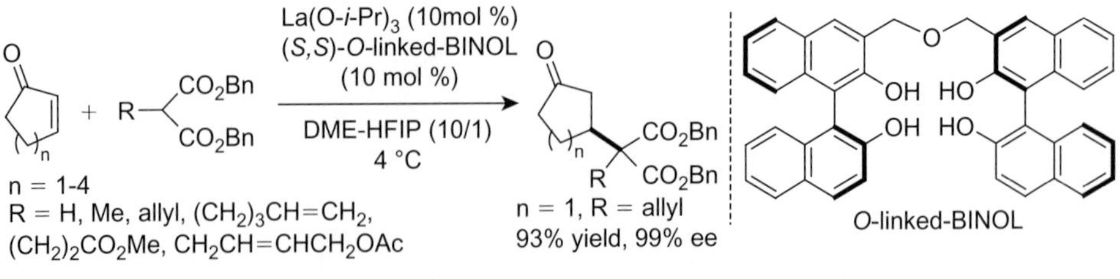

Scheme 8.31 La(O-*i*Pr)$_3$/O-linked-BINOL-catalyzed asymmetric Michael reaction of malonates.

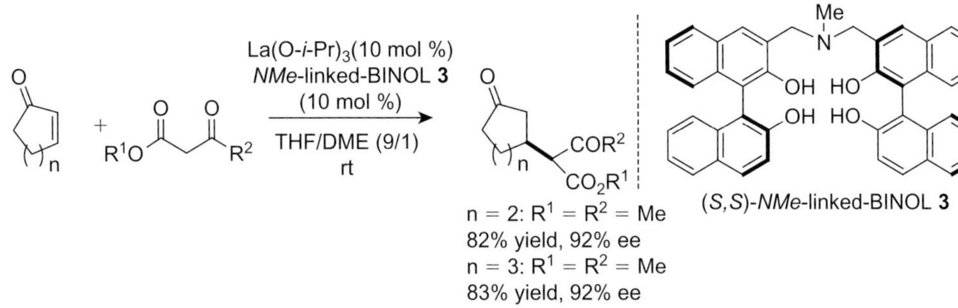

Scheme 8.32 La(O-iPr)$_3$/NMe-linked-BINOL-catalyzed asymmetric Michael reaction of β-keto esters.

confirmed by X-ray crystallographic analysis, coordination of the heteroatom in the linker part of O-linked-BINOL to Zn was important for the formation of the trinuclear Zn$_3$(linked-BINOL)$_2$ complex from Et$_2$Zn and linked-BINOL (Scheme 8.33). The trinuclear Zn catalyst was applied for catalytic asymmetric aldol (Scheme 8.34) [74], Michael (Scheme 8.35) [75], and Mannich-type reaction

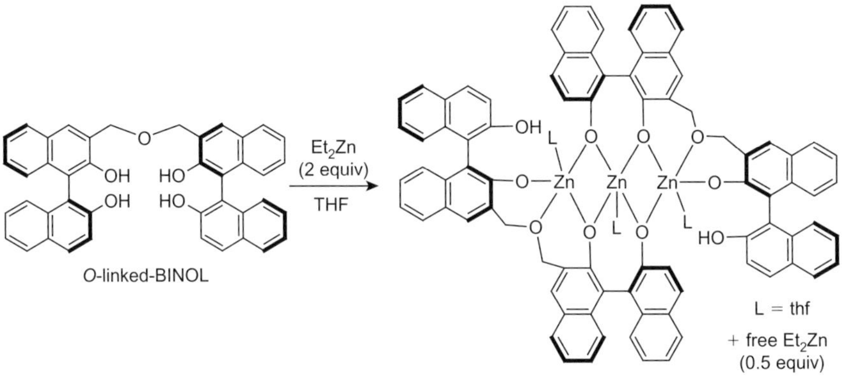

Scheme 8.33 Preparation of trinuclear Zn complex from Et$_2$Zn and O-linked-BINOL.

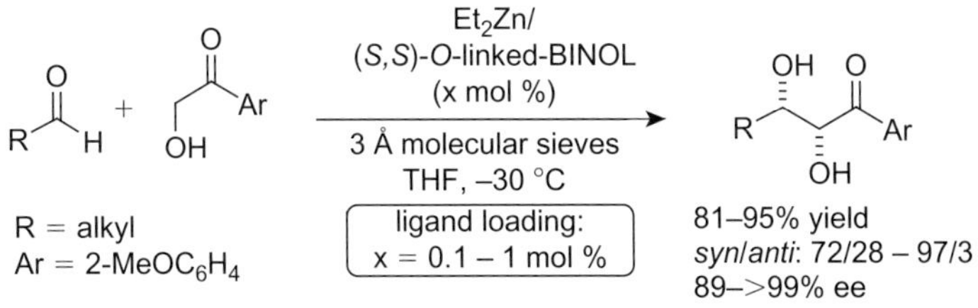

Scheme 8.34 Et$_2$Zn/O-linked-BINOL-catalyzed asymmetric direct aldol reaction.

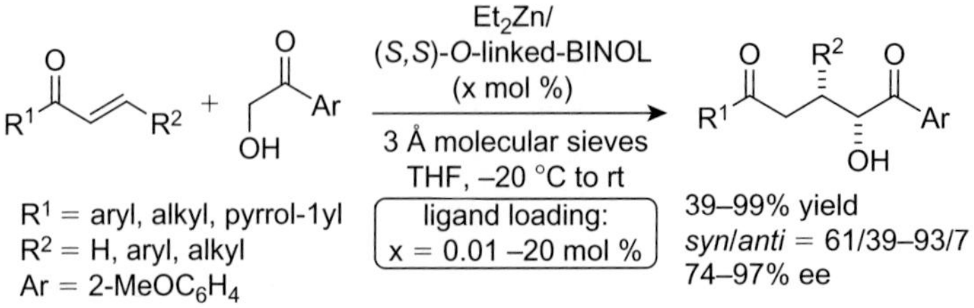
Et$_2$Zn/
(S,S)-O-linked-BINOL
(x mol %)
3 Å molecular sieves
THF, –20 °C to rt
ligand loading:
x = 0.01 –20 mol %
R^1 = aryl, alkyl, pyrrol-1yl
R^2 = H, aryl, alkyl
Ar = 2-MeOC$_6$H$_4$
39–99% yield
syn/anti = 61/39–93/7
74–97% ee

Scheme 8.35 Et$_2$Zn/O-linked-BINOL-catalyzed asymmetric direct Michael reaction.

of hydroxyketones (Schemes 8.36 and 8.37) [76]. High catalytic activity was achieved with 0.1–0.01 mol% catalyst loading in all cases. In all reactions in Schemes 8.34–8.37, the absolute stereochemistry at the α-hydroxy group in the products was the same, because enantiofacial selectivity of the Zn-enolate intermediate is excellent. A cooperative mechanism of the multi-Zn center was proposed for the generation of Zn-enolate *in situ* from hydroxyketones. Linked-BINOLs were also utilized as group XIII metal complexes, such as Ga [71] and In [77].

BINOL is also utilized for alkaline earth-metal and alkali metal-catalyzed reactions. Alkaline earth metal/BINOL catalysts, such as a Ba(O-*i*Pr)$_2$/BINOL, generally

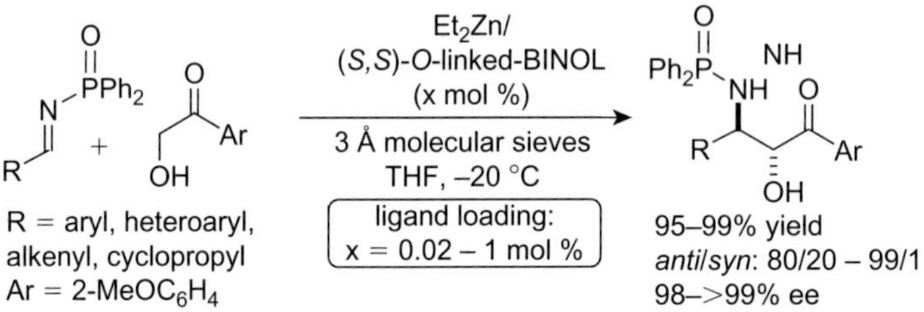
Et$_2$Zn/
(S,S)-O-linked-BINOL
(x mol %)
3 Å molecular sieves
THF, –20 °C
ligand loading:
x = 0.02 – 1 mol %
R = aryl, heteroaryl,
alkenyl, cyclopropyl
Ar = 2-MeOC$_6$H$_4$
95–99% yield
anti/syn: 80/20 – 99/1
98–>99% ee

Scheme 8.36 Et$_2$Zn/O-linked-BINOL-catalyzed asymmetric *anti*-selective direct Mannich-type reaction.

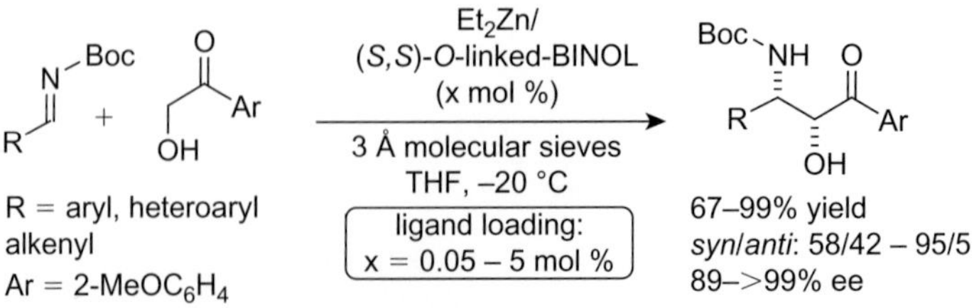
Et$_2$Zn/
(S,S)-O-linked-BINOL
(x mol %)
3 Å molecular sieves
THF, –20 °C
ligand loading:
x = 0.05 – 5 mol %
R = aryl, heteroaryl
alkenyl
Ar = 2-MeOC$_6$H$_4$
67–99% yield
syn/anti: 58/42 – 95/5
89–>99% ee

Scheme 8.37 Et$_2$Zn/O-linked-BINOL-catalyzed asymmetric *syn*-selective direct Mannich-type reaction.

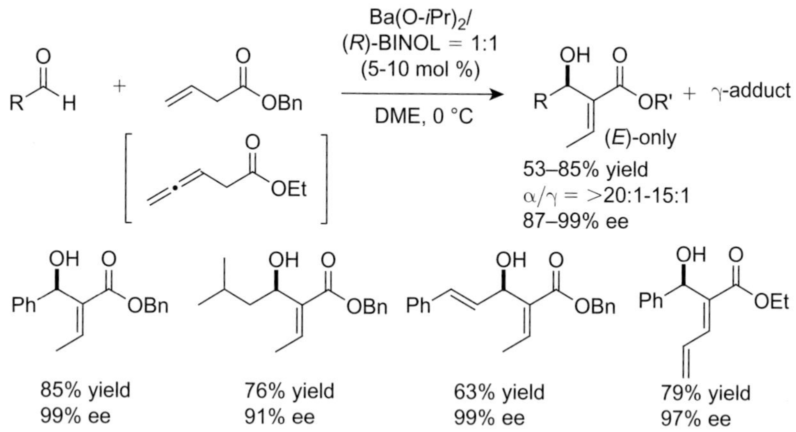

Scheme 8.38 Ba/BINOL-catalyzed direct aldol reaction/isomerization sequence.

have higher Brønsted basicity than rare earth metal/BINOL, group XIII metal/BINOL, and Zn/BINOL complexes. Because the acidity of α-proton of ester donors is much lower than that of ketones and aldehydes, the use of ester donors for *in situ catalytic* generation of metal enolates was rather difficult. For such a purpose, catalysts with high Brønsted basicity are required to catalytically generate a metal enolate *in situ*. The Ba(O-*i*Pr)$_2$/BINOL (1 : 1) catalyst was suitable for the nucleophilic activation of β,γ-unsaturated ester donors. The Ba-catalyst promoted a direct aldol reaction/isomerization sequence to afford α-alkylidene-β-hydroxy esters in 87–99% ee (Scheme 8.38) [78]. The Ba-catalyst was also applicable to β-allenyl ester, giving the product in 97% ee. Mechanistic studies indicated that the first direct aldol reaction gave aldol adducts in low diastereo- and enantioselectivity. The Ba-catalyst selectively isomerized one of enantiomers to afford α-alkylidene-β-hydroxy ester adduct in excellent enantioselectivity (up to 99% ee). Because the Ba-catalyst rapidly promoted the retro-aldol reaction, four isomers exist in rapid equilibrium. Thus, a dynamic kinetic asymmetric transformation (DYKAT) process is established to afford product in up to 85% yield and 99% ee (Scheme 8.39). The DYKAT process is synthetically useful because the α-alkylidene-β-hydroxy esters with β-substituents are rather difficult to synthesize via a conventional asymmetric aza-Morita–Baylis–Hillman reaction.

Chiral Li-BINOL complexes are among the simplest metal/BINOL complexes. In early reports using *dry* Li-BINOL complexes, however, the ability of Li/BINOL complexes to induce enantioselectivity was rather limited compared with other metal/BINOL complexes [9a, 79]. Ishihara and coworkers reported the utility of *wet or alcoholic* Li-BINOL catalysts to drastically improve enantioselectivity in asymmetric cyanosilylation of aldehydes [80]. *Dry* Li-BINOL complex prepared from BuLi/BINOL (1 : 1 ratio) promoted the cyanosilylation of benzaldehyde in 99% yield, but in only 58% ee. By adding a tiny amount of H$_2$O (3 mol%), enantioselectivity was improved to 95% ee. After screening of lithium metal sources,

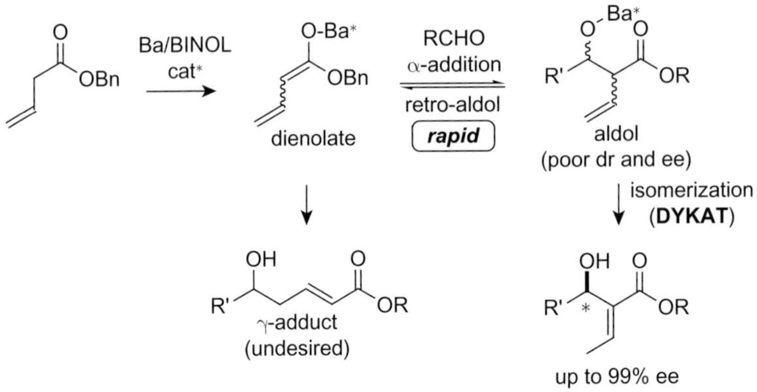

Scheme 8.39 Reaction mechanism of a Ba/BINOL-catalyzed dynamic kinetic asymmetric transformation (DYKAT) process.

LiO-i-Pr was the best in terms of enantioselectivity. With the *alcoholic* Li-BINOL catalyst prepared from LiO-i-Pr/BINOL, cyanosilylation of various aryl and heteroaryl aldehydes proceeded in good to excellent enantioselectivity (81–98% ee, Scheme 8.40). Ishihara also reported a Li/3,3′-substituted-BINOL complex as a Lewis acid/Brønsted base catalyst for direct Mannich-type reactions of 1,3-dicarbonyl compounds [81]. BINOL without 3,3′-substituent gave products in poor enantioselectivity (0–8% ee). Introduction of 3,4,5-F$_3$-C$_6$H$_2$-groups at the 3,3′-positions was the key to strengthening the chiral environment around Li metal center for high enantioselectivity. In a similar manner as observed in asymmetric cyanation, the addition of alcohol (t-BuOH) to generate *alcoholic* Li catalyst was important to achieve the best enantioselectivity. Interestingly, the enantiofacial selectivity of Boc-imines changed depending on the structure of the nucleophiles (Scheme 8.41). A similar direct Mannich-type reaction, using a Zr-BINOL derivative complex, was also reported by Kobayashi [82].

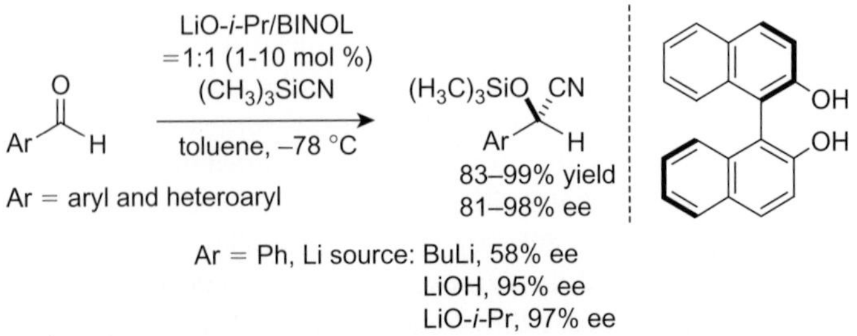

Scheme 8.40 Li/BINOL-catalyzed asymmetric cyanation of aldehydes.

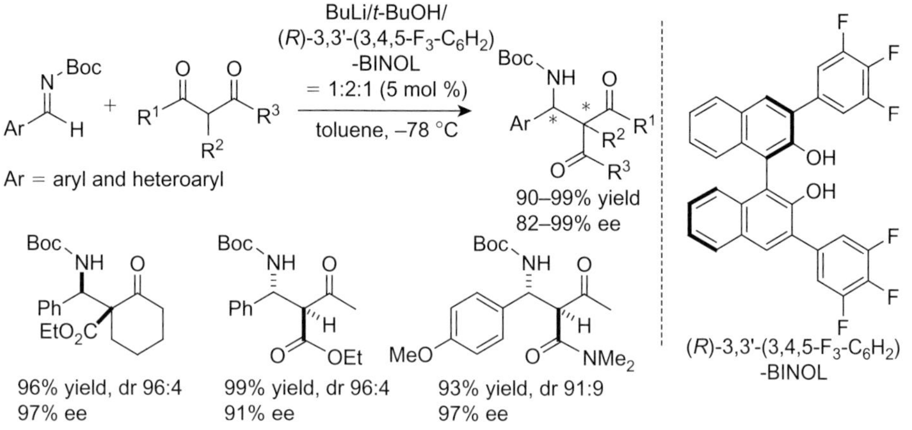

Scheme 8.41 Li/BINOL-catalyzed asymmetric direct Mannich-type reaction.

8.4.4
Lewis Acid/Lewis Base Bifunctional Aluminium-Catalyst

The previous sections described several examples using 3,3′-substituted-BINOL. In those cases, 3,3′-substituents are often introduced as a sterically demanding unit to strengthen the chiral environment around metal centers. The direction of access to reactants (electrophiles) around the Lewis acidic metal center is nicely controlled based on steric factors. In contrast, a totally different strategy to control the direction of nucleophiles' access to electrophiles was developed by introducing Lewis basic functionality at the 3,3′-positions in the BINOL scaffold. A BINOL derivative with a phosphine oxide moiety at 3,3′-positions was designed for catalytic asymmetric cyanosilylation of aldehydes (Al-cat A, Scheme 8.42) [83]. With the Al-cat A, cyanosilylation proceeded to give products in 83–98% ee from broad

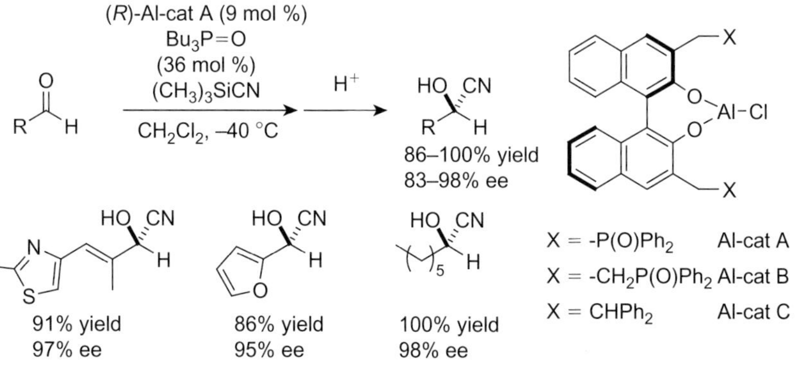

Scheme 8.42 Lewis acid/Lewis base bifunctional Al/BINOL catalyst for asymmetric cyanation of aldehydes.

range of aldehydes. The position of the Lewis basic moiety was critical to avoid undesirable intramolecular Lewis acid/Lewis base interaction. An Al-cat B with a longer linker resulted in poor enantioselectivity. Control experiments using a BINOL with simply sterically hindered diphenylmethyl-substituents, Al-cat C, also resulted in poor selectivity. Thus, a diphenylphosphine oxide moiety at the 3,3'-positions in Al-cat A was speculated to control the orientation of $(CH_3)_3Si\text{-}CN$ through O to Si interaction, while aldehyde is activated via the Lewis acidic Al metal center. A Lewis acid/Lewis base system based on the BINOL scaffold with Lewis basic functionality was nicely applied for the Strecker reaction of aldimines (Scheme 8.43) [84], Reissert-type three-component reaction of quinolines (Scheme 8.44) [85], isoquinolines (Scheme 8.45) [86], and pyridine derivatives (Scheme 8.46)

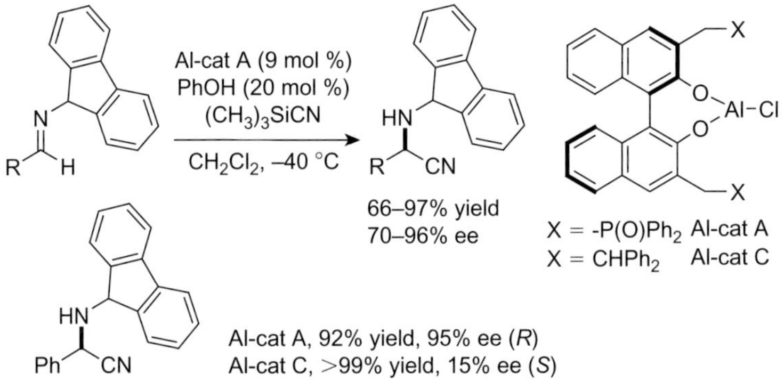

Scheme 8.43 Catalytic asymmetric Strecker-type reaction using a bifunctional Al-catalyst.

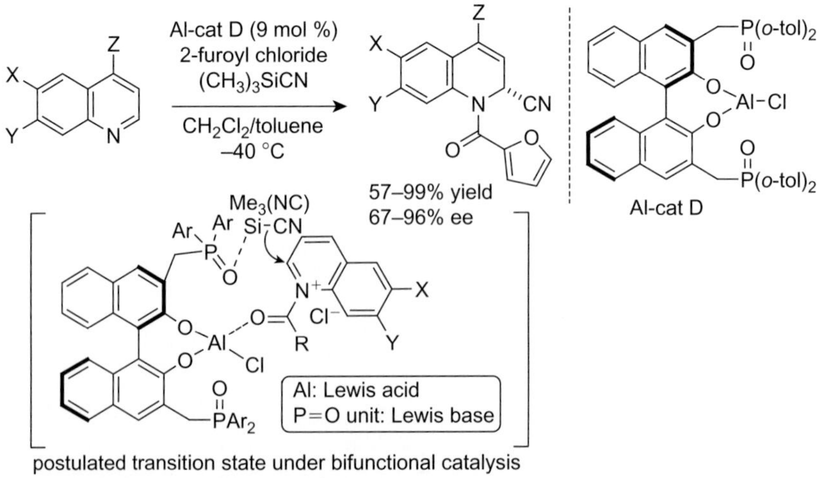

Scheme 8.44 Catalytic asymmetric three-component Reissert-type reaction of quinolines.

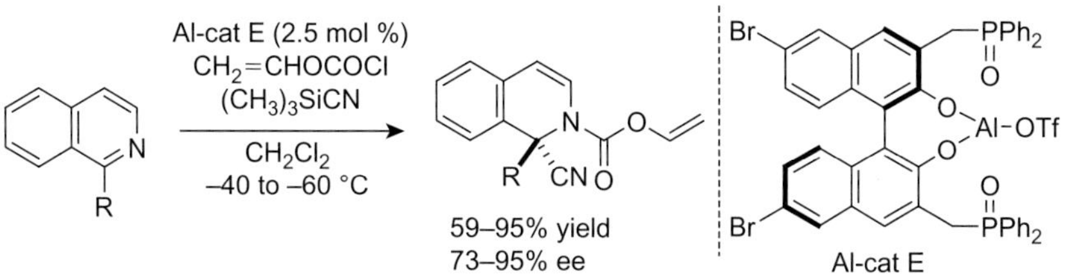

Scheme 8.45 Catalytic asymmetric three-component Reissert-type reaction of isoquinolines.

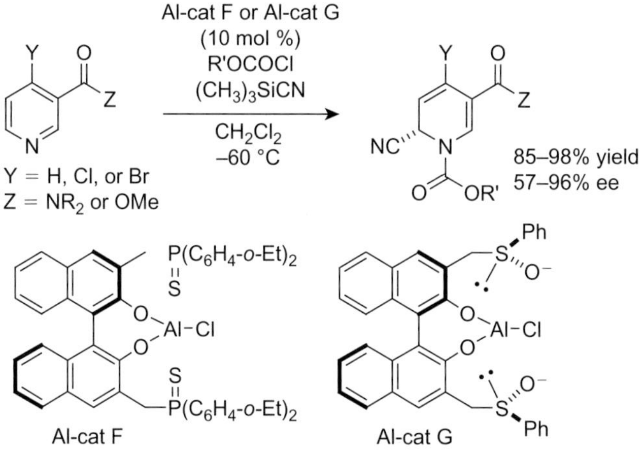

Scheme 8.46 Catalytic asymmetric three-component Reissert-type reaction of pyridine derivatives.

[87], and cyanophosphorylation of aldehydes (Scheme 8.47) [88]. In the Reissert-type three-component reactions, the Lewis acidic Al metal was speculated to control the position of acyliminium intermediate, while the orientation of $(CH_3)_3Si-CN$ was controlled through the Lewis basic moiety at the 3-position. The proposed transition state is shown in Scheme 8.44. Depending on each targeted reaction, the position and structure of the Lewis basic moiety as well as Lewis acidity of Al-metal center was nicely modified by introducing suitable substituents at 3,3'-positions and 6,6'-positions. For example, 6,6'-Br-substituents were important in improving the Lewis acidity of Al-metal for the construction of a quaternary carbon stereocenter (Al-cat E, Scheme 8.45). For the Reissert-type reaction of pyridine derivatives, either a sulfoxide or phosphine sulfide unit as Lewis basic functional group was the best for high enantioselectivity (Al-cat F and Al-cat G, Scheme 8.46). For cyanophosphorylation of aldehydes, a ligand with a *tert*-amine moiety was suitable to generate an active cyanide nucleophile from diethyl cyanophosphonate (Al-cat H, Scheme 8.47).

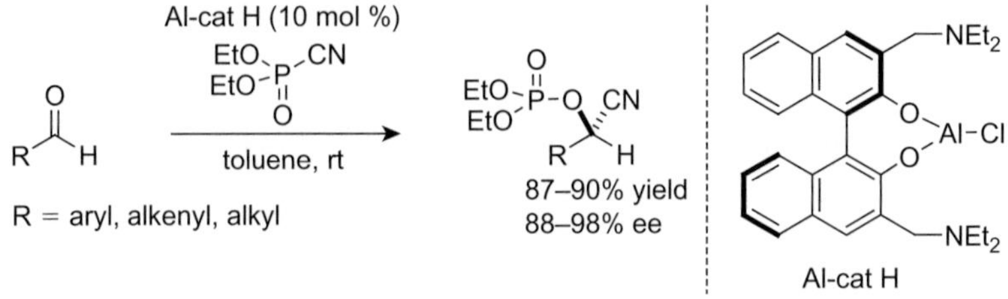

Scheme 8.47 Catalytic asymmetric cyanophosphorylation of aldehydes.

8.5
BINOL in Organocatalysis

Sections 8.2–8.4 have described the use of BINOL as chiral ligand for metal catalysis. On the basis of recent growing interest in organocatalysis, BINOLs have also been utilized as a basic and reliable chiral scaffold in designing organocatalytic reactions. This section introduces a few selected examples to describe the utility of the BINOL unit in organocatalysis.

Schaus and coworkers reported the utility of BINOL derivatives as promoters for the addition of various types of boronates to ketones and imines. Asymmetric allylation of ketones with allyl borane was promoted by catalytic amount of BINOL derivatives. Among various chiral diols screened, 3,3′-Br-BINOL gave the best enantioselectivity. In the proposed catalytic cycle (Scheme 8.48), the key steps to realizing an efficient catalytic process were a ligand exchange process at both the beginning and end of the catalytic cycle. On the basis of mechanistic studies, the structure of boronate reagent was optimized to accelerate the rate-limiting

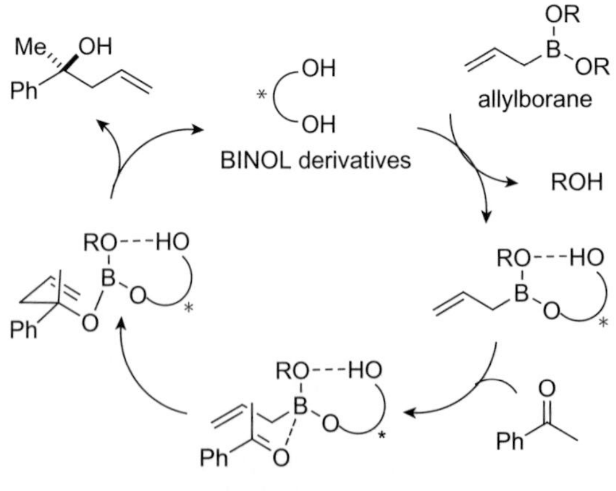

Scheme 8.48 Proposed catalytic cycle of asymmetric allylation promoted by BINOL derivatives.

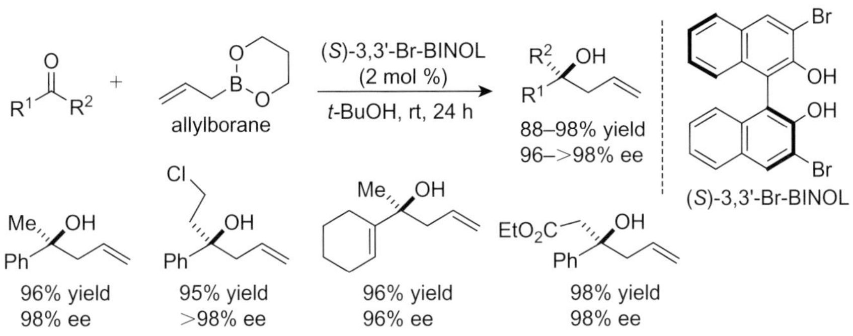

Scheme 8.49 3,3′-Br-BINOL-catalyzed asymmetric allylation of ketones.

catalyst regeneration step. With the optimized system, catalytic asymmetric allylation proceeded with as little as 2 mol% of 3,3′-Br-BINOL, giving products in excellent enantioselectivity (Scheme 8.49) [89]. Asymmetric crotylation of acetophenone also proceeded in excellent diastereo- and enantioselectivity. High diastereoselectivity was accomplished from both (E)- and (Z)-crotylboranes (Scheme 8.50). Schaus and coworkers also expanded the system for the addition of boranes to imines. Catalytic asymmetric allylation, crotylation, arylation, alkenylation, and alkynylation of imines were established (Scheme 8.51) [90]. Depending on the targeted reactions, the structures of BINOL derivatives as well as boronates were optimized to achieve high enantioselectivity. A catalytic asymmetric Petasis reaction between alkenyl boronates, secondary amines, and glyoxylates was also developed by using sterically hindered VAPOL as a promoter (Scheme 8.52) [91].

Shi and Sasai designed bifunctional Brønsted acid/Lewis base organocatalysts by introducing a Lewis basic unit into BINOLs (Figure 8.8) [92, 93]. In a similar manner as observed in Lewis acid/Lewis base bifunctional Al-catalyst in Section 8.4.4, the position of the Lewis basic unit was crucial to realizing efficient catalytic activity and enantioselectivity. The catalyst with a pyridine moiety was suitable for

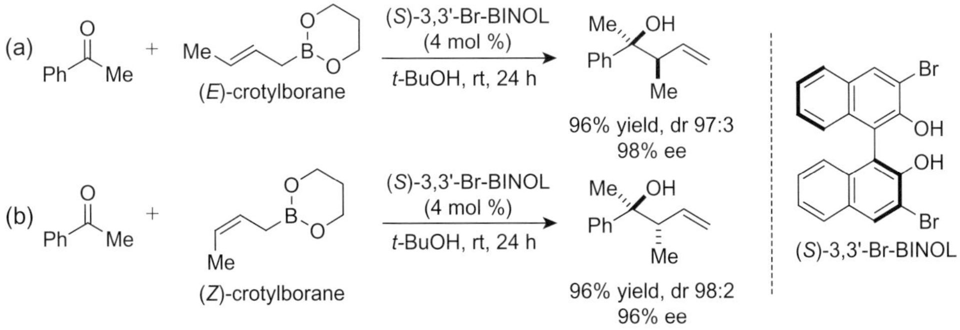

Scheme 8.50 3,3′-Br-BINOL-catalyzed asymmetric crotylation of ketone.

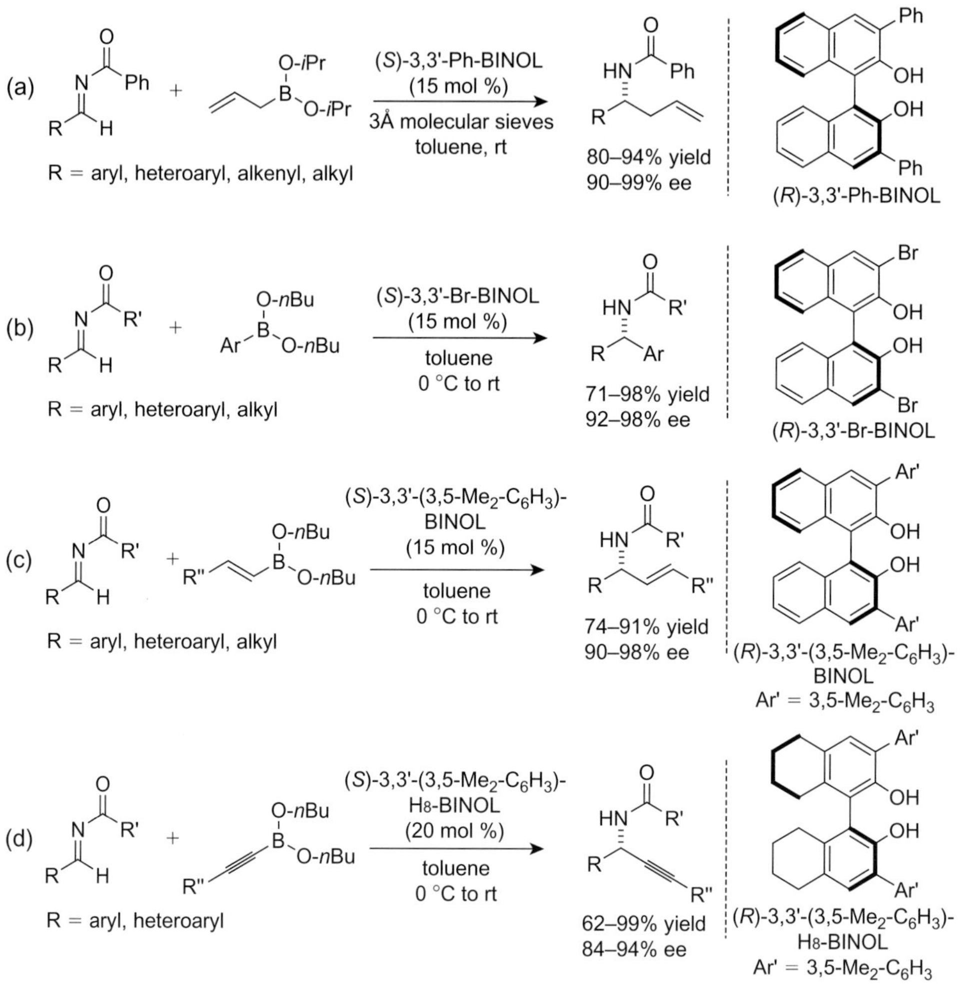

Scheme 8.51 Catalytic asymmetric allylation (a), arylation (b), alkenylation (c), and alkynylation (d) of imines promoted by BINOL derivatives.

the catalytic asymmetric aza-Morita–Baylis–Hillman reaction of enones, giving products in 87–95% ee (Scheme 8.53) [93]. In the proposed mechanism, the Lewis basic pyridine moiety at the 3-position functions as a nucleophilic catalyst to generate nucleophilic species from vinyl ketones and acrolein, and the BINOL unit functions as a chiral Brønsted acid to activate imines. Through this dual activation mechanism, high enantioselectivity was realized.

Chiral phosphoric Brønsted acids using the BINOL scaffold are one of the most successful achievements in organocatalysis (Figure 8.9). Akiyama and Terada independently discovered the utility of chiral phosphoric acid catalysis.

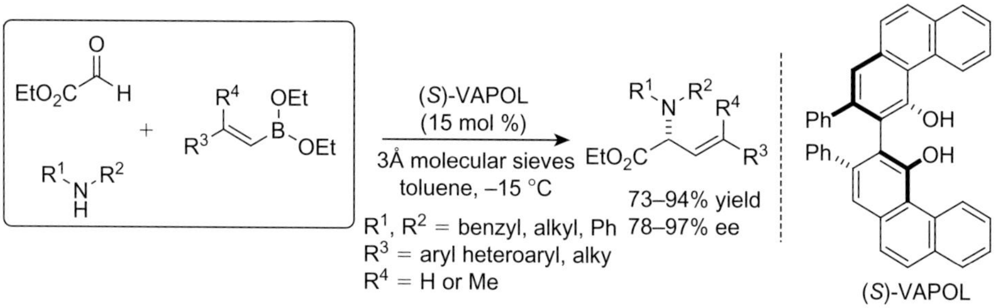

8.5 BINOL in Organocatalysis

Scheme 8.52 Catalytic asymmetric Petasis reaction between alkenyl boronates, secondary amines, and glyoxylates.

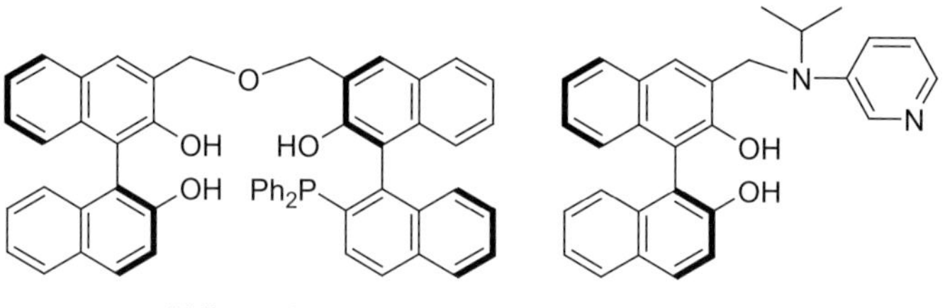

Shi's catalyst

Sasai's catalyst

Figure 8.8 Structures of Brønsted acid/Lewis base bifunctional organocatalysts.

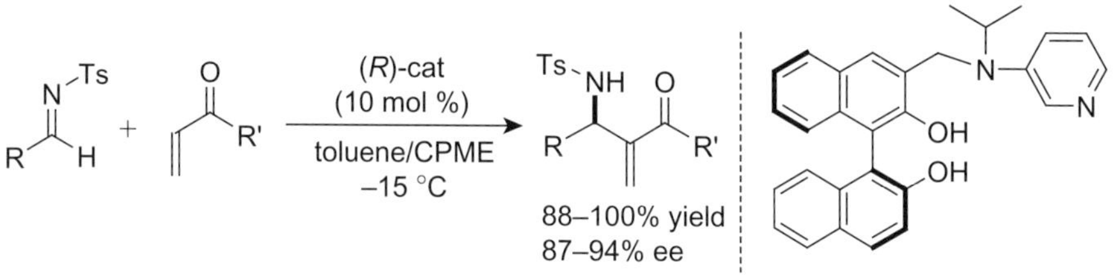

Scheme 8.53 Catalytic asymmetric aza-Morita–Baylis–Hillman reaction promoted by a bifunctional organocatalyst based on the BINOL scaffold.

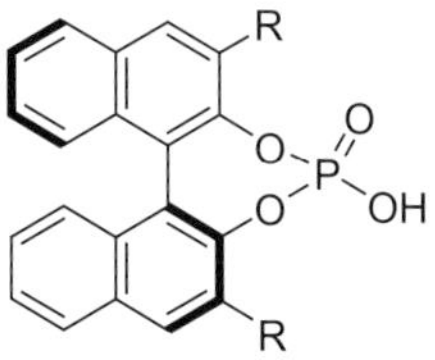

Figure 8.9 BINOL-derived monophosphoric acids as chiral Brønsted acid catalysts.

Introduction of a sterically hindered aromatic ring or silyl group was the key for high enantioselectivity in a similar manner as described in Lewis acid-catalyzed reactions. Related catalysts have been investigated intensively by many researchers and a broad range of enantioselective reactions has been developed. However, these topics are not covered in this chapter due to space limitations. For comprehensive reviews on chiral phosphoric Brønsted acids, see elegant reviews on organocatalysis [94].

8.6
Summary

The utility of BINOL and its derivatives as privileged ligands has been described. BINOLs have been utilized in wide range of reactions. One of the key aspects of BINOLs as privileged ligands and catalysts is the diversity of available BINOL derivatives. As discussed in each section, the electronic and steric tuning of BINOLs as well as the modification of dihedral angles often played a key role in achieving synthetically useful stereoselectivity. In the early stages of research, BINOLs were used as chiral ligands for Lewis acid-catalyzed reactions. The utility of BINOLs was greatly expanded by using them as ligands for bimetallic Lewis acid/Lewis base bifunctional catalysts. Introduction of additional functionality at the 3,3′-positions was also useful in establishing Brønsted acid-assisted-Lewis acid catalysis as well as Lewis acid/Lewis base catalysis. Most recently, the utility of BINOL is rapidly expanding in the rational design of organocatalysis.

References

1 Cram, D.J. and Cram, J.M. (1978) *Acc. Chem. Res.*, **11**, 8–14.

2 Meca, L., Reha, D., and Havlas, Z. (2003) *J. Org. Chem.*, **68**, 5677–5680.

3 Brunel, J.M. (2007) *Chem. Rev.*, **107**, PR1–PR45.

4 Chen, Y., Yekta, S., and Yudin, A.K. (2003) *Chem. Rev.*, **103**, 3155–3211.

5 Kocovsky, P., Vyskocyl, S., and Smrcina, M. (2003) *Chem. Rev.*, **103**, 3213–3246.

6 Noyori, R., Tomino, I., and Tanimoto, Y. (1979) *J. Am. Chem. Soc.*, **101**, 3129–3131.

7 Graves, C.R., Scheidt, K.A., and Nguyen, S.T. (2006) *Org. Lett.*, **8**, 1229–1232.

8 Lin, Y.-M., Fu, I.-P., and Uang, B.-J. (2001) *Tetrahedron: Asymmetry*, **12**, 3217–3221.

9 (a) Schiffers, R. and Kagan, H.B. (1997) *Synlett*, 1175–1178; (b) Ushio, H. and Mikami, K. (2005) *Tetrahedron Lett.*, **46**, 2903–2906.

10 Bougauchi, M., Watanabe, T., Arai, T., Sasai, H., and Shibasaki, M. (1997) *J. Am. Chem. Soc.*, **119**, 2329–2330.

11 (a) Daikai, K., Kamaura, M., and Inanaga, J. (1998) *Tetrahedron Lett.*, **39**, 7321–7322; (b) Nemoto, T., Ohshima, T., Yamaguchi, K., and Shibasaki, M. (2001) *J. Am. Chem. Soc.*, **123**, 2725–2732.

12 (a) Nemoto, T., Kakei, H., Gnanadesikan, V., Tosaki, S.-y., Ohshima, T., and Shibasaki, M. (2002) *J. Am. Chem. Soc.*, **124**, 14544–14545; (b) Tosaki, S.-y., Tsuji, R., Ohshima, T., and

Shibasaki, M. (2005) *J. Am. Chem. Soc.*, **127**, 2147–2155.

13 Nemoto, T., Ohshima, T., and Shibasaki, M. (2001) *J. Am. Chem. Soc.*, **123**, 9474–9475.

14 (a) Kinoshita, T., Okada, S., Park, S.-R., Matsunaga, S., and Shibasaki, M. (2003) *Angew. Chem. Int. Ed.*, **42**, 4680–4684; (b) Matsunaga, S., Qin, H., Sugita, M., Okada, S., Kinoshita, T., Yamagiwa, N., and Shibasaki, M. (2006) *Tetrahedron*, **62**, 6630–6639.

15 Chen, Z., Morimoto, H., Matsunaga, S., and Shibasaki, M. (2006) *Synlett*, 3529–3532.

16 (a) Kakei, H., Tuji, R., Ohshima, T., and Shibasaki, M. (2005) *J. Am. Chem. Soc.*, **127**, 8962–8963; (b) Kakei, H., Tsuji, R., Ohshima, T., Morimoto, H., Matsunaga, S., and Shibasaki, M. (2007) *Chem. Asian J.*, **2**, 257–264.

17 Hara, K., Park, S.-Y., Yamagiwa, N., Matsunaga, S., and Shibasaki, M. (2008) *Chem. Asian J.*, **3**, 1500–1504.

18 (a) Guo, Q.-X., Wu, Z.-J., Luo, Z.-B., Liu, Q.-Z., Ye, J.-L., Luo, S.-W., Cun, L.-F., and Gong, L.-Z. (2007) *J. Am. Chem. Soc.*, **129**, 13927–13938; (b) Luo, Z., Liu, Q., Gong, L., Cui, X., Mi, A., and Jiang, Y. (2002) *Chem. Commun.*, 914–915.

19 Takizawa, S., Katayama, T., Kameyama, C., Onitsuka, K., Suzuki, T., Yanagida, T., Kawai, T., and Sasai, H. (2008) *Chem. Commun.*, 1810–1812.

20 Yamamoto, H. and Ishihara, K. (eds) (2008) *Acid Catalysis in Modern Organic Synthesis*, Wiley-VCH Verlag GmbH, Weinheim.

21 Mikami, K. and Shimizu, M. (1992) *Chem. Rev.*, **92**, 1021–1050.

22 (a) Mikami, K., Terada, M., and Nakai, T. (1989) *J. Am. Chem. Soc.*, **111**, 1940–1941; (b) Mikami, K., Terada, M., and Nakai, T. (1990) *J. Am. Chem. Soc.*, **112**, 3949–3954.

23 Keck, G.E., Tarbet, K.H., and Geraci, L.S. (1993) *J. Am. Chem. Soc.*, **115**, 8467–8468.

24 (a) Mikami, K. and Matsukawa, S. (1994) *J. Am. Chem. Soc.*, **116**, 4077–4078; (b) Keck, G.E. and Krishnamurthy, D. (1995) *J. Am. Chem. Soc.*, **117**, 2363–2364.

25 Motoyama, Y., Terada, M., and Mikami, K. (1995) *Synlett*, 967–968.

26 Long, J., Hu, J., Shen, X., Ji, B., and Ding, K. (2002) *J. Am. Chem. Soc.*, **124**, 10–11.

27 (a) Kano, T., Hashimoto, T., and Maruoka, K. (2005) *J. Am. Chem. Soc.*, **127**, 11926–11927; (b) Kano, T., Hashimoto, T., and Maruoka, K. (2006) *J. Am. Chem. Soc.*, **128**, 2174–2175.

28 Review: Yuan, Y., Ding, K., and Chen, G. (2008) in *Acid Catalysis in Modern Organic Synthesis*, Vol. 2 (eds H. Yamamoto and K. Ishihara), Wiley-VCH Verlag GmbH, Weinheim, pp. 721–823.

29 Terada, M., Matsumoto, Y., Nakamura, Y., and Mikami, K. (1999) *Inorg. Chim. Acta*, **296**, 267–272.

30 (a) Ishitani, H., Ueno, M., and Kobayashi, S. (1997) *J. Am. Chem. Soc.*, **119**, 7153–7154; (b) Ishitani, H., Ueno, M., and Kobayashi, S. (2000) *J. Am. Chem. Soc.*, **122**, 8180–8186.

31 Kobayashi, S., Ishitani, H., and Ueno, M. (1998) *J. Am. Chem. Soc.*, **120**, 431–432.

32 Ishitani, H., Komiyama, S., Hasegawa, Y., and Kobayashi, S. (2000) *J. Am. Chem. Soc.*, **122**, 762–766.

33 (a) Ishitani, H., Yamashita, Y., Shimizu, H., and Kobayashi, S. (2000) *J. Am. Chem. Soc.*, **122**, 5403–5404; (b) Yamashita, Y., Ishitani, H., Shimizu, H., and Kobayashi, S. (2002) *J. Am. Chem. Soc.*, **124**, 3292–3302.

34 Gastner, T., Ishitani, H., Akiyama, R., and Kobayashi, S. (2001) *Angew. Chem. Int. Ed.*, **40**, 1896–1898.

35 Yamashita, Y., Saito, S., Ishitani, H., and Kobayashi, S. (2002) *Org. Lett.*, **4**, 1221–1223.

36 (a) Maruoka, K., Itoh, T., Shirasaka, T., and Yamamoto, H. (1988) *J. Am. Chem. Soc.*, **110**, 310–312; (b) Maruoka, K., Hoshino, Y., Shirasaka, T., and Yamamoto, H. (1988) *Tetrahedron Lett.*, **29**, 3967–3970.

37 Yamamoto, H. and Futatsugi, K. (2005) *Angew. Chem. Int. Ed.*, **44**, 1924–1942.

38 (a) Ishihara, K. and Yamamoto, H. (1994) *J. Am. Chem. Soc.*, **116**, 1561–1562; (b) Ishihara, K., Kurihara, H., and Yamamoto, H. (1996) *J. Am. Chem. Soc.*, **118**, 3049–3050; (c) Ishihara, K.,

Kurihara, H., Matsumoto, M., and Yamamoto, H. (1998) *J. Am. Chem. Soc.*, **120**, 6920–6930.
39 Teo, Y.-C. and Loh, T.-P. (2005) *Org. Lett.*, **7**, 2539–2541.
40 Takita, R., Yakura, K., Ohshima, T., and Shibasaki, M. (2005) *J. Am. Chem. Soc.*, **127**, 13760–13761.
41 Review: Kobayashi, S., Sugiura, M., Kitagawa, H., and Lam, W.W.-L. (2002) *Chem. Rev.*, **102**, 2227–2302.
42 Kobayashi, S. and Kawamura, M. (1998) *J. Am. Chem. Soc.*, **120**, 5840–5841.
43 Reviews: (a) Ogawa, C., Gu, Y., Boudou, M., and Kobayashi, S. (2008) in *Acid Catalysis in Modern Organic Synthesis*, Vol. 2 (eds H. Yamamoto and K. Ishihara), Wiley-VCH Verlag GmbH, Weinheim, pp. 589–634; (b) Shibasaki, M., Matsunaga, S., and Kumagai, N. (2008) in *Acid Catalysis in Modern Organic Synthesis*, Vol. 2 (eds H. Yamamoto and K. Ishihara), Wiley-VCH Verlag GmbH, Weinheim, pp. 635–720.
44 Reviews: (a) Shibasaki, M. and Yoshikawa, N. (2002) *Chem. Rev.*, **102**, 2187–2209; (b) Matsunaga, S. and Shibasaki, M. (2008) *Bull. Chem. Soc. Jpn.*, **81**, 60–75.
45 Shibasaki, M., Sasai, H., and Arai, T. (1997) *Angew. Chem. Int. Ed. Engl.*, **36**, 1236–1256.
46 (a) Sasai, H., Suzuki, T., Itoh, N., Tanaka, K., Date, T., Okamura, K., and Shibasaki, M. (1993) *J. Am. Chem. Soc.*, **115**, 10372–10373; (b) Sasai, H., Tokunaga, T., Watanabe, S., Suzuki, T., Itoh, N., and Shibasaki, M. (1995) *J. Org. Chem.*, **60**, 7388–7389.
47 (a) Yoshikawa, N., Yamada, Y.M.A., Das, J., Sasai, H., and Shibasaki, M. (1999) *J. Am. Chem. Soc.*, **121**, 4168–4178; (b) Yoshikawa, N., Kumagai, N., Matsunaga, S., Moll, G., Ohshima, T., Suzuki, T., and Shibasaki, M. (2001) *J. Am. Chem. Soc.*, **123**, 2466–2467.
48 (a) Sasai, H., Arai, T., Satow, Y., Houk, K.N., and Shibasaki, M. (1995) *J. Am. Chem. Soc.*, **117**, 6194–6198; (b) Sasai, H., Emori, E., Arai, T., and Shibasaki, M. (1996) *Tetrahedron Lett.*, **37**, 5561–5564; (c) Funabashi, K., Saida, Y., Kanai, M., Arai, T., Sasai, H., and Shibasaki, M. (1998) *Tetrahedron Lett.*, **39**, 7557–7558.
49 Emori, E., Arai, T., Sasai, H., and Shibasaki, M. (1998) *J. Am. Chem. Soc.*, **120**, 4043–4044.
50 (a) Gröger, H., Saida, Y., Sasai, H., Yamaguchi, K., Martens, J., and Shibasaki, M. (1998) *J. Am. Chem. Soc.*, **120**, 3089–3103; (b) Sasai, H., Arai, S., Tahara, Y., and Shibasaki, M. (1995) *J. Org. Chem.*, **60**, 6656–6657; (c) Yokomatsu, T., Yamagishi, T., and Shibuya, S. (1993) *Tetrahedron: Asymmetry*, **4**, 1783–1784; (d) Rath, N.P. and Spilling, C.D. (1994) *Tetrahedron Lett.*, **35**, 227–230.
51 (a) Yamagiwa, N., Tian, J., Matsunaga, S., and Shibasaki, M. (2005) *J. Am. Chem. Soc.*, **127**, 3413–3422; (b) Tian, J., Yamagiwa, N., Matsunaga, S., and Shibasaki, M. (2002) *Angew. Chem. Int. Ed.*, **41**, 3636–3638; (c) Abiko, Y., Yamagiwa, N., Sugita, M., Tian, J., Matsunaga, S., and Shibasaki, M. (2004) *Synlett*, 2434–2436.
52 Tosaki, S.-y., Hara, K., Gnanadesikan, V., Morimoto, H., Harada, S., Sugita, M., Yamagiwa, N., Matsunaga, S., and Shibasaki, M. (2006) *J. Am. Chem. Soc.*, **128**, 11776–11777.
53 Yamada, K.-i., Harwood, S.J., Gröger, H., and Shibasaki, M. (1999) *Angew. Chem. Int. Ed.*, **38**, 3504–3506.
54 Wooten, A.J., Carroll, P.J., and Walsh, P.J. (2008) *J. Am. Chem. Soc.*, **130**, 7407–7419.
55 Chavarot, M., Byrne, J.J., Chavant, P.Y., and Vallee, Y. (2001) *Tetrahedron: Asymmetry*, **12**, 1147–1150.
56 (a) Gnanadesikan, V., Horiuchi, Y., Ohshima, T., and Shibasaki, M. (2004) *J. Am. Chem. Soc.*, **126**, 7782–7783; (b) Horiuchi, Y., Gnanadesikan, V., Ohshima, T., Masu, H., Katagiri, K., Sei, Y., Yamaguchi, K., and Shibasaki, M. (2005) *Chem. Eur. J.*, **11**, 5195–5204.
57 (a) Yamagiwa, N., Matsunaga, S., and Shibasaki, M. (2003) *J. Am. Chem. Soc.*, **125**, 16178–16179; (b) Yamagiwa, N., Qin, H., Matsunaga, S., and Shibasaki, M. (2005) *J. Am. Chem. Soc.*, **127**, 13419–13427.

58 Kakei, H., Sone, T., Sohtome, Y., Matsunaga, S., and Shibasaki, M. (2007) *J. Am. Chem. Soc.*, **129**, 13410–13411.
59 Sone, T., Yamaguchi, A., Matsunaga, S., and Shibasaki, M. (2008) *J. Am. Chem. Soc.*, **130**, 10078–10079.
60 Sone, T., Lu, G., Matsunaga, S., and Shibasaki, M. (2009) *Angew. Chem. Int. Ed.*, **48**, 1677–1680.
61 Arai, T., Sasai, H., Aoe, K., Okamura, K., Date, T., and Shibasaki, M. (1996) *Angew. Chem. Int. Ed. Engl.*, **35**, 104–106.
62 Shimizu, S., Ohori, K., Arai, T., Sasai, H., and Shibasaki, M. (1998) *J. Org. Chem.*, **63**, 7547–7551.
63 Xu, Y., Ohori, K., Ohshima, T., and Shibasaki, M. (2002) *Tetrahedron*, **58**, 2585–2588.
64 Yamada, K., Arai, T., Sasai, H., and Shibasaki, M. (1998) *J. Org. Chem.*, **63**, 3666–3672.
65 (a) Ohshima, T., Xu, Y., Takita, R., Shimizu, S., Zhong, D., and Shibasaki, M. (2002) *J. Am. Chem. Soc.*, **124**, 14546–14547; (b) Mihara, H., Sohtome, Y., Matsunaga, S., and Shibasaki, M. (2008) *Chem. Asian J.*, **3**, 359–366; (c) Nara, S., Toshima, H., and Ichihara, A. (1997) *Tetrahedron*, **53**, 9509–9524.
66 (a) Arai, T., Bougauchi, M., Sasai, H., and Shibasaki, M. (1995) *J. Org. Chem.*, **60**, 2926–2927; (b) Arai, T., Sasai, H., Yamaguchi, K., and Shibasaki, M. (1998) *J. Am. Chem. Soc.*, **120**, 441–442; (c) Yamagishi, T., Suemune, K., Yokomatsu, T., and Shibuya, S. (2002) *Tetrahedron*, **58**, 2577–2583.
67 Iida, T., Yamamoto, N., Sasai, H., and Shibasaki, M. (1997) *J. Am. Chem. Soc.*, **119**, 4783–4784.
68 Arai, T., Yamada, Y.M.A., Yamamoto, N., Sasai, H., and Shibasaki, M. (1996) *Chem. Eur. J.*, **2**, 1368–1372.
69 Iida, T., Yamamoto, N., Matsunaga, S., Woo, H.-G., and Shibasaki, M. (1998) *Angew. Chem. Int. Ed.*, **37**, 2223–2226.
70 Review: Matsunaga, S. and Shibasaki, M. (2006) *Chem. Soc. Rev.*, **35**, 269–279.
71 Synthesis of linked-BINOL: Matsunaga, S., Das, J., Roels, J., Vogl, E.M., Yamamoto, N., Iida, T., Yamaguchi, K., and Shibasaki, M. (2000) *J. Am. Chem. Soc.*, **122**, 2252–2260.
72 Kim, Y.S., Matsunaga, S., Das, J., Sekine, A., Ohshima, T., and Shibasaki, M. (2000) *J. Am. Chem. Soc.*, **122**, 6506–6507.
73 Majima, K., Takita, R., Okada, A., Ohshima, T., and Shibasaki, M. (2003) *J. Am. Chem. Soc.*, **125**, 15837–15845.
74 Kumagai, N., Matsunaga, S., Kinoshita, T., Harada, S., Okada, S., Sakamoto, S., Yamaguchi, K., and Shibasaki, M. (2003) *J. Am. Chem. Soc.*, **125**, 2169–2178.
75 (a) Harada, S., Kumagai, N., Kinoshita, T., Matsunaga, S., and Shibasaki, M. (2003) *J. Am. Chem. Soc.*, **125**, 2582–2590; (b) Matsunaga, S., Kinoshita, T., Okada, S., Hrada, S., and Shibasaki, M. (2004) *J. Am. Chem. Soc.*, **126**, 7559–7570.
76 (a) Matsunaga, S., Kumagai, N., Harada, S., and Shibasaki, M. (2003) *J. Am. Chem. Soc.*, **125**, 4712–4713; (b) Matsunaga, S., Yoshida, T., Morimoto, H., Kumagai, N., Matsunaga, S., and Shibasaki, M. (2004) *J. Am. Chem. Soc.*, **126**, 8777–8785.
77 Harada, S., Handa, S., Matsunaga, S., and Shibasaki, M. (2005) *Angew. Chem. Int. Ed.*, **44**, 4365–4368.
78 Yamaguchi, A., Matsunaga, S., and Shibasaki, M. (2009) *J. Am. Chem. Soc.*, **131**, 10842–10843.
79 (a) Loog, O. and Mäeorg, U. (1999) *Tetrahedron: Asymmetry*, **10**, 2411–2415; (b) Nakajima, M., Orito, Y., Ishizuka, T., and Hashimoto, S. (2004) *Org. Lett.*, **6**, 3763–3765.
80 Hatano, M., Horibe, T., and Ishihara, K. (2010) *J. Am. Chem. Soc.*, **132**, 56–57.
81 Hatano, M., Ikeno, T., Miyamoto, T., and Ishihara, K. (2005) *J. Am. Chem. Soc.*, **127**, 10776–10777.
82 Kobayashi, S., Salter, M.M., Yamazaki, Y., and Yamashita, Y. (2010) *Chem. Asian J.*, **5**, 493–495.
83 Hamashima, Y., Sawada, D., Kanai, M., and Shibasaki, M. (1999) *J. Am. Chem. Soc.*, **121**, 2641–2642.
84 Takamura, M., Hamashima, Y., Usuda, H., Kanai, M., and Shibasaki, M. (2000) *Angew. Chem. Int. Ed.*, **39**, 1650–1652.
85 (a) Takamura, M., Funabashi, K., Kanai, M., and Shibasaki, M. (2000)

J. Am. Chem. Soc., **122**, 6327–6328; (b) Takamura, M., Funabashi, K., Kanai, M., and Shibasaki, M. (2001) *J. Am. Chem. Soc.*, **123**, 6801–6808.

86 Funabashi, K., Ratni, H., Kanai, M., and Shibasaki, M. (2001) *J. Am. Chem. Soc.*, **123**, 10784–10785.

87 Ichikawa, E., Suzuki, M., Yabu, K., Albert, M., Kanai, M., and Shibasaki, M. (2004) *J. Am. Chem. Soc.*, **126**, 11808–11809.

88 Baeza, A., Casa, J., Nájera, C., Sansano, J.M., and Saá, J.M. (2003) *Angew. Chem. Int. Ed.*, **42**, 3143–3146.

89 (a) Lou, S., Moquist, P.N., and Schaus, S.E. (2006) *J. Am. Chem. Soc.*, **128**, 12660–12661; (b) Barnett, D.S., Moquist, P.N., and Schaus, S.E. (2009) *Angew. Chem. Int. Ed.*, **48**, 8679–8682.

90 (a) Lou, S., Moquist, P.N., and Schaus, S.E. (2007) *J. Am. Chem. Soc.*, **129**, 15398–15404; (b) Bishop, J.A., Lou, S., and Schaus, S.E. (2009) *Angew. Chem. Int. Ed.*, **48**, 4337–4340.

91 Lou, S. and Schaus, S.E. (2008) *J. Am. Chem. Soc.*, **130**, 6922–6923.

92 Review: Shi, Y.-L. and Shi, M. (2007) *Eur. J. Org. Chem.*, 2905–2916.

93 Matsui, K., Takizawa, S., and Sasai, H. (2005) *J. Am. Chem. Soc.*, **127**, 3680–3681.

94 Reviews: (a) Akiyama, T. (2008) in *Acid Catalysis in Modern Organic Synthesis*, Vol. **1** (eds H. Yamamoto and K. Ishihara), Wiley-VCH Verlag GmbH, Weinheim, pp. 62–107; (b) Terada, M. (2008) *Chem. Commun.*, 4097–4112.

9
TADDOLate Ligands
Hélène Pellissier

9.1
Introduction

Apart from the success of BINOL [1], TADDOL is one of the most efficient chiral backbones for asymmetric synthesis. Tetraaryl-1,3-dioxolane-4,5-dimethanols (TADDOLs), containing two adjacent diarylhydroxymethyl groups in a trans relationship on a 1,3-dioxolane ring (Figure 9.1), were introduced in 1982 by Seebach *et al.* [2]. TADDOLs are easily prepared from chiral acetals or ketals of tartrate esters by reaction of the latter with aromatic Grignard reagents.

Within the molecule of TADDOL, one of the hydrogen atoms of the alcohol group participates in an intramolecular hydrogen bond and the other is free for intermolecular interactions. These unique features were confirmed by X-ray crystallographic analysis, which revealed that the heteroatoms on the diarylmethyl groups are almost always in close proximity to each other, joined together by H-bonds, and predisposed to form chelate complexes in which the metallic centers reside in propeller-like chiral environments. In the vast majority of cases, the TADDOL units are present in conformations with near-C_2-symmetry, which feature perfect staggering about the exocyclic C–C bonds and staggering about the endocyclic C–C bonds that is as near ideal as possible for a dioxolane ring. TADDOLs also display an antiperiplanar arrangement of endo- and exocyclic C–O bonds, and quasi-axial and quasi-equatorial placement, respectively, of the two members constituting each pair of aryl groups, where in the former case an "edge-on" conformation is preferred and in the latter a "face-on" conformation is preferred when the molecule is observed along the C_2 axis. An intramolecular bond forms between the OH groups, so that one OH proton remains available for intermolecular hydrogen bonding. The dioxolane ring and the seven-membered ring that results from hydrogen bonding are disposed in a *trans*-fused bicyclo[5.3.0] decane-like arrangement, whereby the bridging hydrogen atom falls nearly along the C_2 axis – that is, at the very place where a chelate-bonded metal ion would be situated in a seven-membered ring chelate. The similarity between this system and a bis(diphenylphosphanyl)metal complex is quite striking. A series of X-ray crystallographic analyses has shown that TADDOL derivatives bearing XH/Y rather

Privileged Chiral Ligands and Catalysts. Edited by Qi-Lin Zhou
Copyright © 2011 WILEY-VCH Verlag GmbH & Co. KGaA, Weinheim
ISBN: 978-3-527-32704-1

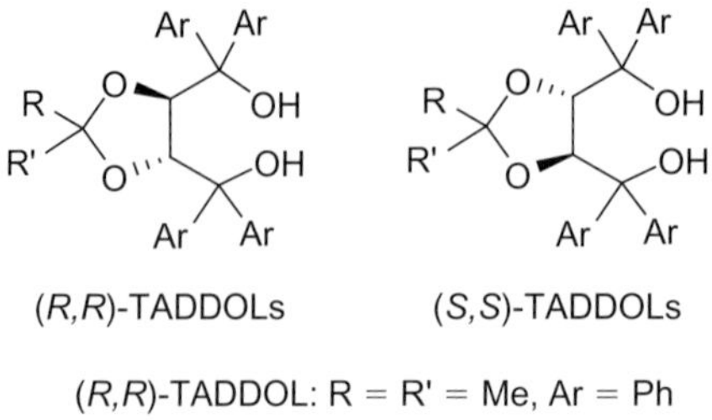

(R,R)-TADDOLs   (S,S)-TADDOLs

(R,R)-TADDOL: R = R' = Me, Ar = Ph

Figure 9.1 Structure of TADDOLs.

than OH/OH groups on the diphenylmethyl substituents of the dioxolane rings also assume the same conformation, although in some cases the hydrogen bonds are much weaker. Furthermore, TADDOL derivatives in which the heteroatoms at the diphenylmethyl groups participate in six- or seven-membered rings show precisely the same orientation of the aryl groups as titanium TADDOLates, and they conform almost exactly to the structures of the TADDOL precursors. Structures of analogs in which other heterocyclic, carbocyclic, or carbobicyclic ring systems replace the dioxolane ring fall nicely into place as well, as do systems in which all or some of the aryl groups are replaced by cyclohexyl or methyl groups. TADDOLs and their derivatives are extraordinarily versatile chiral reagents, which can be used as stoichiometric chiral auxiliaries, chiral ligands for both stoichiometric and catalytic asymmetric reactions, as well as, more recently, chiral organocatalysts. The goal of this chapter is to cover the principal uses of TADDOL derivatives in asymmetric synthesis, which were previously reviewed by Seebach et al., in 2001 [3]. This chapter is subdivided into six sections, according to the different types of asymmetric reactions based on the use of TADDOLs and their derivatives, namely, nucleophilic additions to C=O bonds, nucleophilic conjugate additions to electron-deficient C=C double bonds, nucleophilic substitutions, cycloaddition reactions, oxidation and reduction reactions, and miscellaneous reactions.

9.2
Nucleophilic Additions to C=O Double Bonds

9.2.1
Organozinc Additions to Aldehydes

One of the most powerful methods for the catalytic asymmetric generation of C−C bonds is the enantioselective addition of organometallic reagents to aldehydes or ketones. Since the initial report of Oguni and Omi in 1984 on the reaction of diethylzinc with benzaldehyde in the presence of a catalytic amount of (S)-leucinol with moderate enantioselectivity (49% ee) [4], research on asymmetric organozinc

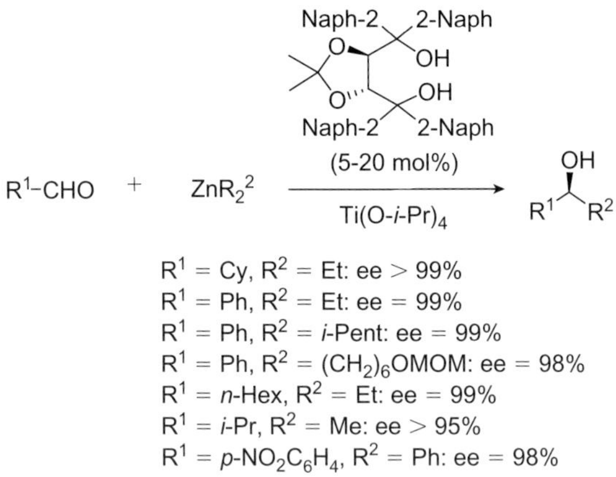

R^1-CHO + ZnR_2^2 →[Naph-2, 2-Naph TADDOL (5-20 mol%) / Ti(O-i-Pr)$_4$] R^1-CH(OH)-R^2

R^1 = Cy, R^2 = Et: ee > 99%
R^1 = Ph, R^2 = Et: ee = 99%
R^1 = Ph, R^2 = i-Pent: ee = 99%
R^1 = Ph, R^2 = (CH$_2$)$_6$OMOM: ee = 98%
R^1 = n-Hex, R^2 = Et: ee = 99%
R^1 = i-Pr, R^2 = Me: ee > 95%
R^1 = p-NO$_2$C$_6$H$_4$, R^2 = Ph: ee = 98%

Scheme 9.1 Organozinc additions to aldehydes.

additions to carbonyl compounds has grown dramatically [5]. In particular, Seebach et al. carried out an extensive study on the use of titanium complexes of TADDOLs for the asymmetric organozinc addition [6]. A selection of the best results obtained by these authors for the addition of dialkylzinc reagents to aldehydes catalyzed by (i-PrO)$_2$Ti-TADDOLate bearing 2-naphthyl groups in the presence of an excess of Ti(Oi-Pr)$_4$ are collected in Scheme 9.1 [6b]. More recently, Seebach's conditions were used by Takemoto et al. in developing a total synthesis of macrolactin A, a 24-membered polyene macrolide antibiotic having strong activity against B16-F10 murine tumor cells and HIV-1 [7].

In 2004, Gau and Sheen reported the synthesis of new titanium TADDOLate complexes containing bidentate nitrogen donors such as 2,2′-bipyridine or 1,10-phenanthroline [8]. These complexes showed good reactivities in ethylations of benzaldehyde with low to moderate enantioselectivities (≤88% ee). The first fluorous analogs of TADDOL were recently synthesized by Ando et al. by replacing the aryl groups of TADDOL with three bulky perfluoroalkyl groups [9]. When applied to the addition of ZnMe$_2$ to aldehydes, these catalysts showed an unprecedented activity combined with an excellent enantioselectivity of up to 98% ee. Finally, notably, excellent results were also obtained by using various heterogeneous TADDOL catalysts [10–20].

9.2.2
Allylations

The allylmetallation of aldehydes and ketones, leading to products with a maximum of two new stereocenters and versatile functionalities, is an important example of acyclic stereocontrol. In this context, TADDOLs were demonstrated to be excellent ligands in asymmetric allylations of aldehydes [21]. Thus, the enantioselective transfer of allyl groups to aldehydes with allylcyclopentadienyltitanium

TADDOLates provided the corresponding homoallylic alcohols with excellent enantioselectivities of up to 99% ee. An example was reported by Cossy et al. with an enantioselective allyltitanation applied to a polyfunctionalized aldehyde derived from *(S)*-glycidol [22, 23]. In 2004, Floreancig and Wang extended the scope of this methodology to the crotylation of aromatic aldehydes [24]. On the other hand, Ghosez et al. developed asymmetric reactions between unsaturated aldehydes and a silyl-substituted allyltitanate reagent, generated from allyldimethylphenylsilane, *n*-BuLi and *(R,R)*-Cl-TiCpTADDOL [25]. These reactions gave the corresponding *anti*-β-hydroxyallylsilanes in good yields and high diastereo- and enantioselectivities (>98% de, 90–96% ee). This methodology could also be applied to the allyltitanations of an acetonide-protected aldehyde and to an aldehyde derived from solketal, providing the corresponding β-hydroxyallylsilanes in 98% and 97% ee, respectively [26]. In 2004, Studer et al. reported the synthesis of a TADDOL-derived cyclohexadienyltitanium complex that was involved in highly diastereo- and enantioselective allylations of aldehydes, providing the corresponding *syn*-cyclohexadienes as the major products [27]. This methodology was applied to the syntheses of (+)-nephrosteranic acid, (+)-*trans*-whisky lactone, and (+)-*trans*-cognac lactone.

9.2.3
Aldol-Type Reactions

The asymmetric aldol reaction is one of the most important topics in modern catalytic synthesis. A new class of catalytic asymmetric aldol reactions has been developed in the last few years with the use of chiral organocatalysts [28], becoming one of the most advanced types of synthesis in the field of organocatalysis. In comparison to the well-studied application of chiral diols as ligands, the emergence of TADDOLs and their derivatives as effective organocatalysts is a very recent development. An indirect aldol reaction such as the vinylogous Mukaiyama aldol reaction has proved to be a powerful method for complex molecule synthesis. In 2005, Rawal et al. reported the use of 1-naphthyl-TADDOL as organocatalyst in the vinylogous Mukaiyama aldol reaction of silyl dienol ethers with a range of aldehydes, affording regiospecifically the products in good to excellent yields and with ee values as high as 90% [29]. Among the many permutations of Mukaiyama aldol reactions, the reaction of silylated enolates of amides with aldehydes is of particular interest since the presence of the enamine unit is expected to render these compounds highly nucleophilic. In 2006, Rawal et al. developed highly diastereo- and enantioselective Mukaiyama aldol reactions of *O*-silyl-*N,O*-ketene acetals mediated by a cyclohexylidene-TADDOL derivative [30]. The reaction was effective for a range of aldehydes, giving the corresponding amide aldol products in useful yields and selectivities (Scheme 9.2).

In 2007, the enantioselective vinylogous aldol reaction of Chan's diene [31] with a range of aldehydes was developed by Villano et al. using TADDOLs as organocatalysts, providing the corresponding vinylogous aldols in complete γ-selectivity in moderate enantioselectivity (≤61% ee) [32]. In addition to organocatalytic

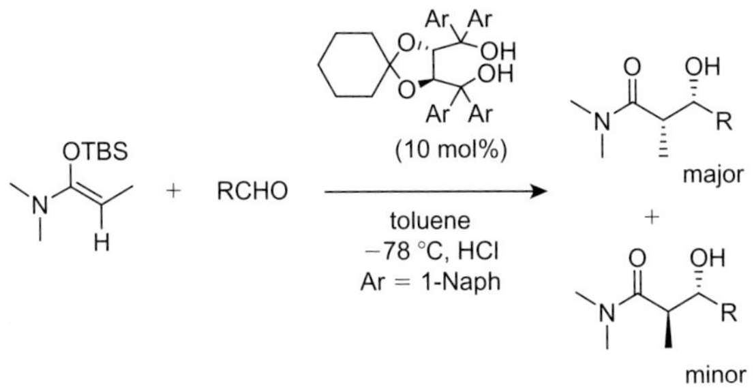

9.2 Nucleophilic Additions to C=O Double Bonds | 337

R = Ph: 94%, 88% de, 98% ee (major)
R = p-O$_2$NC$_6$H$_4$: 93%, 86% de, 94% ee (major)
R = p-ClC$_6$H$_4$: 86%, 90% de, 97% ee (major)
R = p-CF$_3$C$_6$H$_4$: 84%, > 92% de, 96% ee (major)
R = p-Tol: 77%, 88% de, 97% ee (major)
R = m-Tol: 77%, 80% de, 96% ee (major)
R = m-BrC$_6$H$_4$: 84%, 80% de, 80% ee (major)
R = 3,4-(Cl)$_2$C$_6$H$_3$: 71%, 80% de, 80% ee (major)
R = m-MeOC$_6$H$_4$: 81%, 86% de, 97% ee (major)
R = o-MeOC$_6$H$_4$: 50%, 78% de, 91% ee (major)
R = 2-Naph: 50%, 82% de, 97% ee (major)
R = n-Pr: 47%, 80% de, 91% ee (major)

Scheme 9.2 Mukaiyama aldol reactions.

aldol-type reactions, Schneider and Hansch reported the first Zr-catalyzed enantioselective aldol-Tishchenko reactions involving ketone aldols as enol equivalents [33]. Thus, a chiral zirconium TADDOLate complex was found to promote the aldol-Tishchenko reaction of ketone aldol adducts with a range of aldehydes, providing 1,3-*anti*-diol monoesters in good to excellent yields, complete diastereocontrol, and with up to 57% ee. The reductive coupling of α,β-unsaturated carbonyl compounds to aldehydes and ketones, termed the "reductive aldol reaction," has become a topic of intensive investigation [34]. In 2008, Krische et al. reported the first enantioselective reductive aldol couplings of vinyl ketones, which were achieved through the design of a new class of TADDOL-like phosphonite ligands [35]. As shown in Scheme 9.3, the Rh-catalyzed hydrogenative aldol coupling of methyl vinyl ketone or ethyl vinyl ketone with aldehyde allowed the

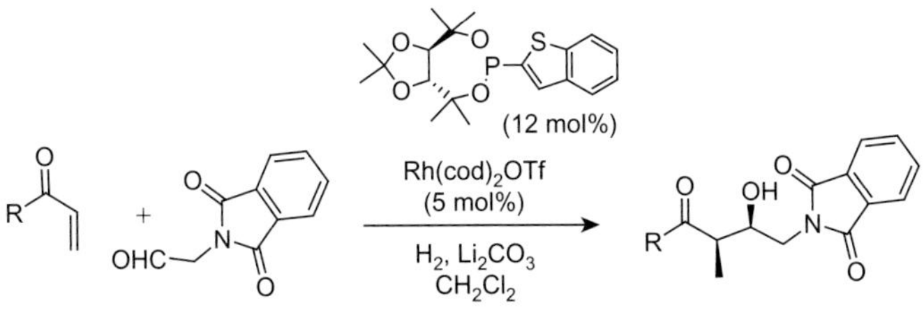

R = Me: 88%, 96% de, 96% ee
R = Et: 94%, 96% de, 94% ee

Scheme 9.3 Reductive aldol reactions.

corresponding linear aldol adducts to be obtained with high diastereo- and enantioselectivities of up to 96% de and 96% ee, respectively.

9.2.4
Miscellaneous Reactions

In 1992, Seebach *et al.* showed that a stoichiometric amount of magnesium TADDOLate behaved as a chiral Lewis acid in Grignard additions to aromatic, heteroaromatic, and α,β-unsaturated ketones, providing the corresponding enantiopure alcohols (Scheme 9.4) [36].

In 1987, Narasaka *et al.* reported highly enantioselective cyanohydrin syntheses mediated by a stoichiometric amount of a titanium TADDOLate [37]. As shown in Scheme 9.5, both aliphatic and aromatic aldehydes led to the corresponding products, in 80–90% yields and with excellent enantioselectivities (≥97% ee).

In more recent years, there has been considerable interest in asymmetric catalytic addition reactions of terminal alkynes to aldehydes [39], since chiral propargylic alcohols represent versatile precursors for the synthesis of many important products. The asymmetric addition of alkynylzinc reagents to aldehydes is an important method of producing such compounds [38–40]. As an example, *(R,R)*-TADDOL was employed as ligand by Singh and Kamble for the phenylacetylene addition to benzaldehyde, providing the corresponding propargylic alcohol in good yield (85%) and enantioselectivity of 60% ee [41].

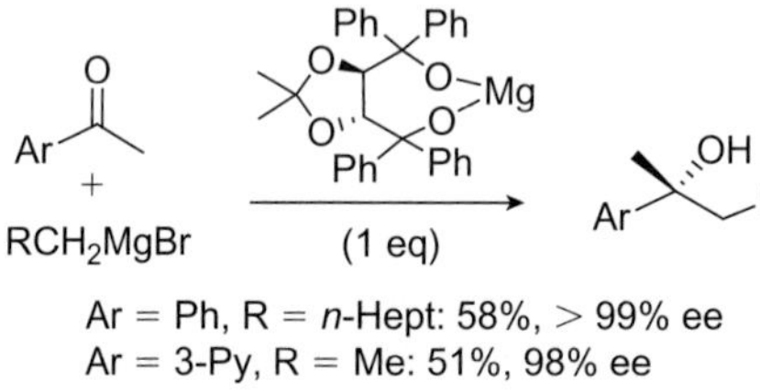
Ph
Ph
O
O
Mg
O
O
Ph
Ph
Ar
+
RCH2MgBr
(1 eq)
OH
R
Ar
Ar = Ph, R = n-Hept: 58%, > 99% ee
Ar = 3-Py, R = Me: 51%, 98% ee

Scheme 9.4 Additions of Grignard reagents to ketones.

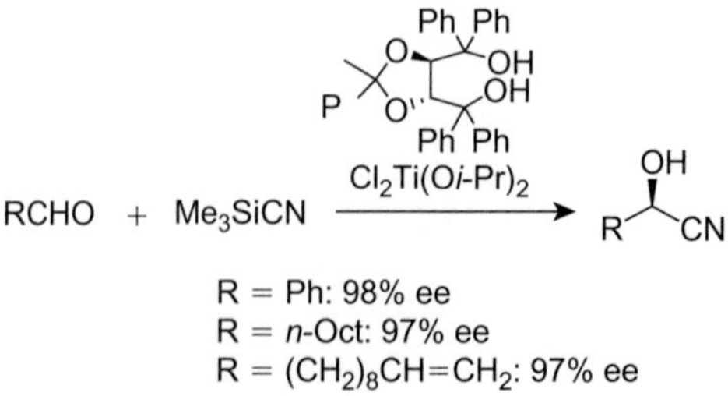
Ph Ph
OH
OH
P
Ph Ph
Cl2Ti(Oi-Pr)2
RCHO + Me3SiCN
OH
R
CN
R = Ph: 98% ee
R = n-Oct: 97% ee
R = (CH2)8CH=CH2: 97% ee

Scheme 9.5 Synthesis of cyanohydrins.

In 2001, Mikami et al. reported the synthesis of novel chiral titanium complexes, which were composed of 3,3′-modified biphenolate (BIPOLate) ligands atropisomerically controlled by (R,R)-TADDOLs [42]. This type of complexes proved to be highly efficient catalysts for the methylation of aldehydes, giving enantioselectivities of up to 100% ee. In another context, Goldfuss et al. reported an enantioselectivity of 90% ee for the methylation of benzaldehyde performed with methylaluminum TADDOLate, albeit in low yield (12%) [43]. Vinyloxy ethoxides, in the presence of a Lewis acidic and/or a coordinating counterion, react with aldehydes to yield the corresponding β-hydroxy-1,3-dioxolanes. In 2005, Redlich et al. developed this type of reaction in the presence of TADDOL-modified vinyloxides, providing the corresponding β-hydroxy-1,3-dioxolanes in good yields and moderate diastereoselectivities (≤64% de) [44].

A new procedure for asymmetric cyanohydrin synthesis was developed by Maruoka et al., employing (R,R)-TADDOL as ligand and acetone cyanohydrin as cyanide source [45]. In the presence of this ligand, $Zr(Ot\text{-}Bu)_4$ could serve as an effective promoter for the Meerwein–Ponndorf–Verley cyanation of a range of aldehydes, providing the corresponding cyanohydrins in moderate to good yields and enantioselectivities (29–85% ee). On the other hand, Kim et al. developed a highly efficient double activation catalysis by TADDOL/Ti(Oi-Pr)$_4$ and Ph$_3$PO for the cyanosilylation of various aldehydes [46]. This cyanosilylation reaction yielded the corresponding trimethylsilyl ethers in excellent yields and moderate enantioselectivities (≤60% ee).

In 2004, Johnson et al. demonstrated that TADDOL-derived metallophosphites could be useful umpolung catalysts for the enantioselective cross silyl benzoin reaction [47]. In this reaction, the chiral lithium phosphite formed in situ via the deprotonation of the corresponding phosphite with n-BuLi, catalyzed the cross silyl benzoin reaction between an acyltriethylsilane and an aldehyde to afford the corresponding silyloxy benzoin product. The tetra(o-fluorophenyl)-TADDOL phosphite provided the best combination of enantiocontrol and reactivity, while the presence on the catalyst of more powerful electron-withdrawing groups caused a dramatic drop in reactivity, and all other variations on the TADDOL phosphite backbone led to a decrease in selectivity relative to the fluorinated catalyst. This catalyst was applied to the cross benzoin reaction between several acylsilanes and aldehydes, providing high yields and enantioselectivities (Scheme 9.6). In 2007, these workers extended the scope of this methodology to the 1,3-silylacylation of a range of nitrones, providing access to the corresponding N-aryl α-amino ketones in good yields and high enantioselectivities of up to 97% ee [48].

9.3
Nucleophilic Conjugate Additions to Electron-Deficient C=C Double Bonds

The enantioselective conjugate addition is a fundamentally important transformation in asymmetric synthesis [49]. In this context, Alexakis and Benhaim reported the use of a series of TADDOL-derived chiral phosphorus ligands for the

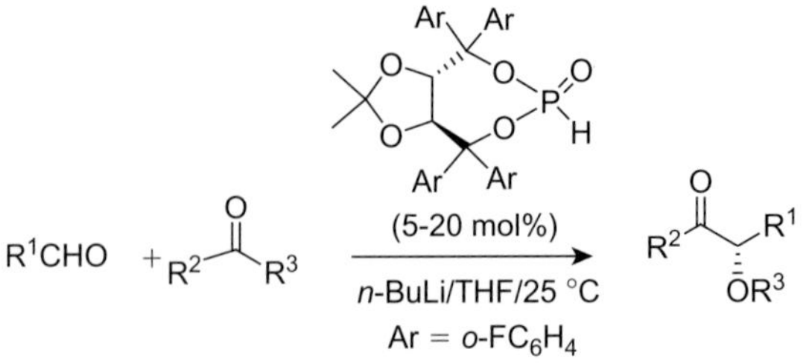

R¹CHO + R²C(O)R³ → (catalyst, 5-20 mol%), n-BuLi/THF/25 °C, Ar = o-FC$_6$H$_4$ → R²C(O)CH(R¹)(OR³)

Catalyst: TADDOL-derived phosphite (Ar groups = o-FC$_6$H$_4$)

R^1 = R^2 = Ph, R^3 = SiEt$_3$: 84%, 82% ee
R^1 = Ph, R^2 = p-ClC$_6$H$_4$, R^3 = SiEt$_3$: 75%, 82% ee
R^1 = p-ClC$_6$H$_4$, R^2 = Ph, R^3 = SiEt$_3$: 82%, 87% ee
R^1 = Ph, R^2 = p-MeOC$_6$H$_4$, R^3 = SiEt$_3$: 87%, 91% ee
R^1 = p-MeOC$_6$H$_4$, R^2 = Ph, R^3 = SiEt$_3$: 83%, 88% ee
R^1 = p-ClC$_6$H$_4$, R^2 = p-MeOC$_6$H$_4$, R^3 = SiEt$_3$: 83%, 90% ee
R^1 = p-MeOC$_6$H$_4$, R^2 = p-ClC$_6$H$_4$, R^3 = H: 79%, 83% ee
R^1 = Ph, R^2 = p-(Me$_2$N)C$_6$H$_4$, R^3 = H: 80%, 81% ee
R^1 = p-(Me$_2$N)C$_6$H$_4$, R^2 = Ph, R^3 = SiEt$_3$: 86%, 86% ee
R^1 = Ph, R^2 = 2-Fu, R^3 = SiEt$_3$: 65%, 85% ee
R^1 = Ph, R^2 = i-Pr, R^3 = SiEt$_3$: 78%, 73% ee
R^1 = Ph, R^2 = n-Hex, R^3 = SiEt$_3$: 88%, 41% ee
R^1 = n-Hex, R^2 = Ph, R^3 = SiEt$_3$: 72%, 67% ee

Scheme 9.6 Cross silyl benzoin reactions.

conjugate addition of dialkylzinc reagents to alkylidene malonates in the presence of copper triflate as catalyst [50]. The best enantioselectivities were obtained in the presence of ligands in which the TADDOL moiety was not the sole chiral center, since these ligands provided an additional chirality at the exocyclic moiety. As an example, the use of ligands containing a chiral substituted cyclohexanol skeleton allowed the corresponding products to be obtained in quantitative yields and moderate enantioselectivities (≤73% ee). In 2001, Feringa et al. reported the synthesis of new chiral bidentate phosphoramidites derived from TADDOL and involved these dimeric ligands in the Cu-catalyzed addition of ZnEt$_2$ to cyclic α,β-enones [51]. In 2008, Schmalz et al. reported the synthesis of a novel class of chiral phosphine-phosphite ligands suitable for the Cu-catalyzed 1,4-addition of Grignard reagents to cyclohexenone [52]. Among a library of ligands, those derived from TADDOL gave the highest enantioselectivities of up to 92% ee combined with a good regioselectivity (Scheme 9.7).

In 2006, Palacios et al. developed the Cu-catalyzed asymmetric conjugate addition of ZnEt$_2$ to α,β-unsaturated imines derived from α-amino acids. The reaction was performed in the presence of a TADDOL-derived phosphoramidite complex, affording the corresponding chiral α-dehydroamino esters bearing a stereogenic center in the γ-position with enantioselectivities of up to 88% ee [53]. With the aim of developing a novel route to amino acids, Seebach et al. reported the synthesis of

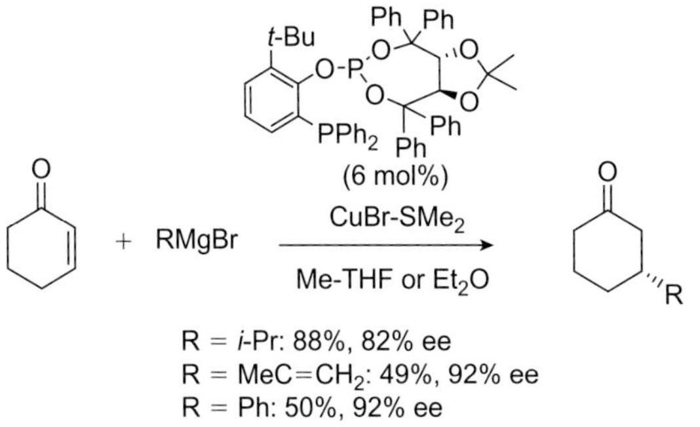

9.3 Nucleophilic Conjugate Additions to Electron-Deficient C=C Double Bonds | 341

Scheme 9.7 Conjugate additions of Grignard reagents to cyclohexenone.

a new type of substrate based on an achiral Ni(II) complex of a Schiff base of dehydroalanine [54]. An efficient catalytic method for asymmetric conjugate addition of CH acids to these novel Michael acceptors was successfully developed using TADDOLs as catalysts and provided the corresponding chiral amino acids after hydrolysis of the intermediate nickel complexes. A series of TADDOL derivatives were tested as catalysts, showing that the derivative bearing 1-naphthyl groups invariably led to the best enantioselectivities, of up to 80% ee, which were obtained in the presence of nucleophiles such as malonic ester derivatives, whereas nucleophiles such as thiophenol or amines reacted with the Schiff base of dehydroalanine without enantioselectivity.

In parallel with the development of aldehydes as acyl donors for conjugate additions, acylsilanes have shown promise as suitable pronucleophiles in conjugate addition reactions. As an example, Johnson et al. achieved a metallophosphite-catalyzed conjugate addition of an acylsilane such as benzoyldimethylphenylsilane to an unsaturated amide, providing the corresponding 1,4-diketone with high diastereoselectivity and moderate enantioselectivity ($\leq 82\%$ ee) [55]. The scope of this methodology was extended by these workers who selected a menthone-derived TADDOL phosphite as the most efficient catalyst for the addition of p-methoxybenzoylcyclohexyldimethylsilane to various aryl and alkyl N,N-dimethylacrylamide derivatives [56]. The corresponding chiral diketones were obtained in good yields and high enantioselectivities after desilylation (Scheme 9.8).

In addition, the ready access to, and the easy recovery of, TADDOLs allow their use as stoichiometric additives or reagents. In this area, Enders et al. described the diastereoselective Fe_2O_3-mediated Michael addition of a chiral TADDOL-derived phosphite to α,β-unsaturated malonates [57]. This reaction constituted the first example of an asymmetric P—C bond formation under heterogeneous conditions. The easy cleavage of the chiral auxiliary from the addition products led to the corresponding chiral β-substituted β-phosphonomalonates in good yields and high enantioselectivities. This inexpensive and readily accessible Fe_2O_3/KOH system could be successfully applied under mild conditions to a range of aromatic α,β-unsaturated malonates, providing high enantioselectivities (90–94% ee). Extension

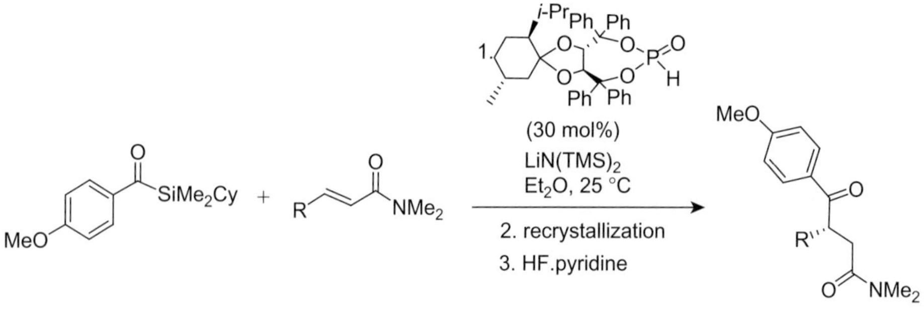

R = Ph: 68%, 99% ee
R = p-MeOC$_6$H$_4$: 63%, 92% ee
R = p-Tol: 78%, 90% ee
R = m-Tol: 67%, 99% ee
R = 2-Fu: 15%, 24% ee
R = p-ClC$_6$H$_4$: 66%, 98% ee
R = p-CF$_3$C$_6$H$_4$: 80%, 90% ee
R = 2-Naph: 66%, 97% ee
R = Me: 56%, 86% ee
R = Et: 82%, 71% ee
R = N-tosylindol-3-yl: 60%, 97% ee

Scheme 9.8 Conjugate additions of p-methoxybenzoylcyclohexyldimethylsilane to unsaturated amides.

of this methodology to the use of nitroalkenes as Michael acceptors provided the corresponding α-substituted β-nitrophosphonates with enantioselectivities of up to 95% ee [57].

9.4
Nucleophilic Substitutions

9.4.1
Allylic Substitutions

Asymmetric allylic substitution is a fundamental transformation in organic synthesis [58]. The vast majority of the studies reported apply palladium as the metal catalyst of choice, involving a plethora of ligands [59], mainly with phosphorus and/or nitrogen donor atoms [60]. As an example, Pfaltz and Hilgraf reported the synthesis of a series of novel P,N-ligands containing a chiral oxazoline ring and a cyclic phosphite group derived from TADDOL as a second chiral unit [61]. These TADDOL-phosphite-oxazoline ligands were successfully employed in Pd-catalyzed allylic alkylations of unsymmetrical substituted allyl substrates with dimethyl malonate, providing a mixture of the corresponding branched and linear regioisomers in high yields. In most cases, the reaction occurred with good regiocontrol and gave enantioselectivities of up to 94% ee. In contrast, the number of chiral monodentate ligands that have been studied in Pd-catalyzed asymmetric allylic transformations is still rather limited. Reports on the application of chiral monodentate ligands in allylic alkylations deal almost without exception with chiral phosphines and, especially, chiral phosphoramidite and phosphite ligands. In this context, van Leeuwen *et al.* showed that bulky monodentate phosphoramidite ligands based on a TADDOL backbone could be successfully employed in the Pd-catalyzed allylic alkylation reaction with carbon nucleophiles such as dimethyl malonate [62]. Owing to the mono-coordination of these ligands, high enantioselectivities of up to 92% ee could be obtained for the allylic alkylation of 1,3-diphenyl-2-propenyl acetate with dimethyl malonate (Scheme 9.9).

In 2006, Alexakis *et al.* reported the use of chiral monodentate TADDOL-derived phosphites in the allylic alkylation of a range of symmetrical allyl acetates with dimethyl malonate, providing variable enantioselectivities (4–79% ee) [63]. In addition, the Pd-catalyzed allylic amination reaction of allyl carbonates was developed by Takacs *et al.* by using self-assembled chiral TADDOL-derived bidentate P,P-ligands [64]. These diphosphites were prepared starting from a racemic mixture of monosubstituted *(R,R)*- and *(S,S)*-bisoxazoline (box) ligands, which was reacted with $Zn(OAc)_2$, affording the corresponding neutral $(box)_2Zn$ complex. A series of monosubstituted box derivatives were prepared, each member bearing a pendant TADDOL-derived monophosphite. As shown in Scheme 9.10, excellent yields and enantioselectivities were obtained for the allylic amination

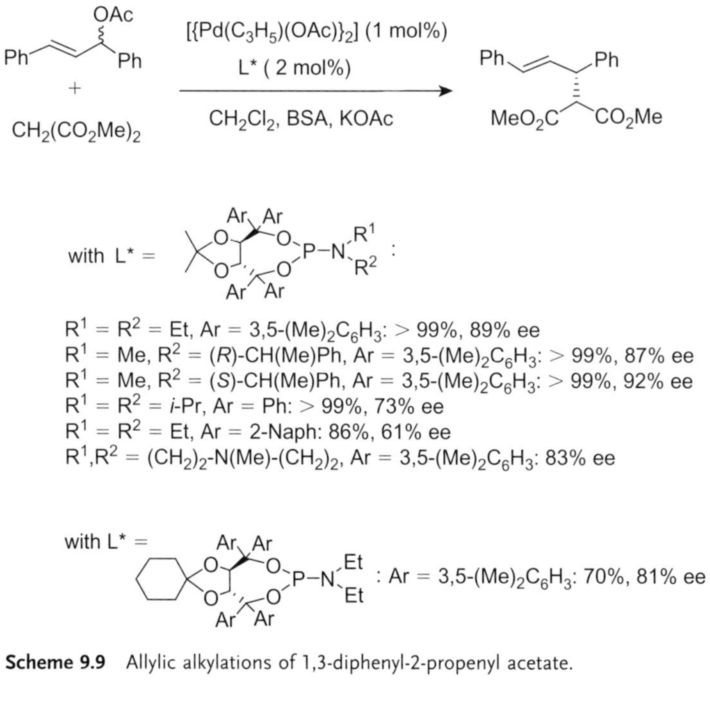
OAc
Ph
Ph
+
CH2(CO2Me)2
[{Pd(C3H5)(OAc)}2] (1 mol%)
L* (2 mol%)
CH2Cl2, BSA, KOAc
Ph
Ph
MeO2C
CO2Me
with L* =
Ar Ar
O O P–N R1 R2
:
R1 = R2 = Et, Ar = 3,5-(Me)2C6H3: > 99%, 89% ee
R1 = Me, R2 = (R)-CH(Me)Ph, Ar = 3,5-(Me)2C6H3: > 99%, 87% ee
R1 = Me, R2 = (S)-CH(Me)Ph, Ar = 3,5-(Me)2C6H3: > 99%, 92% ee
R1 = R2 = i-Pr, Ar = Ph: > 99%, 73% ee
R1 = R2 = Et, Ar = 2-Naph: 86%, 61% ee
R1,R2 = (CH2)2-N(Me)-(CH2)2, Ar = 3,5-(Me)2C6H3: 83% ee
with L* =
Et
Et
: Ar = 3,5-(Me)2C6H3: 70%, 81% ee

Scheme 9.9 Allylic alkylations of 1,3-diphenyl-2-propenyl acetate.

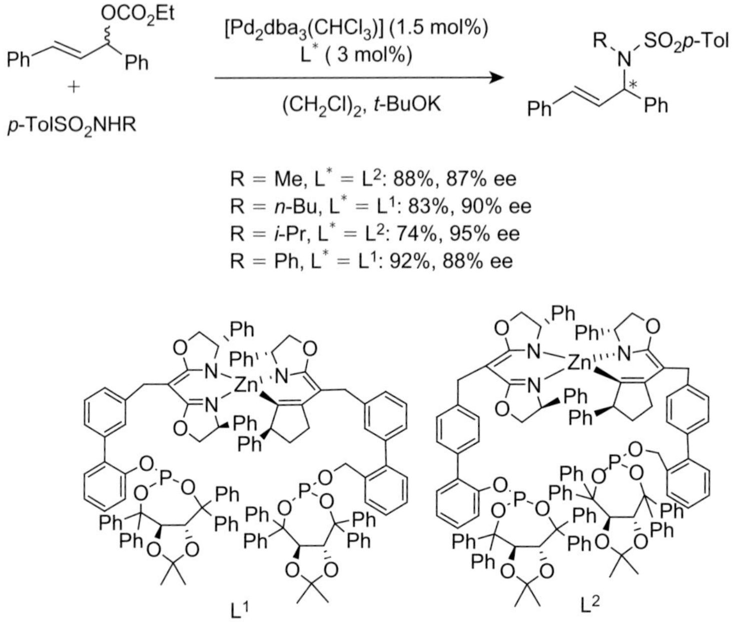

[Scheme 9.10 depicting allylic amination reaction:]

Ph–CH=CH–CH(OCO₂Et)–Ph + p-TolSO₂NHR → [Pd₂dba₃(CHCl₃)] (1.5 mol%), L* (3 mol%), (CH₂Cl)₂, t-BuOK → Ph–CH=CH–C*H(N(R)SO₂p-Tol)–Ph

R = Me, L* = L²: 88%, 87% ee
R = n-Bu, L* = L¹: 83%, 90% ee
R = i-Pr, L* = L²: 74%, 95% ee
R = Ph, L* = L¹: 92%, 88% ee

[Structures of ligands L¹ and L²]

Scheme 9.10 Allylic aminations.

reaction of 1,3-diphenyl-2-propenyl ethyl carbonate with a range of N-substituted sulfonamides.

Metals other than palladium were also involved in the allylic substitution reaction. Thus, Alexakis et al. reported efficient Cu-catalyzed S_N2' allylic substitutions of cinnamyl chloride with Grignard reagents using a series of chiral phosphorus ligands [65]. Among these, a ligand derived from (R,R)-TADDOL and (−)-N-methylephedrine showed a remarkably increased asymmetric induction over the other ligands tested, allowing enantioselectivities of up to 73% ee to be achieved.

9.4.2
α-Halogenations of Carbonyl Compounds

Although organic molecules containing fluorine are extremely rare in nature [66], organofluoro-compounds are becoming increasingly important in medicinal chemistry and as crop-protection agents [67]. In this context, TADDOLs have been demonstrated to be excellent chiral ligands in asymmetric halogenations as fluorinations [68]. The first catalytic enantioselective electrophilic fluorination, described by Togni et al. in 2000, concerned the fluorination of β-keto esters performed with Selectfluor [1-chloromethyl-4-fluoro-1,4-diazoniabicyclo[2.2.2]octane bis(tetrafluoroborate)] as fluorinating agent in the presence of a [TiCl₂(TADDOlato)] catalyst, providing up to 90% ees along with high yields (85–95%) [68a, 69]. In addition, these workers studied the chlorination of β-keto esters under similar conditions using *N*-chlorosuccinimide as chlorinating agent, which gave comparable enantioselectivities of up to 88% ee combined with high yields (83–97%) [68b]. Furthermore, Togni *et al.* developed catalytic fluorination/chlorination competition experiments of a β-keto ester in the presence of a [TiCl$_2$(TADDOLato)] catalyst [70]. Owing to their structure and versatile functional groups, α-acyl-γ-lactams are extremely interesting since they may act as valuable building blocks for the synthesis of molecules with potentially useful biological activity. In this context, Togni *et al.* developed the enantioselective fluorination of α-acyl-γ-lactams performed in the presence of *N*-fluorobenzenesulfonimide as fluorinating agent and with a [TiCl$_2$(TADDOLato)] catalyst [71]. The reaction provided the corresponding fluorinated products in variable enantioselectivities (6–87% ee). In addition, this methodology was further extended to the α-heterodihalogenation of β-keto esters, providing the corresponding α-chloro-α-fluoro-β-keto esters with moderate enantioselectivities ($\leq$65% ee) [72].

9.4.3
Miscellaneous Substitutions

The formation of a quaternary stereogenic center coinciding with a hydroxylation process is still a very rare reaction. In this context, Togni *et al.* expanded the scope of the methodology previously applied to the α-fluorination of β-keto esters to the hydroxylation of these compounds [73]. Therefore, a [TiCl$_2$(TADDOLato)] catalyst was used to promote the α-hydroxylation of a range of β-keto esters in the presence of 2-(phenylsulfonyl)-3-(4-nitrophenyl)oxaziridine as the oxidizing agent, leading to the corresponding chiral hydroxylated products in high yields and variable enantioselectivities (9–94% ee).

Chiral compounds containing sulfur functionalities have been recognized as important auxiliaries as well as ligands [74]. In this area, Togni and Jereb developed a highly efficient direct enantioselective catalytic sulfenylation of β-keto esters using a similar methodology to that applied for the analogous hydroxylation and fluorination reactions [75]. Indeed, the use of a [TiCl$_2$(TADDOLato)] catalyst in the presence of phenylsulfenyl chloride allowed the corresponding chiral α-sulfenylated β-keto esters to be obtained in high yields and moderate enantioselectivities of 53–88% ee.

Moreover, TADDOLs have been demonstrated to be excellent chiral ligands in asymmetric alkylations [76]. Even though transition-metal-catalyzed enantioselective reactions will certainly continue to play a central role in synthetic organic chemistry in the future, the last few years have seen an increasing trend in the use of metal-free catalysts [28]. In comparison to the well-studied application of chiral diols as ligands, the emergence of TADDOLs as effective general organocatalysts is very recent. In 2006, Gonzalez-Muniz *et al.* studied the intramolecular base-promoted alkylation of *N*-chloroacetyl derivatives of several amino acids performed in the presence of TADDOL, providing the corresponding chiral β-lactams with moderate enantioselectivities ($\leq$70% ee) [77].

9.5
Cycloaddition Reactions

9.5.1
Diels–Alder reactions

Chiral diols such as TADDOLs have been demonstrated to be excellent chiral ligands in enantioselective Diels–Alder reactions [78]. The most extensive efforts were dedicated to Diels–Alder reactions of enoyloxazolidinones of fumaric, acrylic, and crotonic acids with cyclopentadiene, which led almost quantitatively to a single stereoisomer by using (R,R)-TADDOLs as ligands of $TiCl_2(Oi\text{-}Pr)_2$ [78c, 79]. Enoyloxazolidinones were also cycloadducted to other dienes under similar conditions, providing the corresponding cycloadducts with excellent enantioselectivities of up to 99% ee [80]. In addition, cycloadditions involving other α,β-unsaturated carbonyl compounds, such as quinines [81], ene-1,2-diones [82], phenylsulfonylmethyl enones [83], maleimides [84], and amidoacrylates [85] were also accomplished with Cl_2Ti-TADDOLate catalysis, leading in many cases to enantiopure products. More recently, Renaud and Corminboeuf demonstrated that N-alkoxyacrylates were suitable substrates for enantioselective Diels–Alder reactions performed in the presence of TADDOLs [86]. High enantioselectivities of up to 92% ee were achieved for a range of N-alkoxyacrylates by using a simple chiral Lewis acid prepared from (i-Bu)$_3$Al and 1-naphthyl-TADDOL. For a long time, it was not known that organocatalysts could be used to catalyze the Diels–Alder reactions. In recent years, however, several types of organocatalysts have been developed. In this area, Rawal et al. reported an elegant example of using TADDOLs as organocatalysts for the Diels–Alder reaction of aminosiloxydienes with acrolein dienophiles to afford the corresponding products in good yields and high enantioselectivities of up to 92% ee (Scheme 9.11) [87].

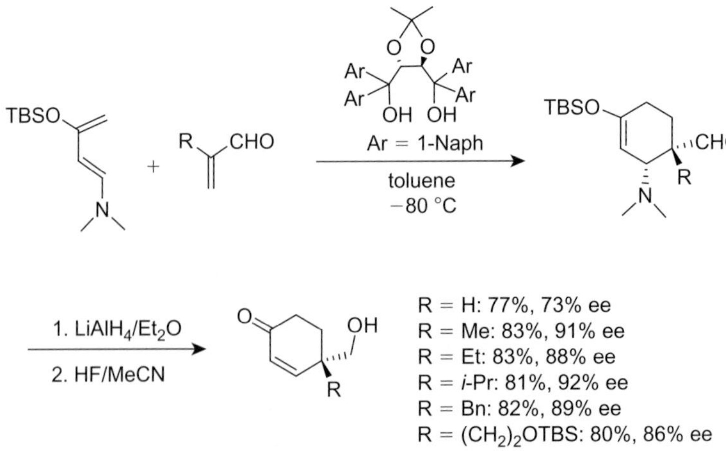

Scheme 9.11 Organocatalytic Diels–Alder reactions.

9.5.2
Hetero-Diels–Alder Reactions

Although less studied than the all-carbon version, the hetero-Diels–Alder reaction has attracted greater attention over the past decades and constitutes nowadays an extremely useful reaction. It provides a versatile regio- and stereoselective approach towards important six-membered partly saturated heterocycles [88]. In this context, Gautun *et al.* reported the hetero-Diels–Alder reactions of cyclohexadiene with *N*-sulfinyl dienophiles promoted by chiral Ti(IV)-based Lewis acids prepared from Me_2TiCl_2 and TADDOLs [89]. In almost all cases, the major diastereomer resulted from an *endo* approach of the reagents and was obtained with up to 83% yield and 69% ee. In recent years, several organocatalytic enantioselective hetero-Diels–Alder reactions have been developed. Of these reactions, one of the most exciting developments is the 1-naphthyl-TADDOL-promoted reaction of 1-amino-3-siloxybutadiene with aldehydes, which was reported by Rawal *et al.* [90]. These cycloadditions provided the cycloadducts highly enantioselectively after treatment with AcCl (Scheme 9.12). In 2004, this TADDOL derivative was employed by Ding *et al.* to induce the hetero-Diels–Alder reaction of an aldehyde with Brassard's diene, affording the corresponding chiral δ-lactone derivative with enantioselectivities of up to 98% ee. [91]. The usefulness of this methodology was demonstrated by its application in the total synthesis of the natural product (+)-dihydrokawain.

In 2006, Ding *et al.* investigated hetero-Diels–Alder reactions of Danishefsky's diene and its analogs with various aldehydes in the presence of a range of TADDOL derivatives [92]. The use of naphth-1-TADDOL as organocatalyst allowed high enantioselectivities of 76–98% ee to be obtained. While investigating the viability

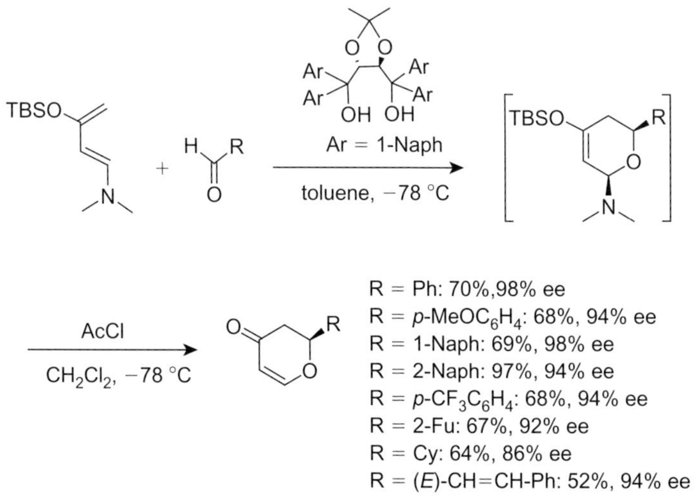
TBSO
H
R
O
N
O O
Ar
Ar
Ar
Ar
OH OH
Ar = 1-Naph
toluene, −78 °C
TBSO
R
O
N
AcCl
CH2Cl2, −78 °C
O
R
O
R = Ph: 70%,98% ee
R = p-MeOC6H4: 68%, 94% ee
R = 1-Naph: 69%, 98% ee
R = 2-Naph: 97%, 94% ee
R = p-CF3C6H4: 68%, 94% ee
R = 2-Fu: 67%, 92% ee
R = Cy: 64%, 86% ee
R = (E)-CH=CH-Ph: 52%, 94% ee

Scheme 9.12 Organocatalytic hetero-Diels–Alder reactions of aminosiloxydiene.

of the enantioselective vinylogous aldol reaction of Chan's diene in the presence of this catalyst, Villano et al. found that the involvement of electron-poor aromatic aldehydes in this reaction enhanced the reactivity, and made a competing hetero-Diels–Alder reaction take place in comparable (or higher) yields and enantioselectivities (21–59% ee) under solvent-free conditions [32].

9.5.3
Miscellaneous Cycloadditions

TADDOLs have also been demonstrated to be excellent chiral ligands in asymmetric 1,3-dipolar cycloadditions [93]. In 2006, Porco et al. developed a synthesis of chiral rocaglamides on the basis of a 1,3-dipolar cycloaddition of an oxidopyrylium species derived from excited-state intramolecular proton transfer of a 3-hydroxyflavone, using specifically functionalized TADDOL derivatives as organocatalysts [94]. The 1,3-dipolar photocycloaddition of a 3-hydroxyflavone in the presence of methyl cinnamate provided the corresponding cycloadduct in 61% yield, which was further converted into a mixture of the expected *endo* methyl rocaglate with 82% ee and its corresponding *exo* isomer by employing a base-mediated α-ketol rearrangement/hydroxyl-directed reduction sequence. In 2007, these workers reported the conversion of *endo* methyl rocaglate into the complex natural rocaglate (−)-silvestrol, exhibiting a very potent cytotoxic activity against human lung cancer cells [95]. In addition, a 1,3-dipolar cycloaddition of azomethine ylides, using alkenylboronic esters equipped with TADDOLs as chiral auxiliaries, was developed by Zong [96]. The cycloaddition provided the corresponding cycloadduct intermediates, in moderate diastereoselectivities (42–60% de), which were then converted into the more stable corresponding (+)-pinanylboronic ester-substituted pyrrolidine derivatives through transesterification with (+)-pinanediol.

In 2006, Rovis and Yu developed a regio- and enantioselective Rh-catalyzed [2+2+2]-cycloaddition of alkenyl isocyanates with terminal alkynes, affording the corresponding bicyclic lactams and/or vinylogous amides, using TADDOL-derived phosphoramidites as chiral ligands [97]. The cycloaddition generally proceeded cleanly to furnish the cycloadducts in high yields and enantioselectivities of up to 94% ee and in a highly regioselective manner. The synthetic utility of this methodology was demonstrated in a total synthesis of (+)-lasubine II [98]. Furthermore, this strategy was successfully applied to the synthesis of chiral bicyclic amidines by using carbodiimides [99]. Indeed, the Rh-catalyzed [2+2+2]-cycloaddition of alkenyl carbodiimides with terminal alkynes afforded the corresponding bicyclic amidines as the major products with excellent enantioselectivities of up to 99% ee (Scheme 9.13).

In 1995, Charette et al. reported the use of a Ti-TADDOLate complex for the asymmetric cyclopropanation of allylic alcohols, providing excellent enantioselectivities of up to 98% ee [100]. In 2001, these authors extended the scope of this methodology to the use of other Ti-TADDOLate complexes, which were applied to the cyclopropanation of a wide variety of allylic alcohols, providing modest

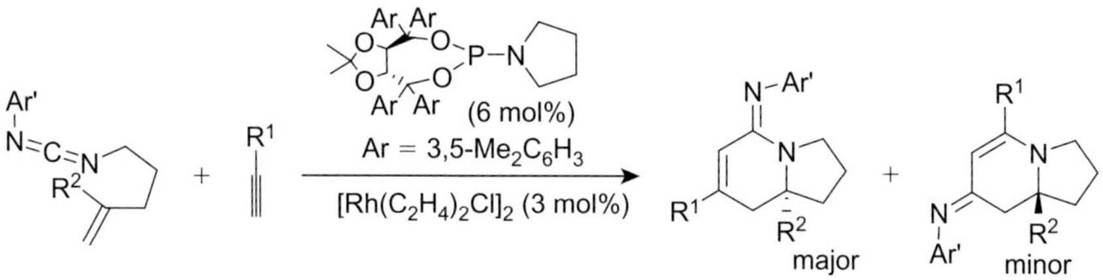

Scheme 9.13 [2 + 2 + 2]-Cycloadditions of alkenyl isocyanates.

Ar' = o-MeOC$_6$H$_4$, R^1 = p-BrC$_6$H$_4$, R^2 = H:
75%, major:minor > 95:5, 98% ee (major)
Ar' = o-MeOC$_6$H$_4$, R^1 = m-FC$_6$H$_4$, R^2 = H:
77%, major:minor > 95:5, 99% ee (major)
Ar' = o-MeOC$_6$H$_4$, R^1 = 3,5F$_2$C$_6$H$_3$, R^2 = H:
66%, major:minor > 95:5, 99% ee (major)
Ar' = o-MeOC$_6$H$_4$, R^1 = p-AcC$_6$H$_4$, R^2 = H:
78%, major:minor > 95:5, 99% ee (major)
Ar' = o-MeOC$_6$H$_4$, R^1 = 3,5F$_2$C$_6$H$_3$, R^2 = Me:
79%, major:minor > 95:5, 98% ee (major)
Ar' = o-CF$_3$C$_6$H$_4$, R^1 = m-Tol, R^2 = H:
74%, major:minor = 89:11, 98% ee (major)

to excellent yields and enantioselectivities (48–94% ee) [101]. To extend the scope of the cyclopropanation reaction, the same group developed a new family of chiral phosphates derived from TADDOL [102]. The use of these ligands in the Simmons–Smith cyclopropanation of both functionalized and unfunctionalized olefins led to the formation of the desired cyclopropanes in good yields and moderate enantioselectivities (31–75% ee).

9.6
Oxidation and Reduction Reactions

The asymmetric epoxidation of olefins is one of the most useful and challenging reactions in modern organic chemistry [103] due to the fact that chiral epoxides constitute versatile building blocks and that many biologically active compounds and natural products contain epoxide functionalities [104]. In this area, Seebach and Aoki reported the preparation of a TADDOL-derived hydroperoxide, TADOOH, from H$_2$O$_2$ and TADDOL by replacement of one OH group in TADDOL by an OOH group [3, 105]. This stable hydroperoxide was tested as a chiral oxidant in several asymmetric oxidations, such as the epoxidation of enones in base catalysis. The treatment of a range of enones with TADOOH in the presence of n-BuLi furnished the corresponding epoxy ketones in high yields and variable enantioselectivities (40–97% ee). In 2003, Vogl et al. investigated the Vo(V)-catalyzed epoxidation of allylic alcohols in the presence of TADOOH [106]. Thus, treatment of a range of allylic alcohols with an achiral vanadium catalyst and

TADOOH allowed the corresponding chiral epoxides to be obtained in moderate enantioselectivities (41–72% ee). Similar reactions were also performed by Zhang et al. in the presence of an achiral oxovanadium(IV)-substituted polyoxometallate (POM) and TADOOH [107]. The use of this resistant oxovanadium sandwich-type POM, $[ZnW(VO)_2(ZnW_9O_{34})_2]^{12-}$, achieved chemo-, regio-, and stereoselective epoxidations of allylic alcohols by TADOOH with very high catalytic efficiency [up to 42 000 TON (turnover number)], allowing moderate to high enantioselectivities (18–90% ee) to be obtained.

Chiral sulfoxides belong to the class of chiral organosulfur compounds that are most widely used in asymmetric synthesis [74b, 108]. In this area, TADOOH was applied as oxidant by Seebach and Aoki to an enantioselective oxidation of methyl phenyl sulfide [3, 105]. Thus, TADOOH oxidized methyl phenyl sulfide without any catalysis, providing the corresponding (S)-sulfoxide in 73% yield and 86% ee.

In addition, this oxidant was applied to the asymmetric Baeyer–Villiger oxidation of bicyclic and tricyclic cyclobutanones with kinetic resolution [3, 105]. In the presence of DBU and LiCl, the cyclobutanone was oxidized into the normal Baeyer–Villiger product, along with the enantiomerically enriched unreacted cyclobutanone and, in some cases, the abnormal Baeyer–Villiger product. Thus, these results, providing enantioselectivities of up to 99% ee, demonstrated that TADOOH turned out to be an enantiomer-differentiating, and also regioselective, stoichiometric oxidant in the Baeyer–Villiger reaction of cyclobutanones.

In contrast, the enantioselective hydrogenation of olefins with chiral rhodium or ruthenium catalysts is the best-established and most widely used method in asymmetric catalysis [109]. Meanwhile, unfunctionalized olefins are still particularly difficult substrates, because, in general, a polar group adjacent to the C=C bond, which can coordinate to the metal catalyst, is required for high catalyst activity and enantioselectivity. Indeed, there are very few examples of highly enantioselective hydrogenations of olefins lacking such a polar group. To overcome these limitations, Pfaltz et al. found a new class of hydrogenation catalysts, showing exceptionally high activity with unfunctionalized olefins and giving, in many cases, excellent enantioselectivity of up to 95% ee [61, 110]. These highly efficient catalysts are iridium complexes of chiral P,N-ligands derived from TADDOL. Under these conditions, several types of olefins could be hydrogenated with moderate to high enantioselectivity (58–95% ee) and full conversion in almost all cases by using these iridium complexes as their BAr_F salts (BAr_F=tetrakis [3,5-bis(trifluoromethyl)phenyl]borate). On the other hand, high enantioselectivities of up to 94% ee combined with good activities were obtained by van Leeuwen et al. in the Rh-catalyzed hydrogenation of olefins, such as dimethyl itaconate and methyl α-acetamidoacrylate (Scheme 9.14) [111]. In this work, the chiral ligands were calyx[4]arene-based TADDOL-containing diphosphites.

Excellent enantioselectivities were obtained by Seebach and Beck for the reduction of arylketones using a lithium aluminum hydride derivative of TADDOL (Scheme 9.15) [112].

In 1999, Seebach and Heldmann reported the Rh-catalyzed hydrosilylation of ketones with Ph_2SiH_2 performed in the presence of TADDOL derivatives [113].

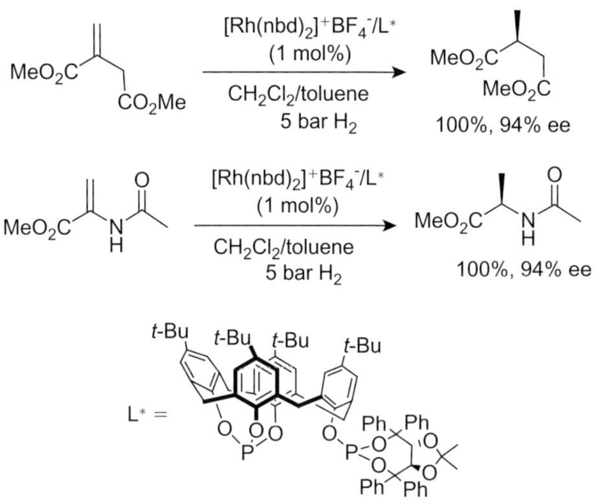

Scheme 9.14 Rhodium-catalyzed hydrogenations of olefins.

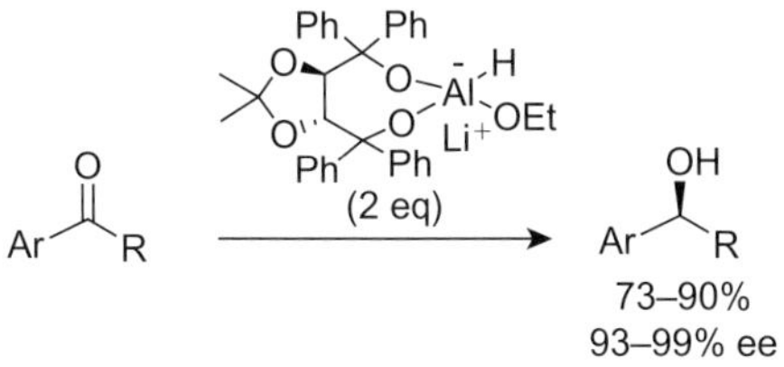

Scheme 9.15 Reduction of ketones.

The better enantioselectivities of up to 98% ee were obtained by using a phosphorus derivative of TADDOL bearing an additional chelating unit such as an oxazolinyl group. In the same context, Finn et al. reported the Rh-catalyzed hydrosilylation of ketones based on the use of a chiral phosphite P,N-ligand derived from TADDOL and (+)-N-methylephedrine [114]. A high level of asymmetric induction was observed for various ketones (15–89% ee).

9.7
Miscellaneous Reactions

Chiral $(i\text{-PrO})_2$Ti-TADDOLates were used by Seebach et al. for transesterifications involving differentiation between enantiomers for ring openings of lactones [115] and azalactones [116], and for openings of *meso* five-membered cyclic anhydrides [117] and *meso*-N-sulfonylimides, providing the corresponding isopropyl esters with enantioselectivities of up to >99% ee [118]. On the other hand, these authors exploited the acidity of TADDOLs to promote the asymmetric protonation of achiral enolates, allowing the conversion of racemic ketones into their enantiomerically forms with enantioselectivities of up to 99% ee [3].

Among the already existing methods for the synthesis of chiral α-hydroxy carbonyl compounds, the direct organocatalytic enantioselective α-aminoxylation of carbonyl compounds is one of the most important strategies. Nitroso compounds such as nitrosobenzene are useful electrophiles for performing this type of reaction although the nitrogen versus oxygen reactivity should be carefully controlled through the selection of appropriate catalysts and reaction conditions [119]. In this context, Yamamoto and Momiyama reported regio- and enantioselective nitroso aldol reactions in which a piperidine enamine of cyclohexanone reacted with nitrosobenzene with an enantioselectivity of up to 91% ee in the presence of *(S,S)*-1-naphthyl-TADDOL as catalyst, leading selectively to the *N*-nitroso aldol reaction product [120]. The scope of this reaction was extended by Greck *et al.*, confirming the exclusive formation of the N-regioisomers in a highly enantioselective manner with up to 93% ee [121].

Chiral α-branched amines are common substructures within biologically active materials and hence attract broad interest. Additions of carbon fragments to C=N bonds of imines and related compounds build up the carbon framework in the same operation as asymmetric induction, so that this approach is one of the more attractive entries to chiral amines [122]. In this context, the Mannich reaction is a widely applied means of producing β-amino carbonyl compounds starting from cheap and readily available substrates, an aldehyde, an amine, and a ketone [123]. As an alternative strategy, the reaction can also be performed as a nucleophilic addition of a C-nucleophile to a preformed imine that is prepared starting from the aldehyde and an amine source. In recent years, several organocatalytic Mannich reactions were successfully developed. As an example, Akiyama *et al.* reported Mannich-type reactions of a ketene silyl acetal with a range of aldimines catalyzed by a chiral phosphate having the TADDOL scaffold, which provided the corresponding β-amino acid esters with high enantioselectivities (73–92% ee) [124].

The Strecker reaction, starting from an aldehyde, ammonia, and a cyanide source, is an efficient method for the preparation of α-amino acids and their derivatives. Interestingly, several completely different types of chiral organocatalysts were found to have catalytic hydrocyanation properties. Among these molecules is TADDOL, which was recently implicated in the hydrocyanation of aldimines by Rueping *et al.* [125]. Although the enantioselectivities obtained remained moderate (8–56% ee), the results demonstrated the feasibility of chiral diols such as TADDOLs as promising chiral catalysts for the Strecker reaction.

In 2002, Schmalz *et al.* reported the preparation of chiral bidentate P,P ligands derived from TADDOL, which were further tested for the Rh-catalyzed hydroboration of styrene to give 1-phenylethanol in enantioselectivities of up to 91% ee [126]. In 2007, this methodology was applied to an efficient and highly stereoselective synthetic entry to a *trans*-7,8-dimethoxycalamenene, a projected intermediate for the total synthesis of marine biologically active serrulatane and amphilectane diterpenes [127]. Furthermore, Takacs *et al.* extended the scope of this reaction to a wide range of either electron-donating or -withdrawing substituted styrenes and by using very simple TADDOL-derived monodentate ligands [128]. In this case, the best results (Scheme 9.16) were obtained by using pinacolborane. Furthermore, a TADDOL-derived (4′-*tert*-butyl)phenylphosphite was applied as ligand to the efficient Rh-catalyzed hydroboration of a β,γ-unsaturated amide, providing the corresponding β-hydroxycarbonyl derivative in good yield (79%) and excellent enantioselectivity of 97% ee (Scheme 9.16) [129]. In 2008, these authors reported the synthesis of a series of TADDOL phosphite-bearing self-assembled ligands, readily prepared in combinatorial arrays by chirality-directed self-assembly upon addition of zinc(II), providing a focused ligand library for Rh-catalyzed hydroboration of *ortho*-substituted styrenes [130]. For example, the application of these chiral heterodimeric self-assembled ligands to the Rh-catalyzed hydroboration of 2-methoxystyrene allowed the corresponding alcohol to be obtained in both excellent yield and enantioselectivity of 94% and up to 99% ee, respectively. In 2005, Morken *et al.* used a closely related ligand to induce chirality in a Pd-catalyzed diboration of allenes, providing a reactive chiral allylboron intermediate that was a versatile reagent for the allylation of aldehydes [131]. On the basis of the high level of enantioselectivity obtained for the diboration reaction (93–98% ee), a one-pot diboration/allylboration/oxidation cascade process was elaborated, achieving β-hydroxy-ketones with high enantioselectivity (84–95% ee).

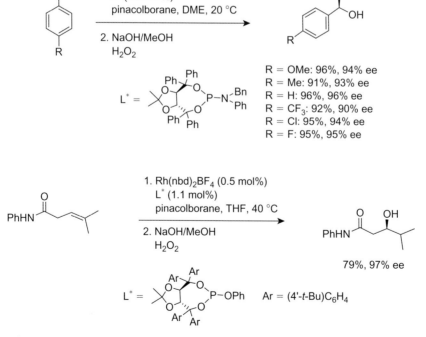

Scheme 9.16 Rhodium-catalyzed hydroborations.

In 2005, a new method for the asymmetric diamination of alkenes was developed by Muniz and Nieger by using a bis(imido)osmium reagent in the presence of a catalytic amount of titanium TADDOLate [132]. In particular, the reaction of various oxazolidinones allowed the corresponding chiral osmaimidazolidinones to be formed in high yields and moderate to high enantioselectivities (68–90% ee). A TADDOL-derived ligand was employed by Shibasaki *et al.* in the Zr-catalyzed conversion of cyclohexene into its corresponding *trans*-β-cyanohydrin with moderate enantioselectivity (62% ee) [133].

The pinacol coupling of aldehydes to give diols is an important method for constructing vicinally functionalized C–C bonds. In recent years, TADDOL was successfully exploited as its corresponding *in situ* generated titanium complex in a pinacol coupling reaction of aldehydes, providing the corresponding chiral diols with variable diastereoselectivities (28–100% de) and moderate enantioselectivities (49–74% ee) [134]. In 2005, similar reactions were developed by Studer and Knoop by using for the first time a chiral Ti(III)-TADDOLate [135]. In this study, it was demonstrated that cyclohexadienyl-Ti(IV) complexes underwent thermal Ti–C bond homolysis to generate the corresponding Ti(III) complexes. These Ti(III) complexes could be used in the reductive dimerization of benzaldehyde, yielding the corresponding diol in 48% yield and moderate stereoselectivities (50% de and 56% ee).

On the other hand, TADDOL was involved in an enantioselective oxidative coupling of the titanium enolate of 3-phenylacetyl-2-oxazolidinone with a ferrocenium cation as oxidant, affording the corresponding chiral dimer with 91% yield and moderate stereoselectivities (50% de, 76% ee) [136].

In the last 28 years, the asymmetric Heck reaction has received considerable attention, providing high enantioselectivities [137]. As an example, Feringa *et al.* developed a highly enantioselective intramolecular Heck reaction of cyclohexadienone monoacetals performed in the presence of a bidentate TADDOL-based phosphoramidite ligand [138]. Moreover, these workers found that monodentate phosphoramidites derived from TADDOL were effective ligands for this reaction, providing enantioselectivities of up to 96% ee.

In a different area, Xue and Jiang reported the Rh-catalyzed hydroformylation of styrene performed in the presence of a diphosphite derived from TADDOL [139]. Thus, the use of a diphosphite derived from a (1*R*,5*S*,6*R*)-*trans,trans*-spirol and *(R,R)*-TADDOL or *(S,S)*-TADDOL provided moderate enantioselectivities ($\leq$49% ee).

9.8
Conclusions

Since its introduction by Seebach in 1982, TADDOL has generated a great diversity of readily accessible derivatives, which have become among the most widely used ligands for catalytic enantioselective reactions. This chapter presents the principal and highly versatile reactions that employ TADDOLs and their derivatives as chiral catalysts and illustrates in depth the diversity of useful products that can be obtained through the use of these powerful versatile chiral reagents [140]. In particular, several enantioselective catalytic reactions involving TADDOLs and their derivatives as organocatalysts have recently appeared in the literature.

References

1. Brunel, J. M. (2005) *Chem. Rev.*, **105**, 857–897.
2. (a) Seebach, D., Beck, A.K., Mukhopadhyay, T., and Thomas, E. (1982) *Helv. Chim. Acta*, **65**, 1101–1133; (b) Seebach, D., Beck, A.K., Schiess, M., Widler, L., and Wonnacott, A. (1983) *Pure Appl. Chem.*, **55**, 1807–1822; (c) Seebach, D., Beck, A.K., Imwinkelried, R., Roggo, S., and Wonnacott, A. (1987) *Helv. Chim. Acta*, **70**, 954–974.
3. Seebach, D., Beck, A.K., and Heckel, A. (2001) *Angew. Chem. Int. Ed.*, **40**, 92–138.
4. Oguni, N. and Omi, T. (1984) *Tetrahedron Lett.*, **25**, 2823–2824.
5. Jacobsen, E.N., Pfaltz, A., and Yamamoto, H. (eds) (1999) *Comprehensive Asymmetric Catalysis*, Vols **I–III**, Springer, Berlin.
6. (a) Seebach, D., Plattner, D.A., Beck, A. K., Wang, Y.M., and Hunziker, D. (1992) *Helv. Chim. Acta*, **75**, 2171–2209; (b) Seebach, D., Beck, A.K., Schmidt, B., and Wang, Y.M. (1994) *Tetrahedron*, **50**, 4363–4384; (c) Ito, Y.N., Ariza, X., Beck, A.K., Bohac, A., Ganter, C., Gawley, R.E., Kühnle, F.N.M., Tuleja, J., Wang, Y.M., and Seebach, D. (1994) *Helv. Chim. Acta*, **77**, 2071–2210; (d) Schmidt, B. and Seebach, D. (1991) *Angew. Chem., Int. Ed. Engl.*, **30**, 99–101.
7. Fukuda, A., Kobayashi, Y., Kimachi, T., and Takemoto, Y. (2003) *Tetrahedron*, **59**, 9305–9313.
8. Sheen, W.-S. and Gau, H.-M. (2004) *Inorg. Chim. Acta*, **357**, 2279–2284.
9. (a) Sokeirik, Y.S., Mori, H., Tarui, A., Sato, K., Omote, M., Ando, A., and Kumadaki, I. (2007) *Org. Lett.*, **9**, 1927–1929; (b) Sokeirik, Y.S., Hoshina, A., Omote, M., Sato, K., Tarui, A., Kumadaki, I., and Ando, A. (2008) *Chem. Asian J.*, **3**, 1850–1856.
10. Altava, B., Burguette, M.I., Escuder, B., Luis, S.V., Salvador, R.V., Fraile, J.M., Mayoral, J.A., and Royo, A.J. (1997) *J. Org. Chem.*, **62**, 3126–3134.
11. Seebach, D. (1997) US Patent Application 959390.
12. Altava, B., Burguette, M.I., Luis, S.V., and Mayoral, J.A. (1994) *Tetrahedron*, **50**, 7535–7542.
13. Altava, B., Burguette, M.I., Escuder, B., Luis, S.V., Salvador, R.V., Fraile, J.M., Mayoral, J.A., Garcia, J.I., and Vincent, M.J. (2000) *Angew. Chem. Int. Ed.*, **39**, 1503–1506.
14. Seebach, D., Marti, R.E., and Hintermann, T. (1996) *Helv. Chim. Acta*, **79**, 1710–1740.
15. Sellner, H. and Seebach, D. (1999) *Angew. Chem. Int. Ed.*, **38**, 1918–1920.
16. Sellner, H., Rheiner, P.B., and Seebach, D. (2002) *Helv. Chim. Acta*, **85**, 352–387.
17. Degni, S., Wilen, C.-E., and Leino, R. (2001) *Org. Lett.*, **3**, 2551–2554.
18. Heckel, A. and Seebach, D. (2002) *Chem. Eur. J.*, **8**, 560–572.
19. Degni, S., Strandman, S., Laari, P., Nuopponen, M., Wilen, C.-E., Tenhu, H., and Rosling, A. (2005) *React. Funct. Polym.*, **62**, 231–240.
20. Altava, B., Burguete, M.I., Garcia-Verdugo, E., Luis, S.V., and Vicent, M.J. (2006) *Green Chem.*, **8**, 717–726.
21. (a) Hafner, A., Duthaler, R.O., Marti, R., Rihs, G., Rothe-Streit, P., and Schwarzenbach, F. (1992) *J. Am. Chem. Soc.*, **114**, 2321–2336; (b) Fürstner, A. and Langemann, K. (1997) *J. Am. Chem. Soc.*, **119**, 9130–9136; (c) Adam, J.M., Ghosez, L., and Houk, K.N. (1999) *Angew. Chem. Int. Ed.*, **38**, 2728–2730.

22 Cossy, J., Pradaux, F., and BouzBouz, S. (2001) *Org. Lett.*, **3**, 2233–2235.
23 Cossy, J., Willis, C., Bellosta, V., and BouzBouz, S. (2002) *J. Org. Chem.*, **67**, 1982–1992.
24 Wang, L. and Floreancig, P.E. (2004) *Org. Lett.*, **6**, 569–572.
25 De Fays, L., Adam, J.-M., and Ghosez, L. (2003) *Tetrahedron Lett.*, **44**, 7197–7199.
26 Adam, J.-M., de Fays, L., Laguerre, M., and Ghosez, L. (2004) *Tetrahedron*, **60**, 7325–7344.
27 (a) Schleth, F. and Studer, A. (2004) *Angew. Chem. Int. Ed.*, **43**, 313–315; (b) Schleth, F., Vogler, T., Harms, K., and Studer, A. (2004) *Chem. Eur. J.*, **10**, 4171–4185.
28 Pellissier, H. (2007) *Tetrahedron*, **63**, 9267–9331.
29 Gondi, V.B., Gravel, M., and Rawal, V.H. (2005) *Org. Lett.*, **7**, 5657–5660.
30 McGilvra, J.D., Unni, A.K., Modi, K., and Rawal, V.H. (2006) *Angew. Chem. Int. Ed.*, **45**, 6130–6133.
31 Chan, T.H. and Brownbridge, P. (1979) *Chem. Commun.*, 578–579.
32 (a) Villano, R., Acocella, M.R., Massa, A., Palombi, L., and Scettri, A. (2007) *Tetrahedron Lett.*, **48**, 891–895; (b) Villano, R., Acocella, M.R., Massa, A., Palombi, L., and Scettri, A. (2009) *Tetrahedron*, **65**, 5571–5576.
33 Schneider, C. and Hansch, M. (2003) *Synlett*, 837–840.
34 Nishiyama, H. and Shiomi, T. (2007) *Top. Curr. Chem.*, **279**, 105–137.
35 (a) Bee, C., Han, S.B., Hassan, A., Iida, H., and Krische, M.J. (2008) *J. Am. Chem. Soc.*, **130**, 2746–2747; (b) Han, S.B., Hassan, A., and Krische, M.J. (2008) *Synthesis*, **17**, 2669–2679.
36 (a) Weber, B. and Seebach, D. (1992) *Angew. Chem., Int. Ed. Engl.*, **31**, 84–96; (b) Weber, B. and Seebach, D. (1994) *Tetrahedron*, **50**, 6117–6128.
37 (a) Minamikawa, H., Hayakawa, S., Yamada, T., Iwasawa, N., and Narasaka, K. (1988) *Bull. Chem. Soc. Chem. Jpn.*, **61**, 4379–4383; (b) Narasaka, K., Yamada, T., and Minamikawa, H. (1987) *Chem. Lett.*, 2073–2076.
38 Cozzi, P.G., Hilgraf, R., and Zimmermann, N. (2004) *Eur. J. Org. Chem.*, 4095–4105.
39 Pu, L. (2003) *Tetrahedron*, **59**, 9873–9886.
40 (a) Pu, L. and Yu, H.B. (2001) *Chem. Rev.*, **101**, 757–824; (b) Frantz, D.E., Fassler, R., Tomooka, R., and Carreira, E.M. (2000) *Acc. Chem. Res.*, **33**, 373–381.
41 Kamble, R.M. and Singh, V.K. (2003) *Tetrahedron Lett.*, **44**, 5347–5349.
42 Ueki, M., Matsumoto, Y., Jodry, J.J., and Mikami, K. (2001) *Synlett*, 1889–1892.
43 Soki, F., Neudörfl, J.-M., and Goldfuss, B. (2008) *J. Organomet. Chem.*, **693**, 2139–2146.
44 Maier, P., Redlich, H., and Richter, J. (2005) *Tetrahedron: Asymmetry*, **16**, 3848–3852.
45 Ooi, T., Miura, T., Takaya, K., Ichikawa, H., and Maruoka, K. (2001) *Tetrahedron*, **57**, 867–873.
46 Kim, S.S., Kwak, J.M., and Rajagopal, G. (2006) *Bull. Korean Chem. Soc.*, **27**, 1638–1640.
47 Linghu, X., Potnick, J.R., and Johnson, J.S. (2004) *J. Am. Chem. Soc.*, **126**, 3070–3071.
48 Garrett, M.R., Tarr, J.C., and Johnson, J.S. (2007) *J. Am. Chem. Soc.*, **129**, 12944–12945.
49 (a) Krause, N. and Hoffmann-Röder, A. (2001) *Synthesis*, **2**, 171–196; (b) Joshi, N.N. and Jha, S.C. (2002) *Arkivoc*, 167–196.
50 Alexakis, A. and Benhaim, C. (2001) *Tetrahedron: Asymmetry*, **12**, 1151–1157.
51 Mandoli, A., Arnold, L.A., de Vries, A.H.M., Salvadori, P., and Feringa, B.L. (2001) *Tetrahedron: Asymmetry*, **12**, 1929–1937.
52 Robert, T., Velder, J., and Schmalz, H.-G. (2008) *Angew. Chem. Int. Ed.*, **47**, 7718–7721.
53 Palacios, F. and Vicario, J. (2006) *Org. Lett.*, **8**, 5405–5408.
54 Belokon, Y.N., Harutyunyan, S., Vorontsov, E.V., Peregudov, A.S., Chrustalev, V.N., Kochetkov, K.A., Pripadchev, D., Sagyan, A.S., Beck, A.K., and Seebach, D. (2004) *Arkivoc*, 132–150.

55 Nahm, M.R., Linghu, X., Potnick, J.R., Yates, C.M., White, P.S., and Johnson, J.S. (2005) *Angew. Chem. Int. Ed.*, **44**, 2377–2379.

56 Nahm, M.R., Potnick, J.R., White, P.S., and Johnson, J.S. (2006) *J. Am. Chem. Soc.*, **128**, 2751–2756.

57 Enders, D., Tedeschi, L., and Förster, D. (2006) *Synthesis*, 1447–1460.

58 (a) Helmchen, G. and Pfaltz, A. (2000) *Acc. Chem. Res.*, **33**, 336–345; (b) Trost, B.M. and Crawley, M.L. (2003) *Chem. Rev.*, **103**, 2921–2943.

59 Seebach, D., Devaquet, E., Ernst, A., Hayakawa, M., Kühnle, F.N.M., Schweizer, W.B., and Weber, B. (1995) *Helv. Chim. Acta*, **78**, 1636–1650.

60 Hilgraf, R. and Pfaltz, A. (1999) *Synlett*, 1814–1816.

61 Hilgraf, R. and Pfaltz, A. (2005) *Adv. Synth. Catal.*, **347**, 61–77.

62 Boele, M.D.K., Kamer, P.C.J., Lutz, M., Spek, A.L., de Vries, J.G., van Leeuwen, P.W.N.M., and van Strijdonck, G.P.F. (2004) *Chem. Eur. J.*, **10**, 6232–6246.

63 Mikhel, I.S., Bernardinelli, G., and Alexakis, A. (2006) *Inorg. Chim. Acta*, **359**, 1826–1836.

64 Takacs, J.M., Reddy, D.S., Moteki, S.A., Wu, D., and Palencia, H. (2004) *J. Am. Chem. Soc.*, **126**, 4494–4495.

65 Alexakis, A., Malan, C., Lea, L., Benhaim, C., and Fournioux, X. (2001) *Synlett*, 927–930.

66 Meyer, M. and O'Hagan, D. (1992) *Chem. Br.*, 785–788.

67 (a) Lowe, K.C. and Powell, R.L. (2001) *J. Fluorine Chem.*, **109**, 1–94; (b) Hiyama, T. (2000) in *Organofluorine Compounds*, Springer, Berlin.

68 (a) Hintermann, L. and Togni, A. (2000) *Angew. Chem. Int. Ed.*, **39**, 4359–4362; (b) Hintermann, L. and Togni, A. (2000) *Helv. Chim. Acta*, **83**, 2425–2435.

69 Togni, A., Mezzetti, A., Barthazy, P., Becker, C., Devillers, I., Frantz, R., Hintermann, L., Perseghini, M., and Sanna, M. (2001) *Chimia*, **55**, 801–805.

70 Toullec, P.Y., Devillers, I., Frantz, R., and Togni, A. (2004) *Helv. Chim. Acta*, **87**, 2706–2711.

71 Perseghini, M., Massaccesi, M., Liu, Y., and Togni, A. (2006) *Tetrahedron*, **62**, 7180–7190.

72 Frantz, L., Hintermann, L., Perseghini, M., Broggini, D., and Togni, A. (2003) *Org. Lett.*, **5**, 1709–1712.

73 Toullec, P.Y., Bonaccorsi, C., Mezzetti, A., and Togni, A. (2004) *Proc. Natl. Acad. Sci. U.S.A.*, **101**, 5810–5814.

74 (a) Pellissier, H. (2007) *Tetrahedron*, **63**, 1297–1330; (b) Pellissier, H. (2006) *Tetrahedron*, **62**, 5559–5601.

75 Jereb, M. and Togni, A. (2005) *Org. Lett.*, **7**, 4041–4043.

76 Schmidt, B. and Seebach, D. (1994) *Tetrahedron*, **50**, 7473–7484.

77 (a) Bonache, M.A., Cativiela, C., Garcia-Lopez, M.T., and Gonzalez-Muniz, R. (2006) *Tetrahedron Lett.*, **47**, 5883–5887; (b) Bonache, M.A., Lopez, P., Martin-Martinez, M., Garcia-Lopez, M.T., Cativiela, C., and Gonzalez-Muniz, R. (2006) *Tetrahedron*, **62**, 130–138.

78 (a) Chapuis, C. and Jurczak, J. (1987) *Helv. Chim. Acta*, **70**, 436–440; (b) Narasaka, K., Iwasawa, N., Inoue, M., Yamada, T., Nakashima, J., and Sugimori, J. (1989) *J. Am. Chem. Soc.*, **111**, 5340–5345; (c) Gothelf, K.V., Hazell, R.G., and Jorgensen, K.A. (1995) *J. Am. Chem. Soc.*, **117**, 4435–4436; (d) Gothelf, K.V. and Jorgensen, K.A. (1996) *Acta Chem. Scand.*, **50**, 652–660.

79 (a) Seebach, D., Dahinden, R., Marti, R.E., Beck, A.K., Plattner, D.A., and Kühnle, F.N.M. (1995) *J. Org. Chem.*, **60**, 1788–1799; (b) Narasaka, K., Iwasawa, N., Inoue, M., Yamada, T., Nakashima, J., and Sugimori, J. (1989) *J. Am. Chem. Soc.*, **111**, 5340–5345; (c) Narasaka, K., Inoue, M., and Okada, N. (1986) *Chem. Lett.*, 1109–1112; (d) Narasaka, K., Inoue, M., and Okada, N. (1986) *Chem. Lett.*, 1967–1968; (e) Corey, E.J. and Matsumura, Y. (1991) *Tetrahedron Lett.*, **32**, 6289–6292; (f) Gothelf, K.V. and Jorgensen, K.A. (1995) *J. Org. Chem.*, **60**, 6847–6851; (g) Yamamoto, I. and Narasaka, K. (1995) *Chem. Lett.*, 1129–1130.

80 (a) Iwasawa, N., Sugimori, J., Kawase, Y., and Narasaka, K. (1989) *Chem. Lett.*, 1947–1950; (b) Narasaka, K., Saitou, M., and Iwasawa, N. (1991) *Tetrahedron: Asymmetry*, **2**, 1305–1318;

(c) Yamamoto, I. and Narasaka, K. (1994) *Bull. Chem. Soc. Jpn.*, **67**, 3327–3333.

81 (a) Engler, T.A., Letavic, M.A., and Takusagawa, F. (1992) *Tetrahedron Lett.*, **33**, 6731–6734; (b) Engler, T.A., Letavic, M.A., Lynch, K.O., and Takusagawa, F. (1994) *J. Org. Chem.*, **59**, 1179–1183; (c) Engler, T.A., Letavic, O., and Reddy, J.P. (1991) *J. Am. Chem. Soc.*, **113**, 5068–5070; (d) Engler, T.A., Letavic, M.A., Iyengar, R., LaTessa, K.O., and Reddy, J.P. (1999) *J. Org. Chem.*, **64**, 2391–2405.

82 (a) Quinkert, G., Del Grosso, M., Bucher, A., Bauch, M., Döring, W., Bats, J.W., and Dürner, G. (1992) *Tetrahedron Lett.*, **33**, 3617–3620; (b) Quinkert, G., Del Grosso, M., Döring, A., Döring, W., Schenkel, R.I., Bauch, M., Dambacher, G.T., Bats, J.W., Zimmermann, G., and Dürner, G. (1995) *Helv. Chim. Acta*, **78**, 1345–1391.

83 (a) Wada, E., Yasuoka, H., and Kanemasa, S. (1994) *Chem. Lett.*, 1637–1640; (b) Wada, E., Pei, W., and Kanemasa, S. (1994) *Chem. Lett.*, 2345–2348.

84 Bienaymé, H. (1997) *Angew. Chem., Int. Ed. Engl.*, **36**, 2670–2673.

85 Cativiela, C., Lopez, P., and Mayoral, J.A. (1991) *Tetrahedron: Asymmetry*, **2**, 1295–1304.

86 Corminboeuf, O. and Renaud, P. (2002) *Org. Lett.*, **4**, 1735–1738.

87 Thadani, A.N., Stankovic, A.R., and Rawal, V.H. (2004) *Proc. Natl. Acad. Sci. U.S.A.*, **101**, 5846–5850.

88 (a) Waldmann, H. (1994) *Synthesis*, 535–551; (b) Pellissier, H. (2009) *Tetrahedron*, **65**, 2839–2877.

89 Bayer, A., Hansen, L.K., and Gautun, O.R. (2002) *Tetrahedron: Asymmetry*, **13**, 2407–2415.

90 Huang, Y., Unni, A.K., Thadani, A.N., and Rawal, V.H. (2003) *Nature*, **424**, 146.

91 Du, H., Zhao, D., and Ding, K. (2004) *Chem. Eur. J.*, **10**, 5964–5970.

92 Zhang, X., Du, H., Wang, Z., Wu, Y.-D., and Ding, K. (2006) *J. Org. Chem.*, **71**, 2862–2869.

93 (a) Gothelf, K.V., Thomsen, I., and Jorgensen, K.A. (1996) *J. Am. Chem. Soc.*, **118**, 59–64; (b) Gothelf, K.V. and Jorgensen, K.A. (1998) *Chem. Rev.*, **98**, 863–909; (c) Pellissier, H. (2007) *Tetrahedron*, **63**, 3235–3285.

94 Gerard, B., Sangji, S., O'Leary, D.J., and Porco, J.A. (2006) *J. Am. Chem. Soc.*, **128**, 7754–7755.

95 Gerard, B., Cencic, R., Pelletier, J., and Porco, J.A. (2007) *Angew. Chem. Int. Ed.*, **46**, 7831–7834.

96 Zong, K. (2005) *Bull. Korean Chem. Soc.*, **26**, 717–718.

97 Yu, R.T. and Rovis, T. (2006) *J. Am. Chem. Soc.*, **128**, 12370–12371.

98 Lee, E.E. and Rovis, T. (2008) *Org. Lett.*, **10**, 1231–1234.

99 Yu, R.T. and Rovis, T. (2008) *J. Am. Chem. Soc.*, **130**, 3262–3263.

100 Charette, A.B. and Brochu, C. (1995) *J. Am. Chem. Soc.*, **117**, 11367–11368.

101 Charrette, A.B., Molinaro, C., and Brochu, C. (2001) *J. Am. Chem. Soc.*, **123**, 12168–12175.

102 Voituriez, A. and Charette, A.B. (2006) *Adv. Synth. Catal.*, **348**, 2363–2370.

103 Xia, Q.-H., Ge, H.-Q., Ye, C.-P., Liu, Z.-M., and Su, K.-X. (2005) *Chem. Rev.*, **105**, 1603–1662.

104 José, M.C., Maria, T.M., and Shazia, A. (2004) *Chem. Rev.*, **104**, 2857–2900.

105 Aoki, M. and Seebach, D. (2001) *Helv. Chim. Acta*, **84**, 187–207.

106 Adam, W., Beck, A.K., Pichota, A., Saha-Möller, C.R., Seebach, D., Vogl, N., and Zhang, R. (2003) *Tetrahedron: Asymmetry*, **14**, 1355–1361.

107 Adam, W., Alsters, P.L., Neumann, R., Saha-Möller, C.R., Seebach, D., and Zhang, R. (2003) *Org. Lett.*, **5**, 725–728.

108 (a) Andersen, K.K. (1988) *The Chemistry of Sulfones and Sulfoxides* (eds S. Patai, Z. Rappoport, and C.J.M. Stirling), John Wiley & Sons, Inc., New York, Ch. 3, pp. 56–94; (b) Walker, A.J. (1992) *Tetrahedron: Asymmetry*, **3**, 961–998; (c) Solladié, G. (1991) *Comprehensive Organic Synthesis*, Vol. 6 (eds B.M. Trost and I. Fleming), Pergamon, Oxford, Ch. 3, pp. 148–170; (d) Kresze, G. (1985) *Methoden der Organischen Chemie* (Houben-Weyl) (ed. D. Klamann), Georg Thieme Verlag, Stuttgart, pp. 669–886; (e) Solladié, G. and Carreno, M.C. (1995) *Organosulfur Chemistry. Synthetic Aspects*

(ed. P.C.B. Page), Academic Press, New York, Ch. 1, pp. 1–47.
109 Noyori, R. (2002) *Angew. Chem. Int. Ed.*, **41**, 2008–2022.
110 Pfaltz, A., Blankenstein, J., Hilgraf, R., Hörmann, E., McIntyre, S., Menges, F., Schönleber, M., Smidt, S., Wüstenberg, B., and Zimmermann, N. (2003) *Adv. Synth. Catal.*, **345**, 33–43.
111 Marson, A., Freixa, Z., Kamer, P.C.J., and van Leeuwen, P.W.N.M. (2007) *Eur. J. Inorg. Chem.*, 4587–4591.
112 (a) Beck, A.K., Dahinden, R., and Kühnle, F.N.M. (1996) *ACS Symp. Ser.*, **641**, 52–69; (b) Seebach, D., Beck, A.K., Dahinden, R., Hoffmann, M., and Kühnle, F.N.M. (1996) *Croat. Chem. Acta*, **69**, 459–484.
113 Heldmann, D.K. and Seebach, D. (1999) *Helv. Chim. Acta*, **82**, 1096–1110.
114 Yao, S., Meng, J.-C., Siuzdak, G., and Finn, M.G. (2003) *J. Org. Chem.*, **68**, 2540–2546.
115 Seebach, D., Jaeschke, G., Gottwald, K., Matsuda, K., Formisano, R., Chaplin, D. A., Breuning, M., and Bringmann, G. (1997) *Tetrahedron*, **53**, 7539–7556.
116 Gottwald, K. and Seebach, D. (1999) *Tetrahedron*, **55**, 723–738.
117 (a) Seebach, D., Jaeschke, G., and Wang, Y.M. (1995) *Angew. Chem., Int. Ed. Engl.*, **34**, 2395–2396; (b) Jaeschke, G. and Seebach, D. (1998) *J. Org. Chem.*, **63**, 1190–1197.
118 Ramon, D.J., Guillena, G., and Seebach, D. (1996) *Helv. Chim. Acta*, **79**, 875–894.
119 Yamamoto, H. and Momiyama, N. (2005) *Chem. Commun.*, 3514–3525.
120 (a) Momiyama, N. and Yamamoto, H. (2005) *J. Am. Chem. Soc.*, **127**, 1080–1081; (b) Akakura, M., Kawasaki, M., and Yamamoto, H. (2008) *Eur. J. Org. Chem.*, 4245–4249.
121 Thomassigny, C., Prim, D., and Greck, C. (2006) *Tetrahedron Lett.*, **47**, 1117–1119.
122 Friestad, G.K. and Mathies, A.K. (2007) *Tetrahedron*, **63**, 2541–2569.
123 (a) Cordova, A. (2004) *Acc. Chem. Res.*, **37**, 102–112; (b) Marques, M.M.B. (2006) *Angew. Chem. Int. Ed.*, **45**, 348–352.
124 Akiyama, T., Saitoh, Y., Morita, H., and Fuchibe, K. (2005) *Adv. Synth. Catal.*, **347**, 1523–1526.
125 Rueping, M., Sugiono, E., and Moreth, S.A. (2007) *Adv. Synth. Catal.*, **349**, 759–764.
126 (a) Blume, F., Zemolka, S., Fey, T., Kranich, R., and Schmalz, H.-G. (2002) *Adv. Synth. Catal.*, **344**, 868–883; (b) Velder, J., Robert, T., Weidner, I., Neudörfl, J.-M., Lex, J., and Schmalz, H.-G. (2008) *Adv. Synth. Catal.*, **350**, 1309–1315.
127 Werle, S., Fey, T., Neudörfl, J.M., and Schmalz, H.-G. (2007) *Org. Lett.*, **9**, 3555–3558.
128 Moteki, S.A., Wu, D., Chandra, K.L., Reddy, D.S., and Takacs, J.M. (2006) *Org. Lett.*, **8**, 3097–3100.
129 Smith, S.M., Thacker, N.C., and Takacs, J.M. (2008) *J. Am. Chem. Soc.*, **130**, 3734–3735.
130 Moteki, S.A. and Takacs, J.M. (2008) *Angew. Chem. Int. Ed.*, **47**, 894–897.
131 Woodward, A.R., Burks, H.E., Chan, L. M., and Morken, J.P. (2005) *Org. Lett.*, **7**, 5505–5507.
132 (a) Muniz, K. and Nieger, M. (2005) *Chem. Commun.*, 2729–2731; (b) Almodovar, I., Hövelmann, C.H., Streuff, J., Nieger, M., and Muniz, K. (2006) *Eur. J. Org. Chem.*, 704–712.
133 Yamasaki, S., Kanai, M., and Shibasaki, M. (2001) *J. Am. Chem. Soc.*, **123**, 1256–1257.
134 Wen, J., Zhao, J., and You, T. (2006) *J. Mol. Catal. A*, **245**, 278–280.
135 Knoop, C.A. and Studer, A. (2005) *Adv. Synth. Catal.*, **347**, 1542–1546.
136 Nguyen, P.Q. and Schäfer, H.J. (2001) *Org. Lett.*, **3**, 2993–2995.
137 Bolm, C., Hildebrand, J.P., Muniz, K., and Hermanns, N. (2001) *Angew. Chem. Int. Ed.*, **40**, 3284–3308.
138 (a) Imbos, R., Minnaard, A.J., and Feringa, B.L. (2002) *J. Am. Chem. Soc.*, **124**, 184–185; (b) Imbos, R., Minnaard, A.J., and Feringa, B.L. (2003) *J. Chem. Soc., Dalton Trans.*, 2017–2023.
139 Xue, S. and Jiang, Y.-Z. (2004) *Chin. Chem. Lett.*, **22**, 1456–1458.
140 Jacobsen, E.N. and Yoon, T.P. (2003) *Science*, **299**, 1691–1693.

10
Cinchona Alkaloids

Hongming Li, Yonggang Chen, and Li Deng

10.1
Introduction

Cinchona alkaloids constitute a class of readily available natural products. As represented by quinine and quinidine, an outstanding feature of these alkaloids is that they exist in pseudoenantiomeric pairs (Scheme 10.1). From the viewpoint of catalysis development, the quinuclidinyl moiety as a tertiary amine constitutes the most important feature of cinchona alkaloids. It could serve as an active center for Brønsted base catalysis, Lewis base catalysis, and nucleophilic catalysis. In addition to providing an active center for these various modes of organocatalysis, it also allows cinchona alkaloids to bind to metals. Consequently, cinchona alkaloids could also serve as ligands to form chiral metal complexes. Upon alkylations of the quinuclidine nitrogen, the resulting ammonium salts could serve as chiral phase-transfer catalysts, another class of chiral organic catalysts. Their availability and the unparalleled possibility to be applied as or converted into various classes of chiral catalysts allow cinchona alkaloids to provide a highly attractive starting point for the development of asymmetric catalysis. Another critical structural feature of cinchona alkaloids is the presence of the 9-OH group and, for quinidine and quinine, also the 6′-OMe group. These functional groups provide a handle for modifications, thereby affording possibilities for catalyst tunings. As demonstrated later in this chapter, these sites of modifications are extremely valuable for the development of bifunctional chiral organic catalysts. These structural features associated with cinchona alkaloids render them, again and again, "the system of choice" for the development of asymmetric catalysis with both transition metals and organic molecules.

It is impossible to present even a survey of the incredibly rich chemistry promoted by cinchona alkaloid-derived catalysts in the format of a book chapter. First, we will limit our discussions to homogeneous asymmetric catalysis. However, the readers are reminded that cinchona alkaloids are utilized in one of the most important transition metal-catalyzed heterogeneous asymmetric hydrogenations [1]. Numerous authoritative and comprehensive reviews covering all aspects of the Os-catalyzed asymmetric dihydroxylations, more commonly known as Sharpless

Privileged Chiral Ligands and Catalysts. Edited by Qi-Lin Zhou
Copyright © 2011 WILEY-VCH Verlag GmbH & Co. KGaA, Weinheim
ISBN: 978-3-527-32704-1

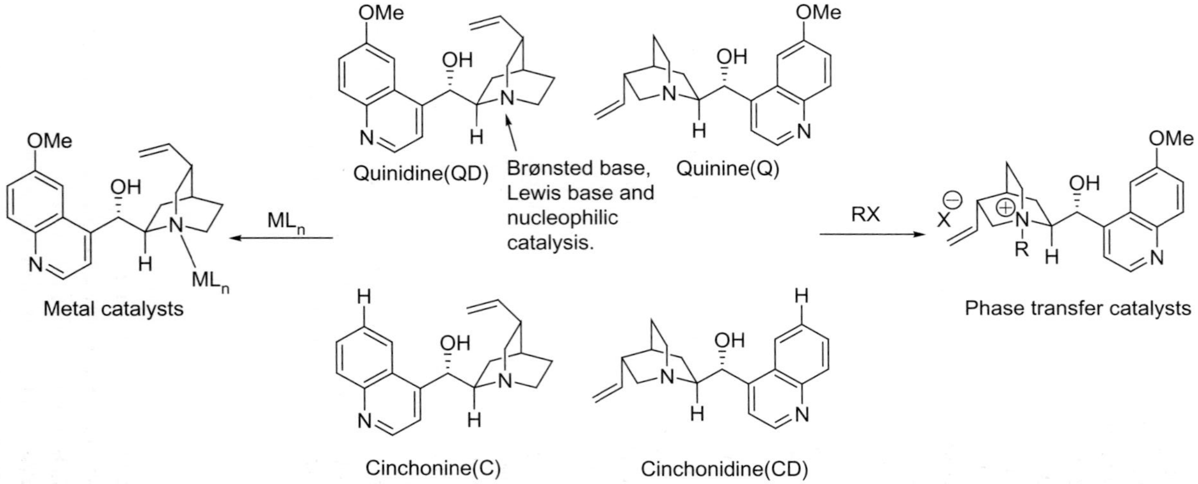
OMe
OH
H
N
Quinidine(QD)
Brønsted base,
Lewis base and
nucleophilic
catalysis.
Quinine(Q)
Metal catalysts
ML_n
RX
X
R
Phase transfer catalysts
Cinchonine(C)
Cinchonidine(CD)

Scheme 10.1

asymmetric dihydroxylation, are available in the literature. In this chapter we only briefly present the initial discovery and development of modified cinchona alkaloids as ligands for the Os-complex, the roles played by these ligands in the dramatic expansion of the scope of Sharpless asymmetric dihydroxylation and the insights into the origin of enantioselectivity revealed by mechanistic studies. Similarly, the development of asymmetric phase-transfer catalysts with cinchona alkaloid-derived ammonium salts has also been covered in great details in several reviews. Thus our coverage focuses on a few key breakthroughs that first revealed and subsequently established cinchona alkaloids as a privileged structural motif for asymmetric phase-transfer catalysis. The topics of nucleophilic catalysis and base catalysis are discussed with a focus on more recent studies [2]. In light of the latest breakthroughs in cinchona alkaloid-promoted cooperative organocatalysis, the successful development of cinchona alkaloids as bifunctional catalysts that simultaneously activate two reactants for an asymmetric transformation is presented as a separate topic. Also included in this topic are the latest advances in the development of novel asymmetric reaction cascades with cinchona alkaloids as a multifunctional organic catalyst to address selectivity issues in more than one bond-forming or breaking event.

The concept of privileged chiral catalysts and ligands is arguably most conspicuously demonstrated by the broad utility of various classes of cinchona alkaloid-derived chiral catalysts in asymmetric synthesis. In our view, synthetic and/or mechanistic studies providing the first precedent(s) of a highly enantioselective reaction mediated by a specific mode of catalysis with cinchona alkaloids constitute the key starting point that sets in motion the process that eventually establishes the broad utility of that particular mode of catalysis in asymmetric synthesis. The second important step in this process is the demonstration that a cinchona alkaloid, with the same mode of catalysis, is able to promote another enantioselective reaction of distinct chemical mechanism. Subsequent studies that further expand the scope and the synthetic utility of that particular mode of catalysis are highly valuable, yet, in terms of impact, they are not as instrumental as earlier studies. Thus the readers are reminded that this chapter is organized following the authors' personal view along with other inevitable biases.

10.2
Metal Catalysis

The Sharpless asymmetric dihydroxylation (AD) with OsO_4-cinchona alkaloid complexes stands as one of the most important asymmetric reactions in organic synthesis [3]. In 1980, Sharpless and Hentges reported that, in the presence of a stoichiometric amount of dihydroquinidine acetate (DHQD-3) and OsO_4, enantioselective osmylation of alkenes followed by hydrolysis afforded the corresponding diols in up to 94% ee (Scheme 10.2) [4, 5]. This result set the stage for the exploration of modified cinchona alkaloids for the development of a highly enantioselective catalytic asymmetric dihydroxylations. Notably, in addition to

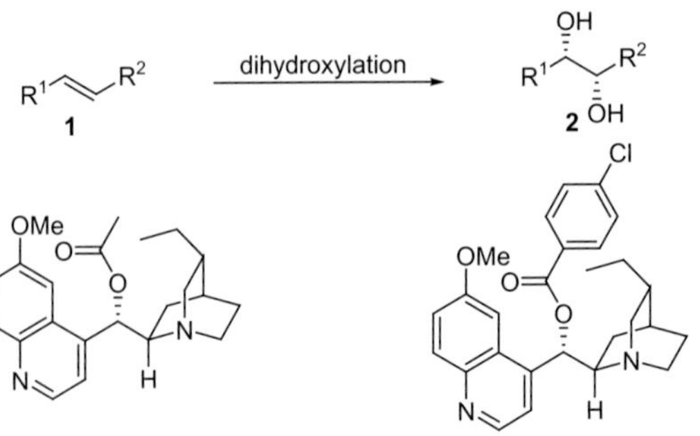

DHQD-3
100 mol% Cat., 100 mol% OsO$_4$
62–90% yield, 25–83% ee
(Sharpless & Hentges, 1980)

DHQD-4
13.4 mol% Cat., 0.2 mol% OsO$_4$
1.2 eq NMO (terminal oxidant)
80–95% yield, 20–88% ee
(Sharpless et. al., 1988)

Scheme 10.2

achieving high enantioselectivity, how to establish oxidative catalytic turnover was also a key issue to be addressed. In 1988, Sharpless and coworkers discovered that, with N-methylmorpholine N-oxide (NMO) as the terminal oxidant and dihydroquinidine p-chlorobenzoate (DHQD-4) as the ligand, catalytic asymmetric dihydroxylations could be accomplished in good yield and high enantioselectivity [5].

Following this stunning breakthrough, extensive screening of numerous modified cinchona alkaloids identified bis-cinchona alkaloids such as (DHQD)$_2$-PHAL [6a], (DHQD)$_2$PYR [6b], and (DHQD)$_2$AQN [6c] as more effective ligands for the Sharpless AD (Figure 10.1). The employment of these ligands, which were made commercially available, allowed the dramatic expansion of the substrate scope for Sharpless AD. High enantioselectivity has been achieved with olefins in each of the six possible substitution patterns. More specifically, terminal olefins, *trans*-disubstituted and trisubstituted olefins can be reliably oxidized in high enantioselectivity. Unsurprisingly, the enantioselectivity for 1,1-disubstituted olefins is more dependent of the electronic and steric differences between the two substituents. The cis-disubstituted olefins and tetrasubstituted olefins represent significantly more challenging classes of substrates. In parallel to ligand development, the Sharpless AD was also improved through optimizations of terminal oxidants and reaction protocols. As a result, a mix of the most general ligand (DHQD)$_2$PHAL or (DHQ)$_2$PHAL with potassium osmate, potassium ferricyanide, and potassium carbonate is now available with the commercial name of AD-mix α or β. Since its discovery, the Sharpless AD has evolved into one of a handful of powerful asymmetric reactions that are relied upon by synthetic chemists in both academic and industrial settings.

As discussed in great detail in various reviews, the remarkable combination of high efficiency in asymmetric induction and extraordinary generality in substrate scope rendered the Sharpless AD a target for mechanistic studies, which were mainly carried out by the Sharpless [3] and Corey groups [3, 7]. These mechanistic studies revealed insights into both the chemical mechanism of the reaction and

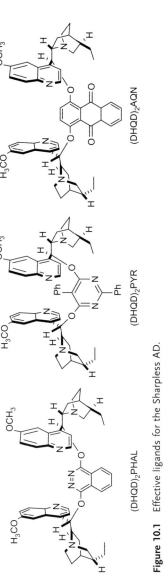

Figure 10.1 Effective ligands for the Sharpless AD.

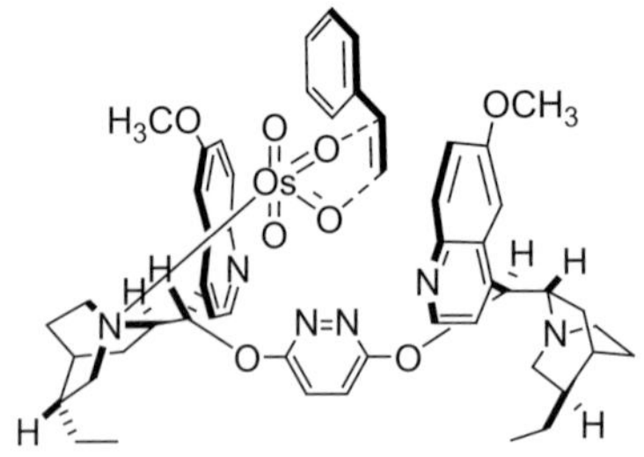

Figure 10.2 A [3 + 2] cycloaddition between the olefin and the cinchona alkaloid-coordinated OsO$_4$ is the stereochemistry-defining step in Sharpless AD.

the origin of enantioselectivity induction by the cinchona alkaloid-derived ligands. The now commonly accepted mechanism for Sharpless AD, first proposed by Corey and coworkers, invokes a [3 + 2] cycloaddition between the olefin and the cinchona alkaloid-coordinated OsO$_4$ as the stereochemistry-defining step (Figure 10.2) [7].

An insightful observation from these studies is that, while the alkaloids themselves are found to be conformationally flexible [8], the binding of the quinuclidine nitrogen to a metal complex reduces this flexibility [9]. The active conformation of the bis-cinchona alkaloid ligands was also elucidated by comparing enantioselectivities obtained from the asymmetric dihydroxylation of various olefins using ligand DHQD-6 and a designed, conformationally locked bis-cinchona alkaloid DHQD-7. The latter contains two linkages (pyridazinone and adipate) joining the two quinidine units (Figure 10.3) [10]. The remarkable similarity between the enantioselectivities obtained with these two ligands in the dihydroxylation of several olefins established that DHQD-6 adopts a rigid conformation upon binding to OsO$_4$ to form the active complex for olefin oxidation. Importantly, these mechanistic understandings proved to be invaluable in guiding the development of new ligands for highly challenging asymmetric dihydroxylations of complex

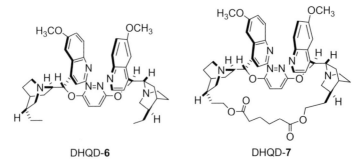

Figure 10.3 Two bis-cinchona alkaloid ligands; bis-cinchona alkaloid DHQD-7 is conformationally locked.

olefins, such as the regioselective and enantioselective dihydroxylations of terpenes [11].

The Sharpless AD is a landmark achievement in synthetic method development. It also, in a truly stunning fashion, brings cinchona alkaloids to the attention of synthetic chemists in the field of asymmetric catalysis. It changed our perception of the potential of cinchona alkaloids in asymmetric catalysis and, consequently, provided the most critical factor that triggered the subsequent studies that established cinchona alkaloids as privileged ligands and catalysts in asymmetric synthesis.

10.3
Phase-Transfer Catalysis

10.3.1
Asymmetric Alkylations

Phase-transfer catalysis with organic ammonium salts represents both a well-recognized and broadly applicable concept for the development of catalytic transformations. Accordingly, cinchona alkaloid-derived chiral ammonium salts were first explored by Wynberg and associates as chiral phase-transfer catalysts in 1978 for a conjugate addition of nitromethane to chalcone [12]. However, a synthetically useful enantioselectivity was not obtained in these early studies. The first example of a highly enantioselective reaction with a chiral phase-transfer catalyst was realized by a group at Merck [13] with their report of the asymmetric methylation of 6,7-dichloro-5-methoxy-2-phenyl-1-indanone (8) catalyzed by C-9 (Scheme 10.3), which stands as a landmark breakthrough in asymmetric phase-transfer catalysis.

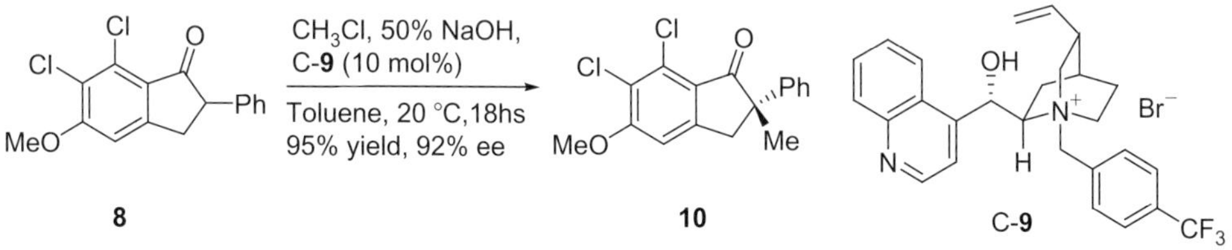

Scheme 10.3

The successful development of a chiral phase-transfer catalyst to afford high efficiency toward a reaction of substantial substrate scope was not accomplished until 1997, when the Corey [14] and Lygo [15] groups independently reported highly enantioselective alkylation of glycine Schiff base 11 using chiral quaternary N-(9-anthracenylmethyl)cinchonidinium salts CD-15 and CD-14, respectively (Scheme 10.4). Notably, O'Donnell's pioneering studies [16] of the same alkylation with N-benzyl cinchonidinium salt CD-13, although they did not attain high

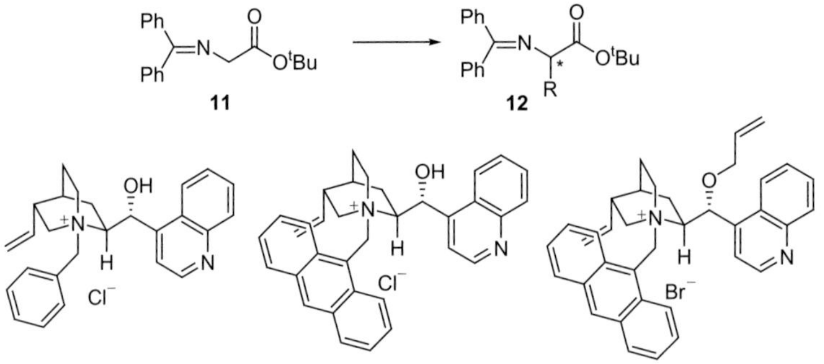

CD-13
O'Donnell (1989)
50% NaOH, rt
60–82% yield
48–66% ee

CD-14
Lygo (1997)
50% KOH, rt
40–86% yield
67–91% ee

CD-15
Corey (1997)
CsOH·H$_2$O, −40 °C
67–91% yield
92–99.5% ee

Scheme 10.4

enantioselectivity, defined conditions for achieving high yield for this reaction with phase-transfer catalysis.

10.3.2
Asymmetric Conjugate Additions

Building on the breakthrough in asymmetric alkylations with catalyst **CD-15**, Corey and coworkers carried out a series of studies to develop a broad range of asymmetric transformations with phase-transfer catalysis. In 1998, cinchonidium salt **CD-15** was found to be an effective catalyst for the promotion of asymmetric conjugate additions of glycine Schiff base **11** to cyclic α,β-enones **16a**, ethyl vinyl ketone **16b**, and α,β-unsaturated esters **16c** (Scheme 10.5) [17]. Their pioneering studies established that cinchona alkaloid-derived ammonium salts could be highly effective catalysts for asymmetric conjugate additions, thereby establishing another milestone in establishing these chiral phase-transfer catalysts as privileged chiral catalysts.

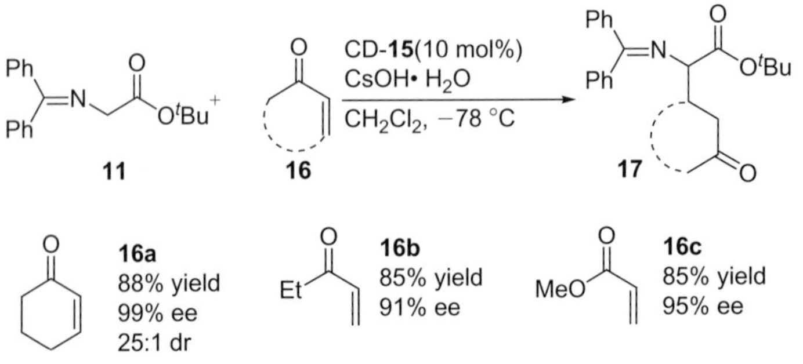

16a
88% yield
99% ee
25:1 dr

16b
85% yield
91% ee

16c
85% yield
95% ee

Scheme 10.5

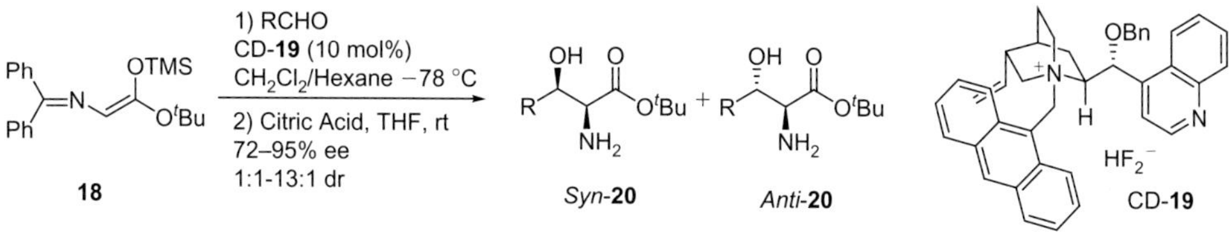

Scheme 10.6

10.3.3
Asymmetric Aldol Reactions

In parallel to the aforementioned studies, Corey and coworkers also demonstrated that CD-**19** could promote highly enantioselective aldol reactions of trimethylsilyl enol ether derivatives of *tert*-butyl glycinate-benzophenone Schiff base **18** with aldehydes (Scheme 10.6) [18]. Following these conceptually important advances, Corey and coworkers further expanded the utility of *N*-(9-anthracenylmethyl) cinchonidinium salts to realize a series of highly enantioselective alkylations [19], conjugate additions [20], and other aldol reactions [21].

10.3.4
Examples of Recent Applications

Corey's studies provided the most critical experimental precedents demonstrating cinchona alkaloid-derived ammonium salts as broadly useful phase-transfer catalysts, thereby establishing asymmetric phase-transfer catalysis as a compelling strategy for asymmetric synthesis. These studies also represent some of the most important developments in the field that is known today as asymmetric organocatalysis. As exemplified by some recent examples (Schemes 10.7–10.9), the invention of new phase-transfer chiral catalysts based on cinchona alkaloids and their applications has been actively pursued for the development of a diverse range of asymmetric transformations [22–24]. Providing one of the most general and

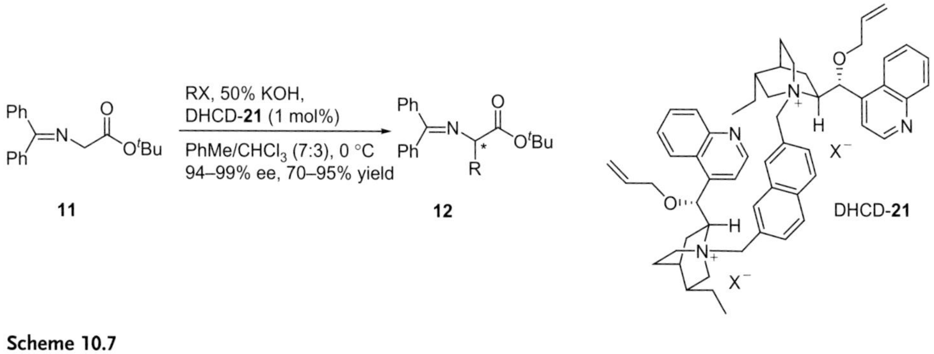

Scheme 10.7

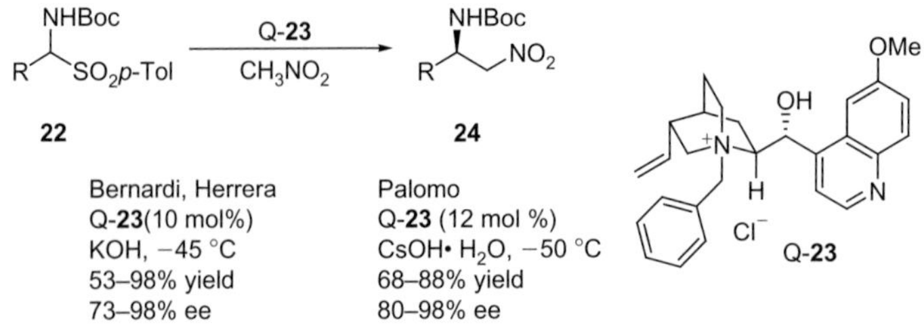
NHBoc
R
SO2p-Tol
Q-23
CH3NO2
NHBoc
R
NO2
22
24
Bernardi, Herrera
Q-23(10 mol%)
KOH, −45 °C
53–98% yield
73–98% ee
Palomo
Q-23 (12 mol %)
CsOH• H2O, −50 °C
68–88% yield
80–98% ee
OMe
OH
N
H
N
Cl⁻
Q-23

Scheme 10.8

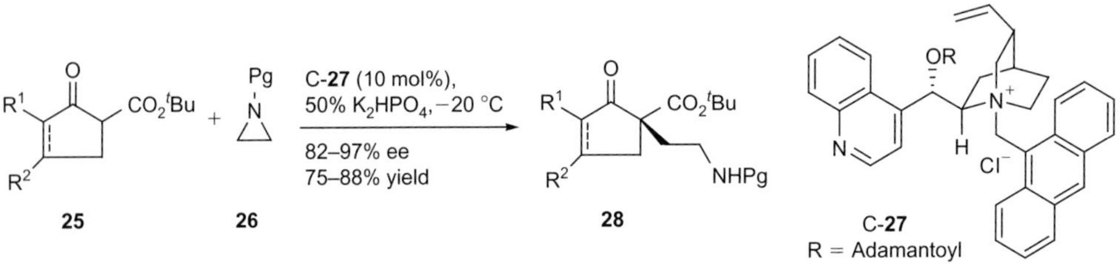
O
R1
CO2tBu
R2
+
Pg
N
C-27 (10 mol%),
50% K2HPO4, −20 °C
82–97% ee
75–88% yield
O
R1
CO2tBu
R2
NHPg
25
26
28
OR
N
H
N
Cl⁻
C-27
R = Adamantoyl

Scheme 10.9

attractive approaches for the development of practical asymmetric reactions, asymmetric phase-transfer catalysis has become one of the most important areas in asymmetric catalysis. As already summarized in a series of excellent reviews [25], cinchona alkaloids continue to play a crucial role in that area as a privileged starting point for the development of new catalysts and reactions.

10.4 Nucleophilic Catalysis

10.4.1 Asymmetric Reactions with Ketenes

In the 1970s to 1980s Wynberg and coworkers carried out extensive studies in exploring cinchona alkaloids as chiral base catalysts, nucleophilic catalysts, and phase-transfer catalysts [26]. These studies resulted in the historically important first documentation of a highly enantioselective reaction promoted by a cinchona alkaloid-derived catalyst. In 1982, Wynberg reported an asymmetric synthesis of *(S)*- and *(R)*-malic acid **32** (Scheme 10.10) [27]. With 1–2 mol% of quinidine, the reaction between ketene **29** and chloral **30** proceeded smoothly to give β-lactone **31** in excellent yield and stereoselectivity. A nucleophilic catalysis mechanism was proposed in which the formation of a chiral ammonium-enolate **34** and the subsequent highly enantioselective aldol reaction between the enolate and the chloral led to the formation of the chiral lactone product (Scheme 10.11). This result

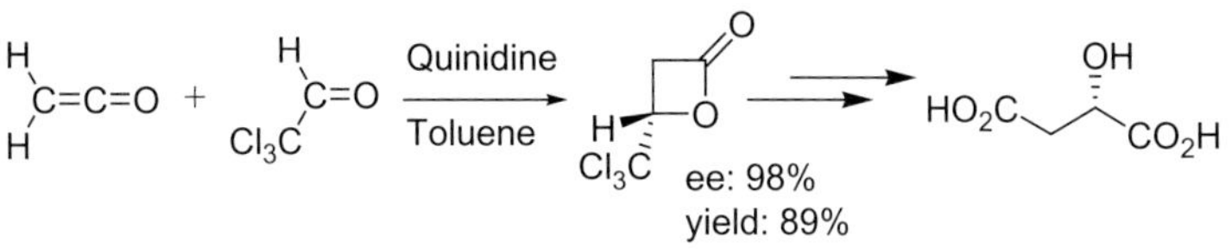

Scheme 10.10

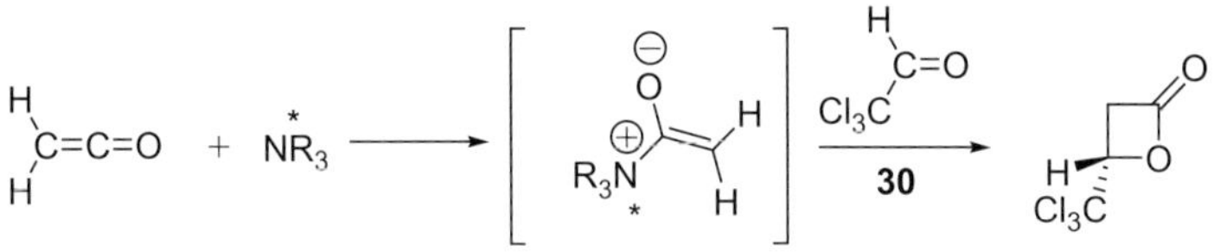

Scheme 10.11

proved to be truly ahead of its time as the next report of a highly enantioselective reaction with a cinchona alkaloid as a nucleophilic catalyst did not appear in the literature until late 1990s.

Remarkably, Wynberg's breakthrough was not followed with further investigations until Calter's report of a highly efficient asymmetric dimerization of methylketene with trimethylsilylquinine (TMSQN, Q-**36**) in 1996 [28]. In Calter's investigation, the methylketene was generated *in situ* from the corresponding acid chloride **35** and one equivalent of *i*-Pr$_2$NEt was utilized as deprotonation agent (Scheme 10.12) [28c]. Furthermore, Calter also demonstrated that TMSQN was superior to quinine and that the scope of the reaction was substantial. In 2001, Romo and coworkers reported an intramolecular aldol-lactonization of oxo-acid **39** using *O*-acetyl quinidine (QD-**40**, AcQD) as the catalyst (Scheme 10.13) [29].

The successful applications of cinchona alkaloids as nucleophilic catalysts for formal [2 + 2] reactions are not limited to ketenes and aldehydes [30]. In 2000, Lectka and coworkers reported a highly enantioselective and diastereoselective reaction of ketenes, generated from acid chloride **42**, and imine **43** to afford β-lactams **45** (Scheme 10.14) [31a]. Benzoylquinine (Q-**44**, BQN) was used both as a nucleophilic catalyst and a dehydrohalogenation agent. In addition, a stoichiometric amount of proton sponge was added to neutralize HCl. The scope of BQN-promoted reactions has been extended to α-halogenation (Schemes 10.15 and 10.16) [31b, c] and [4 + 2] cycloadditions [31d–f]. In these studies Lectka and coworkers showed that, in the presence of catalytic amount of Pd (II) or Ni(II) metal complex, the yield of the reactions were dramatically improved. Experimental evidences indicated that metal-ketene enolates were formed with the metal complex, which afforded improved reactivity (Scheme 10.16). These studies by the Lectka group represent some of the most important progresses in asymmetric nucleophilic catalysis by cinchona alkaloids.

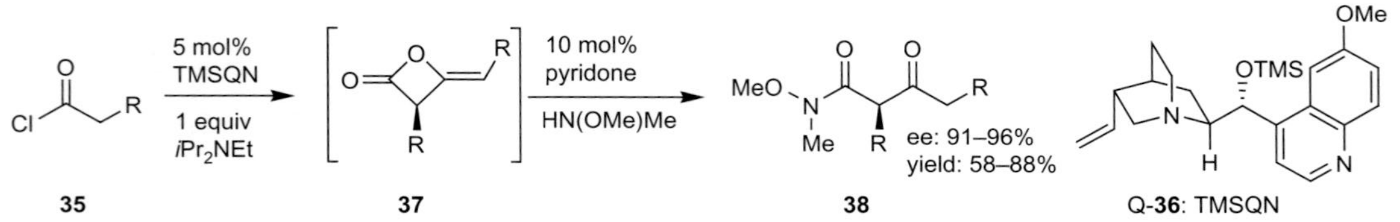

Scheme 10.12

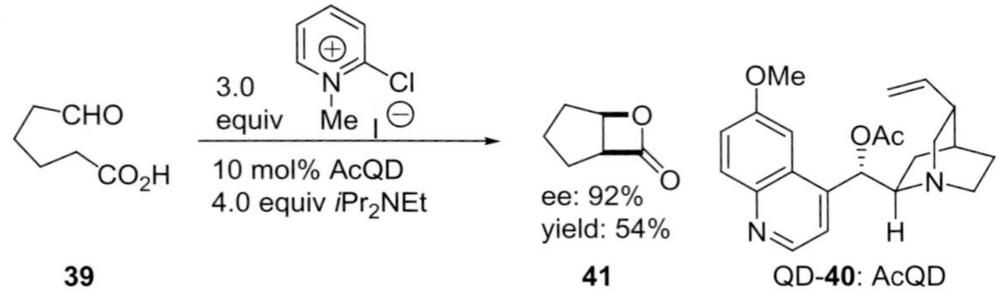

Scheme 10.13

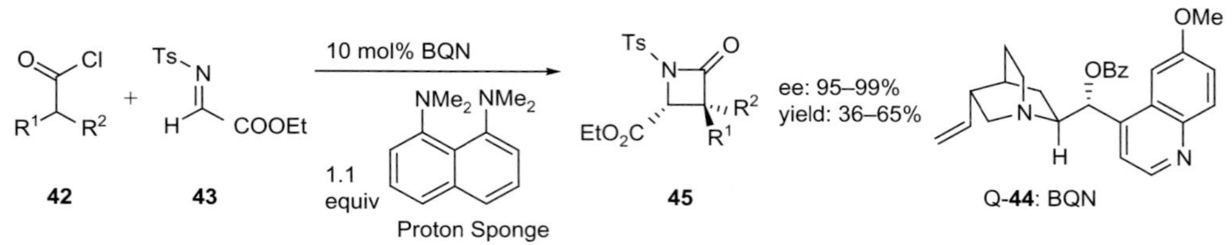

Scheme 10.14

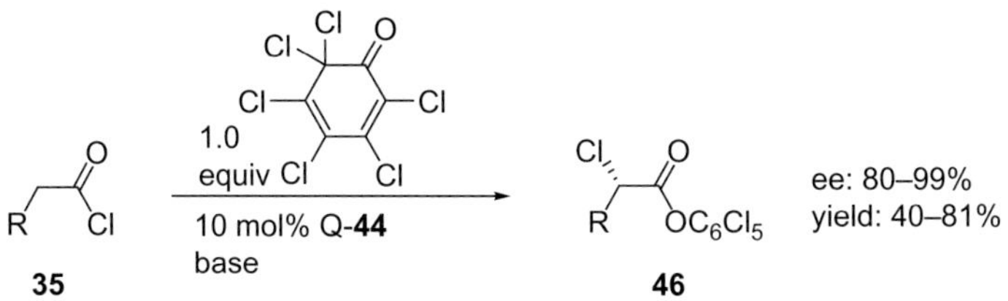

Scheme 10.15

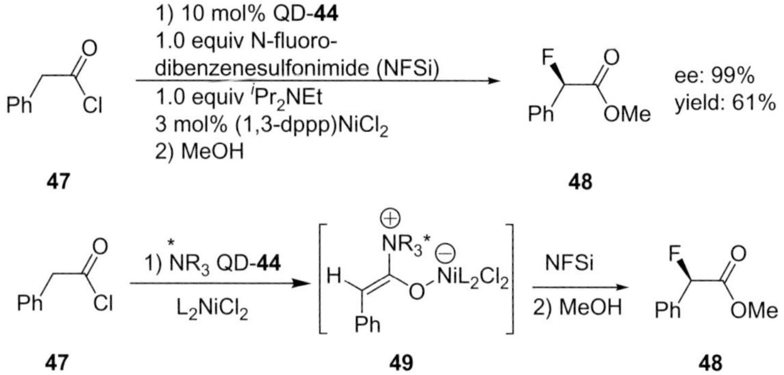

Scheme 10.16

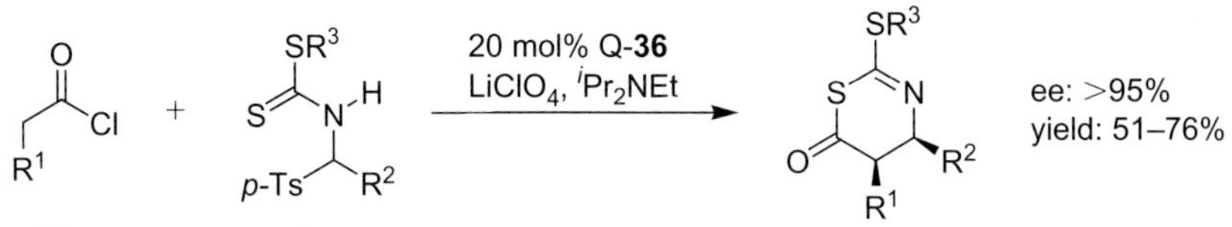

Scheme 10.17

An interesting recent application of asymmetric transformations of ketenes was reported by Nelson and coworkers, who demonstrated that a formal [4+2] cycloaddition of ketenes and N-thioacylimines, generated *in situ* from acid chloride 35 and α-amido sulfone 50, can be promoted efficiently by cinchona alkaloid Q-36 (Scheme 10.17) [32].

10.4.2
Asymmetric Morita–Baylis–Hillman Reactions

In 1999, Hatakeyama and coworkers developed an iso-cuperidine (β-ICD, QD-54) promoted Morita–Baylis–Hillman (MBH) reaction of aldehyde 52 with acylates 53 (Scheme 10.18) [33]. The synergistic cooperation of nucleophilic catalysis, through

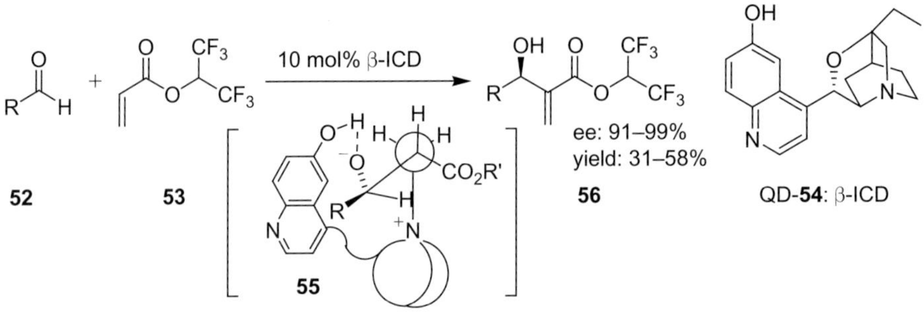

Scheme 10.18

the tertiary amine moiety of β-ICD, and acid catalysis, through phenolic OH moiety of β-ICD, was suggested to explain the high levels of stereochemical induction. Hatakeyama's report constitutes a particularly important development in asymmetric catalysis for several reasons. Firstly, it is the first example of a highly enantioselective asymmetric MBH reaction. Secondly, utilizing a non-ketene precursor for the formation of an enolate, it provides a significant expansion of the scope for nucleophilic catalysis by cinchona alkaloids. In addition, Hatakeyama's report provides the first experimental evidence of efficient cooperative catalysis afforded by a cinchona alkaloid. Later, Shi and Xu extended the utility of β-ICD to promote the aza-MBH reactions of imines **57** with methyl vinyl ketone (**58**) (Scheme 10.19) [34].

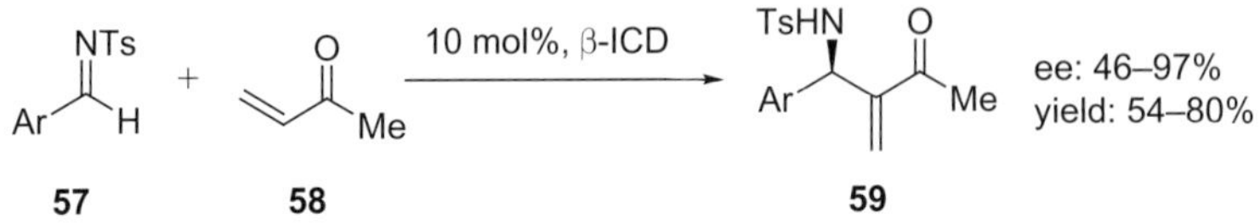

Scheme 10.19

10.4.3
Asymmetric Cyanation of Simple Ketones

During extensive and systematic studies aiming to establish cinchona alkaloids as broadly useful chiral Brønsted base, nucleophilic, and Lewis base catalysts, Deng and Tian reported in 2001 a highly enantioselective cyanation of ketones catalyzed by DHQD-PHN or (DHQD)$_2$AQN (Scheme 10.20) [35]. The enantioselectivity of the reaction is postulated to come from the addition of the cyanide, as part of a chiral ammonium complex **63**, generated from the addition of the cinchona alkaloid catalyst to ethyl cyanocarbonate, to ketone **60** (Scheme 10.21). This form of nucleophilic catalysis by cinchona alkaloids is distinct from that in the aforementioned asymmetric ketene enolate chemistry and MBH reactions. These

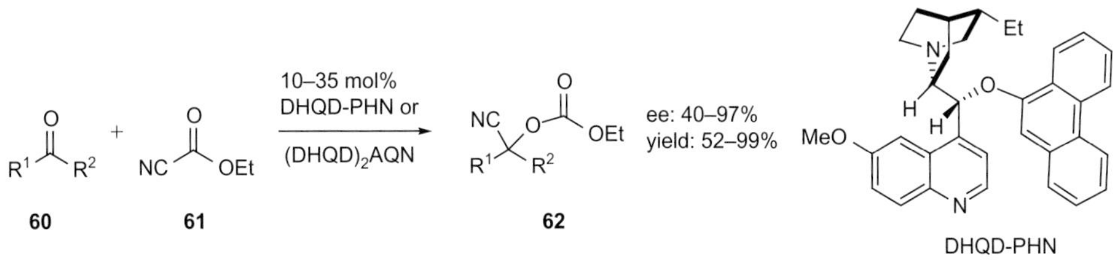

Scheme 10.20

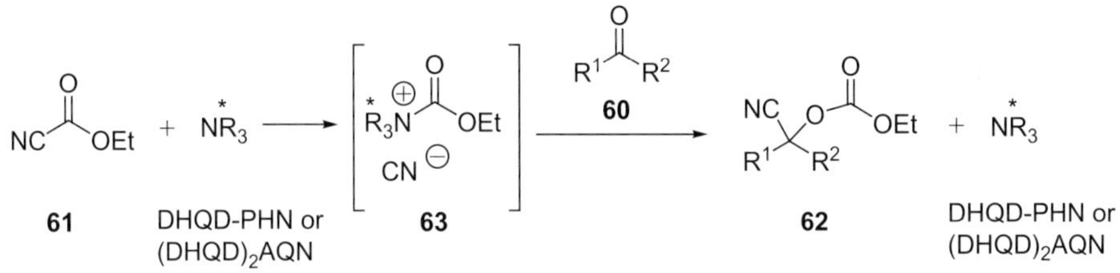

Scheme 10.21

results also provide the first example of a highly enantioselective nucleophilic addition to ketones with an organic catalyst.

10.4.4
Recent Applications of Nucleophilic Catalysis by Cinchona Alkaloids

10.4.4.1 Asymmetric Conjugate Additions

Gaunt and coworkers developed a novel enantioselective approach to prepare chiral cyclopropanes (Scheme 10.22) [36]; 9-O-methylquinidine (QD-66) and (DHQD)$_2$PYR were identified as optimal catalysts. This particular organocatalytic cyclopropanation may proceed through an ammonium ylide pathway with the cinchona alkaloid as a nucleophilic catalyst.

Chen and coworkers reported an asymmetric N-allylic alkylation of indoles 68 with Morita–Baylis–Hillman carbonates 69 catalyzed by (DHQD)$_2$PHAL

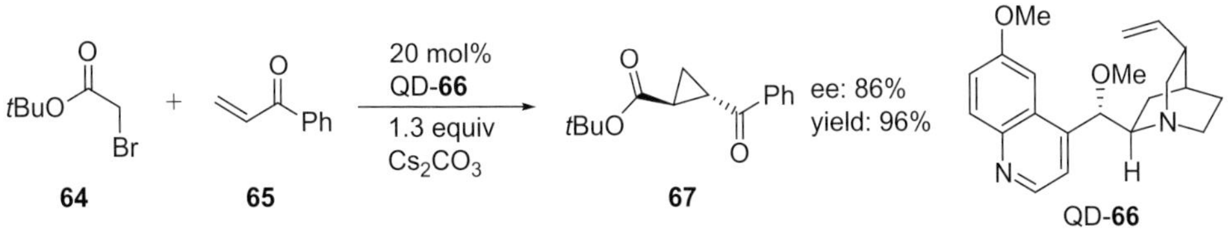

Scheme 10.22

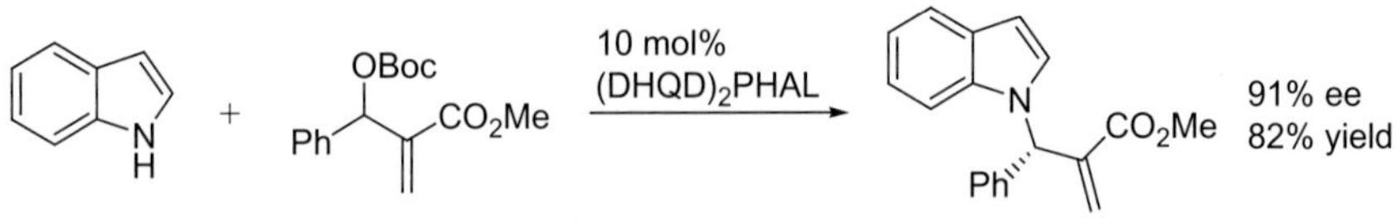

Scheme 10.23

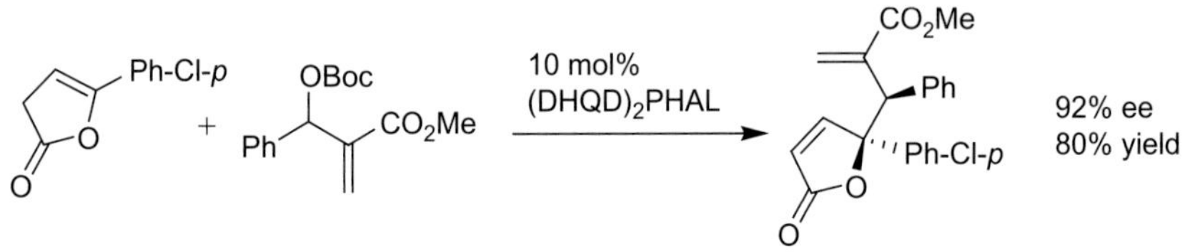

Scheme 10.24

(Scheme 10.23) [37]. The scope of the nucleophiles for this reaction was further extended to butenolides **71** in later studies (Scheme 10.24) [38].

10.4.4.2 Asymmetric Electrophilic Halogenations of Olefins

Some of the most interesting applications of cinchona alkaloid-mediated nucleophilic catalysis are directed toward the development of asymmetric electrophilic halogenations of alkenes. In 2010, Borhan reported a (DHQD)$_2$PHAL-mediated chlorolactonization of 4-pentenoic acids **73** (Scheme 10.25) [39]. Simultaneously, Tang reported QD-**76**-catalyzed bromolactonization of (Z)-enynes **75** with NBS (Scheme 10.26) [40].

Wynberg's landmark study provided the first indication that cinchona alkaloids could act as efficient nucleophilic catalysts, although the initial scope of the electrophile was limited to one particular aldehyde. Recent developments, especially those reported by the Lectka group, have greatly expanded the scope of the electrophile for cinchona alkaloid-mediated ketene-enolate chemistry. In addition,

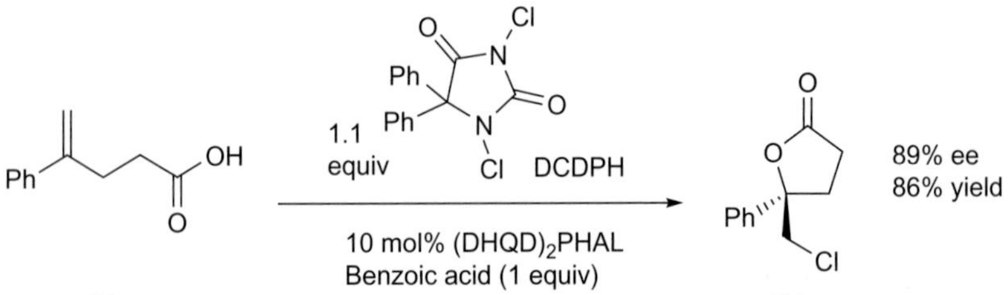

Scheme 10.25

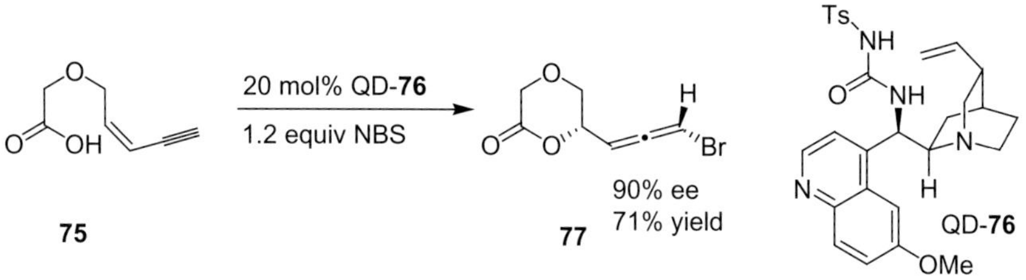

Scheme 10.26

the *in situ* generation of the ketene from the corresponding acid chlorides dramatically improved the practicality of these reactions. The studies by Hatakeyama and Deng provided early precedents indicating the potential of cinchona alkaloids as a broadly useful class of chiral nucleophilic catalysts. The latest reports of highly enantioselective halogenations of simple olefins have extended enantioselective organocatalysis to olefins, a brand new class of synthetically important substrates, thereby providing new applications of nucleophilic catalysis by cinchona alkaloids.

10.5
Base Catalysis

In contrast to the remarkable progress made by Wynberg and Hatakeyama in demonstrating the potential of cinchona alkaloids as effective enantioselective nucleophilic catalysts, early attempts to explore cinchona alkaloids as chiral base catalysts seemed to show that cinchona alkaloids were not able to fulfill this tantalizing promise [41]. However, over the last decade, base catalysis mediated by cinchona alkaloids has been transformed into one of the most widely practiced strategies in asymmetric organocatalysis.

10.5.1
Asymmetric Alcoholysis of Cyclic Anhydrides

Asymmetric alcoholysis of various cyclic anhydrides constitutes a transformation of considerable synthetic potential by providing an attractive approach toward important chiral carboxylic acid derivatives such as hemiesters, α-amino and α-hydroxy acids. Not surprisingly, extensive efforts involving both chiral metal and organic catalysts have been devoted to the development of an effective asymmetric alcoholysis [42]. Pioneering studies by Oda, Aitkin, and Bolm employing natural cinchona alkaloids showed that a highly enantioselective alcoholysis was possible only when these alkaloids were employed in more than stoichiometric amounts (Scheme 10.27) [43].

In 2000 Deng and coworkers reported that even a catalytic amount (5–30 mol%) of modified cinchona alkaloids, such as those bis-cinchona alkaloids applied as ligands in Sharpless asymmetric dihydroxylations [3], is highly enantioselective for

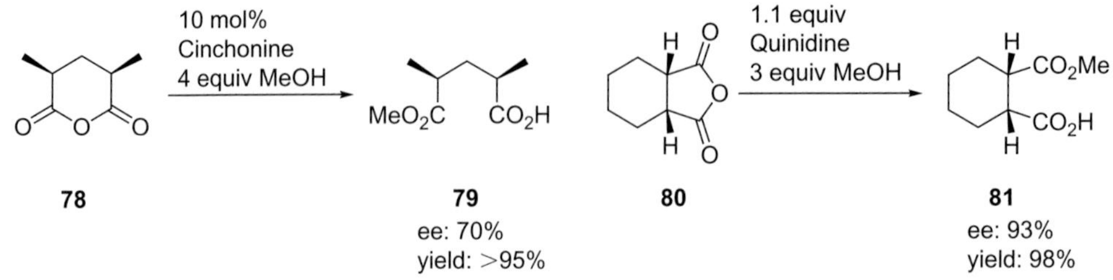

Scheme 10.27

the promotion of asymmetric alcoholysis of both succinic and glutaric anhydrides (Scheme 10.28) [44]. At the same time, Bolm and coworkers found that the combination of a stoichiometric amount of pempidine and a catalytic amount of quinidine (10 mol%) afforded high enantioselectivity for the alcoholysis of succinic anhydrides (Scheme 10.29) [45].

Deng's group further extended the cinchona alkaloid-catalyzed alcoholysis to parallel kinetic resolutions of monosubstituted succinic anhydrides (Scheme 10.30) [46], kinetic resolutions [47], and also dynamic kinetic resolutions [48] of UNCAs **89** (Scheme 10.31) and 1,3-dioxolane-2,4-diones **92** [49] (Scheme 10.32). These modified cinchona alkaloid-catalyzed alcoholyses might proceed by either a

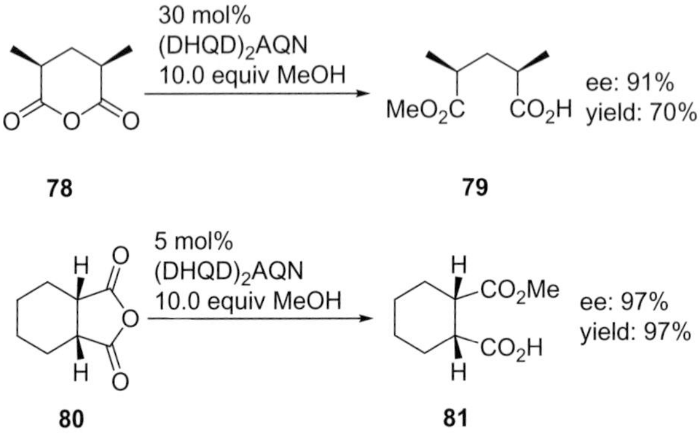

Scheme 10.28

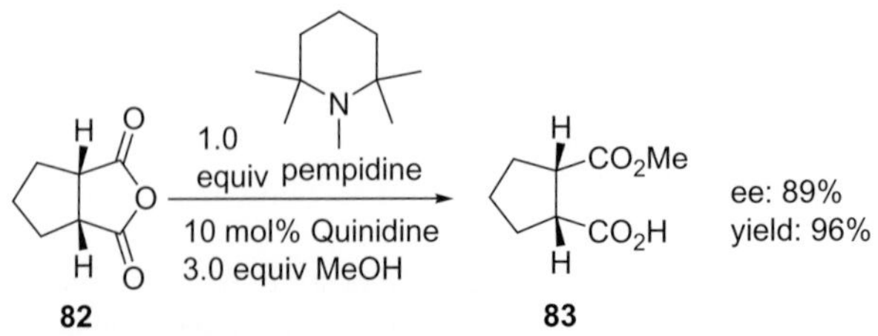

Scheme 10.29

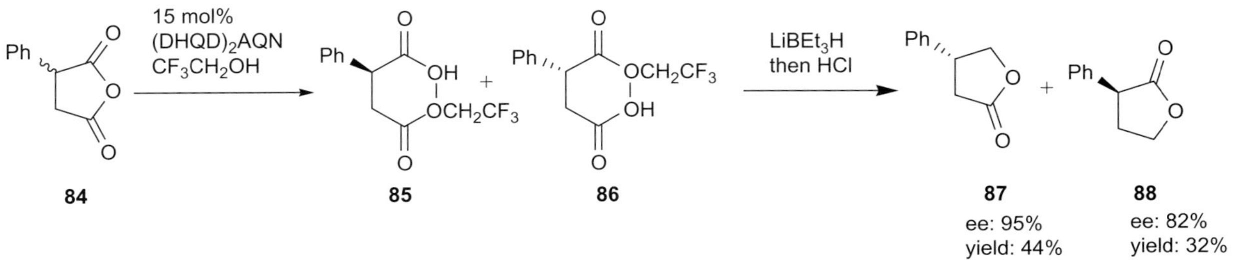

Ph
O
O
O
84
15 mol%
(DHQD)$_2$AQN
CF_3CH_2OH
Ph
O
OH
OCH_2CF_3
O
85
+
Ph
O
OCH_2CF_3
OH
O
86
LiBEt$_3$H
then HCl
Ph
O
O
87
ee: 95%
yield: 44%
+
Ph
O
O
88
ee: 82%
yield: 32%

Scheme 10.30

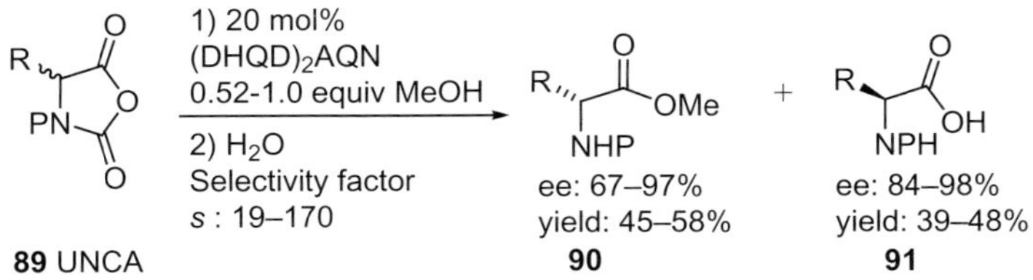

R
O
O
PN
O
89 UNCA
1) 20 mol%
(DHQD)$_2$AQN
0.52-1.0 equiv MeOH
2) H_2O
Selectivity factor
s : 19–170
R
O
OMe
NHP
ee: 67–97%
yield: 45–58%
90
+
R
O
OH
NPH
ee: 84–98%
yield: 39–48%
91

Scheme 10.31

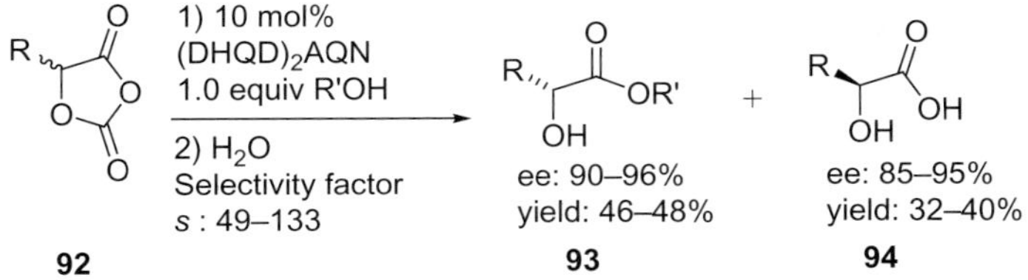

R
O
O
O
O
92
1) 10 mol%
(DHQD)$_2$AQN
1.0 equiv R'OH
2) H_2O
Selectivity factor
s : 49–133
R
O
OR'
OH
ee: 90–96%
yield: 46–48%
93
+
R
O
OH
OH
ee: 85–95%
yield: 32–40%
94

Scheme 10.32

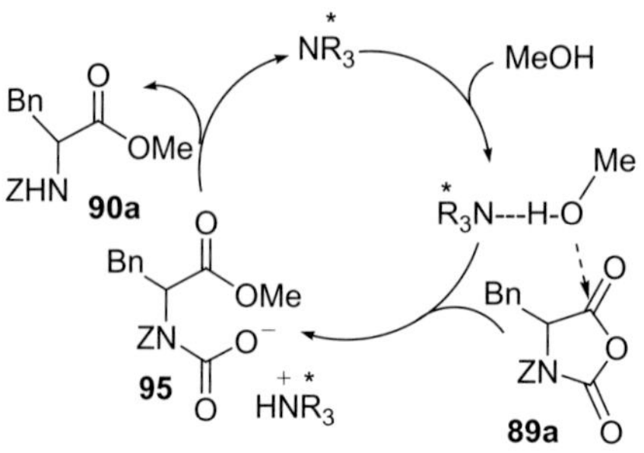

Scheme 10.33

nucleophilic or a general base catalysis mechanism. However, although chiral amines, including cinchona alkaloids, had only been shown as effective chiral nucleophilic catalysts, kinetic studies by Deng and coworkers indicated that the highly enantioselective alcoholysis of UNCA **89a** with DHQD-PHN proceeded by a general base catalysis mechanism (Scheme 10.33). These studies demonstrated for the first time that cinchona alkaloids are able to catalyze highly efficient and robust asymmetric reactions via general base catalysis, thereby providing the first documentation of a chiral amine as an effective Brønsted base catalyst.

10.5.2
Conjugate Additions

The discovery of cinchona alkaloids as effective Brønsted base catalysts for asymmetric alcoholysis raised the exciting possibility of applying this then new mode of catalysis as a broadly applicable approach to activate nucleophiles for asymmetric reactions. Guided by these considerations, Deng and coworkers investigated the ability of modified cinchona alkaloids to promote highly enantioselective conjugate additions of thiols to cyclic enones. Wynberg and coworkers had previously reported that natural cinchona alkaloids promoted these conjugate additions in modest enantioselectivity [50]. In 2002, Deng and coworkers reported (DHQD)$_2$PYR as a highly effective catalyst for the sulfa-conjugate addition, which afforded excellent enantioselectivity for a broad range of cyclic enones (Scheme 10.34) [51]. The successful application of base catalysis by modified cinchona alkaloids to the development of a highly enantioselective reaction with a mechanism distinct from the

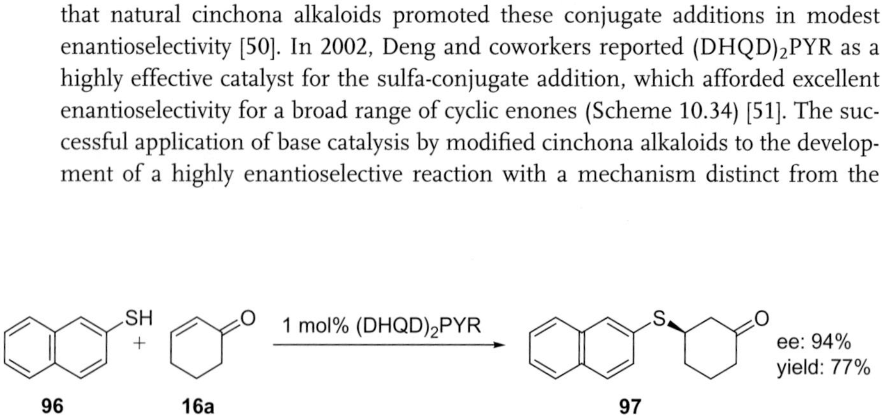

Scheme 10.34

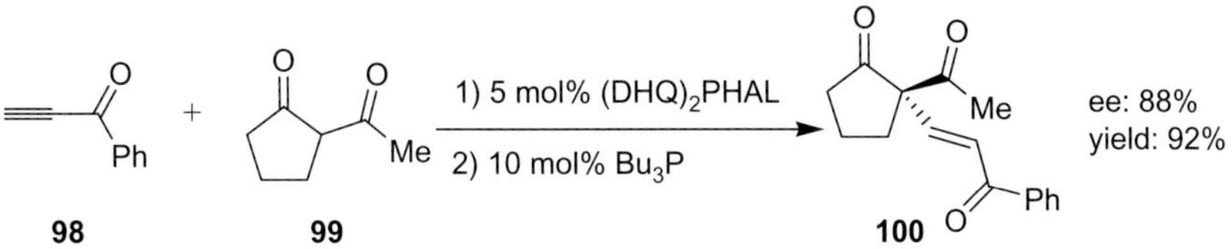

Scheme 10.35

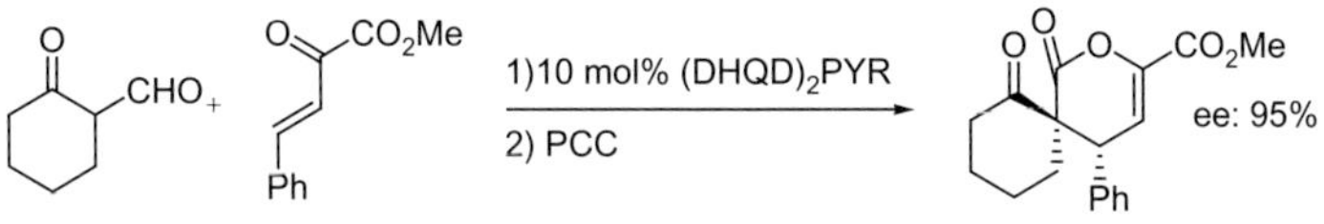

Scheme 10.36

asymmetric alcoholysis provided a strong indication of the considerable potential of cinchona alkaloids to act as Brønsted base catalysts.

The scope of the modified cinchona alkaloid-catalyzed conjugate additions has also been extended to carbon nucleophiles. In 2004, Jørgensen and coworkers reported an efficient (DHQ)$_2$PHAL-catalyzed conjugate addition of β-ketoesters to alkynones (Scheme 10.35) [52]. More recently, Ma and coworkers reported an asymmetric synthesis of spiro-3,4-dihydropyrans utilizing a (DHQD)$_2$PYR-catalyzed conjugate addition for asymmetric induction (Scheme 10.36) [53].

10.5.3
Asymmetric Mannich and Aldol Reactions

In recent years, cinchona alkaloids acting in the Brønsted catalysis mode have also been applied to the activation of various enolizable nucleophiles for asymmetric 1,2-additions. In 2005, Jørgensen and coworkers reported a highly selective, (DHQD)$_2$PYR-catalyzed Mannich reaction of α-substituted cyanoacetates (Scheme 10.37) [54]. In 2007, Shibata, Toru, and coworkers reported an enantioselective aldol reaction of oxindoles to trifluoropyruvate with (DHQD)$_2$PHAL (Scheme 10.38) [55].

The discovery and development of cinchona alkaloids as effective and versatile chiral Brønsted base catalysts stands as one of the most important advances in

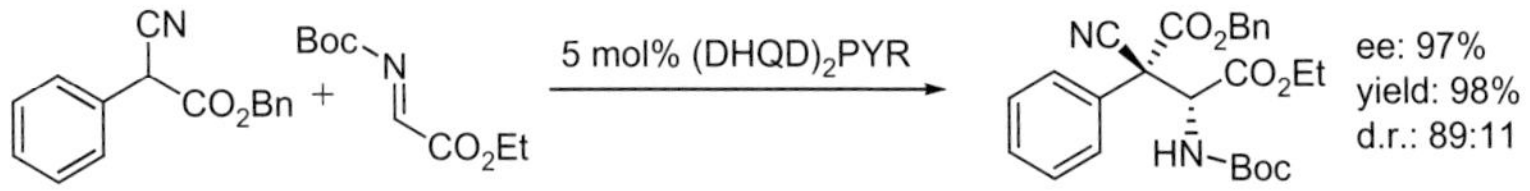

Scheme 10.37

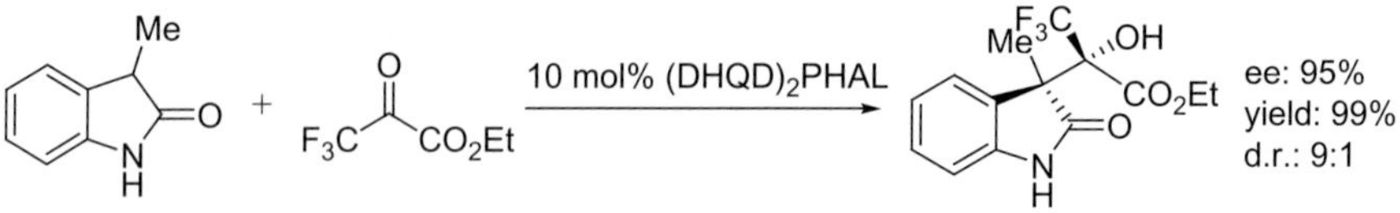

Scheme 10.38

the field of organocatalysis. Apart from providing a general strategy to promote asymmetric catalysis based on the activation of nucleophiles, this new development also pointed to the possibility of developing powerful bifunctional asymmetric catalysis in which the Brønsted base catalysis by a chiral tertiary amine couples with various electrophile-activation strategies such as Brønsted acid or iminium catalysis. As illustrated below, these highly exciting premises have already become a widely applied strategy in the field of asymmetric organocatalysis.

10.6
Cooperative and Multifunctional Catalysis

10.6.1
Acid–Base Cooperative Catalysis

Side chains of natural amino acids that bear either hydrogen bond donors or acceptors are weakly acidic or basic. These amino acid residues, each as an individual unit, are only moderately active catalytic centers. Consequently, much of the remarkable catalytic activity demonstrated by enzymes arises from cooperative catalysis, which utilizes multiple amino acid residues to achieve highly orchestrated, simultaneous activations of multiple reactants participating in a reaction [56]. As fundamentally important progress had been made in establishing small organic molecules, such as chiral thioureas and modified cinchona alkaloids, as highly enantioselective yet versatile chiral Brønsted acids and bases, respectively, the stage was set to tackle the challenge of developing powerful acid–base bifunctional chiral catalysts based on small organic molecules. Although the studies carried out almost three decades ago by Wynberg and coworkers with natural cinchona alkaloids did not result in a highly enantioselective reaction [41], the discovery of modified cinchona alkaloids as highly effective chiral Brønsted base catalysts indicated the possibility of developing powerful acid–base bifunctional catalysts using the cinchona alkaloid skeleton.

10.6.1.1 Asymmetric Conjugate Additions
In 2004, two different highly enantioselective asymmetric conjugate additions with bifunctional cinchona alkaloids were reported by the Jørgensen and Deng groups simultaneously. Jørgensen and coworkers described an asymmetric amination of α-substituted β-cyanoesters, β-ketoesters, and 1,3-diketones with dialkyl

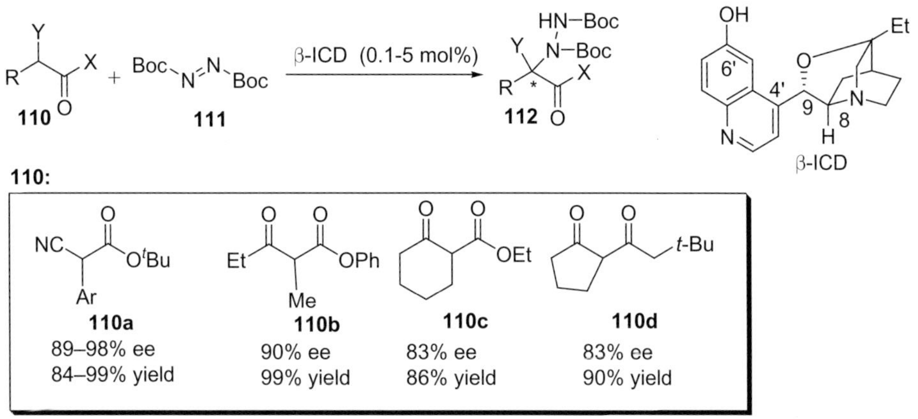

10.6 Cooperative and Multifunctional Catalysis

Scheme 10.39

110a: Ar, NC-C(=O)-O'Bu, 89–98% ee, 84–99% yield
110b: Et-C(=O)-C(Me)-C(=O)-OPh, 90% ee, 99% yield
110c: cyclohexanone-C(=O)-OEt, 83% ee, 86% yield
110d: cyclopentanone-C(=O)-t-Bu, 83% ee, 90% yield

azodicarboxylates **111** mediated by the rigid 6′-OH cinchona alkaloid β-ICD (Scheme 10.39) [57], which is the same catalyst that was reported previously by Hatakeyama as an effective nucleophilic-acid bifunctional catalyst for an enantioselective Morita–Baylis–Hillman reaction [33].

Deng and coworkers found that cupreidine (QD-**113a**), cupreine (Q-**113a**), and their derivatives (Figure 10.4) can catalyze highly enantioselective conjugate additions of methyl malonates and β-ketoesters to either aryl or alkyl nitroalkenes **115** (Scheme 10.40) [58]. These authors also reported that a much faster reaction rate and higher enantioselectivity were observed with cupreidine (C6′-OH) compared with those observed with quinidine (C6′-OMe), thereby demonstrating the critical role played by the 6′-OH group. This is the first example demonstrating that conformationally flexible cinchona alkaloids such as cupreidine or cupreine could act as effective bifunctional catalysts.

Cupreine and cupreidine, unlike β-ICD, bearing a rigid C8–C9 bond, feature a rotational C8–C9 bond as the connection between the basic quinuclidine unit and the acidic 6′-OH quinoline unit. The authors envisioned that this critical feature should enable these conformationally flexible cinchona alkaloids to adopt an optimal conformation for mechanistically distinguished asymmetric reactions,

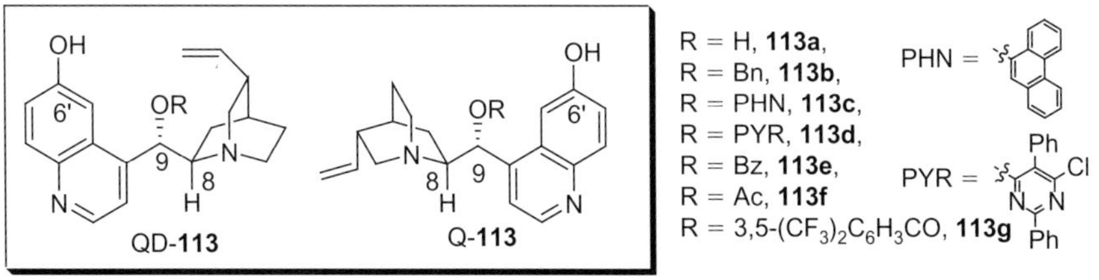

R = H, **113a**,
R = Bn, **113b**,
R = PHN, **113c**,
R = PYR, **113d**,
R = Bz, **113e**,
R = Ac, **113f**
R = 3,5-(CF$_3$)$_2$C$_6$H$_3$CO, **113g**

Figure 10.4 Cupreidine (QD-**113a**) and cupreine (Q-**113a**) and their derivatives.

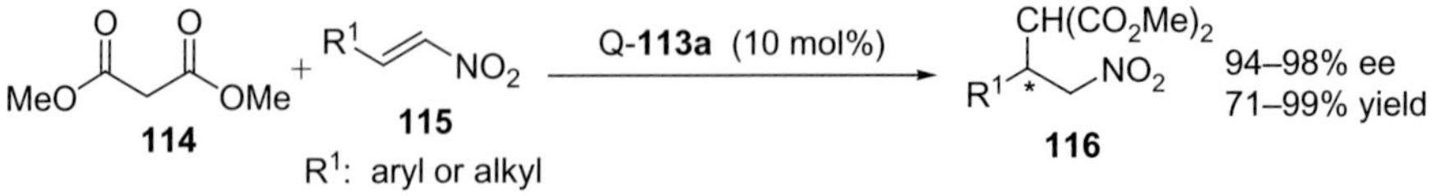

Scheme 10.40

thereby providing a key element for the design of highly selective yet broadly applicable chiral catalysts. Furthermore, in the framework of cinchona alkaloids bearing a rotational C8–C9 bond, the easily modifiable C9-substitutuent provides either a handle for catalyst tuning or additional sites for the introduction of hydrogen bond donor moieties for the development of new bifunctional catalysts. Another advantage of these cinchona alkaloids is that they can be derived from either *pseudo*-enantiomer of the natural cinchona alkaloids. Indeed, these exciting possibilities, raised by the first report from the Deng group of an efficient chiral bifunctional organic catalyst derived from quinidine and quinine, have been turned into reality at an extraordinarily pace.

In a subsequent paper published later in 2004, Deng and coauthors established that the nucleophiles for the conjugate addition to nitroalkenes could be extended to a wide range of trisubstituted carbon nucleophiles bearing various functionalities, thereby allowing the direct creation of adjacent quaternary–tertiary stereocenters in excellent *enantio*- and *diastereo*-selectivities (Scheme 10.41) [59]. Based on the elucidation of the active conformer of the catalyst and other experimental results, these authors also proposed a transition state model that featured multipoint hydrogen-bonding interactions as the origin of the remarkably efficient

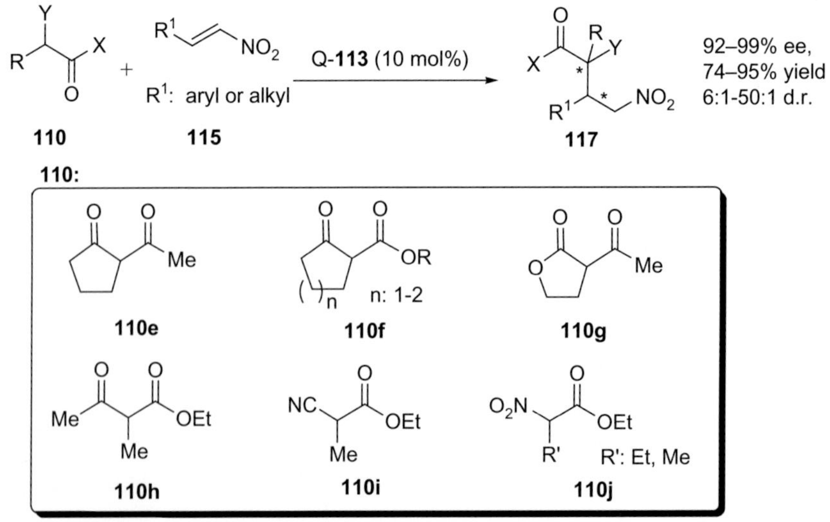

Scheme 10.41

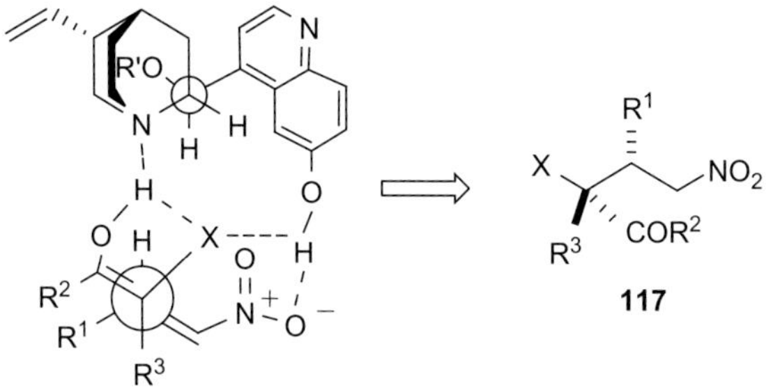

Figure 10.5 Transition state model featuring multipoint hydrogen-bonding interactions as the origin of the remarkably efficient catalysis demonstrated by QD-113

catalysis demonstrated by QD-113 or Q-113 (Figure 10.5) [59]. In this stereochemical model, the corresponding enol of the enolizable nucleophiles was postulated to be the active form of the nucleophile, which is activated and oriented through a two-point hydrogen-bonding interaction with the catalyst while the catalyst also simultaneously activates the Michael acceptor through the acidic 6′-OH.

Subsequent studies by Deng and coworkers demonstrated that these 6′-OH cinchona alkaloids could efficiently promote highly enantioselective and diastereoselective Michael additions of trisubstituted nucleophiles to α,β-unsaturated sulfones [60], enones [61], and enals [62] (Scheme 10.42). Furthermore, cinchona alkaloids QD-113 and Q-113 also promote the highly enantioselective amination of α-cyanoacetates with dialkyl azodicarboxylate to afford both enantiomers of the synthetically valuable chiral amines bearing a tetrasubstituted stereocenter [63].

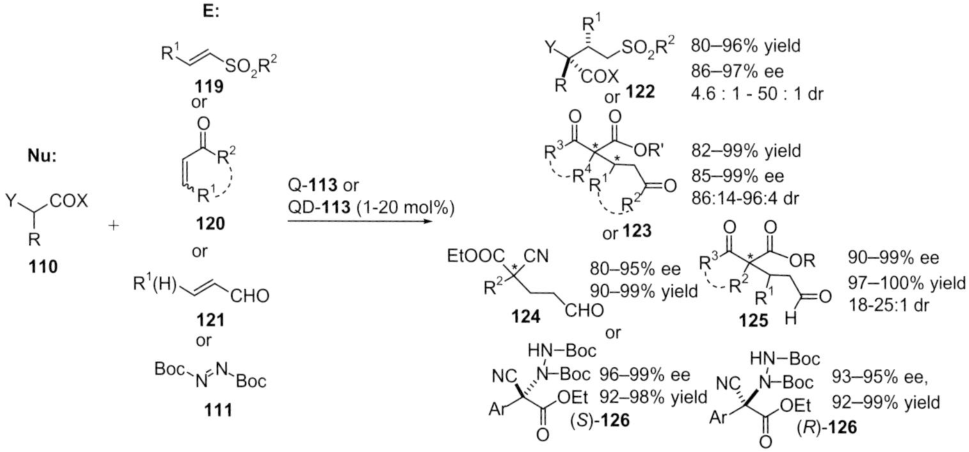

Scheme 10.42

From the viewpoint of synthetic method development, these asymmetric reactions provide practical access to a broad range of valuable chiral building blocks, including those containing all-carbon substituted quaternary stereocenters. They have already found applications in asymmetric syntheses of natural products [64] and drug candidates [65].

As the scope of the 6'-OH cinchona alkaloid catalysts expanded in asymmetric conjugate additions [66], new cinchona alkaloid-derived bifunctional catalysts designed for this important class of asymmetric transformations began to emerge. For example, by replacing the 6'-OH group with a 9-thiourea as the hydrogen bond donor moiety (Figure 10.6) [67], Chen and Soós independently reported 9-thiourea cinchona alkaloids as bifunctional catalysts for the conjugate additions of thiophenol to α,β-unsaturated imide [68] and nitromethane to chalcones [69], respectively (Scheme 10.43). Although the 9-thiourea cinchona alkaloids only afforded high enantioselectivity for the latter in these initial studies, subsequent investigations by several groups established them as another class of highly efficient and general bifunctional catalysts for asymmetric conjugate addition reactions (Scheme 10.43) [70]. For the asymmetric conjugate addition to nitroalkenes, the scope of the nucleophiles was extended to 5-aryl-1,3-dioxolanone heterocycles **145** [71] and naphthol derivatives [72] (Scheme 10.44). A particularly interesting application of these catalysts is the biomimetic decarboxylative conjugate addition of β-keto-acid (**149**) to nitroalkenes reported by Wennemers and coworkers (Scheme 10.45) [73].

Jørgensen and coworkers developed an asymmetric conjugate addition of oxime **151** to nitroalkenes with the thiourea cinchona alkaloids Q-**133**, thus documenting a C–O bond forming enantioselective Michael addition (Scheme 10.46) [74]. Wang and coworkers demonstrated that, by careful design of nucleophiles bearing multiple functional groups, tandem Michael/Michael additions could be initiated by the Q-**133**-catalyzed addition of *trans*-3-(2-mercaptophenyl)-2-propenoic acid ethyl ester **153** to nitroalkenes to create six-membered cyclic chiral building blocks [75]. Similarly, the same group applied a Q-**133**-catalyzed addition of *o*-formyl thiophenols **155** to α,β-unsaturated imides to initiate a Michael/aldol addition cascade [76]. Scheidt and coworkers developed an intramolecular asymmetric conjugate addition of phenol to β-ketoester alkylidene (**158**) with C6'-thiourea quinidine derivative QD-**135** [77]. The 1,4-adducts were transformed into chiral flavanones and chromanones upon decarboxylation.

10.6.1.2 Asymmetric 1,2-Additions to Carbonyls

Deng and coworkers reported in 2005 that 6'-OH cinchona alkaloids QD-**113e** and Q-**113e** are also highly effective in the promotion of asymmetric nitroaldol (Henry) reactions between nitromethane and alkyl, alkenyl, and aryl α-ketoesters **160** (Scheme 10.47) [78]. They also demonstrated that the adducts can be transformed into other tetrasubstituted chiral building blocks such as β-lactam, aziridine, and α-methylcysteine derivatives. Recently, this method has been successfully applied in the synthesis of a drug candidate [79]. In 2006, 6'-thiourea catalyst QD-**135** was reported by Hiemstra and coworkers as an effective catalyst for an

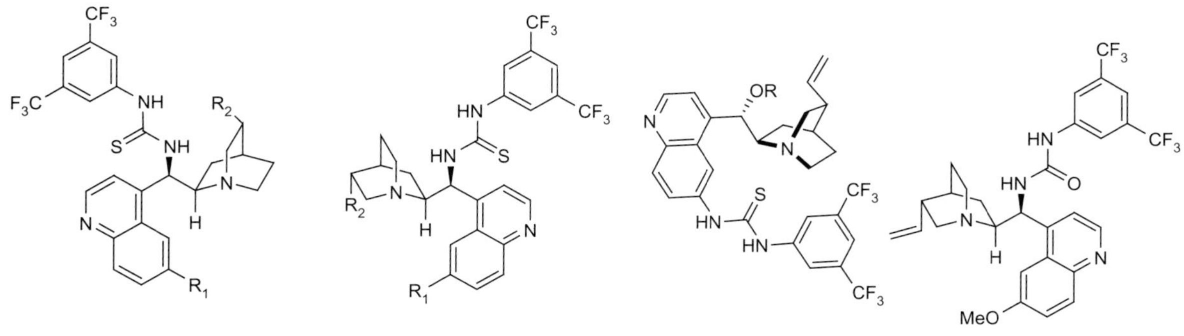

CN-127, R^1 = H, R^2 = CH=CH$_2$
CN-128, R^1 = H, R^2 = CH$_2$CH$_3$
QD-129, R^1 = OMe, R^2 = CH=CH$_2$
QD-130, R^1 = OMe, R^2 = CH$_2$CH$_3$

CD-131, R^1 = H, R^2 = CH=CH$_2$
CD-132, R^1 = H, R^2 = CH$_2$CH$_3$
Q-133, R^1 = OMe, R^2 = CH=CH$_2$
Q-134, R^1 = OMe, R^2 = CH$_2$CH$_3$

QD-135, R = Bn
QD-136, R = TIPS

Figure 10.6 9-Thiourea and -urea cinchona alkaloids that can act as bifunctional catalysts.

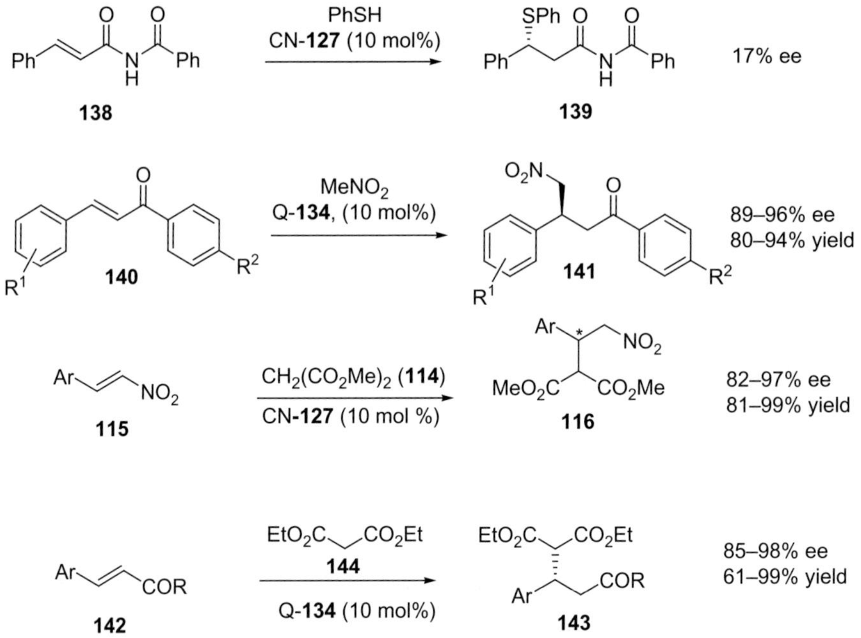

Scheme 10.43

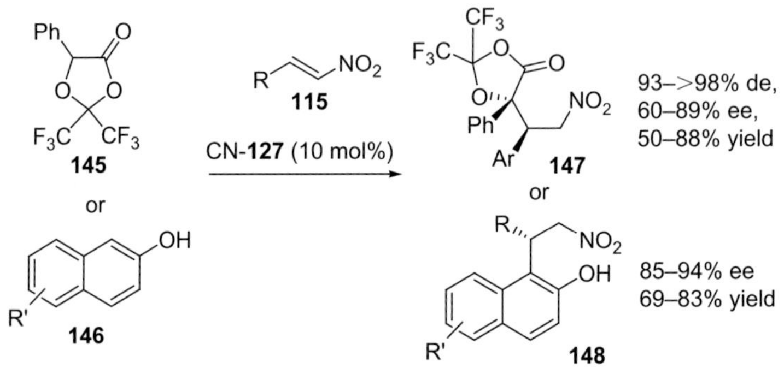

Scheme 10.44

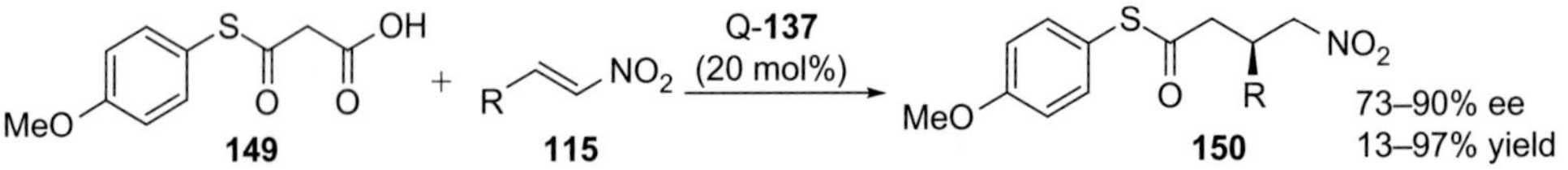

Scheme 10.45

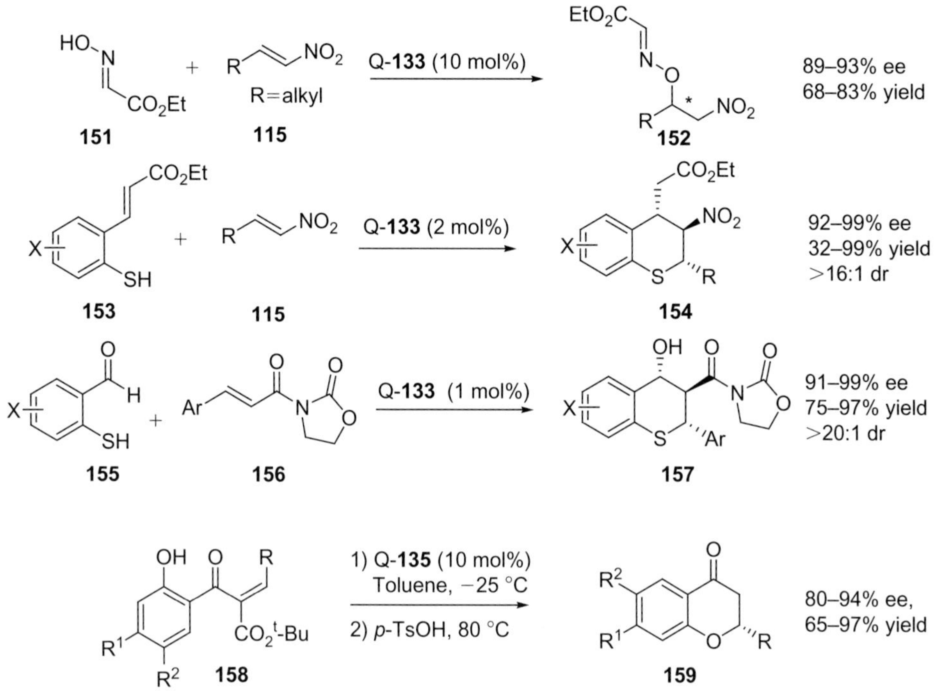

Scheme 10.46

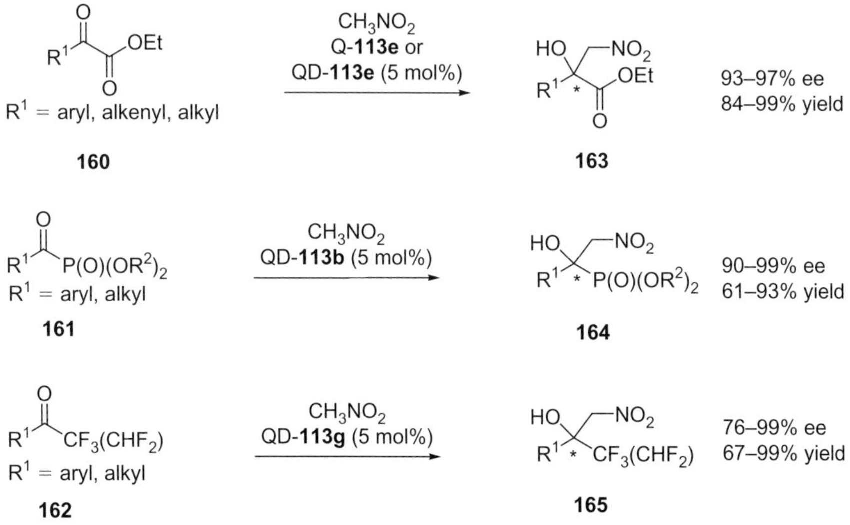

Scheme 10.47

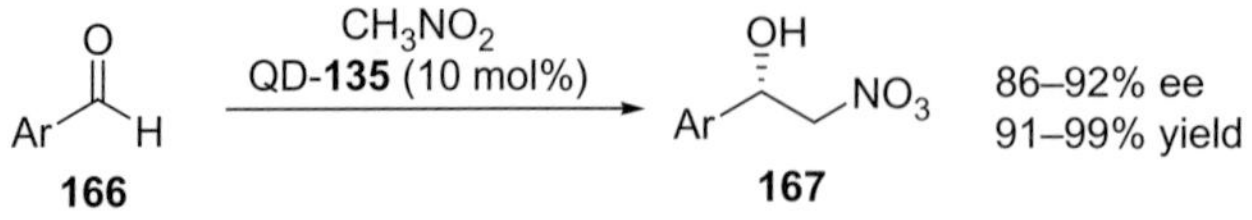

Scheme 10.48

enantioselective Henry reaction of nitromethane and aromatic aldehydes (Scheme 10.48) [80]. These studies provided the first successful application of cinchona alkaloids as acid–base bifunctional catalysts for asymmetric 1,2-addition reactions to carbonyls. Recent studies by the Zhao [81] and Bandini [82] groups further extended the scope of 6′-OH cinchona alkaloid-catalyzed Henry reactions to α-ketophosphonates **161** and fluoromethyl ketones **162**, respectively (Scheme 10.47).

10.6.1.3 Asymmetric 1,2-Additions to Imines

The catalytic asymmetric Mannich reaction provides an attractive approach for the construction of chiral amine synthons. In 2005, the Schaus group reported a highly enantioselective Mannich reaction of β-ketoesters with acyl imines catalyzed by cinchonine (Scheme 10.49) [83]. The scope of the nucleophiles was subsequently extended to cyclic 1,3-dicarbonyl compounds [84].

In a series of studies, the 9-thiourea cinchona alkaloids were found to be effective in promoting the additions of various nucleophiles to imines. In 2005, Ricci and coworkers reported a highly enantioselective aza-Henry reaction of nitromethane to N-Boc or N-Cbz aryl imines with C9 thiourea cinchona alkaloids Q-**133** (Scheme 10.50) [85]. In 2006, the Dixon [86] and Deng [87] groups

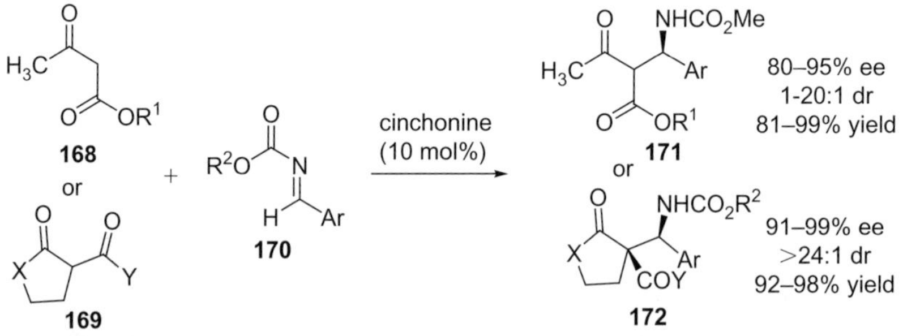

Scheme 10.49

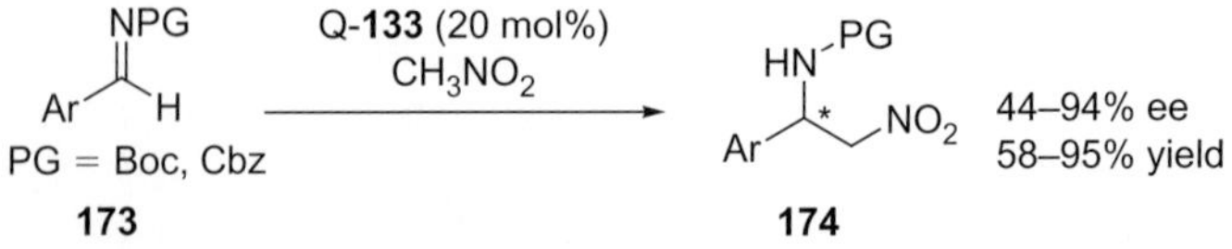

Scheme 10.50

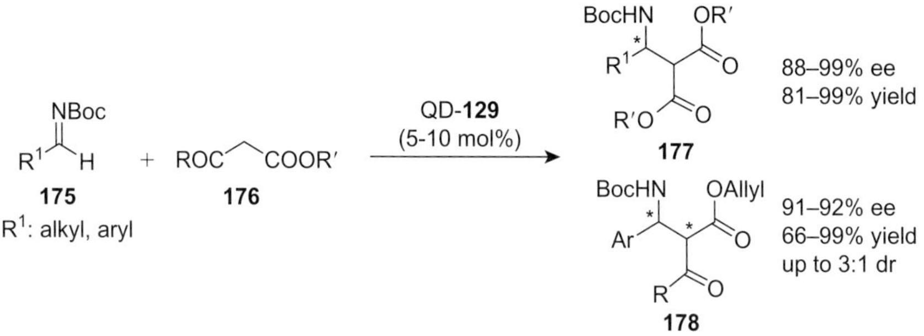

Scheme 10.51

independently reported highly enantioselective additions of malonates and β-ketoesters to N-carbamate imines catalyzed by 9-thiourea cinchona alkaloid QD-129 (Scheme 10.51). The Mannich adducts could be readily converted into important chiral building blocks such as β-amino acids and β-amino ketones. Deng and coworkers reported that even N-Boc alkyl imines, a class of highly sensitive imines toward either acid or base, could be utilized as substrates, thereby demonstrating the mild yet effective nature of hydrogen-bonding mediated cooperative catalysis.

More recently the Seidel group reported an asymmetric Mannich reaction of α-isothiocyanato imides and sulfonyl protected imines with a 6′-OH cinchona alkaloid QD-113f [88]. This reaction afforded α,β-diamine acid derivatives in excellent enantio- and diastereo-selectivities with 0.25–1.0 mol% of catalyst (Scheme 10.52).

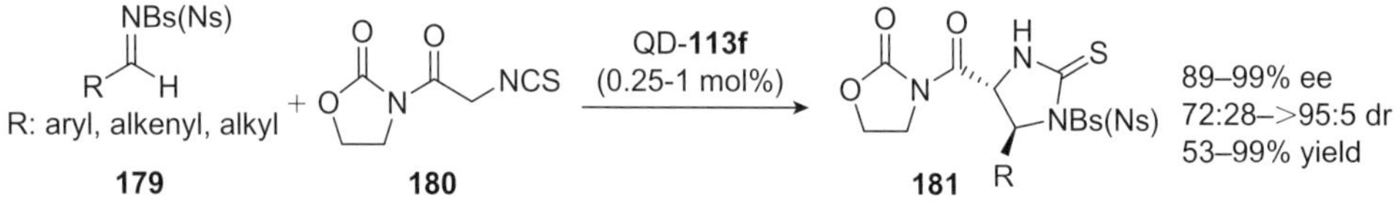

Scheme 10.52

10.6.1.4 Asymmetric Friedel–Crafts Reactions

The asymmetric Friedel–Crafts reaction of indoles with prochiral electrophiles constitutes an important approach to enantiomerically enriched indole derivatives. In 2005, Török and coworkers reported a highly enantioselective Friedel–Crafts addition of indoles with cinchonine and cinchonidine (Scheme 10.53) [89]. Presumably, cinchonine and cinchonidine promote the reaction in a bifunctional fashion by activating indole and trifluoropyruvate 108 via hydrogen bonding with the quinuclide nitrogen and the 9-OH, respectively. However, the scope of the reaction with respect to the electrophile is limited to a single substrate, the highly reactive ethyl trifluoropyruvate. In 2006, the Deng group demonstrated that 6′-OH cinchona alkaloids Q-113c and QD-113c are significantly more effective catalysts

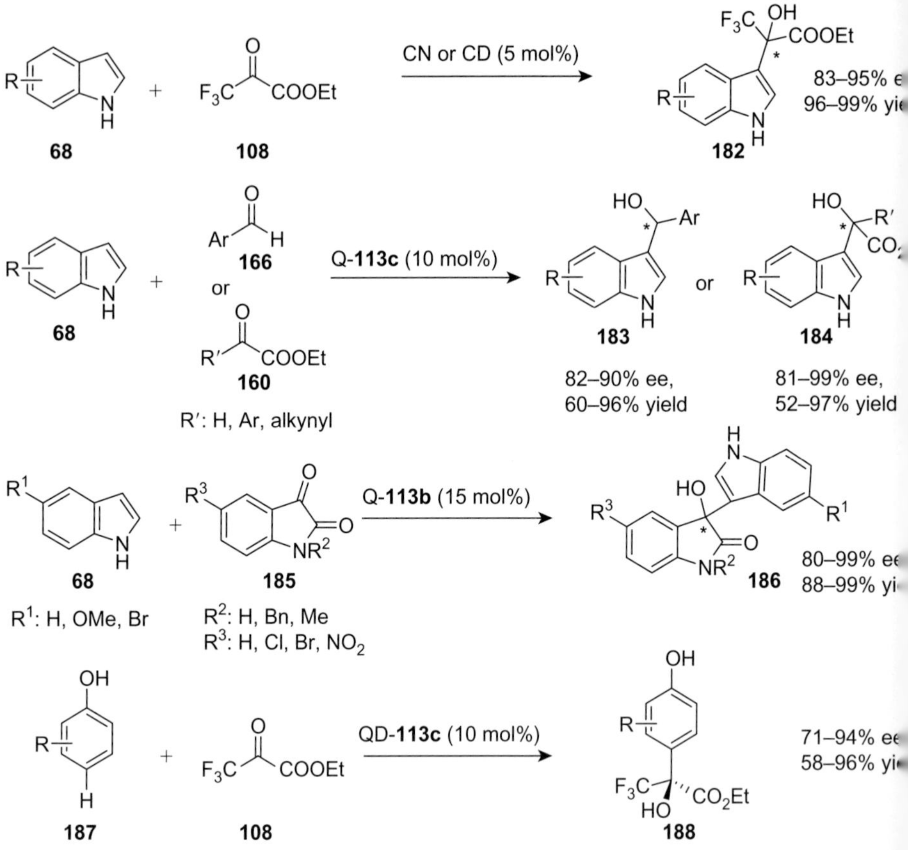

Scheme 10.53

for this reaction, as they promoted highly enantioselective Friedel–Crafts additions of indoles to aryl and alkynyl pyruvates, glyoxalate, and aryl aldehydes [90]. In recent years, the scope of the asymmetric Friedel–Crafts reaction mediated by the 6′-OH cinchona alkaloids has been further expanded. Specifically, Chimni reported a highly enantioselective Friedel–Crafts reaction of isatins **185** [91], and Chen reported a *para*-regioselective, asymmetric addition of simple phenol to ethyl trifluoropyruvate **108** [92].

The asymmetric Friedel–Crafts reaction between indoles and alkyl imines constitutes a synthetically valuable transformation. However, due to the sensitivity of the alkyl imines and the resulting Friedel–Crafts adducts toward acid and base, the development of this reaction presented a challenging goal in asymmetric synthesis. In 2006, the Deng group reported an efficient enantioselective Friedel–Crafts reaction of indoles and sulfonyl-protected aryl and alkyl imines with 9-thiourea cinchona alkaloids QD-**129** and Q-**133** (Scheme 10.54) [93]. Notably, with some highly sensitive alkyl imines QD-**129** or Q-**133** proved to be uniquely effective catalysts for this transformation. Once again, this study demonstrated the

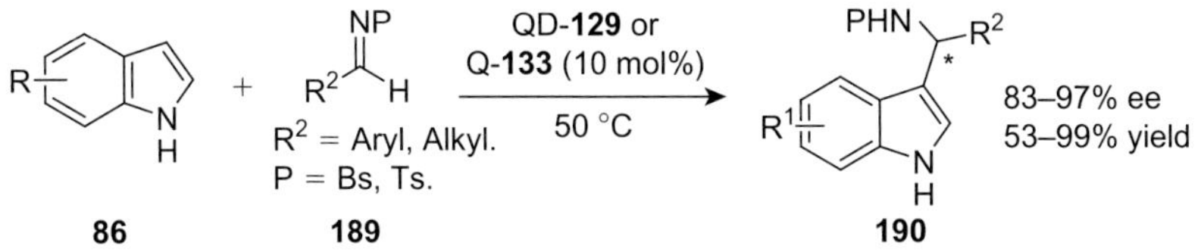

10.6 Cooperative and Multifunctional Catalysis

Scheme 10.54

remarkable ability of hydrogen bonding-mediated cooperative catalysis as an effective approach for the promotion of reactions involving acid- and/or base-sensitive reactants or products.

10.6.1.5 Asymmetric Diels–Alder Reactions

The asymmetric Diels–Alder reaction with 2-pyrones as a diene provides useful bicyclic chiral building blocks containing multiple functionalities. Owing to its electron-deficient and partially aromatic character, 2-pyrones are particularly challenging dienes for normal-electron demanding Diels–Alder reactions. With bifunctional 6′-OH cinchona alkaloid QD-**113c**, the Deng group developed the first highly enantioselective Diels–Alder reaction with 3-hydroxy-2-pyrones **191** (Scheme 10.55) [94]. Furthermore, the 9-thiourea cinchona alkaloid QD-**129** was found to be effective for reactions with acrylonitriles, thereby providing the first example of a highly enantioselective Diels–Alder reaction with a α,β-unsaturated

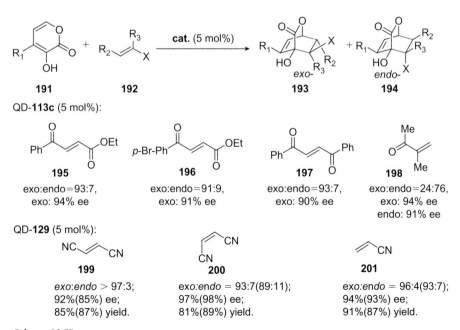

Scheme 10.55

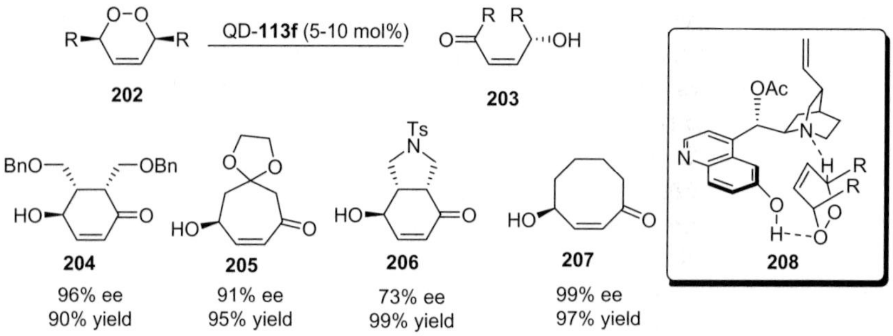

Scheme 10.56

nitriles. These studies illustrate the ability of acid–base bifunctional organic catalysts to promote asymmetric Diels–Alder reactions.

10.6.1.6 Asymmetric Fragmentation

With 6'-OH cinchona alkaloid QD-**113f**, Toste and coworkers reported an asymmetric Kornblum–DeLaMare reaction to produce optically active γ-hydroxyl-enones from cyclic meso peroxides (Scheme 10.56). They proposed that in the transition state (**208**), the C6'-OH of QD-**113f** activates the peroxide bond as an acid while the basic quinuclidine deprotonates the acidic C–H while the bulky substituents of meso peroxides point to an open space [95]. Similarly, with thiourea-cinchona alkaloid QD-**129**, the Jørgensen group disclosed an enantioselective fragmentation for the desymmetrization of *meso* cyclopropane-cyclopentanone and epoxycyclopentanones to give chiral γ-substituted cyclopentenones **210** (Scheme 10.57) [96].

As illustrated above, great strides have been made in the development of asymmetric acid–base bifunctional organic catalysts based on cinchona alkaloids. Although β-ICD and cupreidine and cupreine derivatives were simultaneously discovered to be highly enantioselective acid–base bifunctional catalysts, the latter

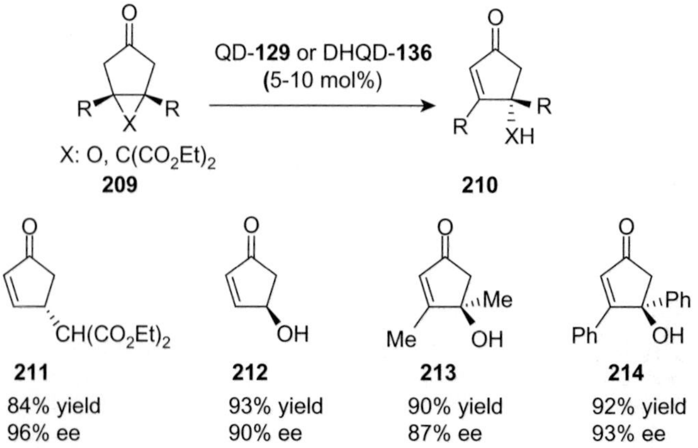

Scheme 10.57

6′-OH cinchona alkaloids, bearing a rotational C8–C9 bond, have been successfully applied to numerous asymmetric reactions and have therefore proved to be a far more general class of catalysts. These studies demonstrated that the conformational flexibility resulting from the rotational C8–C9 bond linking the two active centers is a key feature of these powerful bifunctional catalysts. The impact of this and other lessons learned with 6′-OH cinchona alkaloids is vividly illustrated by the surge in reports of designer acid–base bifunctional catalysts based on the cinchona alkaloid skeleton, such as the 9- and 6′-thiourea cinchona alkaloids. The list of such catalysts is still growing (Scheme 10.58) [97].

Cinchona alkaloid-derived bifunctional catalysts have played a leading role in transforming acid–base bifunctional organocatalysis from a dubious proposition into one of the most widely applied concepts for catalyst design and reaction development in asymmetric synthesis. The remarkable activity of these catalysts also raises the possibility of exploring the cinchona alkaloid skeleton for the design and development of other types of bifunctional catalysts with the two active centers associated with the quinuclidine moiety and one of the C9-substituents, respectively. As outlined below, this possibility is already turning into reality.

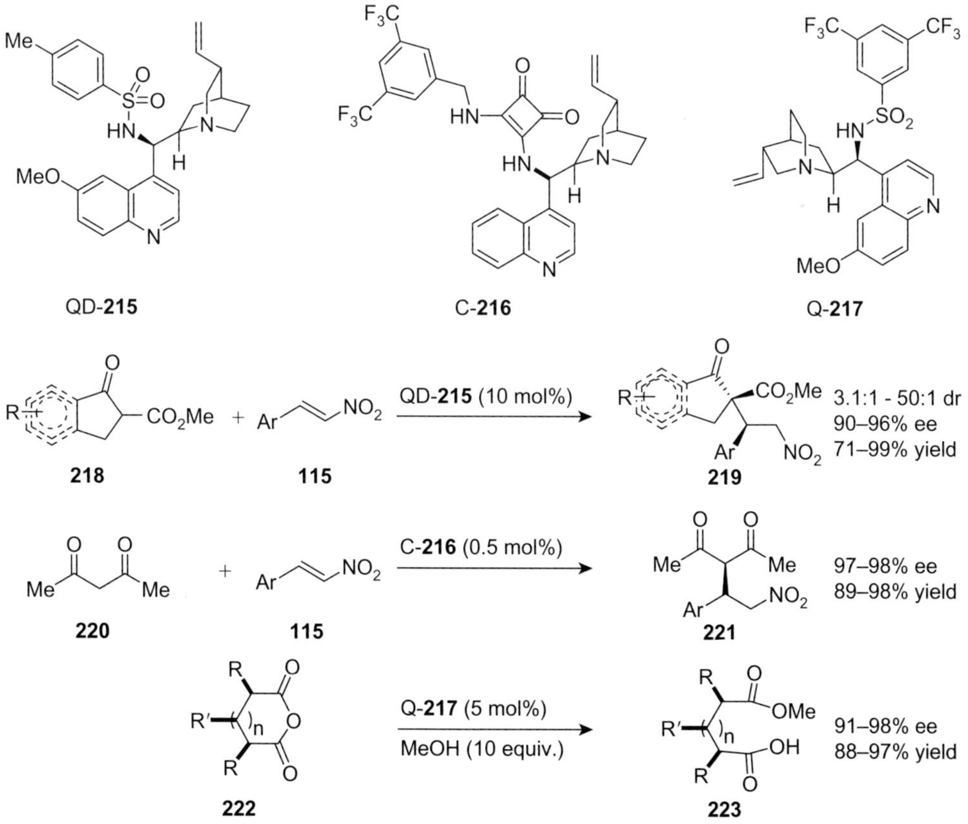

Scheme 10.58

10.6.2
Base–Iminium Cooperative Catalysis

The enantioselective iminium catalysis with chiral amines pioneered by the MacMillan group has been developed into a versatile approach to activate α,β-unsaturated carbonyls [98]. Although tremendous success has been achieved for the asymmetric reaction with α,β-unsaturated aldehydes via iminium catalysis by pyrrolidine- and proline-based chiral secondary amines, successful applications of these catalysts for the activation of the sterically more hindered α,β-unsaturated ketones were rare [99]. Recently, chiral primary amines have been demonstrated to be effective for the promotion of asymmetric reactions with α,β-unsaturated ketones. These important precedents in asymmetric iminium catalysis in combination with the advance in acid–base bifunctional catalysis with cinchona alkaloids suggested the possibility of developing cinchona alkaloid-derived bifunctional catalysts based on base–iminium cooperative catalysis.

10.6.2.1 Asymmetric Conjugate Additions

The J.-G Deng group reported a highly enantioselective and diastereoselective vinylogous Michael addition of α,α-dicyanoalkenes **224** to cyclic and acyclic enones using 9-NH$_2$ quinine Q-**227** (Scheme 10.59) [100, 101]. This catalyst is designed to activate the enones by iminium catalysis with the 9-NH$_2$ and the α,α-dicyanoalkenes **224** by base catalysis with the quinuclidine, respectively. Similarly to that in other asymmetric reactions with iminium catalysis, an acid such as TFA was applied as a co-catalyst. This report provides the first indication of the 9-NH$_2$ cinchona alkaloid as an effective catalyst to mediate base–iminium cooperative catalysis.

The scope of the nucleophiles of this conjugate addition to enones was extended to benzo-cyclic 1,3-dicarbonyl **228** (Scheme 10.60), which allowed Deng's group to develop a highly attractive asymmetric synthesis of the chiral anticoagulant drug *(S)*-warfarin (**229a**) [102]. In subsequent reports, the scope of the nucleophiles was further extended to malononitrile [103] and α-nitroesters [104]. More recent studies showed that 9-NH$_2$ cinchona alkaloids are also effective in promoting asymmetric reactions to enones with nitrogen nucleophiles **230** (Scheme 10.61) [105] and nucleophilic epoxidations of cyclic [106] and acyclic [107] enones (Scheme 10.62).

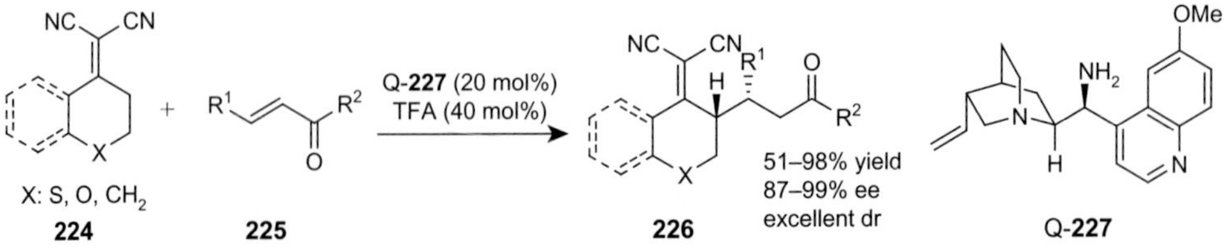

Scheme 10.59

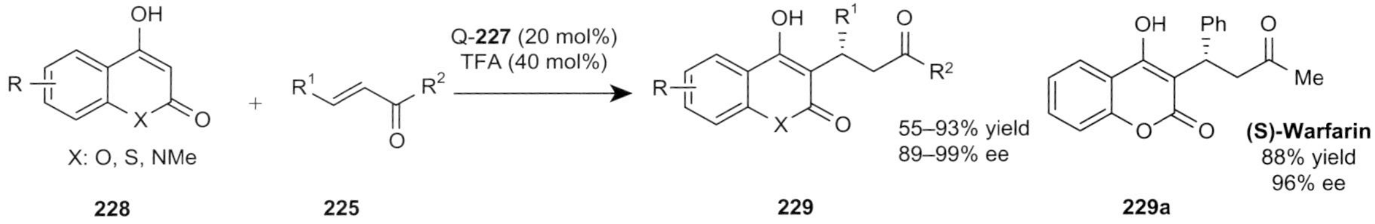

Scheme 10.60

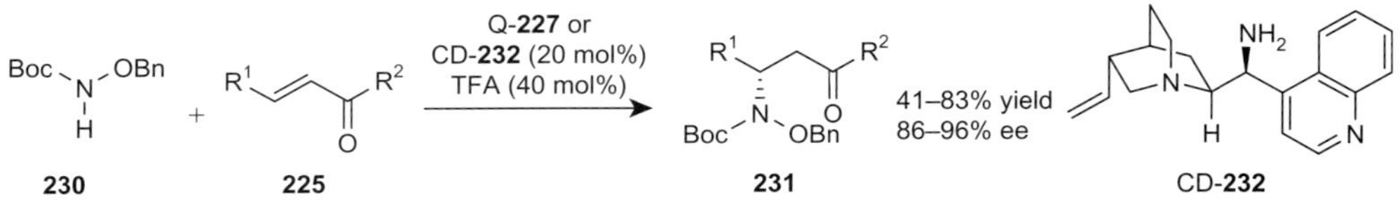

Scheme 10.61

10.6.2.2 Asymmetric Fridel–Crafts Additions

Highly enantioselective Friedel–Crafts additions of indoles to α,β-unsaturated aldehydes catalyzed by the chiral imidazolidinone were first disclosed by the MacMillan group in 2002 [108]. However, poor enantioselectivity was obtained for reactions with α,β-unsaturated ketones [109]. In 2006, two groups developed asymmetric Friedel–Crafts addition of indole to α,β-unsaturated ketones, independently, with C9-NH$_2$ cinchona alkaloids derivatives (Scheme 10.63). With CN-**240**, the Chen group developed an enantioselective indole alkylation to various enones in moderate to good enantioselectivity [110]. Shortly after, Melchiorre reported a more efficient protocol by using Q-**241** along with N-Boc phenylglycine (Boc-D-Phg-OH) as the acid co-catalyst to attain good to excellent enantioselectivity for this Friedel–Crafts reaction [111]. The poor reactivity and selectivity of N-Me indole are consistent with a base-iminium dual activation mechanism for the reaction.

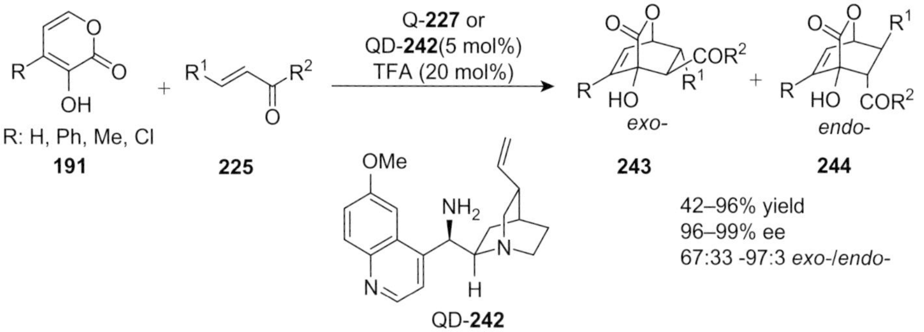

Scheme 10.64

10.6.2.3 Asymmetric Diels–Alder Reactions

As mentioned above, Deng's group found that acid–base bifunctional cinchona alkaloids QD-113c and QD-129 are able to promote a highly enantioselective Diels–Alder reaction of 2-pyrone with activated dienophiles. However, these catalysts are ineffective for the Diels–Alder reaction of 2-pyrone (191) with simple α,β-unsaturated ketones, a class of less reactive but synthetically valuable dienophiles [99]. In 2008, the Deng group disclosed an asymmetric Diels–Alder reaction between these two challenging classes of Diels–Alder reactants. With the readily available 9-NH$_2$ quinidine derivative QD-242 and quinine derivative Q-227, excellent enantioselectivity and good to excellent endo/exo selectivity can been secured for a wide range of α,β-unsaturated ketones (Scheme 10.64) [112]. The catalyst can also accept various 4-substituented 2-pyrones (191).

10.6.2.4 Semipinacol-Type 1,2-Carbon Migration

Recently, Tu and coworkers demonstrated that 9-NH$_2$ cinchona alkaloid Q-246 is an efficient catalyst for the asymmetric vinylogous α-ketol rearrangement of cyclic hydroxy enones to afford chiral spirocyclic diketones bearing all-carbon substituted quaternary stereocenters (Scheme 10.65) [113]. N-Boc-L-phenylglycine (NBLP) turns out to be the optimal acid co-catalyst. Good to excellent enantioselectivity and diastereoselectivity can be obtained with a significant range of cyclic hydroxy enones. This study is particularly noteworthy in extending the utility of cinchona alkaloid-derived catalysts to a rare class of asymmetric rearrangement reactions.

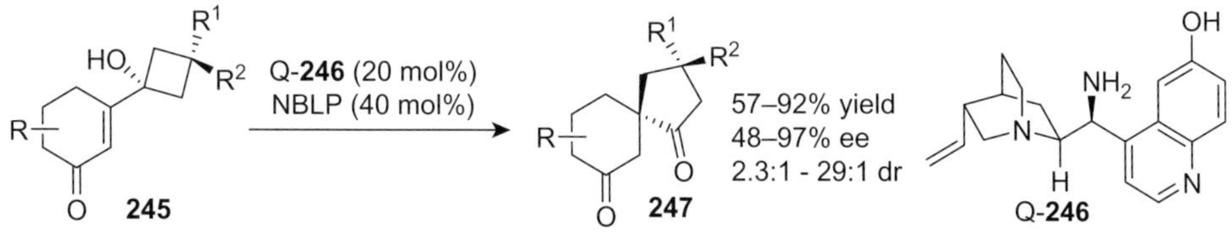

Scheme 10.65

In summary, the successful development of the 9-NH$_2$ cinchona alkaloids as base–iminium bifunctional catalysts for a wide range of mechanistically distinctive reactions with α,β-unsaturated ketones represents important recent progress in asymmetric synthesis [114]. These studies further expanded the range of asymmetric cooperative catalysis that can be mediated by cinchona alkaloid-derived, bifunctional organic catalysts.

10.6.3
Multifunctional Cooperative Catalysis

The development of dual-functional cooperative organocatalysis with cinchona alkaloids to promote asymmetric reaction cascades has become a new topic of interests. In these reaction cascades, the cinchona alkaloid promotes each of the bond-breaking and -forming steps as a bifunctional catalyst by activating and orienting the two participating reactants or intermediates. In contrast, other asymmetric reaction cascades utilize a cinchona alkaloid as either a mono- or bifunctional catalyst to promote only the first stereochemistry-generating step to initiate the asymmetric reaction cascades.

10.6.3.1 Tandem Conjugate Addition–Protonation Reactions

In 2006, Deng and coworkers reported that cinchona alkaloid QD-113c/Q-113c is an efficient catalyst for the highly enantioselective and diastereoselective reactions of trisubstituted carbon nucleophiles and α-chloroacrylonitrile (**249**) (Scheme 10.66) [115]. These reactions create two nonadjacent stereocenters: the first in the nucleophilic addition and the second in the protonation of the enol(ate) intermediate, thereby affording a one-pot construction of 1,3-tertiary-quaternary stereocenters. Importantly, with DABCO this reaction proceeded with hardly any diastereoselectivity, which confirms that QD-113c was responsible for both the enantioselectivity and the diastereoselectivity. In 2007, Deng and coworkers further established that the 9-thiourea cinchona alkaloids QD-129/Q-133, catalysts presenting the hydrogen bond donor and acceptor in a different spatial relationship compared to that presented by 6′-OH cinchona alkaloids, could promote the reaction with a sense of diastereoselectivity complementary to that afforded by QD-113c/Q-113c [116]. The synthetic implication of such catalyst-controlled diastereo- and enantioselective reaction cascades was illustrated in the total syntheses of a diastereomeric pair of natural products, manzacidins A (**251a**) and C (**251b**) via a unified route. In these reaction cascades cinchona alkaloids QD-113c/Q-113c and QD-129/Q-133 served a dual role, facilitating the enantioselective addition to generate the quaternary stereocenter and then the stereoselective protonation of the transient enolate intermediate as a chiral donor (**253**, Scheme 10.67).

10.6.3.2 Catalytic Asymmetric Peroxidations

Numerous peroxy natural products are potent antitumor, anticancer, and antiparasite agents [117]. Some peroxy natural products have found wide clinical use as antimalarial drugs, with artemisinin being one of the most important [118].

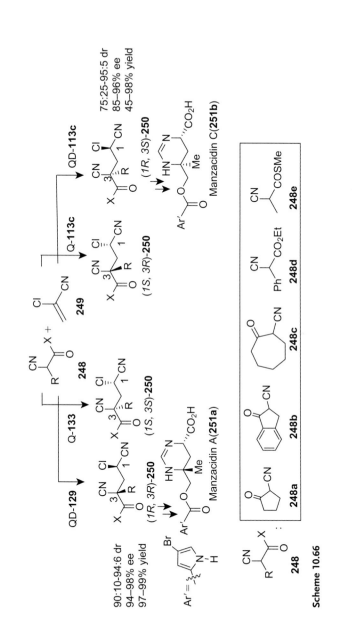

Scheme 10.66

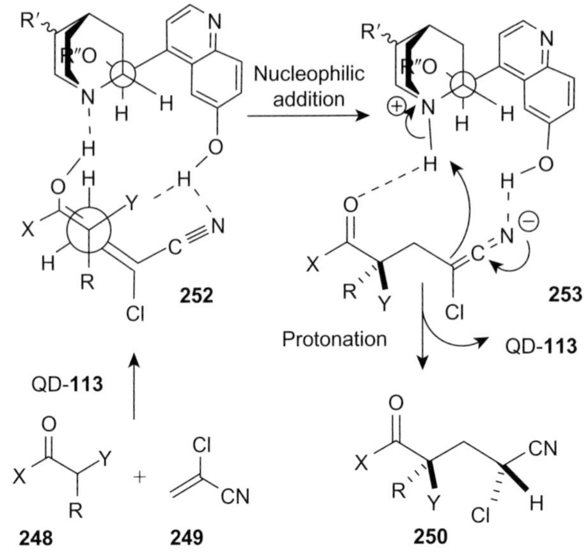

Scheme 10.67

However, there is a general lack of enantioselective methods for the preparation of optically active chiral peroxides.

The base-promoted reactions of α,β-unsaturated ketones **225** with hydroperoxides **254** represent a classic epoxidation reaction. It is well established that the epoxide **238** is formed via a two-step mechanism (Scheme 10.68): nucleophilic addition of the hydroperoxide **254** to **225** followed by an intramolecular nucleophilic substitution by the resulting enolate **257** that breaks the peroxide bond. In principle this epoxidation pathway (**225** → **238**) could be converted into a

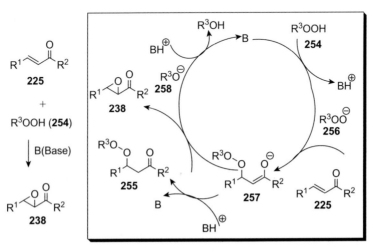

Scheme 10.68

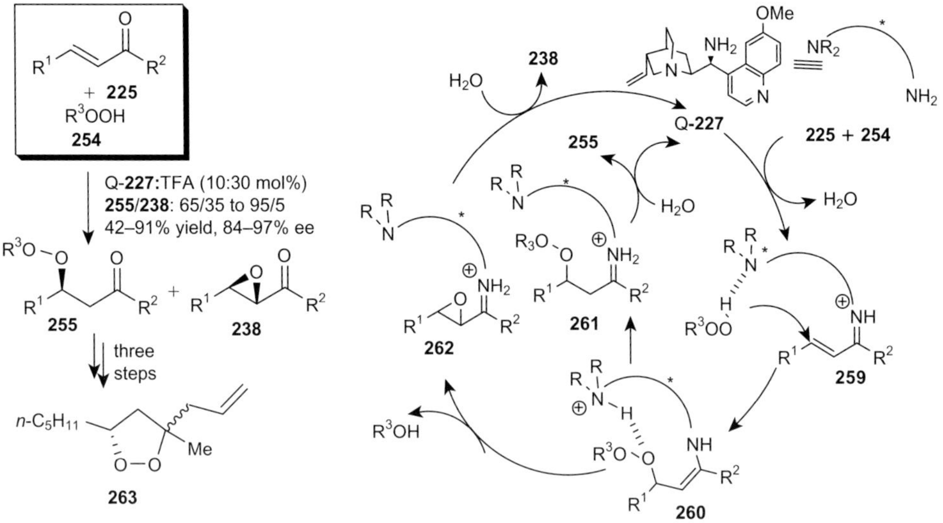

Scheme 10.69

peroxidation pathway (**225** → **255**) if **257** could be trapped by protonation. However, the overwhelming preference of the enolate **257** for intramolecular nucleophilic substitution is evident from the lack of reported peroxidations of α,β-unsaturated carbonyl compounds.

Deng and coworkers hypothesized that a cinchona alkaloid derivative such as Q-**227** could not only render the nucleophilic addition of the hydroperoxide **254** to the iminium intermediate **259** enantioselective (Scheme 10.69), but could also influence the partitioning of the peroxyenamine intermediate **260** between the epoxidation (**260** → **262**) and the peroxidation (**260** → **261**) pathways. In particular, with the protonated quinuclidine as a proton source nearby to facilitate the protonation of the peroxyenamine, the peroxidation might be accelerated. Presumably, due to steric crash and/or multipoint interactions between the peroxyenamine intermediate and the covalently linked cinchona alkaloid, the bond-rotational freedom of the peroxyenamine intermediate should be hampered in complex **260** (Scheme 10.69). This conformational rigidity imposed by catalyst Q-**227** on the peroxyenamine would diminish its ability to adopt the active conformation by which the nucleophilic enamine moiety is optimally aligned relative to the O–O bond for the nucleophilic attack, which in turn would decelerate the epoxidation. Deng and coworkers reported in 2008 that the 9-NH$_2$ cinchona alkaloid Q-**227** could promote an unprecedented, highly enantioselective catalytic peroxidation [107]. Instead of affording stereoselective protonation, the cinchona alkaloid catalyst served the function of trapping the peroxyenamine intermediate via protonation. These results presented a novel application of dual-functional cooperative catalysis, namely the alteration of the partitioning of a reaction intermediate to create a new asymmetric peroxidation pathway.

These studies demonstrated the possibility of developing multifunctional cooperative catalysis with cinchona alkaloids not only to exercise stereoselectivity

10.7
Conclusion

Since the documentation of the first catalytic enantioselective reaction, a quinine-catalyzed addition of hydrogen cyanide to benzaldehyde reported by Bredig and Fiske in 1913 [119], studies of cinchona alkaloid-mediated asymmetric catalysis have been defining the frontier of asymmetric catalysis. Wynberg's visionary investigations constitute the first systematic effort aiming to establish asymmetric catalysis as a broadly applicable strategy for asymmetric synthesis [41]. The Sharpless asymmetric dihydroxylation represents one of the pinnacle achievements in the development of a highly enantioselective and general catalytic reaction. Corey's studies in chiral phase-transfer catalysis established the first class of highly versatile chiral organic catalysts. Over the last decade a series of breakthroughs made by several groups, as outlined in this chapter, have redefined the scope of asymmetric nucleophilic and Brønsted base catalysis. Most recently, cinchona alkaloid-derived bifunctional catalysts have been playing a leading role in establishing cooperative organocatalysis as a powerful strategy for asymmetric synthesis. To the present authors, the already uncovered potential of cinchona alkaloids as versatile chiral catalysts truly exceeded our previous highest expectation. Nonetheless, we anticipate that much new chemistry remains to be discovered and developed with these marvelous natural products, as they share several key features with proteins in having hydrogen-bond donor, hydrogen-bond acceptor and aromatic motifs linked by a flexible yet easily tunable framework.

Acknowledgments

We are grateful for the financial support from National Institute of Health (National Institute of General Medical Sciences, R01-GM61591). We sincerely thank our colleagues, Mr. Brian Provencher and Mr. Keith Bartelson, for their help during the preparation of the manuscript.

References

1 For a recent review, see: Blaser, H.-U. and Studer, M. (2007) *Acc. Chem. Res.*, **40**, 1348–1356.
2 For an early review, see: Kacprzak, K. and Gawronski, J. (2001) *Synthesis*, 961–998.
3 For two outstanding reviews, see (a) Kolb, H.C., Vannieuwenhze, M.S., and Sharpless, K.B. (1994) *Chem. Rev.*, **94**, 2483–2547; (b) Noe, M.C., Letavic, M.A., Snow, S.L., and McCombie, S.W. (1994) *Org. React.*, **66**, 109–625.
4 Hentges, S.G. and Sharpless, K.B. (1980) *J. Am. Chem. Soc.*, **102**, 4263–4265.

5 Jacobsen, E.N., Marko, I., Mungall, W.S., Schroder, G., and Sharpless, K.B. (1988) *J. Am. Chem. Soc.*, **110**, 1968–1970.
6 (a) Sharpless, K.B., Amberg, W., Bennani, Y.L., Crispino, G.A., Hartung, J., Jeong, K.S., Kwong, H.L., Morikawa, K., Wang, Z.M., Xu, D.Q., and Zhang, X.L. (1992) *J. Org. Chem.*, **57**, 2768–2771; (b) Crispino, G.A., Jeong, K.S., Kolb, H.C., Wang, Z.M., Xu, D.Q., and Sharpless, K.B. (1993) *J. Org. Chem.*, **58**, 3785–3786; (c) Becker, H. and Sharpless, K.B. (1996) *Angew. Chem. Int. Ed.*, **35**, 448–451.
7 Noe, M.C. (1996) PhD dissertation, Harvard University. The Development of a Mechanistic Model and Applications for the Sharpless Asymmetric Dihydroxylation of Olefins.
8 Dijkstra, G.D.H., Kellogg, R.M., Wynberg, H., Svendsen, J.S., Marko, I., and Sharpless, K.B. (1989) *J. Am. Chem. Soc.*, **111**, 8069–8076.
9 Corey, E.J. and Noe, M.C. (1996) *J. Am. Chem. Soc.*, **118**, 319–329.
10 Corey, E.J. and Noe, M.C. (1993) *J. Am. Chem. Soc.*, **115**, 12579–12580.
11 Corey, E.J., Noe, M.C., and Lin, S.Z. (1995) *Tetrahedron Lett.*, **36**, 8741–8744.
12 Colonna, S., Hiemstra, H., and Wynberg, H. (1978) *Chem. Commun.*, 238–239.
13 Dolling, U.H., Davis, P., and Grabowski, E.J.J. (1984) *J. Am. Chem. Soc.*, **106**, 446–447.
14 Corey, E.J., Xu, F., and Noe, M.C. (1997) *J. Am. Chem. Soc.*, **119**, 12414–12415.
15 Lygo, B. and Wainwright, P.G. (1997) *Tetrahedron Lett.*, **38**, 8595–8598.
16 (a) O'Donnell, M.J., Bennett, W.D., and Wu, S. (1989) *J. Am. Chem. Soc.*, **111**, 2353–2355; (b) O'Donnell, M.J., Wu, S., and Huffman, J.C. (1994) *Tetrahedron.*, **50**, 4507–4518; (c) O'Donnell, M.J., Wu, S., Esikova, I., and Mi, A. (10 September 1996) Catalytic enantioselective synthesis of α-amino acid derivatives by phase-transfer catalysis. U.S. Patent 5,554,753.
17 Corey, E.J., Noe, M.C., and Xu, F. (1998) *Tetrahedron Lett.*, **39**, 5347–5350.
18 Horikawa, M., Busch-Petersen, J., and Corey, E.J. (1999) *Tetrahedron Lett.*, **40**, 3843–3846.
19 Corey, E.J., Bo, Y., and Busch-Petersen, J. (1998) *J. Am. Chem. Soc.*, **120**, 13000–13001.
20 (a) Corey, E.J. and Zhang, F.-Y. (1999) *Org. Lett.*, **1**, 1287–1290; (b) Corey, E.J. and Zhang, F.-Y. (2000) *Org. Lett.*, **2**, 4257–4259; (c) Zhang, F.-Y. and Corey, E.J. (2000) *Org. Lett.*, **2**, 1097–1100; (d) Zhang, F.-Y. and Corey, E.J. (2001) *Org. Lett.*, **3**, 639–641; (e) Lygo, B. and To, D.C.M. (2002) *Chem. Commun.*, 2360–2361.
21 (a) Corey, E.J. and Zhang, F.-Y. (1999) *Angew. Chem. Int. Ed.*, **38**, 1931–1934; (b) Zhang, F.-Y. and Corey, E.J. (2004) *Org. Lett.*, **6**, 3397–3399.
22 Park, H.-G., Jeong, B.-S., Yoo, M.-S., Lee, J.-H., Park, M.-K., Lee, Y.-J., Kim, M.-J., and Jew, S.-S. (2002) *Angew. Chem. Int. Ed.*, **41**, 3036–3038.
23 (a) Fini, F., Sgarzani, V., Pettersen, D., Herrera, R.P., Bernardi, L., and Ricci, A. (2005) *Angew. Chem. Int. Ed.*, **44**, 7975–7978; (b) Palomo, C., Oiarbide, M., Laso, A., and Lopez, R. (2005) *J. Am. Chem. Soc.*, **127**, 17622–17623.
24 Moss, T.A., Fenwick, D.R., and Dixon, D.J. (2008) *J. Am. Chem. Soc.*, **130**, 10076–10077.
25 (a) Nelson, A. (1999) *Angew. Chem. Int. Ed.*, **38**, 1583–1585; (b) Ooi, T. and Maruoka, K. (2007) *Angew. Chem. Int. Ed.*, **46**, 4222–4266; (c) Maruoka, K. (ed.) (2008) *Asymmetric Phase Transfer Catalysis*, Wiley-VCH Verlag GmbH, Weinheim; (d) Jew, S.-S. and Park, H.-G. (2009) *Chem. Commun.*, 7090–7103.
26 (a) Wynberg, H. and Helder, R. (1975) *Tetrahedron Lett.*, **46**, 4057–4059; (b) Smaardijk, Sb.A. and Wynberg, H. (1987) *J. Org. Chem.*, **52**, 135–137.
27 Wynberg, H. and Staring, E.G.J. (1982) *J. Am. Chem. Soc.*, **104**, 166–168.
28 (a) Calter, M.A. (1996) *J. Org. Chem.*, **61**, 8006–8007; (b) Calter, M.A. and Guo, X. (1998) *J. Org. Chem.*, **63**, 5308–5309; (c) Calter, M.A., Orr, R.K., and Song, W. (2003) *Org. Lett.*, **5**, 4745–4748.
29 Cortez, G., Tennyson, R.L., and Romo, D. (2001) *J. Am. Chem. Soc.*, **123**, 7945–7946.
30 For reviews, see: (a) France, S., Guerin, D.J., Miller, S.J., and Lectka, T. (2003) *Chem. Rev.*, **103**, 2985–3012; (b) France,

S., Weatherwax, A., Taggi, A.E., and Lectka, T. (2004) *Acc. Chem. Res.*, **37**, 592–600.

31. (a) Taggi, A.E., Hafez, A.M., Wack, H., Young, B., Drury, W.J., and Lectka, T. (2000) *J. Am. Chem. Soc.*, **122**, 7831–7832; (b) Wack, H., Taggi, A.E., Hafez, A.M., Drury, W.J., and Lectka, T. (2001) *J. Am. Chem. Soc.*, **123**, 1531–1532; (c) Paull, D.H., Scerba, M.T., Alden-Danforth, E., Widger, L.R., and Lectka, T. (2008) *J. Am. Chem. Soc.*, **130**, 17260–17261; (d) Bekele, T., Shah, M.H., Wolfer, J., Abraham, C.J., Weatherwax, A., and Lectka, T. (2006) *J. Am. Chem. Soc.*, **128**, 1810–1811; (e) Abraham, C.J., Paull, D.H., Scerba, M.T., Grebinski, J.W., and Lectka, T. (2006) *J. Am. Chem. Soc.*, **128**, 13370–13371; (f) Abraham, C.J., Paull, D.H., Bekele, T., Scerba, M.T., Dudding, T., and Lectka, T. (2008) *J. Am. Chem. Soc.*, **130**, 17085–17094.

32. Xu, X., Wang, K., and Nelson, S.G. (2007) *J. Am. Chem. Soc.*, **129**, 11690–11691.

33. Iwabuchi, Y., Nakatani, M., Yokoyama, N., and Hatakeyama, S. (1999) *J. Am. Chem. Soc.*, **121**, 10219–10220.

34. Shi, M. and Xu, Y.-M. (2002) *Angew. Chem. Int. Ed.*, **41**, 4507–4510.

35. Tian, S.-K. and Deng, L. (2001) *J. Am. Chem. Soc.*, **123**, 6195–6196.

36. Papageorigiou, C., Cubillo de Dios, M.A., Ley, S.V., and Gaunt, M.J. (2004) *Angew. Chem. Int. Ed.*, **43**, 4641–4644.

37. Cui, H.-L., Feng, X., Peng, J., Lei, J., Jiang, K., and Chen, Y.-C. (2009) *Angew. Chem. Int. Ed.*, **48**, 5737–5740.

38. Cui, H.-L., Huang, J.-R., Lei, J., Wang, Z.-F., Chen, S., Wu, L., and Chen, Y.-C. (2010) *Org. Lett.*, **12**, 720–723.

39. Whitehead, D.C., Yousefi, R., Jaganathan, A., and Borhan, B. (2010) *J. Am. Chem. Soc.*, **132**, 3298–3300.

40. Zhang, W., Zheng, S., Liu, N., Werness, J.B., Guzei, I.A., and Tang, W. (2010) *J. Am. Chem. Soc.*, **132**, 3664–3665.

41. Wynberg, H. (1986) *Top. Stereochem.*, **16**, 87–130.

42. For reviews of enantioselective alcoholysis, see: (a) Chen, Y., McDaid, P., and Deng, L. (2003) *Chem. Rev.*, **103**, 2965–2984; (b) Tian, S.-K., Chen, Y., Hang, J., Tang, L., McDaid, P., and Deng, L. (2004) *Acc. Chem. Res.*, **37**, 621–631.

43. (a) Hiratake, J., Yamamoto, Y., and Oda, J. (1985) *J. Chem. Soc., Chem. Commun.*, 1717–1719; (b) Hiratake, J., Inagaki, M., Yamamoto, Y., and Oda, J. (1987) *J. Chem. Soc., Perkin. Trans.* **1**, 1053–1058; (c) Aitken, R.A., Gopal, J., and Hirst, J.A. (1988) *J. Chem. Soc., Chem. Commun.*, 632–634; (d) Aitken, R.A. and Gopal, J. (1990) *Tetrahedron: Asymmetry*, **1**, 517–520; (e) Bolm, C., Gerlach, A., and Dinter, C.L. (1999) *Synlett*, 195–196.

44. Chen, Y., Tian, S.-K., and Deng, L. (2000) *J. Am. Chem. Soc.*, **122**, 9542–9543.

45. Bolm, C., Schiffers, I., Dinter, C.L., and Gerlach, A. (2000) *J. Org. Chem.*, **65**, 6984–6991.

46. Chen, Y. and Deng, L. (2001) *J. Am. Chem. Soc.*, **123**, 11302–11303.

47. Hang, J., Tian, S.-K., Tang, L., and Deng, L. (2001) *J. Am. Chem. Soc.*, **123**, 12696–12697.

48. Hang, J., Li, H., and Deng, L. (2002) *Org. Lett.*, **4**, 3321–3324.

49. Tang, L. and Deng, L. (2002) *J. Am. Chem. Soc.*, **124**, 2870–2871.

50. Hemistra, H. and Wynberg, H. (1981) *J. Am. Chem. Soc.*, **103**, 417–430.

51. McDaid, P., Chen, Y., and Deng, L. (2002) *Angew. Chem., Int. Ed.*, **41**, 338–340.

52. Bella, M. and Jørgensen, K.A. (2004) *J. Am. Chem. Soc.*, **126**, 5672–5673.

53. Yao, W., Pan, L., Wu, Y., and Ma, C. (2010) *Org. Lett.*, **12**, 2422–2425.

54. Poulsen, T.B., Alemparte, C., Saaby, S., Bella, M., and Jørgensen, K.A. (2005) *Angew. Chem., Int. Ed.*, **44**, 2896–2899.

55. Ogawa, S., Shibata, N., Inagaki, J., Nakamura, S., Toru, T., and Shiro, M. (2007) *Angew. Chem. Int. Ed.*, **46**, 8666–8669.

56. Silverman, R.B. (2000) *The Organic Chemistry of Enzyme-Catalyzed Reactions*, Academic Press, San Diego, CA.

57. Saaby, S., Bella, M., and Jørgensen, K.A. (2004) *J. Am. Chem. Soc.*, **126**, 8120–8121.

58. Li, H., Wang, Y., Tang, L., and Deng, L. (2004) *J. Am. Chem. Soc.*, **126**, 9906–9907.

59. Li, H., Wang, Y., Tang, L., Wu, F., Liu, X., Guo, C., Foxman, B.M., and Deng, L.

60. (a) Li, H., Song, J., Liu, X., and Deng, L. (2005) *J. Am. Chem. Soc.*, **127**, 8948–8949; (b) Li, H., Song, J., and Deng, L. (2009) *Tetrahedron*, **65**, 3139–3148.
61. Wu, F., Li, H., Hong, R., and Deng, L. (2006) *Angew. Chem. Int. Ed.*, **45**, 947–950.
62. Wu, F., Hong, R., Khan, J., Liu, X., and Deng, L. (2006) *Angew. Chem. Int. Ed.*, **45**, 4301–4305.
63. Liu, X., Li, H., and Deng, L. (2005) *Org. Lett.*, **7**, 167–169.
64. Duvall, J.R., Wu, F., and Snider, B.B. (2006) *J. Org. Chem.*, **71**, 8579–8590.
65. Xu, F., Corley, E., Zacuto, M., Conlon, D.A., Pipik, B., Humphrey, G., Murry, J., and Tschaen, D. (2010) *J. Org. Chem.*, **75**, 1343–1353.
66. For a mini-review on cuperidine and cuperine see: Marcelli, T., van Maarseveen, J.H., and Hiemstra, H. (2006) *Angew. Chem. Int. Ed.*, **45**, 7496–7504.
67. For a recent review, see: Connon, S.J. (2008) *Chem. Commun.*, 2499–2510.
68. Li, B.-J., Jiang, L., Liu, M., Chen, Y.-C., Ding, L.-S., and Wu, Y. (2005) *Synlett*, 603–606.
69. Vakulya, B., Varga, S., Csámpai, A., and Soós, T. (2005) *Org. Lett.*, **7**, 1967–1969.
70. (a) Wang, J., Li, H., Zu, L., Jiang, W., Xue, H., Duan, W., and Wang, W. (2006) *J. Am. Chem. Soc.*, **128**, 12652–12653; (b) Ye, J., Dixon, D.J., and Hynes, P.S. (2005) *Chem. Commun.*, 4481–4483; (c) McCooey, S.H. and Connon, S.J. (2005) *Angew. Chem. Int. Ed.*, **44**, 6367–6370.
71. Hynes, P.S., Stranges, D., Stupple, P.A., Guarna, A., and Dixon, D.J. (2007) *Org. Lett.*, **9**, 2107–2110.
72. Liu, Y.-Y., Cui, H.-L., Chai, Q., Long, J., Li, B.-J., Wu, Y., Ding, L.-S., and Chen, Y.-C. (2007) *Chem. Commun.*, 2228–2230.
73. Lubkoll, J. and Wennemers, H. (2007) *Angew. Chem. Int. Ed.*, **46**, 6841–6844.
74. DinÅr, P., Neilsen, M., Bertelsen, S., Niess, B., and Jørgensen, K.A. (2007) *Chem. Commun.*, 3646–3648.
75. Zu, L., Wang, J., Li, H., Xe, H., Jiang, W., and Wang, W. (2007) *J. Am. Chem. Soc.*, **129**, 1036–1037.
76. Wang, J., Xie, H., Li, H., Zu, L., and Wang, W. (2008) *Angew. Chem. Int. Ed.*, **47**, 4177–4179.
77. Biddle, M.M., Lin, M., and Scheidt, K.A. (2007) *J. Am. Chem. Soc.*, **129**, 3830–3831.
78. Li, H., Wang, B., and Deng, L. (2006) *J. Am. Chem. Soc.*, **128**, 732–733.
79. Cochi, A., MÅtro, T., Pardo, D.G., and Cossy, J. (2010) *Org. Lett.*, **12**, 3693–3695.
80. Marcelli, T., van der Hass, R.N.S., van Maarseveen, J.H., and Hiemstra, H. (2006) *Angew. Chem. Int. Ed.*, **45**, 929–931.
81. Mandal, T., Samanta, S., and Zhao, C.-G. (2007) *Org. Lett.*, **9**, 943–945.
82. Bandini, M., Sinisi, R., and Umani-Ronchi, A. (2008) *Chem. Commun.*, 4360–4362.
83. Lou, S., Taoka, B.M., Ting, A., and Schaus, S. (2005) *J. Am. Chem. Soc.*, **127**, 11256–11257.
84. Ting, A., Lou, S., and Schaus, S.E. (2006) *Org. Lett.*, **8**, 2003–2006.
85. Bernardi, L., Fini, F., Herrera, R.P., Ricci, A., and Sagarzani, V. (2006) *Tetrahedron*, **62**, 375–380.
86. Tillman, A.L., Ye, J., and Dixon, D.J. (2006) *Chem. Commun.*, 1191–1193.
87. (a) Song, J., Wang, Y., and Deng, L. (2006) *J. Am. Chem. Soc.*, **128**, 6048–6049; (b) Song, J., Shih, H.-W., and Deng, L. (2007) *Org. Lett.*, **9**, 603–606.
88. Li, L., Ganesh, M., and Seidel, D. (2009) *J. Am. Chem. Soc.*, **131**, 11648–11649.
89. Török, B., Abid, M., London, G., Esquibel, J., Török, M., Mhadgut, S.C., Yan, P., and Prakash, G.K.S. (2005) *Angew. Chem. Int. Ed.*, **44**, 3086–3089.
90. Li, H., Wang, Y.-Q., and Deng, L. (2006) *Org. Lett.*, **8**, 4063–4065.
91. Chauhan, P. and Chimni, S.S. (2010) *Chem. Eur. J.*, **16**, 7709–7713.
92. Zhao, J.-L., Liu, L., Gu, C.-L., Wang, D., and Chen, Y.-J. (2008) *Tetrahedron Lett.*, **49**, 1476–1479.
93. Wang, Y.-Q., Song, J., Hong, R., Li, H., and Deng, L. (2006) *J. Am. Chem. Soc.*, **128**, 8156–8157.

94 Wang, Y., Li, H., Wang, Y.-Q., Liu, Y., Foxman, B.M., and Deng, L. (2007) *J. Am. Chem. Soc.*, **129**, 6364–6365.

95 Staben, S.T., Linghu, X., and Toste, F. D. (2006) *J. Am. Chem. Soc.*, **128**, 12658–12659.

96 Dickmeiss, G., Sio, V.D., Udmark, J., Poulsen, T.B., Marcos, V., and Jørgensen, K.A. (2009) *Angew. Chem. Int. Ed.*, **48**, 6650–6653.

97 (a) Luo, J., Xu, L.-W., Hay, R.A.S., and Lu, Y. (2009) *Org. Lett.*, **11**, 437–440; (b) Malerich, J.P., Hagihara, K., and Rawal, V.H. (2008) *J. Am. Chem. Soc.*, **130**, 14416–14417; (c) Oh, S.-H., Rho, H.-S., Lee, J.-W., Lee, J.-E., Youk, S.-H., Chin, J., and Song, C.-E. (2008) *Angew. Chem. Int. Ed.*, **47**, 7872–7875.

98 For reviews of asymmetric iminium catalysis, see: (a) Lelais, G. and MacMillan, D.W.C. (2006) *Aldrichimica Acta*, **39**, 79–87; (b) Erkkilä, A., Majamder, I., and Pihko, P.M. (2007) *Chem. Rev.*, **107**, 5416–5470.

99 For a discussion of challenging issues in the development of catalytic asymmetric Diels–Alder reactions with simple α,β-unsaturated ketones, see: Northrup, A.B. and MacMillian, D.W. C. (2002) *J. Am. Chem. Soc.*, **124**, 2458–2460.

100 Xie, J.-W., Chen, W., Li, R., Zeng, M., Du, W., Yue, L., Chen, Y.-C., Wu, Y., Zhu, J., and Deng, J.-G. (2007) *Angew. Chem., Int. Ed.*, **46**, 389–392.

101 Brunner, H., Bugler, J., and Nuber, B. (1995) *Tetrahedron: Asymmetry*, **6**, 1699–1702.

102 Xie, J.-W., Yue, L., Chen, W., Du, W., Zhu, J., Deng, J.-G., and Chen, Y.-C. (2007) *Org. Lett.*, **9**, 413–415.

103 Li, X., Cun, L., Lian, C., Zhong, L., Chen, Y., Liao, J., Zhu, J., and Deng, J. (2008) *Org. Biomol. Chem.*, **6**, 349–353.

104 Liu, C. and Lu, Y. (2010) *Org. Lett.*, **12**, 2278–2281.

105 Lu, X. and Deng, L. (2008) *Angew. Chem., Int. Ed.*, **47**, 7710–7713.

106 Wang, X., Reisinger, C.M., and List, B. (2008) *J. Am. Chem. Soc.*, **130**, 6070–6071.

107 Lu, X., Liu, Y., Sun, B., Cindric, B., and Deng, L. (2008) *J. Am. Chem. Soc.*, **130**, 8134–8135.

108 Austin, J.F. and MacMillan, D.W.C. (2002) *J. Am. Chem. Soc.*, **124**, 1172–1173.

109 Li, D.-P., Guo, Y.-C., Ding, Y., and Xiao, W.-J. (2006) *Chem. Commun.*, 799–801.

110 Chen, W., Du, W., Yue, L., Li, R., Wu, Y., Ding, L.-S., and Chen, Y.-C. (2007) *Org. Biomol. Chem.*, **5**, 816–821.

111 Bartoli, G., Bosco, M., Carlone, A., Pesciaioli, F., Sambri, L., and Melchiorre, P. (2007) *Org. Lett.*, **9**, 1403–1405.

112 Singh, R.P., Bartelson, K., Wang, Y., Su, H., Lu, X., and Deng, L. (2008) *J. Am. Chem. Soc.*, **130**, 2422–2423.

113 Zhang, E., Fan, C., Tu, Y., Zhang, F., and Song, Y. (2009) *J. Am. Chem. Soc.*, **130**, 14626–14627.

114 For a recent review of asymmetric catalysis with primary amines, see: Chen, Y.-C. (2008) *Synlett*, 1919–1930.

115 Wang, Y., Liu, X., and Deng, L. (2006) *J. Am. Chem. Soc.*, **128**, 3928–3930.

116 Wang, B., Wu, F., Wang, Y., Liu, X., and Deng, L. (2007) *J. Am. Chem. Soc.*, **129**, 768–769.

117 For reviews, see: (a) Casteel, D.A. (1999) *Nat. Prod. Rep.*, **16**, 55–73; (b) Dembitsky, M.V. (2008) *Eur. J. Med. Chem.*, **43**, 223–251.

118 Mutabingwa, T.K. (2005) *Acta Tropica*, **95**, 305–315.

119 Bredig, G. and Fiske, P.S. (1913) *Biochem. Z.*, **46**, 7–23.

11
Proline Derivatives
Shilei Zhang and Wei Wang

11.1
Introduction

In the past decade, organocatalysis has enjoyed great success and become an important branch of catalysis [1]. The seminal enamine and iminium catalysis (termed "aminocatalysis") works, independently reported by List and Barbas [2–4] and by MacMillan [5–7] have conceptualized the field. Notably, among the organocatalysts developed, amino acid proline occupies a "privileged" status [8].

In the initial stage of organocatalysis, L-proline (**1**) has played a key and phenomenal role (Figure 11.1) [8]. The work on L-proline-catalyzed intermolecular aldol reaction by List and Barbas trigged intense interest [3]. The enamine chemistry was quickly extended to a handful of new organic transformations, including Mannich, Michael, Morita–Baylis–Hillman, α-aminations, and α-aminoxylation reactions of carbonyl compounds [8]. L-Proline has been recognized as a simple enzyme [9]. The observed high catalytic activity and stereo-control capacity are attributed to its unique structure. L-Proline is the sole DNA encoded amino acid bearing a secondary amine moiety. The rigid ring structure plays a special role in the formation of the secondary and tertiary structures of peptides and proteins and their biological functions. As an extensively studied organocatalyst, this bifunctional molecule with a carboxylic acid and an essential secondary amine displays remarkable catalytic activity in facilitating the formation of iminium- and enamine species from corresponding aldehyde/ketone precursors. The high activity arises from the enhanced acidity of the carboxylic acid ($pK_a = 1.99$ in water at room temperature) compared to a primary amino acid such as leucine ($pK_a = 2.36$ in water) due to the rigid structure. More importantly the pyrrolidine portion with the "envelope" conformation is critical for facile formation of an enamine and high stereocontrol. This observation is demonstrated in the structure–activity relationship studies of amino acid catalysts including acyclic and four- and six-membered systems in an aldol reaction [10].

It is also realized that L-proline as a catalyst has several drawbacks. First, the major problem is that, in numerous instances, poor catalytic activities and low

Privileged Chiral Ligands and Catalysts. Edited by Qi-Lin Zhou
Copyright © 2011 WILEY-VCH Verlag GmbH & Co. KGaA, Weinheim
ISBN: 978-3-527-32704-1

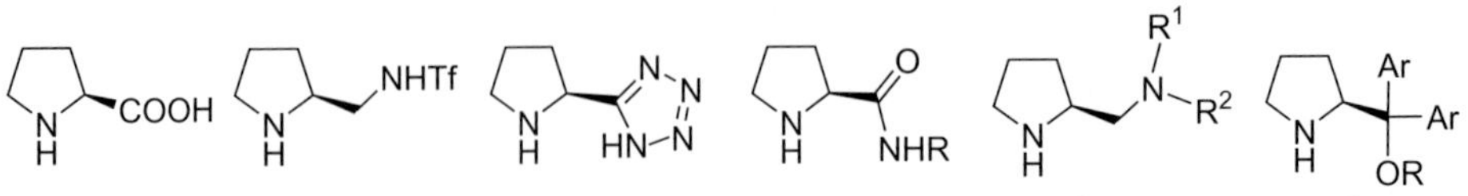

Figure 11.1 Structures of some representative pyrrolidine-derived catalysts.

enantioselectivities are observed. Second, a relatively high catalyst loading is usually required to effect the desired reaction on a reasonable timescale; commonly L-proline is used at levels of around 20–30 mol%. Third, L-proline has limited solvent compatibility; often reactions are performed in polar solvents such as DMSO, DMF, MeOH, or H_2O. Finally, and most importantly, it is plagued by difficulties in terms of structural modification and tuning to improve its catalytic activity and stereoselectivity. These limitations connected with proline as an organocatalyst prompted organic chemists to seek more effective alternatives. Analogs of proline have been developed, accordingly, to overcome these problems. Notable examples are pyrrolidine trifluoromethanesulfonamide (2) [11], 5-pyrrolidin-2-yltetrazole (3) [12, 13], prolinamides (4), pyrrolidine amines (5), and diarylprolinol silyl ethers (6) [14–16]. It is noted that they share the common structural features. They all posses an essential pyrrolidine architecture and are considered as proline derivatives. Significantly, they have proved to be more efficient in several catalytic reactions than proline. It is estimated that more than 800 articles have dealt with applications of this type of catalyst [17]. These facts clearly manifest the "privileged" status of proline and its pyrrolidine derivatives in organocatalyst design and development.

Because of space constraints, it is impossible to discuss every single case. We apologize for works that are not discussed and refer to the reader to several excellent reviews for details. This chapter will focus only on the organic reactions promoted by pyrrolidine-derived catalysts with representative examples. It is noted that MacMillan's chiral imidazolidinones are not included due to space limitations and the structural difference from pyrrolidines [6, 7].

11.2
Proline as Organocatalyst

11.2.1
Aldol Reactions

11.2.1.1 Intermolecular Aldol Reactions
Although the proline-catalyzed synthesis of Wieland–Miescher ketone (the Hajos–Parrish–Eder–Sauer–Wiechert reaction) was discovered as early as the 1970s [2], further study on the catalytic potential of proline nearly stopped until List and Barbas reported the asymmetric intermolecular aldol reaction between unmodified ketones and aldehydes [3, 10, 18]. The reaction was conducted in DMSO at

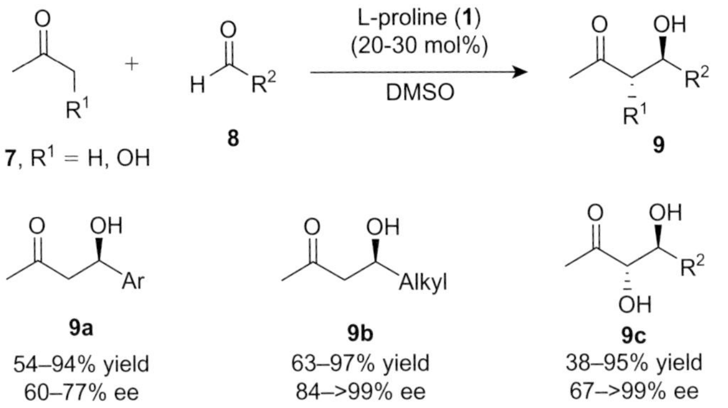

11.2 Proline as Organocatalyst

Scheme 11.1 The first intermolecular aldol reaction catalyzed by L-proline.

9a 54–94% yield, 60–77% ee
9b 63–97% yield, 84–>99% ee
9c 38–95% yield, 67–>99% ee

room temperature with L-proline (20–30 mol%) as promoter. A wide range of substrates proceeded smoothly to provide the aldol products with moderate to excellent enantioselectivities and in moderate to high yields. When acetone was used as aldol donor, α-branched aliphatic aldehydes gave better ees than aromatic aldehydes (**9b** versus **9a**). Hydroxyacetone is also a good aldol donor, affording the anti-diols (**9c**) (Scheme 11.1).

A formidable synthetic challenge is the cross-aldol reaction of nonequivalent aldehydes. It is not easy to obtain the desired aldol products from the reaction of two different aldehydes because of the tendency for the aldehyde to undergo self-dimerization and polymerization. In 2002, the MacMillan group successfully solved this problem by employing a syringe pump to slowly add an aldehyde donor **10** to the solution of aldehyde acceptor **11** (Scheme 11.2) [19]. In this manner, the desired cross-aldol products **12** were obtained in high yields and with excellent enantioselectivities. Notably, almost no other byproducts, such as dehydration or self-dimerization products, were observed.

The cross-aldol products are very important chiral building blocks that have been applied for the synthesis of natural products prelactone B by Pihko's group and trichostatin A by Wang and Duan and colleagues and for the intermediate of callipeltoside C by MacMillan and coworkers [20–22].

The utilization of functionalized aldehydes and ketones in L-proline-catalyzed aldol reactions has been hotly pursued as a result of their broad synthetic utility.

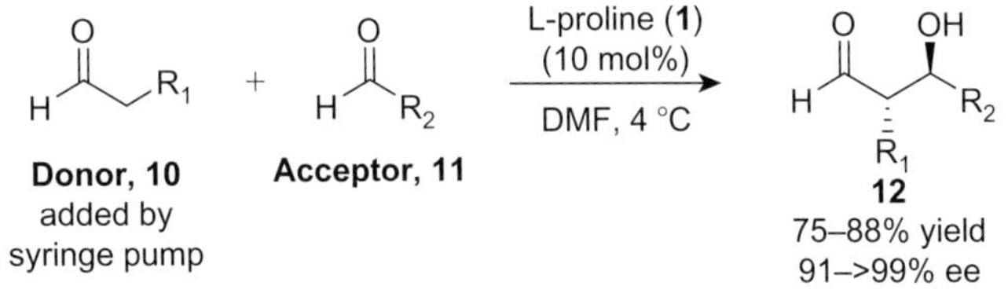

12 75–88% yield, 91–>99% ee

Scheme 11.2 MacMillan's L-proline-catalyzed cross-aldol reaction of nonequivalent aldehydes.

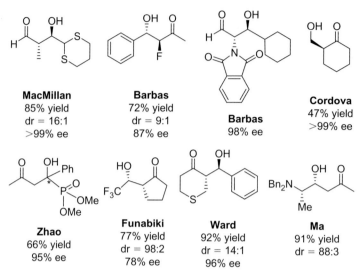

Figure 11.2 Structurally diverse aldol adducts obtained by L-proline-catalyzed aldol reactions.

Figure 11.2 shows some of the aldol products possessing functional groups obtained by aldol reactions of ketones/aldehydes with aldehydes/ketones [23–30]. In most cases, the products exhibit an *anti* configuration (in cases of two newly-created stereogenic centers). Significantly, the interesting desymmetrization [31] and dynamic kinetic resolution (DKR) [32] were also realized by L-proline-catalyzed aldol reactions.

11.2.1.2 Intramolecular Aldol Reactions

The first intramolecular L-proline aldol reaction is the well-known Hajos–Parrish–Eder–Sauer–Wiechert reaction (Scheme 11.3) [2, 33]. This ketone–ketone cross-aldol reaction generated the very useful Wieland–Miescher ketone **14**, an intermediate in steroid synthesis, with a high enantioselectivity. Notably, the reaction inspired List and Barbas to carry out the intermolecular aldol reaction of acetone to aldehydes nearly 30 years later, and then began the organocatalysis era [3].

In 2003, List and coworkers reported the first proline-catalyzed intramolecular aldehyde–aldehyde aldol reaction (Scheme 11.4) [34]. The reaction provides β-hydroxy cyclohexane carbonyl derivatives **16** that are of potential widespread

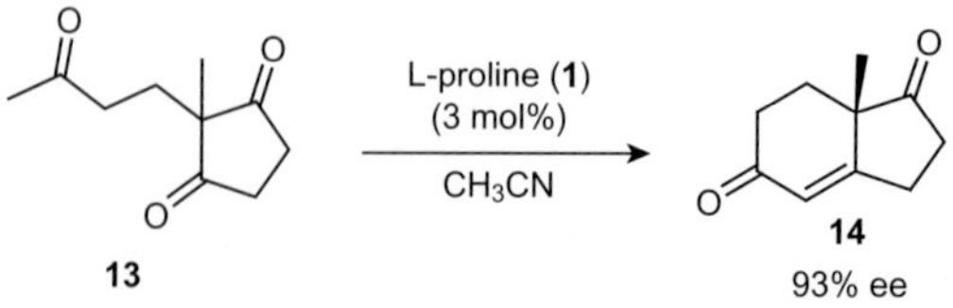

Scheme 11.3 Intramolecular aldol Hajos–Parrish–Eder–Sauer–Wiechert reaction.

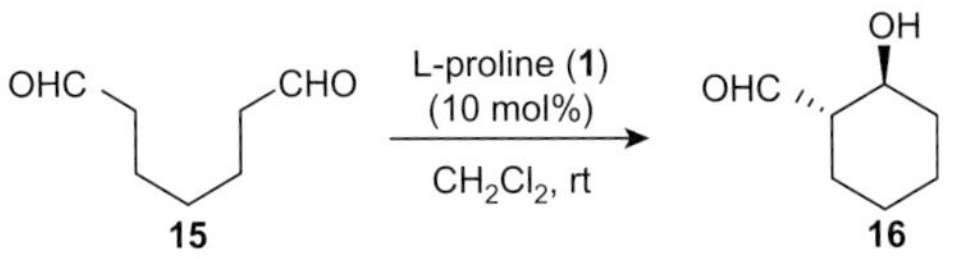

Scheme 11.4 The first intramolecular aldehyde–aldehyde aldol reaction.

usage in target-oriented synthesis. For example, Pearson used this chemistry in the total synthesis of the natural product (+)-cocaine [35].

11.2.1.3 Synthesis of Carbohydrates by Proline-Catalyzed Aldol Reactions

The synthesis of biologically important carbohydrates with defined configuration has been a long-standing topic in chemistry [36]. Organocatalyzed enantioselective aldol reactions provide a simple and direct entry to carbohydrates from commercially available dihydroxyacetone (DHA), glyceraldehyde, glycolaldehyde, or their derivatives.

In 2004, MacMillan and colleagues reported an elegant $C_2 + C_2 + C_2$ strategy to prepare carbohydrates in two steps (Scheme 11.5) [37, 38]. An organocatalyzed aldol addition/Mukaiyama aldol addition reaction sequence was implemented. The proline-catalyzed aldol provided the *anti*-configured aldol adducts (**18**). In the second Mukaiyama aldol reaction, the configuration of products **19** depended on the reaction conditions (Lewis acids and solvents). Thus, this allowed for the generation of three different products – glucose (**19a**), mannose (**19b**), or allose (**19c**) derivatives with a high degree of stereoselectivities.

Enders and coworkers investigated the aldol addition of protected DHA **20** to glyceraldehyde derivatives **21** ($C_3 + C_3$ strategy) (Scheme 11.6) [39]. After deprotection of aldol product **22** with Dowex, D-psicose (**23**) was obtained in quantitative yield. This new protocol represents a simple, biomimetic approach to selectively and differently protected simple carbohydrates and related compounds in only one

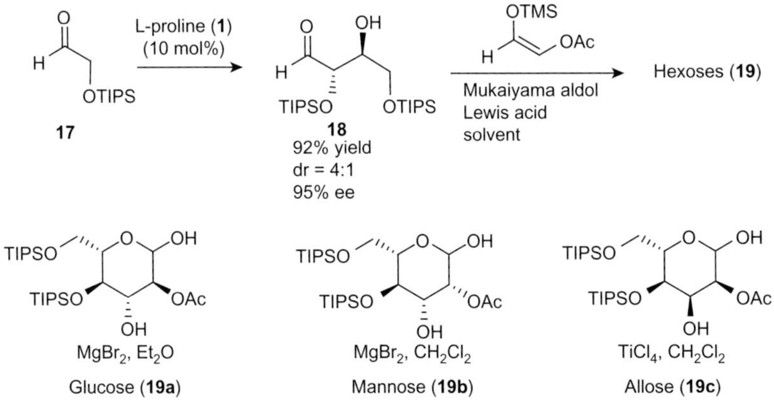

Scheme 11.5 MacMillan's $C_2 + C_2 + C_2$ approach to carbohydrates.

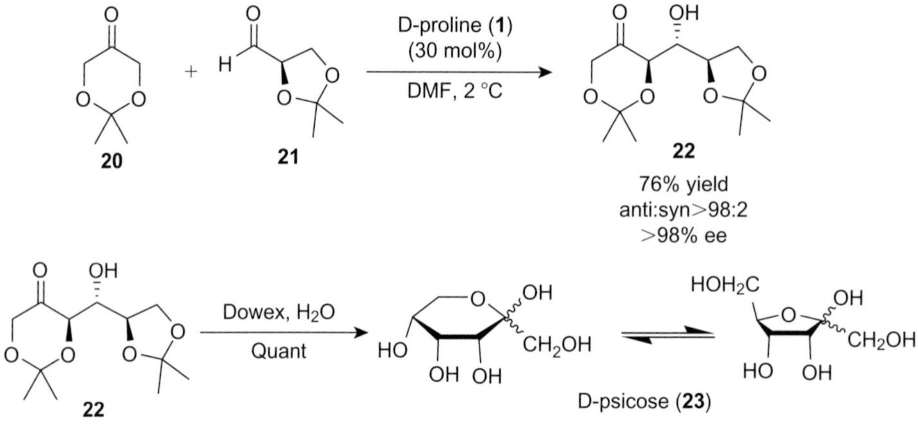

Scheme 11.6 Enders' $C_3 + C_3$ strategy to prepare carbohydrates.

step. Similarly, the same authors reported a $C_3 + C_2$ strategy to selectively obtain protected aldopentoses and derivatives [40].

11.2.2
Mannich Reactions Catalyzed by Proline

The direct, asymmetric Mannich reaction catalyzed by small organic molecules is one of the most powerful and convenient methods for the construction of chiral α- or β-amino acid derivatives. The resulting Mannich products are of particular interest due to their broad utilities as building blocks in the synthesis of pharmaceutically valuable compounds and peptidomimetics.

In 2000, List reported the first proline-mediated asymmetric three-component Mannich reaction (Scheme 11.7) [41]. The simple protocol of mixing a catalytic amount of L-proline, p-anisidine, and p-nitrobenzaldehyde in acetone/DMSO (1 : 4) gave the corresponding Mannich product **25a** after 12 h in 50% yield and with 94% ee. Moreover, aliphatic aldehydes **24** gave high yield and enantioselectivity as well. The PMP protecting group can be readily removed under oxidative conditions to give rise to chiral primary amines.

Shortly after List's report, the use of aldehydes as donors in proline-catalyzed Mannich reaction was disclosed by Hayashi and Córdova, respectively [42, 43]. As with List's results, the products have *syn*-configuration.

In a related study, Barbas reported a proline-catalyzed Mannich reaction of N-PMP-protected α-imino ethyl glyoxylate **27** with various unmodified ketones **26** (Reaction 1) or aldehydes **29** (Reaction 2, Scheme 11.8) [44, 45]. Importantly, the process afforded valuable functionalized α-amino acids **28** and **30**, respectively, with excellent regio-, diastereo-, and enantioselectivities. The method has been employed by Merck for the synthesis of a DPP-IV inhibitor [46].

The applications of functionalized imines and carbonyl compounds in proline-catalyzed Mannich reactions in organic synthesis have been explored intensively.

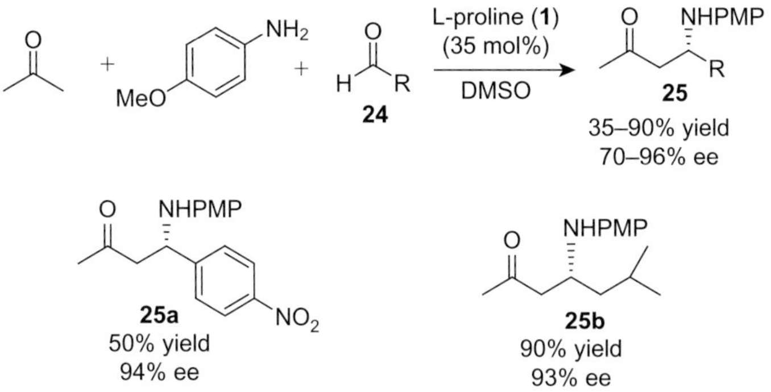

Scheme 11.7 List's proline-mediated asymmetric three-component Mannich reaction.

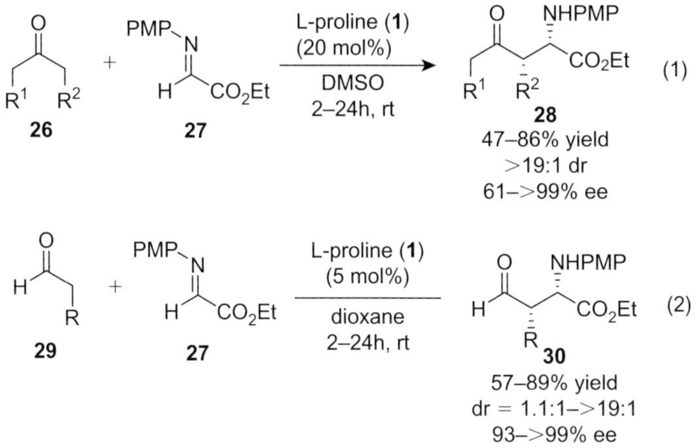

Scheme 11.8 N-PMP-protected α-imino ethyl glyoxylate as substrates in L-proline-catalyzed Mannich reactions.

Figure 11.3 shows some examples [47–54]. The products serve as useful building blocks in diversity-oriented synthesis of nitrogen-containing molecules.

11.2.3
Michael Addition Reactions Catalyzed by Proline

Catalytic asymmetric Michael addition reactions are one of the most fundamental reactions for C—C bond formation, second only to the aldol reaction in importance in organocatalysis. However, very few examples can be identified of the application of proline as catalyst, probably due to its low catalytic efficiency in this type of reaction. One example, is the Michael addition of nitroalkanes **31** to cycloalkenones **32** reported by Hanessian (Scheme 11.9) [55]. This reaction used *trans*-2,5-dimethylpiperazine (**33**) as additives and provided the Michael products **34** with

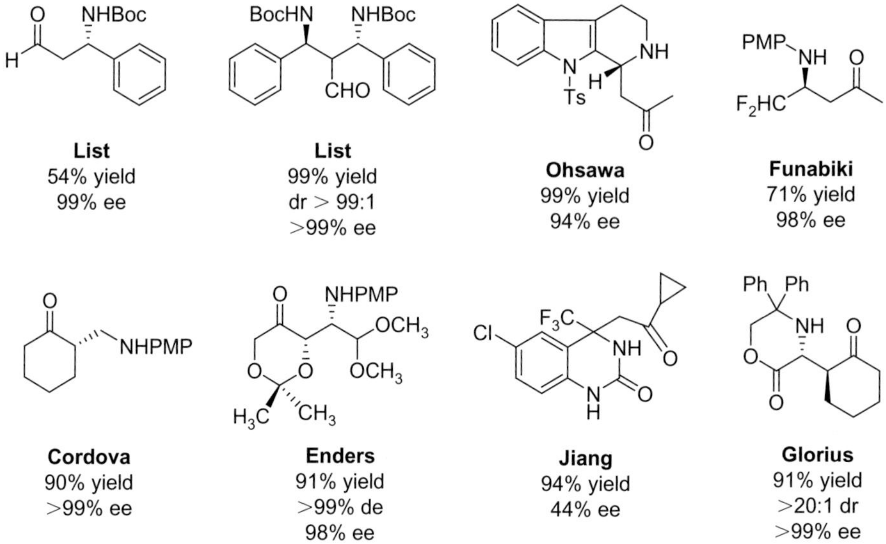

Figure 11.3 Structurally diverse products obtained by L-proline-catalyzed Mannich reactions.

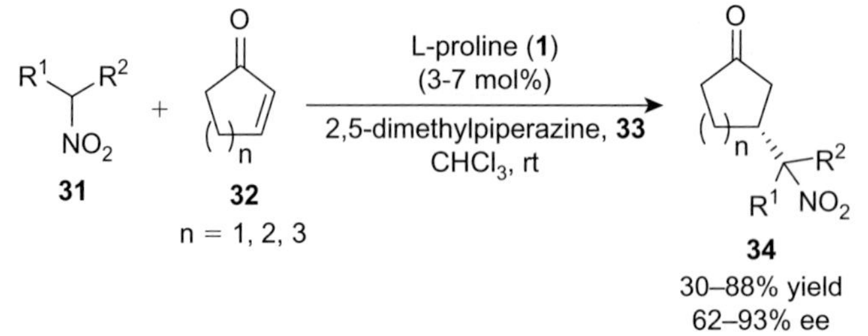

List
54% yield
99% ee

List
99% yield
dr > 99:1
>99% ee

Ohsawa
99% yield
94% ee

Funabiki
71% yield
98% ee

Cordova
90% yield
>99% ee

Enders
91% yield
>99% de
98% ee

Jiang
94% yield
44% ee

Glorius
91% yield
>20:1 dr
>99% ee

Scheme 11.9 L-Proline-catalyzed Michael addition of nitroalkanes to cycloalkenones.

30–88% yield
62–93% ee

moderate to high enantioselectivities. List reported a Michael addition of unmodified ketones to nitro olefins in good yields, but the enantioselectivities were very poor (7–23% ee) [56].

11.2.4
Morita–Baylis–Hillman (MBH) Reactions Catalyzed by Proline

The Morita–Baylis–Hillman reaction is a versatile carbon–carbon bond-forming reaction for the synthesis of densely functionalized compounds from aldehydes and electron-deficient alkenes in the presence of Lewis bases such as tri-substituted phosphines and tertiary amines. Generally, proline is not a good promoter in the MBH reactions. In reported studies, only moderate enantioselectivities were observed [57–59]. In contrast, an aza-Morita–Baylis–Hillman reaction provided products with good to excellent ee values. As shown in Scheme 11.10, Barbas

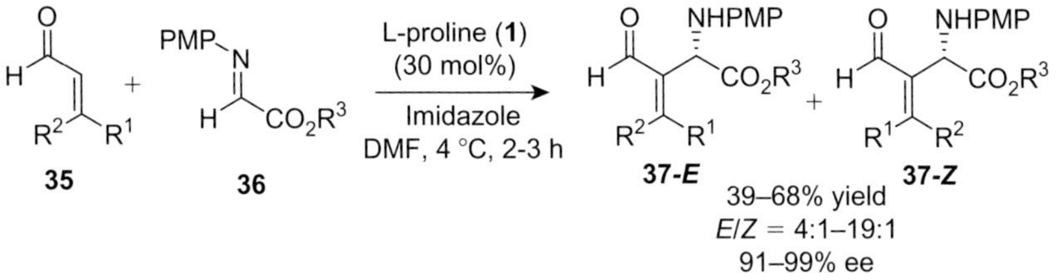

Scheme 11.10 Aza-Morita–Baylis–Hillman reaction catalyzed by L-proline and imidazole.

proved that imines **36** derived from glyoxylic acids were good acceptors in the aza-Morita–Baylis–Hillman reaction, affording 91–99% ee, albeit moderate yields [60].

11.2.5
α-Amination, α-Aminoxylation, and α-Alkylation of Carbonyl Compounds Catalyzed by Proline

Carbon–heteroatom, especially C–N and C–O bonds, are highly valuable in organic chemistry because numerous biologically active compounds contain such moieties. α-Functionalization of carbonyl compounds presents a direct approach. α-Amination and α-aminoxylation of carbonyl compounds catalyzed by L-proline have been demonstrated to be efficient methods. Activation of the α-position of carbonyl compounds by forming enamines with proline allows them to attack the electrophilic N=N or N=O double bond to form C–N and C–O bonds.

In 2002, List and Jørgensen, respectively, reported the first α-amination of aldehydes **38** and ketones **41** catalyzed by L-proline (Scheme 11.11) [61, 62]. These reactions gave products **40** and **42**, respectively, in high yields and with excellent enantioselectivities. The products can be applied to prepare 2-oxazolidinones and other natural and non-natural α-amino and α-hydrazino acids or ketone derivatives in short steps.

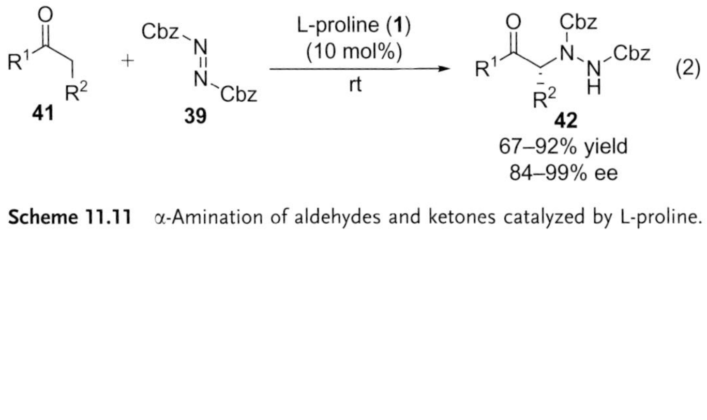

Scheme 11.11 α-Amination of aldehydes and ketones catalyzed by L-proline.

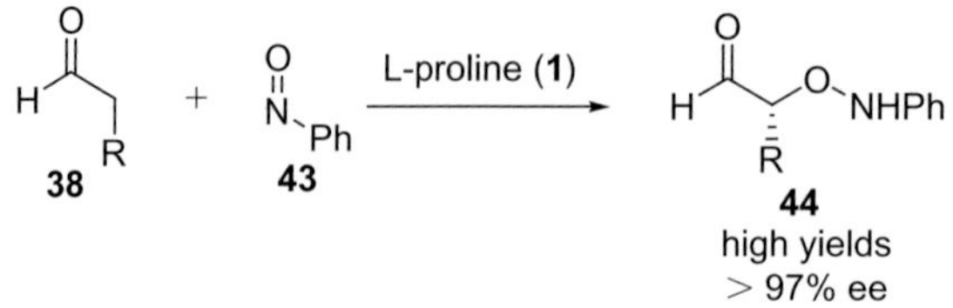

Scheme 11.12 α-Aminoxylation of aldehydes and ketones.

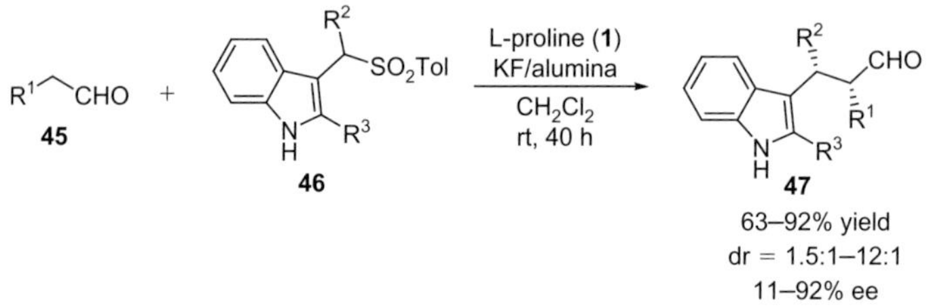

Scheme 11.13 Intermolecular α-alkylation of aldehydes.

Shortly afterwards, the α-aminoxylation of aldehydes **38** with nitroso **43** was realized by three groups almost at the same time (Scheme 11.12) [63–65]. The reaction was very efficient and afforded the products **44** bearing new C–O bonds with excellent ee values and the reaction time was only 10–20 min in Zhong's study. These valuable intermediates can be transformed into multiple synthetic blocks, especially 1,2-diols. The ketone version of this reaction was subsequently reported by Córdova's group [66].

Catalytic asymmetric α-alkylation of carbonyl compounds is a great challenge in organocatalysis. In 2004, List et al. presented the first catalytic asymmetric intramolecular α-alkylation of aldehydes catalyzed by proline derivative [67]. In contrast, the more challenging intermolecular version was seldom documented because of deactivation of the amine catalyst by N-alkylation with alkyl halides. Recently, Melchiorre et al. reported a proline-catalyzed intermolecular formal α-alkylation of aldehydes **45** (Scheme 11.13) [68]. The electrophiles were in fact a vinylogous iminium ion generated in situ from arylsulfonyl indoles **46** under base conditions.

11.2.6
Cascade/One-Pot Reactions Catalyzed by Proline

Catalytic asymmetric cascade/one-pot reactions are powerful synthetic tools for the facile construction of complex chiral molecular architectures from simple achiral materials in a single operation under mild conditions with high stereocontrol [69]. The combination of two or more reaction steps that are compatible under the same reaction conditions makes organocatalytic cascade/one-pot reactions possible.

In 2003, Barbas and colleagues reported a one-pot α-amination-aldol cascade (Scheme 11.14) [70]. In the proline-catalyzed three-component one-pot reaction,

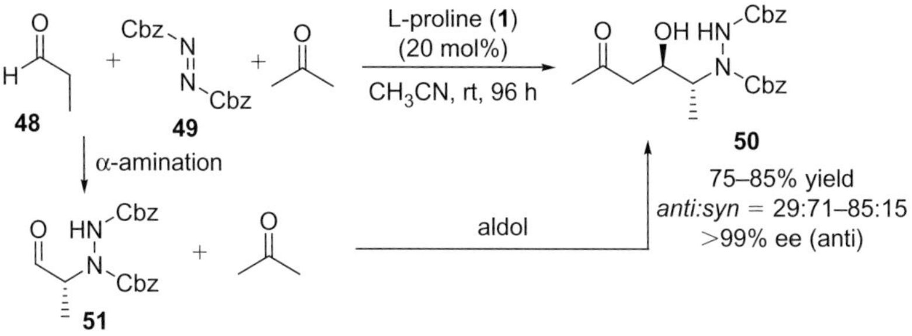

Scheme 11.14 Proline-catalyzed α-amination–aldol cascade reaction.

critically the reactivity to α-amination of propionaldehyde **48** was higher than other possible reactions such as α-amination of acetone and aldol reaction between acetone and propionaldehyde, so a successive α-amination-aldol reaction took place orderly in good yields and with excellent enantioselectivities. It was also found that relatively low diastereoselectivities were observed due to proline-catalyzed racemization of the amino aldehyde intermediates **51** during the reaction.

Watanabe *et al.* reported a self-condensation of α,β-unsaturated aldehydes **52** to give cyclohexadiene products **53** (Scheme 11.15) [71]. It was proposed that the reaction underwent a Michael-like imine addition sequence. A stoichiometric amount of L-proline was necessary. Notably, a rare γ-activation of α,β-unsaturated aldehydes was involved in this reaction.

In two related studies, Zhong *et al.* described two α-aminoxylation/aza-Michael reactions [72, 73]. In the cascade reactions, the resulting amines from α-aminoxylation of aldehydes underwent a subsequent intramolecular conjugate addition and multifunctionalized tetrahydro-1,2-oxazines (THOs) were produced with excellent stereoselectivities.

11.3
Proline Analogs as Organocatalysts

As shown above, proline has been demonstrated to be one of the most successful organocatalysts. However, it is impossible to rely on proline to do everything well. As discussed above, it is realized that in many cases poor catalytic activities and low enantioselectivities along with limited solvent compatibility are observed due to its intrinsic structural limitations. Therefore, structural modification and derivation of proline are necessary to improve catalytic activity and reaction enantioselectivity in certain reactions.

11.3.1
4-Hydroxyproline as Organocatalyst

As an analog of proline, the commercially available 4-hydroxyproline has been found some useful applications in organocatalysis. The hydroxyl group in

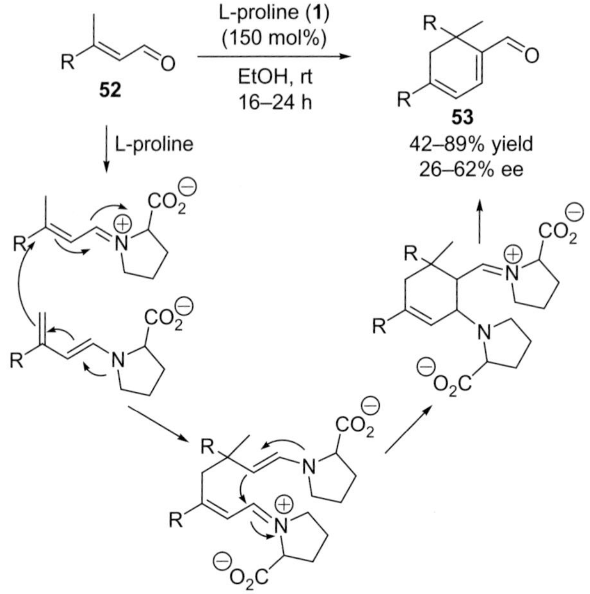

Scheme 11.15 Self-condensation of α,β-unsaturated aldehydes.

4-hydroxyproline allows its easy modifications, such as esterification, silicification, or immobilization to polymers for various purposes.

In 2005, Iwabuchi et al. reported an intramolecular aldolization catalyzed by 4-hydroxyproline derived silyl ether *cis*-**54** and *trans*-**55** to afford chiral bicyclo[3.3.1] alkanones **57** and **58**, respectively, with high diastereo- and enantioselectivities (Scheme 11.16) [74]. In contrast, only moderate ee (77%) was obtained when L-proline was employed.

One of the major drawbacks in using organocatalyst-catalyzed reactions is the high catalyst loadings (10–30 mol%) generally required to complete the transformations on reasonable timescales. This raises a cost concern when a large amount of chiral materials are used for a large scale of synthesis in industrial

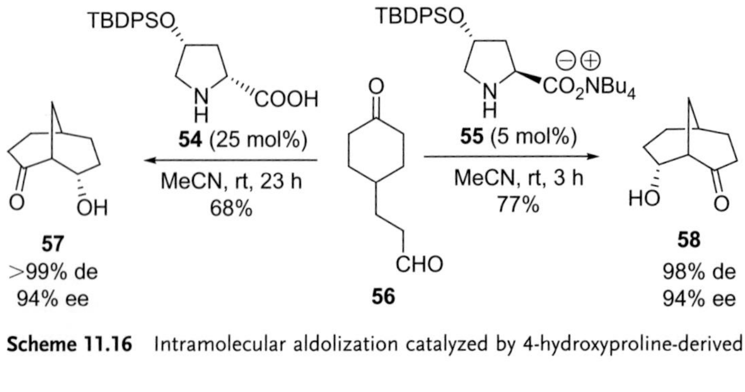

Scheme 11.16 Intramolecular aldolization catalyzed by 4-hydroxyproline-derived silyl ethers.

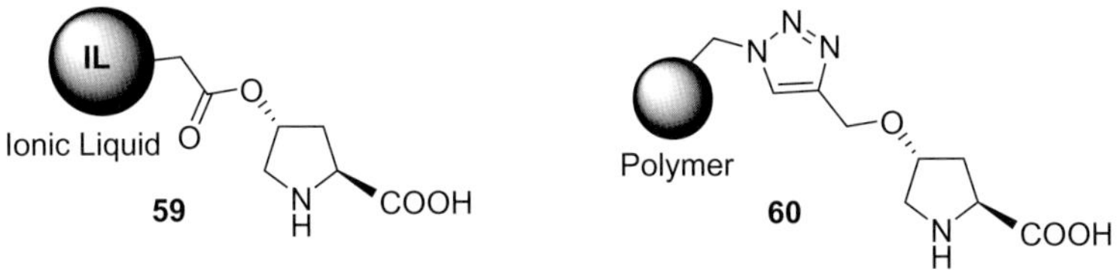

Figure 11.4 Ionic-liquid-supported and polymer-supported 4-hydroxyproline catalysts.

applications. An alternative strategy is to design recyclable and subsequently reusable versions of organocatalysts. The 4-OH group in 4-hydroxyproline can provide a handle to tether functionalities in the design of recyclable catalysts. Ionic-liquid-supported [75, 76] and polymer-supported [77–79] 4-hydroxyproline organocatalysts **59** and **60** have been reported (Figure 11.4). The merits of these catalysts are (i) the catalytic reactions can be carried out in water and (ii) the catalysts are separable and recyclable.

11.3.2
Other Proline Analogs as Organocatalysts

In 2004, List *et al.* reported the first catalytic asymmetric intramolecular α-alkylation of aldehydes **61** (Scheme 11.17) [67]. Only moderate ee (68%) was observed with L-proline. Nevertheless, they found that *(S)*-α-methylproline (**62**) could dramatically improve the reaction rate and enantioselectivity.

In 2005, MacMillan reported an enantioselective organocatalytic cyclopropanation between α,β-unsaturated aldehydes **64** and dimethylphenylacyl sulfonium ylide **65** (Scheme 11.18) [80]. With L-proline as catalyst, only 46% ee was obtained. This low enantioselectivity is attributed to the formation of both *(E)*- and *(Z)*-iminium isomers with L-proline, a configurational equilibrium that typically leads to diminished enantiocontrol. However, with the catalyst dihydroindole 2-carboxylic acid (**66**), the van der Waals force resulting from the phenyl ring will lead to favorable *(Z)*-iminium **68** formation and thus good enantiocontrol can be induced. Under optimal reaction conditions, the cyclopropanes **67** were obtained with high ees.

It is established that the proline-catalyzed Mannich reaction usually provides products with *syn* configuration. However, the generation of *anti* stereoisomers in

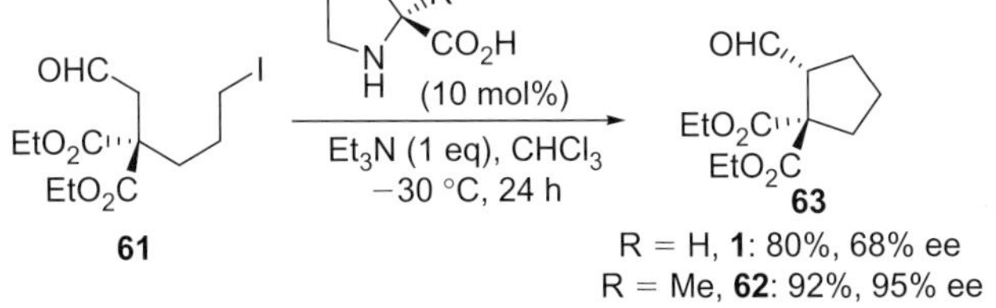

Scheme 11.17 Organocatalytic enantioselective intramolecular α-alkylation of aldehydes.

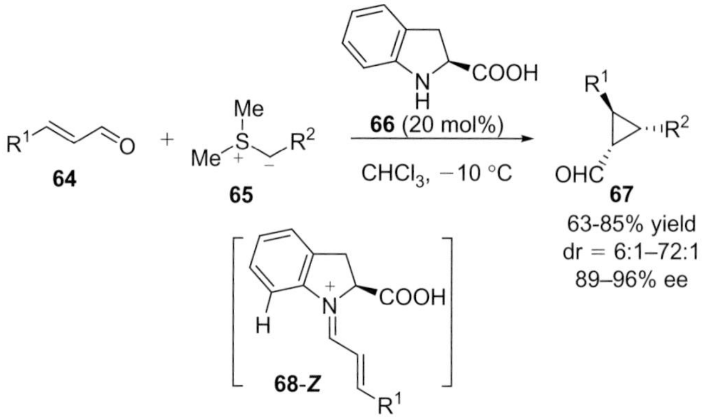

Scheme 11.18 Cyclopropanations of α,β-unsaturated aldehydes and dimethylphenylacyl sulfonium ylides.

organocatalytic Mannich reactions has plagued chemists for a while. In 2006, based on the catalytic mechanism, Barbas designed two new proline analogs, 5-methyl-3-pyrrolidinecarboxylic acid (**71**) and 3-pyrrolidinecarboxylic acid (**77**) (Scheme 11.19) [81–83]. With these two catalysts, the *anti*-selective Mannich reactions of aldehydes **69** or ketones **75** and N-PMP-protected α-imino ethyl glyoxylates **70** and **76**, respectively, were realized with good diastereoselectivities and excellent enantioselectivities. Critically, formation of the less sterically crowded favorable *(S)-trans* conformation **73** of *(E)*-enamine that reacts with **70** at the *Re* face, controlled by the ionic interaction from the carboxylate and the N-atom (**74**), delivers the *anti*-product **72**.

11.4
5-Pyrrolidin-2-yltetrazole as Organocatalyst

Tetrazoles are generally used in medicinal chemistry as bioisosteres for carboxylic acids to increase the solubility of the drug while retaining the properties of the acid. The strategy is also applied for the design of its analog pyrrolidine-tetrazole (**3**) with improved catalyst properties in some aspects.

In 2004, three groups, Yamamoto [84], Ley [13], and Arvidsson [85], independently reported the synthesis and application of *(S)* 5-pyrrolidin-2-yltetrazole (**3**) in the α-aminoxylation of carbonyl compounds, Mannich, and aldol reactions. These reactions furnished similar results to those of proline. Nevertheless, a notable improvement is, for example, in Ley's case, that the Mannich reaction can be conducted in the nonpolar solvent CH_2Cl_2, in which no reaction occurred when L-proline was used.

Further investigations revealed that *(S)*-pyrrolidine-tetrazole **3** was superior to proline in several cases. For example, Yamamoto *et al.* reported the aldol reactions

Scheme 11.19 anti-Selective Mannich reactions catalyzed by 3-pyrrolidinecarboxylic acids.

of ketones to chloral [12] Barbas reported Mannich reactions [86] and α-amination of α-branched aldehydes (Figure 11.5) [87]. In all of these reactions, L-proline displayed lower catalytic activities and enantiocontrol than that of *(S)*-pyrrolidine-tetrazole 3.

Cascade reactions were also reported with pyrrolidine-tetrazole 3 as catalyst. Yamamoto and colleagues disclosed an *O*-nitroso aldol/Michael cascade between α, β-unsaturated cycloketones 79 and nitrosobenzene 80, affording endocyclic adducts 81 with excellent ees and in moderate yields (Scheme 11.20) [88]. Ley *et al.* illustrated an organocatalytic nitrocyclopropanation reaction with a Michael–alkylation sequence [89]. The product contains an interesting bicyclo[4.1.0]heptane structure.

Pyrrolidine-tetrazole 3 catalyzed aldol reactions [90, 91], Michael reactions [92–96], and Biginelli reactions [97] were also reported. Notably, in general better results were observed than those of L-proline.

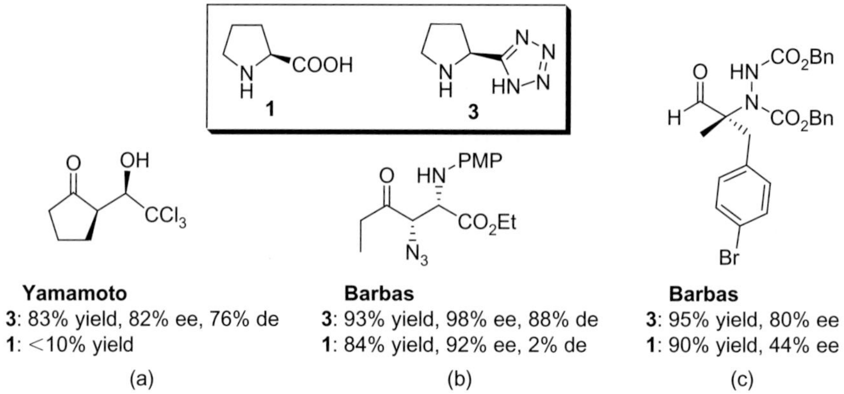
COOH
1
3
Yamamoto
3: 83% yield, 82% ee, 76% de
1: <10% yield
(a)
PMP
CO2Et
Barbas
3: 93% yield, 98% ee, 88% de
1: 84% yield, 92% ee, 2% de
(b)
CO2Bn
CO2Bn
Br
Barbas
3: 95% yield, 80% ee
1: 90% yield, 44% ee
(c)

Figure 11.5 Improved results obtained in aldol, Mannich, and α-amination reactions catalyzed by pyrrolidine-tetrazole **3**.

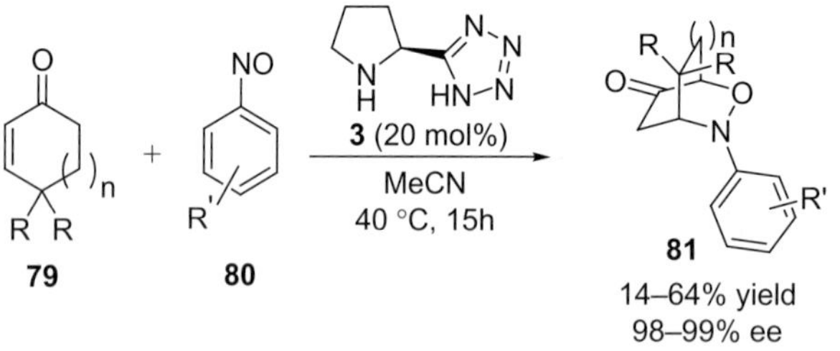
NO
3 (20 mol%)
MeCN
40 °C, 15h
79
80
81
14–64% yield
98–99% ee

Scheme 11.20 Cascade reactions catalyzed by *(S)*-pyrrolidine-tetrazole **3**.

11.5 Pyrrolidine-Based Sulfonamides as Organocatalysts

In 2004, Wang and coworkers devised a new pyrrolidine sulfonamide organocatalyst (**2**, Figure 11.6). Trifluoromethanesulfonamide has a comparable pK_a with acetic acid in water (6.3 versus 4.76) and DMSO (9.7 versus 12.3). The new catalyst displays similar or higher catalytic activity and enantiocontrol than proline in many cases.

The efficiency of catalyst **2** in promoting various organic asymmetric transformations has been examined, for example, in aldol reactions of α,α-dialkyl aldehydes or aryl methyl ketones to aryl aldehydes [98, 99], Mannich reactions of ketones with α-imino esters [100], α-aminoxylation reactions of aldehydes and ketones [11], Michael addition of aldehydes or ketones to nitrostyrenes [101, 102] or chalcones [103]. Notably, in some cases, significant improved results are obtained [98, 99, 101–103]. For instance, in the Michael addition of aldehydes or ketones to nitrostyrenes, good to excellent ee values were obtained. In contrast, the same reaction only afforded low reaction yields and/or poor enantioselectivities with proline.

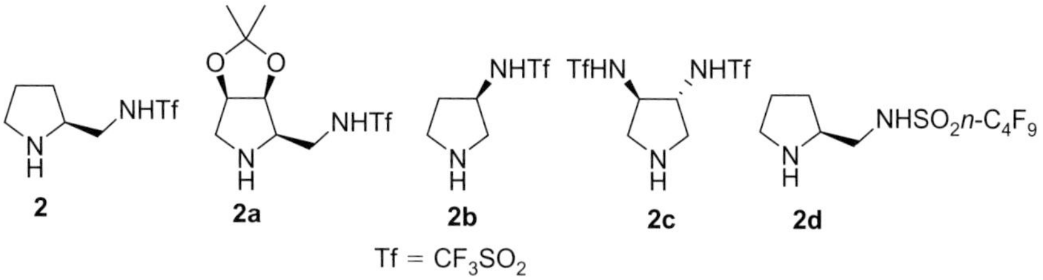

Figure 11.6 Pyrrolidine-sulfonamide catalysts.

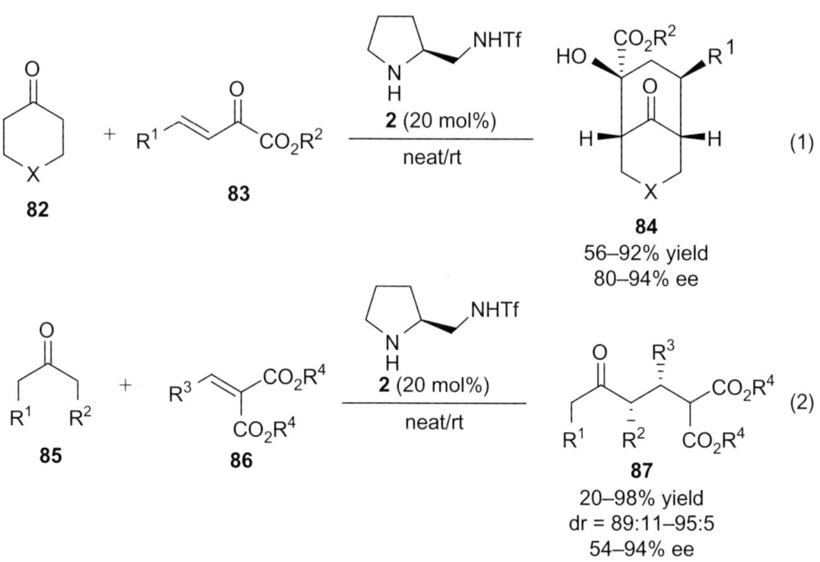

Scheme 11.21 Annulation and Michael addition reactions catalyzed by (S)-pyrrolidine-sulfonamide **2**.

Tang and coworkers reported the **2**-catalyzed enantioselective formal [3 + 3] annulation of ketones **82** and enones **83** [104], and the Michael addition of ketones **85** to alkylidene malonates **86** (Scheme 11.21) [105]. Among the catalysts screened, (S)-pyrrolidine-sulfonamide **2** was the best one in promoting these reactions to give good yields and high enantioselectivities.

11.6
Pyrrolidine-Based Amides as Organocatalysts

Among reported pyrrolidine-based catalysts, pyrrolidine amides constitute a significant portion because amides are readily prepared from diverse amines and they may exhibit various and adjustable catalytic activities. Notably, more than 90% of reported reactions mediated by these amide catalysts are aldol reactions.

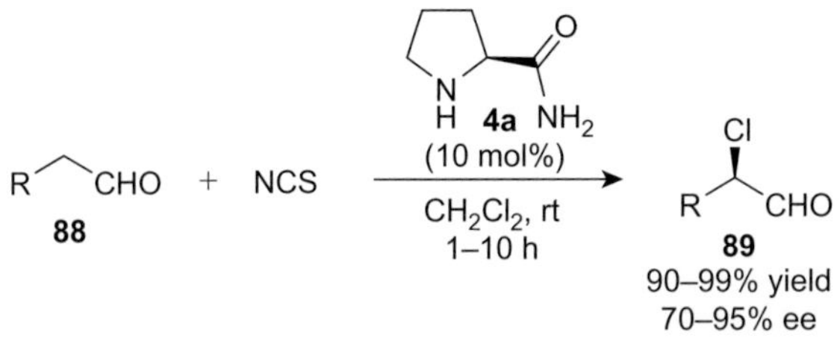

Scheme 11.22 α-Chlorination of aldehydes catalyzed by L-prolinamide **4a**.

Jørgensen has reported the use of L-prolinamide **4a** for asymmetric α-chlorination of aldehydes **88**, affording optically active α-chloro aldehydes **89** in excellent yields and moderate to good enantioselectivities (Scheme 11.22) [106]. Meanwhile, Wang reported the first organocatalytic α-selenenylation reaction of aldehydes and ketones catalyzed by L-prolinamide [107, 108].

Other applications of L-prolinamide **4a** include aldol reactions of ketones to diethyl formylphosphonate hydrate [109], ketones to 1,2-diketones [110], and acetone to pyruvaldehyde [111]. In these instances, L-prolinamide shows better catalytic activity than L-proline.

In L-proline-catalyzed aldol reactions of acetone to aromatic aldehydes, generally less than 90% ee values were obtained. Gong and coworkers designed a series of pyrrolidine-based amino alcohol amide catalysts (Scheme 11.23) [112, 113]. They found catalyst **4b** exhibited excellent catalytic capacity in the aldol reaction of acetone to a wide scope of aldehydes **90**. The reaction gave products **91** with excellent enantioselectivities in all cases using even as low as 2 mol% of catalyst. The high catalytic activity of **4b** may come from the strong hydrogen bond interactions between an aldehyde, an amide, and a hydroxyl group, as shown in the proposed model **92**.

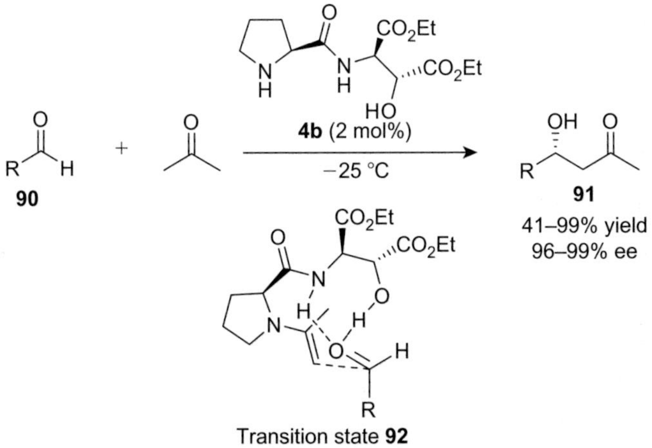

Scheme 11.23 Aldol reactions catalyzed by Gong's catalyst **4b**.

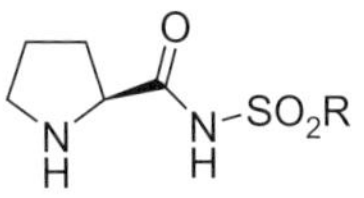

Figure 11.7 An unusual N-sulfonylamide catalyst.

An unusual N-sulfonylamide (**4c**) was also employed by several groups in exploring aldol reactions [114], Mannich reaction [115], Michael reactions [116], and Diels-Alder reactions [117] (Figure 11.7).

11.7
Pyrrolidine Diamine Catalysts

Pyrrolidine diamines are also an important class of organocatalysts and widely applied in organocatalysis (Figure 11.8). They have been used in aldol, Mannich, Michael, and Diels–Alder reactions. In most cases, they are used in salt forms in catalysis and the counter anions are generally trifluoroacetate or tosylate.

Catalyst **5a** (called the "Barbas catalyst") is the earliest one, first explored by Barbas and coworkers. They found that it was a good promoter for several organic transformations such as aldol [118] and Michael [119] of α-branched aldehydes to benzaldehydes or nitrostyrenes. Hayashi demonstrated that it was an optimal promoter for an intramolecular version of an aldol reaction (Scheme 11.24) [120]. Jørgensen et al. developed the first organocatalytic enantioselective Mannich reaction of ketimines and unmodified aldehydes using **5a** [121].

Some cycloaddition examples catalyzed by diamines were developed. Karlsson and colleagues disclosed a 1,3-dipolar cycloaddition reaction of 1-cycloalkene-1-carboxaldehyde (**95**) with nitrones **96** [122]. The process was facilitated by diamine **5b**, furnishing exo-bicyclic isoxazolidines **97** in low to good yields and variable enantioselectivities (Scheme 11.25).

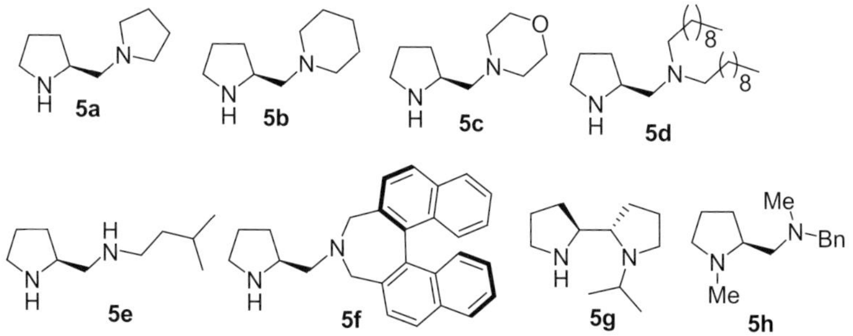

Figure 11.8 Pyrrolidine diamine catalysts.

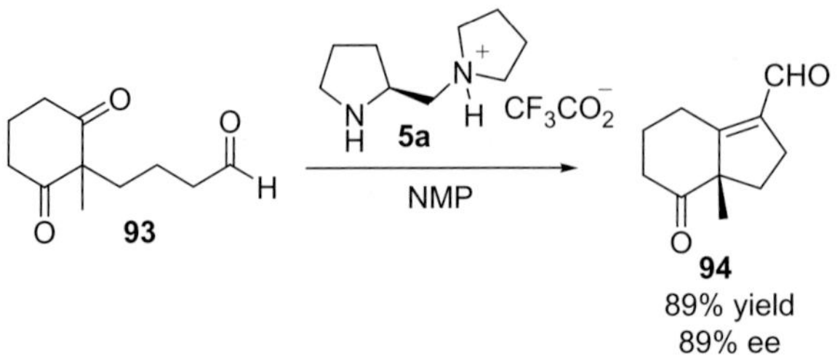

Scheme 11.24 Intramolecular aldol reaction catalyzed by **5a**.

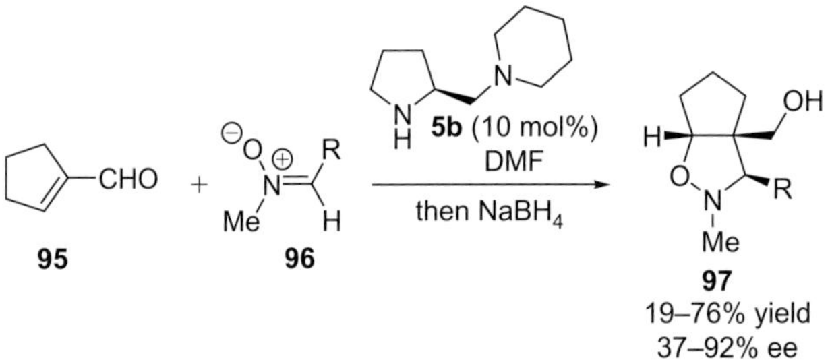

Scheme 11.25 1,3-Dipolar cycloaddition of nitrones with 1-cycloalkene-1-carboxaldehyde.

Catalyst **5a** was also employed as a promoter in cascade reactions. For example, Dondoni et al. developed an ethyl pyruvate (**98**) homoaldol reaction (Scheme 11.26) [123]. Córdova reported a Michael–aldol cascade reaction between salicylic aldehyde derivatives and α,β-unsaturated cyclic ketones [124] or 2-mercaptobenzaldehyde and α,β-unsaturated cyclic ketones [125].

Catalysts **5c–g** were applied in aldol reactions [126], Michael addition of aldehydes or ketones to nitroolefins [127–130], Michael addition of aldehydes to vinyl sulfones [131, 132]. Catalysts **5h** bearing a ditertiary-amine moiety was used for a kinetic resolution of racemic primary alcohols [133].

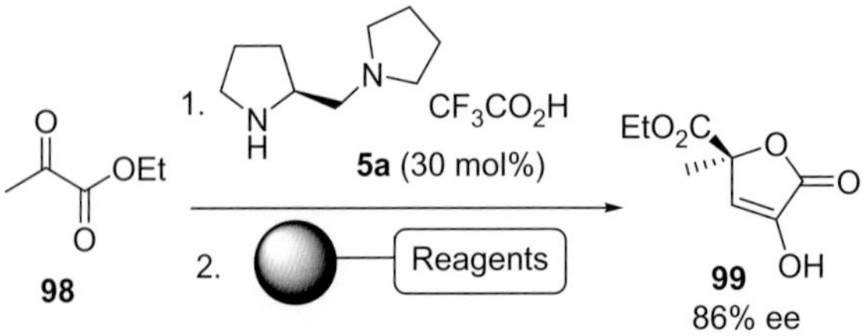

Scheme 11.26 Ethyl pyruvate homoaldol reaction.

11.8
Diarylprolinols or Diarylprolinol Ether Catalysts

In addition to L-proline, chiral diarylprolinols [134] and diarylprolinol ethers [135] are another class of widely used organocatalysts and occupy an important position in aminocatalysis (Figure 11.9). They are involved in catalyzing a wide range of organic reactions through enamine/iminium chemistry. In addition to catalyzing single-step transformations, remarkably they have proved to be one of the most successful catalysts for cascade processes.

11.8.1
Aldol Reactions, Mannich Reactions, and Other α-Functionalizations of Aldehydes Catalyzed by Diarylprolinols or Diarylprolinol Silyl Ethers

The title reactions have been widely explored with L-proline and it derivates, as discussed above. However, the limitations of these catalysts in some reactions were also realized. Diarylprolinol or diarylprolinol silyl ether catalysts provide better or unexpected results with these reactions.

Acetaldehyde, a difficult substrate, is rarely used in organocatalysis. Hayashi and coworkers demonstrated that diarylprolinol **6e** was an efficient promoter in catalyzing the aldol reaction of acetaldehyde with various aldehydes (**100**) (Scheme 11.27) [136]. This represents the first example of the use of diarylprolinol in an enantioselective aldol reaction. A related Mannich reaction was also uncovered by the same group [137].

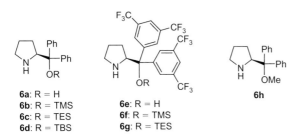

Figure 11.9 Diarylprolinol and diarylprolinol ether catalysts.

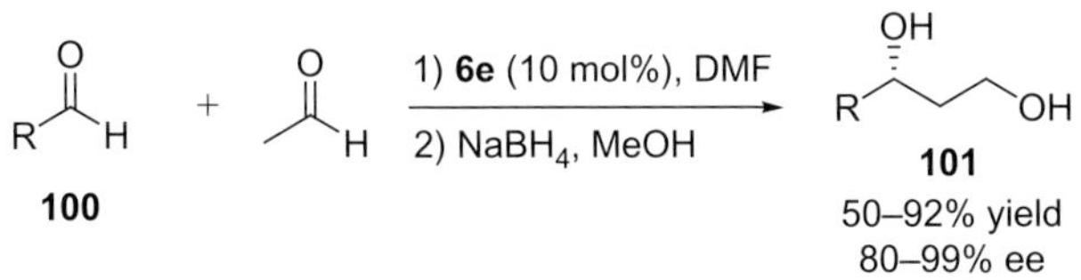

Scheme 11.27 Aldol reactions with acetaldehyde.

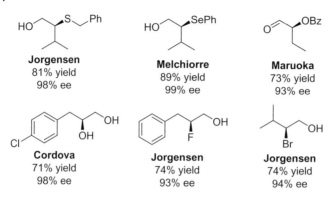

Figure 11.10 Products of α-functionalizations of aldehydes catalyzed by diarylprolinol silyl ethers.

The α-functionalizations of aldehydes, such as α-sulfenylation [14], α-selenenylation [138], α-benzoyloxylation [139, 140], α-oxidation [141], or α-halogenation [142] were achieved with good to excellent enantioselectivities by employing diarylprolinol silyl ethers as catalysts (Figure 11.10).

11.8.2
Michael Addition Reactions Catalyzed by Diarylprolinols or Diarylprolinol Silyl Ethers.

11.8.2.1 Michael Additions through an Enamine Pathway

As shown above, diarylprolinols and ethers are effective promoters for the formation of enamines from carbonyl compounds. Accordingly, the chemistry has also been applied in asymmetric Michael addition reactions with electron-deficient olefins to generate highly functionalized molecules with one or two chiral centers. These compounds are highly versatile building blocks in organic synthesis (Scheme 11.28). Hayashi et al. first applied diarylprolinol TMS ether **6b** (Figure 11.9) for a highly enantioselective Michael reaction of aldehydes to nitroalkenes with excellent levels of enantioselectivities (in most cases, ≥98% ees obtained) [15]. Soon after this, related studies were extensively carried out to extend the substrate scope [143–150].

The Michael addition of aldehydes to enones was reported by Gellman [16], Ma [151], and Gaunt [152], affording 1,5-dicarbonyl compounds with excellent enantioselectivities.

The diarylprolinol silyl ethers **6** catalyzed Michael additions of aldehydes go beyond nitroolefins and enones as acceptors. Maleimides [153], vinyl phosphonates [154], alkylidine malonates [155], and vinyl sulfones [156, 157] are also possible for the process, generating adducts with functional diversity.

11.8.2.2 Michael Additions through an Iminium Mechanism

In addition to facilitating the generation of enamines, diarylprolinol ethers are highly effective activators for the formation of iminium ions from corresponding

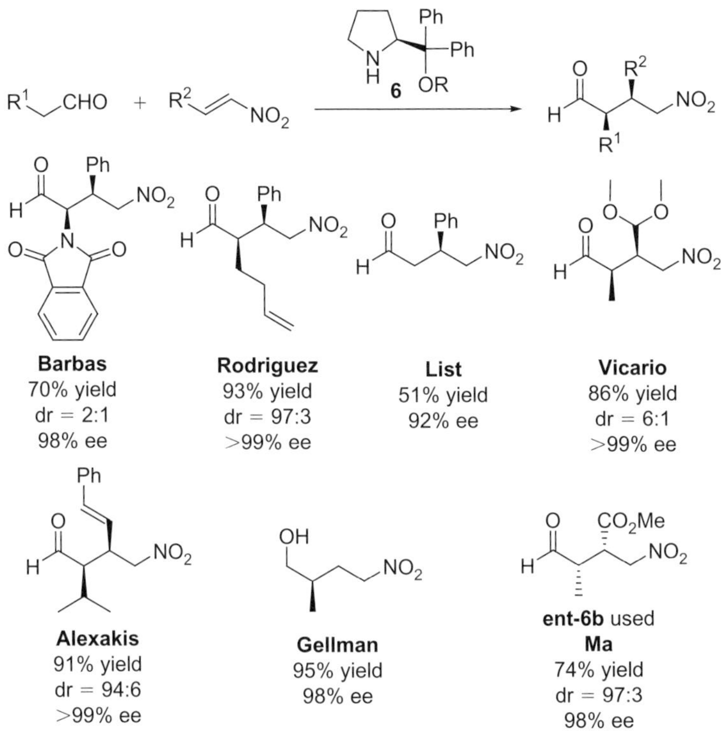
Ph
Ph
N
H
6
OR
R1 CHO
R2
NO2
O
R2
NO2
H
R1
Barbas
70% yield
dr = 2:1
98% ee
Rodriguez
93% yield
dr = 97:3
>99% ee
List
51% yield
92% ee
Vicario
86% yield
dr = 6:1
>99% ee
Alexakis
91% yield
dr = 94:6
>99% ee
Gellman
95% yield
98% ee
CO2Me
ent-6b used
Ma
74% yield
dr = 97:3
98% ee

Scheme 11.28 Products of Michael reactions of aldehydes and nitroalkenes catalyzed diarylprolinol ethers.

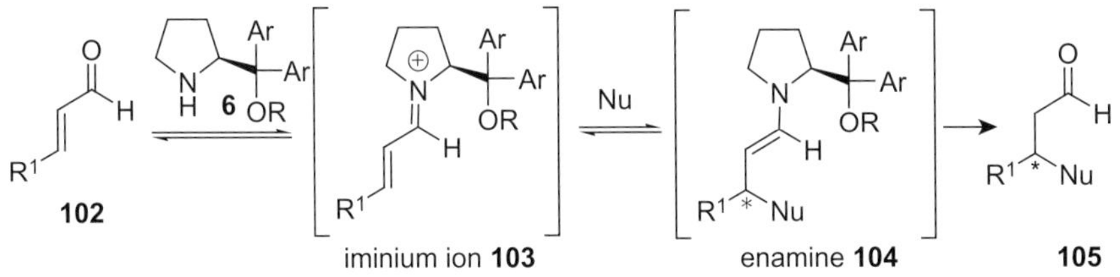
Ar
Ar
N
H 6 OR
R1
102
iminium ion 103
Nu
enamine 104
105

Scheme 11.29 Efficient iminium catalysis mediated by diarylprolinol ethers.

α,β-unsaturated aldehydes (**102**). The resulting iminium ions **103** render the β-carbons more electrophilic than their carbonyl precursors for nucleophilic attack (Scheme 11.29). A wide range of nucleophiles have been examined with this activation mode, including carbon- and heteroatom-centered nucleophiles such as nitrogen, oxygen, sulfur, and phosphorus. Remarkably, in general, excellent enantioselectivities and good yields are observed. These beautiful chemistries clearly manifest the "privileged" status of this class of catalysts.

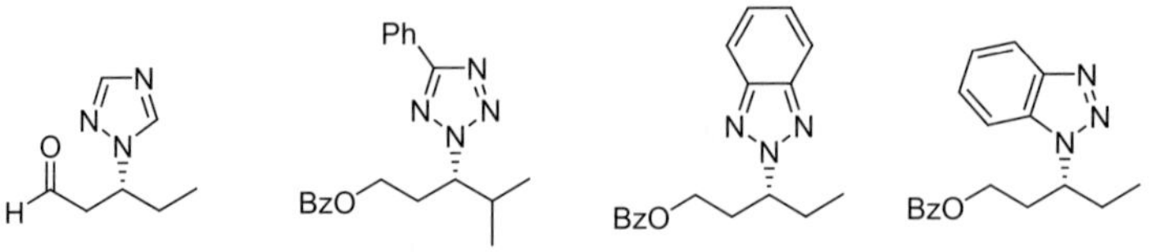

Figure 11.11 Products of the Michael addition of N-heterocyclic compounds to α,β-unsaturated aldehydes.

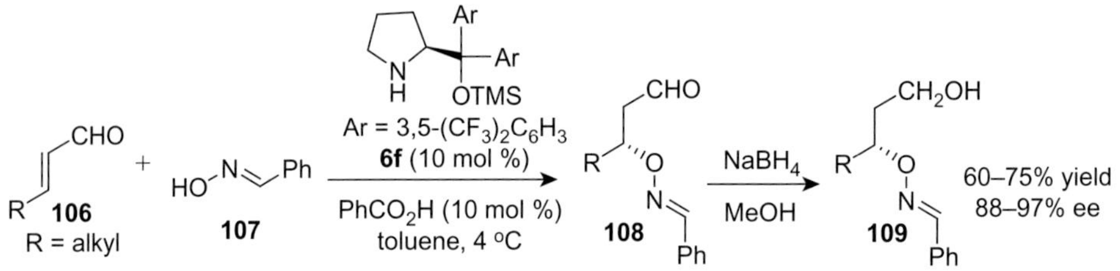

Scheme 11.30 Oxa-Michael reaction of enals with oximes.

In 2007, Jørgensen et al. reported a Michael addition of N-heterocyclic compounds to α,β-unsaturated aldehydes (Figure 11.11) [158]. In the presence of catalyst **6f**, 1,2,4-triazole, 5-phenyltetrazole, and 1,2,3-benzotriazole were successfully introduced into the β-position of α,β-unsaturated aldehydes in good yields and with high enantioselectivities. The heterocyclic molecules displayed interesting biological activities [159, 160].

Michael addition of oxygen-centered nucleophiles to α,β-unsaturated aldehydes was less documented due to its weak nucleophilicity. In 2007, Jørgensen successfully developed an oxa-Michael addition to enals **106** by using active oximes **107** as nucleophile (Scheme 11.30) [161]. The reaction was promoted by catalyst **6f** and gave moderate yields and good to excellent ees. The resulting aldehydes **108** were reduced to give the corresponding alcohols **109** and, after removal of oxime moiety by hydrogenation with Pd(OH)$_2$/C, gave synthetically useful 1,3-diols. The limitation of this methodology is that only aliphatic substituted enals are suitable substrates (no aromatic ones were reported).

Conjugate addition of thiols **111** to α,β-unsaturated aldehydes **110** was disclosed by Jørgensen in 2005 (Scheme 11.31) [162]. Various alkyl thiols could effectively participate in the process in good yields and with good to excellent enantioselectivities. Notably, in the process of optimization, the authors observed the formation of a stable enamine species between the catalyst and adduction product. This contributes to the slow reaction rate because of slow turnover.

Chiral phosphines are very important ligands in organometallic catalysis, and usually they are prepared by resolutions or by using a stoichiometric amount of chiral auxiliaries. Melchiorre [163] and Córdova [164] independently reported organocatalytic hydrophosphination of α,β-unsaturated aldehydes (Scheme 11.32).

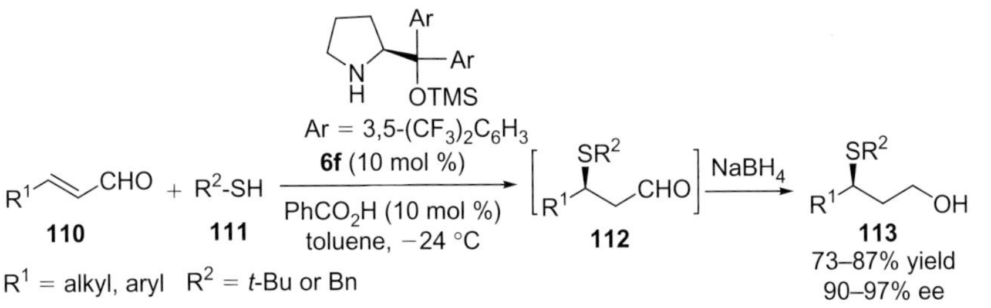

Scheme 11.31 Conjugated addition of thiols to α,β-unsaturated aldehydes.

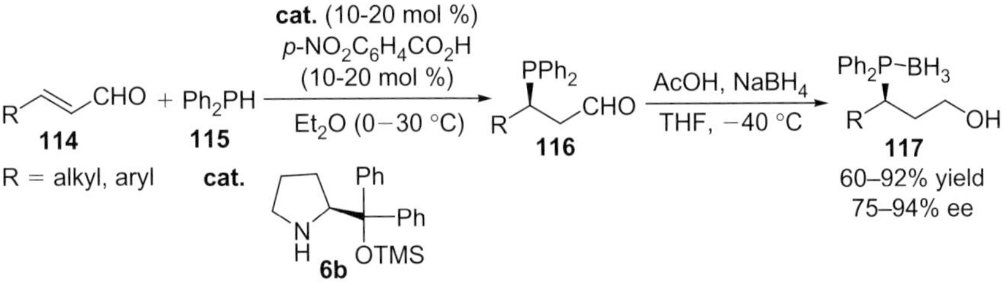

Scheme 11.32 (S)-Diphenylprolinol TMS ether catalyzed conjugate addition of hydrophosphines to enals.

The use of diarylprolidinol silyl ethers as promoters for the conjugate addition of diphenylphosphine **115** to α,β-unsaturated aldehydes **114** led to highly enantioselective adducts, which were reduced *in situ* by NaBH$_4$ to give air-stable phosphine-borane-alcohol derivatives **117**.

Organocatalytic Michael additions of carbon-centered nucleophiles to α,β-unsaturated aldehydes have been studied extensively. These nucleophiles include activated methylene compounds, nitroalkane, and aromatic rings. Scheme 11.33 shows some representative examples using chiral diarylprolinol ethers to catalyze conjugate addition processes [165–169].

11.8.3
Cycloaddition Reactions Catalyzed by Diarylprolinols or Diarylprolinol Silyl Ethers

In 2003, Jørgensen *et al.* reported the first organocatalytic enantioselective inverse-electron-demand hetero-Diels–Alder reaction (Scheme 11.34) [170]. Notably, a chiral pyrrolidine bearing a bulky side chain (**121**) was used for the process. Oxidation of the intermediates with pyridine chlorochromate (PCC) afforded lactones **124** as single diastereomer.

Three diarylprolinol silyl ethers promoted [3 + 2] cycloaddition reactions with α,β-unsaturated aldehydes were reported (Scheme 11.35). They are Chen's azomethine imines **125** (Reaction 1) [171], Nevalainen's nitrones **128** (Reaction 2)

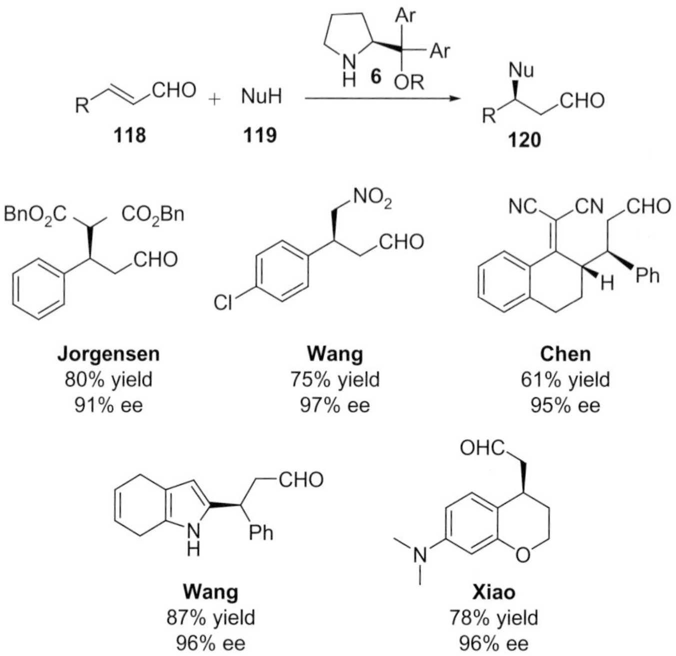

Scheme 11.33 Michael additions of carbon-centered nucleophiles to α,β-unsaturated aldehydes.

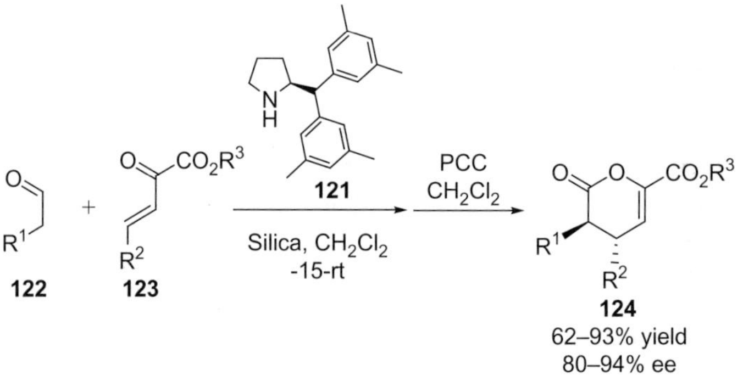

Scheme 11.34 A hetero-Diels–Alder reaction.

[172], and Vicario's azomethine ylides **131** (Reaction 3) [173]. These reactions gave five-membered heterocycles. Notably, in Chen's case, the *exo* **127** were major products, while other two cases gave predominantly *endo* **130** and **133**, respectively.

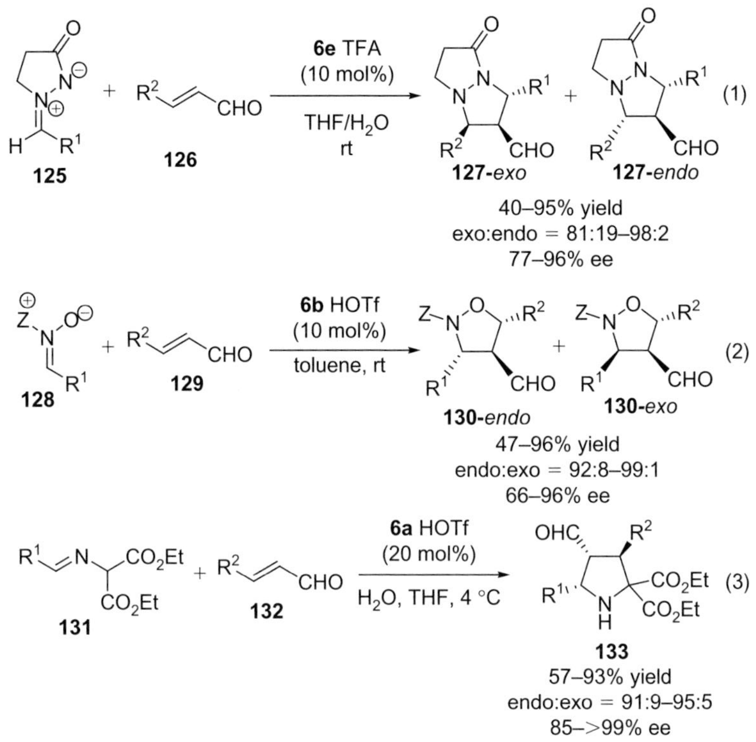

Scheme 11.35 [3 + 2] Cycloaddition reactions.

11.8.4
Cascade Reactions Catalyzed by Diarylprolinol Silyl Ethers

Diarylprolinol silyl ether catalysts can activate both the α-position of aldehyde by forming an enamine and the β-position of α,β-unsaturated aldehyde by forming an iminium ion. The combination of two activation modes in a one-pot operation has led to impressive powerful cascade reactions for facile construction of complex molecular architectures [174, 175]. By rational design of appropriate amphiprotic species as reaction partners of α,β-unsaturated aldehydes, three-, five-, or six-membered rings can be obtained by diarylprolinol silyl ether promoted cascade reactions.

11.8.4.1 Three-Membered Rings Formed by a [1 + 2] Strategy

Jørgensen and colleagues described the first iminium-catalyzed epoxidation reaction of α,β-unsaturated aldehydes **134** (Scheme 11.36) [176]. The process, catalyzed by diarylprolinol silyl ether **6f**, gave epoxides **135** with excellent levels of enantioselectivity and in good yields when H_2O_2 was used as the oxidant. The epoxidation process involved an oxa-Michael–substitution cascade. Using a similar strategy, Córdova developed an aziridination of α,β-unsaturated aldehydes [177].

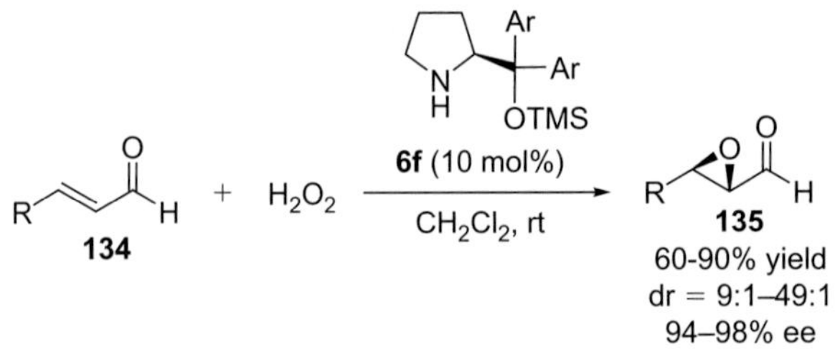

Scheme 11.36 Epoxidation of α,β-unsaturated aldehydes.

Michael–alkylation reaction of alkyl halides **137** with α,β-unsaturated aldehydes **136** is an attractive process for synthesis of cyclopropanes **138**. Nevertheless, it is a challenging task because of the high tendency for N-alkylation of the secondary amino group of the catalyst by alkyl halides, leading to poisoning of the catalyst. Wang [178] has developed diarylprolinol silyl ether-catalyzed Michael–alkylation cascade reactions to generate highly functionalized chiral cyclopropanes in a one-pot transformation (Scheme 11.37).

11.8.4.2 Five-Membered Rings Formed by a [3 + 2] Strategy

Jørgensen et al. reported an organocatalytic domino Michael–aldol reaction between enals **139** and 2-mercapto-1-phenylethanone (**140**) (Scheme 11.38) [179]. Additives were very important for the formation of the tetrahydrothiophene products. If benzoic acid was used, the reaction gave rise to 2,3,4,4-tetrasubstituted tetrahydrothiophene carbaldehydes **141** as single isomer in good yields and with excellent ees (Reaction 1); the use of $NaHCO_3$ as additive led to (tetrahydrothiophen-2-yl)phenyl methanones **142** as main products in moderate yields and with moderate ees (Reaction 2).

Wang has designed a new cascade process involving a Michael–Michael sequence (Scheme 11.39) [180]. The transformation, which results in the formation of two new C–C bonds and three contiguous stereogenic centers, enables the facile assembly of tetrasubstituted, highly functionalized cyclopentanes **145** from simple achiral molecules **143** and **144** with high levels of enantio-and diastereocontrol in a single operation.

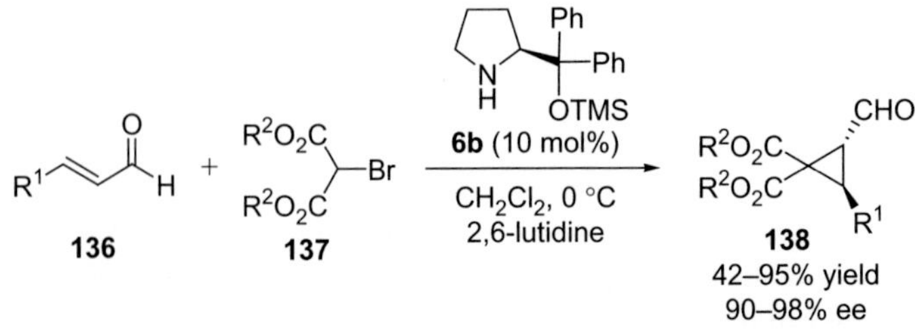

Scheme 11.37 Michael–alkylation of alkyl halides with α,β-unsaturated aldehydes.

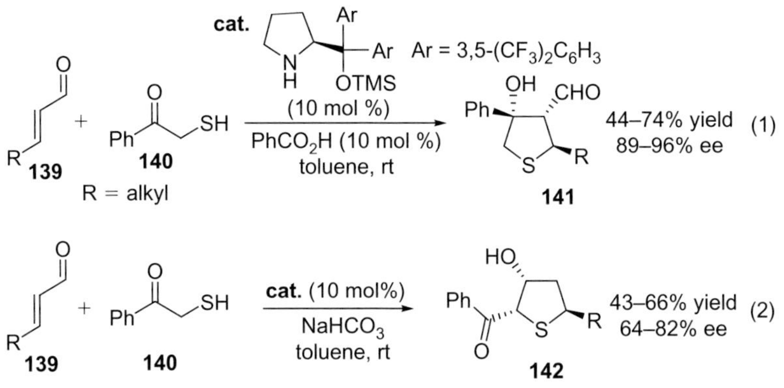

Scheme 11.38 Michael–aldol reactions between enals and 2-mercapto-1-phenylethanone to form five-membered rings.

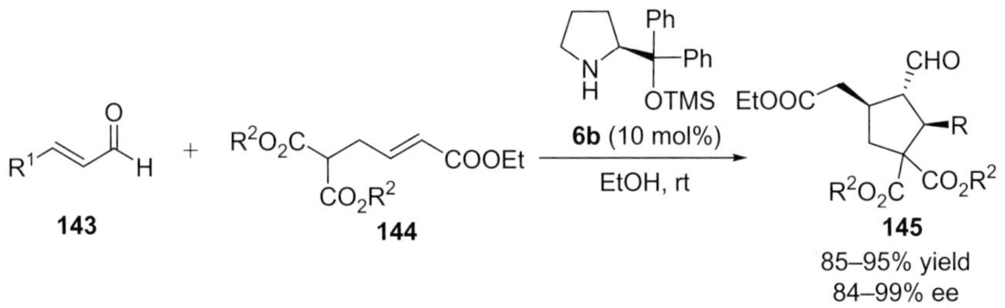

Scheme 11.39 Michael–Michael reactions catalyzed by a chiral diarylprolinol silyl ether.

This cascade catalytic strategy also proved to be a highly efficient and facile approach to synthetically useful, highly functionalized chiral trisubstituted tetrahydrothiophenes [181], pyrrolidines [182], and cyclopentanes with the generation of four stereogenic centers [183].

11.8.4.3 Six-Membered Rings Formed by a [4+2] Strategy

The Wang [184–186] and Córdova groups [187–189] have each developed enantioselective cascade hetero-Michael–aldol–dehydration processes, where S, O, and N served as nucleophiles for the initial conjugate addition reaction (Scheme 11.40). These cascade processes provided an efficient approach to the preparation of biologically significant benzo(thio)pyrans and hydroquinolines **148**.

11.8.4.4 Six-Membered Rings Formed by a [3+3] Strategy

It is recognized that, as discussed above, in the [4+2] strategies the enals (formed iminium ions) act as electrophiles (β-carbon) in the initial Michael addition and then as nucleophiles (α-carbon, enamine forms) in the following aldol/Michael

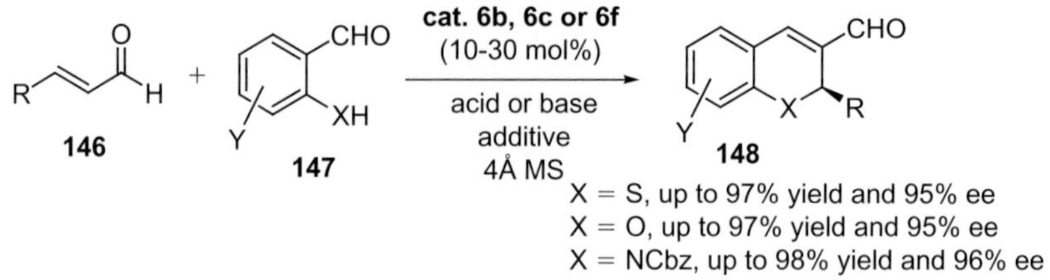

Scheme 11.40 Hetero-Michael–aldol–dehydration reactions.

reaction. In such a way, the α- and β-carbons of α,β-unsaturated aldehydes are integrated into the final cyclohexanes.

Jørgensen and colleagues developed a series of [3 + 3] cascade reactions via the rational design of substrates. In these reactions, the β-position and the aldehyde moiety of enals serve as electrophiles. The first example is a cascade Michael–Darzens-type reaction with enals **149** using γ-chloro-β-ketoesters **150** as bifunctional nucleophilic molecules (Scheme 11.41) [190]. In the presence of organocatalyst **6f**, the ketoester attacked the **6f**–enal iminium complex to give a conjugate addition adduct **154**, which underwent an intramolecular aldolization under basic conditions (NaOAc). Treatment of the cyclic cyclohexanone **151** with K_2CO_3 gave rise to an epoxide (**152**). Notably, the final products after saponification and decarboxylation were obtained with high levels of enantio- and diastereoselectivity.

Similar cascade reactions were also reported by Jørgensen and coworkers, including Michael–aldol [191], Michael–Knoevenagel [192], Michael–MBH [193], and Michael-Henry reactions [194].

11.8.4.5 Six-Membered Rings Formed by a [2 + 2 + 2] Strategy

A milestone in organocatalyzed cascade reactions resulted from work by Enders and coworkers (Scheme 11.42), who developed a powerful three-component triple

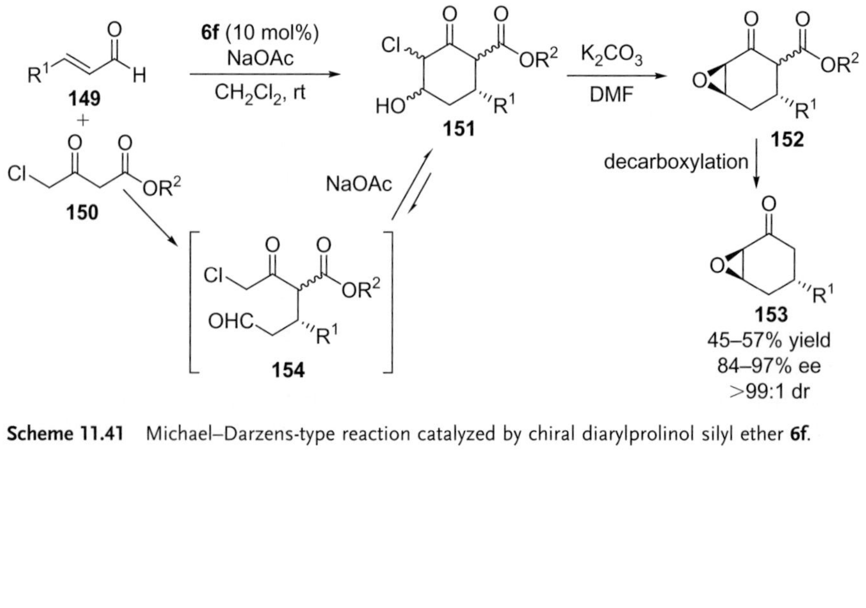

Scheme 11.41 Michael–Darzens-type reaction catalyzed by chiral diarylprolinol silyl ether **6f**.

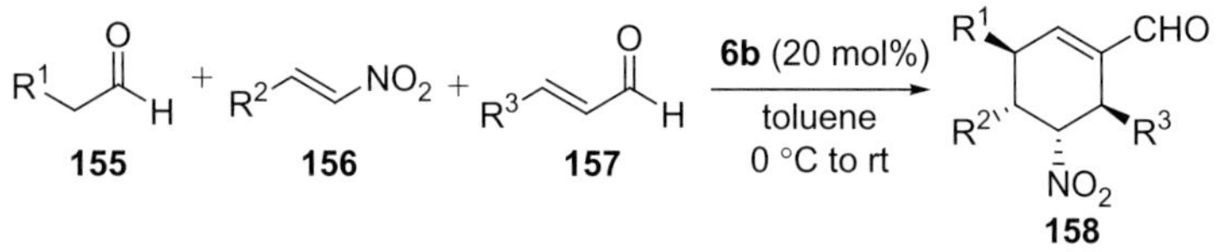

Scheme 11.42 Triple cascade Michael–Michael–aldol condensation reaction catalyzed by chiral diphenylprolinol silyl ether **6b**.

cascade process [195]. The process involves a Michael–Michael–aldol condensation sequence (enamine–imine–enamine), catalyzed by diphenylprolinol TMS ether **6b**, to form tetrasubstituted cyclohexene carbaldehydes **158** with high chemo-, regio-, and stereocontrol from readily available aldehydes **155**, nitroolefins **156**, and enals **157**. Remarkably, three C–C bonds and four new stereogenic centers are efficiently created in a one-pot transformation. The strategy has been extended by Melchiorre [196] and by Gong [197].

11.8.4.6 Other Cascade Reactions

Chiral diarylprolinol ethers have also been demonstrated as general and useful catalysts in both [1+2+3] [198, 199] and [1+4] [200] fashions, affording the corresponding six- and five-membered rings with excellent enantiocontrol. Moreover, Jørgensen [201] and Hong [202] also disclosed multicomponent cascade reactions.

11.9
Concluding Remarks

This chapter describes "privileged" proline and its derivatives as effective organocatalysts in asymmetric catalysis. Impressively, since 2000, a relatively short period of time, proline has quickly established its "privileged" status in organocatalysis. This declaration has been supported by the fact that proline itself is a general promoter in various organic transformations. Moreover, drawing inspiration from the catalyst, chemists have developed an array of proline analogs and derivatives as catalysts that can dramatically improve and expand the scope of aminocatalysis. As demonstrated, several unprecedented transformations have been developed and they serve as useful approaches in organic synthesis. It is expected that new pyrrolidine catalysts and new organic reactions will continue to be disclosed. In addition, there will increasing applications of these methods in the syntheses of natural products and biologically interesting molecules. There are, however, still some problems to be overcome, for example, in general, high catalyst loadings are required for effective transformations. Efforts to elucidate mechanistic aspects may assist in the development of new more efficient catalysts and new activation modes, but this is a challenging task for organic chemists.

References

1 Berkessel, A. and Groger, H. (eds) (2005) *Asymmetric Organocatalysis-From Biomimetic Concepts to Applications in Asymmetric Synthesis*, Wiley-VCH Verlag GmbH, Weinheim.
2 Hajos, Z.G. and Parrish, D.R. (1974) *J. Org. Chem.*, **39**, 1615–1621.
3 List, B., Lerner, R.A., and Barbas, C.F., III (2000) *J. Am. Chem. Soc.*, **122**, 2395–2396.
4 Mukherjee, S., Yang, J.W., Hoffmann, S., and List, B. (2007) *Chem. Rev.*, **107**, 5471–5569.
5 Ahrendt, K.A., Borths, C.J., and MacMillan, D.W.C. (2000) *J. Am. Chem. Soc.*, **122**, 4243–4244.
6 Lelais, G. and MacMillan, D.W.C. (2006) *Aldrichim. Acta*, **39**, 79–87.
7 Erkkilä, A., Majander, I., and Pihko, P.M. (2007) *Chem. Rev.*, **107**, 5416–5470.
8 List, B. (2002) *Tetrahedron*, **58**, 5573–5590.
9 Movassaghi, M. and Jacobsen, E.N. (2002) *Science*, **298**, 1904–1905.
10 Sakthivel, K., Notz, W., Bui, T., and Barbas, C.F., III (2001) *J. Am. Chem. Soc.*, **123**, 5260–5267.
11 Wang, W., Wang, J., Li, H., and Liao, L.-X. (2004) *Tetrahedron Lett.*, **45**, 7235–7238.
12 Torii, H. Nakadai, M. Ishihara, K. Saito, S. Yamamoto, H. (2004) *Angew. Chem. Int. Ed.*, **43**, 1983–1986.
13 Cobb, A.J.A., Shaw, D.M., and Ley, S.V. (2004) *Synlett*, 558–560.
14 Marigo, M., Wabnitz, T.C., Fielenbach, D., and Jorgensen, K.A. (2005) *Angew. Chem. Int. Ed.*, **44**, 794–797.
15 Hayashi, Y., Gotoh, H., Hayashi, T., and Shoji, M. (2005) *Angew. Chem. Int. Ed.*, **44**, 4212–4215.
16 Chi, Y.-G. and Gellman, S. H. (2005) *Org. Lett.*, **7**, 4253–4256.
17 MacMillan, D.W.C. (2008) *Nature*, **455**, 304–308.
18 Notz, W. and List, B. (2000) *J. Am. Chem. Soc.*, **122**, 7386–7387.
19 Northrup, A.B. and MacMillan, D.W.C. (2002) *J. Am. Chem. Soc.*, **124**, 6798–6799.
20 Pihko, P.M. and Erkkila, A. (2003) *Tetrahedron Lett.*, **44**, 7607–7609.
21 Zhang, S.-L., Duan, W.-H., and Wang, W. (2006) *Adv. Synth. Catal.*, **348**, 1228–1234.
22 Carpenter, J., Northrup, A.B., Chung, D., Wiener, J.J.M., Kim, S.–G., and MacMillan, D.W.C. (2008) *Angew. Chem. Int. Ed.*, **47**, 3568–3572.
23 Samanta, S. and Zhao, C.-G. (2006) *J. Am. Chem. Soc.*, **128**, 7442–7443.
24 Pan, Q., Zou, B., Wang, Y., and Ma, D. (2004) *Org. Lett.*, **6**, 1009–1012.
25 Thayumanavan, R., Tanaka, F., and Barbas, C.F., III (2004) *Org. Lett.*, **6**, 3541–3544.
26 Storer, R.I. and MacMillan, D.W.C. (2004) *Tetrahedron*, **60**, 7705–7714.
27 Zhong, G., Fan, J., and Barbas, C.F., III (2004) *Tetrahedron Lett.*, **45**, 5681–5684.
28 Casas, J., Sunden, H., and Córdova, A. (2004) *Tetrahedron Lett.*, **45**, 6117–6119.
29 Ward, D.E. and Jheengut, V. (2004) *Tetrahedron Lett.*, **45**, 8347–8350.
30 Funabiki, K., Yamamoto, H., Nagaya, H., and Matsui, M. (2006) *Tetrahedron Lett.*, **47**, 5507–5510.
31 Companyó, X., Valero, G., Crovetto, L., Moyano, A., and Rios, R. (2009) *Chem. Eur. J.*, **15**, 6564–6568.
32 Wang, Y.-J., Shen, Z.-X., Li, B., Zhang, Y., and Zhang, Y.-W. (2007) *Chem. Commun.*, 1284–1286.
33 Eder, U., Sauer, G., and Wiechert, R. (1971) *Angew. Chem. Int. Ed. Engl.*, **10**, 496–497.
34 Pidathala, D., Hoang, L., Vignola, N., and List, B. (2003) *Angew. Chem. Int. Ed.*, **42**, 2785–2788.
35 Mans, D.M. and Pearson, W.H. (2004) *Org. Lett.*, **6**, 3305–3308.
36 Markert, M. and Mahrwald, R. (2008) *Chem. Eur. J.*, **14**, 40–48.
37 Northrup, A.B., Mangion, I.K., Hettche, F., and MacMillan, D.W.C. (2004) *Angew. Chem. Int. Ed.*, **43**, 2152–2154.
38 Northrup, A.B. and MacMillan, D.W.C. (2004) *Science*, **305**, 1752–1755.

39 Enders, D. and Grondal, C. (2005) *Angew. Chem. Int. Ed.*, **44**, 1210–1212.
40 Grondala, C. and Enders, D. (2007) *Adv. Synth. Catal.*, **349**, 694–702.
41 List, B. (2000) *J. Am. Chem. Soc.*, **122**, 9336–9337.
42 Hayashi, Y., Tsuboi, W., Ashimine, I., Urushima, T., Shoji, M., and Sakai, K. (2003) *Angew. Chem. Int. Ed.*, **42**, 3677–3680.
43 Córdova, A. (2004) *Chem. Eur. J.*, **10**, 1987–1997.
44 Córdova, A., Notz, W., Zhong, G., Betancort, J.M., and Barbas, C.F., III (2002) *J. Am. Chem. Soc.*, **124**, 1842–1843.
45 Córdova, A., Watanabe, S., Tanaka, F., Notz, W., and Barbas, C.F., III (2002) *J. Am. Chem. Soc.*, **124**, 1866–1867.
46 Janey, J.M., Hsiao, Y., and Armstrong, J.D.I. (2006) *J. Org. Chem.*, **71**, 390–392.
47 Yang, J.W., Chandler, C., Stadler, M., Kampen, D., and List, B. (2008) *Nature*, **452**, 453–455.
48 Chandler, C., Galzerano, P., Michrowska, A., and List, B. (2009) *Angew. Chem. Int. Ed.*, **48**, 1978–1980.
49 Itoh, T., Yokoya, M., Miyauchi, K., Nagata, K., and Ohsawa, A. (2003) *Org. Lett.*, **5**, 4301–4304.
50 Funabiki, K., Nagamori, M., Goushi, S., and Matsui, M. (2004) *Chem. Commun.*, 1928–1929.
51 Ibrahem, I., Casas, J., and Córdova, A. (2004) *Angew. Chem. Int. Ed.*, **43**, 6528–6531.
52 Enders, D., Grondal, C., Vrettou, M., and Raabe, G. (2005) *Angew. Chem. Int. Ed.*, **44**, 4079–4083.
53 Jiang, B., Dong, J.J., Si, Y.G., Zhao, X.L., Huang, Z.G., and Xua, M. (2008) *Adv. Synth. Catal.*, **350**, 1360–1366
54 Hahn, B.T., Fröhlich, R., Harms, K., and Glorius, F. (2008) *Angew. Chem. Int. Ed.*, **47**, 9985–9988.
55 Hanessian, S. and Pham, V. (2000) *Org. Lett.*, **2**, 2975–2978.
56 List, B., Pojarliev, P., and Martin, H.J. (2001) *Org. Lett.*, **3**, 2423–2425.
57 Imbriglio, J.E., Vasbinder, M.M., and Miller, S.J. (2003) *Org. Lett.*, **5**, 3741–3743.
58 Vasbinder, M.M., Imbriglio, J.E., and Miller, S.J. (2006) *Tetrahedron*, **62**, 11450–11459.
59 Tang, H., Zhao, G., Zhou, Z., Zhou, Q., and Tang, C. (2006) *Tetrahedron Lett.*, **47**, 5717–5721.
60 Utsumi, N., Zhang, H., Tanaka, F., and Barbas, C.F.I. (2007) *Angew. Chem. Int. Ed.*, **46**, 1878–1880.
61 List, B. (2002) *J. Am. Chem. Soc.*, **124**, 5656–5657.
62 Kumaragurubaran, N., Juhl, K., Zhuang, W., Bogevig, A., and Jorgensen, K.A. (2002) *J. Am. Chem. Soc.*, **124**, 6254–6255.
63 Zhong, G. (2003) *Angew. Chem. Int. Ed.*, **42**, 4247–4250.
64 Brown, F.J., Brochu, M.P., Sinz, C.J., and MacMillan, D.W.C. (2003) *J. Am. Chem. Soc.*, **125**, 10808–10809.
65 Hayashi, Y., Yamaguchi, J., Hibino, K., and Shoji, M. (2003) *Tetrahedron Lett.*, **44**, 8293–8296.
66 Bogevig, A., Sunden, H., and Córdova, A. (2004) *Angew. Chem. Int. Ed.*, **43**, 1109–1112.
67 Vignola, N. and List, B. (2004) *J. Am. Chem. Soc.*, **126**, 450–451.
68 Shaikh, R.R., Mazzanti, A., Petrini, M., Bartoli, G., and Melchiorre, P. (2008) *Angew. Chem. Int. Ed.*, **47**, 8707–8710.
69 Tietze, L.F., Brasche, G., and Gericke, K. M. (2006) *Domino Reactions in Organic Synthesis*, Wiley-VCH Verlag GmbH, Weinheim.
70 Chowdari, N.S., Ramachary, D.B., and Barbas, C.F., III (2003) *Org. Lett.*, **10**, 1685–1688.
71 Bench, B.J., Liu, C., Evett, C.R., and Watanabe, C.M.H. (2006) *J. Org. Chem.*, **71**, 9458–9463.
72 Lu, M., Zhu, D., Lu, Y., Hou, Y., Tan, B., and Zhong, G. (2008) *Angew. Chem. Int. Ed.*, **47**, 10187–10191.
73 Zhu, D., Lu, M., Chua, P.J., Tan, B., Wang, F., Yang, X., and Zhong, G.-F. (2008) *Org. Lett.*, **10**, 4585–4588.
74 Itagaki, N., Kimura, M., Sugahara, T., and Iwabuchi, Y. (2005) *Org. Lett.*, **7**, 4185–4188.
75 Miao, W. and Chana, T.H. (2006) *Adv. Synth. Catal.*, **348**, 1711–1718.

76 Lombardo, M., Pasi, F., Easwar, S., and Trombinia, C. (2007) *Adv. Synth. Catal.*, **349**, 2061–2065.

77 Font, D., Jimeno, C., and Pericas, M.A. (2006) *Org. Lett.*, **8**, 4653–4655.

78 Kehat, T. and Portnoy, M. (2007) *Chem. Commun.*, 2823–2825.

79 Kristensen, T.E., Vestli, K., Fredriksen, K.A., Hansen, F.K., and Hansen, T. (2009) *Org. Lett.*, **11**, 2968–2971.

80 Kunz, R.K. and MacMillan, D.W.C. (2005) *J. Am. Chem. Soc.*, **127**, 3240–3241.

81 Mitsumori, S., Zhang, H., Cheong, P.H.-Y., Houk, K.N., Tanaka, F., and Barbas, C.F., III (2006) *J. Am. Chem. Soc.*, **128**, 1040–1041.

82 Zhang, H., Mifsud, M., Tanaka, F., and Barbas, C.F., III (2006) *J. Am. Chem. Soc.*, **128**, 9630–9631.

83 Zhang, H., Mitsumori, S., Utsumi, N., Imai, M., Garcia-Delgado, N., Mifsud, M., Albertshofer, K., Cheong, P.H.-Y., Houk, K.N., Tanaka, F., and Barbas, C.F. I. (2008) *J. Am. Chem. Soc.*, **130**, 875–886.

84 Momiyama, N., Torii, H., Saito, S., and Yamamoto, H. (2004) *Proc. Natl. Acad. Sci. U.S.A.*, **101**, 5374–5378.

85 Hartikka, A. and Arvidsson, P.I. (2004) *Tetrahedron: Asymmetry*, **15**, 1831–1834.

86 Chowdari, N.S., Ahmad, M., Albertshofer, K., Tanaka, F., and Barbas, C.F., III (2006) *Org. Lett.*, **8**, 2839–2842.

87 Chowdari, N.S. and Barbas, C.F., III (2005) *Org. Lett.*, **7**, 867–870.

88 Yamamoto, Y., Momiyama, N., and Yamamoto, H. (2004) *J. Am. Chem. Soc.*, **126**, 5962–5962.

89 Hansen, H.M., Longbottom, D.A., and Ley, S.V. (2006) *Chem. Commun.*, 4838–4840.

90 Jheengut, V. and Ward, D.E. (2007) *J. Org. Chem.*, **72**, 7805–7808.

91 Ward, D.E., Jheengut, V., and Beye, G.E. (2006) *J. Org. Chem.*, **71**, 8989–8992.

92 Mitchell, C.E.T., Brenner, S.E., and Ley, S.V. (2005) *Chem. Commun.*, 5346–5348.

93 Mitchell, C.E.T., Cobb, A.J.A., and Ley, S.V. (2005) *Synlett*, 611–614.

94 Knudsen, K.R., Mitchell, C.E.T., and Ley, S.V. (2006) *Chem. Commun.*, 66–68.

95 Mitchell, C.E.T., Brenner, S.E., Garcia-Fortanet, J., and Ley, S.V. (2006) *Org. Biomol. Chem.*, **4**, 2039–2049.

96 Wascholowski, V., Knudsen, K.R., Mitchell, C.E.T., and Ley, S.V. (2008) *Chem. Eur. J.*, **14**, 6155–6165.

97 Wu, Y.-Y., Chai, Z., Liu, X.-Y., Zhao, G., and Wang, S.-W. (2009) *Eur. J. Org. Chem.*, 904–911.

98 Wang, W., Li, H., and Wang, J. (2005) *Tetrahedron Lett.*, **46**, 5077–5079.

99 Mei, K., Zhang, S., He, S., Li, P., Jin, M., Xue, F., Luo, G., Zhang, H., Song, L., Duan, W., and Wang, W. (2008) *Tetrahedron Lett.*, **49**, 2681–2684.

100 Wang, W., Wang, J., and Li, H. (2004) *Tetrahedron Lett.*, **45**, 7243–7246.

101 Wang, W., Wang, J., and Li, H. (2005) *Angew. Chem. Int. Ed.*, **44**, 1369–1371.

102 Wang, J., Li, H., Lou, B.-S., Zu, L.-S., Guo, H., and Wang, W. (2006) *Chem. Eur. J.*, **12**, 4321–4332.

103 Wang, J., Li, H., Zu, L.-S., and Wang, W. (2006) *Adv. Synth. Catal.*, **348**, 425–428.

104 Cao, C.-L., Sun, X.-L., Kang, Y.-B., and Tang, Y. (2007) *Org. Lett.*, **9**, 4151–4154.

105 Cao, C.-L., Sun, X.-L., Zhou, J.-L., and Tang, Y. (2007) *J. Org. Chem.*, **72**, 4073–4076.

106 Halland, N., Braunton, A., Bachmann, S., Marigo, M., and Jorgensen, K.A. (2004) *J. Am. Chem. Soc.*, **126**, 4790–4791.

107 Wang, W., Wang, J., and Li, H. (2004) *Org. Lett.*, **6**, 2817–2820.

108 Wang, J., Li, H., Mei, Y.-J., Lou, B., Xu, D.-G., Xie, D.-Q., Guo, H., and Wang, W. (2005) *J. Org. Chem.*, **70**, 5678–5687.

109 Dodda, R. and Zhao, C.-G. (2006) *Org. Lett.*, **8**, 4911–4914.

110 Samanta, S. and Zhao, C.-G. (2006) *Tetrahedron Lett.*, **47**, 3383–3386.

111 Alberg, D.G., Poulsen, T.B., Bertelsen, S., Christensen, K.L., Birkler, R.D., Johannsen, M., and Jørgensen, K.A. (2009) *Bioorg. Med. Chem. Lett.*, **19**, 3888–3891.

112 Tang, Z., Jiang, F., Yu, L.-T., Cui, X., Gong, L.-Z., Mi, A.-Q., Jiang, Y.-Z., and Wu, Y.-D. (2003) *J. Am. Chem. Soc.*, **125**, 5262–5263.
113 Tang, Z., Yang, Z.-H., Chen, X.-H., Cun, L.-F., Mi, A.-Q., Jiang, Y.-Z., and Gong, L.-Z. (2005) *J. Am. Chem. Soc.*, **127**, 9285–9289.
114 Silva, F., Sawicki, M., and Gouverneur, V. (2006) *Org. Lett.*, **8**, 5417–5419.
115 Yang, H. and Carter, R.G. (2009) *J. Org. Chem.*, **74**, 2246–2249.
116 Zhang, Q., Ni, B., and Headley, A.D. (2008) *Tetrahedron*, **64**, 5091–5097.
117 Yang, H. and Carter, R.G. (2009) *J. Org. Chem.*, **74**, 5151–5156.
118 Mase, N., Tanaka, F., and Barbas, C.F., III (2004) *Angew. Chem. Int. Ed.*, **43**, 2420–2423.
119 Mase, N., Thayumanavan, R., Tanaka, F., and Barbas, C.F., III (2004) *Org. Lett.*, **6**, 2527–2530.
120 Hayashi, Y., Sekizawa, H., Yamaguchi, J., and Gotoh, H. (2007) *J. Org. Chem.*, **72**, 6493–6499.
121 Zhuang, W., Saaby, S., and Jorgensen, K.A. (2004) *Angew. Chem. Int. Ed.*, **43**, 4476–4478.
122 Karlsson, S. and Högberg, H.-E. (2003) *Eur. J. Org. Chem.*, 2782–2791.
123 Dambruoso, P., Massi, A., and Dondoni, A. (2005) *Org. Lett.*, **7**, 4657–4660.
124 Rios, R., Sunden, H., Ibrahem, I., and Córdova, A. (2007) *Tetrahedron Lett.*, **48**, 2181–2184.
125 Rios, R., Sunden, H., Ibrahem, I., Zhao, G.-L., and Córdova, A. (2006) *Tetrahedron Lett.*, **47**, 8679–8682.
126 Mase, N., Nakai, Y., Ohara, N., Yoda, H., Takabe, K., Tanaka, F., and Barbas, C.F., III (2006) *J. Am. Chem. Soc.*, **128**, 734–735.
127 Betancort, J.M. and Barbas, C.F., III (2001) *Org. Lett.*, **3**, 3737–3740.
128 Mase, N., Watanabe, H., Yoda, H., Takabe, K., Tanaka, F., and Barbas, C.F., III (2006) *J. Am. Chem. Soc.*, **128**, 4966–4967.
129 Pansare, S.V. and Pandya, K. (2006) *J. Am. Chem. Soc.*, **128**, 9624–9625.
130 Vishnumaya and Singh, V.K. (2007) *Org. Lett.*, **9**, 1117–1119.
131 Sulzer–Mosse, S., Alexakis, A., Mareda, J., Bollot, G., Bernardinelli, G., and Filinchuk, Y. (2009) *Chem. Eur. J.*, **15**, 3204–3220.
132 Mossé, S. and Alexakis, A. (2005) *Org. Lett.*, **7**, 4361–4364.
133 Terakado, D., Koutaka, H., and Oriyama, T. (2005) *Tetrahedron: Asymmetry*, **16**, 1157–1165.
134 Lattanzi, A. (2009) *Chem. Commun.*, 1452–1463.
135 Mielgo, A. and Palomo, C. (2008) *Chem. Asian. J.*, **3**, 922–948.
136 Hayashi, Y., Itoh, T., Aratake, S., and Ishikawa, H. (2008) *Angew. Chem. Int. Ed.*, **47**, 2082–2084.
137 Hayashi, Y., Okano, T., Itoh, T., Urushima, T., Ishikawa, H., and Uchimaru, T. (2008) *Angew. Chem. Int. Ed.*, **47**, 9053–9058.
138 Tiecco, M., Carlone, A., Sternativo, S., Marini, F., Bartoli, G., and Melchiorre, P. (2007) *Angew. Chem. Int. Ed.*, **46**, 6882–6885.
139 Gotoha, H. and Hayashi, Y. (2009) *Chem. Commun.*, 3083–3085.
140 Kano, T., Mii, H., and Maruoka, K. (2009) *J. Am. Chem. Soc.*, **131**, 3450–3451.
141 Ibrahem, I., Zhao, G.-L., Sunden, H., and Córdova, A. (2006) *Tetrahedron Lett.*, **47**, 4659–4663.
142 Franzån, J., Marigo, M., Fielenbach, D., Wabnitz, T.C., Kjærsgaard, A., and Jorgensen, K.A. (2005) *J. Am. Chem. Soc.*, **127**, 18296–18304.
143 Albertshofer, K., Thayumanavan, R., Utsumi, N., Tanaka, F., and Barbas, C.F., III (2007) *Tetrahedron Lett.*, **48**, 693–696.
144 Bonne, D., Salat, L., Dulcere, J.-P., and Rodriguez, J. (2008) *Org. Lett.*, **10**, 5409–5412.
145 Garcia-Garcia, P., Ladepeche, A., Halder, R., and List, B. (2008) *Angew. Chem. Int. Ed.*, **47**, 4719–4721.
146 Hayashi, Y., Itoh, T., Ohkubo, M., and Ishikawa, H. (2008) *Angew. Chem. Int. Ed.*, **47**, 4722–4724.
147 Ruiz, N., Reyes, E., Vicario, J.L., Badia, D., Carrillo, L., and Uria, U. (2008) *Chem. Eur. J.*, **14**, 9357–9367.

148 Belot, S., Massaro, A., Tenti, A., Mordini, A., and Alexakis, A. (2008) *Org. Lett.*, **10**, 4557–4560.

149 Chi, Y., Guo, L., Kopf, N.A., and Gellman, S.H. (2008) *J. Am. Chem. Soc.*, **130**, 5608–5609.

150 Zhu, S., Yu, S., and Ma, D. (2008) *Angew. Chem. Int. Ed.*, **47**, 545–548.

151 Wang, J., Ma, A., and Ma, D. (2008) *Org. Lett.*, **10**, 5425–5428.

152 Vo, N.T., Pace, R.D.M., O'Hara, F., and Gaunt, M.J. (2008) *J. Am. Chem. Soc.*, **130**, 404–405.

153 Zhao, G.-L., Xu, Y., SundÅn, H., Eriksson, L., Sayah, M., and Córdova, A. (2007) *Chem. Commun.*, 734–735.

154 Sulzer-Mosse, S., Tissot, M., and Alexakis, A. (2007) *Org. Lett.*, **9**, 3749–3752.

155 Zhao, G.-L., Vesely, J., Sun, J., Christensen, K.E., Bonneau, C., and Córdova, A. (2008) *Adv. Synth. Catal.*, **350**, 657–661.

156 Zhu, Q. and Lu, Y. (2008) *Org. Lett.*, **10**, 4803–4806.

157 Landa, A., Maestro, M., Masdeu, C., Puente, A., Vera, S., Oiarbide, M., and Palomo, C. (2009) *Chem. Eur. J.*, **15**, 1562–1565.

158 Diner, P., Nielsen, M., Marigo, M., and Jørgensen, K.A. (2007) *Angew. Chem. Int. Ed.*, **46**, 1983–1987.

159 Ming, Z.-H., Xu, S.-Z., Zhou, L., Ding, M.-W., Yang, J.-Y., Yang, S., and Xiao, W.-J. (2009) *Bioorg. Med. Chem. Lett.*, **19**, 3938–3940.

160 Lin, Q., Meloni, D., Pan, Y., Xia, M., Rodgers, J., Shepard, S., Li, M., Galya, L., Metcalf, B., Yue, T.-Y., Liu, P., and Zhou, J. (2009) *Org. Lett.*, **11**, 1999–2002.

161 Bertelsen, S., Diner, P., Johansen, R.L., and Jørgensen, K.A. (2007) *J. Am. Chem. Soc.*, **129**, 1536–1537.

162 Marigo, M., Schulte, T., Franzen, J., and Jørgensen, K.A. (2005) *J. Am. Chem. Soc.*, **127**, 15710–15711.

163 Carlone, A., Bartoli, G., Bosco, M., Sambri, L., and Melchiorre, P. (2007) *Angew. Chem. Int. Ed.*, **46**, 4504–4506.

164 Ibrahem, I., Rios, R., Vesely, J., Hammar, P., Eriksson, L., Himo, F., and Córdova, A. (2007) *Angew. Chem. Int. Ed.*, **46**, 4507–4510.

165 Brandau, S., Landa, A., Franzen, J., Marigo, M., and Jorgensen, K.A. (2006) *Angew. Chem. Int. Ed.*, **45**, 4305–4309.

166 Zu, L.-S., Xie, H.-X., Li, H., Wang, J., and Wang, W. (2007) *Adv. Synth. Catal.*, **349**, 2660–2664.

167 Hong, L., Liu, C., Sun, W., Wang, L., Wong, K., and Wang, R. (2009) *Org. Lett.*, **11**, 2177–2180.

168 Xie, J.-W., Yue, L., Xue, D., Ma, X.-L., Chen, Y.-C., Wu, Y., Zhu, J., and Deng, J.-G. (2006) *Chem. Commun.*, 1563–1565.

169 Lu, H.-H., Liu, H., Wu, W., Wang, X.-F., Lu, L.-Q., and Xiao, W.-J. (2009) *Chem. Eur. J.*, **15**, 2742–2746.

170 Juhl, K. and Jorgensen, K.A. (2003) *Angew. Chem. Int. Ed.*, **42**, 1498–1501.

171 Chen, W., Yuan, X.-H., Li, R., Du, W., Wu, Y., Ding, L.-S., and Chen, Y.-C. (2006) *Adv. Synth. Catal.*, **348**, 1818–1822.

172 Chow, S.S., Nevalainen, M., Evans, C.A., and Johannes, C.W. (2007) *Tetrahedron Lett.*, **48**, 277–280.

173 Vicario, J.L., Reboredo, S., Badía, D., and Carrillo, L. (2007) *Angew. Chem. Int. Ed.*, **46**, 5168–5170.

174 Yu, X.-H. and Wang, W. (2008) *Org. Biomol. Chem.*, **6**, 2036–2047.

175 Enders, D., Grondal, C., and Huttl, M.R.M. (2007) *Angew. Chem. Int. Ed.*, **46**, 1570–1581.

176 Marigo, M., Franzen, J., Poulsen, T.B., and Jorgensen, K.A. (2005) *J. Am. Chem. Soc.*, **127**, 6964–6965.

177 Vesely, J., Ibrahem, I., Zhao, G.-L., Rios, R., and Córdova A. (2007) *Angew. Chem. Int. Ed.*, **46**, 778–781.

178 Xie, H.X., Zu, L.-S., Li, H., Wang, J., and Wang, W. (2007) *J. Am. Chem. Soc.*, **129**, 10886–10894.

179 Brandau, S., Maerten, E., and Jorgensen, K.A. (2006) *J. Am. Chem. Soc.*, **128**, 14986–14991.

180 Zu, L.S., Li, H., Xie, H.-X., Wang, J., Jiang, W., Tang, Y., and Wang, W. (2007) *Angew. Chem. Int. Ed.*, **46**, 3732–3734.

181 Li, H., Zu, L.S., Xie, H.X., Wang, J., Jiang, W., and Wang, W. (2007) *Org. Lett.*, **9**, 1833–1835.

182 Li, H., Zu, L.S., Xie, H.X., Wang, J., and Wang, W. (2008) *Chem. Commun.*, 5636–5638.

183 Zhao, G.L., Ibrahem, I., Dziedzic, P., Sun, J., Bonneau, C., and Córdova, A. (2008) *Chem. Eur. J.*, **14**, 10007–10011.

184 Wang, W., Li, H., Wang, J., and Zu, L.-S. (2006) *J. Am. Chem. Soc.*, **128**, 10354–10355.

185 Li, H., Wang, J., E-Nunu, T., Zu, L.-S., Jiang, W., Wei, S.-H., and Wang, W. (2007) *Chem. Commun.*, 507–509.

186 Li, H., Wang, J., Xie, H.-X., Zu, L.-S., Jiang, W., Duesler, E.N., and Wang, W. (2007) *Org. Lett.*, **9**, 965–968.

187 Rios, R., Sunden, H., Ibrahem, I., Zhao, G.-L., Eriksson, L., and Córdova, A. (2006) *Tetrahedron Lett.*, **47**, 8547–8551.

188 Sunden, H., Ibrahem, I., Zhao, G.-L., Eriksson, L., and Córdova, A. (2007) *Chem. Eur. J.*, **13**, 574–581.

189 Sunden, H., Rios, R., Ibrahem, I., Zhao, G.-L., Eriksson, L., and Córdova, A. (2007) *Adv. Synth. Catal.*, **349**, 827–832.

190 Marigo, M., Bertelsen, S., Landa, A., and Jorgensen, K.A. (2006) *J. Am. Chem. Soc.*, **128**, 5475–5479.

191 Bolze, P., Dickmeiss, G., and Jørgensen, K.A. (2008) *Org. Lett.*, **10**, 3753–3756.

192 Albrecht, Ł., Richter, B., Vila, C., Krawczyk, H., and Jørgensen, K.A. (2009) *Chem. Eur. J.*, **15**, 3093–3102.

193 Cabrera, S., Aleman, J., Bolze, P., Bertelsen, S., and Jorgensen, K.A. (2008) *Angew. Chem. Int. Ed.*, **47**, 121–125.

194 Reyes, E., Jiang, H., Milelli, A., Elsner, P., Hazell, R.G., and Jorgensen, K.A. (2007) *Angew. Chem. Int. Ed.*, **46**, 9202–9205.

195 Enders, D., Huttl, M.R.M., Grondal, C., and Raabe, G. (2006) *Nature*, **441**, 861–863.

196 Penon, O. Carlone, A. Mazzanti, A. Locatelli, M. Sambri, L. Bartoli, G., and Melchiorre, P. (2008) *Chem. Eur. J.*, **14**, 4788–4791.

197 Zhang, F.-L., Xu, A.-W., Gong, Y.-F., Wei, M.-H., and Yang, X.-L. (2009) *Chem. Eur. J.*, **15**, 6815–6818.

198 Carlone, A., Cabrera, S., Marigo, M., and Jorgensen, K.A. (2007) *Angew. Chem. Int. Ed.*, **47**, 1101–1104.

199 Ruano, J.L.G., Marcos, V., Suanzes, J. A., Marzo, L., and Aleman, J. (2009) *Chem. Eur. J.*, **15**, 6576–6580.

200 Enders, D., Wang, C., and Bats, J.W. (2008) *Angew. Chem. Int. Ed.*, **47**, 7539–7542.

201 Bertelsen, S., Johansen, R.L., and Jørgensen, K.A. (2008) *Chem. Commun.*, 3016–3018.

202 Kotame, P., Hong, B.-C., and Liao, J.-H. (2009) *Tetrahedron Lett.*, **50**, 704–707.

Index

a

acid/base bifunctional metal/BINOL
  catalyst 308ff
1,2-addition
– to carbonyl compounds 25f, 287, 386
– to imino compounds 390
1,4-addition, *see also* conjugate addition
– of α-alkoxylamines 311
– of boron nucleophiles 196
– of carbon nucleophiles to α,β-unsaturated
  acceptors 186
– of heteroatom nucleophiles to
  α,β-unsaturated acceptors 193ff
– of nitrogen nucleophiles 193
– of oxygen nucleophiles 195
– of sulfur nucleophiles 195
ALB catalyst, *see* aluminum/lithium/
  (BINOL)$_2$ catalyst
alcoholysis of cyclic anhydrides 377
aldehyde
– addition of allyltrimethoxysilane to 25
– addition of organozinc compounds
  to 287, 334f
– addition of terminal alkynes to 307
– α-alkylation of 418
– allylation of 335f
– α-amination of 417
– α-aminoxylation of 418
– arylation of 25, 155, 253
– α-chlorination of 426
– cross-aldol reaction of nonequivalent 411
– cyanation of 286, 320f
– cyanophosphorylation of 324
– cyanosilylation of 27
– dienylation of 26
– intramolecular α-alkylation of 421
– organozinc addition to 334
– racemic 2-substituted, hydrogenation
  of 153

– self-condensation of α,β-unsaturated
  aldehydes 420
– umpolung allylation of 158
– vinylation of 25
aldol reaction 176, 317, 336, 368, 381, 410,
  429
– aldol-Tishchenko reaction 310, 337
– application in carbohydrate
  synthesis 413
– BINAP-transition metal catalyzed
  21ff
– cross-aldol reaction of nonequivalent
  aldehydes 411
– direct intermolecular 310
– ethyl pyruvate homoaldol reaction
  428
– intramolecular 412, 420, 428
– Mukaiyama 177, 301, 303ff,
  337, 413
– nitroaldol reaction 180f
– of acetone 426
– reductive 337
– with acetaldehyde 429
– with enolsilanes 176ff
– with silyl dienolates 23
– with silylketene acetals 176ff
aldol-type reaction 21, 301, 336
alkenylation
– of carbonyl compounds 25
– of imino compounds 25
alkylation 367
– allylic 18
– of α-heteroatom stabilized
  nucleophile 182
– of tetrasubstituted tin enolate with alkyl
  halide 288
– orthoester 30
alkyne
– [2+2+2] cycloadditon of 43

Privileged Chiral Ligands and Catalysts. Edited by Qi-Lin Zhou
Copyright © 2011 WILEY-VCH Verlag GmbH & Co. KGaA, Weinheim
ISBN: 978-3-527-32704-1

allene
- hydroamination of 17
- in allylic cyclization reactions 166
- in [4 + 1] cycloaddition 79
- in hydroacylations 78
allyl enol carbonate
- Tsuji allylation of 232
allylalcohol
- isomerization of 13
allylamine
- isomerization of 13, 125
allylation 301, 335
- catalytic cycle of BINOL promoted 324
- decarboxylative Tsuji 231ff, 250
- electronic discrimination of bonds in Pd (PHOX)allyl complexes 222
- of carbonyl compounds 24, 84, 325
- of imino compounds 24, 84
- of ketones 84
- umpolung, of aldehydes 158
allylether
- isomerization of 13
allylic alkylation 18, 87, 120, 343
allylic amination 228f, 344
allylic functionalization, see also allylation 127, 196
allylic oxygenation 196
allylic substitution 126, 342
- application in total synthesis 230
- on cyclic acetates 227
- palladium-catalyzed nucleophilic 224ff
- tungsten-catalyzed 229
- with nitromethane 228
aluminum-catalyzed asymmetric reaction
- using chiral Al/Li-BINOL complex 313f
- using chiral BINOL-Al complex 305, 321f
- using chiral BINOL-Al(III) complex 300
- using chiral salen-Al(III) complex 278, 289
aluminum/lithium/(BINOL)$_2$ catalyst (ALB catalyst), see also BINOL/Al/Li 312
amination 196, 385
- conjugate 193
- of carbonyl compounds 28
- of α-cyanoacetate 385
annulation reaction 425
arylation 253
- copper-catalyzed intramolecular 83
- of carbonyl compounds 25, 29, 156, 253
- of imino compounds 25
- rhodium-catalyzed 156
asymmetric catalysis
- acid-base cooperative 382ff
- base 377ff
- Brønsted acid-assisted Lewis acid (BLA) 304ff
- metal 363ff
- multifunctional cooperative 400
- nucleophilic 370ff
- phase-transfer 367ff
aza-Claisen rearrangement 251
aza-Henry reaction, see nitro-Mannich reaction
aza-Michael reaction 32
aza-Morita-Baylis-Hillman reaction 327, 416
aziridination 208

b

backbone modification of phosphacyclic ligand 60
Barbas catalyst, see pyrrolidine diamine as organocatalyst
BASPHOS, see also DuPhos-type chiral ligand, backbone modified 62
Bayer-Villiger oxidation 86
(Z)-α-(benzamido)cinnamic acid
- hydrogenation of 3
benzoquinine (BQN) 371
biaryldiphosphine 2
biaryldiphosphine as chiral ligand, see BINAP as chiral ligand
BINAL-H 296
- in reduction of ketones 296
- synthesis of 296
BINAP (2,2'-diphenylphosphino-1,1'-binaphthyl) 1
- derivatives of 2
BINAP as chiral ligand
- in aldol-type reactions 21
- in alkenylations of carbonyl compounds 25
- in alkenylations of imino compounds 25
- in allylations of carbonyl compounds 24
- in allylations of imino compounds 24
- in allylic alkylations 18
- in aminations of carbonyl compounds 28
- in arylations of carbonyl compounds 25, 29
- in arylations of imino compounds 25
- in aza-Michael reactions 32
- in conjugate addition using organoboron reagents 32
- in conjugate additions using Grignard reagents 32
- in cyanations of carbonyl compounds 27
- in cyanations of imino compounds 27
- in [2 + 2 + 2] cycloadditions 42
- in [3 + 2] cycloadditions 39

– in [5 + 2] cycloadditions 39
– in Diels-Alder reactions 35
– in dienylation of imino compounds 26
– in dienylations of carbonyl compounds 26
– in ene reactions 38
– in fluorinations of carbonyl compounds 28
– in hydroacylations 14
– in hydroaminations 14
– in hydroborations 14
– in hydrogenations of functionalized ketones 6
– in hydrogenations of olefines 3
– in hydrogenations of simple ketones 9
– in hydrosilylations 14
– in intermolecular Heck reactions 19
– in intramolecular Heck reactions 18
– in intramolecular reactions of enynes 38
– in isomerizations of allylalcohols 13
– in isomerizations of allylamines 13
– in Mannich-type reactions 21
– in Michael-type reactions 30
– in nucleophilic additions to carbonyl compounds 24
– in nucleophilic additions to imino compounds 24
– in orthoester alkylations of carbonyl compounds 29
– in Phauson-Kand-type reactions 43
– in ring-opening reactions 45
– in α-substitutions of carbonyl compounds 28
Binaphane, see also DuPhos-type chiral ligand, backbone modified 62, 75, 79f
1,1′-binaphthyl-2,2′-diol (BINOL) 295
– ligand structures derived from 296
Binapine, see also bisphosphacycle-based chiral ligand, P-chiral 65, 67, 72
BINAP-transition metal complex 2
BINAP-type ligands, see BINAP as chiral ligand 2f
BINOL, see 1,1′-binaphthyl-2,2′-diol
– chiral 7,7′-alkoxy-substituted 299
– in carbon-carbon forming reactions 300ff, 325ff
– in crotylation of ketones 325
– in organocatalysis 324
– in oxidation reactions 296
– in Petasis reaction 327
– in reduction reactions 296
– linked BINOL-catalysts, see also O-linked BINOL 316f
BINOL/aluminum
– in cyanation of aldehydes 321

– in cyanophosphorylation of aldehydes 324
– in hetero-Diels-Alder reaction 305
– in Strecker-type reaction 322
– in three-component Reissert-type reactions of N-heteroaromatic compounds 322f
BINOL/aluminum/lithium (ALB catalyst)
– in Michael reactions 313f
BINOL/barium
– in direct aldol reaction/isomerization reaction forming α-alkylidene-β-hydroxy esters 319
BINOL/indium
– in Diels-Alder reactions 306
– in direct addition of terminal alkynes to aldehydes 307
BINOL/lanthanum
– in epoxidation of α,β-unsaturated N-acylimidazoles 298
– in epoxidation of α,β-unsaturated N-acylpyrroles 298
– in expoxidation of enones 297
BINOL/lanthanum/lithium in Corey-Chaykovsky cyclopropanation 311
– in direct aldol/Tishchenko sequential reactions 310
– in intermolecular direct aldol reactions 310
– in one-pot epoxidation/ring-expansion reaction 312
BINOL/lithium
– in cyanation of aldehydes 320
– in direct Mannich-type reaction 321
BINOL/samarium
– in epoxidation of α,β-unsaturated amides 297
BINOL/titanium
– in 1,3-dipolar cycloadditions of nitrones 303
– in allylations 301
– in Diels-Alder reactions 302
– in ene reactions yielding glyoxylate esters 301
– in Mukaiyama aldol reactions 301
BINOL/vanadium
– in oxidative coupling of naphthols 300
BINOL/ytterbium
– in 1,3-dipolar cycloadditions of nitrones 307
BINOL/yttrium
– in epoxidaton of α,β-unsaturated esters 299
– in epoxidaton of α,β-unsaturated phosphine oxides 299

BINOL/yttrium/lithium
- in 1,4-addition of α-alkoxylamines 311
BINOL/zirconium
- in *anti*-selective Mukaiyama aldol reactions 305
- in Mannich-type reactions 303
BIPHEMP, *see also* BINAP-type ligands 2
bisbenzodioxanPhos, *see* SYNPHOS 2
bis-cinchona alkaloid chiral catalyst 364
- (DHQ)$_2$PHAL 364, 381
- (DHQD)$_2$AQN 364f, 374f, 378f
- (DHQD)$_2$PHAL 364f, 375f, 381f
- (DHQD)$_2$PHN 374f, 380
- (DHQD)$_2$PYR 364f, 380f
bisoxazoline scaffold 173f
bisoxazoline-based chiral ligand (Box)
- in addition of activated carbon nucleophiles to carbonyl electrophiles 182
- in addition of activated carbon nucleophiles to imino compounds 183
- in 1,4-additions of carbon nucleophiles to α,β-unsaturated acceptors 186
- in 1,4-additions of heteroatom nucleophiles to α,β-unsaturated acceptors 193ff
- in aldol reactions 176
- in allylic functionalization reactions 196
- in aminations 196
- in aziridinations 208
- in carbon-carbon bond formations 176ff
- in carbon-heteroatom bond formation 193ff
- in carbonylative cyclizations 199
- in chiral Lewis-acids 176ff
- in cyclizations 191, 199
- in cyclopropanations 205
- in desymmetrization reactions 201
- in Diels-Alder cycloadditions 202, 209, 211
- in ene reactions 184
- in Friedel-Crafts reactions 184
- in halogenations 197
- in hydroaminations 201
- in insertion reactions into heteroatom-hydrogen bonds 198
- in kinetic resolution reactions 201
- in Mannich-type reactions 177
- in nitroaldol (Henry) reactions 179
- in nitro-Mannich (aza-Henry) reactions 181
- in oxygenations 197
- in reactions of radicals alpha to carbonyls 190
- in rearrangement reactions 191
- in Wacker-type cyclizations 199
- molecular structures of transition metal complexes with 174f
- structures of 172, 178, 180ff, 188, 195, 199ff, 203, 207, 210
bisphosphacycle-based chiral ligand
- application of 65
- design of new 63
- development of 55
- in conjugate additions 85
- in cyclizations 78
- in cycloadditions 78
- in cycloisomerizations 78
- in hydroacylations 77
- in hydroformylations 74
- in hydrogenation of C=C bonds 65
- in hydrogenation of C=N bonds 73
- in hydrogenation of C=O bonds 72
- in hydrosilylations 76
- in nucleophilic additions to ketimines 82
- in nucleophilic additions to ketones 82
- in phosphinations 81
- P-chiral 65
- synthesis of 56ff
Bitianp, *see also* BINAP-type ligands 2
Bn-Box 177
BoPhoz, *see also* Josiphos-type chiral ligand 98
BPE 55ff
- structural features of 55
BPE-4, *see also* DuPhos-type chiral ligand, backbone modified 62
BPM 56f
Butiphane, *see also* DuPhos-type chiral ligand 60

c

C$_{1-6}$TunaPhos, *see also* BINAP-type ligands 2
candoxatril intermediate 67
carbapenem 8
carbohydrate 413
carbon dioxide
- addition to propylene oxide 271
- alternating copolymerization with racemic epoxides 272
carbon-carbon bond formation 120, 156, 176
- addition of dicarbonyl compounds to dienes 123
- three-component 159
- using metal/BINOL chiral Lewis acid catalysts 300ff
carbon-heteroatom bond formation 163ff, 193ff

carbonyl compound
– 1,2-addition to   386
– α-alkylation of   417
– amination of   28, 417
– α-aminoxylation of   352, 417
– arylation of   29
– fluorination of   28
– halogenation of   197, 344
– α-heteroatom functionalization   196
– nucleophilic addition to   24
– orthoester alkylation of   29
– rhodium-catalyzed arylation   156
– α-substitution reactions of   28
– synthesis of chiral β-substituted   32
(+)-carissone   237
cascade reaction   423, 435ff
– direct aldol/isomerization   319
– epoxidation/ring-expansion   312
– hetero-Michael-aldol-dehydration   437
– Michael-aldol reaction   436
– Michael-alkylation reaction   436
– Michael-Darzens-type reaction   438
– Michael-Michael reaction   437
– Michael-Michael-aldol condensation   439
– of α,β-unsaturated aldehyde catalyzed by diarylprolinol silyl ether   435ff
– oxo-Michael-substitution   435
cascade/one-pot reaction   418f, 428
(+)-cassiol   237
CatSium® ligand, *see also* DuPhos-type chiral ligand   61
chiral amine ligand
– dmapen   12
– iphan   12
– pica   12
chiral amplification   312
chiral spiro ligand
– design of   138
– in allene-based allylic cyclization reactions   166
– in arylation of carbonyl compounds   156
– in arylation of imino compounds   156
– in carbene insertion into heteroatom-hydrogen bonds   164
– in carbon-carbon bond formation   156ff
– in conjugate additions   158
– in hydrogenation of imines   154
– in hydrogenations of aldehydes   151
– in hydrogenations of enamides   144
– in hydrogenations of enamines   146
– in hydrogenations of ketones   151
– in hydrogenation of 2-substituted quinolines   154

– in hydrogenations of α,β-unsaturated acids   146
– in hydrosilylation   163
– in hydrosilylation/cyclization reactions   161
– in oxidative cyclization   162f
– in ring expanding cycloisomerization   163
– in ring opening   158
– in three-component coupling reaction   159
– in umpolung allylation of aldehydes   158
– preparation of   139
– spiro dinitrogen ligand   143
– with 1,6-disubstituted spiro[4.4]nonane backbone   138, 141, 143
– with 1,1'-spirobiindane backbone   138, 140
chlorolactonization   376
chromium-catalyzed asymmetric reaction
– using chiral salen-Cr(III) complexes   266, 281, 287
cinchona alkaloid
– application in total synthesis   369, 375
– chiral bifunctional catalysts based on   388ff
– chiral metal catalysts based on   362
– chiral phase transfer catalysts based on   362
– in 1,2-additions to carbonyl compounds   386
– in 1,2-additions to imino compounds   390
– in alcoholysis of cyclic anhydrides   377
– in aldol reactions   368, 381
– in alkylations   367
– in asymmetric acid-base cooperative catalysis   382ff
– in asymmetric base catalysis   377ff
– in asymmetric base-iminium cooperative catalysis   396ff
– in asymmetric metal catalysis   363
– in asymmetric nucleophilic catalysis   370ff
– in asymmetric phase-transfer catalysis   367ff
– in catalytic peroxidations   400
– in conjugate additions   368, 375, 380, 382, 396
– in cyanations of simple ketones   374
– in Diels-Alder reactions   393, 399
– in electrophilic halogenations of olefines   376
– in fragmentations   394
– in Friedel-Crafts reactions   391, 398
– in Mannich reactions   381

cinchona alkaloid (*Continued*)
- in Morita-Baylis-Hillman reactions 373
- in multifunctional catalysis 400
- in reactions with ketenes 370
- in semipinacol-type 1,2-carbon migration 399
- in tandem conjugate addition-protonation reactions 400

cinchonidine (CD) as chiral catalyst 362, 368f

cinchonine (C) as chiral catalyst 362, 378, 383f, 390

Cl,MeO-BIPHEP, *see also* BINAP-type ligands 2

CnrPHOS, *see also* DuPhos-type chiral ligand, backbone modified 62

cobalt-catalyzed asymmetric reaction
- using chiral biaryldiphosphine-Co complexes 44
- using chiral Co-PHOX complexes 249
- using chiral salen-Co(II) complexes 276, 283
- using chiral salen-Co(III) complexes 266, 270ff

conjugate addition, *see also* 1,4-addition 85, 277, 375, 380, 382, 396
- copper-catalyzed asymmetric 158
- of amines 193
- of carbamates 194
- of glycine Schiff bases to carbonyl compounds 368f
- of hydrazine 194
- of hydrogen cyanide 278
- of hydrophosphines to α,β-unsaturated aldehydes 433
- of nitrile to α,β-unsaturated imides 280
- of radicals 190
- of substituted olefine to nitroalkene 384
- of weakly acidic nitrogen nucleophiles 277f
- sulfa- 380
- to electron-deficient olefines 339
- using Grignard reagents 32, 121, 341
- using organoboron reagents 32

conjugate reduction
- of activated C=C bonds 110

copolymerization
- alternating 272

copper-catalyzed asymmetric reaction
- carbene insertion into heteroatom-hydrogen bonds 164
- conjugate addition 158
- conjugate reduction of activated C=C bonds 110

- ring opening with Grignard reagents 158
- using chiral BINAP-Cu(I) complexes 22f, 34f
- using chiral Cu(I)/Box complexes 196, 207
- using chiral Cu(II)-PHOX complexes 247
- using chiral DuPhos-Cu complexes 83, 87
- using chiral Josiphos-Cu complexes 111, 118, 120, 123
- using chiral Lewis-acidic Cu(II)/Box complexes 176ff, 192, 196, 201ff, 210
- using chiral SEGPHOS-Cu complexes 16, 25, 36f
- using chiral spiro ligands 158, 166
- using chrial Cu(I)/spirobisoxazoline complexes 198f
- using Cu complexes with chiral spiro ligands 158, 164f

cross silyl benzoin reaction 340
cross-aldol reaction 411
- ketone-ketone 412

cupreidine 383, 394
cupreine 383, 394

cyanation
- of aldehydes 320f
- of imino compounds 27
- of simple ketones 374

cyanohydrin
- asymmetric synthesis of 284, 338

cyanophosphorylation 324
cyanosilylation 27
(−)-cyanthiwigin F 236

cyclic anhydrid
- alcoholysis of 377

cyclization, *see also* cycloaddition 38, 78, 191
- allylic allene based 166
- carbonylative 199
- for carbon-heteroatom bond formation 199
- intramolecular aldol reaction 412
- intramolecular Heck reaction 18, 122, 238
- intramolecular hydroacylation 77
- intramolecular reactions of enynes 38
- intramolecular α-alkylation of aldehydes 421
- Nazarov 191f
- of alkynes 43
- of dieneynes via [4 + 2] cycloisomerization 79
- of enynes 39, 42f
- of silyloxy-1,6-enynes 80
- oxidative 162f
- Passerini three-component reaction 288
- Phauson-Kand-type reaction 43
- via cascade reaction 435ff

– Wacker-type cyclization 199
cycloaddition 78, 202ff, 211, 345ff, 433
– [1+2] 435
– [2+2+2] 42, 348, 438
– [3+2] 39, 246, 435f
– [3+3] 210, 437
– [4+1] 79
– [4+2], see also Diels-Alder reaction 123, 437
– [5+2] 39
– 1,3-dipolar 209, 303, 307, 348, 428
– aziridination 208
– cyclopropanation 205, 207, 274, 311, 205, 348, 375
– Diels-Alder 35, 202, 247, 281, 346, 393, 399
– hetero-Diels-Alder 36, 204, 347, 434
– of vinylallene and carbon monoxide 79
– of vinyloxiranes and carbodiimides 40
cyclodimerization of oxabenzonorbornadienes 39
1,2-cyclohexadiamine 258
cycloisomerization 78ff
– of imidazoles 81
– ring expanding 163, 205, 207, 274, 311, 375
– chiral Cu(I)-bisoxazoline complexes in 207
– Corey-Chaykovsky 311
– intramolecular 208
– of α,β-unsaturated aldehydes 421, 436

d
DBFOX 172, 188, 197, 203, 210
decarboxylative protonation, see also Tsujii allylation 249
dehydroamino acid
– hydrogenation of derivatives of 65, 67f, 104ff, 144
(R)-desmethylsibutramine 183
desymmetrization 201, 251
– of anhydrides 123, 252
– of diols 202
– of meso-epoxides 266
– of meso-epoxy diols 268
dextromethorphane 116
diarylprolinol
– as chiral catalyst in aldol reactions with acetaldehyde 429
– as chiral catalyst in cycloadditions 433, 435ff
– as chiral catalyst in α-functionalization reactions of aldehydes 429
diarylprolinol silyl ether

– as chiral catalyst in cascade reactions 435ff, 439
– as chiral catalyst in cycloadditions 433
– as chiral catalyst in α-functionalization reactions of aldehydes 429
(+)-dichroanone 235
Diels-Alder reaction 202, 204, 247, 281, 302, 306, 346f, 393, 399
– application in kinetic resolution of α, β-unsaturated pyrazolidinone imides 205
– BINAP-transition metal catalyzed 35f
– hetero- 36, 204, 281f, 302, 305, 347, 434
– nitroso 36
– with 2-pyrones as diene 393, 399
dienylation
– of carbonyl compounds 26, 123
– of imino compounds 26
Difluorphos 2, 25
dihydroquinidine (DHQD) as chiral catalyst 363, 366
2,2'-diphenylphosphino-1,1'-binaphthyl, see BINAP
DiSquareIP*, see also bisphosphacycle ligand, P-chiral 65
DKR, see kinetic resolution, dynamic
DTB-BINAP 2
DTBM-BINAP 2
DuanPhos, see also bisphosphacycle-based chiral ligand, P-chiral 65, 67, 70, 72f, 86
(S)-duloxetine 72f
DuPhos 55
DuPhos-type chiral ligand
– backbone modified 62
– structural features of 55
– synthesis of 56, 60
duthixantphospholane, see also DuPhos-type chiral ligand 60
duxantphospholane, see also DuPhos-type chiral ligand 60
dynamic kinetic asymmetric transformation (DYKAT) 320

e
enamide
– hydrogenation of 67, 104, 144
β-enamido phosphonate
– hydrogenation of 107
enamine
– hydrogenation of 146
enantiofacial selection 11f, 29
ene reaction 38
– catalyzed by chiral BINOL-Ti(IV) complex 301
– glyoxylate- 184f

enol ester
- hydrogenation of 67
enone
- annulation with ketone 425
- conjugate addition of 2-thienylzinc reagents to 85
- conjugate addition of Grignard reagents to 35, 341
- conjugate addition of organoboron reagents to 33, 85
- conjugate additon of alkenyl metal reagents to 35
- 1,4-disilylation of 15
- epoxidation of 297
- hydrosilylation of 16
- sulfa-conjugate addition to 380
epoxidation 262
- chiral salen-Mn catalysts for 263f
- Corey-Chaykovsky, of ketones 312
- of enones 297, 349
- of α,β-unsaturated amides 297f
- of α,β-unsaturated esters 299
- of α,β-unsaturated phosphines 299
epoxidation/ring-expansion reaction 312
epoxide
- alternating copolymerization with carbon dioxide 272
- desymmetrization of meso- 266
- formed by asymmetric epoxidation 264
- homopolymerization of 273
- kinetic resolution of racemic 269

f
f-Binaphane, see also DuPhos-type chiral ligand, backbone modified 62, 73
FerroTANE, see also DuPhos-type chiral ligand, backbone modified 62, 70
fluorination
- of carbonyl compounds 28
(S)-fluoxetine 72f
fragmentation 394
Friedel-Crafts reaction 26, 184, 391
- of aromatic compounds with C=O and C=N bounds 184
- of indoles 391f, 398
α-functionalization of carbonyl compounds
- alkylation 288, 417
- amination 417
- aminoxylation 352, 417
- substitution 28
FuP(spiro phosphinite ligand) 140, 145f

g
gold-catalyzed asymmetric reaction
- ring expanding cycloisomerization 163

- using Au complexes with chiral spiro ligands 163
- using SEGPHOS-Au complexes 40
Grignard reagent 338
- in conjugate additions 32, 121, 341
- in copper-catalyzed ring opening 158
group IV metal/BINOL chiral Lewis acid catalyst 300
group XIII metal/alkali metal/BINOL catalys 312
group XIII metal/BINOL chiral Lewis acid catalyst 304

h
H_4-BINOL 296, 302
H_8-BINAP 2, 4, 43
H_8-BINOL 296, 302
Hajos-Parrish-Eder-Sauer-Wiechert reaction 412
halogenation
- of carbonyl compounds 197, 344
- of olefines, asymmetric electrophilic 376
Heck reaction 122
- intermolecular 19, 237
- intramolecular 18, 122, 238f
Henry reaction, see nitroaldol reaction
hetero-Michael-aldol-dehydration process 437f
homopolymerization of epoxides 273
hydroacylation 14
- intramolecular 77
hydroamination 14, 201
hydroboration 14, 87, 118, 352f
hydrocarboxylation 120
hydroformylation 74
hydrogenation
- of aryl alkyl ketones 73
- of C=C bonds 65, 104, 111
- of C=N bonds 73, 114
- of C=O bonds 72, 111
- of α-dehydroamino acid derivatives 65, 67, 105
- of enamides 67, 144
- of β-enamido phosphonates 107
- of enamines 146
- of enol esters 67
- of functionalized ketones 6, 111ff
- of heteroarenes 117
- of imino compounds 154, 240
- of β-keto esters 7, 72
- of nitroalkenes to chiral nitro compounds 111
- of non-functionalized ketones 113, 151
- of racemic 2-substituted aldehydes via DKR 153

– of racemic 2-substituted ketones via DKR 151
– of simple ketones 9, 152, 244, 297
– of 2-substituted quinolines 154
– of tetrasubstituted olefines 109, 243
– of trisubstituted olefines 240
– of α,β-unsaturated carboxylic acid derivatives 4, 107, 146, 149, 351
– of α,β-unsaturated ketones 244
– of α,β-unsaturated nitriles 111f
– of vinyl phosphonates 244
– transfer hydrogenation 113, 246
– using chiral PHOX ligands 240ff
– using chiral spiro ligands 144ff
hydrophosphonation 119
hydrosilylation 14, 76, 163, 248, 350
– of enones 16
hydrosilylation/cyclization 161
hydrovinylation 161
4-hydroxy-2-cyclopentone
– isomerization of 14
4-hydroxyproline as organocatalyst 419
– supported 421

i
imino compound
– addition of arylboronic acids to 155
– addition of carbon nucleophiles to 183
– addition of organolithium compounds to 183
– 1,2-addition to 390
– allylation of 326
– arylation of 25
– hydrogenation of 74, 154, 240
– hydrosilylation of 16
– nucleophilic addition to 24, 82
– radical addition to 184
– rhodium-catalyzed arylation 156
– vinylation of 25
indium-catalyzed asymmetric reaction
– using chiral BINOL-In catalyst 306
insertion
– carbene 164
– into heteroatom-hydrogen bonds 198
iridium-catalyzed asymmetric reaction
– allylic substitution 229
– using chiral BINAP-Ir(I) complexes 16, 44
– using chiral bisphosphacycle-Ir complexes 73
– using chiral Ir-PHOX complexes 229f, 240ff, 248
– using chiral Josiphos-Ir complexes 114, 120

– using chiral salen-Ir(III) complexes 276
– using Ir complexes with chiral spiro ligands 146ff, 154f
iron-catalyzed asymmetric reaction
– using chiral Fe(III)/Box complexes 202
iso-cuperidine (β-ICD) 373, 383
isomerization 125
– of allylalcohol 13
– of allylamine 13, 125
– of allylether 13
– of 4,7-dihydro-1,3-dioxepines 87
itaconic acid
– hydrogenation of 104

j
Jacobsen's catalyst 263
Josiphos-type chiral ligand
– application of 95
– CAS registry number and name of 96
– conformational space of
– development of 94
– discovery of 94
– functionalized 95
– geometrical parameters of transition metal complexes with
– in allylic alkylations 120
– in allylic substitutions 126
– in carbon-carbon bond formations 120
– in conjugate reductions of activated C=C bonds 110
– in Heck reactions 122
– in hydroaminations 119
– in hydroborations 118
– in hydrogenations 104ff
– in hydrophosphonations 120
– in isomerization of allylamines 125
– in Michael additions 121
– in non-enantioselective reactions 127
– in ring-opening of oxabicycles 125
– structure-reactivity relationship in catalytic metal complexes 97f
– technical synthesis of 95

k
Kephos, *see also* DuPhos-type chiral ligand 60
KetalPhos, *see also* DuPhos-type chiral ligand, backbone modified 62
ketene
– [2 + 2] reaction of 370ff
– [4 + 2] reaction of 373
ketimine, *see* imino compound
α-ketoester
– cyanosilylation of 27

β-ketoester
– α-amination of 28
– conjugate addition to alkynones 381
– hydrogenation of 7f, 72
– hydrogenation of α-substituted 113
– Tsuji allylation of 232f
– Tsuji allylation of α-fluorinated 235
ketone
– addition of allylcyanid to 84
– addition of allyltrimethoxysilane to 25
– addition of arylboronic acid to 155
– addition of Grignard reagents to 338
– allylation of 325
– annulation with enone 425
– Chorey-Chaykovsky epoxidation of 312
– crotylation of 325
– cyanation of 374
– hydrogenation of 6ff, 151f, 296
– hydrogenation of α,β-unsaturated 244
– hydrosilylation of 16
– intramolecular addition of enamide to 288
– nucleophilic addition to 82
– reduction of 296, 351
– transfer hydrogenation of 114, 246
– α-amination of 417
– α-aminoxylation of 418
Kharasch-Sosnovsky reaction 196
kinetic resolution 201
– by catalytic, asymmetric Diels-Alder reaction 205
– by cinchona-alkaloid catalyzed alcoholysis 378
– by pyrrolidine diamine organocatalysts 428
– dynamic 151f
– of ene-epoxides 124
– of racemic 2-substituted aldehydes 153
– of racemic 2-substituted ketones 151
– of racemic primary alcohols 428
– of racemic terminal epoxides 269
Kornblum-DeLaMare reaction 394

l

Lewis acid/Lewis base bifunctional aluminium catalyst 321
ligand design
– backbone variation 56
– of bisphosphacycle ligands 63
– strategies of 56
LLB complex, *see* rare-earth/alkali metal-BINOL complex
L-proline 409
– properties of proline-based catalysts 409f

m

magnesium-catalyzed asymmetric reaction
– using chiral Lewis-acidic Mg(II)/Box complexes 183, 188, 193, 202f, 210
– using chiral salen-Mn(III) complexes 261, 263
Mannich reaction 381, 414, 422, 429f
– BINAP-Pd catalyzed 24
– direct *anti*-selective 318, 421ff
– direct *syn*-selective 318
– Mukaiyama- 302
– using *N*-tosyl-α-imino ester 179
Mannich-type reaction 23, 83f, 177, 303, 318, 321
manzacidin 401
mechanism
– of asymmetric allylation catalyzed by BINOL derivatives 324
– of asymmetric epoxidation catalyzed by salen-Mn(III) complexes 265f
– of base-promoted epoxidation of α,β-unsaturated ketones 402f
– of BINAP-Rh(I) catalyzed asymmetric [3 + 2] cycloaddition 41
– of BINAP-Rh(I) catalyzed asymmetric 1,4-addition 33f
– of BINAP-Ru(II) catalyzed hydrogenation of methyl (Z)-α-acetamidocinnamate 4f
– of BINAP-Ru(II) catalyzed hydrogenation of β-keto esters 7f
– of cinchona-alkaloid catalyzed asymmetric alcoholysis 382, 384
– of hydrolytic kinetic resolution catalyzed by chiral salen-Co(III) complex 270
– of the coupling of propylene oxide and carbon dioxide catalyzed by salen-Co(III) complex 271
– of TolBINAP/DPEN-Ru(II) catalyzed hydrogenation of simple ketones 10f
– of trimethylsilylcyanation of aldehyde catalyzed by chiral salen-Ti complex 286
Me-f-KetalPhos, *see also* DuPhos-type chiral ligand, backbone modified 62
Meldrum's acid 250
(–)-menthol 13
MeO-BiPHEP, *see also* BINAP-type ligands 2, 26, 44
metal/BINOL complex as acid/base bifunctional catalyst 316
methyl (Z)-α-acetamidocinnate
– hydrogenation of 5, 58
(+)-*cis*-methyl dihydrojasmonate 109
(S)-metolachlor (Dual Magnum®) 115
Michael addition 121, 313, 318, 415

- BINAP-transition metal catalyzed 31
- Mukaiyama- 186f
- of aldehydes 430
- of aldehydes and nitroalkenes 431
- of electron-rich aromatic compounds to α,β-unsaturated acceptors 189
- of N-heterocycles to α,β-unsaturated aldehydes 432
- of nitroalkanes to cycloalkenones 416
- of trisubstituted nucleophiles to α,β-unsaturated acceptors 385
- oxa-Michael addition of enals with oximes 432
- through an iminium pathway 430
- through enamine pathway 430

Michael-aldol reaction 437
Michael-Darzens-type reaction 438
Michael-Michael reaction 437
Michael-Michael-aldol condensation reaction 439
Michael-type reaction 30
modification of phosphacycles 60
Morita-Baylis-Hillman reaction 373, 416
MPL-SegPhos, *see also* DuPhos-type chiral ligand 60

n

Nazarov cyclization 191f
nickel-catalyzed asymmetric reaction
- hydrovinylation 161
- three-component coupling reaction 159
- using chiral BINAP-Ni(II) complexes 29, 31
- using chiral DuPhos-Ni complexes 86
- using chiral Josiphos-Ni complexes 113
- using chiral Ni complexes with chiral spiro ligands 159ff
- using chiral Ni(II)/Box complexes 197f, 203, 210
- using chiral Ni-PHOX complexes 253
- using chiral spiro ligands 159, 161

nigellamine A_2 231
nitroaldol reaction 179, 390
nitro-Mannich reaction 181
nitrone 209, 303
nucleophilic addition 82
- to carbonyl compounds 24, 334ff
- to imino compounds 24
- to ketimines 82
- to ketones 82
nucleophilic substitution 342ff

o

olefin
- conjugate additions to electron-deficient 339
- electrophilic halogenation of 376
- Heck-type alkenylation of 20
- hydroamination of 17
- hydroboration of 87, 118, 353
- hydrogenation of functionalized 4
- hydrogenation of tetrasubstituted 109, 243
- hydrogenation of trisubstituted 240
- hydrosilylation of 15, 163 + C71

O-linked chiral BINOL
- in direct aldol reactions 317
- in Mannich-type reactions 318
- in Michael reactions 316ff

organoboron reagent
- addition of arylboronic acids to carbonyl compounds 155f
- in BINOL-promoted carbon-carbon bond formations 326f
- in conjugate additions 32

organozinc compound
- nucleophilic addition to aldehydes 287, 334ff

orthoester alkylation
- of carbonyl compounds 29

osmium-catalyzed asymmetric reaction
- using chiral Josiphos-Os complexes 114
- using chiral Os-cinchona alkaloid complexes 363

oxidation
- catalytic peroxidation 400
- enolate 197
- using chiral BINOL 296
- using chiral TADDOLate ligands 349

oxidative coupling 300
oxygenation
- allylic 196f
- of carbonyl compounds 197

p

palladium-catalyzed asymmetric reaction
- allylic substitution 224
- hydrosilylation 163
- oxidative cyclization 162
- umpolung allylation of aldehydes 158
- using chiral biaryldiphosphine-Pd(II) complexes 28, 38f
- using chiral BINAP-Pd(II) complexes 15, 17ff, 21, 23, 29f, 32, 35, 40
- using chiral bisphosphacyle-Pd(II) complexes 74, 80
- using chiral DuPhos-Pd(II) complexes 81, 85ff
- using chiral Josiphos-Pd complexes 120, 123
- using chiral Pd(II)/Box complexes 191, 199f, 203

palladium-catalyzed asymmetric reaction (*Continued*)
– using chiral Pd-PHOX complexes 224ff, 231ff, 247, 250
– using chiral Pd-TADDOLate complex 342f
– using chiral PHOX ligands 224
– using chiral spiro ligands 158, 162f
– using Pd complexes with chiral spiro ligands 162f, 166
– Wacker-type oxidative cyclization 163
palladium(PHOX)allyl complex 222
– electronic bond discrimination in 222
Passerini three-component reaction 288
P-chiral phospholane ligand 65f
PennPhos, *see also* DuPhos-type chiral ligand, backbone modified 62, 69, 71, 73
peroxidation 400
Petasis reaction 327
phase-transfer catalysis
– of asymmetric aldol reaction 369
– of asymmetric α-alkylation of carbonyl compounds 367
– of asymmetric α-alkylation of imino compounds 368
– of conjugate addition 368
Phauson-Kand-type reaction 43, 248
Ph-Box 177, 182ff, 192, 201ff
1,2-phenylethylene-1,2-diamine 258, 261
phosphination 81
2-(2-phosphinoaryl)oxazoline (PHOX) 223
PHOX, *see* 2-(2-phosphinoaryl)oxazoline
PHOX-type chiral ligand
– in asymmetric arylations 253
– in [3 + 2] cycloadditions 246
– in decarboxylative protonations 249
– in decarboxylative Tsuji allylations 231ff
– in Diels-Alder reactions 247
– in hydrogenation of imines 240
– in hydrogenation of ketones 244f
– in hydrogenation of tetrasubstituted olefines 240
– in hydrogenation of trisubstituted olefines 240
– in hydrogenation of vinyl phosphonates 243
– in hydrosilylations 248
– in intermolecular Heck reactions 238
– in intramolecular Heck reactions 237
– in nucleophilic allylic substitutions 224ff
– in Phauson-Kand-type reactions 248
– in sigmatropic rearrangements 250
– in transfer hydrogenation 246
– NeoPHOX-Ir catalyst for olefine hydrogenation 243

– SimplePHOX-Ir catalyst for olefine hydrogenation 242
– structural features of 223
– structure of 224, 227, 242f
– synthesis of 222f
– ThrePHOX-Ir catalyst for olefine hydrogenation 242
Ph-pyrazine, *see also* DuPhos-type chiral ligand 60
Ph-quinoxaline, *see also* DuPhos-type chiral ligand 60
platinum-catalyzed asymmetric reaction
– using chiral BINAP-Pt(II) complex 21
– using chiral DuPhos-Pt complexes 82, 86
PMHS, *see* polymethylhydrosiloxane
polymethylhydrosiloxane (PMHS) 16, 110
ppfa 95
*i*Pr-Box 178, 191, 203
pregabalin intermediate 67
proline analogs as chiral organocatalysts 419ff
proline as chiral organocatalyst
– in α-alkylations of carbonyl compounds 417
– in α-aminations of carbonyl compounds 417
– in α-aminoxylations of carbonyl compounds 417
– application in carbohydrate synthesis 413
– in cascade/one-pot reactions 418
– in intermolecular aldol reactions 410
– in intramolecular aldol reactions 412
– in Mannich reactions 414
– in Michael additions 415
– in Morita-Baylis-Hillman reactions 416
propylene oxide
– addition of carbon dioxide to 271
pseudoenantiomeric pair 361f
3-pyrrolidine carboxylic acid
– as organocatalyst for *anti*-selective Mannich reaction 423
pyrrolidine diamine
– as chiral organocatalyst 427
pyrrolidine-based amide
– as chiral organocatalyst 425
pyrrolidine-based sulfonamide
– as chiral organocatalyst 424
pyrrolidine-derived catalyst 410
5-pyrrolidin-2-yltetrazole as organocatalyst 422

q
quinidine (QD) catalyst 362, 370, 375, 377f, 383, 390ff

quinine (Q) 362, 383, 396, 400
quinoline
– 2-substituted, hydrogenation of 154

r
R-5-Fc, *see also* DuPhos-type chiral ligand 60
radical addition
– conjugate 190
– tandem 191
– to imino compounds 184
rare-earth metal/alkali metal-BINOL complex (REMB complex) 308
– catalytic asymmetric reactions using 309ff
– lanthan/lithium-BINOL complex (LLB complex) 308ff
rare earth metal/BINOL chiral Lewis acid catalyst 307
rare-earth-metal-catalyzed asymmetric reaction
– using rare earth metal/alkali metal-BINOL complexes (REMB complex) 308ff
– using rare earth metal-BINOL complexes 297ff, 307
rearrangement reaction
– Claisen- 191f
– semipinacol-type 1,2-carbon migration 399
– [3,3]-sigmatropic 250f
– vinylogous α-ketol 399
– [2,3]-Wittig- 191f
reduction, *see also* hydrogenation
– of α-hydroxy ketone via hydrosilylation 77
– using chiral BINOL 296
– using chiral TADDOLate ligands 349
REMB complex, *see* rare-earth/alkali metal-BINOL complex
rhodium-catalyzed asymmetric reaction
– arylation of carbonyl compounds 156
– arylation of imino compounds 156
– BINAP-Rh(I) catalyzed enantioselective reactions 3, 13ff, 33, 39ff
– enantioselective hydrogenation of α-(acylamino)acrylic acids 3
– hydrosilylation/cyclization reaction 161
– using biaryldiphosphine-Rh(I) complexes 26
– using chiral BPE-Rh(I) complexes 67ff, 75
– using chiral diazaphospholane ligands 75
– using chiral DTBM-SEGPHOS-Rh(I) complex 16
– using chiral DuPhos-Rh(I) complexes 66ff, 77ff, 85
– using chiral Josiphos-Rh complexes 105ff
– using chiral Rh complexes of chiral spiro ligands 144ff, 156, 161
– using chiral Rh-PHOX complexes 248, 252
– using chiral Rh-TADDOLate ligands 337, 350, 353
– using chiral spiro ligands 156, 161
ring formation, *see also* cyclization, cycloaddition, cyclopropanation
– of five-membered rings by a [3+2] strategy 436
– of six-membered rings by a [2+2+2] strategy 413, 438
– of six-membered rings by a [3+3] strategy 437
– of six-membered rings by a [4+2] strategy 436
– of three-membered rings by a [1+2] strategy 435
ring-expanding reaction 163
ring-opening reaction 45, 125
– application to kinetic resolution of ene-epoxides 124
– as desymmetrization reaction 251f, 267f
– copper-catalyzed 158
– of azabicycles 252
– of epoxides 266ff
– of *meso*-epoxides with trimethylsilylazide 266
– of *meso*-oxabicyclic alkenes with DIBAL-H 45
– of oxabicycles 125, 252
– of oxetanes, intramolecular 268
RoPHOS, *see also* DuPhos-type chiral ligand, backbone modified 62
ruthenium-catalyzed asymmetric reaction
– application of chiral BINAP-Ru(II) catalyzed hydrogenation in carbapenem synthesis 8
– hydrogenation of functionalized ketones 6
– hydrogenation of functionalized olefines 4
– hydrogenation of simple ketones 9
– using chiral biaryldiphosphine-Ru(II) complexes 6
– using chiral BINAP-Ru(II) complexes 3ff, 27
– using chiral DuPhos-Ru complexes 73
– using chiral Josiphos-Ru complexes 109, 114
– using chiral Ru-PHOX complexes 246
– using chiral salen-Ru complexes 276f
– using Ru complexes with chiral spiro ligands 147, 151f

salalen as chiral ligand 289
salan as chiral ligand 289
salen metal complex as chiral catalyst
– conformation of 260f
– in additions of carbon dioxide to propylene oxide 271
salen, see N,N'-bis(salicylidine)-ethylenediamine 257
salen-based chiral ligand
– synthesis of 258
salen-metal complex as chiral catalyst
– catalytic activity of 266, 271
– in alternating copolymerization of racemic epoxides and carbon dioxide 272
– in conjugate additions 277
– in cyanohydrin synthesis 284
– in cyclopropanation 274
– in desymmetrization reactions of meso-epoxides 266
– in Diels-Alder reactions 281
– in epoxidations 262ff
– in homopolymerization of epoxides 273
– in kinetic resolution reactions of racemic epoxides 269
– in ring-opening reactions of epoxides 266ff
– structural properties of 259
– structure of 264, 275f, 279f, 284f
– synthesis of 257
– trans-configurated 259f
SDP (spirobiindane diphosphine ligand) 140, 151, 161f
SDPO (spiro diphosphite ligand) 140, 155
SEGPHOS, see also BINAP-type ligands 2, 16, 25, 28, 36, 38
SFDP (spirobifluorene diphosphine ligand) 141, 147
Sharpless asymmetric dihydroxylation (AD) 363
– effective cinchona alkaloid ligands for 365
ShiP (spiro phosphite ligand) 140, 156
silver-catalyzed asymmetric reaction
– using chiral Ag(I)-PHOX complexes 246
– using chiral BINAP-Ag complex 22, 24
silyl enolether 23, 176, 233f
SIPHOS (spiro phosphoramidite ligand) 140, 144f, 158
SIPHOX (spiro phosphine-oxazoline ligand) 140, 148f, 154
SITCP (spiro phosphine ligand) 140, 158f
SPINOL (1,1'-spirobiindane-7,7'-diol) 139
SpinPHOX (spiro phosphine-oxazoline ligand) 149, 154
1,1'-spirobiindane 138f

SpiroBIP (spiro diphosphinite ligand) 143f
SpiroBOX (spiro bis(oxazoline) ligand) 140, 164, 166
spiro[4.4]nonane 138
SpiroNP (spiro bisphosphinamidite ligand) 143f
SpirOP (spiro phosphinite ligand) 138, 144ff
SPRIX (spiro bis(isoxazoline) ligand) 143, 158, 162f
Strecker reaction 322, 352
structure-related property
– of BINAP ligands 1
– of BINAP-transition metal complexes 1f
– of BPE 55
– of chiral salen complexes 259
– of DuPhos 55
– of PHOX ligands 221ff
– of TADDOL derivatives 333
– of transition metal complexes containing Josiphos ligands 99f
SYNPHOS, see also BINAP-type ligands 2
synthesis
– industrial pilot process for dextromethorphane 116
– industrial process for (S)-Metolachlor 115
– of N-acyl-α-amino acid esters 190
– of α-alkylidene-β-hydroxy esters 319
– of Al/Li/(BINOL)$_2$ catalyst (ALB catalyst) 313
– of an HIV integrase inhibitor 117
– of axially chiral biaryls from alkynes 43
– of benzo(thio)pyran derivative 438
– of BPE ligands 56
– of carbapenem intermediate 8
– of carbohydrates by proline-catalyzed aldol reactions 413
– of (+)-carissone 237
– of (+)-cassiol 237
– of (–)-cyanthiwigin F 236
– of chiral alcohols 14, 67, 71
– of chiral alcohols via asymmetric hydrogenation of enol esters 67
– of chiral amines with tetrasubstituted stereocenters 385
– of chiral aryl amine 67ff, 74f, 144ff, 193
– of chiral aryl phosphine 81
– of chiral bisazaphospholane ligand 64
– of chiral cyclopropane 375
– of chiral indanoles 124
– of chiral oxetane 312
– of chiral pyrrolidine 437
– of chiral salen complexes 257
– of chiral salen ligands 258
– of chiral salen-metal complexes 258

- of chiral spiro ligands 139
- of chiral spiro ligands with 1,1'-spirobiindane backbone 139f
- of chiral spiro stereocenters 80
- of chiral tetrasubstituted cyclohexene carbaldehydes 439
- of chiral α-chloro aldehyde 426
- of chiral α-hydroxyesters via asymmetric insertion of α-diazoesters 164
- of chiral β-substituted carbonyl compounds 32
- of chiral γ-lactam with quaternary α-carbon center 287
- of cis-β-salen-metal complexes 262
- of cyanohydrin, asymmetric 284, 338
- of (+)-dichroanone scaffold via asymmetric Tsujii allylation 235
- of (R)-desmethylsibutramine 183
- of (S)-duloxetine 72f
- of DuPhos 56
- of ecteinascidin 743 68
- of enantioenriched tert-alkylamines via desymmetrization 201
- of five-membered N-heterocycle 433ff
- of (S)-fluoxetine 72f
- of hydroquinoline derivative 438
- of Josiphos ligands 95
- of manzacidin 401
- of medium-sized rings by intramolecular hydroacylation 78
- of (−)-menthol 13
- of (+)-cis methyl dihydrojasmonate 109
- of natural products unsing ALB-catalyzed Michael-aldol reactions 315
- of nigellamine A_2 231
- of optically active α-alkyl α-aryl γ-butyrolactones 29
- of P-chiral phospholanes 66
- of PennPhos 63
- of Ph-5-Fc 60
- of Ph-BPE 57
- of Ph-BPM 57
- of PHOX ligands 222f
- of 2-piperazinecarboxylic acid derivatives as crixivan intermediate 105
- of planar-chiral paracyclophanes 80
- of polycarbonates catalyzed by salen-Co(III) complex 272
- of polypropylene 274
- of sanglifehrin A 68
- of silica-containing compounds via carbene insertion 165
- of SPINOL 139
- of steroid intermediate 412
- of α-substituted β-amino acids 279
- of β-substituted γ-aminobutyric acids 279
- of substituted cyclopentanones by intramolecular hydroacylation 77
- of substituted tetralone derivatives via [4 + 2] cycloaddition 123
- of α-trialkylsiloxyketone 340
- of (S)-warfarin 396f
- of Wieland-Miescher ketone 412

t

TADDOL, see tetraaryl-1,3-dioxolane-4,5-dimethanol
TADDOLate as chiral ligand
- in aldol-type reactions 336
- in allylations 335
- in allylic substitutions 342
- in α-aminoxylations of carbonyl compounds 352
- in cycloadditions 345ff, 348
- in Diels-Alder reactions 346
- in α-halogenations of carbonyl compounds 344
- in hetero-Diels-Alder reactions 347
- in nucleophilic additions to C=O bonds 334ff, 338
- in nucleophilic conjugate addtions to electron-deficient C=C bonds 339
- in nucleophilic substitutions 342ff, 345
- in organozinc additions to aldehydes 334
- in oxidation reactions 349
- in reduction reactions 349
- in transesterification 351
TADDOL-derive chiral phosphorus ligand 339ff
TADDOL-phosphite-oxazoline ligand 342
TADOOH ligand 349f
tandem radical reaction 191
TangPhos, see also bisphosphacycle ligand, P-chiral 65, 67, 70f, 74f, 82, 87
Taniaphos, see also Josiphos-type chiral ligand 98
t-Bu-Box 176ff, 183f, 190ff, 197, 203, 207
tetraaryl-1,3-dioxolane-4,5-dimethanol (TADDOL) 333
thiourea cinchona alkaloid 393
tiglic acid 148
titanium-catalyzed asymmetric reaction
- using chiral BINOL-Ti(IV) complexes 300ff
- using chiral salen-Ti(IV) complexes 284
- using chiral Ti-TADDOLate complex 335f, 338, 344f
TolBINAP 2, 10, 16, 22, 25, 29

total synthesis
- application of allylic substitution in   230
- application of decarboxylative Tsuji allylation in   235
- of SM-130686   83
- of vitamin E via citronellol   4
- using cinchona alkaloids   369, 375
transfer hydrogenation   114, 246
transition metal complex
- containing chiral Josiphos ligands   99ff
- containing chiral Josiphos-like ligands   99ff
trimethylsilyl azide   266
Tsuji allylation   231
- application in total synthesis   235
- application to fluorinated derivatives   234
- of allyl enol carbonates   232
- of dioxanone enol ethers   236
- of fluorinated compounds   234f
- of β-keto esters   233
- of silyl enolethers   23, 233f
tungsten-catalyzed asymmetric reaction
- allylic substitution   229
- using chiral W-PHOX complexes   229

u
Ugi amine   95
UlluPhOS, *see also* DuPhos-type chiral ligand   60
umpolung allylation   158
α,β-unsaturated acid
- hydrogenation of   4, 107, 146
α,β-unsaturated amide
- conjugate addition to   342
- epoxidation of   297f
α,β-unsaturated aldehyde
- [3 + 3] cascade reactions of   438

- [3 + 2] cycloaddition of   433
- cyclopropanation of   421, 436
- epoxidation of   436
- Michael-alkylation of   436
α,β-unsaturated carboxylic ester
- epoxidation of   299
- Michael-Addition of   121

v
vanadium-catalyzed asymmetric reaction
- using chiral BINOL-oxovanadium(IV) complexes   299
- using chiral salen-oxovanadium(IV) complexes   285
- using chiral V-TADDOLate complex   349
vinyl phosphonate
- hydrogenation of   243

w
Wacker-type cyclization   199
Walphos, *see also* Josiphos-type chiral ligand   98
(S)-warfarin   396

x
XylBINAP   2, 9, 28

z
zinc-catalyzed asymmetric reaction
- using chiral BINOL-Zn(II) complexes   300
- using chiral Zn(II)/Box complexes   188ff, 197f, 202f
zirconium-catalyzed asymmetric reaction
- using chiral BINOL-Zr(IV) complexes   302 f
- using chiral Zr-TADDOLate complex   337